JOURNAL OF CHROMATOGRAPHY LIBRARY — Volume 34

Polymer Characterization by Liquid Chromatography

Gottfried Glöckner
Technische Universität Dresden, DDR

ELSEVIER
Amsterdam — Oxford — New York — Tokyo 1987

This book is the revised translation of
Polymercharakterisierung durch Flüssigkeitschromatographie
published by VEB Deutscher Verlag der Wissenschaften, Berlin, DDR, 1980
Translated by Bernhardt Simon, Berlin, DDR

Published in coedition with VEB Deutscher Verlag der Wissenschaften, Berlin

Elsevier Science Publishers
Sara Burgerhartstraat 25
P. O. Box 211, 1000 AE Amsterdam, The Netherlands

Distributors for the U.S.A. and Canada
Elsevier Science Publishing Company
52 Vanderbilt Avenue
New York, NY 10017

Library of Congress Cataloging-in-Publication Data

Glöckner, Gottfried, 1925—
Polymer characterization by liquid chromatography.
(Journal of chromatography library; v. 34)
"Revised translation of Polymercharakterisierung
durch Flüssigkeitschromatographie . . . 1980" — T.p. verso.
Bibliography: p.
Includes index.
1. Polymers and polymerization—Analysis. 2. Liquid chromatography. I. Title. II. Series.
QD139.P6G5613 1986   547.7'046   86-6237
ISBN 0-444-99507-2

ISBN 0-444-99507-2 (Vol. 34)
ISBN 0-444-41616-1 (Series)

Printed in the German Democratic Republic

JOURNAL OF CHROMATOGRAPHY LIBRARY — Volume 34

Polymer Characterization by Liquid Chromatography

JOURNAL OF CHROMATOGRAPHY LIBRARY

A Series of Books Devoted to Chromatographic and Electrophoretic Techniques and their Applications

Although complementary to the *Journal of Chromatography*, each volume in the library series is an important and independent contribution in the field of chromatography and electrophoresis. The Library contains no material reprinted from the journal itself.

Volume 16 **Porous Silica.** Its Properties and Use as Support in Column Liquid Chromatography
by K. K. Unger

Volume 17 **75 Years of Chromatography — A Historical Dialogue**
edited by L. S. Ettre and A. Zlatkis

Volume 18 **Electrophoresis.** A Survey of Techniques and Applications
Part A: Techniques
Part B: Applications
edited by Z. Deyl

Volume 19 **Chemical Derivatization in Gas Chromatography**
by J. Drozd

Volume 20 **Electron Capture.** Theory and Practice in Chromatography
edited by A. Zlatkis and C. F. Poole

Volume 21 **Environmental Problem Solving Using Gas and Liquid Chromatography**
by R. L. Grob and M. A. Kaiser

Volume 22 **Chromatography.** Fundamentals and Applications of Chromatographic and Electrophoretic Methods
Part A: Fundamentals
Part B: Applications
edited by E. Heftmann

Volume 23 **Chromatography of Alkaloids**
Part A: Thin-Layer Chromatography
by A. Baerheim Svendsen and R. Verpoorte
Part B: Gas-Liquid Chromatography and High-Performance Liquid Chromatography
by R. Verpoorte and A. Baerheim Svendsen

Volume 24 **Chemical Methods in Gas Chromatography**
by V. G. Berezkin

Volume 25 **Modern Liquid Chromatography of Macromolecules**
by B. G. Belenkii and L. Z. Vilenchik

Volume 26 **Chromatography of Antibiotics**
Second, Completely Revised Edition
by G. H. Wagman and M. J. Weinstein

Volume 27 **Instrumental Liquid Chromatography.** A Practical Manual on High-Performance Liquid Chromatographic Methods
Second, Completely Revised Edition
by N. A. Parris

Volume 28 **Microcolumn High-Performance Liquid Chromatography**
by P. Kucera

Volume 29 **Quantitative Column Liquid Chromatography.** A Survey of Chemometric Methods
by S. T. Balke

Volume 30 **Microcolumn Separations.** Columns, Instrumentation and Ancillary Techniques
edited by M. V. Novotny and D. Ishii

Volume 31 **Gradient Elution in Column Liquid Chromatography.** Theory and Practice
by P. Jandera and J. Churáček

Volume 32 **The Science of Chromatography.** Lectures Presented at the A. J. P. Martin Honorary Symposium, Urbino, May 27—31, 1985
edited by F. Bruner

Volume 33 **Liquid Chromatography Detectors**
Second, Completely Revised Edition
by R. P. W. Scott

Preface

The main subject of this book is the characterization of plastics. To a high degree the properties of these polymers depend on the distribution of the molar mass and of other structural features. Small contributions frequently have a great effect. The characterization of polymers cannot be restricted to the determination of mean values but must yield information on these distributions. Using classical methods, the analytical fractionation of polymer homologues and structurally isomeric polymers is extremely time-consuming. Therefore efficient chromatographic techniques are being increasingly employed in modern polymer characterization.

In the first place, HPLC is applied in the form of size exclusion chromatography (gel permeation chromatography), but it is also possible to use other separation mechanisms. In this volume, more space is devoted to these possibilities than is merited by their current range of application, since the author believes that many a problem of characterization that still exists, besides the determination of the molar mass distribution, will be solved by separation techniques of the non-exclusion types. Nevertheless, the relative importance of size exclusion chromatography will not only be preserved but may even increase because of its use to complement other chromatographic techniques.

The first part of this book is intended as an introduction. For the polymer chemist, these chapters are meant to serve as an aid to the understanding of chromatography; moreover they provide the chromatographer, whose work is extending to the separation of macromolecules, with necessary information about polymers. So the book does not presuppose specialist knowledge and can guide the reader in the challenging borderline area between polymer science and chromatography. For scientists involved in practical work the most important parts are probably sections C "Chromatography under real conditions" and D "Applications".

As already mentioned, size exclusion chromatography (SEC) is without doubt the most common, and therefore also the most important, form of chromatographic polymer characterization. Many explanations, especially when dealing with the experimental characteristics of elution chromatography, are of importance for SEC, and also relevant to adsorption and partition techniques. To avoid repetition, an arrangement was chosen which permits a complete picture of macromolecular chromatography, from the principles to the applications. In this way it was possible to avoid splitting the book into separate parts such as "Exclusion chromatography", "Adsorption chromatography" and "Partition chromatography". At the same time it was possible to present the whole complex of chromatographic mechanisms without difficulty. This order also made it possible to place certain fundamental explanations, for example of the gradient technique or kinetic band broadening, according to their priority. Admittedly this order could cause difficulties for a reader only interested in one particular chromatographic method, because he will not find all the relevant information in successive chapters. However, notes in the text and the index should help in finding any item without too much difficulty.

The manuscript was aimed at helping the analyst or polymer chemist who is looking for information about chromatographic methods for the characterization of polymers. I therefore consciously tried to present the material in the most straightforward way possible. I have also tried to simplify the symbolism as far as possible. Since it is not possible simply to mix the different sets of symbols from chromatography on the one hand and polymer science on the other, because they overlap to some extent, in some cases I have had to deviate from the norm.

The German edition of this book, "Polymercharakterisierung durch Flüssigkeitschromatographie", was published by VEB Deutscher Verlag der Wissenschaften, Berlin, in 1980, and included references to original papers published up to 1977. In places the text has undergone considerable modification as compared with the 1980 edition.

The second part of the book, "Concepts of chromatography: mechanisms and materials", underwent many changes from the German edition. In the present book the packing materials are no longer discussed in connection with an individual separation process, because of the ever-increasing number of materials available which, depending on the conditions, can show separating efficiency due to quite different mechanisms. A well-known example of this is silica.

Chapters 10—12 have been completely rewritten and aim to show a more rounded picture of materials used in liquid chromatography. New sections deal with support characterization, bonded phases and cross-linked organic materials.

However, the increase in length due to these additions made it necessary to omit whole chapters of the German edition. Thus, the description of the apparatus for elution chromatography, of the historical development of chromatographic methods and several other passages have been deleted.

It soon became clear just how difficult it was to translate such a greatly revised manuscript. The fact that the revision, translation and editing of the manuscript had to be carried out more or less simultaneously added further difficulties. I wish to thank all those who have contributed to the preparation this book for their cooperation, not least Fräulein MIEDLICH of the Department of Chemistry of VEB Deutscher Verlag der Wissenschaften, who did a great deal of the detailed editorial work with untiring patience, and her department head, Dr. FICHTE, who had to conduct the difficult concert.

In preparing the revised edition, several thousand new papers were taken into consideration. I would like to thank all colleagues who sent me offprints of their interesting work. I am also grateful for the valuable help received from the library staff and from other departments of the Technical University of Dresden. My special thanks go to Frau CHARLOTTE MEISSNER, who again helped me most conscientiously to cope with the wealth of literature. She also showed remarkable patience in preparing about 800 new entries for the list of references. Dr. ZIMMERMANN, with great enthusiasm, then guaranteed that the advantages of electronic data processing could be utilized in indexing these data.

Now it is for the reader to decide whether he can really benefit from these efforts. Critical comments will be gratefully received.

GOTTFRIED GLÖCKNER

Table of contents

Table of contents

Glossary of symbols and abbreviations

Minor symbols which are used only once are not included.

Glossary of symbols

Symbol	*Definition*	*Units*
${}^{1}A$	Surface area per gram of adsorbent	$m^2 \cdot g^{-1}$
A_2	Second virial coefficient	$cm^3 \cdot g^{-2} \cdot mole$
A_S	Surface area required by an adsorbed molecule	
a	Exponent for the Mark-Houwink relation, $[\eta] = K_\eta \cdot M^a$	
a_I	Activity (of component I in the mixture)	
a_i	Increment of A_S due to the structural element i of a molecule (Section 7.2.)	
b	Number of long chain branches per molecule	
b	Slope of gradient (Section 14.4.)	
b_0	Length of a statistical chain element	
C_1, C_2	Constants in SEC calibration, $V_e = C_1 - C_2 \log M$	
c	Solute concentration	$g \cdot l^{-1}$
c_0	Concentration of sample solution, starting concentration	
c_{eq}	Concentration of solution in adsorption equilibrium	
D	Linear diffusion constant	$cm^2 \cdot s^{-1}$
D	Dielectric constant	
D_1, D_2	Constants in SEC calibration, $M = D_1 \cdot e^{-D_2 \cdot V_e}$	
d	Diameter	
d_C	Internal diameter of the column	mm
d_O	Pore diameter	nm
d_P	Particle diameter of the packing	µm
d_a	Thickness of adsorption layer	nm
d_f	Film thickness of the stationary phase	
d_h	Thickness of the hydrodynamically effective layer	nm
ΔE_λ	Extinction of the solute at wavelength λ	
$F(v)$	Experimental SEC elution curve	
ΔG	Difference in standard free energy	
G_m	Band compression factor in gradient technique	
$G(v - y)$	Instrumental dispersion function	
g	Branching parameter, radius of gyration ratio, $g = \langle S^2\rangle_{0,br}/\langle S^2\rangle_{0,1}$	
g'	Branching parameter, intrinsic viscosity ratio, $g' = [\eta]_{br}/[\eta]_l$	

H	Heterogeneity (dispersity) of a polymer sample, $H = \bar{M}_w/\bar{M}_n$	
ΔH	Enthalpy difference	
$H(M)$	Molar mass distribution (weight distribution function)	
$H(P)$	Chain length distribution	
$h(M)$	Molar mass frequency distribution	
$h(P)$	Chain length frequency distribution	
h	Plate height, height equivalent to a theoretical plate	μm
h_{eff}	Effective plate height	
h^*	Reduced plate height, $h^* = h/d_P$	
h_M	Plate height as determined with a probe polymer of molar mass M	
h_f	Plate height from the front part of a skewed peak	
h_t	Plate height from the rear part of a skewed peak	
h'_a	Longitudinal-diffusion plate height	
h'_e	Eddy-diffusion plate height	
h'_s	Plate height due to stagnant mobile phase	
h'_t	Plate height due to interparticle mobile phase effects	
h''_d	Plate-height contribution caused by resistance to mass transfer	
h''_f	Plate-height contribution due to diffusion in the stationary film	
h''_p	Plate-height contribution due to pore diffusion	
I	Ionic strength	
$I(M)$	Weight-cumulative distribution of molar mass	
K	Distribution constant (activity ratio), $K = a''_S/a'_S$	
K^+	Conventional distribution constant (concentration ratio), $K^+ = c''_S/c'_S$	
K^*	Adsorption constant, $K^* = K \cdot V_a$	$cm^3 \cdot g^{-1}$
K_{av}	Laurent-Killander distribution constant	
K_η	Constant factor in the Mark-Houwink relation, $[\eta] = K_\eta \cdot M^a$	
k	(Column) Capacity factor, mass distribution ratio, $k = m''_S/m'_S = K \cdot q = (t_e - t')/t'$	
k	Degree of coupling (SCHULZ MMD, Section 4.2.4.1.)	
k	Reaction rate constant	
L	Column length, length of separating path	m
L_M	Contour length of a chain molecule, $L_M = P \cdot l_{eff} = Z \cdot b_0$	
l_0	Bond length (for C—C: 154 pm or 1.54 Å units)	
l_{eff}	Effective length of the repeat unit of a chain	
M	Molar mass	$g \cdot mole^{-1}$
M_0	Molar mass of the repeat unit	
M_b	Molar mass of the chain portion between two neighbouring branch points	
M_{lim}	Exclusion limit of SEC column or separating material	
$M_{(t)}$	True molar mass value	
$M_{(u)}$	Molar mass value calculated from an uncorrected chromatogram	
$\bar{M}_n$	Number-average molar mass	
$\bar{M}_v$	Viscosity-average molar mass	
$\bar{M}_w$	Mass-average molar mass	
$\bar{M}_z$	z-Average molar mass	
m_0	Mass of the sample injected	g (mg, μg)
m_I	Mass of component I in a mixture	

m_a	Mass of adsorbed solute per gram of adsorbent	$mg \cdot g^{-1}$
N	Plate number, column plate count	
N_{eff}	Effective plate number, $N_{eff} = N \cdot Q^2$	
N_L	Avogadro's number, $N_L = 6.022 \cdot 10^{23}$	$mole^{-1}$
n	Number of moles	
Δn	Refractive index difference due to the solute	
n_p	Peak capacity, maximum number of peaks detectable in a chromatogram	
P	Degree of polymerization	
$\bar{P}_n, \bar{P}_v, \bar{P}_w, \bar{P}_z$	Average values, see molar-mass averages	
p	Pressure	MPa (kPa)
p_{CO}	Fraction of CO groups adsorbed (from IR measurements)	
p_{ESR}	Fraction of segments adsorbed (fixed, from ESR measurements)	
Q	Q Factor, molar mass per unit chain length	$g \cdot mole^{-1} \cdot Å$
Q	Retention factor, capacity term for the resolution equation, $Q = k_I/(1 + k_I)$	
Q_i	Increment of adsorption energy due to the structural element i	
q	Phase ratio, ratio of stationary phase volume to mobile phase volume, $q_{SEC} = V_P/V_I$, in AC and LLC: $q_{LC} = V''/V_M$	
$\boldsymbol{R}$	Gas constant	
R	Retention ratio, $R = t'/(t' + t'')$	
$\langle R^2 \rangle$	End-to-end displacement (mean-square value)	cm^2
R_M	Logarithmic function of retention rate, logarithmic capacity factor, $R_M = \log k = \log [(1 - R)/R]$	
R_f	Relative distance of migration in TLC, $R_f = s/L$	
R_s	Resolution	
R_{sp}	Specific resolution	
$R_{\Theta, i}$	Rayleigh factor in light scattering	
RI	Resolution index	
r	Radius	
r_O	Pore radius	nm
r_P	Particle radius	μm
r_v	Rate of vaporization	
S	Selectivity factor in SEC, $S = (V_{e, II} - V_{e, I})/\log (M_I/M_{II})$	
S	Slope factor in gradient elution, $S = \log (k_W/k_t)/\varphi_{II}$	
ΔS	Entropy difference	
$\langle S^2 \rangle$	Mean-square radius of gyration	cm^2
S^0	Adsorption energy for the solute in adsorption from a pentane solution onto an adsorbent with standard activity $\alpha_A = 1$	
s	Distance travelled, migration distance of a solute	
sk	Skew parameter, molar-mass correction factor for band-broadening asymmetry	
T	Temperature	K
T	Separation power of SEC	
t	Time	
t'	Observed elution time of a non-retained substance	
t''	Net retention time	
t_M	Eluent hold-up time	
t_e	Total retention time, $t_e = t' + t''$	

U	Non-uniformity coefficient (Uneinheitlichkeit), $U = H - 1$	
u	Linear velocity of the mobile phase	$cm \cdot s^{-1}$
V_I	Molar volume of component I	$cm^3 \cdot mole^{-1}$
V'	Elution volume of a non-retained solute, $V' = V_M$ in AC and LLC, but $V' = V_I$ in SEC	ml
V''	Volume of the stationary phase	ml
V_0	Injected sample volume	ml, µl
V_C	Volume of the empty column	cm^3
${}^1V_G$	Specific gel bed volume, volume per gram of swollen gel	$cm^3 \cdot g^{-1}$
V_I	Interstitial volume (void volume)	cm^3
V_M	Eluent hold-up volume	ml
V_P	Total pore volume of the column	cm^3
${}^1V_P$	Pore volume per gram of gel	$cm^3 \cdot g^{-1}$
V_W	Wall volume of the packing material in the column	cm^3
V_a	Surface volume of the adsorbent, volume of a monomolecular solvent layer on the surface of one gram of adsorbent, $V_a = (n''_E \cdot V_E)/m_A$	$cm^3 \cdot g^{-1}$
V_{av}	Average retention volume	
V_c	Count volume, siphon volume	ml
V_e	Elution volume, $V_e = V_r + V'$	ml
V_h	Hydrodynamic volume, $V_h = M[\eta]$	$l \cdot mole^{-1}$
V_{mix}	Volume of the mixing chamber	cm^3
V_r	Retention volume	ml
v	Flow rate	$ml \cdot min^{-1}$
W	Peak width at base	ml
$W(y)$	Distribution of molar mass or hydrodynamic volume as a function of peak position	
w_I	Mass fraction of component I in a mixture (weight fraction)	
x_I	Mole fraction of component I	
y	Recorder deflection, chromatogram height	
Z	Segment number, number of statistical elements per chain	
z	Number of cycles in recycling	
α	Linear coil expansion coefficient, $\alpha^2 = \langle R^2 \rangle / \langle R^2 \rangle_0$	
α_A	Activity of the adsorbent	
Γ_a	Amount adsorbed per unit surface area	$mg \cdot m^{-2}$
γ	Surface tension	
γ	Skew parameter of distribution function	
δ	Solubility parameter	$cal^{1/2} \cdot cm^{-3/2}$ (Hildebrand units)
$\delta_a, \delta_b, \delta_d, \delta_0, \delta_{in}$	Solubility parameter increments due to proton donor capacity. proton acceptor capacity, dispersion forces, dipole orientation and induction forces, respectively	
ε	Exponent for the relationship between the branching parameters g and g', $g' = g^\varepsilon$	
ε	Extended volume parameter, $\varepsilon = (2a - 1)/3$	
$\varepsilon°$	Solvent strength, adsorption energy of the solvent	
$\varepsilon°(RP)$	Solvent strength in reversed-phase chromatography	

ε_I	Interstitial porosity, $\varepsilon_I = V_I/V_C$	
ε_P	Internal porosity, $\varepsilon_P = {}^1V_P/({}^1V_P + 1/\varrho_W) = V_P/(V_P + V_W)$	
ε_t	Total porosity, $\varepsilon_t = \varepsilon_I + \varepsilon_P(1 - \varepsilon_I)$	
η	Viscosity	mPa · s
$[\eta]$	Intrinsic viscosity	$l \cdot g^{-1}$
θ	Symbol indicating the pseudo-ideal state of polymer solutions (exhibiting $A_2 = 0$, $a = 0.5$, $\alpha = 1$, $\varepsilon = 0$, $\chi = 0.5$ at the Flory temperature)	
ϑ	Temperature	°C
ϑ_0, ϑ_u	Temperature at column inlet and outlet, respectively	
ϑ	Bond angle	
$\varkappa$	Flow parameter in TLC	
$\varkappa$	Permeability of a packed column	
λ	Branching frequency, $\lambda = b/M$	
λ	Retention parameter in FFF, $\lambda = \dfrac{\text{distance from the wall}}{\text{channel width}}$	
μ	Dipole moment	
$\Delta\mu$	Difference in chemical potential	
$\mu_1, \mu_2, \ldots$	Statistical moments of a distribution	
ν	Reduced velocity, $\nu = u \cdot d_P/D'$	
ν_μ	Refractive increment at a constant value of chemical potential	
ν_φ	Refractive increment at a constant solution composition	
ϱ	Density	$g \cdot cm^{-3}$
ϱ_a	Density of an adsorption layer	
ϱ_W	Density of pore wall material	
σ	Standard deviation, half-width of a Gaussian curve at 60.7% of its apex, $\sigma = W/4$	
σ_M	Standard deviation of MMD in a plot vs. M	
σ_a	instrumental dispersion, $\sigma_a^2 = \sigma_c^2 + \sigma_e^2$	
σ_c	σ value due to column dispersion	
σ_d	Increment caused by band broadening in the detector	
σ_e	σ value due to extra-column effects	
σ_f	σ value corresponding to the leading part of a peak	
σ_t	σ value corresponding to the rear part of a peak	
σ_v, σ_h	σ increments caused by the first and second half of a column, respectively	
φ_I	Volume fraction of component I	
χ	Huggins constant, interaction parameter in the Flory-Huggins theory	

Subscripts and general notations

X′	(single prime)	Mobile phase or sol phase
X″	(double prime)	Stationary phase or gel phase
X_I, X_{II}	(roman numerals)	Indication of components in a mixture according to increasing values of a certain quantity, e.g., increasing adsorption energy, molar mass, etc.
X_A, X_B	(capital letters)	Indicating a certain material or chemical group A adsorbent, E eluent (solvent), S solute
X_a, X_b	(small letters)	Indicating a certain state of matter a adsorbed, br branched, l linear
$\bar{X}$ or $\langle X \rangle$		Average values
ΔX		Differences, change of quantity X

Abbreviations

Polymers

CA	Cellulose acetate
CN	Cellulose nitrate
Dextr.	Dextran
EVA	Poly(ethylene-co-vinyl acetate), mass % vinyl acetate indicated in parentheses
PA	Polyamides
PαMS	Poly(α-methylstyrene)
PEMA	Poly(ethyl methacrylate)
PAN	Polyacrylonitrile
PBMA	Poly(butyl methacrylate)
PBd	Polybutadiene
PBn	Poly(but-1-ene)
PC	Polycarbonate
PCS	Poly(p-chlorostyrene)
PDMS	Polydimethylsiloxane
PE	Polyethylene
PEG	Poly(ethylene glycol)
PETP	Poly(ethylene terephthalate)
PIB	Polyisobutene
PMA	Poly(methyl acrylate)
PMMA	Poly(methyl methacrylate)
POE	Poly(ethylene oxide) (IUPAC recommendation: PEO)
POP	Poly(propylene oxide) (IUPAC recommendation: PPO)
PP	Polypropylene
PS	Polystyrene
PTFE	Poly(tetrafluoroethylene)
PVAC	Poly(vinyl acetate)
PVAL	Poly(vinyl alcohol)
PVC	Poly(vinyl chloride)
PVP	Poly(vinyl pyrrolidone)

Solvents

AC	Acetone
AcN	Acetonitrile
Bzn	Benzene
CBzn	Chlorobenzene
CHx	Cyclohexane
CP	Cyclopentane
DCE	Dichloroethane
DCM	Dichloromethane
DMF	Dimethylformamide
DMSO	Dimethyl sulphoxide
Dx	Dioxane
E	Diethyl ether
Eol	Ethanol
EAt	Ethyl acetate
Hp	Heptane
Hx	Hexane
M	Methanol
MEK	Methyl ethyl ketone
MAt	Methyl acetate
NM	Nitromethane
Pn	Pentane
TCB	1,2,4-Trichlorobenzene
TBE	Tetrabromomethane
TCE	Tetrachloroethylene
TCM	Chloroform (trichloromethane)
Tetra	Carbon tetrachloride (tetrachloromethane)
THF	Tetrahydrofuran
Tol	Toluene
Tri	Trichloroethylene
W	Water

Methods/General Terms

AC	Adsorption chromatography
AFC	Affinity chromatography
CCD	Chemical composition distribution
E	Eluent
FFF	Field flow fractionation
GC	Gas chromatography
GF	Gel filtration chromatography (SEC in aqueous media using soft gels)
GPC	Gel permeation chromatography
HDC	Hydrodynamic chromatography
HPLC	High performance LC
HPTLC	High performance TLC
IR	Infrared
L	Solvent (eluent)
LC	Liquid chromatography
LLC	Partition chromatography, liquid-liquid partition chromatography
LSC	Liquid-solid chromatography
MMD	Molar mass distribution
NMR	Nuclear magnetic resonance spectroscopy
PC	Precipitation chromatography
R.I.	Refractive index
RP	Reversed phase
RPC	Reversed phase chromatography
S	Solute
SEC	Size exclusion chromatography
SEC/W	SEC in aqueous media
SFC	Supercritical fluid chromatography
TLC	Thin-layer chromatography
UV	Ultraviolet

A Basic facts

1. Introduction

This chapter is intended to illustrate the kind of problems chromatography is faced with in the characterization of polymers and the chromatographic methods used in this connection. The reader who is already familiar with these topics may pass on to the more specific treatments commenced in Chapter 6 onward. Chapters 2 and 3 are addressed to the polymer scientist who is seeking access to the fundamentals of chromatography, Chapters 4 and 5 to the expert in chromatography who is attempting to extend his field of work to the investigation of polymers. (In this context, polymers are understood to comprise synthetic products and natural products of comparable complexity.)

Mother Nature alone is capable of producing enormous numbers of macromolecular compounds, whose molecules all have the same molar mass and the same configuration. The admirable uniformity of proteins and nucleic acids results from their structure-controlled synthesis in the living cell. On the other hand, synthetic polymers exhibit more or less broad distributions for all structural parameters. This is true for both technical products and polymers made in the laboratory. (Exceptions such as synthetic peptides or products made by the duplication principle may be ignored to a first approximation.) Other natural polymers that do not serve key-and-lock functions but are only required as materials are not monodisperse but similar to the synthetic ones.

This book deals primarily with the problems involved in the investigation of polydisperse polymers, i.e. the characterization of synthetic products and comparable natural products, e.g., of natural rubber. These products exhibit.

(i) a *chain-length distribution* which in most cases is so broad that it must be treated mathematically as a continuous distribution. This will be discussed in Section 4.2. Even the so-called monodisperse standards, used for instance in the calibration of exclusion chromatography, and obtained by special, demanding polymerization processes, are mixtures of components with different degrees of polymerization. Evidence in support of this has been obtained by chromatographic investigations, examples of which will be presented in Chapters 9, 18 and 19.

(ii) The *constitutional features* of synthetic polymers are also found to exhibit a distribution. For homopolymers this relates to the linkage of the monomer units in 1,1- or 1,2-positions, to possible defects as compared with the ideal structure, to the occurrence of short- and long-chain branchings or to differences with respect to the end groups. For copolymers, the proportions of the different monomer units and their arrangement along the chain must also be considered (chemical distribution and sequence-length distribution).

(iii) Polymers of asymmetric monomers may exhibit additional distributions with respect to their configurational features.

(iv) For the sake of completeness, the distributions with respect to the *conformational features* are considered, which in some cases can also be observed in synthetic polymers. These are of greater importance for the constitutionally and configurationally uniform biopolymers whose regular structure offers many more starting points for the formation of stable conformers.

Some of the distributions mentioned above are distributions along the polymer chain. In these cases, for example, spectroscopic methods are used to determine the frequency of certain linkages or the distribution of sequence lengths in copolymers. Other methods for the determination of intramolecular distributions can also be employed to investigate the decomposition of the polymer chain. These methods include the brilliant techniques for sequence analysis of proteins and peptides by stepwise hydrolysis, or pyrolytic gas chromatography.

This book deals with the chromatographic characterization of *polymers*. The investigation of decomposition products is disregarded, as is the analysis of small-molecule auxiliary agents and additives that are, nevertheless, of undisputed importance in the use of polymer products.

This leads us to the question of the features of macromolecular chromatography which may justify the title chosen here. GIDDINGS (1967b) commented as follows: "The difference between macromolecular and small-molecule chromatography, from a fundamental point of view, is considerably greater than the difference between gas and liquid chromatography . . ."

The significant features of macromolecular chromatography are due partly to special properties of the polymers themselves, and partly to effects occurring in the chromatographic process. Let us start with the properties of the polymers.

(i) It is not possible to vaporize them without decomposition. This is why liquid chromatography is the method to choose when a chromatographic separation by molecular features is required. However, it should be noted, that chromatographic analyses are also carried out on solid polymers. These are packed as a stationary phase into columns on which the retention of known gaseous samples is studied. In this inverse gas chromatography, information can be obtained on the interactions between the sample and the polymer as well as some indication of the structure of the polymeric solid.

(ii) As a rule, there are only a rather limited number of solvents for polymers. Solubility is of much greater importance in the liquid chromatography of polymers than in that of small-molecule samples. Therefore the interactions between polymers and solvents will be discussed in detail in Chapter 5.

(iii) The diffusion coefficients of macromolecules in solutions, ranging between 10^{-6} and 10^{-9} $cm^2 s^{-1}$, are smaller than those of dissolved small-molecule substances by at least two orders of magnitude. Therefore the kinetic factors discussed in Chapter 15 have an even stronger influence than elsewhere in liquid chromatography. For example, the well known minimum in plots of the reduced height equivalent to a theoretical plate vs. the reduced velocity has not yet been found with polymers.

(iv) In many cases the number of species occurring simultaneously is so large that it has not yet been found feasible to separate them from one another. This is especially true for the chain-length distribution, the determination of which is dealt with in detail in Chapter 16. The chromatogram of a polymer sample seldom looks like one would expect from HPLC, or like those for the analyses of small-molecule mixtures. The fine chromatograms of protein mixtures are exceptions. Considerable successes with respect to the separation of individual components have also been achieved for oligomers (cf. Sections 18.3. and 19.11.3.). But these systems contain only a small number of components compared to the polymer systems.

Let us proceed to features resulting from the interplay of polymer properties and the chromatographic technique. Of course the principles and laws are the same as in the chromatography of small-molecule substances, but in some cases their effect involves entirely new phenomena. This is true, for example, for (v) the behaviour of polymers in adsorption chromatography. Retention presupposes bonding with a simultaneous decrease in free enthalpy; chromatographic migration requires a reversibility of adsorption under chromatographic conditions. For an individual segment, the equilibrium between adsorption and desorption is not difficult to reach, but this segment cannot leave the surface after desorption if its neighbours in the chain are still adsorbed. As the equilibrium must lie on the side of adsorption, the probability of a free segment being re-adsorbed is higher than that of a simultaneous desorption of all neighbouring segments. This means irreversible adsorption of the macromolecule as a whole, but reversible segment adsorption. This phenomenon, which will be discussed in more detail in Chapters 6 and 18, is of far-reaching importance for the chromatography of flexible polymers and may necessitate the use of displacers. On the other hand, fairly "normal" adsorption chromatography can be carried out with conformationally stable biopolymers.

(vi) The liquid chromatography of small-molecule substances is mainly carried out on porous packing material whose pores are large compared with the size of the dissolved particles. The solute can then enter into the pores as easily as the solvent; usually no exclusion effects are observed. Macromolecules, on the other hand, are much larger than the solvent molecules. It would be necessary for the packing materials to have very large pores in order that neither of the two classes of substances will be affected by the boundary walls. Most of the chromatographic analyses of polymers are, however, carried out by the principle of size exclusion, i.e., with pores of the same order of magnitude as those of the macromolecules. As a consequence of the short distances from the walls, additional exclusion effects caused by electrostatic interactions are likely to occur. This phenomenon is most evident in chromatographic analyses using water as an eluent. It will be dealt with in Section 19.3.3.3. In this context, the tendency of flexible polymer molecules to creep into pores smaller then the coil sizes is also worth mentioning (cf. Section 16.6.1.).

In cases where the polymer molecules are completely excluded from fine-pore supports, it must not be overlooked that the much smaller molecules of the solvent can enter freely into the pores. The volume available in the chromatographic bed for the solvent is much larger than that for the dissolved sample, so that segregation phenomena of different kinds may occur.

(vii) As a result of the factors mentioned above, especially of (v) and (vi), in practice it is rarely possible to realize a separation on the basis of a single mechanism. This is discussed in detail under the heading "Real GPC" in Section 16.6. The superposition of several mechanisms usually raises problems due to the fact that they have opposing directions of separation.

(viii) Finally, the energy input, which also causes disturbances, e.g., by heating, in the case of small-molecule compounds, may lead to decomposition due to the shearing action for substances with molar masses of about 10^7 g $\cdot$ mole^{-1}.

Thus, macromolecular chromatography is not simply a partial field of small-molecule liquid chromatography.

At the beginning of this chapter it was stated that the main concern of this book would be the chromatographic analyses of synthetic polymers and of natural products of comparable complexity. This is because these products, with their large number of components and the flexibility of their molecules, illustrate the difficulties that may be encountered in the investigation of polymers. It is only a slight overstatement to say that, compared with those products, the conformationally stable biopolymers exhibit more "normal" behaviour in liquid chromatography. On the other hand, it should not be overlooked that these biopolymers, with their possibilities of conformational isomerism, their tendency to build up electric charges, etc., raise specific problems that are of little importance in the case of synthetic polymers.

Let us now turn to the question of which of the chromatographic mechanisms of separation are used for the characterization of polymers. The choice is determined by the particular problem.

The most common problem is the determination of the chain-length distribution. For this purpose, gel permeation chromatography (GPC), or size exclusion chromatography (SEC), has been developed, which nowadays is used so widely that one may consider it to comprise the whole of the chromatographic investigation of polymers. Other methods of separation by the degree of polymerization include the different variants of field-flow fractionation and hydrodynamic chromatography (Sections 13.1 and 13.2.), which also make it possible to investigate very large particles such as latex particles or blood cells. A separation in terms of molecular size can also be achieved by adsorption or precipitation chromatography as well as by chromatography in supercritical media. Although so far these methods have had little application, they are included and even described in some detail here. This has been done on the one hand because their separating power is in part quite superior and, on the other hand, in the expectation that their importance and range of application will greatly increase in the future.

The separation of copolymers according to their chemical composition can be achieved by adsorption and precipitation chromatography. Prior to this analysis, it is of advantage to fractionate the copolymers by their chain-length, e.g., by means of SEC, and to determine the chemical distribution in the fractions.

Macromolecules with certain functional or sterically important groups can be separated by adsorption chromatography, hydrophobic chromatography or affinity chromatography. Such methods may offer a selectivity and a sensitivity which can rarely be achieved by other methods of polymer characterization. Affinity chromatography plays an important role in the purification and isolation of biologically active substances possessing groups of a certain structure, for which the material of the chromatographic bed is specially prepared using the key-and-lock principle.

In most cases the technique used is column liquid chromatography, where the high-performance technique predominates. Therefore the greater part of this book will deal with the characterization of polymers by HPLC in columns. The final chapter presents the most interesting results obtained using thin-layer chromatography.

2. Chromatographic techniques

The complex general field of chromatography can be subdivided following different principles according to the chosen point of view:

Column chromatography/planar chromatography. This subdivision is based on the geometry of the chromatographic bed. CVET (1906) used glass columns packed with chromatographically active material, through which the mixture to be separated flowed. This principle of *column chromatography* was further developed in a great number of investigations and has been implemented in many variants today. Besides glass, stainless steel is used for making columns resistant to pressures of more than 10 MPa (100 bar). For laboratory columns the interior diameters range from fractions of a millimetre (in the case of capillary column chromatography) to several centimetres (for preparative columns), and the lengths range from a few centimetres to several metres. In most cases the columns are filled with packings in which the chromatographic process takes place. However, capillary column chromatography is accomplished in very long empty columns, the interior walls of which are coated with the stationary phase.

Planar chromatography includes thin-layer chromatography (TLC) and paper chromatography. In this case the spread-out, chromatographically active phase also interacts with the surrounding gas atmosphere. In the flat bed methods usually the eluent is soaked up by the stationary phase acting like a wick. Column separations are rarely carried out according to this principle. In the most simple case the eluent flows freely through the column under its natural pressure. However, in most cases, the pressure is increased by use of pumps. The higher flow-rates thus obtained enable separations to be completed within a much shorter time. Commercial apparatus for modern column chromatography yield high outputs with a relatively small amount of human effort. In contrast to this, flat bed techniques have a lower initial cost but continuously require a considerably higher amount of labour.

Gas chromatography/liquid chromatography. This subdivision is based on the state of aggregation of the mobile phase.

Gas chromatography (GC) is a column technique, whereas liquid chromatography (LC) may be carried out in columns and on films. The stationary phase may be solid (GSC, LSC) or liquid (GLC, LLC). A liquid stationary phase is usually placed upon a solid carrier, which is porous in most cases. This increases its mechanical stability. Moreover, the extent of the interface with the mobile phase is increased in this way. The mobile phase must not remove the stationary phase from the chromatographic bed. Therefore only stationary liquids having a negligibly low vapour pressure can be used for GLC, while for LLC they must be insoluble in the mobile phase.

Gas chromatography can only be used in the investigation of samples which are volatile at the usual separating temperatures, up to 600 K. If necessary, non-volatile substances of special interest can be chemically converted into volatile derivatives. Liquid chromatography can be used to investigate all substances which are soluble in the mobile phase. When working in closed systems, the state of aggregation of the sample itself is of no importance.

Elution chromatography/development chromatography. This subdivision is based on the distance, s, travelled by a sample in a chromatographic bed of length L until the termination of the experiment. In the case of *development chromatography*, $s < L$, i. e. at the end of the experiment the components are still in the bed. They have travelled different distances from the starting point and are characterized by R_f *values*:

$$R_f = \frac{\text{distance of the substance from the starting point}}{\text{distance of the eluent front from the starting point}}$$

After every analysis in the developing technique the chromatographic bed has to be regenerated. As a rule, development chromatography is carried out as a planar technique, e.g., TLC.

In the case of *elution chromatography* the analysis is continued until all of the components have arrived at the end of the separation path. The stronger the forces retaining the components on the chromatographic bed, the later they will arrive at its end. They are detected in the eluate and identified by their retention time, t_e. The *retention ratio*, R, is independent of the geometry of the chromatographic bed, and hence comparable to the R_f value:

$$R = \frac{\text{eluent hold-up time, t}'}{\text{retention time of the component, } t_e}$$

Usually elution chromatography is realized as a column technique. The chromatographic bed is cleaned by the process itself in every analysis. The next sample can be loaded immediately after the last component has left the chromatographic bed. In the case of a longer experiment in which all the components travel relatively close to each other, the next sample can even be loaded while the preceding one is still on the chromatographic bed. If the intervals between the individual loadings are properly adjusted, the samples will not interfere with one another. Continuous analysis of the eluate can be carried out automatically by means of detectors. Thus elution chromatography, although requiring a higher initial cost than development chromatography, yields much more information with regard to the amount of work required for the execution of a given series of analyses.

Elution chromatography/displacement chromatography/frontal chromatography. Usually the working conditions are chosen so that the separation is accomplished by *elution chromatography*. In this method the components are transferred into the stationary phase at higher concentrations, returning completely into the mobile phase as soon as the concentration in the latter has become low enough. All the components migrate, at a speed dependent on the tendency to be transferred into the stationary phase. After travelling a sufficient distance, the sample is separated into component bands.

Displacement chromatography also offers the possibility of a separation of substances which cannot be removed from the chromatographic bed by diluting the mobile phase, or where this process is too slow. In this case the eluent is changed, i.e., the bed is washed with a solvent which itself has such a strong affinity to the stationary phase that it displaces weaker components from it. Thus it is possible to fractionate the sample using eluents of different strengths.

In *frontal chromatography* a sample solution is continuously added to the chromatographic bed. The component having the highest affinity to the stationary phase is retained directly after the start. The other components follow immediately. The following solution flows through the inlet zone of the bed, which is saturated with the preferentially retained component, so that the composition of the following solution remains unchanged. In the next zone the solution component having the highest affinity displaces the weaker component from the stationary phase and moves it forward. This process occurs repeatedly at all levels of affinity. At the end of the bed, at first only the depleted solvent arrives. The first sample zone contains the least-retained components. The zones of the other components follow. They are distinctly separated from one another by fronts.

Frontal analysis was developed as a result of efforts to eliminate the effect of nonlinearities. Today it is still used for removing small impurities from solutions, e.g., "filtration" on carbon adsorbents or water desalination by means of ion exchangers.

Partition chromatography/precipitation chromatography/exclusion chromatography/adsorption chromatography/affinity chromatography/ligand chromatography. This subdivision is based on the mechanisms which lead to chromatographic separation.

Partition chromatography requires stationary phases in which the components of the sample have different solubilities. Liquid films or polymer layers are used for this purpose. The thinner the stationary layer, the sooner the partition equilibria are reached and the more rapidly the mobile phase can flow.

Precipitation chromatography is a special form of partition chromatography, where the stationary phase is generated from the mobile one.

Exclusion chromatography occurs in solid stationary phases, the pore sizes of which are of the same order of magnitude as those of the particles to be separated. Too large molecules cannot diffuse into the pores; they are excluded and move along the chromatographic bed without being retained. The more deeply the other molecules can penetrate into the pore system, the longer they are retained.

Adsorption chromatography occurs on stationary phases having active centres which adsorb the sample components more or less strongly. Polar intermolecular forces are of special importance for the adsorption.

Affinity chromatography is a technique for separating substances according to their different abilities specifically and reversibly to bind to groups of the stationary phase. In this way it is possible to solve rather difficult problems such as the isolation of enzymes.

Ligand chromatography allows the separation of stereoisomers. Depending on their configurations, the sample components form more or less strongly bonded complexes with ligand groups of the chromatographic bed, thus being differently retained.

Wet-bed/dry-bed chromatography. This subdivision of chromatography is based on the nature of the chromatographic bed at the beginning of the analysis. Planar chromatography is usually started with *dry* beds which take up the eluent by capillary forces. On the other hand, columns are operated using the *wet-bed technique* in most cases, the sample being injected into the already flowing eluent.

Low-pressure/high-pressure chromatography. The development of efficient apparatus has allowed the use of high starting pressures in the liquid chromatography, so that also the high flow resistances encountered in long, high-resolution columns can be overcome. Thus high-pressure liquid chromatography (HPLC) is at the same time high-performance liquid chromatography, but the highest resolving power is achieved by means of carefully optimized techniques without using an extremely high pressure.

Gradient elution/isocratic elution. A gradient is formed when there are different migration conditions at various points of the chromatographic bed. This can be achieved in several ways. One possibility is to modify the eluent during the development. In contrast, in isocratic elution the transport efficiency of the eluent is constant everywhere. Isocratic solvent mixtures have a constant development capacity.

3. Foundations and fundamental concepts of chromatography

In a chromatographic process, a mobile phase moves along a stationary phase. Substances which can be carried along by the mobile phase, and retained by the stationary one, move along the chromatographic bed. They move forward rapidly if they prefer to stay in the mobile phase.

From the aspect of matter, the concept of a *phase* designates a homogeneous region being separated from other regions by a phase boundary. However, it can also mean a period or time interval. In the discussion of chromatographic facts, frequently both interpretations are reasonable.

3.1. Retention time, mobile phase hold-up time, and relative rate of migration

Whether substances travel ahead of or lag behind other ones in a chromatographic separation is not determined by their speed in the mobile phase, but only by the time they spend in the stationary phase. When they travel in the mobile phase they all move forward with the same velocity, i.e., with that of the eluent. The separation results from the fact that the different substances spend different periods of time in the stationary phase while the mobile phase is flowing past. The sum of all these dwelling times is the *net retention time* t'' (cf., the notation proposed by Bayer et al., 1968 and Ettre, 1981).

Substances which are not retained at all appear at the end of the separating path after a time t'. This *mobile phase hold-up time* t' is the time required by the eluent to flow through the chromatographic bed. The same time t' must also be spent in the mobile phase by those substances which are temporarily retained by the stationary phase. Their retention time is the sum of $t' + t''$. This *total retention time* depends, via t'', on the chromatographic behaviour of the substances and, via t', on external parameters such as the rate of elution and the length of the separation path.

In order to compare data obtained from different devices one defines the quotient

$$R = \frac{t'}{t' + t''} \tag{3-1}$$

which is called the *retention ratio*. It indicates the relative speed of a substance having the retention time t'' as a fraction of the flow rate of the eluent. An equivalent formulation is based on the distribution of the substance between the two phases

$$R = \frac{m'_S}{m'_S + m''_S} \tag{3-2}$$

where m'_S is the mass of the substance in the mobile phase and m''_S is the mass of the substance in the stationary phase. Both quotients yield the same value. R is zero for a substance occurring only in the stationary phase ($m'_S = 0$). Its net retention time, t'', is infinite. R approaches the upper limit 1 as t'' tends to zero. This is the case for substances which occur only in the mobile phase ($m''_S = 0$).

3.2. Distribution constants

The *distribution constant K* determines how a substance, on a time average, is distributed between the mobile and the stationary phase. The thermodynamic potential at equilibrium yields

$$K = \frac{a''_S}{a'_S} \tag{3-3}$$

where a''_S is the activity of the substance S in the stationary phase and a'_S is the activity of S in the mobile phase. Eqn. (3-3) indicates that the activity in the stationary phase is proportional to that in the mobile phase. The higher the value of K, the more strongly the substance is retained by the stationary phase, and the more slowly it travels forward.

The exact form of eqn. (3-3) will be easier to handle if the molar fractions, x, are substituted for the activities

$$a'_S \approx x'_S = \frac{n'_S}{n'_S + n'_E} \tag{3-4a}$$

$$a''_S \approx x''_S = \frac{n''_S}{n''_S + n''_E} \tag{3-4b}$$

where n_S is the number of moles of the substance and n_E is the number of moles of the solvent. If both phases are highly dilute, n_S can be neglected against n_E:

$$K = \frac{a''_S}{a'_S} \approx \frac{x''_S}{x'_S} \approx \frac{n''_S/n''_E}{n'_S/n'_E} = \frac{m''_S/n''_E}{m'_S/n'_E} \tag{3-5}$$

Also in this form K is called the *thermodynamic distribution constant* in chromatographic literature.

For polymers, the values of thermodynamic quantities are frequently expressed on a molar basis (see also the discussion of eqn. 4-10). This is irrelevant for the distribution constant because this dimension appears in the numerator as well as in the denominator of eqn. (3-3) to the same degree, and hence cancels. However, if a molar activity is stipulated, then also for polymers the distribution constant can be approximately represented as a quotient of volume fractions (cf. eqn. 9-2).

The *conventional* or *chromatographic distribution constant* is defined as the ratio

$$K^+ = \frac{c''_S}{c'_S} = \frac{m''_S/V''}{m'_S/V'} = \frac{n''_S/V''}{n'_S/V'} \tag{3-6}$$

where $c''_S = m''_S/V''$ is the concentration in the stationary phase; V'' is the volume of the stationary phase; $c'_S = m'_S/V'$ is the concentration in the mobile phase; and V' is the volume of the mobile phase. There is the following relationship between the conventional distribution constant, K^+, and the retention rate, R:

$$\frac{1-R}{R} = K^+(V''/V') \tag{3-7}$$

According to BATE-SMITH and WESTALL (1950), the term $\log \frac{1-R}{R}$ is denoted by R_M. Thus eqn. (3-7) can be rewritten as:

$$R_M = \log K^+ + \log(V''/V') \tag{3-8}$$

The R_M value increases with increasing distribution constant, whereas the retention rate, R, decreases.

The term $K^+(V''/V')$ is called the (*mass*) *distribution ratio* or *capacity factor* and denoted by the symbol k. The capacity factor indicates the ratio of the quantity of substance in the stationary phase to that in the mobile phase,

$$k = n''/n' = m''/m' \tag{3-9}$$

From eqns. (3-1) and (3-7) and the definition $t' + t'' = t_e$ one obtains the relationship

$$t_e = t'(1 + k) \tag{3-10}$$

This equation shows that the capacity factor can quite simply be determined from the experimental values of t_e and t'.

In dilute solutions, the volume of the mobile phase is proportional to the number of moles of the eluent. The more exactly this also holds for the stationary phase, the smaller will be the difference between K and K^+.

3.3. The formation of bands

Chromatographic techniques are continuous ones. The transport of the fluid phase and the exchange of substance between the mobile and the stationary phase take place simultaneously and continuously. Both processes require a driving force, a potential gradient. For the transport, the pressure differential occurring along the chromatographic bed provides this gradient, while the activity difference between the mobile and the stationary phases plays a corresponding role in the transfer of matter. At the leading edge of the sample band, the activity in the mobile phase exceeds that in the stationary one, while at the trailing edge the activity in the stationary phase is greater; the equilibrium of partition is not reached, and would only be so at the band centre.

The classical model of theoretical plates as developed by MARTIN and SYNGE (1941) does not take into consideration these deviations from equilibrium, but allows a simple mathematical description of chromatography. Here the equalization of concentrations and the transport are considered to be cycles proceeding one after another, and the balance of matter in the individual sections of the separation path is investigated. The model differs from reality insofar as it splits up the simultaneous, continuous processes into small steps occurring one after another. In this way it facilitates the understanding of what really happens. It must, however, be stressed that the contradiction between the model and reality will only be eliminated in the limit of an infinitely large number of infinitely small separating stages and infinitely short cycle times.

Consider the model of theoretical plates, assuming the simplest case: let the values of the partition coefficient and the phase ratio be such that at equilibrium in each separating stage the amount of substance, m', in the mobile phase is equal to the amount, m'', in the stationary phase. Then the process can be represented as shown by Fig. 3-1. In the starting condition all of the separating stages are free of sample. At the starting point the sample is fed into the first separating stage via the mobile phase. The sample contains a certain quantity of substance, $m'_{(1)}$. We could set $m'_{(1)} = 1$, but to avoid calculating with fractions, we set $m'_{(1)} = 2^{10} = 1024$. Initially, the stationary phase still does not contain any

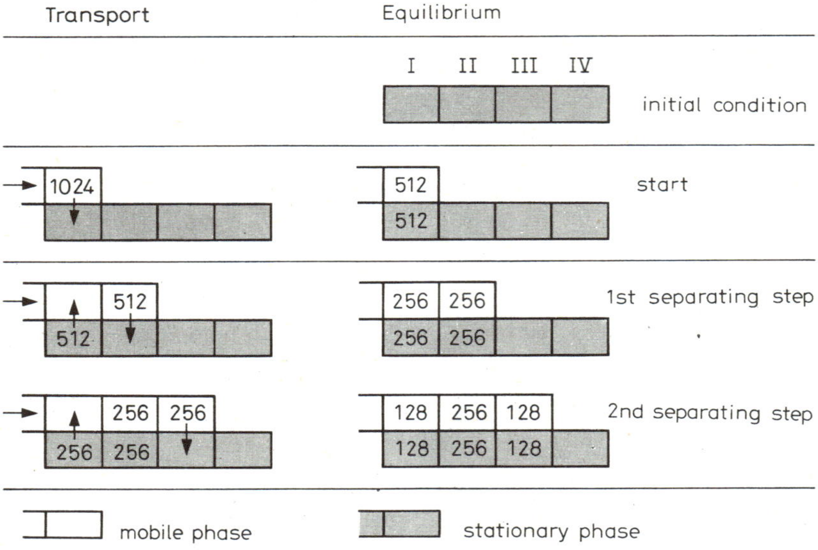

Fig. 3-1
Formation of a chromatographic band according to the model of theoretical plates, shown for the case where $V' = V''$ and $K^+ = 1$

The sample, containing $m' = 1024$ mass units, is in the mobile phase and is transported to the first separating stage after the analysis has been started. Under the chosen conditions the sample is uniformly divided among the stationary and the mobile phases of the first stage. The arrow points to the direction of the matter transfer.

1st separating step: The eluent displaces the mobile phase to the second separating stage. The amount of substance $m''_{(1)} = 512$ remains in the stationary phase of the first stage. After equilibrium has been established, half of this quantity each resides in the mobile and half in the stationary phase; thus $m'_{(1)} = 0$ and $m''_{(1)} = 512$ have changed into $m'_{(1)} = 256$ and $m''_{(1)} = 256$. From the first separating stage, the amount of substance $m''_{(2)} = 512$ has been transferred to the second stage with the mobile phase. After establishment of equilibrium in the 2nd stage, $m'_{(2)} = 512$ and $m''_{(2)} = 0$ have changed into $m'_{(2)} = m''_{(2)} = 256$ by the transfer of substance from the mobile phase into the stationary one. The total amount of all substance fractions, i.e., $m'_{(1)} + m''_{(1)} + m'_{(2)} + m''_{(2)}$, is equal to 1024. The higher separating stages are free of sample.

2nd separating step: All the fractions of the mobile phase are transported to the next stage. Pure eluent is fed into the first stage. After the equalization of concentration in the direction of the arrows, the first stage contains 128 + 128 units, the second stage 256 + 256 units and the third stage 128 + 128 units.

Table 3-1
Continuation of the separation shown by Fig. 3-1
The numbers indicate the total amounts of substance contained in each stage.

Separating step	Separating stage I	II	III	IV	V	VI	VII	VIII	IX	X	XI
Start	1024										
1st	512	512									
2nd	256	512	256								
3rd	128	384	384	128							
4th	64	256	384	256	64						
5th	32	160	320	320	160	32					
6th	16	96	240	320	240	96	16				
7th	8	56	168	280	280	168	56	8			
8th	4	32	112	224	280	224	112	32	4		
9th	2	18	72	168	252	252	168	72	18	2	
10th	1	10	45	120	210	252	210	120	45	10	1

substance, i.e., $m''_{(1)} = 0$. After the partition equilibrium has been reached, one has $m'_{(1)} = m''_{(1)} = 512$. All other separating stages are still free of substance.

The transport of the mobile phase creates the starting condition for the next separating step. Fig. 3-1 shows the first three steps. The subsequent procedure is indicated in Table 3-1. The figures indicate the total quantities contained in the respective stages. They are the binomial coefficients of the expression $(m'' + m')^n$.

Fig. 3-2 shows the stepped curve after $n = 5$ and after $n = 10$ steps, respectively (a and b). The bands become broader with increasing distance travelled. However, the latter increases more rapidly, so that the ratio of the band width to the distance travelled, called the *reduced band width*, decreases (Fig. 3-2c). This relative contraction allows the chromatographic separation of substances having different partition coefficients. As $n \to \infty$, the symmetrical binomial distribution approaches the normal distribution.

Fig. 3-2d shows the Gaussian curve resulting from the stepped curve for large numbers of steps.

In practice, already at the beginning of the analysis the sample naturally occupies a finite volume, which usually corresponds to a square pulse. Carberry and Bretton (1958) investigated mathematically the variation of such a pulse shape and found that for separation paths of normal lengths the sample attains a distribution corresponding to a normal distribution around the centre of the pulse after only a relatively short time of travelling. In chromatographic beds with only a modest resolution the width of the starting pulse has a minor effect on the width of this Gaussian curve.

For Gaussian curves the relationship between the migration distance and the band width can be described as follows: after the first stage, i.e., after a migration through a (sufficiently large) number of steps n, the band has the form of a bell-shaped curve with a *standard deviation* σ_n. This curve shows how the substance is distributed over the separating stages. For the amount of substance contained in each stage, the migration over a distance with n additional steps is now considered. The amount of substance again is spread over many neighbouring stages according to a Gaussian curve. All of the curves have the same

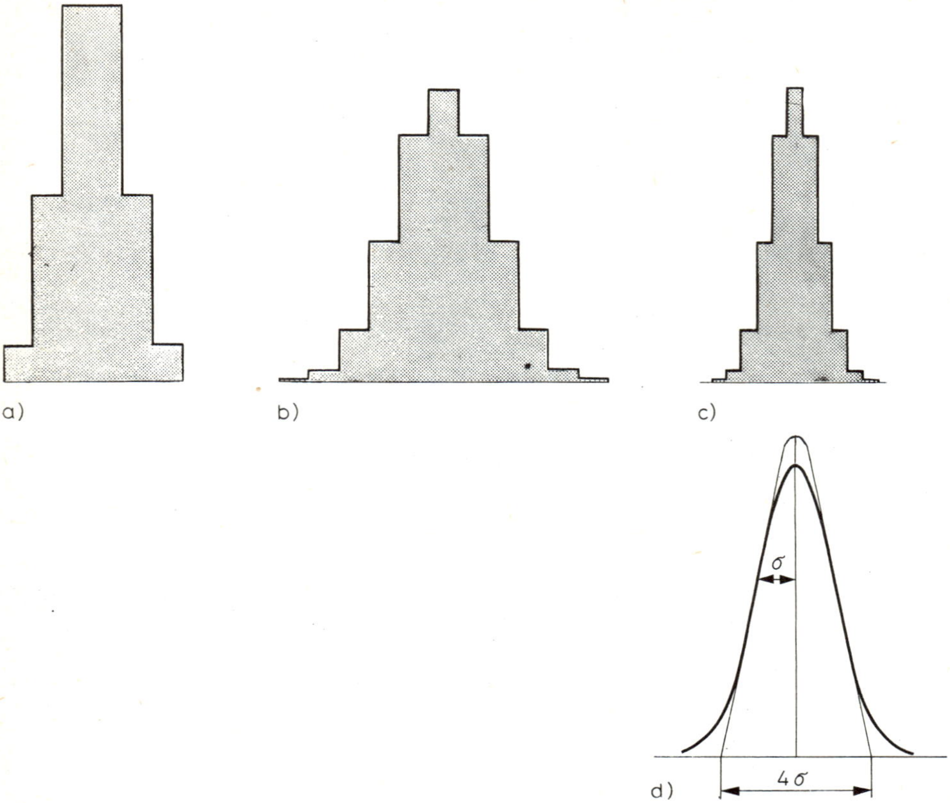

Fig. 3-2
Transition from the stepped curve to the Gaussian curve

a) Stepped curve after five separating steps
b) Stepped curve after ten separating steps
c) Stepped curve b, referred to the same distance travelled as for curve a; the reduced band width decreases with increasing number of steps
d) Curve of the normal distribution according to GAUSS (bell-shaped curve)

$$f(x) = \frac{1}{\sigma\sqrt{2\pi}}\, e^{-\frac{(x-x_0)^2}{2\sigma^2}}$$

which is approached by the binomial distribution for a large number of steps. The standard deviation is the half-width of the curve at the point of inflection. This point lies at an ordinate value of 60.7 if the vertex ordinate of the curve is set equal to 100. From the area A and the mode $f(0)$, the standard deviation can be calculated:

$$A = \sqrt{2\pi} \cdot \sigma \cdot f(0)$$

The tangents at the points of inflection intersect the baseline at a distance of $W = 4\sigma$ from each other. W is called the *peak width* at base (band intercept). The *variance* of the Gaussian distribution is the square of the standard deviation, i.e., σ^2.

standard deviation σ_n, because for each part of the sample the process fully corresponds to the first stage from zero to n. The sum of this large number of distribution curves yields the band of the total substance at the point $2n$ with the standard deviation σ_{2n}. As the chromatographic transport over the distance $0 \rightarrow n$ and that over $n \rightarrow 2n$ are independent of each other, the variance, σ_{2n}^2, at the point $2n$ can be calculated from the variances obtained after each distance:

$$\sigma_{2n}^2 = \underset{0 \rightarrow n}{\sigma_n^2} + \underset{n \rightarrow 2n}{\sigma_n^2} = 2\sigma_n^2 \tag{3-11}$$

From this it follows that:

$$\sigma_{2n} = \sigma_n \sqrt{2} \tag{3-12}$$

Doubling the distance of migration increases the band width by a factor of $\sqrt{2}$. A similar relationship also exists between the number of separating stages (or theoretical plates), N, and the standard deviation:

$$\sigma = \text{const} \sqrt{N} \tag{3-13}$$

However, since the *number of theoretical plates*, N, increases linearly with the distances travelled by the substance, it follows that:

$$\frac{s}{\sigma} = \frac{\text{const } N}{\text{const} \sqrt{N}} = \sqrt{N} \tag{3-14}$$

This is the quantitative expression for the decrease in the reduced band width as mentioned above.

The stochastic theory of chromatographic processes as developed by GIDDINGS and EYRING (1955), GIDDINGS (1957) and MCQUARRIE (1963) likewise leads to the result that the standard deviation for a single hand increases in proportion to the square root of the number of theoretical plates, $\sqrt{N}$, whereas the relative band width, σ/s, decreases with $1/\sqrt{N}$. According to this derivation the shape of the curve, however, slightly deviates from the Gaussian distribution. The elution curves obtained have a positive skew, i.e., a tail. KUBIN's theory (1965) leads to a similar result.

Eqn. (3-14) enables the number of theoretical plates to be determined from the standard deviation of a band, the centre of which has travelled through the distance s. The number of theoretical plates for the entire separating distance can be calculated using the ratio of s to the total length L.

When the band is determined in the eluate, its net retention time, t'', or its *retention volume*, V_r, determine its position in the chromatogram. In elution chromatography, these quantities correspond to the migration distance, s, in development chromatography. If the column is filled with eluent before the start of separation, each quantity of liquid fed into the column displaces an equal volume out of the column. When the *dead volume*, V' (volume of the mobile phase), has passed through the column, then the centre of the band has travelled a fraction, R, of the total column length, L. (R corresponds to the R_f value in thin-layer chromatography, being calculated from the latter by means of a correction factor.) The following relationships exist (see also Fig. 3-3):

$$R = \frac{s}{L} = \frac{t'}{t' + t''} = \frac{V'}{V' + V_r} = \frac{V'}{V_e} \tag{3-15}$$

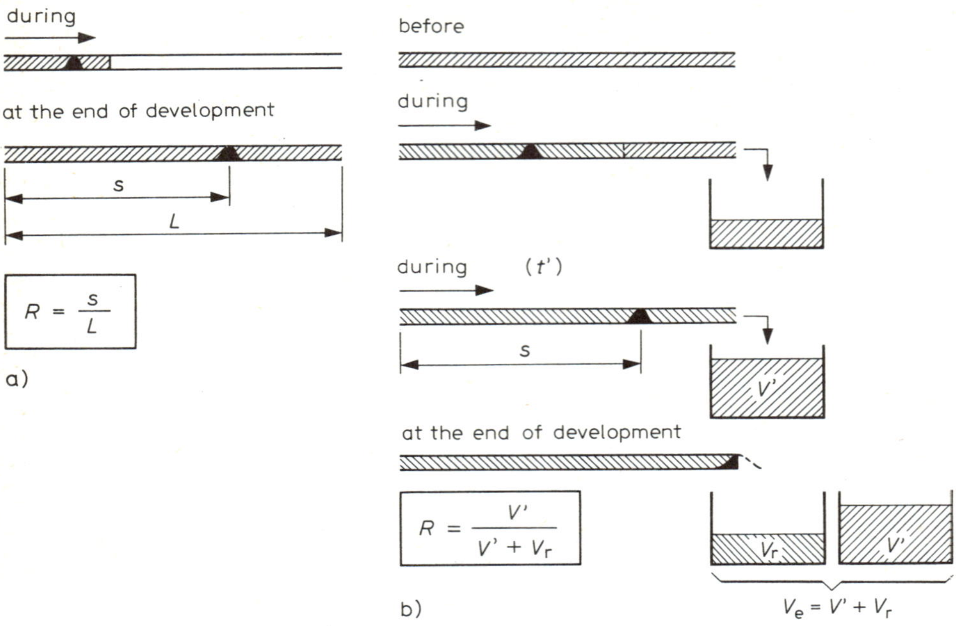

Fig. 3-3
Determination of the retention ratio

a) in development chromatography (dry bed technique)
b) in elution chromatography (wet bed technique).

When the front of the eluent, by means of which the sample has been fed, arrives at the end L of the separation path, the substance is at the point s in the interior of the column. The discharge volume, V', is the *eluent hold-up volume (dead volume)* of the column. The time required to arrive at this point is the dead time, t'. The time required from this point until the discharge of substance is the net retention time, t'', during which the *retention volume* V_r flows out of the column.

V_e is the *elution volume* or *break-through volume*, which is the total volume leaving the column until the outflow of the substance. V_r, the *net retention volume*, is the volume passed during the net retention time:

$$V_r = V_e - V' \qquad (3\text{-}16)$$

If, unlike the usual technique in elution chromatography, the substance is transported through a dry column, then a portion, V', will remain in the voids and the volume collected at the outlet is the net retention volume V_r. The problems encountered in the exact determination of V' will be dealt with in Sec. 16.7.

The *standard deviation*, σ, is the half-width of the chromatographic band at the inflection point (cf., Fig. 3-2). It can be calculated from the elution volume V_e at peak maximum and the elution volume $V_{c(y)}$ at the point y where the chromatographic band reaches 60.7% of its maximum intensity:

$$\sigma = V_{e(y)} - V_e \qquad (3\text{-}17)$$

(Here the slight additional widening of the band over this relatively short distance is neglected.)

The *number of theoretical plates* is found to be

$$N = \left(\frac{V_e}{\sigma}\right)^2 \tag{3-18}$$

If the peak width W is determined instead of σ (Fig. 3-2d), then the associated formula is

$$N = 16\left(\frac{V_e}{W}\right)^2 \tag{3-19}$$

This number of theoretical plates of a column of the length L/mm is inversely proportional to the *height h of a theoretical plate* (h or HETP):

$$h = L/N \tag{3-20}$$

Using this quantity h, the efficiencies of separating paths of different lengths can be compared.

The concepts of the number of theoretical plates and the height equivalent to a theoretical plate are derived from the technique of distillation. There the number of plates indicates how many distribution steps are required to separate a binary mixture. In contrast, in chromatography the number of theoretical plates indicates the width of the band of an individual substance after travelling through a certain distance.

To avoid any misunderstanding, let us once again stress that the theoretical plates are only intellectual devices in the description of the separating efficiency. The height equivalent to a theoretical plate is a kinetic quantity having the dimension of length, but by no means represents the size of an actual segment of the separation path. The latter is homogeneous, and not subdivided into cells, stages or plates.

Distillation columns having the same number of plates show the same separating efficiency for a certain mixture — this follows from the definition. In contrast, chromatographic columns having the same number of plates need not show the same separating power. Thus, the number of theoretical plates alone does not suffice to characterize the resolving power of a chromatographic technique.

3.4. Chromatographic resolution

A chromatographic separation is considered successful if successive bands are so distinctly separated from each other that the corresponding retention volumes and the peak areas can be determined without any difficulty. The smaller the peak width and the greater the distance between the peaks, the better this determination can be accomplished. The distance between the peaks of the two components I and II can be expressed as the difference between the elution volumes, V_e, or between the distances, s, travelled. The following quotient

$$R_s = \frac{V_{e,II} - V_{e,I}}{2\sigma_{II} + 2\sigma_I} = \frac{s_I - s_{II}}{2\sigma_I + 2\sigma_{II}} \tag{3-21}$$

characterizes the resolution of a chromatographic process. For $R_s \geqq 1$, adjacent bands are well separated. If the peak-to-peak distance is 1.5 times the peak width, a "6σ separation" exists. In this case the detector signal drops to the zero line between successive bands. The more the value of the quotient decreases below 1, the worse the resolution will be, especially if the concentration of the components varies. For incompletely resolved peaks having a Gaussian distribution, the resolution can also be measured by the ratio of the peak height to the valley depth, which represents a very sensitive measure of the degree of separation (CHRISTOPHER, 1971). For example, when $R_s = 1$ the valley depth is 25% of the peak height (SNYDER, 1970).

Eqn. (3-21) is the starting point for deriving the important equation controlling chromatographic resolution, eqn. (3-25), which offers a deeper understanding of chromatographic processes. This equation can be derived by means of the elution volumes or the distances travelled. For neighbouring bands, $\sigma_I = \sigma_{II}$. Substitution into eqn. (3-21) yields

$$R_s = \frac{V_{e,II}/V_{e,I} - 1}{4\sigma_I/V_{e,I}} = \frac{1}{4}\sqrt{N}\,(V_{e,II}/V_{e,I} - 1) \qquad (3\text{-}22)$$

The number of theoretical plates, N, introduced by means of eqn. (3-18), is the same for all of the neighbouring bands, so that the subscript can be omitted.

From eqns. (3-2, 5, 15) it follows that

$$V_e = V'\left(1 + K\frac{n''_E}{n'_E}\right) \qquad (3\text{-}23)$$

For a sufficient dilution, n''_E/n'_E indicates the relative amounts of the stationary and mobile phase. For this quotient we introduce the *phase ratio*, q:

$$V_e = V'(1 + Kq) \qquad (3\text{-}24)$$

The phase ratio always expresses the amount of the stationary phase compared to that of the mobile phase. For the different types of chromatography, the question of which information is most useful in each particular case can be answered in different ways. The introduction of the phase ratio makes it possible for the general eqn. (3-25) to be subsequently modified for special cases.

Using eqn. (3-24) we obtain the resolution:

$$\boxed{R_s = \underbrace{\frac{1}{4}\cdot\sqrt{N}}_{a)}\cdot\underbrace{\frac{K_{II} - K_I}{K_I}}_{b)}\cdot\underbrace{\frac{K_I}{K_I + 1/q}}_{c)}} \qquad (3\text{-}25)$$

This equation shows that the resolving power depends on three factors:

(a) the *column efficiency factor*, which increases with the root of the number of theoretical plates, and hence with the root of the length of the separating path,

(b) the *relative distribution factor*, which increases with the difference in the distribution constants, and

(c) the *retention factor*, which is determined by the phase ratio.

A chromatographic bed has a high retention factor if it can take up a large amount of substance. This depends on the quality and amount of the stationary phase. The retention factor increases with increasing q. Of course the phase ratio cannot vary arbitrarily. Therefore, the smaller is K_I the more the retention factor fall short of its limiting value 1. This *effect of too small distribution coefficients* necessarily leads to an unsatisfactory resolution. If the K values are too high there is also a detrimental effect. The sample remains in the stationary phase for a long time and hardly moves forward. The separation takes too long.

By means of extreme value investigations, SNYDER (1968) found that an optimum is reached at $K = 2/q$. This holds for elution chromatography if the optimization is accomplished by varying the pressure drop along the column and the length of the column while the flow rate is kept constant. For a constant length (i.e. pressure and flow rate being variable), the optimum lies between $3/q$ and $6/q$, whereas it is found to occur at (or closely below) $3/q$ if the pressure is kept constant (SNYDER et al., 1979).

The resolution can most efficiently be improved through the relative distribution factor. In the case of unsatisfactory separations, one can look for conditions which increase the differences between the distribution coefficients of the components. Increasing the length of the separation path will only be the last resort. Because of the square root term in the column efficiency factor, doubling the resolving power in this way requires the length of the separating path, and hence also the duration of separation, to be increased by a factor of 4.

3.5. Separation of multicomponent mixtures

In the preceding section it has been shown that an optimum resolution requires K values which are within rather narrow limits. The operating conditions have to be chosen so as to make the *distribution coefficients* fall into the most favourable range. The means of attaining this differ according to the various chromatographic techniques.

In gas chromatography on solid adsorbents it is possible to vary the stationary phase and/or the temperature. With increasing temperature adsorption decreases, i.e., the distribution coefficients decrease. Temperature has just the same effect in gas-liquid chromatography, where the choice of the liquid stationary phase offers a further possibility of adjusting the K values.

In liquid-liquid chromatography, both the stationary and the mobile liquid phases determine the partition coefficients of the substances to be separated. True, the requirement that the two liquids must not be miscible and, what is more, that their mutual solubilities should be negligible, restricts the choice to a very limited number of combinations. Temperature has quite a minor effect on the partition coefficients. Usually the K values of the different substances vary in the same direction, so that the separability is not improved.

Macromolecular substances are investigated mainly by precipitation chromatography or gel chromatography. In precipitation chromatography solvents, nonsolvents and the temperature have a considerable effect on the partition coefficients. In gel chromatography it is mainly the pore width which determines the distribution of the macromolecules between the stationary and the mobile phases. Temperature has almost no effect, whereas the quality of the solvent may have a limited effect.

In adsorption chromatography at a liquid/solid interface, K depends on the activity of the adsorbent, the temperature and the eluent. Chromatographically strong eluents suppress the adsorption of the sample components and lead to low K values, whereas in weak eluents the same substances have high distribution coefficients.

If a mixture of quite different substances has to be investigated, the conditions for optimum K values cannot be satisfied for all components at the same time. Fig. 3-4 illustrates this situation for a mixture of six components, the retention of which by the stationary phase increases from I to VI.

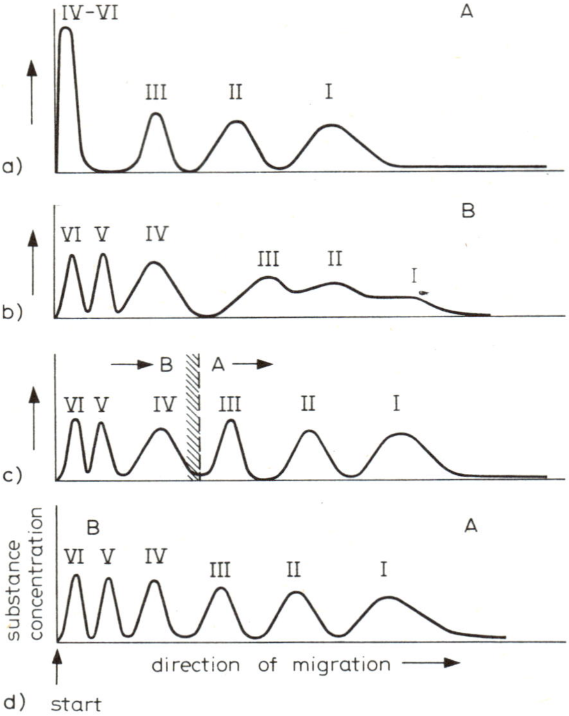

Fig. 3-4
Separation of a mixture of six components having different distribution coefficients

a) The weak eluent A develops the components I, II and III, but not the more strongly retained components IV—VI, which move far too slowly.
b) The strong eluent B separates the components IV—VI. At the start of the chromatogram the resolution is, however, unsatisfactory.
c) Replacement of the weak eluent A by the strong eluent B after the development of peaks I—III produces a good separation of all the components.
d) Gradient elution with a continuous change from A to B also produces a good separation.

A good separation within a certain range of distribution coefficients always implies a poor separation in other ranges. This is the *fundamental problem of chromatographic separation*, the general elution problem. Its effect increases with increasing differences between the individual components of the mixture. Special techniques are required to overcome this *general elution problem*. For example, the development can be started with a weak solvent and completed with a stronger one.

3.6. Non-linear concentration relationships

The statements made in Sections 3.2., 3.3. and 3.4. were based on the fact that the amount of substance in the stationary phase is proportional to that in the mobile phase and that the partition equilibrium is reached rapidly. In reality the situation may be different from this ideal behaviour. This section will deal with effects of *non-linear distribution isotherms*.

The partition coefficient is constant if the concentration of the substance in the stationary phase is proportional to that in the mobile phase. This is the case only in the lower range

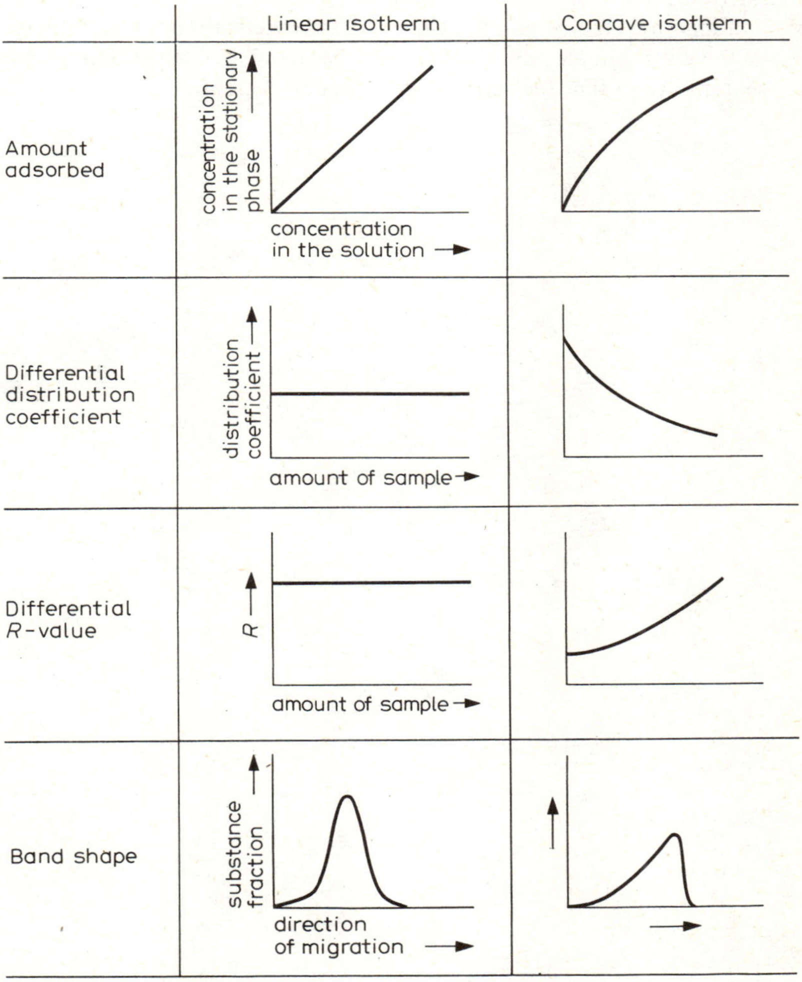

Fig. 3-5
Effect of the loading isotherms on the distribution coefficients, the R values and the band shapes

Non-linear isotherms yield concentration-dependent R values and skewed bands.

of concentration and might be expected in liquid-liquid partition chromatography. However, very often the amount of substance in the stationary phase increases less rapidly than that in the mobile phase. As a consequence of such a concave isotherm (see Fig. 3-5), the band becomes asymmetric. Starting from a relatively high initial value, for infinite dilution, the differential distribution coefficient decreases as the concentration of the mobile phase increases. This makes the R value concentration-dependent, i.e., the trailing edge of a travelling band undergoes a stronger retention by the stationary phase than the main part of the substance, which is moved through the chromatographic bed at the leading edge of the band. A long tail develops on the trailing edge. This effect may worsen to such a degree that successive bands are no longer separated. Difficulties of this kind must be expected in preparative separations.

For small loadings, the height equivalent to a theoretical plate (cf., eqn. (3-20)) is independent of the given sample amount. On the other hand, above the limit of linear retention h increases rapidly, so that the separating efficiency decreases.

4. Macromolecules: size, constitution, configuration, conformation

The molar mass of macromolecules is greater than 10000 g · mole^{-1}. Sometimes values of several million are found. In comparison with small molecules, macromolecules have huge structures, however, as in the former their atoms and atomic groups are held together by normal chemical bonds.

Since carbon plays a predominant rôle in macromolecular structures, in most cases the bonds in question are ordinary homopolar σ-bonds.

4.1. Molar mass and degree of polymerization

Macromolecules are built up of relatively simple units which are based on the monomer structures. The number of the monomer molecules which form a macromolecule by polymerization, polycondensation, etc., is called the degree of polymerization. Thus, a PVC molecule with a degree of polymerization of $P = 2000$ is built up of 2000 vinyl chloride molecules and could be represented by the symbol

$$\alpha - \left(CH_2 - \underset{\displaystyle Cl}{\underset{|}{C}}H - \right)_{2000} - \omega$$

Whether or not the end groups α or ω are mentioned depends, last but not least, on whether they play a demonstrable rôle in the context discussed. For many problems they can be neglected to a good approximation.

The following relationship exists between the molar mass, M, of the macromolecule, its degree of polymerization, P, and the molar mass, M_0, of the repeat unit (the end groups being neglected):

$$P \cdot M_0 = M \tag{4-1}$$

For macromolecules produced by polymerization or polyaddition, M_0 is equal to the molar mass of the monomer. Synthetic reactions with low-molecular-weight by-products (polycondensation, polyrecombination) lead to M_0 values which are smaller than the molar mass of the starting materials.

In most cases the numerical values of M_0 are about 100. Consequently the above limit of $M \geqq 10^4$ g · mole^{-1} yields $P \geqq 10^2$ for the degree of polymerization. Oligomers have a structure based on the same principle as that of polymers, but they contain only a small number of basic units.

4.2. Distribution of the degrees of polymerization

In many biopolymers the macromolecules have uniform size and structure. According to present knowledge this holds for proteins and nucleic acids among other things. On the other hand, synthetic polymers contain macromolecules of different sizes. They are polymer homologues if they differ from each other only in their degrees of polymerization.

4.2.1. Mean values

Naturally the degree of polymerization in a homologous series can only vary step by step (from 100 to 101, from 101 to 102, etc.). Calculations of mean values carried out on this basis lead to sum expressions such as those for the *number average*

$$\bar{P}_n = \frac{\Sigma\, n_i P_i}{\Sigma\, n_i} = \frac{\Sigma\, m_i}{M_0\, \Sigma\, (m_i/M_i)} \tag{4-2a}$$

$$\bar{M}_n = \frac{\Sigma\, n_i M_i}{\Sigma\, n_i} = \frac{\Sigma\, m_i}{\Sigma\, (m_i/M_i)} \tag{4-3a}$$

Number averages are determined by osmotic measurements, vapour pressure methods or end-group determinations and are of special importance for kinetic problems in polymer science. The corresponding expressions for *weight averages* (which are obtained, say, by scattered-light measurements) are

$$\bar{P}_w = \frac{\Sigma\, n_i P_i^2}{\Sigma\, n_i P_i} = \frac{\Sigma\, m_i \cdot M_i}{M_0\, \Sigma\, m_i} \tag{4-4a}$$

or

$$\bar{M}_w = \frac{\Sigma\, n_i M_i^2}{\Sigma\, n_i M_i} = \frac{\Sigma\, m_i \cdot M_i}{\Sigma\, m_i} \tag{4-5a}$$

In some cases the so-called z-average is additionally required (cf. Section 16.2):

$$\bar{P}_z = \frac{\Sigma\, n_i \cdot P_i^3}{\Sigma\, n_i \cdot P_i^2} = \frac{\Sigma\, m_i \cdot M_i^2}{M_0\, \Sigma\, m_i \cdot M_i} \tag{4-6a}$$

$$\bar{M}_z = \frac{\Sigma\, n_i \cdot M_i^3}{\Sigma\, n_i \cdot M_i^2} = \frac{\Sigma\, m_i \cdot M_i^2}{\Sigma\, m_i \cdot M_i} \tag{4-7a}$$

If a large number of summands are involved, such calculations are rather troublesome. Since in most cases the molar-mass distribution (MMD) is so broad that its graphical representation as a sequence of points practically converges into a smooth curve, and since for limits far enough from each other the EULER-MCLAURIN formula can be used to set, with a negligible error,

$$\sum_{x=x_1}^{x_2} x = \int_{x_1}^{x_2} x\, \mathrm{d}x \tag{4-8}$$

it is customary to treat the MMD as continuous distribution with the limits 0 and ∞. Then Eqns. (4-2a) through (4-7a) can be written as:

$$\bar{P}_n = \frac{\int_0^\infty P h(P)\, \mathrm{d}P}{\int_0^\infty h(P)\, \mathrm{d}P} \tag{4-2 b}$$

$$\bar{M}_n = \frac{\int_0^\infty M h(M)\, \mathrm{d}M}{\int_0^\infty h(M)\, \mathrm{d}M} \tag{4-3 b}$$

$$\bar{P}_w = \frac{\int_0^\infty P^2 h(P)\, \mathrm{d}P}{\int_0^\infty P h(P)\, \mathrm{d}P} \tag{4-4 b}$$

$$\bar{M}_w = \frac{\int_0^\infty M^2 h(M)\, \mathrm{d}M}{\int_0^\infty M h(M)\, \mathrm{d}M} \tag{4-5 b}$$

$$\bar{P}_z = \frac{\int_0^\infty P^3 h(P)\, \mathrm{d}P}{\int_0^\infty P^2 h(P)\, \mathrm{d}P} \tag{4-6 b}$$

$$\bar{M}_z = \frac{\int_0^\infty M^3 h(M)\, \mathrm{d}M}{\int_0^\infty M^2 h(M)\, \mathrm{d}M} \tag{4-7 b}$$

$\bar{M}_n$, $\bar{M}_w$ and $\bar{M}_z$ or data derived from them give first information about the width of the distribution.

In Chapter 16 we shall use the *heterogeneity* $H = \bar{M}_w/\bar{M}_n$. The *non-uniformity* coefficient, U (Uneinheitlichkeit according to SCHULZ) is $U = H - 1$.

Because of the relationship in eqn. (5-8), viscosity measurements yield a viscosimetric mean value:

$$\bar{M}_v = \left(\frac{\sum n_i M_i^{a+1}}{\sum n_i M_i}\right)^{1/a} = \left[\frac{\int_0^\infty M^{a+1} h(M)\, \mathrm{d}M}{\int_0^\infty M h(M)\, \mathrm{d}M}\right]^{1/a} \tag{4-9}$$

Normally ($a < 1$) the *viscosity average* defined in this way is somewhat smaller than the weight average (eqn. (4-5b)), but much greater than the number average. For $a = 1$ the viscosity average coincides with the weight average.

4.2.2. Frequency and mass distribution functions

The function $h(P)$ is the *frequency distribution function*. The integral $\int_0^\infty Ph(P)\,\mathrm{d}P$ indicates the total number of moles of the basic structural units in the system considered. The number of moles of the molecules having the degree of polymerization P is $h(P)$. It is customary to refer this quantity to a system obtained from one mole of monomer molecules. Then

$$\int_0^\infty Ph(P)\,\mathrm{d}P = 1 \tag{4-10}$$

and, with the use of Eqn. (4-2b), it follows that

$$\int_0^\infty h(P)\,\mathrm{d}P = \frac{1}{\bar{P}_n} \tag{4-11}$$

If, by analogy with statistics, the integral in eqn. (4-11) is normalized to 1, i.e., if $\int_0^\infty h_1(P)\,\mathrm{d}P = 1$, then $h_1(P)$ indicates the number of moles of the chains having the degree of polymerization P in a system containing $\bar{P}_n$ moles of repeat units. This makes the accuracy with which the numerical values of $h_1(P)$ can be stated depend on the accuracy of $\bar{P}_n$. For the mathematical treatment $h_1(P)$ offers advantages over $h(P)$, but in most cases the practical problems arising in the experimental determination of distribution functions can be more easily controlled by the normalization according to eqn. (4-10), which we shall also use.

The *mass distribution function* $H(P)$ of the degree of polymerization is defined as

$$H(P) = Ph(P) \tag{4-12}$$

By analogy:

$$H(M) = Ph(M) \tag{4-13}$$

The mass distribution function is normalized to 1, i.e., we can write

$$\int_0^\infty H(P)\,\mathrm{d}P = 1 \qquad \text{and} \tag{4-14}$$

$$\int_0^\infty H(M)\,\mathrm{d}M = 1 \tag{4-15}$$

Consequently 1 g of polymer contains $H(P)$ grams with the polymerization degree P, and $H(M)$ grams with the relative molar mass M.

The numerical values of $h(P)$ and $h(M)$ are not identical. If the argument P is replaced by the argument M, which is greater than P by a factor of M_0, the function value reduces correspondingly:

$$h(P) = M_0 \cdot h(M) \tag{4-16}$$

By analogy, the mass distribution function is

$$H(P) = M_0 \cdot H(M) \tag{4-17}$$

This situation can also be explained as follows:

The graphical representation of $H(P)$ vs. P yields a curve which, together with the abscissa, encloses an area of size 1, as is shown by eqn. (4-14). If H is plotted vs. M instead of vs. P, then the coefficients of measurement of the abscissa are greater by a factor M_0. To keep the area unchanged, the scale of the ordinate must be reduced correspondingly. The transformation $H(P) \rightarrow H(M)$ has to compensate for the change caused by the substitution $P \rightarrow M$ in the abscissa scale.

4.2.3. Determination of the distribution from fractionation data

Distribution functions can be experimentally determined by dividing the polymer to be investigated into fractions, the M_i (or P_i) of which, as well as the mass fractions m_i are to be estimated. To derive the distribution of the starting material from these data, first the mass fractions m_i are plotted as a cumulative bar graph vs. the respective M_i value. (The graph starts with the fraction having the smallest value of the molar mass.)

These mass bars make it possible to construct the curve of the *integral mass distribution function*

$$I(P) = \int_0^P H(P)\,\mathrm{d}P \tag{4-18}$$

or

$$I(M) = \int_0^M H(M)\,\mathrm{d}M \tag{4-19}$$

$I(P)$ and $I(M)$, respectively, indicate the fraction of a gram corresponding to the macromolecules with polymerization degrees between 0 and P (or molar masses between 0 and M) in 1 g of the starting material. (If the upper limit of integration approaches infinity, eqns. (4-18) and (4-19) tend to (4-14) and (4-15), respectively.)

The curve of the integral mass distribution function passes through the mass centres of the bars for the inner fractions. The lowermost and the uppermost mass bars, however, are only touched by the curve, because, for these marginal fractions, the difference between the actual distribution and that assumed in the construction of the integral curve is not compensated by overlapping. (As a matter of principle curves having several points of inflection should not be plotted until the latter have been unambiguously confirmed by multiple fractionations.) The upper part of Fig. 4-1 shows the bar graph and the integral curve, while the lower part shows the mass distribution curve, $H(M)$, obtained by a point-by-point differentiation, and the associated frequency distribution curve, $h(M)$, as calculated by eqn. (4-13).

4.2.4. Theoretical functions for the distribution of degrees of polymerization

Statistical derivations of the molar mass distribution of polycondensates have been carried out independently by SCHULZ (1935) and FLORY (1936). Starting from the probability p that a functional group, e.g., —COOH in polyesters or polyamides, reacts in a chain-

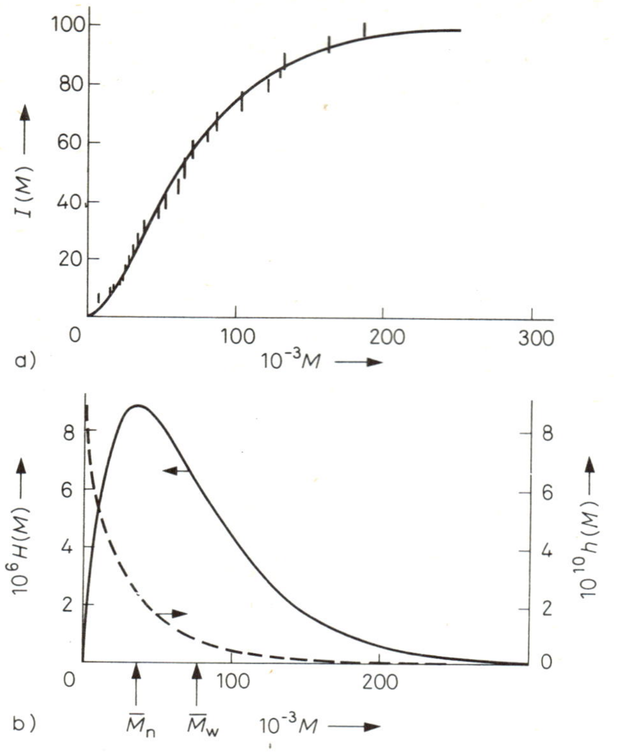

Fig. 4-1
Evaluation of experimental fractionation (polycarbonate in methylene chloride, methanol as a precipitant)

a) Bar diagram. The mass percentages of the fractions are plotted vs. the relative molar mass in a cumulative representation according to eqn. (4—19). The curve plotted is a Schulz distribution (eqn. (4-21)), with $k = 0.9$ and $c = 39000/M_0$.
b) Differential mass distribution (solid line) and frequency distribution (dashed curve, ordinate on the right).

forming bond, and the probability $1 - p$ that this group occurs as a free end group, equations for $h(P)$ and $H(P)$ were derived.

Analogous results were obtained for polymers in the case of chain termination by a disproportionation of free radicals. In this case every macromolecule stems from a macroradical with the same degree of polymerization, i.e., the distribution function of the "dead" polymer and that of the "living" chains are identical. These distributions can also be calculated on the basis of balance equations.

4.2.4.1. The generalized SCHULZ distribution

A substantial step forward was the investigation about the effect of coupling reactions on the molar mass distribution (SCHULZ (1939)).

An example of a coupling reaction is termination by recombination in free radical polymerization, where in each case two living macroradicals of the primary distribution combine into one particle of the final distribution. For the coupling degree k, the generalized Schulz distribution is

$$h(P) = \frac{1}{c^2\Gamma(k+1)}\left(\frac{P}{c}\right)^{k-1} \cdot \exp(-P/c) \tag{4-20}$$

$$H(P) = \frac{1}{c\Gamma(k+1)}\left(\frac{P}{c}\right)^{k} \cdot \exp(-P/c) \tag{4-21}$$

where c is a constant (cf. eqn. (4-26)).

These equations in the generalized form shown above, being properly derived on the basis of straightforward kinetic assumptions, give a surprisingly good representation of experimentally determined distributions (GLÖCKNER, 1964; TUNG, 1966c). Of course only integral values of k are reasonable in a physical sense: only two or three or four particles, etc., of the primary distribution can couple with each other. Accordingly SCHULZ formulated the relationships using the factorial $k!$, which is defined only for integral numbers k in $\Gamma(k+1) = k!$. The generalization achieved by passing from $k!$ to $\Gamma(k+1)$ extends the range of application of the equation enormously (ZIMM, 1948), but k can no longer be considered as a degree of coupling.

For $k = 1$, eqns. (4-20) and (4-21) yield expressions which can directly be derived for termination by disproportionation:

$$h(P) = c^{-2} \exp(-P/c) \tag{4-22}$$

$$H(P) = c^{-2} P \exp(-P/c) \tag{4-23}$$

The graphical representation of eqn. (4-22) gives an exponentially decreasing curve, whereas $H(P)$ vs. P, plotted according to eqn. (4-23), passes through a maximum. On the other hand, for $k > 1$ the frequency distribution shows a maximum as well.

Substitution of eqn. (4-20) for $h(P)$ in eqns. (4-2b) and (4-4b), respectively, gives:

$$\bar{P}_n = k \cdot c \tag{4-24}$$

$$\bar{P}_w = (k+1)\, c \tag{4-25}$$

For the parameters c and k this gives

$$c = \bar{P}_w - \bar{P}_n \tag{4-26}$$

$$k = \left(\frac{\bar{P}_w}{\bar{P}_n} - 1\right)^{-1} = U^{-1} \tag{4-27}$$

The parameter k, which was introduced as a degree of coupling in the derivation, turns out to be the reciprocal of non-uniformity U. Thus the meaning of the generalization becomes most obvious: the non-uniformity will only by chance coincide with the reciprocal of an integer. Intermediate values are much more probable. Moreover, for technical products the distribution is frequently so broad that $U > 1$. Using the generalized SCHULZ formula, such distributions can also be represented very well — but of course the interpretation of k as a degree of coupling has to be dispensed with.

4.2.4.2. Additional functions for the description of the chain length distribution

As an ideal case it can be imagined that a certain kinetic model will be reflected by the distribution of polymerization degrees of the product. A great variety of pertinent relationships have been derived, in which almost real conditions, such as transfer reactions, the decrease of the initiator concentration or the consumption of the monomer, have also appropriately been taken into account [B 4].

Frequently the actual distribution of a polymer is formed by the superposition of basic distributions which differ from one another to a greater or lesser extent. This certainly holds for technical products and, very likely, also for most of the lab-scale polymers. Heating-up periods, inhomogeneities in the reaction system, a step-by-step variation of the temperature, etc., cause a successive formation of polymers having different distributions. The final product is a mixture of these polymers. From this point of view it is justified to use distribution functions also without derivation from an assumed mechanism of polymerization, thus following the aim of achieving as close a fit as possible to the experimental results, so that missing data can be obtained by interpolation, and parameters of the molecular distribution can be rapidly determined.

A three-parameter function of the form

$$H(P) = \frac{m}{\frac{k+1}{cm}\,\Gamma\left(\frac{k+1}{m}\right)} P^k \cdot e^{-\frac{P^m}{c}} \tag{4-28}$$

performs very well in this respect (STOCKMAYER and MUUS, 1959; KUBIN, 1967).

A large number of experimental investigations have shown that frequently even functions having two parameters are sufficient to represent the results. Such functions can be considered special cases of eqn. (4-28). For example, when $m = 1$ we have the generalized Schulz distribution as given by eqn. (4-21). For $m = k + 1$ and $c = t^{k+1}$ one obtains (TUNG, 1956)

$$H(P) = \frac{k+1}{t}\left(\frac{P}{t}\right)^k e^{-(P/t)^{k+1}} \tag{4-29}$$

When $m = 2$ and $c = u^2$ one has the Maxwell distribution:

$$H(P) = \frac{2}{u\Gamma\left(\frac{k+1}{2}\right)}\left(\frac{P}{u}\right)^k e^{-(P/u)^2} \tag{4-30}$$

Another set of two-parameter distribution functions can be reduced to the logarithmic normal distribution (LANSING and KRAMER, 1935; WESSLAU, 1956),

$$H(P) = \frac{1}{\sigma_M\sqrt{2\pi}}\,\frac{1}{P}\exp\left[-\frac{(\ln P/P_0)^2}{2\sigma_M^2}\right] \tag{4-31}$$

When plotted on a cumulative probability paper with a logarithmic abscissa, this distribution gives straight lines. P_0 is the abscissa value corresponding to the ordinate value of 0.5, and σ_M is the slope. In Cartesian coordinates, for a logarithmic abscissa, eqn. (4-31) yields a bell-shaped curve. So this distribution bears a close relation to gel chromatography, where, due to the logarithmic calibration relationship, the elugrams are curves when plotted vs. lg M. P_0 is the degree of polymerization of the peak maximum. For number and weight averages we obtain the relationships:

$$\bar{P}_n = P_0 \exp(-\sigma_M^2/2) \quad (4\text{-}32)$$

$$\bar{P}_w = P_0 \exp(+\sigma_M^2/2) \quad (4\text{-}33)$$

$$\bar{P}_w/\bar{P}_n = U + 1 = \exp(\sigma_M^2) \quad (4\text{-}34)$$

The maximum lies at

$$P_0 = (\bar{P}_n \bar{P}_w)^{0,5} \quad (4\text{-}35)$$

For very high degrees of polymerization, experimentally determined distributions tend to zero more rapidly than the logarithmic normal distribution (KOTLIAR, 1964).

4.3. Constitution of the macromolecules

The constitution of chemical compounds depends on how the atoms in the molecules are bonded to one another. Differences in the constitution of macromolecules may already be caused by the structure of the primary structural units, e.g. $H_2C{=}CHO{-}CO{-}CH_3$ (vinyl acetate) and $H_2C{=}CH{-}CO{-}OCH_3$ (methyl acrylate). Such differences in constitution lead to differences in properties and behaviour, which are as obvious as those between ethanal and ethylene oxide (both C_2H_4O) or, more closely, those between methyl acetate and ethyl formate (both $C_3H_6O_2$). For the protein macromolecules of uniform compositions the investigation of constitution by sequential analysis has reached a very high standard.

Besides constitutional characteristics, the modification of which may lead to entirely different polymers, very slight deviations may also play an important rôle. To a first approximation they appear only as a fault in the perfect structure. They may however be the causes of quality fluctuations or differences between products which seem unexplainable at first sight. Some of these characteristics will be explained below.

Chain structure of the polymers. In vinyl polymers, the majority of the basic molecules are linked together in a regular 1,2-pattern ("head-to-tail structure"), but a few of them are inserted in the reverse order. The fraction of the 1,1-additions depends on the substituent and increases with increasing polymerization temperature.

Diene polymers such as polybutadiene or polyisoprene contain the monomer units linked in a 1,2- or 1,4-pattern. Both constitutions still contain 50% of the double bonds (Fig. 4-2a, b). Frequently 1,2- and 1,4-additions simultaneously occur in one and the same molecule. Dienes with isolated double bonds can form cyclopolymers, which no longer contain any reactive double bonds (Fig. 4-2c).

Chain branchings. Truly linear macromolecules are rare in polymers, because in the process of formation transfer reactions also take place, which lead to branched structures. Long-chain branchings are produced by intermolecular chain transfer reactions which increase with increasing polymer concentration, i.e., with the degree of conversion. Short side chains containing from three to five carbon atoms result from intramolecular transfer reactions, which to a first approximation are independent of the degree of conversion. Branched molecules have a more compact structure and a smaller hydrodynamic volume

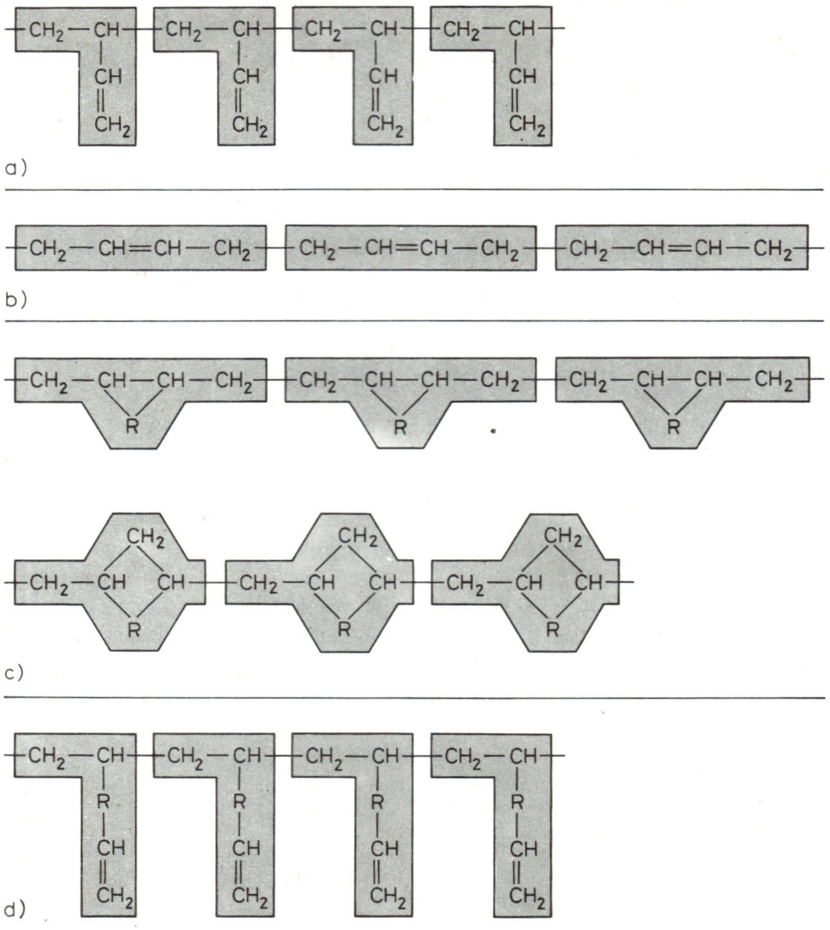

Fig. 4-2
Constitutional isomerism for polydienes
a) polybuta-1,2-diene;
b) polybuta-1,4-diene;
c) cyclopolymers of a diene with isolated double bonds;
d) 1,2-polymer derived from the same diene as a starting material.

than linear coiled molecules. Among other things, this has consequences in gel chromatography.

End groups and chemically modified chain units. Generally the end groups can be neglected because they have little influence. However, the degree of branching can be determined from the number of end groups per macromolecule. Moreover there are problems, e.g., in connection with graft reactions, where it has to be clarified whether or not the existing chains have a certain end group. If the end group in question is a relatively polar group added to non-polar molecules, then this question can readily be answered by means of adsorption chromatography. Chemical modifications, e.g., by oxidation, which affect only a few structural units in macromolecules, can be detected from the varying adsorption behaviour.

4.4. Configuration of the macromolecules

Configuration isomers have the same constitution but different steric structures. They can be converted into one another by cleavage of bonds and formation of new valence bonds. Examples of low-molecular-weight configuration isomers are maleic acid and fumaric acid or aldoses such as glucose, mannose and galactose. In macromolecules the cis-trans isomerism of the first example and the position isomerism of the second example may occur. Poly-1,4-dienes have a cis- or trans-configuration with respect to the double bonds in the chain. Natural rubber is 1,4-cis-polyisoprene, whereas 1,4-trans-polyisoprene forms the thermoplastic guttapercha. Special polymerization techniques also make it possible to prepare the isomers *artificially* in an almost pure form.

Position isomerism is of importance for polymers with two different substituents. In vinyl polymers the substituents in question are the side group R originating from the monomer molecule, and the hydrogen atom bonded to the same carbon atom. Also for a regular 1,2-constitution there are several possibilities for the sterical position of the substituents, which are shown schematically in Fig. 4-3. Every hooklet symbolizes a structural unit, e.g.,

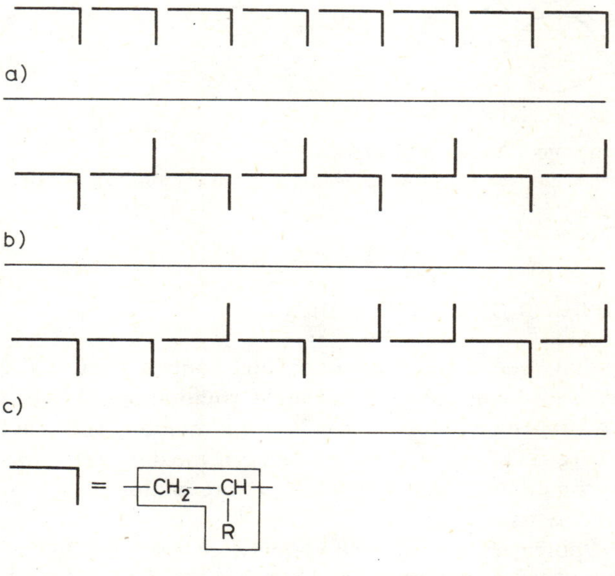

Fig. 4-3
Differences in configuration of vinyl polymers

The hooklets symbolize monomeric units in 1,2-addition (see Fig. (4-2a)). The carbon backbone is straightened, extending in a plane normal to the plane of the drawing.
a) Isotactic configuration
b) Syndiotactic configuration
c) Atactic configuration

$-CH_2-\underset{\displaystyle Cl}{\underset{|}{C}}H$ in PVC. The H atoms are not indicated. The carbon atoms of the straightened backbone form a zigzag sequence in a plane perpendicular to the plane of the drawing. Ordinary polymers have an approximately atactic structure, with a slight excess of syndiotactic diads. Stereoregular isotactic or syndiotactic polymers can be prepared in an almost pure form by special syntheses.

While for vinyl polymers the type of taxis simply results from the alternating or non-alternating position of the substituents, a more precise definition must be used if the structural units contribute an odd number of atoms to the structure of the backbone. HUGGINS et al. (1962) found quite an amusing formulation: In an isotactic polymer, a "hypothetical observer, advancing along the bonds constituting the main chain, finds each of the unsymmetrical carbon atoms with all of its substituents arranged in the same steric order". In a syndiotactic polymer, in each of the successive (conventional) structural units this arrangement is opposite to that in the preceding unit (see Fig. 4-4).

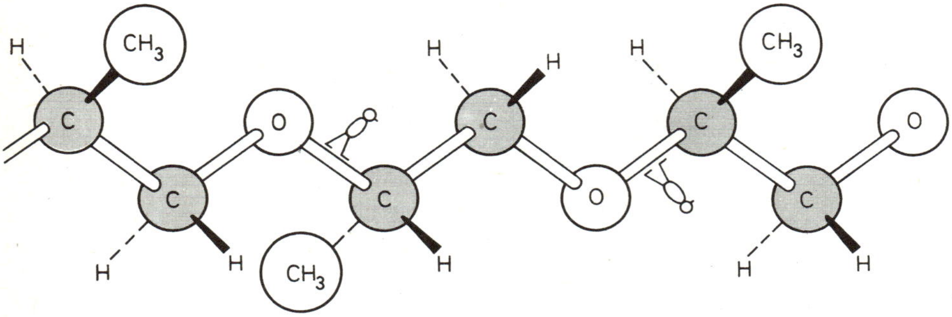

Fig. 4-4
Isotactic polypropylene oxide $[CH(CH_3)CH_2O]_P$
The "hypothetical observer" walks from an "oxygen peak" to a "carbon valley", and the methyl group is on his left.

4.5. Conformation of the molecules

Conformation isomers have the same constitution and configuration and can be interconverted without any bonds being split. For example, rotation of a CH_2Cl group in 1,2-dichloroethane yields two mirror-symmetric gauche conformations and one trans conformation having lower energy. The energy barriers between the different conformations are low enough to be overcome by thermal energy. Therefore conformation isomers cannot usually be isolated.

In syndiotactic vinyl polymers the interactions between the substituents additionally favour the lower-energy all-trans conformation. Therefore the chains of syndiotactic polymers have a planar zigzag structure in ordered regions. In isotactic polymers the substituents favour other microconformations, e.g., the regular alternation of gauche and trans, which leads to a helix arrangement.

In solutions, the trans/gauche ratio is influenced by the medium. Good solvents produce higher proportions of the trans isomer, and hence a further loosening of the coils.

For a mathematical description of the randomly coiled arrangement, a macromolecule of polymerization degree P can be replaced by a model consisting of Z segments of length b_0. This segment model has to satisfy the following conditions:

- Its straightened length $Z \cdot b_0$ must be equal to the length of the straightened molecule, $P \cdot l_{eff}$. For vinyl polymers the effective length of a monomer unit is $l_{eff} = 2.52 \times 10^{-8}$ cm $= 0.252$ nm. This results from the valence angle of the C—C chain, $\vartheta = 109.5°$, and the bond length $l_0 = 0.154$ nm.
- The end points of the model and the coiled molecule are identical.
- The angles between successive segments may assume any value between 0 and 180°. All of the values have the same probability.

To such a model the statistics of random flight can be applied. For the root mean square of the end-to-end distance R, this yields

$$\langle R^2 \rangle = Z \cdot b_0^2 \tag{4-36}$$

With the assumption of free rotation about all of the bonds of the backbone, the ideal value of the chain end-to-end distance can be calculated from the bond length l_0, the number $2P$ of such bonds in the backbone and a function of the valence angle ϑ:

$$\langle R^2 \rangle_{0,f} = 2P \cdot l_0^2 \cdot \frac{1 - \cos \vartheta}{1 + \cos \vartheta} \tag{4-37}$$

Substitution of the numerical values for l_0 and ϑ into this relationship yields

$$\langle R^2 \rangle_{0,f}^{0.5} = (3.08 \cdot 10^{-8})\, P^{0.5} \text{ cm} \tag{4-38}$$

$\langle R^2 \rangle_{0,f}$ is smaller than $\langle R^2 \rangle$. This already results from the assumption of free rotation, which of course is not valid. Even hydrogen as a substituent forms certain energy barriers, which together with the above-mentioned trans preference make the *unperturbed dimensions*, $\langle R^2 \rangle_0^{0.5}$, of polyethylene 1.6 times greater than the value calculated by eqn. (4-38). The *hindrance parameter*, represents a quantitative measure of this effect:

$$\sigma^2 = \langle R^2 \rangle_0 / \langle R^2 \rangle_{0,f} \tag{4-39}$$

For vinyl polymers, σ is approximately proportional to the molar volume of the substituent (Kuwahara et al., 1966).

Since polymers cannot be evaporated without being decomposed, the dimensions of macromolecules can be determined only in solution. Here the θ state is of special interest, because in this state the coils gain their unperturbed dimensions. In good solvents they are additionally expanded by solvation and an increase of the trans fraction. The expansion coefficient

$$\alpha = (\langle R^2 \rangle / \langle R^2 \rangle_0)^{0.5} \tag{4-40}$$

increases with increasing degree of polymerization and, besides σ, contributes to the fact that $\langle R^2 \rangle > \langle R^2 \rangle_{0,f}$.

Besides the chain end-to-end distance, R, the radius of gyration, S, is also suitable for the description of molecular dimensions. For the θ state we obtain

$$\langle R^2 \rangle_0 = 6 \langle S^2 \rangle_0 \tag{4-41}$$

4.6. Associates

Especially in the neighbourhood of the θ point, the interactions between the macromolecules are so strong, as compared to those between solvent molecules and macromolecules, that they prevent the occurrence of genuine, molecular-disperse solutions. The solutions contain associated particles consisting of several macromolecules; these associates are decomposed by an increase in temperature or dilution. Association phenomena interfere with many measurements because they make the results depend, at least partially, on supermolecular assemblies which change with the external conditions. This effect may become so drastic that even the age and the history of the solutions influence the results.

5. Interactions between polymers and solvents

A substance dissolves in another one if the chemical potential of the mixture is lower than that of the starting system. Since generally the entropy increases upon dissolution, this process will take place if it is not opposed by energetic interactions. This may be the case with non-electrolytes if the cohesion energy of the pure substances is much greater than that of the mixture. Dissolution will occur if the interactions between all of the substances involved are approximately equal. This is most likely to be the case for molecules having similar structures. The old rule of solubility, "similia similibis solventur" (similar substances are dissolved by similar ones) expresses pertinent experience. It also holds for polymers, illustrating in a most simple way why, say, polyvinyl alcohol dissolves in water, polystyrene in aromatic hydrocarbons and polyisobutene in aliphatic hydrocarbons. For polymer solutions the requirement for similarity is indeed even more stringent than for low-molecular-weight substances, because the entropy of mixing is proportional to the number of moles, and hence it is necessarily relatively low for macromolecular systems. On the other hand, the enthalpy is referred to the number of segments, i.e., it is largely independent of the molecular size. Thus the solubility of polymers depends on the enthalpy conditions much more than that of low-molecular-weight substances. (This is most distinctive of polymer mixtures. Here the *incompatibility* is more or less the rule, whereas the corresponding low-molecular-weight substances are miscible.)

5.1. Solubility parameters

According to HILDEBRAND [B 1] the enthalpy of mixing is given by

$$\Delta H_M = V_M \left[\left(\frac{\Delta E_I}{V_I} \right)^{0.5} - \left(\frac{\Delta E_{II}}{V_{II}} \right)^{0.5} \right]^2 \varphi_I \varphi_{II} \tag{5-1}$$

where V_M is the total volume of the mixture, $V_{I,\,II}$ are the molar volumes of the components, $\Delta E_{I,\,II}$ are the values of the internal heat of evaporation and $\varphi_{I,\,II}$ are the volume fractions of I and II, respectively. The *solubility parameters* are given by the following expressions in eqn. (5-2):

$$\delta_I = \left(\frac{\Delta E_I}{V_I} \right)^{0.5} \qquad \delta_{II} = \left(\frac{\Delta E_{II}}{V_{II}} \right)^{0.5} \tag{5-2}$$

The enthalpy of mixing, ΔH_M, is lowest for $\delta_I = \delta_{II}$. This defines the optimum solubility, since the entropy contributes directly to the variation of the chemical potential.

The solubility parameters of volatile substances can be directly determined from the enthalpy of evaporation and the molar volume. For polymers they can be indirectly determined (e.g. from the interfacial tension; cf. BONN and VAN AARTSEN, 1972) or calculated from the increments, G, by means of a relationship given by SMALL (1953)

$$\delta = \varrho \, \Sigma \, G/M_0 \tag{5-3}$$

where ϱ is the density and M_0 is the molar mass of the repeat unit. The increments G are listed in Table 5-1 for several structural units.

Table 5-1
Increments G for calculation of the solubility parameters according to eqn. (5-3)
The solubility parameters are obtained in Hildebrand units.

Structural element	G	Structural element		G
$-CH_2-$	133	$>CO$	(ketones)	275
$-\overset{\vert}{C}H-$	28	$-COO-$	(esters)	310
$-\overset{\vert}{\underset{\vert}{C}}-$	−93	$-O-$	(ethers)	70
$-CH_3$	214	H	(variable)	80 ... 100
$CH_2=$	190	$-CN$		410
$-CH=$	111	$-Cl$		270
$>C=$	19	$-CCl_2-$		427
6-membered ring	95 ... 105	$-CCl_3$		657
5-membered ring	105 ... 115	$-CF_2-$		150
Conjugation	20 ... 30	$-CF_3$		274
Phenyl-	735	$-S-$		225
Phenylene- (o-, m-, p-)	658	$-SH$		315
		$-\overset{\vert}{\underset{\vert}{Si}}-$	(in silicones)	−38

For copolymers and mixed solvents the solubility parameters are determined from the values δ_A and δ_B for the pure components,

$$\delta = \varphi_A \cdot \delta_A + \varphi_B \cdot \delta_B \tag{5-4}$$

The original concept of solubility parameters was restricted to dispersion interactions between non-polar liquids. Naturally the quotient of the internal heat of evaporation and the molar volume can also be established for polar liquids. In the further development of the original concept, this "total solubility parameter" was decomposed into a polar contribution and a dispersion contribution, which can be determined according to the principle of homomorphism (BLANKS and PRAUSNITZ, 1964). Here it is assumed that the contribution of dispersion to the interactions between polar molecules corresponds to the interactions between non-polar molecules having analogous structures and the same size. Thus, for instance, toluene is homomorphic to chlorobenzene. BLANKS and PRAUSNITZ (1964), HANSEN (1965, 1967, 1969), HOY (1969) and others established tables in which the polar as well as the nonpolar contributions to the δ values are listed. Using the Lorentz-Lorenz equation, KELLER et al. (1970), TIJSSEN et al. (1976), and KARGER et al. (1976) calculated the dispersion contribution from the refractive index by means of the polarizability of the molecules. Then they further subdivided the polar contribution by means of the dipole moment and of physical data, for the proton donor (δ_a) acceptor (δ_b) interactions. Table 5-2 contains data of this kind as given by KARGER et al., which are interrelated by the equation:

$$\delta_t^2 = \delta_d^2 + 2\delta_{in}\delta_d + \delta_0^2 + 2\delta_a\delta_b \qquad (5\text{-}5)$$

The *Hildebrand units*, in which the numerical values of the solubility parameters are stated, are the values obtained by inserting in eqn. (5-2) the internal heat of evaporation in cal (1 cal = 4.19 J), and the molar volume, in cm^3.

The concept of solubility parameters is helpful in finding potential solvents rapidly and at a low expense, and in understanding many facts of the solubility behaviour, but it cannot explain all the phenomena and occasionally fails in quantitative conclusions.

Table 5-2
Partial and total solubility parameters
Data given in Hildebrand units, according to KARGER et al. (1976). Conversion into $(J \cdot cm^{-3})^{0.5}$ by factor 2.05. Example: n-pentane, δ_t = 7.1 $(cal \cdot cm^{-3})^{0.5}$ or 14.5 $(J \cdot cm^{-3})^{0.5}$.

Solvent	δ_t	δ_d	δ_0	δ_{in}	δ_a	δ_b	V
n-pentane	7.1	7.1	—	—	—	—	115
n-hexane	7.3	7.3	—	—	—	—	131
Diisopropyl ether	7.1	6.9	1.0	0.1	—	3.0	102
Diethyl ether	7.5	6.7	2.4	0.5	—	3.0	105
Triethylamine	7.5	7.5	—	—	—	4.5	140
Cyclohexane	8.2	8.2	—	—	—	—	108
Propyl chloride	8.4	7.3	2.9	0.6	—	0.7	88
Carbon tetrachloride	8.6	8.6	—	—	—	0.5	97
Diethyl sulphide	8.6	8.2	1.7	0.25	—	2.6	108
Ethyl acetate	8.9	7.0	4.0	1.0	—	2.7	98
Propylamine	8.9	7.3	1.7	0.2	1.8	5.5	82
Ethyl bromide	8.9	7.8	3.1	0.6	—	0.8	77
Toluene	8.9	8.9	—	—	—	0.6	107
Tetrahydrofuran	9.1	7.6	3.5	0.8	—	3.7	82
Benzene	9.2	9.2	—	—	—	0.6	89
Chloroform	9.3	8.1	3.0	0.5	6.5	0.5	81
Ethyl methyl ketone	9.5	7.1	4.7	1.2	—	3.2	90
Acetone	9.6	6.8	5.1	1.5	—	3.0	74
1,2-dichloroethane	9.7	8.2	4.2	0.5	—	0.7	79
Anisole	9.7	9.1	2.1	0.4	—	1.7	109
Chlorobenzene	9.7	9.2	1.9	0.3	—	1.0	102
Bromobenzene	9.9	9.6	1.5	0.2	—	1.0	105
Methyl iodide	9.9	9.3	2.5	0.3	—	0.7	62
Dioxane	10.1	7.8	5.2	1.0	—	4.6	86
Hexamethylphosphoramide	10.5	8.4	3.4	1.7	—	4.0	176
Pyridine	10.6	9.0	3.8	1.0	—	4.9	81
Acetophenone	10.6	9.6	2.7	0.7	—	3.3	117
Benzonitrile	10.7	9.2	3.4	1.0	—	2.3	103
Propionitrile	10.8	6.9	6.6	1.8	—	2.1	71
Quinoline	10.8	10.3	1.8	0.3	—	4.2	118

Table 5-2 (continued)

Solvent	δ_t	δ_d	δ_0	δ_{in}	δ_a	δ_b	V
N,N-dimethylacetamide	10.8	8.2	4.7	1.6	—	4.5	92
Nitroethane	11.0	7.3	6.0	2.2	—	1.0	71
Nitrobenzene	11.0	9.5	3.6	1.1	—	1.0	103
Tricresyl phosphate	11.3	9.6	2.5	1.5	—	(?)	316
Dimethylformamide	11.8	7.9	6.2	2.4	—	4.6	77
Propanol	12.0	7.2	2.6	0.4	6.3	6.3	75
Dimethyl sulphoxide	12.0	8.4	6.1	2.1	—	5.2	71
Acetonitrile	12.1	6.5	8.2	2.8	—	3.8	53
Phenol	12.1	9.5	2.3	0.4	9.3	2.3	92
Ethanol	12.7	6.8	3.4	0.5	6.9	6.9	59
Nitromethane	12.9	7.3	8.3	3.0	—	1.2	54
γ-butyrolactone	12.9	8.0	7.2	3.2	—	(?)	77
Propylene carbonate	13.3	9.8	5.9	2.4	—	(?)	85
Diethylene glycol	14.3	8.2	4.0	0.6	5.3	5.3	96
Methanol	14.5	6.2	4.9	0.8	8.3	8.3	41
Ethylene glycol	17.0	8.0	6.8	1.1	6.1	6.1	56
Formamide	19.2	8.3	(?)	(?)	large	large	40
Water	23.4	6.3	(?)	(?)	large	large	18
Alumina[1])	16.4	10.8	9.8	—	11.4	2.5	—

δ_t = total solubility parameter from vaporization energy; δ_d = dispersion solubility parameter; δ_0 = orientation solubility parameter; δ_{in} = induction solubility parameter; δ_a = proton-donor solubility parameter; δ_b = proton-acceptor solubility parameter; V = molar volume (in ml · mole^{-1})

[1]) KARGER et al. (1978)

5.2. Thermodynamic quality of a solvent

The thermodynamic quality of a solvent with respect to a given polymer determines the value of the second virial coefficient, A_2 (SCHULZ and <OLL, 1952). The *virial coefficients* describe the deviations from the relationhips holding for ideal, dilute mixtures. The ideal equation for the osmotic pressure, π, of a solution of concentration, c

$$\pi = \frac{\boldsymbol{R} \cdot T}{M} \cdot c \tag{5-6}$$

is extended by the introduction of virial coefficients to become:

$$\pi = \boldsymbol{R}T\left(\frac{1}{M}c + A_2c^2 + A_3c^3 + \dots\right) \tag{5-7}$$

In most cases the series can be terminated as early as after the second term. To what extent this is justified is shown by the graphical representation of the reduced osmotic pressure, π/c, vs. the concentration (see Fig. 5-1): if the higher virial coefficients (from A_3) can be neglected, one obtains straight lines. The steeper the slope of these lines, the higher A_2, and the better is the solvent.

Eqn. (5-7) shows that the A_2 term depends on the square of the polymer concentration. It represents that contribution to the measured quantity which is caused by the contact of two

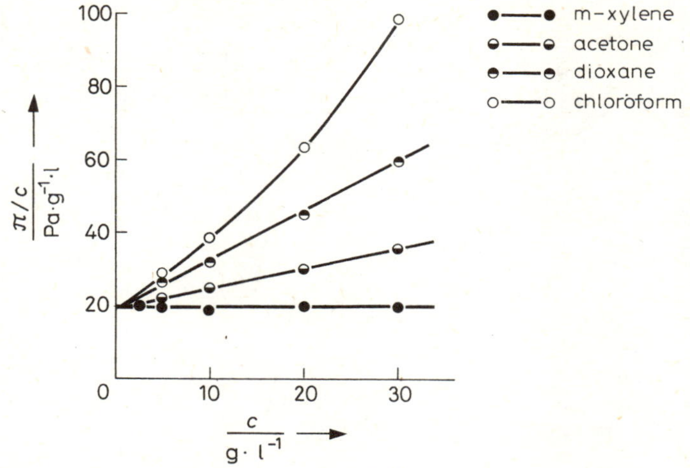

Fig. 5-1
Reduced osmotic pressure vs. concentration
Polymethyl methacrylate (M_n = 128000 g · mole^{-1}) in different solvents; temperature: 20 °C (SCHULZ and DOLL, 1952).

dissolved molecules. For a constant degree of polymerization and in solvents of constant concentration, such binary contacts increase in frequency with increasing expansion of the coils, i.e., with increasing intensity of the interactions between the solute and the solvent. This leads to the relationship between A_2 and the solvent quality.

In a representation such as that in Fig. 5-1, measurements in poor solvents yield curves with a gentle slope. In the limiting case they are parallel to the abscissa. For these pseudo-ideal solutions, again eqn. (5-6) holds, i.e., the virial coefficients are eliminated. For each solvent this limiting case is associated with a certain temperature, which (for the degree of polymerization approaching infinity) is called the *θ temperature* or *Flory temperature*. As the temperature is increased the system passes from the θ state into that of an ordinary solution, with $A_2 > 0$. As the temperature is decreased, phase separation occurs.

The coil expansion can also be used as a measure of the solvent quality. For macromolecules of equal molar mass the expansion coefficient, α, as given by eqn. (4-40) increases with improving quality of the solvent. In the θ state, $\alpha = 1$. Expansion coefficients can be determined by viscosity measurements or from the angular dependence of scattered-light data. Usually the amount of work required exceeds that for the determination of A_2. (Virial coefficients are automatically obtained in the determination of absolute molar masses by light scattering or osmosis.)

The quality of the solvent can be further observed from the exponent a in the relationship between the intrinsic viscosity, $[\eta]$, and the molar mass

$$[\eta] = K_\eta \cdot M^a \tag{5-8}$$

$$[\eta] = \lim_{\substack{c \to 0 \\ \tau \to 0}} (\eta_{sp}/c)$$

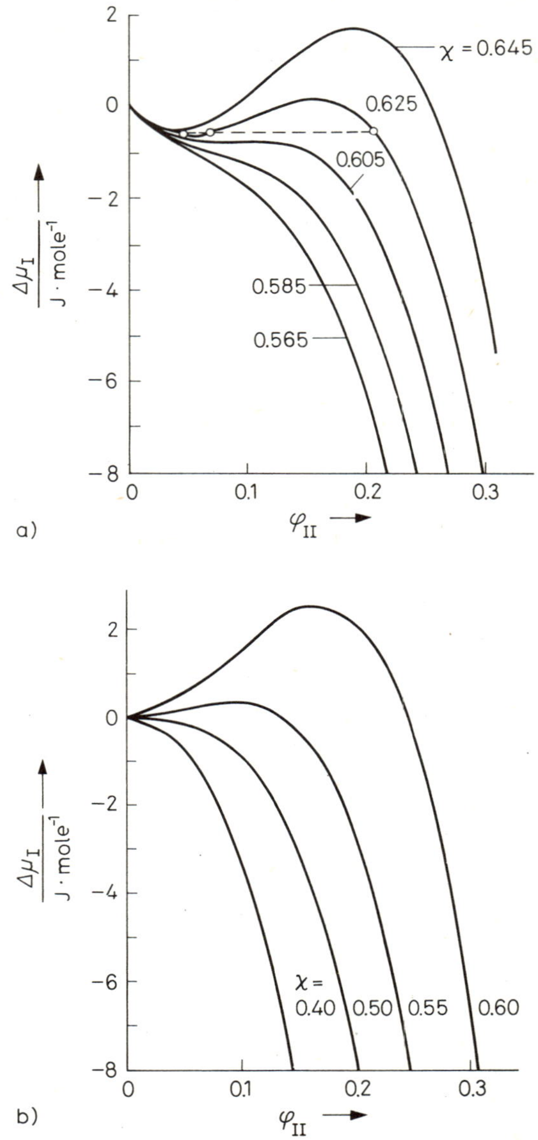

Fig. 5-2
Dependence of the chemical potential of the solvent on the volume fraction of the dissolved polymer according to the Flory-Huggins equation:

$$\Delta\mu_1 = \boldsymbol{RT}[\ln(1 - \varphi_{II}) + (1 - 1/P)\,\varphi_{II} + \chi\varphi_{II}^2]$$

a) $P = 100$; $T = 298.2$ K; $0.565 \leqq \chi \leqq 0.645$
For $\chi \geqq 0.605$, a sol phase with a small φ_{II} and a gel phase with a high φ_{II} value are developed. Both phases are at the same $\Delta\mu_I$ value.

b) $P = \infty$; $T = 298.2$ K; $0.40 \leqq \chi \leqq 0.60$
Systems with $\chi > 0.50$ precipitate a gel phase which is in equilibrium with a sol phase of infinite dilution.

where τ is the shear stress. In the θ state, $a = 0.5$. In good solvents, values of about $a = 0.8$ are possible for coiled molecules. Even higher exponents occur with rigid, rod-shaped molecules.

The *Huggins constant*, χ, in the *Flory-Huggins equation* also depends on the quality of the solvent

$$\ln a_{\mathrm{I}} = \frac{\Delta \mu_{\mathrm{I}}}{RT} = \ln (1 - \varphi_{\mathrm{II}}) + \left(1 - \frac{1}{P}\right) \varphi_{\mathrm{II}} + \chi \varphi_{\mathrm{II}}^2 \quad (5\text{-}9)$$

where a_{I} is the activity of the solvent, $\Delta \mu_{\mathrm{I}}$ is the chemical potential of the solvent in the mixture and φ_{II} is the volume fraction of the dissolved macromolecules having the (uniform) degree of polymerization, P. Fig. 5-2a, b shows the behaviour of $\Delta \mu_{\mathrm{I}}$ as calculated by eqn. (5-9) as a function of φ_{II} for different χ values. For the higher values of χ one obtains curves with extrema. Between the minimum and the maximum, a_{I} should increase with increasing φ_{II}. This contradicts every experience. In this range the solution divides into two phases having the same ordinate value $\Delta \mu_{\mathrm{I}}$: a *gel phase* with a higher polymer concentration (φ''_{II}) is in equilibrium with a *sol phase* (φ'_{II}). On the other hand, steadily decreasing curves which have no extrema are associated with single-phase, homogeneous systems. The curve with the horizontal inflectional tangent separates the two families of curves. This curve is obtained with the critical value for χ, which can be calculated from eqn. (5-9) using the conditions for this point of inflection, $\mathrm{d}\ln a_{\mathrm{I}}/\mathrm{d}\varphi_{\mathrm{II}} = 0$ and $\mathrm{d}^2\ln a_{\mathrm{I}}/\mathrm{d}\varphi_{\mathrm{II}}^2 = 0$. For the θ system ($P \rightarrow \infty$) one has $\chi_{\mathrm{crit}} = 0.5$. The better the solvents, the smaller is χ.

5.3. Polymers in single solvents

5.3.1. Phase equilibria in binary systems

For mixtures of a solvent and a homogeneous polymer, the behaviour can be derived from Fig. 5-2: if the interaction parameter increases beyond the limit χ_{crit}, then the mixture decomposes into a gel phase with the volume fraction φ''_{II} and a sol phase with φ'_{II}. The composition of these phases depends on the value of χ; their quantities depend on the initial concentration of the solution.

For a given polymer-solvent combination, χ has a certain value, but the temperature has an effect as well. The lower the temperature, the greater χ will usually be. In many systems the increase beyond the critical temperature which is required for a phase separation occurs in an experimentally accessible temperature range. Phase diagrams can then be obtained as shown schematically in Fig. 5-3.

The curve defines an extremely asymmetric miscibility gap: the gel phase contains polymer and solvent in comparable quantities, whereas the sol phase is an extremely dilute solution which, in the graphical representation, can hardly be distinguished from a pure solvent. In contrast, miscibility gaps in low-molecular-weight mixed systems are largely symmetrical.

For a genuine binary mixture, the curve shown in Fig. 5-3 can simply be determined by cooling solutions of different initial concentrations to their turbidity points and plotting the turbidity temperatures vs. the concentration.

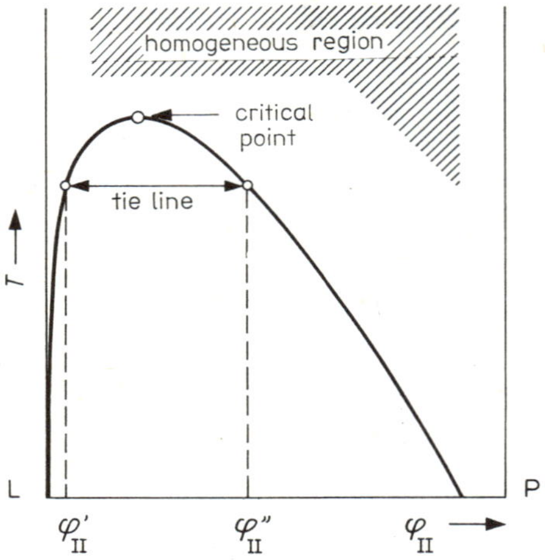

Fig. 5-3
Phase diagram for a polymer having a high and uniform degree of polymerization in a pure solvent

In this case the curve (binodal) can be determined by turbidimetric measurements. The end points of the tie line indicate the compositions φ'_{II} of the sol phase and φ''_{II} of the gel phase, which are in equilibrium at this temperature. The abscissa ranges from a pure solvent L with $\varphi_{II} = 0$ to a pure polymer P with $\varphi_{II} = 1$.

The critical point is characterized by the condition $\varphi'_{II} \rightarrow \varphi''_{II}$. Its position can be estimated from the behaviour of the *tie lines* which in the phase diagram connect the pairs of points corresponding to the phases which are in equilibrium with one another. As the critical value is approached, the pairs of corresponding points draw closer and closer together. The tie line cuts off a shorter and shorter section of the curve and finally degenerates into a tangent. The critical point is identical with the vertex of the turbidity curve only for genuine binary solutions (homogeneous polymer, homogeneous solvent), and the compositions of all the coexistent phases lie on one and the same turbidity curve, independently of the initial concentration. Such a curve is called a *binodal*. It marks off the range of stable solutions from the miscibility gap. Inside the binodal, a second, narrower curve, called a *spinodal*, separates the range of the absolutely unstable systems from the metastable range. In the metastable range the solution may remain homogeneous for a short time, but fluctuations in concentration will already lead to an increased light scattering (see Fig. 5-4). The scattering data for mixtures in the range between the binodal and the spinodal curve depend on the distance to the spinodal as well as on time. In the PICS method (pulse-induced critical scattering; GORDON et al., 1973, 1977; DERHAM et al., 1974), a few μl of polymer solution are periodically cooled very rapidly, in a glass tube 1 mm in diameter, from a temperature anything above that of the turbidity curve to the measuring temperature. The latter lies in the metastable range, dropping in small steps (circ. 0.05 K) from one temperature cycle to another. What is measured is the light scattering at different angles of observation (I_{30}, I_{90}). The representation of $1/I_{30}$ vs. the measuring temperature can be extrapolated to $1/I_{30} = 0$. The associated temperature determines one point of the spinodal. Further points are determined by means of solutions of different concentrations. The spinodal touches the turbidity curve at the critical point with a common tangent. This is also valid for the case

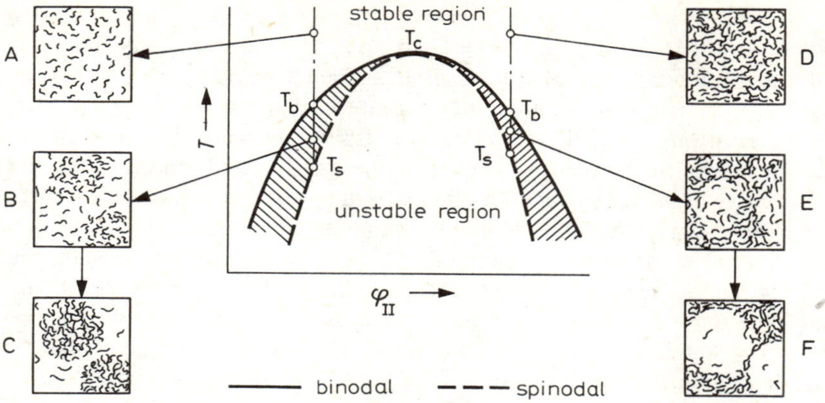

Fig. 5-4
Schematic phase diagram for a solution of a homodisperse polymer

Dilute solutions: A — stable; B — metastable, with fluctuations in concentration; C — phase separation, precipitation of the gel phase

Concentrated solutions: D — stable; E — metastable, with fluctuations in concentration; F — phase separation by segregation of a sol phase (according to DERHAM, GOLDSBROUGH and GORDON, 1974).

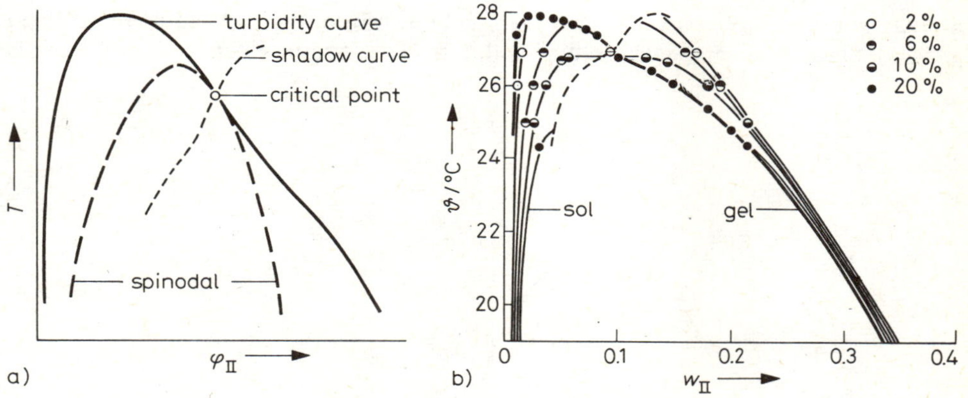

Fig. 5-5
Phase diagram for quasi-binary polymer/solvent systems

a) Schematic representation indicating the spinodal. The latter touches the turbidity curve at the critical point, which at the same time is the point of intersection with the shadow curve.

b) Experimental turbidity curve for polystyrene ($\bar{M}_n$ = 210000 g · mole^{-1}) in cyclohexane, with the equilibrium curves for four different initial concentrations (according to REHAGE, MÖLLER and ERNST, 1965).

to be discussed below, where the turbidity curve is not the curve of coexistence (the binodal) (Fig. 5-5a).

5.3.2. Phase equilibrium for polymolecular samples in a single solvent

While in true binary systems the turbidity curve and the curve of phase coexistence are identical, the *quasi-binary representation of multicomponent systems* leads to an entirely different picture (Fig. 5-5b): the concentrations of the phases being in equilibrium lie on different curves, which depend on the initial concentration. Each curve consists of one branch reflecting the polymer concentration in the gel phase and another branch for the sol phase. The two branches continuously merge into one another only if the initial concentration is, by chance, equal to the *critical concentration*. In all other cases there is a gap between the two branches of the curve, which increases with increasing difference between the initial concentration and the critical one. The curve branches facing each other terminate either at the turbidity curve or at the so-called *shadow curve*. The latter defines the geometrical locus of the composition of the phases precipitating at the very beginning of turbidity. The amount of substance in the shadow phases is far too small for a direct analysis. The first measurable points are located well behind the shadow curve, so that the latter can be determined only by extrapolation. On the other hand, the turbidity curve itself is experimentally accessible. It results from the turbidity temperatures of solutions of graduated concentrations. In quasi-binary systems the vertex of the turbidity curve, i.e., the maximum turbidity temperature (*precipitation threshold*), is not identical with the critical point. The latter is lower, being determined by the point of intersection of the turbidity curve and the shadow curve.

The question why quasi-binary systems are so different from genuine binary systems can be answered by inspecting Figs. 5-6 and 5-7, where the situation is shown for a model system consisting of the solvent L and the two homologous polymers I and II ($P_{\mathrm{I}} < P_{\mathrm{II}}$): on the plane surface extending between the temperature axis and the side $P_{\mathrm{II}}L$ of the triangle, the phase diagram of the true binary system L-P_{II} as known from Fig. 5-3 can be represented (Fig. 5-7a). Since P_{I} is completely soluble within the temperature interval shown, there is

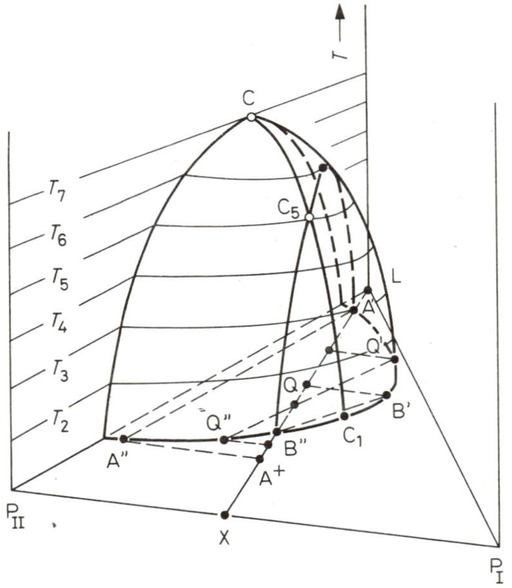

Fig. 5-6
Miscibility gap in a three-component system, $L-P_{\mathrm{I}}-P_{\mathrm{II}}$, at different temperatures
By means of quasi-binary plotting, the parameters of the three-dimensional representation are projected upon the plane normal to LX (according to KONINGSVELD, 1969).

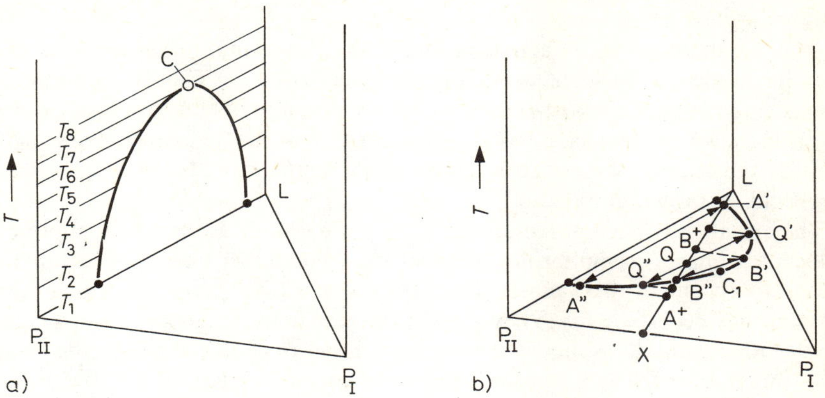

Fig. 5-7
Details of Fig. 5-6

a) Phase diagram of the genuine binary system L-P_{II} (binodal)
b) Triangular diagram of the system L-P_I-P_{II} at the temperature T_1, where the tie lines are indicated, e.g., the tie line between the sol phase Q' and the gel phase Q'', which develop from a solution with the initial concentration Q after cooling to T_1.

no corresponding curve on the right-hand face of the prism ($P_I LT$). Hence the miscibility gap covers only part of the bottom surface, on which the composition of the ternary system can be represented. For the temperature T_1 (origin of the temperature axis) the triangular diagram has the form shown in Fig. 5-7b. The curve A″Q″B″C_1B′Q′A′ encloses the *miscibility gap*. Its points of intersection with the P_{II}L axis are identical with the lowest points of the curve in Fig. 5-7a. The straight line LX connects all systems containing P_I and P_{II} in the same proportion. The greater the distance from L, the higher is the total polymer concentration. Thus LX corresponds to the abscissa in the quasi-binary representation (Fig. 5-5). At the temperature T_1 the miscibility gap is already wide enough to cause a phase separation in a considerable concentration interval between L and X. If the polymer concentration reaches the point A′, the solution will become turbid. The precipitating shadow phase is enriched with the higher-molecular-weight P_{II} (point A″). With the onset of turbidity of a gel of the composition B″, a more dilute phase (sol phase) is separated, which, in comparison with X, contains more of the better soluble P_I (point B′). The total polymer concentration existing at A″ is greater than that at the gel turbidity point B″. The total polymer concentration at B′ and B^+, respectively, is greater than that at A′. While A′ and B″ can be determined experimentally as points on the turbidity curve, A″ as well as the point A^+ projected upon LX are above the turbidity curve on the concentration scale. A^+ and B^+ are points on the shadow curve.

The quantities of the phases in equilibrium with A′ or B″ are too small for a direct determination of the data for A″ and B′. Experimentally it is possible to decompose systems having the initial concentration, Q, into the coexisting phases Q' and Q'', the polymer content of which can be determined and projected upon LX. By shifting the starting point Q along the straight line between A′ and B″, one has to determine a sequence of values so that the extrapolation to A″ and B′ becomes possible.

At the higher temperature T_2 the miscibility gap is smaller (Fig. 5-6). At T_7 (the precipitation threshold for the binary system L-P_{II}) the gap vanishes completely. For mixtures of

P_I and P_{II} with P_I/P_{II} = X, the *precipitation threshold* is equal to T_6. This value is not identical with the *critical temperature*.

At the critical point the compositions of the coexisting phases must be equal. Hence it must lie on the curve which, starting from the critical point C of the genuine binary system, connects the critical concentrations at the individual temperatures. For T_1 this is the point C_1, as is shown by the secant-to-tangent transition of the tie lines. For the quasi-binary system LX the critical point is at C_5. The critical temperature is lower than the precipitation threshold, as shown in Fig. 5-5.

The quasi-coexistence curves in Fig. 5-5 are not binodals; they do not define the geometrical locus of phases in equilibrium with each other (since, because of the mentioned separation, the compositions of these phases lie outside of the LXT plane, on which the quasi-binary diagram has been plotted). They merely connect fictive points projected upon this plane. On the other hand the points on the *turbidity curve* are really located on the plane of drawing. The spinodal for the L-X system, being determined by PICS or the like, is also located on this plane. Therefore the spinodal and the turbidity curve may have a common tangent. (The shape of the spinodal curve depends on $\bar{M}_w$ and $\bar{M}_z$, but not on the specific shape of the molar mass distribution.)

Although the quasi-coexistence curve with the critical concentration also passes through the critical point, it intersects the turbidity curve and the spinodal. The compositions of the equilibrium phases which lead to the construction of this quasi-coexistence curve, too, are known to lie, as in every other case, outside the quasi-binary LX*T* plane. Quite analogously, one may explain why the shadow curve passes through the critical point at yet another angle.

5.4. Polymers in mixed solvents

For solvent/non-solvent mixtures the theoretical treatment of phase equilibria is even more complicated. They are, however, indispensible for some problems of polymer characterization, being frequently used regardless of all difficulties. If a non-solvent is poured into a polymer solution, the latter becomes turbid at a certain point, because a gel phase precipitates. Further addition increases the turbidity until all of the polymer is precipitated. The onset of turbidity depends on the type of the polymer, the solvent and the precipitant, the temperature and the molar mass of the polymer. These quantities can be determined one at a time by turbidimetric titration, if the necessary calibrations are given.

5.4.1. Selective solvation

In mixtures of solvents having different thermodynamic qualities, the better solvent prevails in the solvation sphere of the macromolecules. This has no effect on the refractive index increment, $\nu = dn/dc$, if the latter is measured in the usual way. However, if the solution is allowed to exchange matter with the pure solvent mixture in a dialyzer, then selective sorption causes permeation of the preferred solvent into the solution chamber, and loss of the other solvent. The dialysis is complete when the components capable of permeation have reached equal values of chemical potential in both chambers. The higher the coefficient γ^* of selective sorption (given in ml of the preferred solvent per g of polymer) and the greater the difference in refractivity between the two solvents, the more the refractive increment at constant chemical potential, ν_μ, deviates from the increment at constant chemical composition, ν_φ:

$$\nu_\mu = \nu_\varphi + \gamma^* (\mathrm{d}n/\mathrm{d}\varphi_\mathrm{I}) \tag{5-10}$$

The differential quotient, $\mathrm{d}n/\mathrm{d}\varphi_\mathrm{I}$, gives the variation of the refractive index of the mixed solvent as the content of the component I is varied. The selective sorption can be quantitatively determined according to eqn. (5-10) (TUZAR and KRATOCHVIL, 1967) or from measurements of the nuclear magnetic relaxation (LÜTJE, 1970).

Frequently a mixture of a certain composition shows an even better solvency than the two pure components. At this point the coefficient of selective sorption passes through zero. (It is referred to the same solvent over the whole range.)

5.4.2. Solvent segregation during precipitation

In Section 5.3. it was shown how polymer solutions can be divided into a gel phase and a sol phase by decreasing the temperature. As already stated, a corresponding effect also occurs if a precipitant is added. Usually the solvent-to-precipitant ratio in the gel phase is markedly higher than in the sol phase (see Fig. 5-8).

Multicomponent systems containing solvent, non-solvent and macromolecules of different chain lengths are of practical importance. The four-component system $L/F/P_\mathrm{I}/P_\mathrm{II}$

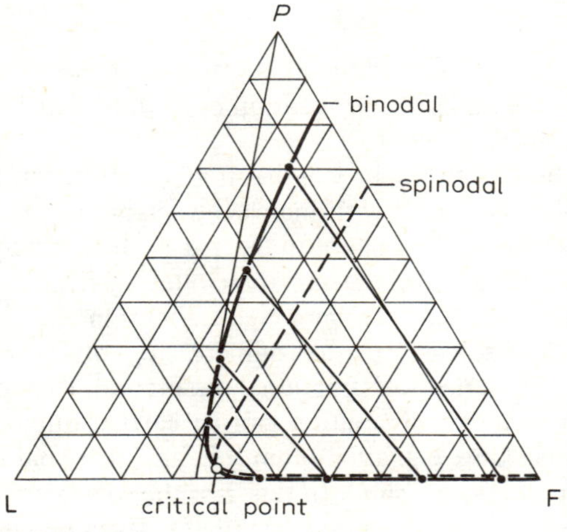

Fig. 5-8
Phase diagram for a system consisting of a solvent L, a precipitant F and a homodisperse polymer ($P = 100$)

The tie lines plotted connect the points of the coexisting phases. In each case the polymer-rich phases contain less precipitant than the sol phases. The dashed spinodal touches the binodal, the solid outer curve, at the critical point. The precipitation threshold depends on the initial concentration of the polymer. The indicated value x is obtained if one starts with concentrated solutions. Highly dilute solutions require even less precipitant than the amount corresponding to the critical point (according to TOMPA, 1949).

represents the simplest case. As the number of components increases, the considerations concerning the phase behaviour grow more and more complex. However, the findings derived from the model system $L/P_I/P_{II}$ (cf., Section 5.3.2.) have always been confirmed in more extensive considerations (HUGGINS and OKAMOTO, 1967; KONINGSVELD, 1977). The following statements are also valid for multicomponent systems:

— The turbidity curve is not a binodal.
— The true composition of the phases in equilibrium lies outside the quasi-binary plane.
— Different initial concentrations lead to different quasi-coexistence curves in the quasi-binary diagram.
— The critical point lies below the precipitation threshold.

As in the systems discussed in Section 5.3.2., the position of the critical point depends on $\bar{M}_w$ and $\bar{M}_z$. Theoretically, the critical polymer concentration is found to be:

$$\varphi_c = C(\bar{M}_z^{0.5}/\bar{M}_w) \qquad (5\text{-}11)$$

OKAMOTO (1958) determined the values of the constant C for a number of solvent mixtures.

5.4.3. Characterization of polymers on the basis of solubility differences

The structure-dependent solubility differences (cf. Section 5.1.) can be used in the identification of polymers. A series of standardized solvents would enable something like analytical schemes to be carried out. An even finer differentiation may be achieved by means of turbidity titration, which also reveals the differences in the molar mass of homologous polymers. Empirically, the following relationship was found to exist for many systems

$$\varphi^* = A' + B/M^{0.5} \qquad (5\text{-}12)$$

where φ^* is the volume fraction of the precipitant at the turbiditv point and A, B are constants. This relationship is suitable for the determination of the molar mass in mg quantities of narrow-cut fractions (GLÖCKNER, 1965b).

The relationship between solubility and chain length offers a significant approach to *fractionation*. If a solution is cooled, or mixed with increasing quantities of a precipitant, then it is initially the components having the highest molar mass which precipitate. These components are isolated, and so are the subsequently precipitating fractions (fractional precipitation, fractionation from the high-molecular-weight side). On the other hand, if a polymer is extracted step by step with solvent mixtures of increasing solvency, then the first fraction has the lowest degree of polymerization (fractional dissolution, fractionation from the low-molecular-weight side). Fractional dissolution can be carried out with polymers which have been applied to a flat backing, e.g. aluminium foil, as a thin coating, or in columns with polymers on granular carriers, e.g. sand. In this case the solvency of the extractant can also be increased continuously. This technique has finally developed into precipitation chromatography (cf., Section 9.5.).

5.5. Resorption and desorption of a solvent

The dissolution of a polymer starts with swelling by the penetration of a solvent. Since the diffusion coefficients in a polymer are small, the process proceeds slowly, but yet much more rapidly than the outward diffusion of macromolecules. The latter have even smaller diffusion coefficients, being most immobile in their mutual penetration (see Fig. 15—10). Even in favourable cases (the sample being finely pulverized and continuously agitated) the dissolution takes almost an hour. If the polymer adheres to the bottom of the container, the process will take a very long time.

In cross-linked molecular assemblies the resorption of a solvent leads only to a *limited swelling*. The quantity of swelling agent in the gel at the equilibrium of swelling depends on the intermolecular interactions and the mesh size of the network. A limited swelling is also observed for non-cross-linked polymers in poor solvents. An *unlimited swelling*, i.e., a homogeneous distribution of the macromolecules in the available liquid volume, can only take place for a solvent of sufficient thermodynamic quality.

The desorption of a solvent also requires a relatively long time. To remove the last parts of solvent, drying is carried out at the maximum permissible temperature and in as good a vacuum as possible (<100 Pa or 1 torr). Nevertheless the sample may contain up to several per cent of solvent even after careful drying for several days. This solvent inclusion occurs especially for good solvents with high boiling points (GLÖCKNER et al., 1975).

The fact that coil molecules in solution exhibit a surprisingly strong retention of the solvent volume contained in their interior is called *solvent immobilization*. Although the coils in dilute solutions contain hardly more than 1 % of polymer, the solvent content being as high as 99 %, they carry along the solvent in both sedimentation and flowing, as if it were enclosed in a sphere. Despite their extremely loose structure in solution, the coils are almost impermeable.

6. Adsorption of polymers

The behaviour of macromolecules on solid surfaces is of interest here because it has a significant influence on chromatographic processes. However, it also has a direct or indirect effect on essential technical processes and problems including the stabilization of particle dispersions, the adhesion of paint coats and adhesives or the reinforcement of rubber and other polymers by fillers, as well as coagulation processes and certain problems in friction and lubrication processes. The complex implications of the interaction between polymers and solid phases, and the rather bewildering variety of experimental findings which cannot readily be reconciled with one another, have stimulated intensive theoretical treatment (FRISCH et al., 1953; SIMHA et al., 1953; SILBERBERG, 1962; FORSMAN and HUGHES, 1963; DIMARZIO, 1965; HOEVE, 1965; MEIER, 1967; BIRSHTEIN, 1979; JOANNY et al., 1979). Summaries are given by HUGHES and v. FRANKENBERG (1963); PATAT et al. (1964); KIPLING (1965); HELLER (1966); SILBERBERG (1970) and ROE (1974). In the present context it is desirable to have a somewhat simplified picture of the real situation, which might be helpful in understanding chromatographic phenomena. The question to be answered in this chapter is that of the conformation of macromolecules in adsorption layers. This also requires some brief statements on experimental techniques and essential results. Supplementary aspects for better understanding of adsorption chromatography of polymers, e.g., desorption behaviour or mutual displacement, will be dealt with in Chapter 18. Treating these aspects here would require anticipating too much from the chapters to come.

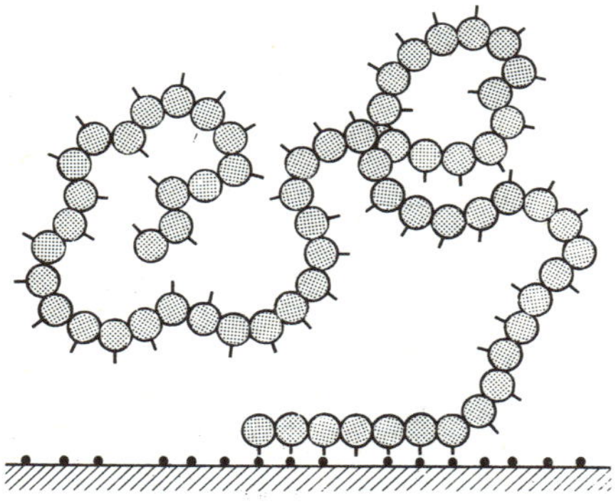

Fig. 6-1
Schematic representation of the start of adsorption
The macromolecule is fixed to groups on the surface using a part of its segments.

The interactions between the flexible molecular coils in solution and the phase boundary, which can be considered to be a solid wall, lead to different results depending on the specific conditions (Fig. 6-1). If the *energy of adsorption* per segment is too small, coils arriving at the wall by diffusion are reflected like balls. On the other hand, if the adsorption energy exceeds a certain threshold value, the coils are retained. If the change in enthalpy due to adsorption exceeds the entropy loss associated with the collapse of the three-dimensional coils to two-dimensional formations, the macromolecules are deformed and adsorbed with almost all of their segments.

6.1. Experimental methods and results

6.1.1. Adsorption isotherms

A known quantity of adsorbent is mixed with the dissolved sample, the mixture is allowed to reach equilibrium and finally the quantity adsorbed is determined, e.g., from the change in concentration. Moisture and polar impurities may greatly interfere with the measurement; therefore the chemicals and, above all, the adsorbent have to be pretreated most carefully. For example, Aerosil is heated to 300 °C under vacuum (1 Pa or 0.01 torr) over a period of 3 to 48 h. If possible, the experiments are carried out in sealed ampoules. The adsorption isotherms indicate the quantity adsorbed either per gram of adsorbent or per square metre of surface as a function of the equilibrium concentration. Their shape corresponds to that of a *Langmuir isotherm:* a limit is even reached at relatively low concentrations (Fig. 6-2). For almost all polymers this limit is of the order of magnitude of 1 mg $\cdot$ m^{-2} (Table 6-1). PATAT and NITSCHMANN (1964) obtained values up to 3 g $\cdot$ m^{-2} by multilayer adsorption from concentrated solutions (up to 90 g $\cdot$ l^{-1}). In this case the isotherms were determined gravimetrically by means of metal sheets hanging in the polymer solutions. The very weak adsorption on chromium foils from extremely dilute solutions was investigated by means of labelled polystyrenes (STROMBERG et al., 1964). In this case a reversible adsorption with a dynamic equilibrium was found to occur.

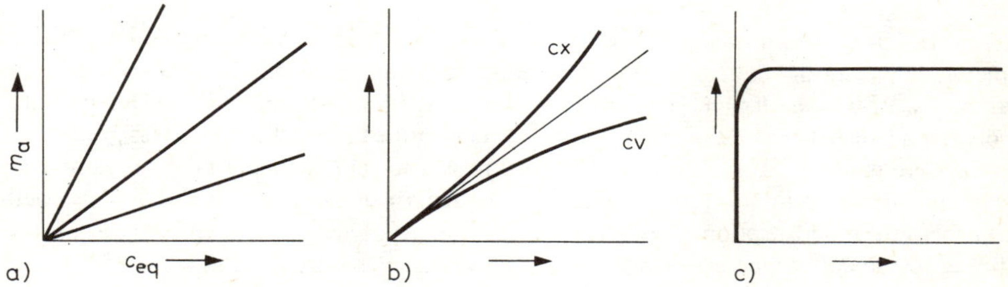

Fig. 6-2
Schematic representation of adsorption isotherms

a) Linear isotherms for three different distribution coefficients (ideal case, which may occur for low-molecular-weight adsorbates in high dilutions)
b) Convex (cx) and concave (cv) course of the isotherms (real case for low-molecular-weight substances)
c) Extremely non-linear isotherm of the concave type with a total coverage (plateau formation) for very low concentrations (typical for the adsorption of polymers).

Table 6-1
Maximum loading, Γ_a, in the adsorption of polymers from dilute solutions

Polymer	$10^{-3}M$	Adsorbent	Solvent	$\frac{\Gamma_a}{mg \cdot m^{-2}}$	Authors
PS	40	carbon black	Tetra	0.7	ELTEKOV (1975)
	300		Tetra	0.8	ELTEKOV (1975)
	300	rutile	Tetra	0.7	ELTEKOV (1975)
	300	Aerosil	Tetra	0.9	ELTEKOV (1975)
	32	carbon black	Bzn	0.5	HOWARD (1972)
	105		Bzn	0.6	THIES (1966)
	135		Bzn	0.6	HOWARD (1972)
	105	Cab-O-sil M5	Tri	0.7	BOTHAM (1970)
PMMA	300	Aerosil	TCM	0.4	ELTEKOV (1975)
	330	carbon black	MEK	0.6	ELTEKOV (1975)
	33		Bzn	0.9	HERD (1971)
	290		Bzn	0.9	HERD (1971)
	822		Bzn	0.9	THIES (1966)
	395(i)	Cab-O-sil M5	TCM	0.7	MIYAMATO et al. (1974)
	340(s)		TCM	0.6	MIYAMATO et al. (1974)
	822		Tri	1.0	BOTHAM (1970)
PVAC	250	iron powder	TCM	1.5	KORAL et al. (1958)
			DCE	2.4	KORAL et al. (1958)
			Bzn	3.1	KORAL et al. (1958)
			Tetra	6.9	KORAL et al. (1958)
POE	30	carbon black	W	0.6	ELTEKOV (1975)
	15		W	0.7	ELTEKOV (1975)
	5		W	0.6	HOWARD (1967)
	5	carbon black	Bzn	0.8	HOWARD (1967)
	18		W	0.8	HOWARD (1967)
	18	Aerosil	Bzn	0.9	HOWARD (1967)
PDMS	520	Cab-O-sil M5	Hxn	0.9	BREBNER et al. (1980)
			Tetra	0.7	BREBNER et al. (1980)

6.1.2. Viscosimetric investigations

The adsorption of macromolecules on the inside surface of capillaries decreases their inside diameter. The resulting increase of the flow resistance can be measured for tubes with very small bores. ROWLAND et al. (1965) carried out investigations with glass frits. The quantity of polymer adsorbed per unit of surface area was determined using the same glass grade that the frits were made of. Control experiments using stearic acid yielded a pore restriction of $\Delta r = 3$ nm, which is in good agreement with the length of molecules, reflecting the well known brush-type adsorption of this substance. Using polymers, layer thicknesses corresponding to the coil diameters of dissolved macromolecules were found.

From similar measurements using capillary viscosimeters, ÖHRN (1956), however, derived layer thicknesses of 120 and 150 nm for polyvinyl acetate and polystyrene, respectively, in toluene. For polystyrene in decalin, investigations by FENDLER et al. (1955) with Ubbelohde viscosimeters yielded 22 ± 8 nm in the 0a capillary ($r = 0.265$ mm) and 32 ± 8.5 nm in the 00 capillary ($r = 0.155$ mm). These layer thicknesses were calculated from the flow times of calibration liquids before and after contact of the capillary with a polystyrene solution ($\bar{M}_w = 620000$ g · mole^{-1}). They are in good agreement with the coil radius, i.e., 32 nm in decalin. The apparent restriction of the capillary can also be calculated by comparing the passage times determined with the two viscosimeters for the polystyrene solution itself. However, this yields a layer thickness of 148 nm. Obviously the hydrodynamic effect of the adsorption layer has a wider range in the polymer solution than in calibration oils. CRAUBNER (1965) found that the thickness, d_h, of the hydrodynamically active layer of polymethyl methacrylate on glass in benzene increases with increasing molar mass and decreases exponentially with increasing concentration, c (in g · dm^{-3}), according to:

$$d_h(c, M) = 0.254 \cdot M^{0.47}\, e^{-0.76c}/\text{nm} \qquad (6\text{-}1)$$

By means of extremely careful measurements using a closed capillary viscosimeter, PRIEL and SILBERBERG (1978) were able to determine the adsorption behaviour of polystyrenes in toluene at extremely low concentrations. The temperature was kept constant within $\pm 1.5 \cdot 10^{-4}$ K, the flow times of about 1000 s were determined with an accuracy of $\pm 10^{-3}$ s and the capillary was always kept filled with the solution (except for a very short time at the end of a measurement) in order to avoid structural changes of the adsorbed layer. The measurements yielded a linear increase in the thickness of the adsorption layers with increasing concentration in the range from 0.03 to 0.15 mg · l^{-1}. For all of the samples with $M > 5 \cdot 10^5$ g · mole^{-1} this increase was practically independent of the molar mass. With a maximum of 15 nm, the layer thickness reached, however, only $^1/_3$ to $^1/_6$ of a coil diameter. The first increase was followed by a plateau in the layer thickness, which extended to a concentration of 1 mg · l^{-1} for a sample with $M = 1.8 \cdot 10^6$ g · mole^{-1}. Further increases in concentration again led to an increase in the layer thickness, which at about 10 mg · l^{-1} was followed by the plateau described by ÖHRN (1956). For this sample the plateau value of 140 nm distinctly exceeded the hydrodynamic coil diameter of 92 nm.

6.1.3. Ellipsometry

Since 1963 (STROMBERG and GRANT), this technique, developed by DRUDE as early as 1889, has also been used for the elucidation of the structure of polymeric adsorption layers. It involves the use of polarized light impinging obliquely upon a reflecting surface. The components of light which are polarized in a plane parallel to the surface are reflected differently from those normal to the surface. If the surface carries a layer of substance, the light is repeatedly reflected within this layer. The layer thickness and the concentration within the layer can be calculated from the amplitude ratio and the phase difference between the horizontal and the vertical wave. For layers with a density profile, the thickness determined has to be associated with a certain point of the profile.

STROMBERG et al. (1965) as well as KILLMANN and v. KUZENKO (1974) investigated the adsorption of polystyrene from θ solutions (cf., Section 5.2.) in cyclohexane (Fig. 6-3). The layer thickness determined increased with the molecular size, approximately corresponding to the coil dimensions (Fig. 6-4). KILLMANN and v. KUZENKO interpreted the difference in the results of the two papers as a consequence of the surface roughness of the chromium base. It might also be due to the slightly different temperatures of measurement, because the θ-temperature is about 35 °C. Below this temperature, an adsorption layer may exhibit an excessive growth due to *phase separation* (HOEVE, 1970). STROMBERG et al. obtained adsorption layers of approximately equal thicknesses on different metals (Cr, Au, Cu, Ag, steel).

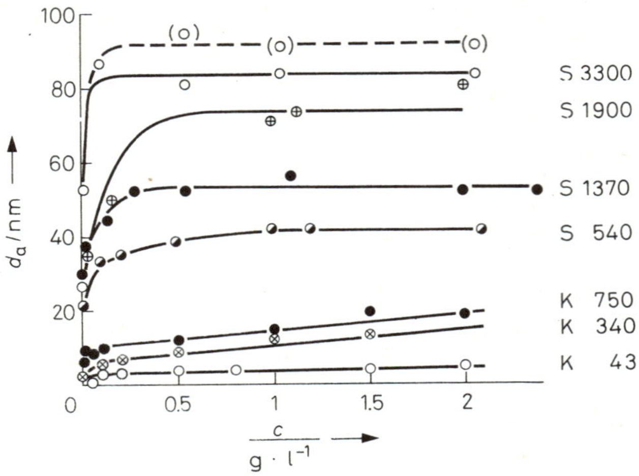

Fig. 6-3
Dependence of the layer thickness on the concentration in the adsorption of polystyrene on chromium

Solvent: cyclohexane. Ellipsometric measurements by STROMBERG et al. (1965) at 34 °C with samples of $10^{-3}\ M$ = 3300; 1900; 1370 and 540 g · mole^{-1}, and measurements by KILLMANN and v. KUZENKO (1974) at 36 °C with samples of $10^{-3}\ M$ = 750; 340 and 43 g · mole^{-1}, respectively. The broken curve above that of sample S 3300 indicates the temporarily occurring maximum of the layer thickness for this substance. The solid curve is obtained after a longer waiting time.

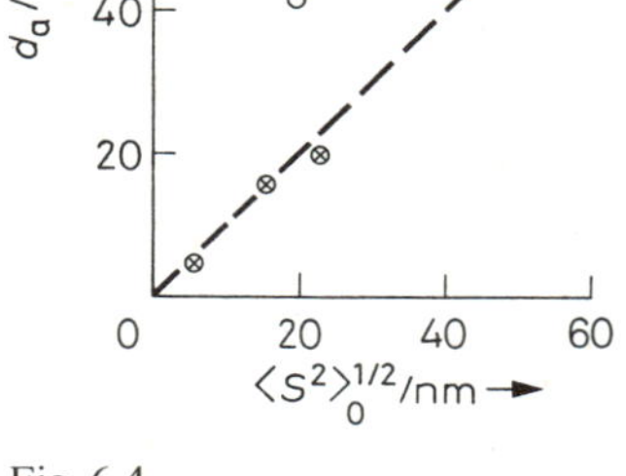

Fig. 6-4
Ellipsometric layer thickness, d_a, as a function of the radius of gyration for polystyrene in cyclohexane on chromium surfaces

⊗ Measurements by KILLMANN and v. KUZENKO (1974) at 36 °C
○ Measurements by STROMBERG et al. (1965) at 34 °C
(As $\langle S^2\rangle_0^{1/2} = 2.63 \cdot 10^{-2}\ M^{1/2}$ nm, proportionality to $\sqrt{M}$ was obtained for both series of measurements.)
The value marked by an arrow decreased to the next lower one in the course of time.

The effect of the material was smaller than that of the surface pretreatment. In good solvents such as dioxane and methyl ethyl ketone it was not possible by means of ellipsometry to detect any polystyrene adsorption on bright chromium surfaces. Polyvinyl pyrrolidone ($M = 1.78 \times 10^6$ g · mole^{-1}) was adsorbed on chromium from an aqueous solution, forming layers 45 nm thick within 10 min at a concentration of 10 g · dm^{-3}. On the other hand, polyethylene glycol (M = 40000 g · mole^{-1}) was adsorbed in films less than 3 nm thick on highly cleaned surfaces (KILLMANN and v. KUZENKO). Polyethylene terephthalate (M = 7400 g · mole^{-1}) was adsorbed from ethyl acetate on chromium at 34 °C, forming layers 7 nm thick (PEYSER et al., 1967). To obtain ellipsometric information about the adsorption on silica, FLEER and SMITH (1976) used oxidized silicon polymers as a base. They investigated the behaviour of polyvinyl alcohol, one sample containing 2% and the other 12% of residual acetate, and found layer thicknesses of 23 nm for both polymers on hydrophobized surfaces. On hydrophilic silica, the more saponified product formed thicker films than the partially saponified one.

In addition to the layer thickness, ellipsometry also yields information about the density of the adsorption layer. The highest values measured were 0.45 g · cm^{-3} for polyethylene glycol layers only 3 nm thick. Usually the values range between 0.025 and 0.250 g · cm^{-3}.

6.1.4. Electrosorption analysis

A. c. polarography at a mercury drop electrode can be employed to derive, among other things, information about the *area required* by molecules which have penetrated into the double layer. Data compiled by JEHRING (1974) show that for polyethylene glycol (I) as well as polyvinyl alcohol (II) and polyvinyl pyrrolidone (III) the area, A_S, required by one macromolecule increases with the molar mass. The proportionality factors in the relationship A_S = const · M are 0.49; 0.36 and 0.18 · 10^{-16} cm^2, respectively. The measured area requirements approximately agree with those necessary for the contact of all the molecular segments with the surface. For the molar mass, M_0, of the repeat unit the factors yield values of 21.6 (I), 15.9 (II) and 20.0 · 10^{-16} cm^2 (III), respectively, for the areas required per basic unit, which are in the same order of magnitude as those obtained from model considerations. A brush-type adsorption is entirely inconsistent with the observed M dependence.

6.1.5. IR spectroscopy

FONTANA and THOMAS (1961) have shown that infrared spectroscopy may yield information about the interactions between the adsorbent and the adsorbate. As in subsequent investigations the adsorbent preferably used was Aerosil. Especially polymers with CO groups and polystyrene were investigated (THIES, 1966; KISELEV et al., 1968; BOTHAM and THIES, 1970; THIES, 1971; HERD et al., 1971; SCHULZ et al., 1977; KUNATH et al., 1978; DAY and ROBB, 1980).

The carbonyl frequency at 1730 cm^{-1} of polymethyl methacrylate or at 1775 cm^{-1} of polycarbonate is shifted by approximately 20 cm^{-1} towards lower wave numbers due to *hydrogen bridges*. If both components of the carbonyl band can be measured, e.g., by compensation, the fraction, p of the CO groups which is adsorbed through such bridges can be determined. DIETZ (1976) supplemented these measurements by investigations of the Si—OH band

Table 6-2
Bonding of polyesters to Aerosil surfaces in the adsorption from chloroform at 25 °C (according to Dietz, 1976)

Fraction p_{CO} of the hydrogen-bonded carbonyl groups, decrease ΔN_{OH} of the quasi-free, superficial SiOH groups and fraction p_{SiOH} forming hydrogen bridges with the carbonyl groups, for a partial and a total coverage of the surface.
Time setting: 24 h; a 24 h pretreatment of the Aerosil at 200 °C and 0.01 Pa (10^{-4} mbar) yields ΔN_{OH} = 890 µval of quasi-free SiOH groups per gram.
In the systematic notation of the polyesters, the first figure indicates the number of methylene groups in the diol, while the second one indicates the number of methylene groups in the acid; 2. Ph means polyethylene orthophthalate.

Characterization		m_a = 20 mg/g_{SiO_2}			Maximum coverage			
Polyester	$10^{-3} M$	p_{CO}	N_{OH} (µval · g^{-1})	p_{SiOH}	m_a (mg · g^{-1})	p_{CO}	N_{OH} (µval · g^{-1})	p_{SiOH}
2.4	6.24	0.43	200	0.22	91	0.35	463	0.52
6.8	2.99	0.67	160	0.18	90	0.48	400	0.45
10.12	4.59	0.81	142	0.16	[illegible]0	0.64	332	0.37
2.Ph	2.43	0.38	150	0.17	86	0.35	415	0.47
PC	27	0.56	88	0.10	95	0.31	229	0.26

at 3660 cm^{-1} (in chloroform), the intensity of which is reduced by a hydrogen bridge with CO groups in favour of a new band at 3450 cm^{-1}.

For a very low loading of the adsorbent the fraction p reaches rather high values (circ. 0.8). Of course the structure of the polymer also has some effect. Flexible, aliphatic chain segments between the carbonyl groups facilitate their contact with the surface Si—OH groups. In this way such polymers have a higher p value than rigid ones (see Table 6-2). For ethylene copolymers containing 7.7—20.5 mole-% vinyl acetate, Botham and Thies (1970) obtained p_{CO} = 0.78, 0.76 and 0.88, respectively, while for pure polyvinyl acetate, p_{CO} = 0.43. The data obtained for polyvinyl pyrrolidone are equally low (Day and Robb, 1980).

The value of p decreases as the loading increases. In the state of saturation it amounts to ca. 70% of the initial value. Extremely low p values were observed in the adsorption of associate particles of relatively rigid molecules (Lipatov et al., 1975).

6.1.6. Electron spin resonance (ESR)

Fox et al. (1974) used magnetic electron spin resonance in the investigation of macromolecular adsorption layers. The polymers were spin-labelled by nitric oxide groups. Labelled groups linked to loops and chain ends which protrude into the solution are more mobile than those linked to trains which are rigidly fixed to the surface. The immobile groups yield an ESR spectrum which can also be observed from frozen polymer solutions. The spectrum of the labelled groups linked to mobile segments is considered equivalent to that of dissolved molecules. Using this technique, Robb and Smith (1974) characterized polyvinyl pyrrolidone containing 3 mole-% of allylamine components on porous silica gel with a surface area of 250 $m^2 \cdot g^{-1}$. The adsorbent was submerged in polymer solutions of graduated concentrations and allowed to stand over 24 hours with occasional shaking. After centrifugation, the polymer concentration was determined both in the solution and on the adsorbent, and after washing the adsorbent with water the ESR spectra were recorded. The latter were assigned by means of the mentioned correspondences, thus determining the fractions, p_{ESR}, of the

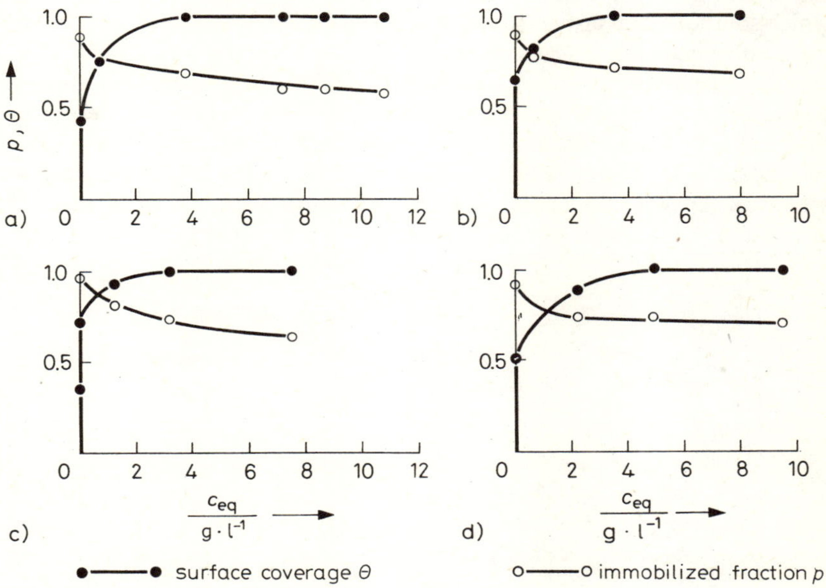

Fig. 6-5
Increase of the relative surface coverage, θ, and decrease of the rigidly bonded fraction, p, with increasing equilibrium concentration c_{eq} (according to ESR measurements by Robb and Smith, 1974)

Adsorption of polyvinyl pyrrolidone ($M = 40000$ g · mole^{-1}) on Aerosil:
a) from water
b) from 0.1 N NaCl
c) from chloroform
Adsorption of polyvinyl pyrrolidone ($M = 18000$ g · mole^{-1}) on Aerosil:
d) from water.

immobilized segments. They are somewhat higher than the p_{CO} values determined by IR spectroscopy. This is reasonable in view of the fact that p_{CO} indicates only those carbonyl groups which have met partner groups at the surface for hydrogen bridge formation. As the Si—OH groups are located 0.7—1 nm from each other, and are certainly not always present just at the locations where the chains lying on the adsorption layer have their carbonyl groups, the fraction of the rigidly adsorbed segments should always be greater than the fraction of the bonded anchor groups. Fig. 6-5 shows the immobilized fraction, p_{ESR}, together with the surface coverage. The fraction decreases from about 0.90 for a partially covered surface to 0.71 ($M = 18000$) and 0.60 ($M = 40000$) for a total coverage.

6.1.7. Calorimetry

Calorimetric measurements of the enthalpy of adsorption, of immersion, and of wetting should give an insight into the energy conditions of polymer adsorption. KILLMANN (1976) reported on this kind of investigation with non-porous Aerosil having a surface area of 200 $m^2 \cdot g^{-1}$ and solutions of polyethylene glycol in water, methanol, benzene or carbon tetrachloride. The temperature effect induced by breaking an ampoule with concentrated polymer solution in a suspension of the adsorbent yields the *enthalpy of adsorption*. The enthalpy of immersion is measured by putting together dry adsorbent and a solution. An analogous experiment using pure solvent yields the *enthalpy of wetting*. In all cases exothermic adsorption enthalpies were found, the values of which increased more and more slowly as the loading increased. The intensity of an IR band at 3300 cm^{-1} (in carbon tetrachloride), which indicates hydrogen bridges originating from SiOH, also increased more and more slowly as the coverage increased. A linear relationship was found between the adsorption enthalpy and the intensity of the IR absorption of the hydrogen bridges (Fig. 6-6).

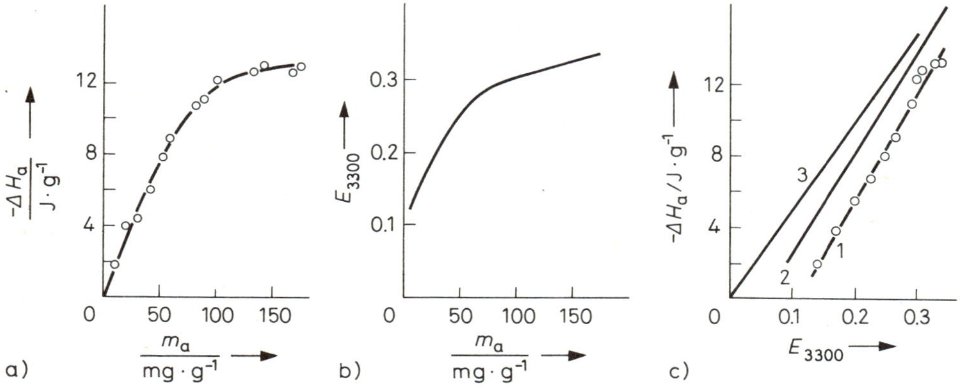

Fig. 6-6
Adsorption of polyethylene glycol ($M = 6000$ g · mole^{-1}) at 25 °C from carbon tetrachloride on Aerosil (according to KILLMANN, 1976)

a) Enthalpy of adsorption per g of Aerosil as a function of the quantity adsorbed
b) Extinction of the IR band at 3300 cm^{-1} as a function of the quantity adsorbed
c) Relationship between the enthalpy of adsorption and the IR absorption at 3300 cm^{-1}
1 for PEG 6000, derived from Figs. a) and b); 2 analogous representation for PEG 600; 3 analogous representation for ethylene glycol (monomer).

6.1.8. Magnetic birefringence

In 1978, SCHOLTEN reported on the application of magnetic birefringence in the investigation of the adsorption behaviour of cellulose esters, polyvinyl formal, gelatine and other polymers on γ-iron oxide and chromium dioxide using suspensions of particles of suitable shapes and sizes. The orientation of such magnetic particles in an external field may cause the suspension to become birefringent. The variation of birefringence induced by a variation of the magnetic field depends on the hydrodynamic friction of the particles in the surrounding liquid. As early as 1910, CORBINO used this effect to determine the size of suspended particles. In the apparatus developed by SCHOLTEN, the measuring cell had a width of 1.75 mm and an optical path length of 10 mm. It was arranged between the poles of an electromagnet and was irradiated with the focused light of a He-Ne laser. A polarizer was arranged before the cell, and a strained glass plate (for producing a suitable additional phase shift), as well as the analyzer and the photodiode, were placed behind the cell. What was observed was the decrease in birefringence obtained after the total compensation of a permanent field by the opposite field of the electromagnet. Polymer addition caused a change in the rate of this optical effect. The thickness of the adsorption layer was derived from the delay, which, for example, was 38 nm for an infinitesimal concentration of cellulose nitrate in i-amyl acetate. Analogous investigations in an alternating field enabled the kinetics of adsorption to be observed.

6.2. Discussion of the experimental results

Let us attempt to arrange the wealth of results indicated above into a more organized picture. Here experiments with porous absorbents should be evaluated with due care if pores and coil dimensions are of the same order of magnitude. To begin with, let us consider the interaction between macromolecules and flat boundary faces. The effect of pores will be discussed later in Section 6.2.5.

6.2.1. The structure of the adsorption layer

The conformation of the adsorbed macromolecules varies within a range whose boundaries are marked by several extreme models (ULLMAN, 1964). They are shown in Fig. 6-7 together with their implications for the quantities adsorbed and the layer thicknesses.

Polymer molecules have a large number of similar, adsorbable groups, but without abandoning the coil conformation only a few of them can be adsorbed on a flat surface (see Fig. 6-1). However, the coil structure is changeable and can be altered by external forces. Conformations having a low thermodynamic probability (such as the flat arrangement of the macromolecule on a smooth surface) can be achieved in spite of the entropy loss if there is an appropriate gain in energy. However, the conversion into such an extreme conformation becomes more and more difficult depending on the degree to which the molecule already differs from the stable coil conformation. Contributing to this are also the stresses acting upon the loops which, being jammed between already fixed trains, still project into the solution and tend to reach the surface. Which conformation is finally reached also depends on the degree of coverage. As long as sufficient space is available, flexible macromolecules are adsorbed in a flat, supported arrangement (SILBERBERG, 1962). This is indicated by the high portion of fixed segments observed by IR and ESR measurements for low degrees of coverage, as well as by the results of electrosorption analysis.

A flat arrangement enables the highest gain of energy, because the enthalpy of adsorption is liberated for a maximum number of segments. Clearly, the adsorption enthalpy per segment must be so high that the molecule approaching by diffusion as a coil is retained rather than being repelled into the solution due to the change in entropy. On the other hand, the adsorption enthalpy must be balanced with the interactions between the solvent and the polymer chain in such a way that the coil does not at once collapse under the action of the surface

Model	Relationship between molar mass and quantity adsorbed, m_a	layer thickness, d_a	portion of adsorbed groups p
flat imposed conformation 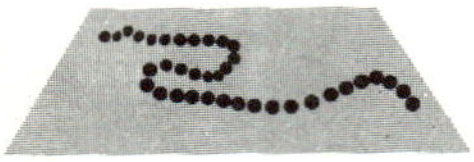	independent of M	independent of M	1
coil 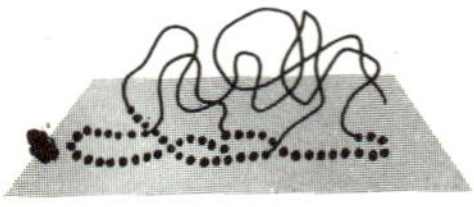	independent of M	$\sim M^{0.5}$	$\ll 1$
coil, collapsed 	$\sim M^{1/3}$	$\sim M^{1/3}$	≈ 0.1
brush type 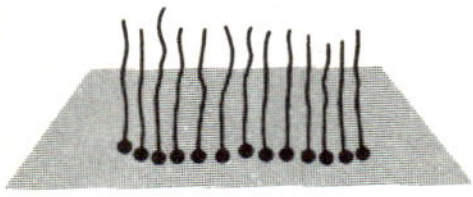	$\sim M$	$\sim M$	$\ll 1$
loops and trains 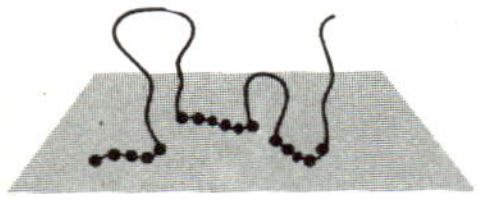	$\sim M^a$ $0 \leq a \leq 0.5$	independent of M	≈ 0.5

Fig. 6-7
Model for the adsorption of macromolecules

forces and remains lying in tangled layers. (For instance, this can be expected in the case of adsorption from θ-solutions.) To enable a higher fraction of the anchor groups to be bonded, the coil must be supported by the solvent until the chain has unfolded and attached to the surface train by train. Moreover, segments already adsorbed must also be able to disengage from the surface temporarily. Obviously this exchange of adsorbed segments is possible in certain cases, e.g., for polyvinyl pyrrolidone on Aerosil in water. This follows from ESR results, which also provide essential information about the conformation in saturated layers. Robb and Smith (1974) found that the fixed fraction $p = 0.90$ produced by spin-labelled molecules on incompletely covered surfaces was reduced to 0.60 by the addition of another, unlabelled polymer. IR measurements likewise indicated a decrease in p with increasing coverage in numerous systems.

The smaller p values for a completely covered surface mean that a greater part of the segments are no longer directly fixed. The saturation itself may occur because there is just no more space on the surface or because all adsorption sites are occupied. According to measurements carried out by DIETZ (1976) for Aerosil, 48% (polyester 2.4) to 74% (PC) of the SiOH groups are still unbonded when the quantity adsorbed reaches its limiting value. Consequently there are still a large number of bondable surface sites under the covering macromolecules. Thus the adsorption ceases because further macromolecules do not find any space on the surface.

In fully occupied adsorption layers, all the molecules are bonded with an average fraction, p, of their segments and projecting into the solution with the fraction $1 - p$. This is reflected by the concept that adsorbed macromolecules exhibit immobile trains and loose loops. However, the p value only provides information about the average fraction of bonded groups, but not about their distribution. If the complete adsorption layer is produced from a sufficiently concentrated solution in a single step, then bonded and non-bonded segments will be uniformly distributed in all macromolecules. However, if in the course of a stepwise adsorption process the quantity adsorbed increases more rapidly than the number of the bonded groups, then p decreases. This does not imply a change in the conformation of the molecules initially adsorbed. (In an extreme case the total quantity might increase by growing onto an adsorbed layer without any contact with the base. Such a multiple-layer adsorption would also lead to a decrease in p.

In this respect the mentioned ESR measurements of ROBB and SMITH are rather interesting: if the unlabelled polymer added in the second experiment were adsorbed without affecting the conformation of the previously adsorbed molecules, then $p = 0.90$ should remain constant (or, in the case of a total coverage, even increase). The decrease of p within 2 minutes after the addition indicates a remarkable mobility of the adsorbed chains. This is all the more surprising as it was not possible to achieve any desorption by water in the investigated polyvinyl pyrrolidone/water/silica system within 24 hours.

The driving force for the conformation balance is the entropy. Further investigations of this kind will have to show whether there are any more systems in which the conformation equilibrium is reached so rapidly. Ellipsometric investigations of polystyrene on metal surfaces, on the other hand, led to the conclusion that this would take about 3 hours (STROMBERG, 1967). Polymethyl methacrylate is adsorbed on Aerosil from dichloroethane (DCE) with a higher p value than from solutions in carbon tetrachloride (T) or butyl chloride (B). KALNINŠ et al. (1976) observed that adsorption layers produced in T only gradually reached the equilibrium value in T/DCE mixtures requiring about one day at 20 °C. The relatively high p value of a layer adsorbed from a DCE solution only slowly decreased to the value corresponding to a solution containing 95% T. Only after several weeks of contact was a final value of p obtained.

Potential measurements with chromium electrodes in contact with polyvinyl acetate even indicated that during the time of observation there is no exchange between the initially adsorbed molecules, which have flat conformation, and those adsorbed finally with large loops (GOTTLIEB, 1960). A high activation energy for the site exchange processes on the surface, e.g., because of too great distances between active surface groups, and a low chain mobility lead to a coexistence of extreme conformations.

Almost all of the spectroscopic investigations reveal 25—50% of bonded segments, whereas almost all of the layer thickness determinations indicate layers 20—50 nm thick. The two results are consistent if it is assumed that part of the molecules are rightly bonded to the surface, while others are loosely adsorbed, with chain ends projecting far into the solution (HESSELINK, 1975). HOEVE (1976) formulated a theory which indicates a density step between the cover close to the surface and an adjacent, diffuse layer, the thickness of which decreases exponentially. The diffuse, outside layer contains only 10% of the segments in the form of a few, very long loops.

6.2.2. Effect of the temperature

The adsorption on solid surfaces is exothermic. For polymers, too, this is confirmed by calorimetric investigations (KILLMANN, 1976) as well as by the evaluation of the band shift in IR spectra.

Following CURTHOYS et al. (1974), DIETZ (1976) calculated the energy of hydrogen bondings involving SiOH groups from the frequency shift to be 21—26 kJ · mole^{-1}, where the upper limit was reached by the flexible 6.8 and 10.12 polyesters (for an explanation, see Table 6-2). The data are in good agreement with the calorimetrically determined value of 21.3 kJ · mole^{-1} for ethyl acetate on Aerosil.

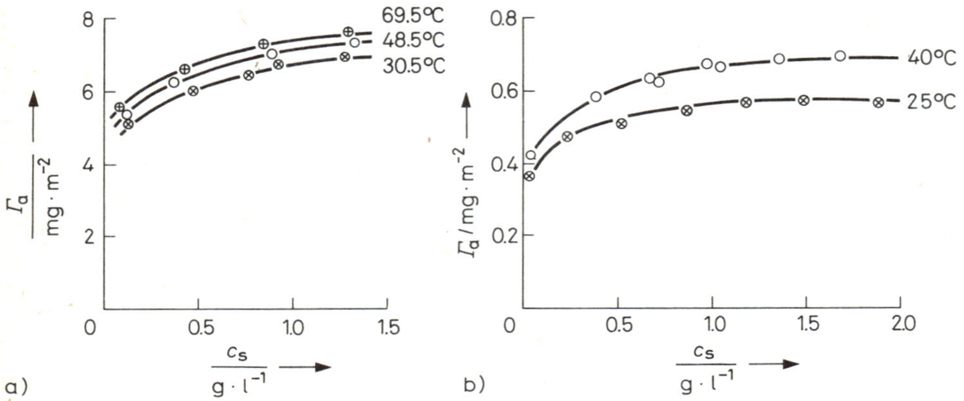

Fig. 6-8
Effect of temperature on the adsorption isotherms

a) Polyvinyl acetate ($\bar{M}_w$ = 250000 g · mole^{-1}) on iron powder from carbon tetrachloride (according to KORAL, ULLMAN and EIRICH, 1958)

b) Polyethylene glycol ($\bar{M}_n$ = 6000 g · mole^{-1}) on Aerosil from benzene (according to KILLMANN, 1976).

A negative (exothermic) adsorption enthalpy means a lower adsorption at a higher temperature. Normally, i.e., for low-molecular-weight substances, this turns out to be true. However, at low temperatures macromolecules usually show a reduced adsorption, e.g., polyethylene glycol on Aerosil (KILLMANN, 1976), polyethylene terephthalate on glass (STROMBERG and GRANT, 1963) and polyvinyl acetate on iron or tin (KORAL et al., 1958) (Fig. 6-8). The Clausius-Clapeyron evaluation of such a temperature dependence would yield an endothermic heat of adsorption, in contradiction to the calorimetric measurements. The Clausius-Clapeyron evaluation fails because the adsorption does not proceed isosterically.

Obviously at different temperatures the entropy effect leads to different conformations having different p values and different maximum coverages. Moreover the abnormal temperature behaviour may also be due to kinetic factors. The conformational rearrangements of partially adsorbed coils, which are required for high coverages, proceed at a sufficient rate only at higher temperatures. A decrease in the maximum coverage with increasing temperature, required by theory, was observed by ELLERSTEIN and ULLMAN (1961) for polymethyl methacrylate from toluene or benzene on iron powder, whereas on Pyrex glass powder there was no temperature dependence at all, possibly due to the mutual cancellation of different effects. The fact that the adsorption of polystyrene from cyclohexane is smaller at 50 °C than at 34.8 °C, the θ-temperature (BURNS and CARPENTER, 1968), can readily be understood in view of the different solvent power (Fig. 6-12).

6.2.3. Effect of the solvent

Like low-molecular-weight substances, polymers are also adsorbed only if the interaction of the solvent with the surface is not too strong. Polar solvents with a high dielectric constant may shield the surface to such a degree that an adsorption does not occur. For preliminary elimination of the solvent quality, let us consider two θ-systems: polystyrene in cyclohexane and polyisobutylene in benzene. While polystyrene is adsorbed on aluminium and alumina, polyisobutylene is not adsorbed (BURNS and CARPENTER, 1968).

Polar solvents from which no adsorption takes place are able to remove adsorption layers produced from other solutions. Thus polyvinyl acetate is removed from iron or tin by acetonitrile (KORAL et al., 1958), polymethyl methacrylate from glass or iron by benzene containing 25% of acetonitrile (ELLERSTEIN and ULLMAN, 1961) and polystyrene from Aerosol by benzene (HERD et al., 1971).

The competition between the dissolved molecules and the solvent determines the adsorption behaviour. For low-molecular-weight solutions the other properties are quite negligible compared with this competition, whereas in polymer solutions the thermodynamic quality of the solvent is decisive.

Thus it is observed from ellipsometric results that chromium does not adsorb polystyrene from methyl ethyl ketone or dioxane solutions, but does adsorb it from cyclohexane (KILLMANN and v. KUZENKO, 1974). The *fraction of fixed carbonyl groups* of polymethyl methacrylate which is adsorbed on Aerosil from carbon tetrachloride, a poor solvent, is $p = 0.55$, much higher than in the case of adsorption from good solvents like chloroform (0.30) or trichloroethylene (0.24) (HERD et al., 1971). The p_{ESR} values determined by ROBB and SMITH (1974) for polyvinyl pyrrolidone show a slight increase (for a 100% coverage) from water (0.69) to 0.1 M NaCl (0.71) and chloroform (0.74).

The *heats of adsorption* for polyethylene glycol ($M = 40000$) on Aerosil increase from methanol (6.6 J/g) through water (17.8) and benzene (69.5) to carbon tetrachloride (71.5) in the same order as the degree of saturation increases (KILLMANN, 1976).

For several polymer systems, Table 6-3 and Fig. 6-9 provide information about the relationship between the adsorption and the thermodynamic quality and displacing power of the solvent.

6.2.4. Effect of the molecular size

The quantity adsorbed usually increases with the molar mass of the solute. Thus the high-molecular-weight polymers formally follow *Traube's rule*, according to which in homologous

Table 6-3
Solvent properties and quantity of polymer adsorbed

The dielectric constant and the ε^0 value of Snyder's adsorption theory (see Section 7.2.) characterize the adsorption tendency of the solvent; the intrinsic viscosity is a measure of the thermodynamic quality for a constant molar mass of the polymer.

Polymer	10^{-3} M	Adsorbent	Solvent	ε^0	Dielectric constant	$\frac{[\eta]}{cm^3 \cdot g^{-1}}$	$\frac{m_a}{mg \cdot g^{-1}}$	Authors
PS	231	Aerosil	Tetra	0.18	2.24	7.8	154	HERD et al.
			Tri	—		84.1	100	(1971)
			Bzn	0.32	2.32	88.6	0	
			MEK	0.51		47.3	0	
PMMA	541	Aerosil	Tetra	0.18	2.24	23.0	>140	
			Bzn	0.32	2.32	125	180	HERD et al.
			Tri	—		135	200	(1971)
			MEK	0.51		71.7	0	
PVAC	250	iron	Tetra	0.18	2.24	33.0	1.54	
			Bzn	0.32	2.32	93.5	0.680	
			TCM	0.40	5.48	138	0.348	KORAL et al.
			DCE	0.49	10.0	110	0.535	(1958)
			AcN	0.65	38.8	81.0	0	
	905	iron	Tetra	0.18	2.24	111	2.74	
			Bzn	0.32	2.32	314	0.698	
			DCE	0.49	10.0	368	0.607	

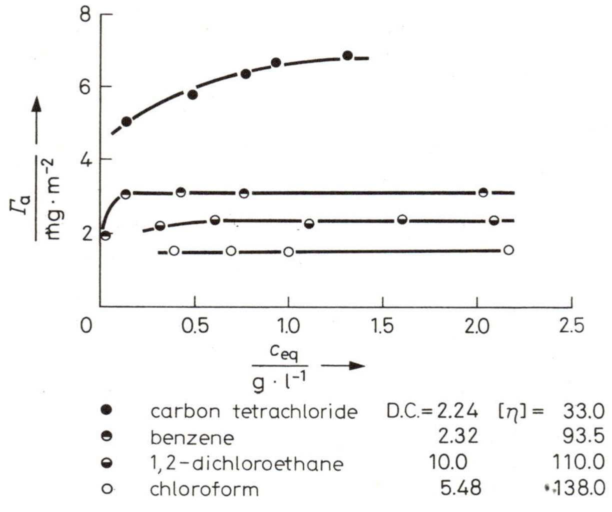

Fig. 6-9
Adsorption isotherms of polyvinyl acetate ($\bar{M}_w = 250\,000$ g · mole^{-1}) on iron powder from four different solvents at 30.4 °C (according to KORAL, ULLMAN and EIRICH, 1958)

The amount adsorbed from a poor solvent such as carbon tetrachloride is greater than that from good solvents. The dielectric constant μ, i.e., the polarity of the solvent, also has a considerable influence. The quantity adsorbed from acetonitrile (μ = 38.8) is nil.

series the adsorption increases with the molecular size. The values can approximately be described by the relationship:

$$m_a = K \cdot M^e \tag{6-2}$$

Table 6-4 contains values of the exponent e. For very high molar masses e approaches zero. Rather high values were observed for the adsorption on silica of oligodimethylsiloxanes having molar masses of less than 3000 g · mole^{-1} (BREBNER et al., 1980).

The poorer the solvent, the stronger is the effect of the molecular size on the amount adsorbed. This influence diminishes with increasing concentration (ROE, 1974). In most cases the layer thickness increases more rapidly with increasing chain length than the amount adsorbed.

Obviously the largest molecules are preferentially adsorbed from mixtures. For the adsorption of polystyrene from toluene on carbon black, FRISCH et al. (1959) observed that the intrinsic viscosity, $[\eta]$, of the supernatant solution increased with increasing concentration of the starting solution. The adsorption of polyvinyl chloride from chlorobenzene on non-porous calcium carbonate depends very much on the chain length, with a preference for particles having molar masses above 100000 g · mole^{-1} (FELTER and RAY, 1970; FELTER, 1971). The molecular size also affects the rate of adsorption. This will be dealt with in greater detail in Chapter 18 because of its immediate impact on chromatographic behaviour. More-

Table 6-4
Values of the exponent e in eqn. (6-2) for the molar-mass dependence of the amount adsorbed

Polymer	Adsorbent	Solvent	e	Authors
PS	glass	CHx	0.26	FURUSAWA et al. (1975)
PS	aluminium (34.7 °C)	CHx	0.13	BURNS and CARPENTER (1968)
PS	aluminium (50 °C)	CHx	0.16	BURNS and CARPENTER (1968)
PMMA	iron	Bzn	0.04	ELLERSTEIN (1961)
		DCE	0.08	ELLERSTEIN (1961)
	Pyrex glass	Bzn	0.00	ELLERSTEIN (1961)
	Aerosil	Bzn	0.00	HERD et al. (1971)
PVAC	iron	Tetra	0.38	KORAL et al. (1958)
		DCE	0.10	KORAL et al. (1958)
		Bzn	0.02	KORAL et al. (1958)
	tin	Tetra	0.20	KORAL et al. (1958)
		DCE	0.10	KORAL et al. (1958)
		Bzn	(0.19)	KORAL et al. (1958)
EVA (14.8)	glass	Bzn	0.33	HARA and IMOTO (1970)
EVA (M > 17800)		Bzn	<0.33	HARA and IMOTO (1970)
PVC	calcium carbonate	CBzn	0.28	FELTER (1971)
PVP	Aerosil	W	~0.5	ROBB and SMITH (1974)
PDMS	glass (30.4 °C)	Bzn	0.40	PERKEL and ULLMAN (1961)
		n-Hp	0.35	PERKEL and ULLMAN (1961)
	iron	Bzn	0.43	PERKEL and ULLMAN (1961)
		n-Hp	0.23	PERKEL and ULLMAN (1961)

over, there is a close relationship between the effect of the molecular size and that of the surface structure.

6.2.5. Effect of the surface structure

Many adsorbents have an extremely large specific surface area because they are porous throughout. The inner surface area can be fully utilized if the pores are large compared to the dimensions of the adsorbate molecules. However, molecules which are too large are excluded from the interior of the pores, and consequently cannot interact at all with most of the total surface. On an adsorbent having pore sizes in the range of the molecular dimensions, the amount of high-molecular-weight fractions of homologous mixtures decreases with increasing molar mass (see Fig. 6-10). If the direct relationship described in Section 6.2.4. is still valid for the other fractions, then the adsorbed amount as a function of the molar mass exhibits a maximum.

FURUSAWA et al. (1975) investigated the adsorption of seven almost uniform polystyrene standards having molar masses between 3700 and 670000 g · mole^{-1} from cyclohexane on porous glass having an average pore size of 17.5 ± 1.7 nm. Fig. 6-11 shows the maximum amounts adsorbed vs. the molecular size. Below 17.5 nm, the pore size, the curve rises normally, according to Traube's rule. Then it passes through a maximum, sloping down steeply at the *exclusion limit*. Without the presence of pores the rise should continue. To test this hypothesis, Fig. 6-12 shows values for adsorption on aluminium and in the adjacent range of molar masses. They confirm the expected behaviour, at the same time demonstrating the special temperature dependence in the vicinity of the θ point as mentioned in Section 6.2.2.

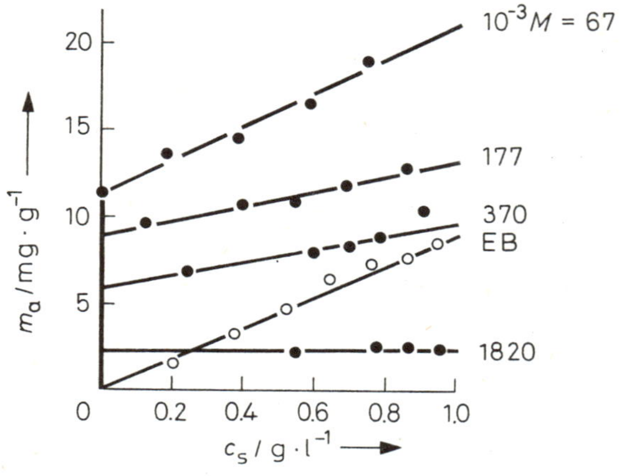

Fig. 6-10
Adsorption of polystyrene from cyclohexane on porous alumina at 50 °C; waiting time: 25 days (according to BURNS and CARPENTER, 1968)

The porous surface (310 m^2 · g^{-1}) is only partially accessible for large-sized molecules. Initially the adsorption isotherms coincide with the ordinate. The reference substance, ethyl benzene (EB), yields a linear isotherm. Since this low-molecular-weight substance reaches a much greater part of the inner surface, the quantity adsorbed (in mg · g^{-1}) is of the same order of magnitude as for polystyrene (on the same surface area, much greater quantities of polymers are adsorbed).

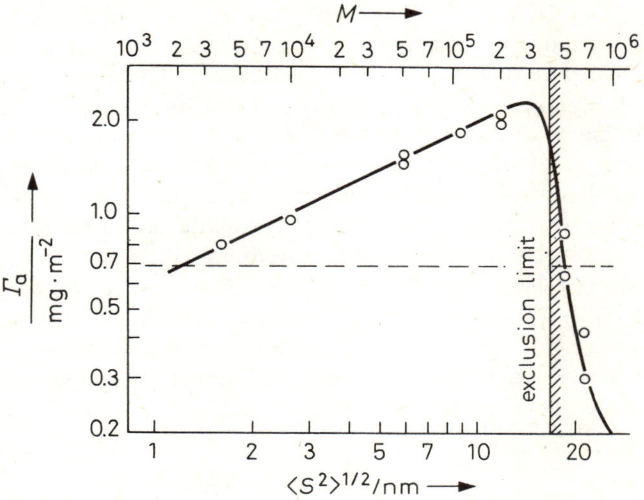

Fig. 6-11
Amount of polystyrene adsorbed from the θ-solvent cyclohexane (35 °C) on porous glass (average pore size 17.5 ± 1.7 nm) as a function of the molar mass and the radius of gyration, respectively, $\langle S^2\rangle_0^{1/2} = 2.63 \cdot 10^{-2} M^{1/2}$ nm (according to FURUSAWA, NAKANISHI and KOTERA, 1975)

Between M = 3700 and 200000 g · mole^{-1} the amount adsorbed increases proportionally to $M^{0.26}$. All of these values lie above the broken horizontal line at Γ_a = 0.69 mg/m^2, which results for a monolayer of undeformed coils.

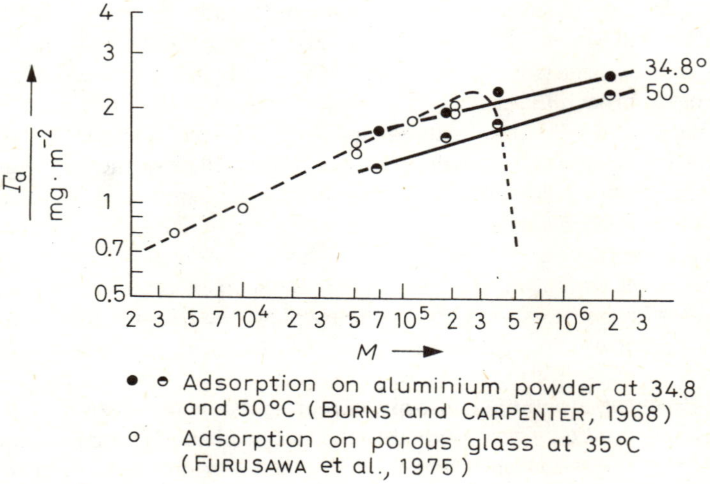

Fig. 6-12
Amount of polystyrene adsorbed from cyclohexane as a function of the molar mass

If the coils were adsorbed in the conformation they have in the solution, then the equilibrium amount should consistently be 0.69 mg $\cdot$ m^{-2} (see Fig. 6-11). However, the measured adsorption is higher for most of the samples. From this it follows that the density of the adsorption layer increases up to three times the coil density. This supports the statement made in Section 6.2.1. about the structure of absorbed macromolecules.

As stated at the beginning of this section, the surface offered by a porous adsorbent can only be fully utilized by molecules which penetrate the pores. If adsorbents of different pore sizes are added to aliquot volumes of a polymer solution, then only a limited adsorption is observed on materials with very narrow pores, whereas the adsorbed amount, Γ, has a high, constant value on adsorbents with very large pore sizes. In the intermediate range, where the pore diameters, d_0, and the values of the end-to-end distance, $\langle R^2 \rangle^{0.5}$, are comparable, Γ increases rapidly with d_0. ŽDANOV et al. (1977) carried out such investigations with polystyrene ($\bar{M}_w = 152000$ g $\cdot$ mole^{-1}; $\langle R^2 \rangle^{0.5} = 39$ nm) in carbon tetrachloride. Using porous glasses with narrow, exactly known pore size distributions, the authors were able to derive the molar mass distribution of polystyrene from the amount adsorbed by the different adsorbents.

6.3. A concluding comparison

In many aspects the adsorption of macromolecules resembles the corresponding behaviour of low-molecular-weight substances. The differences lie above all in the transport to the surface, the extent to which the porous adsorbents can be utilized and the questions relating to the variability of the coils.

The transport to the surface takes place by diffusion. The build-up of the adsorption layer leads to an enrichment with a concentration gradient towards the solution, which has to be overcome by the following molecules. This makes the diffusion towards the interface difficult, but not so much for macromolecules as for low-molecular-weight substances. For the latter, each molecule has to approach the surface to a point where the adsorption forces become effective. Thus each molecule must independently overcome the concentration barrier, which will be steep immediately before the surface. On the other hand, the adsorption of macromolecules starts with the capture of a few segments of the coil periphery. The mass centre must approach the surface by diffusion only up to the distance of a coil radius, which is still relatively large. The subsequent approach is favoured by the adsorption energy, which brings the entire, large-sized molecule close to the surface segment by segment. Moreover, a single macromolecule supplies to the surface the same quantity of adsorbate as hundreds of small molecules which have to diffuse towards the surface separately. Hence polymer solutions can saturate a surface even at low concentrations.

The flexibility of the coil means that the entropy is of considerable importance in polymer adsorption. Nevertheless the adsorption enthalpy plays the decisive rôle in the same way as for low-molecular-weight substances. This follows from the effect of the solvent polarity.

With respect to chromatography, adsorption represents the transition into the stationary phase and must be followed by desorption into the mobile phase. If both processes take place in like manner, reaching a concentration-dependent equilibrium, then all of the chromatographic techniques are applicable. However, here lies an essential difference existing between high- and low-molecular-weight substances. This will be dealt with in Chapter 18.

B Concepts of chromatography: mechanisms and materials

7. Adsorption chromatography

7.1. Adsorption equilibrium (competition model)

Adsorption chromatography utilizes the ability of solid stationary phases to adsorb individual components from mixtures to different extents. Let us consider the simplest case: a substance or sample S in the eluent E. Further, if we suppose that adsorption and desorption are reversible processes and that the active sites of the adsorbent surface are occupied either by substance or by eluent molecules, then:

$$S' + mE'' \rightleftharpoons S'' + mE' \tag{7-1}$$

As in Chapter 3, the mobile phase is indicated by a single prime, the stationary phase by a double one. Thus the formulated equilibrium reaction may be described as follows. A substance molecule S′ adsorbed from the solution displaces *m* solvent molecules E″ from surface positions, the molecule itself becoming part of the stationary phase as S″. The factor *m* indicates the surface area covered by a substance molecule, as referred to the eluent:

$$m = A_S/A_E \tag{7-2}$$

The variation, ΔG, of the thermodynamic potential in the exchange process (7-1) is

$$\Delta G = \Delta H - T\Delta S + RT\Delta \ln a \tag{7-3}$$

at equilibrium $\Delta G = 0$. If the entropy contribution, $T\Delta S$, can be neglected, which is in general permissible for low-molecular-weight substances, then for the adsorption equilibrium

$$\Delta H = -RT\Delta \ln a \tag{7-4a}$$

or, in full:

$$\Delta H''_S + m\Delta H'_E - \Delta H'_S - m\Delta H''_E = -RT \ln \frac{a''_S a'^m_E}{a'_S a''^m_E} \tag{7-4b}$$

This equation can be greatly simplified: the enthalpy difference is determined to a much higher degree by the interactions with the surface than by interactions within the solution, which to a first approximation cancel one another. Moreover, in sufficiently dilute systems the activity of the solvent is approximately equal in the solution and in the adsorption layer. Then the logarithmic expression depends only on a'_S and a''_S, and eqn. (7-4b) can be reduced to:

$$\Delta H''_S - m\Delta H''_E = -RT \ln (a''_S/a'_S) \tag{7-5a}$$

Let us first consider the right-hand side of this equation. The quotient a''_S/a'_S is the *thermodynamic distribution constant* (cf., eqn. (3-3)). For the discussion of adsorption equilibria, the conventional distribution constant which has already been used (eqn. (3-6)), is best stated as an *adsorption constant*

$$K^* = \frac{m''_S/m_A}{m'_S/V'} = \frac{n''_S/m_A}{n'_S/V'} \tag{7-6}$$

where m'_S/V' is the concentration of the substance in the mobile phase or the solution, and m_A is the mass of the solid adsorbent. The adsorption constant describes the relationship detected in recording adsorption isotherms: the amount of substance bonded per gram of adsorbent is referred to the concentration of the solution. While K and K^+ are dimensionless, K^* is expressed in $cm^3 \cdot g^{-1}$. The following relationship exists between the distribution constant, K, and the adsorption constant:

$$K^* = K\frac{n''_E}{n'_E}\frac{V'}{m_A} = KV_a \tag{7-7}$$

$$V_a = \frac{n''_E(V'/n'_E)}{m_A} = \frac{n''_E V_E}{m_A} \tag{7-8}$$

where $V_E = V'/n'_E$ is the molar volume of the eluent; V_a is the "surface volume of the adsorbent", i.e., the volume of a monolayer of eluent covering the adsorbent surface. V_a is proportional to the *accessible surface area*, ^1A, and may serve for the characterization of adsorbents.

With most solvents for chromatography, the following rule connects V_a (in $cm^3 \cdot g^{-1}$) with ^1A (in $m^2 \cdot g^{-1}$):

$$V_a = 3.5 \cdot 10^{-4}\,{}^1A \pm 20\% \tag{7-9a}$$

This simple relationship possible because most eluents do not differ very much with respect to the area covered per molecule and the molar volume.

Using eqn. (7-7), instead of eqn. (7-5a) one obtains:

$$\frac{\Delta H''_S}{2.3RT} - \frac{m\,\Delta H''_E}{2.3RT} = -\log K^* + \log V_a \tag{7-5b}$$

Now let us consider the left-hand side of eqn. (7-5a). The usual symbols employed in the chromatographic literature are obtained if the following stipulations are made:

— The denominators $2.3RT$ are included in the energy values.
— As adsorption enthalpies are generally exothermic, the negative sign is included in the new symbols.
— $\Delta H''_S$ and $\Delta H''_E$ are each sub-divided into two factors, one of which, called α_A, characterizes the *activity of the adsorbent*, and the other one can express a pure substance property of the adsorbate.

We thus obtain:

$$\text{a)}\; -\frac{\Delta H''_S}{2.3RT} = \alpha_A S^0 \qquad \text{b)}\; -\frac{m\,\Delta H''_E}{2.3RT} = A_S\alpha_A\varepsilon^0 \tag{7-10}$$

The factor A_S in eqn. (7-10b) is derived from m by means of eqn. (7-2), taking the molecular area of the eluent as a basis.

This yields the *fundamental equation of adsorption chromatography:*

$$\boxed{\log K^* = \log V_a + \alpha_A(S^0 - A_S \cdot \varepsilon^0)} \qquad (7\text{-}11)$$

Considering the rather far-reaching approximations, this equation derived by SNYDER [A4] provides a remarkably good description of the great variety of phenomena. Extensions of it which may be required will be dealt with in Section 7.5.

The adsorption model proposed by SOCZEWIŃSKI et al. (1969, 1973) is likewise a *competition model*, whereas SCOTT and KUCERA (1973, 1977) developed a *solvent interaction model*. The latter authors assumed that, if multicomponent eluents are used, the retention of the solute on silica is effected by sorption on a layer built up by the eluent component that is adsorbed most strongly. However, the experimental results of SLAATS et al. (1978) suggest that *sorption* may only occur at rather *high* concentrations of the stronger component in the eluent. SNYDER (1974a) evaluated the different models. The ensuing discussion finally induced SNYDER and POPPE (1980) to carry out an extensive investigation, which showed that the "sorption mechanism contains internal inconsistencies and is further contradicted by other evidence."

7.2. Discussion of eqn. (7—11) for adsorption chromatography on polar adsorbents

Eqn. (7-11) shows how the adsorbent, the eluent and the sample interact in adsorption chromatography, thus determining K^* directly and the retention ratio, R, indirectly.

The properties of the adsorbent are described by V_a and α_A. As already mentioned in connection with eqn. (7-8), V_a is the volume of a monolayer of eluent per gram of adsorbent, called the *surface volume* of the adsorbent. It is a measure of the chromatographically utilizable surface, which is reduced for instance by a *preloading* with water. As a water volume of circ. $3.5 \cdot 10^{-4}$ cm^3 likewise yields a monolayer on a surface area of 1 m^2, for partially deactivated adsorbents:

$$V_a = 3.5 \cdot 10^{-4}\, {}^1A - 10^{-4} V_W \qquad (7\text{-}9b)$$

The amount of water added in the deactivation, V_W, is expressed in cm^3 of water per g of adsorbent.

α_A, a dimensionless factor, characterizes the specific *activity of the adsorbent*. The standard value $\alpha_A = 1$ is assigned to highly activated alumina. For deactivated adsorbents α_A is less than 1.

The properties of the sample are described by S^0 and A_S. S^0 is the *energy of adsorption* of a solute passing from a solution in pentane, the standard eluent, to an adsorbent having the standard activity $\alpha_A = 1.0$. It does not depend on the activity of the adsorbent or the eluotropic strength of the solvent, but is determined by the molecular structure and can be calculated as a sum of increments related to the individual structural elements. The *increment addition* is successful if

— the structure of the molecules is such that all the structural elements i can approach the adsorbent surface in like manner,
— the adsorbent surface is so densely covered with equivalent adsorption sites that all the i groups can be attached, and

Table 7-1
Increments Q_i for the additive calculation of adsorption energy according to $S^0 = \Sigma Q_i$ (SNYDER [A 4])

For adsorption on Florisil, using the values which apply to SiO_2 gives sufficient accuracy. The increments for adsorption on MgO may be estimated from the values for Al_2O_3:

$$Q_i(MgO) = 0.77\, Q_i(Al_2O_3)$$

Group	In aliphatic compounds on		In aromatic compounds on		In mixed aliphatic/aromatic compounds on	
	Al_2O_3	SiO_2[1)]	Al_2O_3	SiO_2[1)]	Al_2O_3	SiO_2[1)]
Methyl $-CH_3$	−0.03	0.07	0.06	0.11	—	—
Methylene $-CH_2-$	0.02	−0.05	0.12	0.07	0.07	0.01
Fluorine $-F$	1.64	1.54	0.11	−0.15	—	—
Chlorine $-Cl$	1.82	1.74	0.20	−0.20	—	—
Bromine $-Br$	2.00	1.94	0.33	−0.17	—	—
Iodine $-I$	2.00	1.94	0.51	−0.15	—	—
Ether bridge $-O-$	3.50	3.61	1.04	0.87	1.77	1.83
Sulphide bridge $-S-$	2.65	2.94	0.76	0.48	1.32	1.29
Nitro $-NO_2$	5.40	5.71	2.75	2.77	—	—
Amino $-NH_2$	6.24	8.00	4.41	5.10	—	—
Nitrile $-CN$	5.00	5.27	3.25	3.33	—	—
Carbonyl $-CO-$	5.00	5.27	4.36	4.56	3.74	4.69
Ester $-COO-$	5.00	5.27	4.02	4.18	3.40	3.45
Hydroxyl $-OH$	6.50	5.60	7.40	4.20	—	—
Carboxyl $-COOH$	21	7.6	19	6.1	—	—
Amide $-CONH_2$	8.9	9.6	6.2	6.6	—	—
Phenyl & phenylene	1.86	1.50	1.86	1.50	1.86	1.50
Olefinic or aromatic carbon $-C=$	0.31	0.25	0.31	0.25	0.31	0.25

[1)] narrow-pore silica gel

— the electronic structure of the different groups in the molecule is not changed by the mutual interaction of these groups.

Table 7-1 lists values of the increments, Q_i, of groups which may also be present in macromolecules. The data show that it is not immaterial whether a certain group is bonded aliphatically or aromatically. For example, the value to be associated with the halogen atom would be $Q_i(Al_2O_3) = 1.82$ in PVC, but $Q_i(Al_2O_3) = 0.20$ in poly-p-chlorostyrene. The Q_i values are experimentally determined by means of simple compounds, which show a clear pattern of influences and differ from each other only by the particular group of interest.

A_S is the *effective area covered by the substance molecule* (molecular area). It can be calculated from the monolayer covering the adsorbent surface, and results from the quotient of the specific surface area of the adsorbent and the amount of substance adsorbed per gram. According to SNYDER, the numerical values are referred to benzene (actual molecular area 0.51 nm^2), for which one sets $A_S = 6$. Consequently the unit of A_S corresponds to 0.085 nm^2.

A_S can also be calculated as a sum of increments, a_i, for the individual structural elements of the molecule. If this is done for S^0 as well as for A_S, then the fundamental equation (7-11) yields for the xth member of a homologous series

$$\log K_x^* = \log V_a + \alpha_A \left[\sum^{i=x} (Q_i - a_i\varepsilon^0) \right] \tag{7-12}$$

and for the following $(x + 1)$th member:

$$\log K_{x+1}^* = \log V_a + \alpha_A \left[\sum^{i=x+1} (Q_i - a_i\varepsilon^0) \right] \tag{7-13}$$

For successive homologues this gives

$$\log K_{x+1}^* - \log K_x^* = \alpha_A(Q_i - a_i\varepsilon^0) = \Delta R_M \tag{7-14}$$

found by MARTIN (1949) for partition chromatography, and applied to adsorption chromatography by SPORER and TRUEBLOOD (1959). In this case, for aliphatic samples, it is valid up to $x = 6$.

Table 7-2 shows a_i data. For some structural units one has to assume much higher a_i increments on SiO_2 than on Al_2O_3. The groups in question are most strongly adsorbed, attaching preferentially to sites having a high adsorption energy. The solvent molecules are also more strongly bonded to these sites, so that here the desorption requires more energy, and hence a higher value of $A_S\varepsilon^0$. The *eluotropic strength*, ε^0, is defined as a constant for every chemical compound. Thus the localization of the adsorbate at sites of very high energy yields higher A_S values which, in the sub-division into increments, are associated with the

Table 7-2
Increments, a_i, for aliphatically bonded groups for the calculation of the effective molecular area of adsorbed molecules according to $A_S = \Sigma a_i$ (SNYDER [A 4])

Group	Molecular area, a_i, referred to benzene = 6		
	Calculated from van der Waals' radii	Determined from chromatographic data	
		on Al_2O_3	on SiO_2
Methyl $-CH_3$	1.6	1.6	1.6
Methylene $-CH_2-$	0.9	0.9	0.9
Fluorine $-F$	1.2	1.2	1.2
Chlorine $-C$	1.5	1.5	1.2
Bromine $-Br$	1.8	1.8	1.8
Iodine $-I$	2.1	2.1	2.1
Methyl ether $-OCH_3$	2.1	2.1	9.0
Methyl ester $-COOCH_3$	3.2	3.2	10.5
Acetyl $-COCH_3$	2.6	2.6	9.8
Hydroxy $-OH$	1.3	1.3	8.5
Amide $-CONH_2$	3.1	3.1	10.3
Nitro $-NO_2$	2.3	2.3	9.5
Amino $-NH_2$	1.5	1.5	8.7
Nitrile $-CN$	1.5	1.5	8.7
Phenyl	5.5	4.9	4.9
Phenylene	4.9	4.9	4.9

groups having high adsorption strength. As the surface of silica gel has different types of hydroxyl groups with a distinctive adsorption power, the effect mainly occurs with this adsorbent.

ε^0 is a measure of the *energy of adsorption of the eluent* on an adsorbent having activity $\alpha_A = 1$, referred to the unit of area $A_S = 6$ for benzene and the adsorption energy of pentane on Al_2O_3, which is arbitrarily taken as zero. The adsorption of saturated aliphatic hydrocarbons is mainly effected by dispersion forces. In compounds with polar or polarizable groups the dispersion forces likewise represent the basic value of intermolecular forces, to which further contributions specific to the substance structure are to be added. The difference $S^0 - A_S\varepsilon^0$ increases with these substance-specific interactions. Consequently a reasonable starting position has been chosen by setting $\varepsilon^0_{pentane} = 0$.

Table 7-3 shows ε^0 values which were determined using alumina as an adsorbent. The elution capacity increases with the ε^0 value, in accord with the *eluotropic series* determined

Table 7-3
Eluotropic series of solvents having the strength ε^0 on alumina; the reciprocals of the molar volumes and the relative areas covered by the molecules (according to SNYDER [A 4]) are included

Solvent	$\varepsilon^0(Al_2O_3)$	$10^4 \cdot V^{-1}$	Molecular area A_E[1)]
Fluoroalkane	—0.25		
n-pentane	0.00	87	5.9
i-octane	0.01	61	7.6
Petroleum ether	0.01	~80	6.7
n-decane	0.04	51	10.3
Cyclohexane	0.04	93	6.0
Cyclopentane	0.05	107	5.2
Diisobutene	0.06	64	7.6
Pent-1-ene	0.08	93	5.8
Carbon disulphide	0.15	166	3.7
Carbon tetrachloride	0.18	104	5.0
Amyl chloride	0.26	83	4.2
Xylene	0.26	~82	7.6
i-propyl ether	0.28	71	5.1
i-propyl chloride	0.29	109	3.5
Toluene	0.29	94	6.8
n-propyl chloride	0.30	114	3.5
Chlorobenzene	0.30	98	6.8
Benzene	0.32	113	6.0
Ethyl bromide	0.35	131	3.4
Ethyl ether	0.38	96	4.5
Ethyl sulphide	0.38	93	5.0
Chloroform	0.40[2)]	126	5.0
Methylene chloride	0.42[2)]	157	4.1
Methyl isobutyl ketone	0.43	83	5.3
Tetrahydrofuran	0.57	123	5.0
1,2-ethylene dichloride	0.44	127	4.8
Methyl ethyl ketone	0.51	112	4.6
1-nitropropane	0.53	112	4.5
Acetone	0.56	136	4.2
Dioxane	0.56	117	6.0

Table 7-3 (continued)

Solvent	$\varepsilon^0(Al_2O_3)$	$10^4 \cdot V^{-1}$	Molecular area A_E[1])
Ethyl acetate	0.58	102	5.7
Methyl acetate	0.60	125	4.8
Amyl alcohol	0.61	92	8.0
Dimethyl sulphoxide	0.75	140	4.3
Aniline	0.62	110	6.7
Diethyl amine	0.63	97	7.5
Nitromethane	0.64	185	3.8
Acetonitrile	0.65	191	3.1
Pyridine	0.71	124	5.8
Butyl cellusolve	0.74	77	6.3
Propanol (i- and n-)	0.82	134	4.7
Ethanol	0.88	171	3.8
Methanol	0.95	249	2.9
Ethylene glycol	1.11	180	4.4
Acetic acid	$\gg 1$	175	8.0

1) On silica gel, strong solvents ($\varepsilon^0 \geqq 0.38$) have higher A_E values (≈ 10).
2) For the pure eluent; if stabilized by an alcohol addition, chloroform and methylene chloride show higher values in localized adsorption (cf., Section 7.4.2.).

empirically by TRAPPE (1940). Exceptions are much rarer than in a classification according to the dielectric constants. In Fig. 7-1, ε^0 values determined experimentally on different adsorbents are plotted against one another. Thus, for the calculation of approximate values we obtain:

$$\varepsilon^0_{\text{alumina}} = 1.30\,\varepsilon^0_{\text{silica gel}} = 1.72\,\varepsilon^0_{\text{magnesium oxide}} = 1.92\,\varepsilon^0_{\text{Florisil}} \qquad (7\text{-}15)$$

SNYDER's theory of adsorption was first applied to polymers by KAMIYAMA and INAGAKI (1974). From eqn. (7-11) the authors concluded that a similar chromatographic behaviour on adsorbents of equal activity could be expected if the difference $S^0 - A_S \cdot \varepsilon^0$ has the same value. Like KAMIYAMA et al. (1969) and FONTANA and THOMAS (1958), they assumed that for adsorption of polymers the conditions existing in the related repeat unit might be representative. Thus they calculated S^0 and A_S from the increments Q_i and a_i.

Fig. 7-2 shows results obtained for $R_f = 0.7$ (GLÖCKNER, 1980a). In this case eqn. (7-11) reduces to $S^0 = A_S \cdot \varepsilon^0$. Investigations of adsorption using n-alkanes have shown that the molecular area of larger-sized molecules does not increase proportionally to the chain length (Fig. 7-3). This result is of interest with respect to the conformation of polymer molecules at the adsorption equilibrium (cf., Section 6.2.1.).

7.3. Experimental evaluation of the parameters

In the determination of the numerical values in eqn. (7-11), SNYDER used alumina with a very high activity, setting $\alpha_A = 1.00$ for this substance.

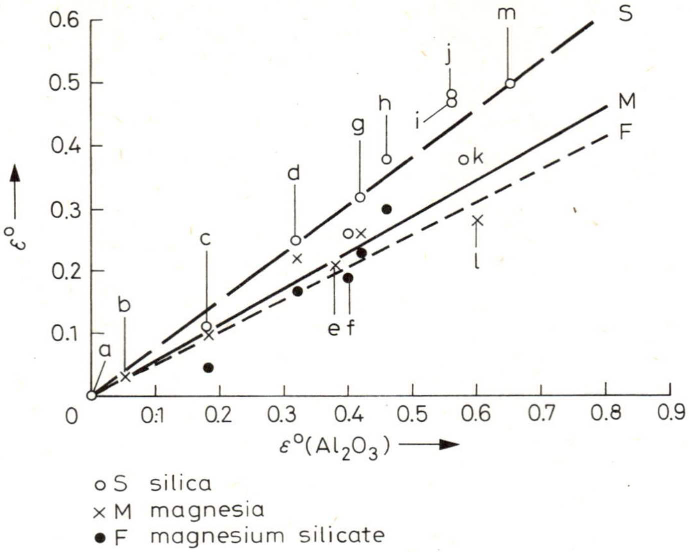

Fig. 7-1
Eluotropic strength, ε^0, of several solvents on silica gel (S), magnesium oxide (M) and magnesium silicate (F) as a function of the value ε^0 (Al_2O_3) on alumina

The following approximate relationships hold:
$\varepsilon^0(S) = 0.77\ \varepsilon^0(Al_2O_3)$
$\varepsilon^0(M) = 0.58\ \varepsilon^0(Al_2O_3)$
$\varepsilon^0(F) = 0.52\ \varepsilon^0(Al_2O_3)$
Solvents:
a P; b CP; c Tetra; d Bzn; e E (for a separation of hydrocarbons); f TCM; g DCM; h E (for a separation of any other compounds); i Ac; j Dx; k EAt; l MAt; n AcN (according to SNYDER [A 4]).

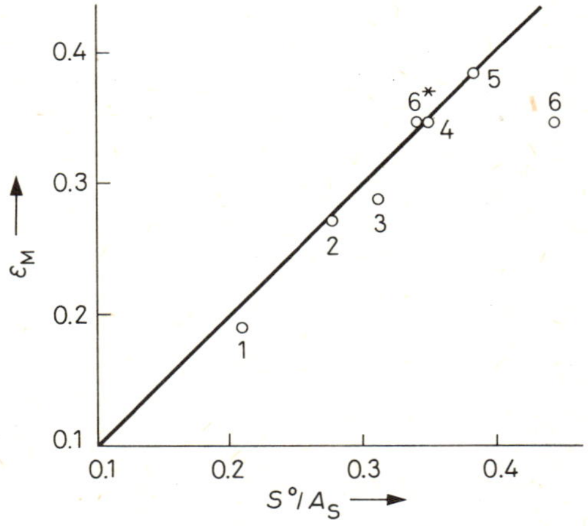

Fig. 7-2
Eluotropic strength giving $R_f = 0.7$ in thin-layer chromatography, plotted vs. the quotients S^0/A_S calculated from increments for the polymers 1 PS; 2 PC; 3 PBMA; 4 styrene/acrylonitrile copolymer; 5 PMMA; 6 CA (for 6* account was taken of the fact that not all of the three acetate groups can be adsorbed simultaneously)

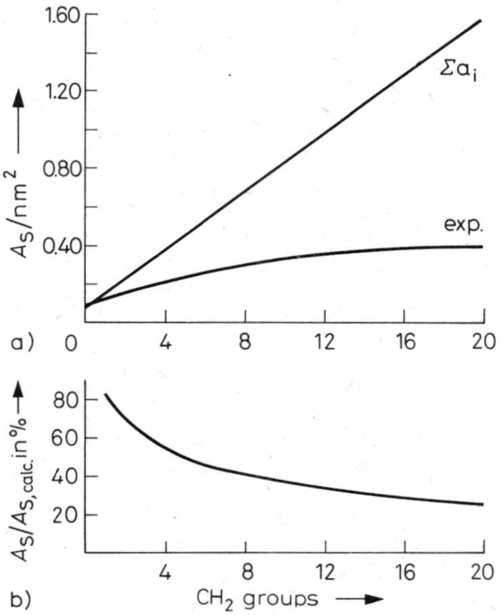

Fig. 7-3
Molecular area of n-alkanes in adsorption chromatography (according to SNYDER [A 4])
The experimentally determined molecular area is much smaller than that calculated from increments.

For this adsorbent the K^* values of a certain sample, if measured in different eluents, yield the following set of equations:

$$\log K_1^* = \log V_a + S^0 - A_S\varepsilon_1^0 \tag{7-16a}$$

$$\log K_2^* = \log V_a + S^0 - A_S\varepsilon_2^0 \quad \text{etc.} \tag{7-16b}$$

From these one obtains by subtraction:

$$\log K_1^* - \log K_2^* = A_S(\varepsilon_2^0 - \varepsilon_1^0) \tag{7-17}$$

The difference on the left-hand side is known from measurement. Using benzene as a sample and pentane as a standard solvent, the ε^0 values of other eluents can be determined, because $A_S = 6$ has been fixed for benzene and $\varepsilon_1^0 = 0$ for pentane. Starting from the assumption that an eluent has the same ε^0 value for all substances, the next step makes it possible to determine the A_S values for other samples by means of the already known ε^0 data, e.g., by plotting $\log K^*$ vs. ε^0 (see Fig. 7-4). Using the new substances, in the third step the determination of ε^0 can again be extended to other solvents. The S^0 data can be obtained in a generally similar way.

Having thus determined the characteristic data for a number of solvents (ε^0) and substance molecules (S^0, A_S) on alumina with the standard activity, it is now possible to determine the parameters V_a and α_A for other adsorbents.

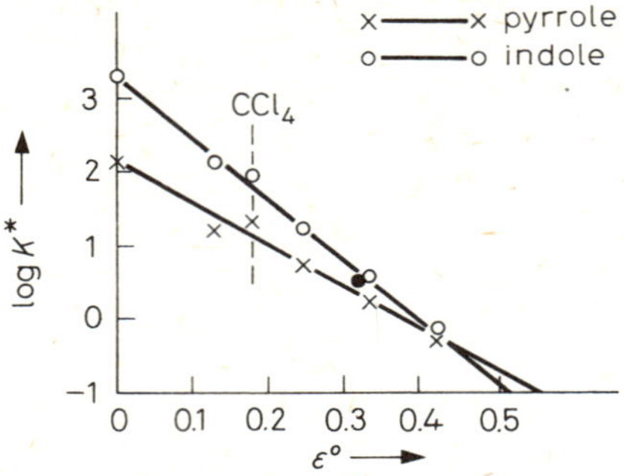

Fig. 7-4
Plot of log K^* values determined by thin-layer chromatography vs. the ε^0 values of the respective eluent employed (mixtures of methylene chloride and pentane)

● Indole developed by benzene
The values of A_s for the substances used can be determined from the slope of the straight line (cf., eqn. (7-17)). The dashed vertical line indicates measured values in carbon tetrachloride ($\varepsilon^0 = 0.18$) (according to SNYDER [A 4]).

7.4. The rôle of the eluent

Among the parameters influencing the separation of a given substance mixture in adsorption chromatography, those of the eluent are of special importance. The following demands are made upon the eluent:

- It should influence the adsorption coefficients in such a way that retention ratios ranging between 0.2 and 0.8 result. (The optimum R value is 0.3.)
- The adsorption coefficients of the sample components must be made to differ so widely that as good a selectivity as possible (see eqn. (3-25)), and hence a high resolution, is achieved. On the other hand, the above condition should be satisfied. Naturally this is only possible if there are not too many components present in the sample.
- The eluent should be a good solvent for the sample, especially in preparative separations. As a rule, the solubility parameters of the substance and of the eluent should differ by one unit at most. (Sometimes, however, a migration in non-solvents is observed.) The adsorbent may alter the solubility. Even a good solvent cannot dissolve an adsorbed substance if it is not at the same time strong enough to displace it from the surface.
- The separation should take place rapidly and without any unnecessary expenditure. Therefore the eluent should have a low viscosity, a favourable boiling point and, for thin-layer chromatography, an optimum flow parameter.
- The properties of the eluent must not affect the detectability of the sample.

7.4.1. Eluent mixtures

Not every separation can be achieved by means of a single-component eluent. As long as complications resulting from different adsorption of the components are ignored, eluent mixtures might be considered chromatographically equivalent to the corresponding pure solvents. For mixtures the properties essential for chromatography, such as the eluotropic strength, ε^0, the viscosity, η, the solubility parameter, δ, etc., range between the values of the component solvents. If the ε^0 value required cannot be realized by a pure solvent but ranges between the values of two liquids suitable for the given problem in respect of their other properties, then the elution desired can usually be performed with a certain mixture of these components. Graduated ε^0 values (with all the other parameters being as constant as possible) can be better realized by a series of mixtures than by pure solvents.

For the eluotropic strength, ε_M, of a binary mixture, SNYDER derived the following relationship

$$\varepsilon_M = \varepsilon_I^0 + \frac{\log\left(x_I + x_{II} \cdot 10^{\alpha_A A_{II}(\varepsilon_{II}^0 - \varepsilon_I^0)}\right)}{\alpha_A A_{II}} \tag{7-18}$$

where ε_I^0 and ε_{II}^0 are the elution parameters of the two components I and II, $x_I = 1 - x_{II}$ is the mole fraction of the weaker component I in the mixture, A_{II} is the molecular area of the more strongly adsorbed component II on the adsorbent and α_A is the activity of the adsorbent. Although in this case the molecular areas, A_S for the substance and A_{II} for the eluent component II, are set approximately equal, eqn. (7-18) nevertheless makes it possible to estimate ε_M with an accuracy of ± 0.02—0.03 units of ε^0.

If ε_I^0 and ε_{II}^0 differ widely from each other, then $x_{II} \cdot 10^{\alpha_A A_{II}(\varepsilon_{II}^0 - \varepsilon_I^0)}$ is much greater than x_I. For an approximate calculation one may use (for $x_{II} > 0.2$ and $\alpha_A(\varepsilon_{II}^0 - \varepsilon_I^0) > 0.2$):

$$\varepsilon_M = \varepsilon_{II}^0 + \frac{\log x_{II}}{\alpha_A \cdot A_{II}} \tag{7-19}$$

This equation has the advantage that it is not necessary to know the eluotropic strength of the weaker solvent.

In most papers the eluent composition is stated in parts by volume or volume percentage. From this the mole fraction x_{II} can be calculated by means of the values V^{-1} listed in Table 7-3. Let us consider, for example, acetonitrile—benzene (1:3 or 25:75, v/v) as an eluent mixture. Acetonitrile is the stronger component, and consequently is given the subscript II. Table 7-3 yields the following values:

Benzene (I)
$\varepsilon_I^0 = 0.32$
$10^4 \cdot V_I^{-1} = 113$
Acetonitrile (II)
$\varepsilon_{II}^0 = 0.65$
$10^4 \cdot V_{II}^{-1} = 191$
$A_{II} = 3.1$

For the mole fraction x_{II} this gives $x_{II} = 191:(191 + 3 \cdot 113) = 0.36$ or, using the % (v/v) data:

$$x_{II} = 25 \cdot 191:(25 \cdot 191 + 75 \cdot 113) = 0.36$$

On an alumina having activity $\alpha_A = 0.7$, the eluotropic strength of the mixture is

$$\varepsilon_M = 0.32 + \{\log [0.36 \cdot 10^{0.7 \cdot 10(0.65-0.32)} + 0.64]\} : (0.7 \cdot 3.1) = 0.50$$

The approximate formula (7-19) yields:

$$\varepsilon_M = 0.65 + (\log 0.36):(0.7 \cdot 3.1) = 0.45$$

Several authors have attempted to calculate the eluotropic strength of binary mixtures by a linear interpolation. This is an over-simplification.

If relationship (7-19) is substituted into the fundamental eqn. (7-11), then using eqn. (3-8) and $A_S/A_{II} \approx 1$ one obtains

$$R_M = \log\left(\frac{V_a m_A}{V'}\right) + \alpha_A S^0 - \alpha_A A_S \varepsilon_{II}^0 - \log x_{II} \tag{7-20}$$

Consequently, in chromatography with solvent mixtures the R_M value varies logarithmically with the mole fraction of the stronger component.

Eqns. (7-18) to (7-20) show that the elution effect of a mixture depends on the activity of the adsorbent, in contrast to the constant ε^0 values of pure solvents. If the activity of the adsorbent is unknown, the strengths of eluent mixtures cannot definitely be stated. This further implies that eluotropic series for mixtures are not equally valid for all adsorbents.

The decrease of retention with increasing concentration of the stronger solvent is not restricted to normal-phase chromatography; it es even more pronounced in reversed-phase chromatography, as formulated in eqn. (7-24). (The interrelation between R_M and k can be deduced from eqn. (3-8).)

7.4.2. Eluent demixing

The flow of an eluent mixture over an adsorbent may itself be considered a chromatographic process in which one of the components is the sample while the other is the eluent. The physico-chemical relationships are the same, being only regarded from a different view-point. If benzene, normally an eluent, is chromatographed as a sample, then one finds $S^0 = 1.86$ (on alumina). As the area occupied by the benzene ring is $A_S = 6$, a value of $\varepsilon^0 = 1.86/6 = 0.31$ is calculated from the properties of the benzene "sample" for the benzene "eluent". The measured value is $\varepsilon^0 = 0.32$. The agreement is so good because there is no localized adsorption of benzene on Al_2O_3 (cf., Section 7.5.3.). If localization occurs, the expression

$$\varepsilon^0 = \left(\frac{S^0 - Q_k \cdot f(Q_k)}{A_S}\right)_{S=E} \tag{7-21}$$

has to be substituted for $\varepsilon^0 = (S^0/A_S)_{S=E}$. Using this equation, SNYDER (1964) found that 23 substances showed a correlation between their behaviour as a sample and that as an eluent, with a standard deviation of ± 0.08 (between the calculated and measured ε^0 data).

In addition to the intended separation of the sample and the eluent, chromatography using eluent mixtures exhibits an unintended separation of the components of the eluent, which obeys the principle of frontal analysis. From the supplied mixture of constant composition, the adsorbent preferentially takes up the stronger component in a zone, the front of which moves forward less rapidly than the remaining eluent. Several components of different eluotropic strengths form a corresponding number of staggered zones in which different elution conditions prevail. Not until the eluent mixture has flowed for some time will the fronts have travelled over the whole bed, which is now in equilibrium with all of the components. From this time onward the elution proceeds isocratically.

For components having extremely different eluotropic strengths, the mixing equation (7-19) is only approximately valid. If the stronger component undergoes localized adsorption, the equation no longer holds. If this component is present in such a low concentration

($x_{II} \ll x_I$) that the molecules of II are just enough to saturate the strong adsorption sites, then the effective eluotropic strength of the mixture is much greater than the calculated value. For that reason commercial chloroform stabilized by addition of alcohol exhibits a much higher ε^0 than the pure product.

The demixing of eluents has consequences mainly in development chromatography. In the flat bed methods, the unintentional development of gradients over the vapour phase has to be taken into account in addition to the chromatographic demixing, especially for components having greater differences in their vapour pressure values.

7.5. Secondary effects

So far the discussion has been based on assumptions which may not be valid in each case. This may cause additional effects which will be discussed in this section.

7.5.1. Interactions in a solution

In the derivation of eqn. (7-11) it was assumed that the enthalpy contributions, $\Delta H'_S$ and $m \cdot \Delta H'_E$, due to the substance and the eluent in the mobile phase, respectively, cancel each other. Obviously this is a rather good approximation for systems with not too strong components, because the relationship (7-11) is satisfied in this case. If the sample is weakly adsorbed, then also the eluents used would never be very strong, because this would lead to rather small distribution coefficients, and hence to a poor resolution (cf., eqn. (3-25)).

The situation is different for strongly adsorbed substances. In this case eluents having high ε^0 values are required. In such systems eqn. (7-11) breaks down. Formally this can be overcome by an additional term:

$$\log K^* = \log V_a + \alpha_A(S^0 - A_S\varepsilon^0) + \Delta_{eas} \qquad (7\text{-}22)$$

Δ_{eas} takes into account the interactions between the sample and the solvent in the mobile phase, e.g. hydrogen bonds. In extreme cases, the two compounds may form a complex adsorbing as a whole. Moreover, Δ_{eas} takes into account a possibly different elution effect of the eluent components on individual sample fractions, which may occur on surfaces having different types of adsorption sites. An eluent component competing with a certain fraction for surface sites of the same type acts upon this fraction as a strong eluent, whereas a component preferring different sites represents a weak eluent for that component (Oščik and Różylo, 1971). The activity of the adsorbent, the structure of its surface and the structure of the adsorbate molecules may also contribute to Δ_{eas}. For example, highly polar eluents may take water from the adsorbent, which was added to adjust the activity. This effect, which greatly influences the sample retention and the selectivity of the separation, depends on the nature of the solvent and on the traces of water which may unintentionally be present in the liquid (Paanakker et al., 1978). If reproducible results are required, the control of the water content is of practical importance. Thomas et al. (1979) define *isohydric solvents* as liquids which, when in contact with a certain adsorbent, adjust the water content of this adsorbent to the same value. This property is ensured by a deliberate pre-moistening of the solvent, which ranges from <0.0005% for isooctane over 0.06% for ethyl acetate to 5.2% for methanol (adsorbent: Spherosil® XOA 600). Empirically a linear relationship has been found between the reciprocal of the mole fraction of this water content and the capacity factor k (cf., eqn. (3-9)) of a solute.

However, water as a deactivation agent exhibits some disadvantages: it is very sparingly soluble in saturated hydrocarbons and several other organic liquids, the concentration in the eluent may undergo unexpected variations due to an exchange with atmospheric moisture, and with silica gel as a stationary phase the equilibrium of activation is not reached until relatively large quantities of eluent have passed through the column. For that reason, SAUNDERS (1976) proposed to perform the adsorbent deactivation by adding acetonitrile to the eluent. The maximum sample capacity, i.e., the largest sample quantity for which the distribution constant does not decrease below 90% of the value for a very small sample quantity, is increased by acetonitrile in a similar ways as with water as a deactivation agent. Both additions effect a considerable extension of the linear range of the adsorption isotherms.

7.5.2. Effects of the adsorbate structure

On adsorbents with pores which are about the size of adsorbate molecules, the normal mechanism is superimposed with steric exclusion. However, on smooth surfaces there are also secondary effects resulting from the molecular structure: the adsorption energy S^0 is exactly equal to the sum of the increments Q_i only in the ideal case. In many cases there are interactions between the adsorbable groups of larger-sized molecules, which make additional contributions, q_{ij}, to the sum of increments. The physical nature of the interactions q_{ij} may be rather varied. The following effects should be considered:

- Disturbance of the smooth contact of the molecule, e.g., a phenyl residue, by the volume of a substituent. Of course *non-planar* molecules are less strongly adsorbed than planar ones. Consequently this effect reduces the adsorption energy ($q_{ij} < 0$)
- *Steric shielding* of an otherwise strongly adsorbable group by substituents in its immediate neighbourhood. This likewise reduces the adsorption energy ($q_{ij} < 0$)
- Reduction of the adsorbability by chemical interactions between two groups of the substance molecule, e.g., *hydrogen bonding* ($q_{ij} < 0$)
- Additional increase of the adsorption energy of an otherwise strongly adsorbable group by other groups which may act as *electron donors*, especially in aromatic systems ($q_{ij} > 0$)
- Simultaneous and equivalent adsorption of neighbouring groups forming *chelate bonds* to one and the same adsorption site of the surface. This effect promotes the adsorption as compared with an isolated attachment ($q_{ij} > 0$)

Thus there are both positive and negative contributions q_{ij}. Therefore, as a rule, the sum of all interactions is so small that the experimental adsorption energy, S^0, still agrees rather well with $\sum^{i} Q_i$.

7.5.3. Localized adsorption

If the adsorption sites on the surface have radii of action which are small compared with the distance between neighbouring centres, as is the case, say, for alumina, then further complications may occur for molecules having several adsorbable groups. The group which has the highest adsorption energy, judged by its Q_i in Table 7-1, is adsorbed locally at one of the distinguished surface sites. Thus the whole molecule is *anchored*, and the other groups can no longer be adsorbed in the same way as with freely movable molecules. The localized adsorption of a group enforces the delocalized adsorption of the other ones.

It is only the strongest group, or one of several equally strong groups, which can contribute the full energy of adsorption as indicated by Q_i; all other groups make a reduced contribution to the total adsorption energy. In such a case the experimental value of S^0 is smaller than the sum calculated from the increments given in Table 7-1. Thus, we can write:

$$S^0 = \sum^{i} Q_i - f(Q_k) \sum^{i-1} Q_i \tag{7-23}$$

The *localization function* $f(Q_k)$ depends on the strength of the group undergoing localized adsorption and, for aliphatic adsorbates, on the number of carbon atoms between the competing groups.

7.6. The rôle of the eluent in reversed-phase chromatography

The higher the probability for a substance to be in the stationary phase, the longer it is retained. Polar components are retained by polar adsorbents, in which case the retention increases with decreasing polarity of the eluent. This yields the series of the $\varepsilon^0(Al_2O_3)$ values, which increase with increasing polarity of medium. On the other hand, on non-polar adsorbents the less polar components undergo a retention which increases with increasing eluent polarity. Howard and Martin (1950) called this technique *reversed-phase chromatography*.

Naturally the $\varepsilon^0(Al_2O_3)$ values, which may, with certain corrections, also be used for chromatography on other polar phases, cannot be employed in the classification of eluents for reversed-phase chromatography. Eluotropic series on non-polar adsorbents exhibit quite a different order (see Table 7-4). The lowest eluotropic strength is shown by water or methanol, the highest by the non-polar hydrocarbons.

Adsorption chromatography on polar adsorbents is based on the attraction between the active sites of the adsorbent and the molecules of the solute, which compete with the solvent molecules for the available adsorption sites. Different views have been taken with regard to the mechanism of reversed-phase chromatography. Partition towards an alkane phase may be true for polymer layers which are capable of swelling, but not for chains which are fixed in a bristle-like arrangement. In many cases retention is mainly due to solvophobic interactions (Horváth et al., 1976; Horváth and Melander, 1978). Fig. 7-5 shows the various possible interactions between a sample molecule and the surface of the stationary phase.

If a non-polar substance from the interior of a polar mobile phase is deposited on the non-polar surface of the adsorbent, this implies a reduction in the number of contacts between the particles of the solute and those of the solvent, i.e., between particles of different polarities. Transferring this from the molecular range to macroscopic dimensions, the deposition of the solute might be characterized as a process leading to a reduction in the interface between polar and non-polar components. This yields a gain in energy which increases with increasing surface tension of the eluent. Water has a high surface tension compared with organic solvents. Consequently, *hydrophobic interactions* lead to a very marked

Table 7-4
Eluotropic series for reversed-phase chromatography

Solvent	Adsorbent: Charcoal [1]	Charcoal A_S [2]	RP 8 relative retention [3]	RP 18 relative retention [3]	Graphitized carbon black $\varepsilon^0(RP)$ [4]
Water	↓	(1.5)	—	—	—
Methanol	↓	(2.9)	1.0	1.0	0
Acetic acid	—	—	2.7	—	—
Acetonitrile	—	—	3.3	3.1	0.039
Ethanol	↓	(3.8)	3.2	3.1	0.051
i-propanol	—	—	8.4	8.3	—
Dimethyl formamide	—	—	9.4	7.6	—
Acetone	↓	(4.2)	9.3	8.8	—
n-propanol	↓	(4.7)	10.8	10.1	—
Ethyl ether	↓	(5.6)	—	—	—
Butanol	↓	(5.6)	—	—	—
Dioxane	—	—	13.5	11.7	—
n-hexane	↓	(6.8)	—	—	0.086
Ethyl acetate	↓	(5.7)	—	—	0.091
Butyl chloride	—	—	—	—	0.112
n-heptane	—	—	—	—	0.119
Dichloromethane	—	—	—	—	0.133
Tetrahydrofuran	—	—	—	—	0.139
n-octane	—	—	—	—	0.139
n-nonane	—	—	—	—	0.161
Benzene	↓	(6.0)	—	—	0.204
n-xylene	—	—	—	—	0.240
References	JERMYN (1957)		KARCH; SEBESTIAN; HALÁSZ and ENGELHARDT (1976a)		COLIN, EON and GUIOCHON (1976)

[1]) Direction of increasing eluotropic strength. (The measure used was the molar concentration of the solvent in water, by means of which the same elution effects were achieved.)
[2]) relative molecular area A_S.
[3]) Relative retention in the elution of the solvent as a sample, $k_{solvent}/k_{CH_3OH}$; eluent: water.
[4]) Eluotropic strength $\varepsilon^0(RP)$, determined chromatographically by means of condensed aromatics used as samples and the solvent used as an eluent.

retention in reversed-phase chromatography. Addition of water-soluble organic liquids such as acetonitrile or methanol results in a reduction in surface tension. The extent of the reduction increases with the concentration, being most dramatic for small additions.

In reversed-phase chromatography, the logarithm of the (isocratic) capacity factor, k (cf., eqn. 3-12), varies almost linearly with the volume fraction, φ, of the organic *modifier*. Over a large range of compositions the data are better described by a quadratic relationship.

$$\ln k = A\varphi^2 + B\varphi + C \qquad (7\text{-}24)$$

Tab 7-4

Solvent[1])	Adsorbent Polystyrene gels ε^0(RP)	Solvent[1])	Adsorbent Polystyrene gels ε^0(RP)
tert. butanol	—0.072	* Dimethyl sulfoxide	0.089
1,1,2-trichloro-1,2,2-trifluoroethane	—0.023	* Acetone	0.089
2,2,4-trimethylpentane	—0.022	1,1,1-trichloroethane	0.105
i-butanol	—0.018	* Methyl ethyl ketone	0.127
* i-propanol	—0.017	Methyl isobutyl ketone	0.144
* Methanol	0	2-pentanone	0.150
sec. butanol	0.010	Diethyl ether	0.151
* n-propanol	0.014	1,2-dichloroethane	0.155
* Ethanol	0.017	* N,N-dimethylformamide	0.160
* n-butanol	0.030	Chloroform	0.185
Propylene carbonate	0.049	Methylene chloride	large
Cyclohexanes	0.053	Ethyl acetate	large
i-hexanes	0.055	* Tetrahydrofuran	large
* Acetonitrile	0.065	1,2-dimethoxyethane	large
Light petroleum	0.066	Benzene	large
n-pentane	0.067	Toluene	large
n-nonane	0.068	o-xylene	large
n-heptane	0.071		
Hexanes	0.074	Reference	ROBINSON, ROBINSON, MARSHALL, BARNES, JOHNSON and SALAS (1980)
Carbon tetrachloride	0.079		
i-propyl ether	0.083		
Cyclohexane	0.088		

[1]) Eluotropic strength ε^0(RP), evaluated from adsorption measurements using benz[α]anthracene or benzo[α]pyrene as a probe.

SCHOENMAKERS et al. (1979) determined the coefficients of this equation for 32 aromatic compounds on a RP 18 column, with methanol, acetonitrile or tetrahydrofuran as organic modifiers. In all cases B is negative, whereas C and generally also A are positive.

Similar investigations were carried out by HENNION et al. (1981) using the modifiers methanol, ethanol, tetrahydrofuran, acetonitrile and 1-propanol. The curves obtained were described by means of the relationship:

$$\ln k = a(1 - \varphi)^n + b \tag{7-25}$$

The exponent n depends on the modifier used. A linear dependence, i.e., $n = 1$, was found only for methanol. With the first four modifiers a was proportional to the molar volume of the aromatic solutes.

Combinations of an organic liquid and water are frequently used in reversed-phase chromatography. In most cases gradient elution is applied, which of course is always started with a high fraction of water but non-aqueous systems are also used. In this case solvophobic interactions may likewise contribute to the separation. The interactions depend on many factors which are in part not yet sufficiently known, including the geometry of the adsorbent surface, the molecular structure of the solvent and the solute and the whole

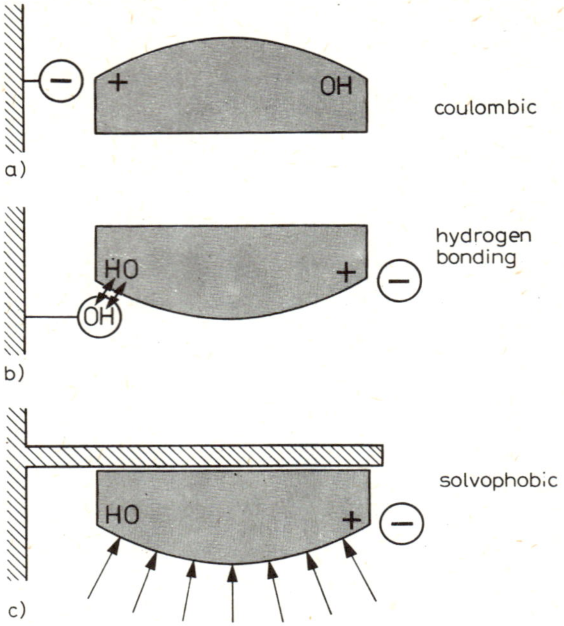

Fig. 7-5
Schematic representation of the interactions in adsorption on polar (a, b) and non-polar (c) stationary phases (according to HORVÁTH, MELANDER and MOLNÁR, 1976)

of the intermolecular forces acting between all the components. Unreacted silanol groups may cause a rather dramatic effect (NAHUM and HORVÁTH, 1981).

Even a layer of alkane chains is not inert to the solvent of the mobile phase. SLAATS et al. (1981) determined the excess concentration of the modifiers acetonitrile and methanol on RP 2, RP 8 and RP 18. The results were interpretable on the basis of a model in which a mixture of constant composition is adsorbed over a limited range of mobile phase compositions. The statement made by other authors that from mixtures with water the organic modifier is exclusively adsorbed, was not sufficient for an explanation of the experimental findings. If the mobile phase contains one component in great excess, this affects the composition of the adsorbed layer, and hence the effective polarity of the bonded phase.

The influence of the solubility of the sample was perceived by LOCKE (1974) and KARGER et al. (1976b). In investigations with ethanol-water mixtures, on RP 18, HENNION et al. (1981) found that the term $k \cdot s/s_0$ remains almost constant (k = capacity factor, s_0 = solubility of the sample in pure ethanol, s = solubility in the eluent mixture).

For members of homologous series, the retention time increases with increasing size of their molecules. This is plausible, since the gain in interfacial energy increases with the contact area.

The fact that reversed-phase chromatography has found such a broad application is, last but not least, due to the possibility of carrying out gradient elutions on non-polar adsorbents almost without any problems. Usually the columns can be rapidly restored to the starting condition required for the next run. This property can also be understood on the basis of the model of solvophobic interactions, according to which attractive forces are only of minor importance. Another practical advantage of reversed-phase chromatography is the fact that traces of water in the eluent have little or no influence on the separations.

On the other hand, the interactions are so specific and so multifarious on the whole that eluotropic series (cf., Table 7-4) do not yet enable any reliable prediction of relative retentions (KARCH et al., 1976). Unlike the $\varepsilon^0(Al_2O_3)$ data, the $\varepsilon^0(RP)$ values are not constants typical of a solvent, but also depend on the samples by means of which they have been determined (COLIN et al., 1976). To get a clear picture of the increasing abundance of data, it has been proposed that the dependence of the capacity factor on the eluent composition (cf., eqn. (7-24)) and on the temperature should be considered simultaneously (MELANDER et al., 1978, 1979).

Gradient elution on reversed phases is usually carried out using binary combinations of water with a modifier, the concentration of which increases during the analysis. Consequently the solvent strength increases. In order to programme the selectivity independently of the polarity, BAKALYAR et al. (1977) generated gradients of ternary mixtures, using two modifiers, the ratio of which was varied in a correspondingly opposite manner to the decrease in the water content.

The importance of the mobile phase in chromatography on non-polar bonded phases was shown by TANAKA et al. (1978). In their investigations on the rôle of organic modifiers in polar group selectivity, these authors used a carefully prepared RP 8 packing material, which did not show any retention in the test with a polar solute in dry heptane (cf., Section 11.10.1.). Using this material, aromatic compounds with different polar groups were analyzed in methanol-water (50:50), acetonitrile-water (30:70) and tetrahydrofuran-water (25:75). These three mixtures resulted in equal methylene group increments of about 2.0, i.e., the mixtures are normalized with respect to their hydrophobic selectivity. (Their surface tensions coincided fairly well.) According to HANSCH and FUJITA (1964), the partition coefficient, P, between water and octanol is a measure of the hydrophobic behaviour of a substance. When log k was plotted vs. log P, the capacity factors measured in the three mobile phases lay on straight lines, having the same slopes, thus confirming the successful normalization of hydrophobic selectivity. When the log k values obtained in THF-W or AcN-W were plotted vs. the corresponding values obtained in M-W, the data points for a homologous series of n-alcohols (which was measured in addition) lay on straight lines of slope 1, which likewise indicated a successful normalization of hydrophobic selectivity. The line parallel to this line which goes through the measuring point for benzene was the line on which the evaluation of group selectivity was based. The latter was low for AcN-W, but very marked for THF-W. For some solutes even the order of elution was reversed when THF-W was used instead of M-W. The variations were very large for small THF additions. The authors concluded that THF and methanol "would constitute an interesting pair to be used with water in a ternary mixture mobile phase for the control of separation of substances with different functional groups."

7.7. The rôle of solubility parameters in chromatographic processes

In connection with the attempts to get generalizable information, the application of the concept of solubility parameters to chromatographic processes should be mentioned. On the basis of this concept it is possible to interpret not only the processes on reversed phases, but also interactions between polar adsorbents and molecules of the mobile phase as well as the partition between two bulk phases.

As usual, however, the generalization involves a loss of precision of the information in an actual case. In addition, the concept of solubility parameters itself has only the character of an approximation, and the numerical values listed in the tables of different authors in part differ considerably from one another. The differences are rather large with the partial solubility parameters δ_d, δ_o, δ_{in}, δ_a and δ_b, but exist also for some of the total values (cf. TIJSSEN et al., 1976).

The advantage lies in the fact that the consideration of solubility parameters gives an insight into the properties of the components involved in the chromatographic process and their interactions, within the framework of general physico-chemical relationships. On the other hand, precise information within narrow ranges is much better obtained by the evaluation and extrapolation of chromatographic measurements (HUBER et al., 1972b, 1973). CHEN and HORVÁTH (1979) reported on *quantitative structure-retention relationships* obtained in this way for some aromatic compounds with different substituents.

The classical concept of solubility parameters was used by ROHRSCHNEIDER (1968) in gas chromatography for the interpretation of relative retentions for non-polar samples on non-polar liquid phases. The further development of this topic, including the use of partial solubility parameters, has been advanced especially by KELLER et al. (1970), KARGER et al. (1976, 1978) and TIJSSEN et al. (1976).

Let us first consider a result of gas chromatography which may also contribute to better understanding of the reversed-phase chromatography discussed above: in 1966, BELYAKOVA et al. investigated the chromatography of low-molecular-weight polar and non-polar compounds on *graphitized carbon black*. This material with its well defined, non-polar surface represents an ideal adsorbent for theoretically relevant studies and may be considered a model for the not so well defined reversed phases. The retention volumes obtained in the above study were plotted by KARGER et al. (1978) as log V_r vs. the product $A_E \cdot \delta_d$ (calculated from the molecular area, A_E, cf., Table 7-3, and the dispersion contribution to the solubility parameter, δ_d, cf., Table 5-2). The result was a fairly good correlation, which shows that the retention on graphitized carbon black is solely due to *dispersion forces*. Even samples as polar as methanol, acetonitrile or nitrobenzene are found exactly where they should lie according to their dispersion forces. It follows that in chromatographic adsorption on carbon black there are no inductive interactions. KARGER et al. (1978) showed that also in the adsorption on polar substances including SiO_2 and Al_2O_3 the inductive forces play a secondary rôle. This fact is explained by the rigid arrangement of the atoms in the adsorbent, which excludes an induction effect of the sample on a non-polar adsorbent such as carbon black. The fact that, on the other hand, a polar adsorbent likewise adsorbs a non-polar sample without any induction effects is related to the flat arrangement of the adsorbed molecule on the surface.

In gas chromatography, the behaviour is determined in an ideal manner only by the interactions between the sample and the adsorbent. The fact that in this case — where no solvophobic contributions can exist — a retention associated with the dispersion interactions occurs on the non-polar material suggests that in RPC these interactions are supplemented by solvophobic effects, and possibly in part cancelled by the corresponding interactions between the solvent and the RP packing material.

In aqueous systems with high values of the surface tension the hydrophobic contributions predominate. As no induction effects occur, the relationship between the adsorption energy, ΔE_{ads}, and the partial solubility parameters (KARGER et al., 1976),

$$\Delta E_{ads} = V(\delta_d \delta_{A,d} + \delta_o \delta_{A,o} + \delta_{in} \delta_{A,d} + \delta_{A,in} \delta_d + \delta_a \delta_{A,b} + \delta_{A,a} \delta_b) \quad (7\text{-}26a)$$

can immediately be reduced to

$$\Delta E_{ads} = V(\delta_d \delta_{A,d} + \delta_o \delta_{A,o} + \delta_a \delta_{A,b} + \delta_{A,a} \delta_b) \quad (7\text{-}26b)$$

where the δ_A values are the partial solubility parameters of the adsorbent. This gives the adsorption energy normalized to the area A_E

$$\varepsilon^0 = C[(\delta_d - \delta_d^0)\, \delta_{A,d} + \delta_o \delta_{A,o} + \delta_a \delta_{A,b} + \delta_{A,a} \delta_b] \quad (7\text{-}27)$$

if A_E can be set proportional to the molar volume, V, which is permissible for a flat arrangement and an equal thickness of the adsorbed layer. For a given adsorbent, C is a constant, and δ_d^0 is the solubility parameter of the eluent to which the ε^0 scale is adjusted. It is assumed that this medium develops only dispersion interactions occur. This is the case with pentane, the reference compound of Table 7-3 ($\varepsilon^0 = 0$, $\delta_d^0 = \delta_t^0 = 7.1$). Eqn. (7-27) shows that ε^0 data may be related to the partial solubility parameters of the liquid concerned, but not to their gross values, δ_t.

In the case of liquid chromatography on a non-polar adsorbent, because $\delta_{A,o}, \delta_{A,a}, \delta_{A,b} = 0$ only the dispersion contributions remain on the right-hand side of eqn. (7-27). KARGER et al. (1978) evaluated data obtained by COLIN et al. (1976) on graphitized carbon black (cf., Table 7-4), and found the expected linear dependence on the dispersion contributions again confirming that induction effects can be neglected.

Eqn. (7-27) allows the calculation of ε^0 values from partial solubility parameters, if the corresponding values of δ_A are known also for the adsorbent. KARGER et al. (1978) estimated these values for Al_2O_3 as $\delta_{A,d} = 10.8$, $\delta_{A,o} = 9.8$, $\delta_{A,a} = 11.4$ and $\delta_{A,b} = 2.5$, and found a good agreement between the measured and the calculated $\varepsilon^0(Al_2O_3)$ values, with a standard deviation of 0.05 units. Using the relationship

$$Q_i = \text{const} \cdot A_S(\delta_o \delta_{A,o} + \delta_a \delta_{A,b} + \delta_{A,a} \delta_b) \quad (7\text{-}28)$$

they were also able to calculate the dimensionless increments, Q_i (cf., Table 7-1), for the contribution of functional groups to the adsorption energy of a monofunctional alkyl derivative from the partial solubility parameters of such types of aliphatic compounds, the dispersion contribution of which was about 7.1 as for the pentane reference. These examples show that obviously the concept of solubility parameters provides a useful instrument for the evaluation of adsorption chromatography. However, it includes too many approximations and simplifications to describe all the phenomena correctly. For instance, it breaks down if localized adsorption occurs (KARGER et al., 1978).

The implications of the concept of solubility parameters for liquid-liquid partition chromatography will be discussed in Chapter 9.

7.8. Other approaches to solvent behaviour in liquid chromatography

The ROHRSCHNEIDER parameters. In 1973, ROHRSCHNEIDER published gas-liquid partition coefficients for six selected solutes in 81 liquids, including commonly used solvents like THF, TCM, Tetra, Tol, AcN and EAt. The coefficients were determined in glass sample flasks (volume 13.4 ml), each of which contained 2 ml of the liquid to be characterized. A 5 μl volume of the test mixture (octane (O) + toluene + ethanol (E) + methyl ethyl ketone + dioxane (D) + nitromethane (N)) was injected by a syringe through a rubber stopper into the solvent. The samples were kept at 25 °C, and after 2 h the first analysis was carried out by means of an automatic headspace analyzer and two different GC columns. The partition coefficients were determined from the measured peak heights, and as a whole reflected the chemical character of the solvents fairly well. For example, the values obtained for benzene were $K_{g,O} = 8310$, $K_{g,E} = 320$, and for propanol $K_{g,O} = 1657$, $K_{g,E} = 4705$.

SNYDER's **classification of solvent properties.** Taking ROHRSCHNEIDER's partition coefficients $K_{g,O}$, $K_{g,E}$, $K_{g,D}$ and $K_{g,N}$ as a basis, SNYDER (1974c; corrected version 1978) developed a classification scheme which we shall discuss in the light of the data for THF. For this solvent ROHRSCHNEIDER obtained $K_{g,O} = 8870$, $K_{g,E} = 2168$, $K_{g,D} = 6141$, and $K_{g,N} = 5379$. SNYDER first multiplied these values by the molar volume of the solvent ($V_X = 81.2$ ml · mole^{-1} for THF) to refer them to an equal number of moles. (The molar volumes, V_S, of the sensors O, E, D and N are, respectively, 163, 58.6, 85.5 and 53.8 ml · mole^{-1}.) The correction

$$K'_g = K_g \cdot V_X \tag{7-29}$$

yields $K'_g = 2168 \cdot 81.2 = 176042$ for ethanol as a sensor. To eliminate the dispersive interaction, these adjusted partition coefficients, K'_g, are divided by K_v:

$$K''_g = K'_g / K_v \tag{7-30}$$

Here, K_v is the K'_g value of the hypothetical n-alkane which has the same molar volume, V_S, as the solute. It results from the adjusted partition coefficient, $K'_{g,O}$, for n-octane in the solvent in question (THF in this case):

$$\log K_v = (V_S/163) \log K'_{g,O} \tag{7-31}$$

For ethanol as a solute this gives $\log K_v = (58.6/163) \cdot \log (8870 \cdot 81.2) = 2.106$. From this one obtains $\log K''_{g,E} = (\log 176042) - 2.106 = 3.140$ and, in the same way, 2.625 for dioxane and 3.707 for nitromethane. If various alkanes are used as solvents, the same procedure with ethanol as a sensor gives $\log K''_{g,E_0} = 1.773$, with rather small deviations from this mean value. The same holds true for dioxane as a sensor, with $\log K''_{g,D_0} = 1.885$, and for nitromethane, with $\log K''_{g,N_0} = 2.190$. These values were supposed to correspond to inductive effects, entropy effects, etc. They are subtracted from the values obtained with the corresponding sensor for the solvent in question. For THF this finally gives $K''_{E,corr} = 3.140 - 1.773 = 1.367$, and in the same way one obtains 0.740 for dioxane and 1.517 for nitromethane. The sum of these values, P' (= 3,624), is defined as the polarity index which indicates the ability of the solvent to interact with the polar test solutes. The lowest value ($P' = 0.1$) belongs to n-hexane, whereas high values are obtained for tetrafluoropropanol, formamide or water. The ratios

$$x_E = \log K''_{E,corr}/P' \tag{7-32a}$$

$$x_D = \log K''_{D,corr}/P' \tag{7-32b}$$

$$x_N = \log K''_{N,corr}/P' \tag{7-32c}$$

were used by SNYDER to characterize the different solvents by points in a triangular diagram. The groupings obtained in this way agree fairly well with experience. This was confirmed by POPPE and SLAATS (1980), who suggested some improvements in the derivation by use of the Flory-Huggins theory.

The TAFT π^* polarity scale. A predictive approach to dipole-dipole interactions based on the dipole moment, μ, and the molar volume, V, of the solute and the dielectric constant of the solvent was given by CARR in 1980. This author was able to show that the π^* scale introduced by TAFT (KAMLET et al., 1977; ABBOUD et al., 1977) as well as the polarity function, $\theta(D)$ (ABBOUD and TAFT, 1979), are of importance in chromatography. The π^* polarity scale is based on the effect of the solvent on the maximum adsorption of the $\pi \rightarrow \pi^*$ or $p \rightarrow \pi^*$ transition. The reference solvent in this system is cyclohexane ($\pi^* = 0$). The empirical π^* values exhibited a linear correlation

$$\theta(D) = \frac{3D \ln D}{D \ln D - D + 1} - \frac{6}{\ln D} - 2 \tag{7-33}$$

which is based on a modification of the equation given by KIRKWOOD (1934) for the interaction of permanent dipoles in solution. From this, CARR derived the equation

$$\ln K = 50.5 \frac{\mu^2}{V} [\theta(D'') - \theta(D')] \tag{7-34}$$

where μ is expressed in Debye units and D'' and D' are the dielectric constants of the stationary and the mobile phase, respectively. This equation predicts the part of the distribution constant that is due to the interactions of permanent dipoles. For non-hydrogen bonding non-aromatic solvents, the function $\theta(D)$ can be used to obtain *a priori* estimates of δ_0, the orientation solubility parameter. CARR obtained a correlation coefficient of 0.97 between measured and calculated values.

7.9. Resolution in adsorption chromatography

In Section 3.4., the resolution equation (3-25) has been derived and discussed in general. Using the relationship (7-7) between the thermodynamic distribution constant K and the adsorption constant K^*, it can readily be rewritten for adsorption chromatography:

$$R_S = \underbrace{\frac{1}{4}\sqrt{N}}_{a)} \cdot \underbrace{\frac{K^*_{II} - K^*_I}{K^*_I}}_{b)} \cdot \underbrace{\frac{K^*_I}{K^*_I + (m_A/V')^{-1}}}_{c)} \tag{7-35}$$

Consequently, the requirements are:

— a high number of plates (*column efficiency factor, a*)

— the greatest possible difference between neighbouring adsorption coefficients (*relative distribution factor, b*)

— not too small a minimum value of K^*_I, or as high a value as possible for the ratio, m_A/V', of the adsorbent to the mobile phase (*retention factor, c*)

These are, mutatis mutandi, the same requirements as in Section 3.4. — of course!

8. Separation by size exclusion

In exclusion chromatography molecules are separated according to their sizes. Particles having the same dimensions migrate with equal speed, even if they differ in chemical structure. The column materials employed are permanently porous gels or gels which become porous upon swelling. Therefore size exclusion chromatography (SEC) is also called gel chromatography. In aqueous systems one speaks of *gel filtration*, in non-aqueous ones, of *gel permeation chromatography* (GPC). These different designations do not represent fundamental differences in the separation mechanism or in the instrumental techniques. A number of gels can be used for separations in water as well as in organic solvents.

In this book, the term *size exclusion chromatography* (SEC) recommended by the ASTM Committees E-19 and D-20 is used (BLY, 1980; ETTRE, 1981) if the separation follows a pure exclusion mechanism, i.e., only according to molecular size. If, however, in addition to the exclusion mechanism there are other interactions between the solute and the packing material, which will be dealt with in detail in Section 16.6., then let us refer to the chromatographic process as *gel permeation chromatography* (GPC).

The *pore structure* of the gels can be visualized by means of an electron microscope (BILLMEYER and ALTGELT, 1971). The pore size is approximately equal to the coil dimensions of the largest molecules separated (HALLER, 1977). For porous glasses, CANTOW and JOHNSON (1967a) found that macromolecules have full access to the inside volume of those pores whose width is at least double the end-to-end distance of the coiled molecules. ENGELHARDT and MATHES (1979) stated that the mean pore diameter of modified silica should be at least 20 nm for the separation of polyvinylpyrrolidones, dextrans or gelatine samples up to 70000 g · mole^{-1}, and at least 50 nm for samples up to 500000 g · mole^{-1}. These dimensions are of the order of magnitude of the end-to-end distances (see eqn. (4-38)).

The diffusion coefficients in the pores are smaller than in free diffusion (Fig. 8-1).

8.1. Distribution equilibrium in SEC

Numerous arguments suggest that in steric exclusion chromatography a distribution equilibrium is reached between the mobile phase V' and the interior of the pores (HEITZ, 1973a). This distribution between stationary and mobile phases in one and the same medium may be treated thermodynamically like a partition between phases (cf., Section 8.5.).

The stationary volume, V'', is the *total pore volume* of the gel bed. The mobile volume, V', is the *interstitial volume*, which depends on the geometry of the gel particles and the packing density.

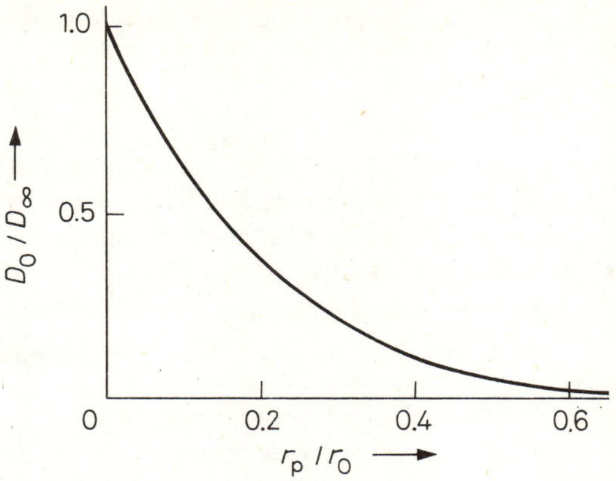

Fig. 8-1
The decrease in diffusion coefficient of ball-shaped particles in pores whose radius, r_O, is of the same order of magnitude as that of the balls, r_P

On the ordinate, the value D_0 for the diffusion within the pore is shown as a fraction of the value, D_∞, for free diffusion (according to ACKERS and STEEVE, 1962).

In SEC the solution flows through the packing of the gel granules, the pores of which contain stagnant solvent. Dissolved particles which are larger than the pores cannot penetrate into the latter. They always remain in the mobile phase, leaving the column with the same volume fraction as that injected. Their elution volume is the dead volume of the packing, i.e., the interstitial volume V'. Small molecules can penetrate into all of the pores. They are distributed among the solvent stagnating there and the solution flowing outside, or the following solvent. While they stay in the pores, the mobile phase continues to flow along. The molecules which diffuse into the pores lag behind more and more, and will be eluted with the maximum elution volume (total permeation volume), $V' + V''$, which exceeds the interstitial volume by the total pore volume V''.

The *total permeation volume*, in which all particles that are smaller than a certain limit are discharged without being separated, is a characteristic quantity of the gel. V'' does not depend on the packing density. For medium-sized molecules, some part of the pores is accessible. The greater this part, the stronger is the retention. This results in a separation according to the molecular size. The structure of the gel defines the limits of the separation range. The *exclusion limit* is the upper mark ("total exclusion"). The lower limit of the range of separation, below which all substance fractions have the same maximum elution volume, is called here the *separation threshold* ("total permeation").

It is usual and practical to indicate the steric exclusion limit and the separation threshold by molar mass values which enable the user immediately to determine the problems for which the various gel types are suitable. The solvent and the polymer used in determining the limits should be stated in addition. For samples having molecular sizes above the exclusion limit, the *distribution constant*, K, is zero. As the molecular size decreases, K increases, reaching the maximum value $K = 1$ at the separation threshold. While in the adsorption mechanism also higher values of K may occur, corresponding to an enrichment of the substance in the stationary phase, in the steric exclusion mechanism the highest possible concentration in the stationary phase is equal to that in the solution. From $c'' = c'$ it follows that $K = 1$. Particles whose sizes lie between the steric exclusion limit and the separation threshold are more or less strongly retained. Their elution volumes range between the dead volume, V', and the maximum retention volume:

$$V_e = V' + K \cdot V'' \quad (8\text{-}1)$$

In size exclusion chromatography the distribution constant, K, at the same time indicates that fraction of the total pore volume, V'', which is accessible for a certain fraction of the sample. The constants in eqn. (8-1) can be determined on a packed column. The elution volume of samples having molecular sizes above the exclusion limit immediately yields the value of V'. The total pore volume, V'', can be determined by means of a sample having molecular sizes below the separation threshold as the difference of its elution volume and the interstitial volume, V', determined previously.

The range $0 \leqq K \leqq 1$ implies that steric exclusion chromatograms are remarkably short compared to adsorption or partition chromatograms. It is rare that the net retention times t'' are longer than the elution time t' for a totally excluded sample.

Between the steric exclusion limit and the separation threshold there is the *separating range*, which includes about two decimal powers. To enable separation beyond these relatively narrow limits for a packing material, combinations of different gel types are used. As a rule, the individual columns are each packed with a different gel type, and the whole chromatographic path is assembled from such columns as required, placing that column with the largest pore size at the column entrance, followed by the others in a stepped, decreasing order of pore sizes. Such a combination first separates the fractions having the highest molar mass from the polymolecular sample, which then flows through the next part of the column at a lower level of viscosity. On the other hand, COOPER et al. (1975a) recommend the opposite order of columns, which separates the sample starting from the small-molecule end. In this case the most difficult separation of the highest molar masses is carried out last. The relationship between the order of the columns and the resolving power was investigated experimentally by JAMES and OUANO (1973). They found that the order of succession appears to be irrelevant. For higher sample concentrations the best resolution was achieved by random combinations.

A few years ago, mixing gels was not recommended because the quality of packing may be affected by particles of different sizes, shapes or densities. However, the progress since made in the production of column materials and in packing techniques has enabled the packing of columns with gel mixtures (KATO and HASHIMOTO, 1974a, b; BASEDOW et al., 1976), which now are also commercially available (linear columns) [F 1].

It is in fact the use of *gel mixtures* which has enabled extraordinary separating efficiencies to be achieved (e.g. MORI, 1979). On the other hand, the composite columns offer the advantage that the combination can rapidly be changed and best fitted to every new problem at low expense. An all-round column must have a broad separating range. However, samples with only a narrow distribution are separated better and more rapidly in columns providing a pore size range which is just sufficient for the problem. In SEC, the modification of the chromatographic bed ist the most important means for increasing the resolving power. The eluent has a much smaller influence than in adsorption or liquid-liquid partition chromatography.

8.2. Relationship between the molar mass and the elution volume

SEC is nothing other than a separating technique which, operated on its own, can say nothing about either the molar mass distribution or the mean values. As a direct result, elution curves are obtained which, at best, show which amount of the sample leaves the column in a certain elution volume. The values M_i and $H(M)$ for calculation of the quantities being of actual interest must be determined by means of *calibration relationships*.

The latter are established using, if possible, monodisperse polymers as calibration standards. The elution volumes of such samples decrease with the logarithm of the molar mass (Moore, 1964):

$$V_e = C_1 - C_2 \log M \tag{8-2}$$

The constants C_1 and C_2 can be taken from the graphical representation of the results of measurement, which at the same time shows the extension of the separating range. The exclusion limit, and the separation threshold, do not sharply terminate the separating range, but with a more or less broad range of transition. Hence the semi-logarithmic relationship (8-2) is not valid over the entire measuring range with constant values of C_1 and C_2.

Eqn. (8-2) describes the relationship as it is established in the calibration: the elution volume is the dependent variable determined as a function of the molar mass. The slope factor, C_2, can be calculated from the positions of two points on the linear part of the calibration curve:

$$C_2 = \frac{V_{e,II} - V_{e,I}}{\log (M_I/M_{II})} \equiv S \tag{8-3}$$

According to Bly et al. (1971), the *selectivity factor* defined in this way is used in the characterization of columns (cf., eqn. (16-53)).

In the evaluation of elugrams, however, the elution volume is the starting quantity and the associated molar mass is the variable desired. Of course this variable can be calculated by the unmodified eqn. (8-2), but it is understandable that some authors use modified forms for this task. Thus Balke and Hamielec (1969), and in a similar way also Provder and Rosen (1971), used the relationship

$$M = D_1 \, e^{-D_2 V_e} \tag{8-4}$$

which is very suitable for the dispersion correction (cf., Sections 16.2. and 16.3.). A comparison of the coefficients gives:

$$C_1 = (2.303 \log D_1)/D_2 \tag{8-5a}$$
$$C_2 = 2.303/D_2 \tag{8-5b}$$

Additional problems arise if the calibrating samples are not ideal. For polydisperse calibration substances, the position of the peak maximum depends on the distribution in the samples. For logarithmic normal distributions it coincides with a mixed average $(\bar{M}_n \cdot \bar{M}_w)^{1/2}$, whereas for Schulz distributions (cf., eqn. (4-21)) the maximum occurs at the weight average, $\bar{M}_w$ (Berger and Schultz, 1965). However, the peak maximum does not always coincide with a directly measurable mean value. Nevertheless, in most cases the peak maximum is found at $\bar{M}_w$ or closely below this value — similarly as the viscosity average, which therefore gives a good measure for the calibrating samples (Coll and Gilding, 1970).

In 1980, Szewczyk suggested that instead of the peak elution volume the average retention volumes should be used, which can be calculated from the elution curve by means of algorithms similar to those for $\bar{M}_n$, $\bar{M}_w$ and $\bar{M}_z$ (see Section 4.2.1.). This eliminates any ambiguity concerning the M value related to the peak maximum and can cope with a heterogeneity in the calibration standards ranging from 1.1 to 2.0. In independent work, van Dijk et al. (1980) also used average elution volumes in calibration.

In this context it is worth mentioning that more accurate investigations of calibration standards revealed a much greater heterogeneity than had been generally assumed (HARTMANN and KLESPER, 1977; PLAZEK and AGARWAL, 1978; LATTIMER et al., 1979).

After a careful calibration, the molar mass of the fractions can in principle be determined from their retention volume alone. Thus the efficiency of SEC approaches that of gas chromatography, which under suitable conditions likewise permits an identification on the basis of retention. In such cases chromatography is not only an efficient separating technique but also a highly productive analytical method for routine problems.

It is true that the elution volumes must be measured with the highest accuracy, because, due to the logarithmic relationship (8-2), even small errors have a great effect on the molar mass. Naturally the calibration relationship must also be known with a high accuracy for the conditions prevailing in the analysis. It need surely not be emphasized that it has to be established using the same polymer and the same solvent. The concentration must also be equal.

However, the relationship between log M and V_e changes in the course of time. Probably traces of some substances are irreversibly retained, for the exclusion limit shifts towards lower values. It is therefore necessary to check the calibration function from time to time. For that purpose it suffices to elute a suitable standard mixture, the multimodal curve of which indicates at a glance whether the full separating efficiency is still maintained. Using this test, HAZELL et al. (1968) continuously monitored columns operated at an elevated temperature. After one year the efficiency of these columns had decreased so much that a replacement was recommendable.

In general, measurements using *internal standards* are not possible for polymers. Of course a higher reliability may be achieved by as little as one additional peak outside of the sample range, which sometimes even results from a peculiarity of the solvent alone (PATEL and STEJNY, 1974). For polymers which do not adsorb in the ultraviolet range, e.g., polyvinyl chloride or polymethyl methacrylate, the internal standard method can, however, be employed if an instrument equipped with two detectors is available (WILLIAMS et al., 1971). To the 1% analytical sample, polystyrene standards are added in such a low concentration (0.003%) that they just give a signal in the UV detector, but are not recognized by the much less sensitive refractive index detector, which records the elugram of the sample. Fluctuations in the concentration of the standard substances are irrelevant, but variations in the sample concentration have a marked effect on the elution volume and the peak shape of the internal standard. Under defined conditions the technique enables the repeated checking of the calibration at low additional expense. Moreover, it immediately signalizes any overloading by the change in the shape of the reference peak.

To be precise, the relationship between the molar mass and the elution volume is still somewhat more complicated than the simple equation (8-2) implies. To sum up:

The separation range of a single gel is rather narrow. The total operation range becomes much wider by means of a combination of suitably graduated gels. For columns operated in series, the elution volumes of the individual sections are added.

The diagram of log M vs. V_e is only approximately linear. This is especially true for single gels. However, combinations may lead to improvements so essential that the evaluation indeed can be carried out using eqn. (8-2). Non-linear calibration curves will be discussed in Section 8.4.

The calibration relationship depends on the concentration and may vary as the age of the packing increases. Initially it is valid only for a certain solvent and for the polymer employed to establish it, because the elution volume is basically related to the particle dimensions, and the molecular dimensions vary if the solvent or the polymer is changed, even if the molar mass is kept constant.

The restriction with respect to the kind of polymer has certain consequences in applying SEC to an unknown system. The simplest method of calibration is by *monodisperse samples*, which are however generally available only in the form of anionic stoichiometrically polymerized polystyrene. How can a calibration be achieved for other polymers by means of these polystyrenes? This question is of great importance for a wider application of gel chromatography and will be dealt with in the following Section.

8.3. Universal calibration of gel chromatography

Macromolecules with different chemical structure generally have different coil volumes even if their molar masses are equal, because on the one hand the molar masses of the repeat units are different, and on the other hand the coil expansion caused by the solvent depends on the chemical nature of both the solvent and the polymer.

8.3.1. The Q value concept

The first attempts to obtain a universal calibration relation led to the Q factors, which are still sometimes used. The starting point was the idea to use the straight length, $P \cdot l_{eff}$, as an index of the molecular size. The effective length of a monomeric unit is $l_{eff} = 2.52 \times 10^{-8}$ cm (cf., Section 4.5.). A chain molecule with molar mass M and molar mass M_0 of the repeat unit has the following length, measured in 10^{-8} cm:

$$P \cdot l_{eff} = 2.52 \cdot P = \frac{M}{M_0/2.52} \tag{8-6}$$

The Q factor

$$Q = M_0/2.52 \tag{8-7}$$

indicates the mass in daltons per 0.1 nm of chain length (1 dalton = 1 g · mole^{-1}). Among the carbon chains, polyethylene has the smallest Q value of $Q_{PE} = 28:2.5 = 11$. For polystyrene, $Q_{PS} = 104:2.5 = 41$.

The calculation with Q factors is based on the hypothesis that polymers of different species might have equal elution volumes if their straight chain lengths coincide. On the basis of the calibration curve established by means of polystyrene standards, the elution volume of the analytical sample first yields an M_{PS} value, from which the value for the sample is calculated. For example, for polyethylene one would have to set $M_{PE} = (11/41)\ M_{PS}$.

This method is not recommended.

8.3.2. Universal calibration by means of the hydrodynamic volume

Plotting V_e values vs. log $(M^{1/2}S)$ or log $(M^{3/2}[\eta])$ yielded a common curve for polystyrene, polymethyl methacrylate and cellulose nitrate, but not by plotting vs. log M or log $\langle S^2 \rangle$, log $[\eta]$ or log $(P \cdot l_{eff})$ (MEYERHOFF, 1965a, b). BENOIT et al. (1966) found that when plotting V_e vs. log M even the curves for linear and branched polystyrenes did not coincide. For equal molar masses, linear samples showed smaller elution volumes than star-shaped ones, which in turn had smaller volumes than samples of a comb-like structure, whereas the viscosities (for equal M) had been found to decrease in the following order:

$$[\eta]_{chain} > [\eta]_{star} > [\eta]_{comb}$$

These observations were interpreted as an effect of the *hydrodynamic volume*, V_h, which according to Einstein's viscosity equation is proportional to the product $[\eta]M$. Plotting log $([\eta]M)$ vs. V_e indeed yielded a common curve, not only for the polystyrene samples of different constitutions, but also for polymethyl methacrylate, polyvinyl chloride, polyphenylsiloxane, polybutadiene and graft copolymers. With the use of log $([\eta]M)$, a universal calibration is possible (see Fig. 8-2).

The behaviour of macromolecules at the entrance to the pores is similar to that under the action of a velocity gradient in viscosity measurements, i.e., the hydrodynamic volume which influences the viscous flow also governs the penetration of the molecules into the pores (GRUBIŠIĆ et al., 1967).

Comparing the elution behaviour of polystyrene and polyisobutylene in 1,2,4-trichlorobenzene at 35, 70, 110 and 150 °C, CANTOW et al. (1967b) also realized that the hydrodynamic dimensions of the polymer molecules are decisive for the separation by gel chromatography.

The hydrodynamic volume has proven successful as a parameter for universal calibration. A great number of experimental studies have confirmed its suitability for many systems such as statistical and block copolymers of styrene and methyl methacrylate (DONDOS

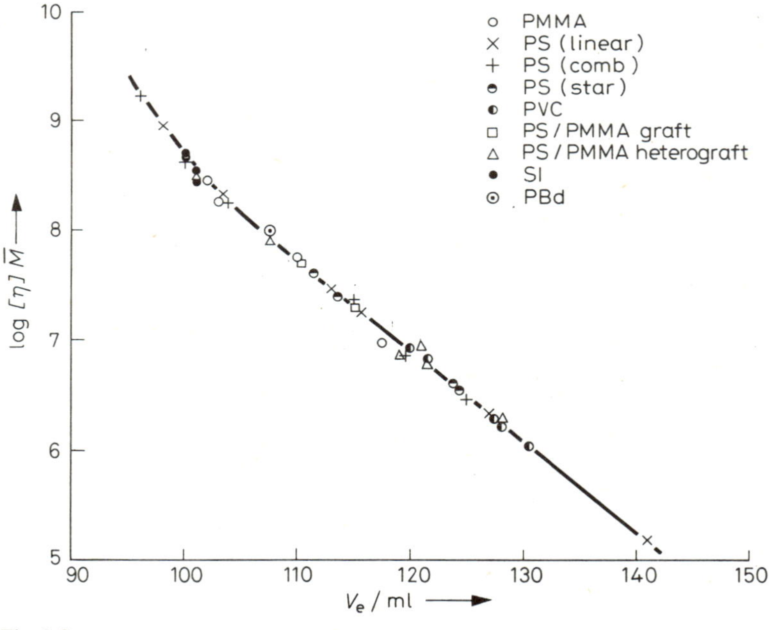

Fig. 8-2
Universal calibration curve (according to GRUBIŠIČ, REMPP and BENOIT, 1967)

et al., 1974), styrene-acrylonitrile copolymers of different compositions (KRANZ et al., 1972), or butadiene-styrene copolymers of different molecular architectures, poly-n-butyl isocyanate and branched copolymers of styrene and divinyl benzene (AMBLER and MCINTYRE, 1975), oligomers down to $M \approx 100$ (AMBLER, 1976), as well as linear polyvinyl acetate (ATKINSON and DIETZ, 1979).

An essential precondition of a universal calibration relation is that in the molar mass range investigated none of the polymers to be compared undergoes a significant variation in molecular conformation. In a paper on poly(n-butyl isocyanate) in THF, AMBLER et al. (1977) explained a deviation of the calibration curve for this polymer from the PS calibration curve by a variation of the molecular geometry.

Whether or not the universal calibration curve is valid for a certain polymer-eluent system can be judged on the basis of the agreement of the log $(M[\eta])$ vs. V_e curve for this system with the corresponding standard curve. However, it is also possible to adopt the course taken by JANČA and KOLÍNSKÝ (1977), using the standard curve to calculate the values $\bar{M}_w$, $\bar{M}_n$ or $[\eta]$ from the elugrams of samples of the system concerned and the $[\eta]$-M relationship of the latter (eqn. (5-8)). If these values agree with those measured directly by light scattering, osmosis or viscosimetry, then the universal calibration is applicable for the system of interest. This method has the advantage that the different distribution in the standards and in the samples of the substance of interest will involve no stronger disturbance than that to be discussed later in Section 16.4.

On the other hand, the Mark-Houwink constants of the system concerned must be known. A checking procedure without that prerequisite is described in Section 19.7.2.

For star-shaped poly-α-methylstyrene, TOSHIO KATO et al. (1975) found that the log $([\eta]M)$ plot is also suitable for the universal calibration, whereas for comb-like polystyrenes TADAYA KATO et al. (1975) observed a slight deviation, which was attributed by them to the variation of the Flory constant, Φ, for a high degree of branching. Using the radii of gyration, these products could be better plotted together with the linear homologues. The authors thus confirmed an analogous result obtained by PANNELL (1972).

For two polymers (I and II) and equal elution volumes, the universal calibration by means of the hydrodynamic volume includes the condition

$$\log (M_{\mathrm{I}} \cdot [\eta]_{\mathrm{I}}) = \log (M_{\mathrm{II}}[\eta]_{\mathrm{II}}) \tag{8-8}$$

Using eqn. (5-8) one obtains for homopolymers:

$$\log M_{\mathrm{II}} = \frac{1}{1 + a_{\mathrm{II}}} \log \frac{K_{\eta,\mathrm{I}}}{K_{\eta,\mathrm{II}}} + \frac{1 + a_{\mathrm{I}}}{1 + a_{\mathrm{II}}} \log M_{\mathrm{I}} \tag{8-9}$$

If K_η and a are known for both polymers, then the calibration curves established by means of polystyrene (I) can be converted for any polymer (II) whatever. The effect of concentration can also be described by means of a semiempirical equation, if the density of the amorphous polymer is additionally known (RUDIN and HOEGY, 1972), or if information about the viscosity in a θ solvent is available (RUDIN and WAGNER, 1976).

For theoretical reasons, COLL and GILDING (1970) used the expression $M[\eta]/f(\varepsilon)$ as a calibrating parameter, together with the following formula derived by PTICYN and EJZNER (1959):

$$f(\varepsilon) = 1 - 2.63\varepsilon + 2.86\varepsilon^2$$
$$\varepsilon = (2a - 1)/3 \tag{8-10}$$

where a is the exponent in eqn. (5-8). Thus in eqn. (8-9) the product $K_\eta \cdot f(\varepsilon)$ appears instead of K_η for both polymers. In most cases, however, the simple relationship is sufficient (COLL, 1971; BAKOŠ et al., 1979).

With cellulose nitrate, OUANO et al. (1973) even found that the agreement of the directly measured $\bar{M}_v$ values with those calculated from polystyrene-calibrated SEC was better if the correction by means of eqn. (8-10) was not carried out. In investigations with polystyrene, polyvinyl chloride and various polymethacrylates, also SAMAY et al. (1978) reached the conclusion that the original Benoit method, where eqn. (8-10) is not taken into account, is preferable.

In some papers, the *unperturbed dimensions* were proposed as a universal calibration parameter (DAWKINS, 1968, 1970; DAWKINS et al., 1969). The unperturbed dimensions are suitable if the solvent employed has exactly the same thermodynamic quality for both polymers, so that the expansion coefficients coincide.

The calibration by the *hydrodynamic volume* is comparatively uncomplicated, and offers the possibility of a universal utilization of SEC. Moreover, it is not even necessary that the $[\eta]$-M relationship (5-8) is explicitly known. The calibration established by means of the polymer I can be transferred to any polymer II for which samples with known $\bar{M}_w$, $\bar{M}_n$ or $[\eta]$ are available (FRANK et al., 1968; WEISS and COHN-GINSBERG, 1970; PROVDER et al., 1971; BELENKIJ and NEFEDOV, 1972; KOLÍNSKÝ and JANČA, 1974; MAHABADI and O'DRISCOLL, 1977). This may be done as follows: From eqns. (4-3b) for $\bar{M}_n$ and (4-5b) for $\bar{M}_w$, using (4-13) and (4-15) one obtains

$$\bar{M}_n = \left[\int_0^\infty \frac{1}{M} H(M)\,\mathrm{d}M\right]^{-1} \tag{8-11}$$

and

$$\bar{M}_w = \int_0^\infty M H(M)\,\mathrm{d}M \tag{8-12}$$

In a similar way it is found that:

$$[\eta] = K_\eta \int_0^\infty M^a H(M)\,\mathrm{d}M \tag{8-13}$$

If the chromatographic curve is *normalized* in the same way as the molar mass distribution, so that one has

$$\int_0^\infty F(V_e)\,\mathrm{d}V_e = 1 \tag{8-14}$$

then

$$H(M)\,\mathrm{d}M = F(V_e)\,\mathrm{d}V_e \tag{8-15}$$

gives the conversion formula

$$H(M) = F(V_e)\frac{\mathrm{d}\,V_e}{\mathrm{d}\log M}\cdot\frac{\log \mathrm{e}}{M} \tag{8-16}$$

The mass fraction with the molar mass M can thus be calculated from the normalized recorder reading, $F(V_e)$, at the point M, the slope of the calibration curve, $\mathrm{d}V_e/\mathrm{dlog}\,M$, and the molar mass M. In a calibration by the *hydrodynamic volume*

$$V_h = M[\eta] = K_\eta \cdot M^{1+a} \tag{8-17}$$

one plots log V_h vs. V_e. Using the normalized distribution of the hydrodynamic volume, by analogy with eqn. (8-16) one obtains

$$H(V_h) = F(V_e)\,\frac{\mathrm{d}\,V_e}{\mathrm{d}\log V_h} \cdot \frac{\log \mathrm{e}}{V_h} \tag{8-18}$$

Substitution of V_h, $H(V_h)$ and dV_h for M, $H(M)$ and dM, respectively, in eqn. (8-11) to (8-13) gives the relationships

$$\bar{M}_n = \left[\int_0^\infty \left(\frac{V_h}{K_\eta}\right)^{-\frac{1}{1+a}} H(V_h)\,\mathrm{d}V_h\right]^{-1} \tag{8-19}$$

$$\bar{M}_w = \int_0^\infty \left(\frac{V_h}{K_\eta}\right)^{\frac{1}{1+a}} H(V_h)\,\mathrm{d}V_h \tag{8-20}$$

$$[\eta] = K_\eta \int_0^\infty \left(\frac{V_h}{K_\eta}\right)^{\frac{a}{1+a}} H(V_h)\,\mathrm{d}V_h \tag{8-21}$$

where it was taken into account that d$V_h = K_\eta M^a \mathrm{d}M$, $H(V_h) = (K_\eta M^a)^{-1}\,H(M)$ and $M = (V_h/K_\eta)^{1/(1+a)}$. Eqns. (8-19) to (8-21) are used to find, for polymer II, the values of K_η and a which make the parameters $\bar{M}_n$, $\bar{M}_w$ or $[\eta]$, as calculated from the gel chromatogram, coincide with the directly measured ones. Two comparisons at least are required for that purpose. Suitable combinations are $\bar{M}_n - \bar{M}_w$ and $\bar{M}_n - [\eta]$. The values of K_η and a obtained in this way depend on the resolution of the gel chromatogram and the accuracy achievable in measuring the reference values $\bar{M}_n$ and $\bar{M}_w$ or $[\eta]$. It is further assumed that the system in question has a common $M[\eta]$ vs. V_e calibration curve with the standard system. The approximate values of K_η and a determined in this way for system II are substituted into eqn. (8-9), by means of which the calibration function is converted. In view of the total amount of computing work required and the necessity to use samples of polymer II with exactly known averages $\bar{M}_n$, $\bar{M}_w$ and $[\eta]$, the method is not more advantageous than the direct calibration with samples of broad distributions, which will be dealt with in the Section 8.3.3.

If the log ($M[\eta]$) vs. V_e calibration relationship is linear, then two comparisons are sufficient; if not, the method must be carried out using several characterized samples, and the calibration curve, constructed from partial sections, must be set up as a polynomial.

In the calibration by the hydrodynamic volume, virtually the effects of both the polymer and the solvent are taken into account. Therefore the calibration should also be applicable if the eluent is changed. This presupposes that this change does not alter those properties of the gel which affect the separation. This may be assumed for the rigid inorganic column materials used in SEC.

Grubišić and Benoit (1968) showed that on porous glass and on silica gel the universal calibration curves for the solvents tetrahydrofuran, dimethylformamide and toluene are identical within the limits of error. In such a case, the universal calibration curve established in a solvent I can indeed be used to evaluate the elugrams obtained in other eluents (OTOCKA and HELLMAN, 1974a). If, however, the separating properties of the gel vary, then a reference polymer which is readily soluble in both eluents is required. The molar mass range of interest must be completely covered, possibly by use of several samples as required. The individual samples need not be characterized. Thus PROVDER et al. (1971) used polymethyl methacrylate to convert the calibration curve established by means of polystyrene standards in tetrahydrofuran (THF) for trifluoroethanol (TFE), which does not dissolve polystyrene. Samples of the transfer polymer were chromatographed in both media, and the normalized chromatograms were compared with each other point by point. Thus it was possible to establish the relationship between the elution volumes in TFE and the corresponding values in THF, which yielded the desired calibration relationship for TFE.

In this connection it should be noted that eqns. (8-19, 8-20, 8-21) make it possible to determine the parameters K_η and a for a given polymer-solvent system from the SEC elution curve of one or two samples in this eluent. For this purpose the equations must be rearranged so that the constant K_η only appears before the integral sign. Then the rearranged equations can be combined in pairs in order to eliminate K_η. This is the case, for example, with the combination of eqns. (8-19) and (8-21), which gives an expression for $\bar{M}_n[\eta]$, or with the combination of (8-20) and (8-21) for $\bar{M}_w[\eta]$, but also for the ratio $[\eta]_I/[\eta]_{II}$, which is obtained using eqn. (8-21). The quantities $[\eta]_I$ and $[\eta]_{II}$ refer to the intrinsic viscosity of two samples (I and II) of the polymer under investigation.

From the elution curves, the paired values V_h and $H(V_h)$ are taken point by point and then summed, which corresponds to the integration. The exponent is assumed to have a value between 0.5 and 1.0, which is varied iteratively in the course of the calculation until the result of the summation agrees with the directly measured value of $\bar{M}_n[\eta]$ (or $\bar{M}_w[\eta]$ or $[\eta]_I/[\eta]_{II}$). Then, using this value for a, the corresponding value of K_η is calculated by any one of the eqns. (8-19, -20, -21). DOBBIN et al. (1980) found that this method, which had already been used by WEISS and COHN-GINSBERG (1969), yielded the best results if *two samples* were measured and the iteration was based on the ratio $[\eta]_I/[\eta]_{II}$.

Finally, there is a very simple approximation method, which is also intended for evaluations without any knowledge of $K_{\eta,II}$ and a_{II} (SMITH, 1974). The usual polystyrene calibration curve and the values of $K_{\eta,I}$ and a_I (for polystyrene in the eluent employed for the calibration) are sufficient for this application. From the elugram of a sample of polymer II, using the ordinary V_e vs. log M_I representation and the constants $K_{\eta,I}$ and a_I, the viscosity average $\bar{M}_{v,I}$ and the associated intrinsic viscosity $[\eta]_I$ are calculated as if the elugram concerned were one of polystyrene. The intrinsic viscosity $[\eta]_{II}$ of the unknown sample is directly measured. By analogy with eqn. (8-8),

$$\bar{M}_{v,II} = \bar{M}_{v,I} \frac{[\eta]_I}{[\eta]_{II}}$$

is a conversion equation for $M_I \rightarrow M_{II}$ for equal elution volumes. The quotient $[\eta]_I/[\eta]_{II}$, which is called the *Benoit factor*, turned out to be surprisingly constant for ethylene-propylene copolymers and polybutadiene in the molar mass ranges from 40 to 470 · 10^3 and from 30 to 810 · 10^3 g · mole^{-1}, respectively. Inspection of eqn. (8-9) shows that the conversion with a constant factor is possible if a_I and a_{II} sufficiently coincide. Moreover, it must be ensured that the samples contain only polymer homologues.

If they also have branched or chemically different molecules, then a certain hydrodynamic volume may possibly apply to different fractions:

$$[\eta]_{\mathrm{I}} \cdot M_{\mathrm{I}} = [\eta]_{\mathrm{II}} \cdot M_{\mathrm{II}} = \ldots = [\eta]_{\mathrm{PS}} \cdot M_{\mathrm{PS}} \tag{8-22}$$

The values with the subscript PS are obtained in the calibration with polystyrene standards. The separation of the sample on the basis of the hydrodynamic volume causes the corresponding fractions I, II, ... to be found simultaneously in the detector. The intrinsic viscosity of this mixture results from the individual contributions $[\eta]_i$ and the mass fractions w_i:

$$[\eta] = w_{\mathrm{I}} \cdot [\eta]_{\mathrm{I}} + w_{\mathrm{II}} \cdot [\eta]_{\mathrm{II}} + \ldots \tag{8-23a}$$

$$= \frac{w_{\mathrm{I}}}{M_{\mathrm{I}}} [\eta]_{\mathrm{I}} \cdot M_{\mathrm{I}} + \frac{w_{\mathrm{II}}}{M_{\mathrm{II}}} [\eta]_{\mathrm{II}} \cdot M_{\mathrm{II}} + \ldots \tag{8-23b}$$

In view of eqn. (8-22) this gives:

$$[\eta] = [\eta]_{\mathrm{PS}} \cdot M_{\mathrm{PS}} [\Sigma(w_i/M_i)] \tag{8-24}$$

On the other hand, because of $w_i = m_i/\Sigma\, m_i$ it is also possible to write eqn. (4-3a) for the number average of the molar mass in the form

$$\bar{M}_n = [\Sigma\,(w_i/M_i)]^{-1}$$

Thus eqn. (8-24) leads to

$$[\eta] \cdot \bar{M}_n = [\eta]_{\mathrm{PS}} \cdot M_{\mathrm{PS}} \tag{8-25}$$

and hence to the important finding that for structurally heterogeneous samples the number average molar masses are required for the calibration by the hydrodynamic volume (Hamielec and Ouano, 1978). The higher mean values such as $\bar{M}_v$ or $\bar{M}_w$ are only permissible for polymer homologues. For branched samples, copolymers or polymer blends, they may cause considerable errors. Probably some of the reported deviations from the universal calibration relation are due to this cause. For example, Williamson and Cervenka (1972) found a common (log $[\eta] \cdot \bar{M}_w$) vs. V_e curve for polyethylene samples of high density and narrow molar mass distribution as well as for polystyrene and copolymers of ethene and butene(I), but not for highly branched low-density polyethylene samples, which deviated markedly from this curve. An analogous universal calibration curve established by Ambler and McIntyre (1977) for ideal and branched polystyrenes showed a quite analogous picture. The highly branched copolymers of styrene and divinylbenzene likewise lay above the curve common to the other samples.

To summarize, the concept of the hydrodynamic volume, if applied correctly, indeed enables the universal calibration of SEC.

Deviations are attributable to adsorption phenomena and other disturbances of the steric exclusion mechanism, which will be discussed in greater detail in Section 16.6.

8.3.3. Calibration by samples with broad distributions

In Section 8.3.2. it was shown that in most cases some effort is involved, although some degree of uncertainty remains, when the calibration relationship established by means of polystyrene standards is transferred to polymer systems for which $K_{\eta,II}$ and a_{II} are unknown. If sufficient narrow fractions were available for each polymer, this problem would not exist. Therefore the attempts to obtain fractions of important polymers with so narrow a distribution that they can be used as standards for a direct calibration are of great interest. For linear polyethylene, PEYROUSET et al. (1975) solved this problem by means of preparative gel chromatography. The samples are commercially available [F 2]. Also polyvinyl chloride and polyvinyl acetate standards may be purchased [F 30]. Owing to the very high production cost, such standards are expensive (£ 35 per gram for PE; JAMIESON, 1973). ATKINSON and DIETZ (1976) used polypropylene fractions obtained by preparative GPC; the non-uniformity of these fractions ranged between 0.25 and 0.91 (mean value: 0.48). The calibration can, however, also be carried out by means of well characterized samples with broad distributions, which are much more readily available. The National Bureau of Standards, USA, has samples of polyethylene which are so sharply defined with respect to their distributions and mean values that they can be used for the calibration of GPC apparatus for the routine investigation of polyethylene samples. Analogous samples of polystyrene and polypropylene exist in the National Physical Laboratory, U. K. (ATKINSON et al., 1978). Standard samples of PVAC, PVC and the acrylics are also available [F 43]. Using a sample whose molar mass distribution is exactly known, the calibration can be carried out according to the principle shown in Fig. 8-3: the elution curve of the standard sample is measured repeatedly, and the

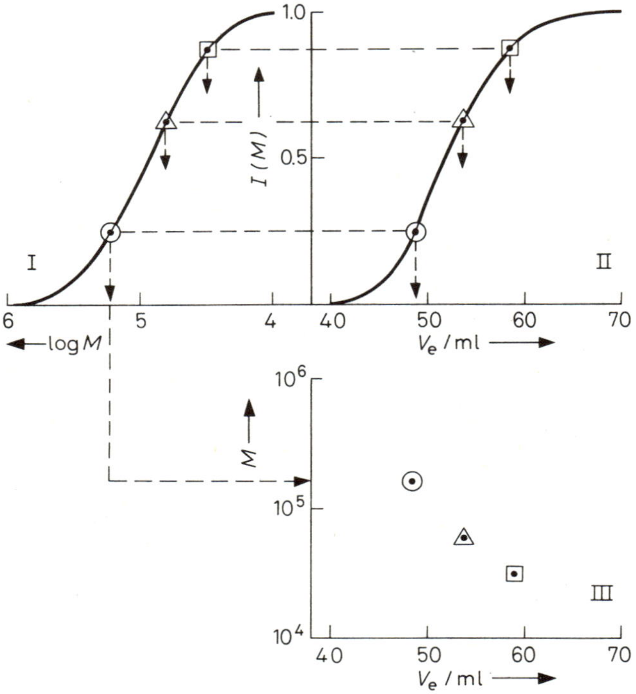

Fig. 8-3
SEC calibration by means of a sample whose broad molar mass distribution is known exactly (diagram I)

Curve II is the integral elution curve, likewise normalized to 1, of this sample. The calibration curve III is constructed from arbitrary pairs of points having equal ordinate values.

area enclosed by the curve is normalized to unity. If a suitable apparatus is used, six repetitions enable the ordinate values, $F(V_e)$, to be reliably associated with the elution volumes. If the detector reading is always proportional to the amount of substance present, then a cumulative plotting of $F(V_e)$ vs. V_e yields an integral curve corresponding to the integral molar mass distribution (cf., Section 4.2.3.). Because of the normalization (8-14), the ordinate can be denoted by $I(M)$ in both cases. As previously stated, the true molar mass distribution is known. In Fig. 8-3 it is plotted as an integral curve with a logarithmic M axis in such a way that the correlation between the fractions eluted first and the highest molar mass values is taken into account. Now the corresponding value of log M can be associated with each value of the elution volume by means of the common ordinate.

This method introduced by WEISS and COHN-GINSBERG (1970) was applied to polyethylene by BARLOW et al. (1977) in which case a secondary standard (PE Marlex® 6009) proved necessary, the distribution of which was broad enough to fully cover the calibration range required. The method can also be used for non-linear calibration curves and yields an arbitrarily dense sequence of points for their construction. However, the essential presupposition is that the molar mass distribution of the calibrating sample is exactly known. Frequently this is uncertain. A final calculation of the mean values of the employed standard by means of the calibration relationship obtained may reveal errors, but it does not suffice for an affirmative answer to the central question.

The calibration can, however, also be carried out using samples with broad distributions, for which only the averages of the molar mass are known. For a linear relationship, one sample is sufficient. Combining eqns. (4-3a), (4-5a) and (8-2) yields:

$$\bar{M}_w/\bar{M}_n = (\Sigma\, m_i\, 10^{-V_i/C_2})\,(\Sigma\, m_i\, 10^{V_i/C_2}) \tag{8-26}$$

V_i and m_i (the elution volume associated with m_i) are taken from the elugram of the sample with a broad distribution at as many points as possible, and substituted into eqn. (8-26). The summation is carried out with the object of finding that value of C_2 which provides the best agreement of the result with the ratio of the directly measured values $\bar{M}_w$ and $\bar{M}_n$, which of course must be known with the highest accuracy for this calibrating sample. LOY (1976) found that 36 iterations are sufficient to determine C_2 with an accuracy of 1/720. Then C_1 of eqn. (8-2) is calculated by iteration according to

$$C_1 \leftarrow C_1 + C_2 \log\,[\bar{M}_w/(\Sigma\, m_i\, 10^{(C_1 - V_i)/C_2})] \tag{8-27}$$

for which purpose as little as two cycles are sufficient in most cases. MALAWER and MONTANA (1980) used a similar algorithmn and gave a graphic interpretation.

The calculations required for this method are simpler than those in the two-parameter iteration developed by BALKE et al. (1969). The equations used in their iteration

$$\bar{M}_w = D_1\, \Sigma\, m_i \cdot e^{-D_2 V_i} \tag{8-28}$$

and

$$\bar{M}_n = \frac{D_1}{\Sigma\, m_i \cdot e^{D_2 V_i}} \tag{8-29}$$

also lead to eqn. (8-26) because of the relationships (8-5a, b) between the parameters C and D.

SZEWCZYK (1976), as well as VRIJBERGEN et al. (1978), developed similar methods by means of which non-linear calibration relationships can also be determined. For a polynomial of degree n (cf., Section 8.4.) it is necessary that at least $(n + 2)$ average values ($\bar{M}_w$, $\bar{M}_n$, $\bar{M}_v$) for suitable calibrating samples are exactly known. VRIJBERGEN et al. used number averages and weight averages of seven dextran samples, determining a calibrating polynomial of degree 3. A minimization method for the determination of calibrating polynomials of degree 2 was described by MCCRACKIN (1977). BELENKIJ et al. (1977) and CVETKOVSKIJ et al. (1977) also made contributions to this set of problems.

The great advantage of all these methods is that the calibrations are carried out directly with the very same polymer system for which they are required. On the other hand, the influence of instrumental spreading is ignored in the methods described here. This implies that the *effective calibration curves* obtained yield correct results only for samples which do not differ too widely from the calibrating samples with respect to the molar mass distribution and the position of the average values. Possible improvements of the calibrating method using samples with broad distributions will be described in Section 16.4.

8.3.4. Normalized calibration curves

In different columns, one and the same polymer-solvent system exhibits different calibration functions, because the column volume, the interstitial volume, V', and the total pore volume, V'', differ from one column to another. Therefore, LAURENT and KILLANDER (1964) worked with normalized calibration curves which represent

$$K_{av} = \frac{V_e - V'}{V_C - V'} \tag{8-30}$$

as a function of the molar mass or of the Stokes radius, which can be derived from the diffusion coefficient. Even for the smallest molecules, which can penetrate into all of the pores, the *Laurent-Killander distribution coefficient* does not reach the limit 1, which applies to the *Wheaton-Bauman distribution coefficient* (1953) employed in eqn. (8-1):

$$K = \frac{V_e - V'}{V''} \tag{8-31}$$

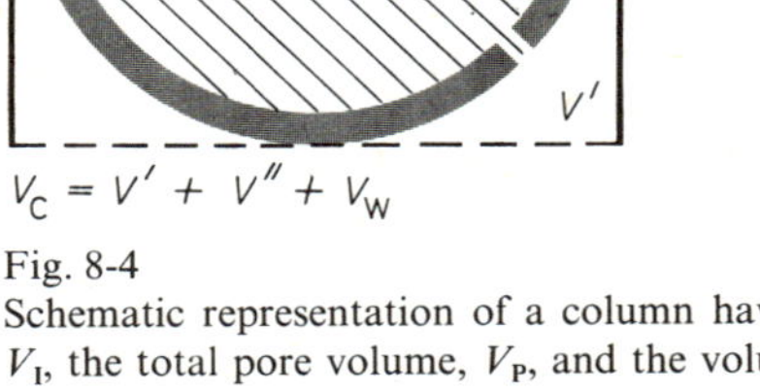

Fig. 8-4
Schematic representation of a column having the total volume V_C, the interstitial volume, V_I, the total pore volume, V_P, and the volume V_W of the wall material (gel matrix)
(The areal proportions shown correspond to an internal porosity $\varepsilon_P = 0.8$, cf., Section 10.4.4.)

This is because the denominator in eqn. (8-30) implicitly exceeds that in eqn. (8-31) by the wall volume, V_W of the gel matrix. Fig. 8-4 illustrates the meaning of the quantities in eqn. (8-30) and (8-31) and their relationships. K is independent of the gel type and the geometry of the column, but can be stated only if the value of the total pore volume, V''', can clearly be determined by experiment. Sometimes this is problematical, e.g., for highly hydrated hydrophilic gels.

8.4. Non-linear calibration relationships

For suitably chosen column combinations, the relationship between the molar mass and the elution volume follows the simple equation (8-2) over several decimal powers. An adequate linearity over four decades is achieved by means of the bimodal pore size distribution approach (YAU et al., 1978a). In this case two columns with different silica packings [F 33] are connected in series, the gels exhibiting logarithmic normal pore-size distributions each with a standard deviation of 0.15, and mean pore sizes differing from each other by a factor of 10.

Naturally the linear range of the calibration curve is always narrower than the interval between the separation threshold and the exclusion limit. For the total separating range the calibration relationship is always non-linear (see Fig. 8-5); see also VANDER LINDEN (1980). In the highly non-linear range in the neighbourhood of the steric exclusion limit, proportionality between V_e and $1/M$ was observed for a Sephadex®-G 25 (MCCLENDON, 1979). Such a relationship may be quite useful for the determination of the interstitial volume, V_I.

If the experimental calibration gives an S-shaped or otherwise bent curve, then a forced linearization may lead to considerable errors in the molar mass values, because of the logarithmic relationship. In such a case it is more correct but also more difficult to evaluate

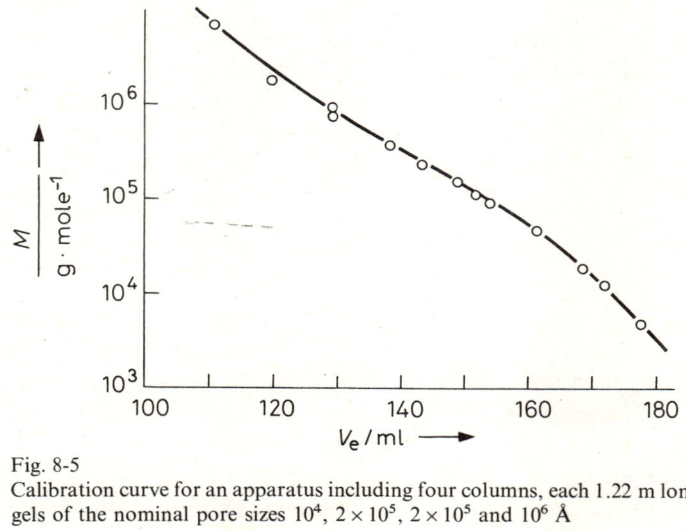

Fig. 8-5
Calibration curve for an apparatus including four columns, each 1.22 m long, with polystyrene gels of the nominal pore sizes 10^4, 2×10^5, 2×10^5 and 10^6 Å

Data points; polystyrene standards in 1,2,4-trichlorobenzene at 145 °C; curve: polynomial of degree 5 ($\log M = a_0 + a_1 V_e + a_2 V_e^2 + a_3 V_e^3 + a_4 V_e^4 + a_5 V_e^5$) (according to MAY and KNIGHT, 1971).

the elugrams by means of a graphical representation of the calibration. For larger series of measurements it is worthwhile to find the function corresponding to the real calibration curve. This is also a prerequisite to the programmed evaluation of non-linear elugrams on computers, and is of special importance in connection with the dispersion correction (TUNG, 1971a; ROSEN and PROVDER, 1971; cf., Chapter 16).

The calibration function can be expressed as a polynomial:

$$\log([\eta]\, M) = a_0 + a_1 V_e + a_2 V_e^2 + \dots a_n V_e^n \qquad (8\text{-}32)$$

Already a polynomial of degree 3 can rather accurately represent the calibration curve, which is S-shaped in most cases. TUNG (1971a) used Legendre polynomials, determining the coefficients by the moment method. From sets of polynomials, ranging in degrees between 3 and 32, in each case that which gave the best fit was selected.

8.5. The principle of separation

In the preceding sections, SEC phenomena have been treated without seeking their causes. This we shall now do.

Other than in adsorption chromatography, the driving force for the separation on gels does not lie in energy effects. The eluotropic strength of the solvent, characterized by ε^0, is irrelevant. If solvent effects occur, they indicate either disturbances (cf., Section 16.6.1.) or influences through the hydrodynamic volume. Also the influence of temperature variations is not nearly as significant as that on the interaction energies. The enthalpies of interaction between the stationary phase and the dissolved particles or the eluent have, if any, quite a minor influence on the separating effect in SEC.

What is decisive is the tendency to an equalization of concentration in communicating spaces (cf., Section 8.1.). As far as the dimensions permit, the dissolved molecules are distributed among the flowing solvent outside the pores and the stationary solvent inside the pores (PORATH, 1963). This distribution among different volume ranges in one and the same medium can be treated like a distribution among phases (CASASSA, 1967; VILENČIK et al., 1974; DOI, 1975). In its normal range, the flow-rate has no influence on the elution volume. This seems to indicate that the distribution reaches an equilibrium rather rapidly.

CASASSA (1967, 1971, 1976) attempted statistical considerations of SEC on the basis of the following assumptions:

— All of the pores have the same size and shape
— Adsorption does not occur
— For solute molecules of the same species, the probability, k_s, of penetrating from the flowing solution into a pore within a certain time interval is equal and constant
— The probability, k_m, referred to the same interval, that the molecules within in pores again return to the mobile phase is equal for particles of the same species. The quantities k_s and k_m are rate constants of the immobilization and mobilization, respectively, as parts of the chromatographic process
— The different elution times are due only to a different retention in the stationary phase. The total retention time, t', in the mobile phase is equal for all of the particles. The molecules in the mobile phase are transported with a uniform velocity

These assumptions as well as the starting point and the procedure in principle correspond to the stochastic treatment of adsorption chromatography by GIDDINGS and EYRING (1955), GIDDINGS (1957) and MCQUARRIE (1963).

If during the migration along the chromatographic bed the number $N = t'k_s$ of delays inside the pores is large, the equation derived initially

$$V_e = V' + KV''(1 - 3/2N) \tag{8-33}$$

becomes the well known relationship $V_e = V' + KV''$. K summarizes the constants

$$K = \frac{k_s/V''}{k_m/V'} \tag{8-34}$$

representing, like the partition coefficient between macroscopic phases, the ratio of concentrations in the mobile and the stationary solvent of a sufficient dilution.

The assumption that $N = t'k_s$ is large is obviously valid, since under normal operating conditions V_e is independent of the rate of elution, and hence of t'. K can be calculated as a function of the geometric proportions of the pores by determining the quotient of the conformation integrals in the stationary and the mobile phase. The pore walls are considered rigid and impenetrable for the solute molecules. Only conformations which do not contact the walls are permitted. Conformations which would penetrate the walls are prohibited. Flatly incumbent conformations are not permitted either, since by assumption there shall be no adsorption. The energy is assumed to be equal for all the permitted structures. For flexible molecules this means that the intramolecular interactions are neglected, and for a sufficiently high degree of polymerization this results in the θ state, for which the conformations can be calculated by random flight statistics.

Thus $\boldsymbol{R} \ln K$ is obtained as a variation of the standard entropy per mole, and the theory confirms what had already been concluded from the phenomena: Size exclusion chromatography is based on *entropy effects*, not on enthalpy effects.

Theoretical considerations on the penetration of rigid spheres of radius r into cavities of different geometries lead to the following results: For fan- or slot-like pores (between flat plates of infinite extension, separated by the distance $2a$) the partition coefficient is $K = (a - r)/a$. For long, cylindrical pores $2a$ in diameter, $K = [(a - r)/a]^2$, and for spherical cavities, $K = [(a - r)/a]^3$.

As an approximation for small values of r/a, one may set $K = 1 - \lambda(r/a)$, where $\lambda = 1$ for fan-shaped cavities, $\lambda = 2$ for cylindrical and $\lambda = 3$ for spherical ones. The factor λ also occurs in the result of Casassa's calculation for flexible macromolecules:

$$K = 1 - \frac{2\lambda}{\sqrt{\pi}} \frac{\langle S^2 \rangle^{1/2}}{a} \tag{8-35}$$

This equation shows how the equilibrium constant of SEC depends on the radius of gyration, S, and the pore dimensions a; it is valid for small values of $\langle S^2 \rangle^{1/2}/a$. As soon as $\langle S^2 \rangle^{1/2}$ reaches the order of magnitude of a, further terms of the power series $(\langle S^2 \rangle^{1/2}/a)^n$ must be taken into account.

Fig. 8-6a shows K as a function of $\langle S^2 \rangle^{1/2}/a$ for various pore types according to Casassa's theory. It is seen that the distribution coefficient tends to zero as $\langle S^2 \rangle^{1/2}/a$ approaches 1. Fig. 8-6b shows the same diagram with interchanged coordinates. Moreover $\langle S^2 \rangle^{1/2}/a$ is plotted logarithmically. Thus the diagram is in a form comparable with SEC calibration

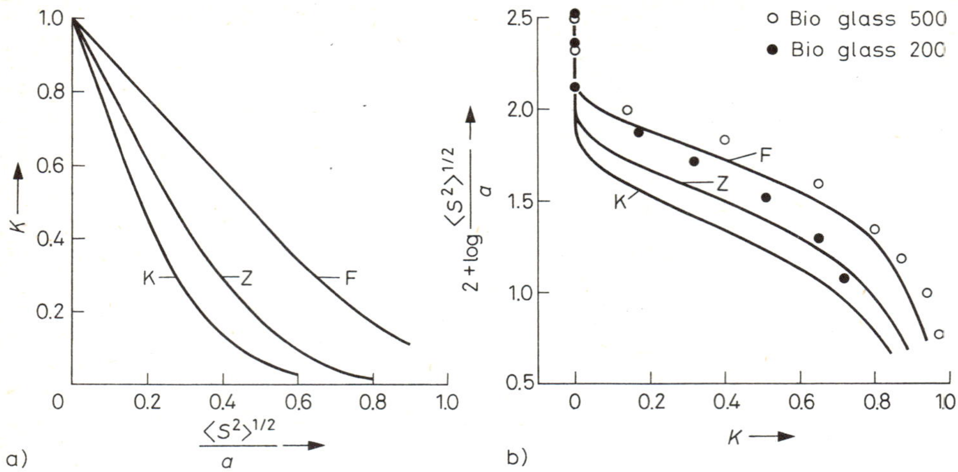

Fig. 8-6
Relationship between the distribution constant, K, in SEC and the ratio of the molecular to the pore size (according to CASASSA, 1967)

a) The relationship according to eqn. (8-35)

b) Semilogarithmic representation of the curves of Fig. 8-6a, with SEC results obtained from columns packed with porous glass (according to YAU, MALONE and SUCHAN, 1971)

F fan-shaped pores, size $2a$; Z cylindrical pores, radius $2a$;
K spherical cavities, radius $2a$.

curves. It is clear that, according to the theory, separation can be expected for a size interval of a little more than one decimal power. Experimental results obtained from homoporous column materials can be compared with this theoretical result. For this purpose packings of porous glass are suitable because they have rather uniform pore sizes, as demonstrated by electron microscopy and mercury porosimetry (HALLER, 1968). The points in Fig. 8-5b indicate the elution volumes of polystyrene standards on two separating materials of this type. In this representation the porosimetric radius has been set equal to the sizing quantity $a(d_p = 2a)$. Naturally this is valid only for cylindrical cavities. If geometrical effects are taken into account, then the experimental data points lie at about twice the values $\langle S^2 \rangle^{1/2}/a$. It should, however, be noted that mercury porosimetry always measures only the pore neck, thus yielding results for the pore diameter which are smaller than those from gas sorption by about a factor of 2. This ink-bottle effect just compensates the first-mentioned influence (CASASSA, 1976).

Fig. 8-6b shows that the positions of the data points and the relationship between the sizing quantities and K are reflected by the theory with surprising accuracy. The separation shown in the diagram takes place in pores of equal sizes and is solely due to the effect described by eqn. (8-35). Thus it is seen from this example that the *slope of the calibration curve* (C_2 in eqn. (8-2)), the selectivity factor, cannot increase arbitrarily in the GPC of flexible polymers. Hopes of achieving particularly sharp separations on gels with very narrow pore size distributions have been abandoned. Also, porous glasses with graduated pore sizes yielded experimentally a better resolution than glasses which were as homoporous as possible (COOPER and JOHNSON, 1971). For branched polymers, Casassa's theory gives an expression analogous to eqn. (8-35), in which a factor ψ depending on the degree of branching, f, is substituted for $\pi^{-1/2}$. Further it follows that the expression $M[\eta]$, the hydrodynamic volume, can be used as a universal calibration parameter, precisely as BENOIT et al. (1966) had proposed on the basis of experimental findings. For hard spheres, impermeable random-flight coils and rigid rods the theory initially gives three different relationships between $M[\eta]$ and $\bar{X}$, the thickness of the forbidden zone along the pore wall. For coils and rods, however, these relationships coincide for an axial ratio of $L/d \leqq 33$. Thus for all non-spherical molecules the universal calibration is applicable, depending only slightly on the internal chain mobility (CASASSA, 1976).

The convincing achievements of the theory clearly indicate that the starting point was chosen correctly. Exclusion chromatography is a normal chromatographic process which can be described as an equilibrium mechanism. Naturally, kinetic factors such as diffusion (KUBIN, 1965; YAU and MALONE, 1967; HERMANS, 1968) and hydrodynamic effects (PEDERSEN, 1962; ACKERS and STEEVE, 1962; ACKERS, 1964; DI MARZIO and GUTTMAN, 1969, 1970; VERHOFF and SYLVESTER, 1970) influence the process of chromatography too, but the decisive influence is exerted by the entropy-based distribution equilibrium (MINDNER and BERGER, 1969). This statement is true for normal GPC. For vacancy permeation chromatography (MALONE et al., 1969) it must obviously be modified. According to investigations carried out by OTOCKA and HELLMAN (1974b), in this case kinetic factors are more important, for the results depend on the flow-rate. Hydrodynamic effects predominate in the separation of large-sized colloidal particles (SMALL et al., 1977), cf., Section 13.2.

8.6. Resolving power of SEC

In exclusion chromatography, the separation efficiency depends on the slope factor C_2 or S, of the calibration curve (cf., eqn. (8-3)). In the linear range, pairs of substances with equal molar mass ratios $M_{\mathrm{I}}:M_{\mathrm{II}}$ have the same distance, ΔV_{e}, in the elugram. In the case of methods which are based on enthalpy differences, the separating conditions can always be optimized for only one pair of substances, whereby the separation for all other constituents of the sample is necessarily impaired. In contrast to this, in steric exclusion chromatography an improvement of the separating efficiency is for the benefit of all sample constituents, and the general elution problem (cf., Section 3.5.) does not exist here.

Improving the resolution by extending the separation path benefits all components lying in the linear range of the calibration. The *general resolution equation* (3-25) holds for exclusion chromatography with $q = V''/V'$.

As in every other case, R_S increases with the number of plates, N, i.e., with the root of the column length. In this connection it is worthwhile remembering that in the derivation the quantity N (without a subscript) was introduced for neighbouring bands. This must be particularly emphasized for the SEC of polymers, because the plate number varies widely in the range $0 \leqq K \leqq 1$ (cf., Section 16.5.2.). Of course the proportionality to the root of the length of the separating path is preserved and, as mentioned above, may be to the benefit of all the components.

For the *relative distribution factor*, $(K_{\mathrm{II}} - K_{\mathrm{I}})/K_{\mathrm{I}}$, we obtain, using eqns. (8-1), (8-2) and the value for the exclusion limit, $\log M_{\mathrm{lim}} = (C_1 - V')/C_2$:

$$\frac{K_{\mathrm{II}} - K_{\mathrm{I}}}{K_{\mathrm{I}}} = \frac{\log (M_{\mathrm{I}}/M_{\mathrm{II}})}{\log (M_{\mathrm{lim}}/M_{\mathrm{I}})} \tag{8-36}$$

Consequently the relative distribution factor depends on the quotient M_I/M_{II} as well as on the parameters C_1 and C_2, which describe specific properties of the apparatus. As regards the parameter M_{lim}, the following remarks are relevant:

In the immediate neighbourhood of the exclusion limit eqn. (8-36) does not hold because here the slope of the calibration curve becomes infinite. It would be a fallacy to expect a particularly high selectivity in this range.

The reason why, nevertheless, M_{lim} is included in eqn. (8-36) is as follows. The further from the exclusion limit the separation takes place, i.e., the greater M_{lim}/M_I, the smaller is the relative distribution factor. If a wide range of samples has to be investigated, it is justified to use columns with high values of the exclusion limit; for if the distribution of the sample extends farther, then the fractions beyond the exclusion limit emerge crowded in the exclusion volume, V'. Thus another maximum may appear, and a part of the distribution curve is cut off. Any maximum occuring near the exclusion limit must be treated with suspicion and artefacts have to be expected. The maximum can be interpreted as a feature of the distribution only if a repetition of the separation using a volume with a higher exclusion limit shows the same behaviour (Fig. 8-7).

However, where routine work is to be done in rather well defined ranges, the column should not be extended "for safety" by an additional unit having a higher steric exclusion limit. This would unnecessarily reduce the relative distribution factor (eqn. (8-36)), broaden

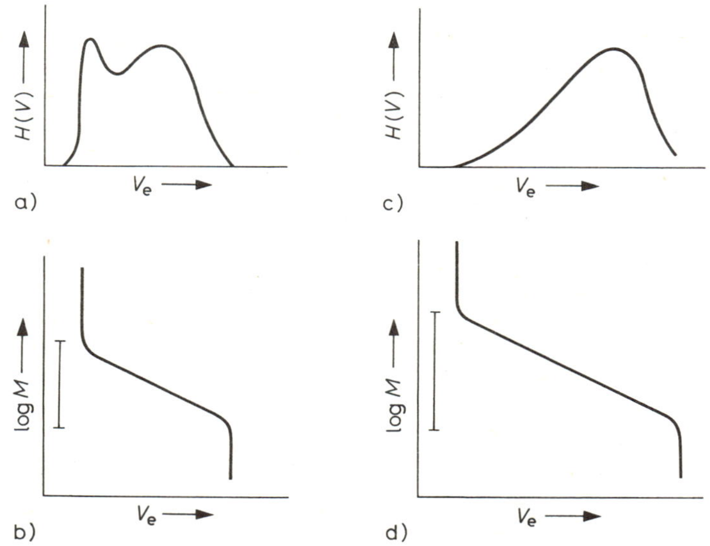

Fig. 8-7
Influence of the separating range on the elugram of a polymer sample with a broad distribution

a) Bimodal elution curve obtained from a column with a separating range which is too narrow
b) Characteristic of this column. The maximum occurring for a small elution volume coincides with the exclusion limit
c) Normal elugram of the same sample, obtained on a column with an adequate separating range
d) Characteristic of this column. The exclusion limit is high enough

The sharp maximum in Fig. 8-7a is an artefact caused by unseparated fractions, which have very high molar mass values.

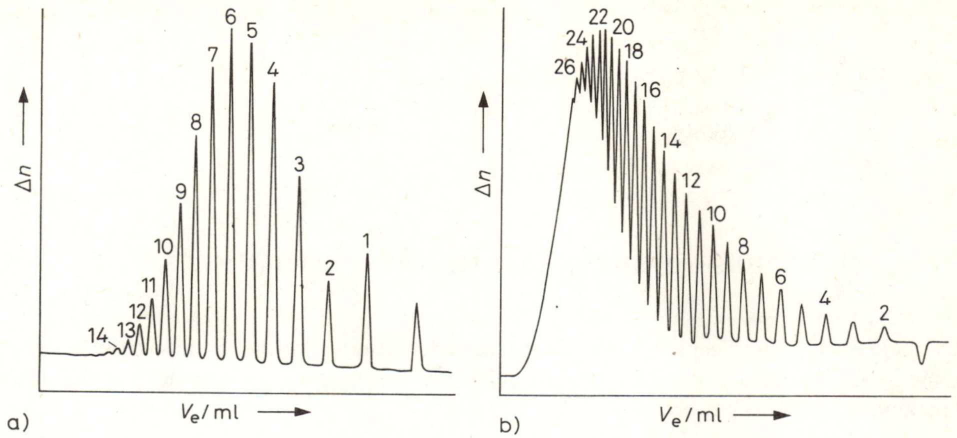

Fig. 8-8
Gel chromatograms of styrene oligomers in dimethyl formamide on Merckogel® 6000
a) $\bar{M}_n$ = 600; 2 mm coiled column, 10 m long, analysis time 40 h
b) $\bar{M}_n$ = 2200; 2 mm column, 20 m long, analysis time 20 days
The numbers indicate the degree of polymerization (according to HEITZ, 1975).

the chromatogram and increase the duration of analysis — three disadvantages which should be avoided. After a systematic investigation of three well characterized polystyrenes (including NBS 705 and NBS 706) by means of various column combinations, MORI (1979) recommended that the steric exclusion limit should be chosen so as to correspond to about ten times the value of $\bar{M}_w$ of the sample to be investigated.

The retention factor would sensibly reduce the resolution if the distribution coefficient is small (cf., Section 3.4.). In chromatographic techniques based on other principles this can be prevented, but not in SEC, which is possible only in the range $0 \leqq K \leqq 1$. This involves a marked *elution-volume* dependence of the resolution. At the starting edge of the elugrams K is only slightly greater than zero. Here, in the range of high molar masses, the retention factor has its lowest value. The larger the elution volume, the higher is the retention factor, and hence the resolution. This contributes to the unusual picture of the exclusion chromatograms in which the peaks appear with approximately the same sharpness at all points, while in all other chromatographic techniques there is a broadening which increases with the elution volume (cf., Section 3.3.). This remarkable behaviour of SEC can be observed most distinctly in the separation of individual homologues (see Fig. 8-8), which can be achieved by means of a highly sophisticated technique.

Naturally the influence of the retention factor has the same impact on all of the methods. Its limiting influence can however be reduced if higher K values are chosen. As this way is barred in SEC, the *phase ratio q* becomes especially important. For high performance, gels with a large total pore volume are required. The interstitial volume must be small; consequently the columns must be packed as densely as possible. The porous layer beads mentioned in Section 10.2. are hardly suitable for use in steric exclusion chromatography, because of their low pore capacity.

9. Chromatographic separation by partition

This chapter deals with separation processes based on the partition of the sample between coexisting liquid phases. The methods are related to some forms of polymer fractionation by means of solubility differences.

9.1. Liquid-liquid partition of low-molecular-weight samples

Frequently the distribution equilibrium between two phases is taken as the starting point for the consideration of chromatographic processes. If, following the example given by MARTIN and SYNGE (1941), substance exchange and phase transport are taken to be successive cycles (cf., Section 3.3.), then the picture obtained would correspond exactly to Craig partition. In this technique a large number of similar vessels, in principle separating funnels, are connected in series. In the transport cycle usually the respective upper phase is transferred to the next vessel. Shaking is then applied until a new equilibrium state has been reached in all of the vessels. As there is no gradient, i.e., one and the same extracting agent is used in all stages, the partition coefficient is constant.

By this principle, SCHWENKE (1964) separated caprolactam oligomers by partitioning between heptane and methanol in an apparatus comprising 100 elements. The mobile upper phase emerging from the outlet was recycled to the input end of the apparatus. Thus it was possible to increase the number of separating steps to 2000. For water-soluble natural macromolecules, multiple partition techniques were widely used (SCHWENKE, 1965) before other methods based on simpler apparatus became available for separation under mild conditions.

These methods are the chromatographic partition techniques. The pioneering work was done by MARTIN and SYNGE (1941). They transferred the partition between chloroform and water from the Craig apparatus into a non-sectionalized column with a continuous eluent flow. Water was employed as a stationary phase on silica gel and chloroform containing the dissolved sample was passed over this column packing.

This principle — an imbibing solid column material with a liquid film having a large surface area and a low-viscosity eluent which neither dissolves nor mechanically removes the stationary film phase — has enabled quite amazing separations.

To prevent mechanical removal of the stationary phase, the *viscosity* and, if possible, also its *adsorption affinity* to the supporting material should be greater than those of the mobile phase. To prevent dissolution, the interaction term in eqn. (5-1) must be at least 2, and hence the *solubility parameters* must differ from each other by at least $\sqrt{2} = 1.4$ Hildebrand units. On the other hand, the sample should dissolve in both phases, which demands the coincidence of the solubility parameters. Thus, the two requirements are in opposition and a compromise is needed. For low-molecular samples this can be achieved in most cases. Here the entropy of mixing is so high that the enthalpy contribution to the free energy of mixing is not critical. Therefore the phase combinations for partition chromatography of low-molecular-weight compounds can be chosen rather freely among immiscible pairs.

9.2. Liquid-liquid partition of macromolecular samples

In the dissolution of macromolecular samples, the entropy contribution is small, and the enthalpy change is decisive (cf., Chapter 5). The solubility parameters of the sample and the solvent must be similar. How can this be reconciled with the condition for immiscible liquids that $\delta_I \geqq (\delta_{II} + \sqrt{2})$ or $\delta_I \leqq (\delta_{II} - \sqrt{2})$? At best the value δ_P for the polymer will lie halfway between δ_I and δ_{II}. However, the difference is still $\sqrt{2}/2 = 0.7$ Hildebrand units. The values for typical polymer solvents are in most cases much closer to the values for the polymers.

In a systematic investigation, DOBRY (1956) found that only 3 of 289 possible combinations of a synthetic polymer and two immiscible liquids in fact contained the polymer in both phases. The remaining 286 combinations proved unsuitable. Thus experimental results as well as the theoretical conclusions will only exceptionally permit the chromatographic separation of polymers by partition between two pure liquids. In multicomponent systems, however, partition equilibria may result. Means of achieving these are the addition of other low-molecular components, or the utilization of the incompatibility between dissimilar polymers.

9.2.1. Fractionation of polymers by partition between immiscible liquids

In fractionation by partition between immiscible liquids the solution of the polymer is extracted by another liquid phase, the solvency of which increases progressively. Systems comprising several components are generally required, which are present in the coexisting phases with different concentrations. Moreover, extreme volume ratios are frequently needed to achieve the desired partition of the polymers between the two phases.

A classical example is the fractionation of polyethylene oxide by partition between water and chloroform-benzene mixtures (SCHULZ and NORDT, 1940; ALMIN et al., 1957, 1959; RING et al., 1966). In a similar way, polyamide 66 (TAYLOR, 1947), polyamide 6 (GORDIJENKO, 1953) and polyethylene terephthalate (UEBERREITER and GÖTZE, 1959; REINISCH et al., 1969) were fractionated. PAILHES et al. (1967) as well as THEIL et al. (1972) used continuous extraction columns. Frequently a coacervate extraction is carried out, in which a very highly swollen, and hence still fluid, gel phase is contacted with the solvent mixtures. Polyalkylene glycols were split into fractions by CASE (1960) using a Craig partition. Using the same technique, polyester oligomers were separated according to size and structural characteristics (SCHÖLLNER et al., 1967, 1968). RIGAMONTI and MEDA (1955) fractionated polyvinyl acetate by means of counter-current distribution in a benzene-methanol-water system.

v. TAVEL and BIERI (1971) found that mixtures with an almost critical behaviour, consisting of about 70 vol.% acetone, 22% hexane and 8% water, formed pairs of phases suitable for the separation of polymethyl methacrylate. It was possible to adjust the partition coefficient of the polymer to any value desired by adding a very small quantity of benzene. With 0.45—0.7% of benzene, various polymers were successfully fractionated (v. TAVEL and RÜFENACHT, 1976). The *partition coefficients* increased with $M^{2/3}$ (RÜFENACHT and v. TAVEL, 1976). Usually the mobile phase was the hexane-rich upper region, while the stationary phase was the lower region containing mainly acetone and water. The density difference between the two phases was as small as 0.03—0.05 $g \cdot cm^{-3}$. To accelerate the precipitation, a *partitioning centrifuge* containing 120 partitioning elements in a drum with a horizontal axle was developed (v. TAFEL and BOLLINGER, 1968). When the drum rotated at a speed of 0.4 Hz (25 rpm) the contents of the 120 glass tubes were stirred thoroughly, so that the partition equilibrium was reached within 3 minutes. At a speed of 16.7 Hz (1000 rpm) the phases separated from each other, so that subsequently the upper phase of each glass tube could be transferred into the next higher one. This was done simultaneously for all of the tubes while the drum was rotating at a speed of 1.7 Hz (100 rpm). The fractions obtained in this way were narrow, with $\bar{M}_w/\bar{M}_n$ = 1.02—1.05. For the starting sample they yielded a chain-length distribution which was markedly narrower than the (uncorrected) GPC curves and agreed with the results of the ultra-centrifuge.

9.2.2. Counter-current fractionation using an auxiliary polymer

ENGLERT and TOMPA (1970) suggested utilizing the *incompatibility* for partition fractionation and extracting the solution of a polydisperse sample by means of a liquid mixture containing graduated concentrations of an auxiliary polymer of a different chemical structure. The procedure starts with a polymer-rich counter-phase which takes up only the shortest chains of the products to be separated. The lower the concentration of the auxiliary polymer in the subsequent batches, the higher the degree of polymerization of the components to be taken up can become. This technique promises an excellent separation efficiency (KONINGSVELD, 1970b). While in aqueous solutions it was already possible to realize separations in a similar way by means of the incompatibility (ALBERTSSON, 1958a/b, 1965; TISELIUS et al., 1963), considerable difficulties were encountered with synthetic polymers in organic solvents. The phases tended to form emulsions, which precipitate poorly due to the minimum density difference. Columns operating on this principle were very sensitive to temperature fluctuations and mechanical vibrations. A solution of this problem obviously might require similar means for phase separation as those described in Section 9.2.1.

The examples show that also polymers can be separated by partition between liquid phases.

The apparatus for multiple partition operates with successive cycles for the transport and the establishment of equilibrium. Now yet another development will be described, which leads from the principle of Craig partition closer to chromatography.

9.3. Counter-current chromatography

While counter-current partition is carried out step by step (also in the apparatus used by v. TAVEL), i.e. with exchange and transport cycles being separated in time, in counter-current chromatography, transport and exchange proceed simultaneously (Ito et al., 1970, 1971, 1972; Tanimura et al., 1970). The mobile upper phase is contacted with the lower phase in many separating elements. A horizontally arranged, helically coiled pipe may serve as a model for such a chromatographic path (Fig. 9-1 a, b). All of the coils contain so much of the lower phase that the upper bends of the coils are separated from one another. The latter are completely filled with the mobile phase, which is slowly pumped through the coiled pipe. When the mobile phase passes from one coil to the next one by overcoming the liquid seal of the lower phase, the phases are moved. This improves the exchange of substance. To prevent the unwanted transport of droplets of the lower phase in the moving upper phase, the coiled pipe is inserted into a centrifuge. In this way dinitrophenyl amino acid derivatives were separated by means of a Teflon® tube with an inside diameter of 2 mm, coiled in 17000 twists.

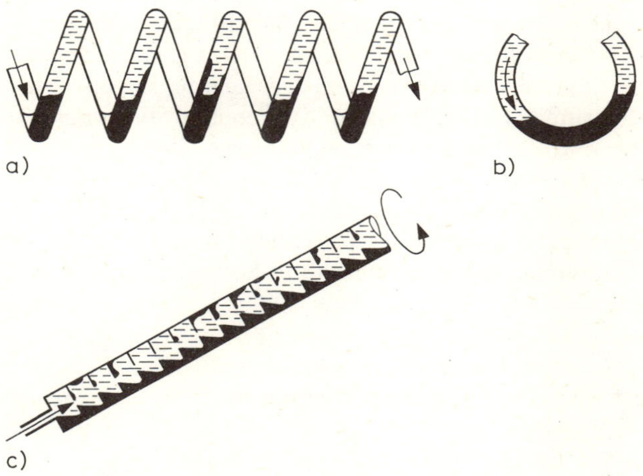

Fig. 9-1
Devices for counter-current chromatography

a) Part of a coiled pipe comprising 17000 coils
The upper phase gradually flows through the coil, which is subdivided into individual chambers by the lower phase portions

b) Schematic representation of the flow process
The lower phase returns to its initial position after part of the mobile phase has flowed through. This is supported by inserting the coil into a centrifuge

c) Rotating chamber column (part of the column comprising 5000 chambers)
The darker-shaded lower phase is lifted along the pipe wall due to the rotation, so that contact with the mobile upper phase is accomplished with a large, continuously changed surface (according to Ito et al., 1970, 1971, 1972; Tanimura et al., 1970).

Further it was possible to perform counter-current chromatography by means of a cylindrical column sectionalized into 5000 small chambers. This column rotated about its axis, which made an angle of 0.52 rad (30°) with the horizontal, at a speed of 3 Hz (180 rpm). The partition walls between the chambers had central bores, through which the mobile upper phase could pass into the next higher chamber (see Fig. 9-1 c). In a different mode of operation, the column axis was moved along a circular path at a speed of 13.3 rps (800 rpm).

In *droplet counter-current chromatography*, the mobile phase falls in single droplets through vertically arranged pipes filled with the stationary liquid. From the sump of the

(n-1)th pipe the mobile phase is transferred to the head of the nth one through a Teflon tube. This variant introduced by TANIMURA et al. (1970) underlies a commercial apparatus which has a battery of 300 pipes each having a length of 400 mm and an inside diameter of 2 mm. Suitable pairs of phases have been given by OGIHARA et al. (1976). HOSTETTMANN et al. (1979a, b) used droplet counter-current chromatography for a careful, preparative isolation of natural materials, determining the optimum solvent combinations by preliminary tests on a thin-layer plate.

In another variant of the counter-current principle (ITO, BOWMAN, 1978) a coil made of a PTFE tube with a horizontal axis was used, which was wound on a metal pipe and moved, together with the latter, by a planetary gearing. The movements were adjusted in such a way that this horizontal flow-through coil planet centrifuge did not require any rotating seals. The tube had a length of 5 m and an inside diameter of 2.6 mm and was wound in about 100 turns on an aluminium tube 1.25 cm in outside diameter. If larger sample volumes (V_0 = 10 ml) had to be passed through the apparatus, then 10 coils of this type were employed in series. The equipment operated on the principle shown in Fig. 9-1a. For a slow rotation, the main effect was due to gravity. When the planetary gearing ran at a high speed (up to 5 Hz, or 300 rpm) the centrifugal force dominated. Consequently two ranges of an optimum separation effect as a function of the speed were found to exist. The efficieny was demonstrated by an example where a mixture of five dipeptides was separated into its constituents within 10 hours, with $v = 1\ \text{ml} \cdot \text{min}^{-1}$ and 5 Hz.

Counter-current chromatography with a *solid* packing material is of importance mainly for preparative fractionation and will be dealt with in Section 17.9.3.

9.4. Chromatography on bonded phases

9.4.1. Low-molecular-weight samples

In its classical form, liquid-liquid partition chromatography is carried out with stationary phases deposited on porous packing materials or cellulose. A column 100 cm long and 2 mm wide normally contains about 0.1 g [D 6] to 0.5 g [D 7] of a stationary phase, i.e., a volume of 0.1—0.5 cm^3. The column has an empty volume of V_C = 3.14 cm^3. About $^2/_3$ of this volume is occupied by the packing material. The total pore volume of the packing is about 1 cm^3. Thus, the stationary liquid by no means fills the pores completely. This might possibly be the case in the highly loaded columns employed by some authors, which contain up to 3 g of a stationary liquid per g of column material (UNGER, 1974).

The amount of the stationary phase, which is usually rather small, has to withstand many separations, in each of which it is flowed over by a volume of the mobile phase which exceeds it by several decimal powers. The required durability can be achieved only if the mobile phase is saturated with the stationary liquid in an *auxiliary column* before it is fed into the actual separating column. From time to time the auxiliary column must be replaced. Nevertheless, the actual liquid-liquid partition chromatography remains experimentally demanding because a gradient in pressure or temperature in the separating column, or the presence of the sample itself may influence the stability of the system. Moreover, it requires a good deal of hard work to apply the stationary liquid to the column material.

Chromatography on bonded phases is much simpler in this respect. It uses column materials which support the stationary phase as a firmly anchored coating (bonded phases cf., Section 11.10.). Typical examples are the octadecyl layers immobilized on porous layer bead material or silica microspheres by Si-C bonds. This anchoring is unaffected by solvents and thermally resistant. The chemical nature of the bonded phases can be modified in a variety of ways, so that many different materials being particularly suitable for particular separation problems are available (cf., Table 11-9). In this respect, chromatography on bonded phases resembles partition chromatography with liquid stationary phases. The selectivity (relative distribution factor) of the phase systems can be varied widely.

As regards the mechanism, however, there are essential differences between the column materials with liquid stationary phases and the chemically bonded layers. The small thickness of the coatings, the high rate of exchange and, above all, the strong influence of functional groups such as -CN and -NH_2 and the like show that in the case of monolayers the chromatographic process takes place at the surface. On the other hand, with polymer coatings on rigid particles the penetration of sample components into the swollen layer must also be taken into account (cf., Section 11.10.3.).

In practice, the bonded phases are superior in some respects: the sample components can be collected as pure components, without traces of the stationary phase; the eluent can easily be exchanged and it is even possible to carry out gradient elutions.

The *phase ratio* is generally small in partition chromatography, being as low as $q = 0.1$, except for highly loaded columns. The *partition coefficients* range from 5 to 500. Due to the almost free choice of the pairs of phases, in this type of chromatography they can be varied within very wide limits more easily than in any other chromatographic technique.

The stability of the stationary phase is the fundamental condition for a constantly efficient and reproducible partition chromatography. In principle it does not matter how the stationary phase is generated — by loading the dry column material outside of the column, by a precipitation technique within the column executed independently of the actual separation or in connection with the latter or by the use of special column materials with chemically bonded layers. For the elution chromatography of low-molecular substances, the bonded phases represent the most convenient variant.

9.4.2. Macromolecular samples

Chromatography on bonded phases is one way to overcome the discrepancy, shown in Section 9.2., in the demands made upon the solubility parameters: a bonded phase may possibly swell, but it does not dissolve in the mobile phase. Now the latter can be chosen freely in this respect, solely on the basis of its reaction to the polymer.

The balanced partition of a polymeric sample between the bonded phase and the eluent can be achieved, though, only under quite specific conditions. Here the compatibility also plays a rôle. A polymeric support must be compatible with the polymer to be separated to such a degree that the sorption does not fail due to incompatibility.

Casper and Schulz (1970, 1971 b) prepared a stationary phase for the phase distribution chromatography of polystyrene by precipitating a high-molecular-weight polystyrene ($M = 8 \cdot 10^6$ g · mole^{-1}) on glass beads. It was tritium labelled, so that its extraction by cyclohexane at 28 °C could be monitored. The treatment was continued until the concentration in the eluate had decreased to less than 1 mg · l^{-1}. Thereafter the column (length 1m, diameter 30 mm) still contained 124.5 mg of the polymer. On the total surface area of 17 m^2 of the packing material, this quantity yielded a coating which swelled to a thickness of 300 to 400 nm in cyclohexane. Polystyrene samples with a narrow distribution and an average degree of polymerization ranging from 300 to 4000 were passed through this column at temperatures between 17 and 25 °C, and the elution volumes were determined. Starting from eqn. (8-1) one obtains

$$K = \left(\frac{V'}{V''}\right)\left(\frac{V_e}{V'} - 1\right) \quad (9\text{-}1)$$

or

$$\log (V_e/V' - 1) = -\log (V'/V'') + \log K \quad (9\text{-}2)$$

$K = \varphi_{gel}/\varphi_{sol}$ is the partition coefficient of the molecules of degree of polymerization P between the stationary gel phase and the mobile sol phase. The experimental data of $\log (V_e/V' - 1)$, when plotted vs. P, yielded lines, the equations of which

$$\log (V_e/V' - 1) = -a + bP^{\nu} \quad (9\text{-}3)$$

if compared with eqn. (9-2) give the following relationships for $\nu = 1$:

$$\log (V'/V'') = a \quad (9\text{-}4)$$

$$\log K = b \cdot P \quad (9\text{-}5)$$

For a an average value of 1.4 was obtained, i.e., V'/V'' is about 30. Table 9-1 gives some partition coefficients as reported by CASPER and SCHULZ. These data show that at a temperature of 10 deg. below the θ temperature (35 °C), the macromolecular product is practically insoluble in the sol phase, and the partition coefficients vary widely in the rather narrow temperature interval between 17 and 25 °C. Increasing the total polymer concentration by three decimal powers had almost no effect on the partition coefficients, whereas the phase ratio, V'/V'', dropped from 28.3 (for 50 mg $\cdot$ l^{-1} and 25 °C) to 14.8 (for 50 g $\cdot$ l^{-1}). Also the effect on the partition coefficients of the mass ratio of two components having different degrees of polymerization was amazingly low. The investigations were continued by GRESCHNER (1979a), who constructed an automated apparatus (1979b) which enabled investigations of a very high accuracy. The column made of V4A steel pipe, with a length of 5.84 m

Table 9-1
Chromatographically determined coefficients for the partition of nearly uniform polystyrene samples between the sol phase and the gel phase in cyclohexane (according to CASPER and SCHULZ, 1971b)

ϑ/°C	Degree of polymerization		
	500	4000	80000
17	0.28	$4.3 \cdot 10^{-5}$	0
21	0.46	$2.2 \cdot 10^{-3}$	0
23	0.57	$1.2 \cdot 10^{-2}$	10^{-39}
25	0.69	$5.1 \cdot 10^{-2}$	10^{-26}

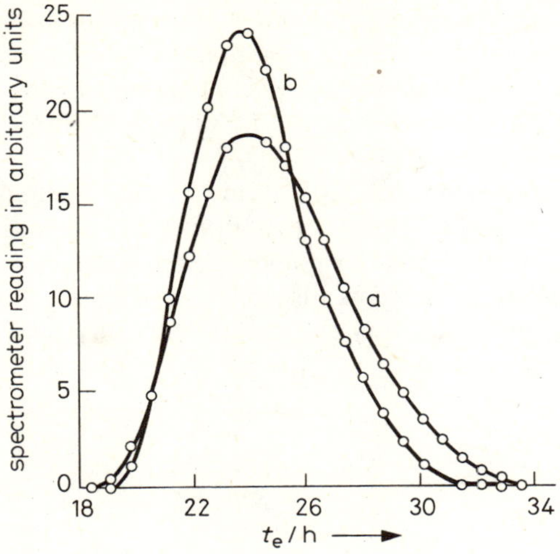

Fig. 9-2
Elution curves of polystyrene samples in phase partition chromatography in cyclohexane at 15 °C

a) Curve of sample K 110000 ($\bar{P}_w$ = 1080; H = 1.009$_5$)
b) Curve for a fraction from sample K 110000, which had been prepared by precipitation chromatography ($\bar{P}_w$ = 1070; H = 1.005$_2$)
Flow rate: 15 ml · h^{-1}
(according to GRESCHNER, 1979b).

(4 × 1.46) and an inside diameter of 10 mm, was packed with glass beads (d_P = 70—80 μm) supporting a macromolecular polystyrene film with a swollen thickness of 305 nm. For a flow-rate of v = 0.25 ml · min^{-1}, calibration by polystyrene standards with narrow distributions yielded plots of ln (V_e/V' — 1) vs. $\bar{P}_w$, which for lower temperatures (below 21 °C) showed particularly marked S shapes. To the left of the point of inflection, which occurred at about M = 20000 g · mole^{-1}, and shifting towards higher molar masses as the temperature increased, the partition process reached thermodynamic equilibrium. Here it was possible to apply eqn. (9-3). v ranged between 0.5 and 1, corresponding to the values given by WOLF et al. (1967, 1978) and KLEINTJENS et al. (1976). For samples with higher molar masses, the establishment of equilibrium did not proceed rapidly enough, but it was in just this dynamic range of operation that excellent separation efficiencies were achieved. Fig. 9-2 may serve as an example, showing the elution curve (a) of a polystyrene standard with $\bar{M}_w$ = 112500 g · mole^{-1} (H = 1.0095) and that (b) of a Baker-Williams fraction of this standard ($\bar{M}_w$ = 111400 g · mole^{-1}, H = 1.0052). The small difference in the distribution of the two samples, the effect of which on the heterogeneity does not even reach 0.5%, can be observed from the two curves with a clearness which in these high-molecular-weight ranges has not yet been achieved by any other method for the determination of the chain-length distribution.

An interesting alternative to the "linear" gel used hitherto in phase partition chromatography, which adheres to the supporting material by physical interactions, is offered by polymer layers fixed chemically on silica gel surfaces (LECOURTIER et al., 1978b, c) (cf., Sect. 11.10.3.)

The fractionating effect observed by LANGHAMMER and QUITZSCH (1961) for polystyrene in toluene, which was passed through a column containing highly cross-linked polystyrene, can likewise be interpreted as phase partition chromatography. In their experiments, the components having the lowest degree of polymerization were the first to leave the column, i. e., the separation was not caused by steric exclusion. The same holds for the analogous work of VAUGHAN and GREEN (1963) (see Fig. 9-3).

Perhaps surface effects are also involved in such separations. The chromatography of optically active poly-α-olefins on poly(S)-3-methylpent-1-ene (PINO 1962, 1966) as well as that of stereoregular block copolymers of propylene on polypropylene supported by silica gel (NATTA et al., 1958, 1960), which can be classified as adsorptive separations, show a remarkable similarity to the examples of phase partition chromatography discussed here.

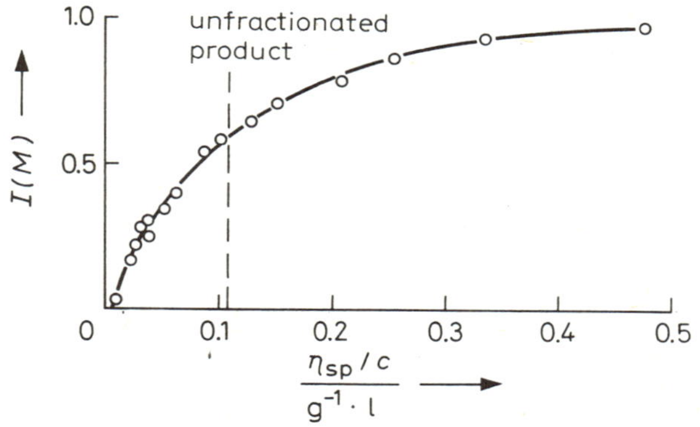

Fig. 9-3
Column fractionation of polystyrene (M_n = 38700 g · mole^{-1}, U = 8.7) by retention on a cross-linked polystyrene (2% of divinylbenzene)
Eluent: benzene-methanol; time required: 2—3 h (according to VAUGHAN and GREEN, 1963).
The small-molecule components leave the column before the macromolecular ones, hence steric exclusion chromatography is not involved (representation as in Fig. 4-1 a).

9.5. Precipitation chromatography

Decreasing the temperature or addition of precipitants causes homogeneous polymer solutions to split into a sol phase and a gel phase. The gel phase contains a higher concentration of polymer and, if solvent mixtures are used, a greater amount of the thermodynamically better solvent than the sol phase (cf., Section 5.4.). The long-chain components of polydisperse samples concentrate in the gel phase, the short-chain ones in the sol phase. In precipitation chromatography the gel acts as the stationary phase, the sol phase as the mobile one. As compared with the standard techniques of chromatography some differences are conspicuous:

a) At the start of the process, the stationary gel phase is not yet fully developed. The separation may take place with polymer-coated supporting materials, the coating of which swells to the gel phase only under the action of the eluent. It may also be that the stationary phase develops continuously from solutions, which are on the verge of phase separation, due to cooling or variations in concentration.

b) The equilibrium between the sol phase and the gel phase is characterized by a coupling between the concentration ratio, $K^+ = c''/c'$, and the phase ratio, $q = V''/V'$. This does not exclude the exchange of components of the polymolecular mixture between the sol and the gel phases. Indeed narrower fractions than in batch fractionation are produced if the sol-gel contact can take place in many partial steps.

c) The stationary phase essentially consists of components of the sample. If the separation is carried out by elution chromatography, then the stationary phase will disappear by the end of the process.

9.5.1. Chromatographic sol-gel fractionation without a temperature gradient

Good separations were obtained if the polymer sample, applied as a very thin layer to the total packing material of the column (sand or glass beads) was extracted by a slow-flowing eluent of increasing solvency (DESREUX et al., 1949, 1950, 1952; KRIGBAUM and KURZ, 1959). MENCER and KUNST (1978) lowered the temperature continuously during the elution; they stressed that in this way the solvent gradient can most easily be adjusted to the solubility characteristics of a sample. KLEIN and FRIEDEL (1970) filled the column completely with a heated polystyrene solution, from which during a gradual cooling process the polymer precipitated on the beads in a prefractionated condition: first the macromolecular components with the lowest solubility, followed by the more easily soluble small-molecule components as an outer layer. The gel film was extracted gradually by butyl acetate-n-propanol (70:30), with increasing temperature. The distribution curves of the fractions were determined by sedimentation analysis; their unusually steep slopes indicated a high uniformity of the fractions. Likewise, by a batchwise extraction of prefractionated layers, FERRIER (1967) fractionated polyethylene, polypropylene and polyvinyl chloride. The fractionating effect of the extraction depends very much on the grain size of the polymer particles deposited on the supporting material (OGAWA et al., 1973).

The batchwise elution has the advantage that the dead volumes at the column outlet, which in most cases are rather large in the simple apparatus used in precipitation chromatography, are negligible in this case. In a continuous elution these volumes may cause a loss of resolution under otherwise equal conditions (KLEIN and WEINHOLD, 1970).

An *automatic fractionation apparatus* for a batchwise extraction of the polymer film has been described by BLAIR (1970): The sample is applied as a concentrated solution to both sides of a continuously running, endless belt of polyethylene terephthalate foil, and fixed by drying. The belt, 70 mm in width, is perforated like a cinematograph film. The edges are left uncoated in order that no parts of the sample are mechanically rubbed off. The belt runs at a speed of 0.4 mm · s^{-1}, guided by pairs of polyamide rollers, through 20 glass vessels arranged in series, which contain an eluent of increasing solvency (see Fig. 9-4). Thus it was possible, within 10 hours, automatically to separate 3 g of synthetic rubber (Neoprene W) into sixteen fractions of approximately equal sizes, the Staudinger indices of which increased from 0.25 to 4.21. This required 20 × 600 ml of eluent. Theoretically, it should be possible to prepare fractions with $\bar{M}_w/\bar{M}_n \leqq 1.06$ by means of this apparatus; the best value obtained was 1.07. The efficiency for a preparative fractionation was demonstrated by the separation of 348 g of Neoprene W into ten fractions in a 25 day continuous operation.

As in a chromatographic process, in Blair's automatic fractionation apparatus, two phases between which matter transfer processes occur are moved against each other. In normal

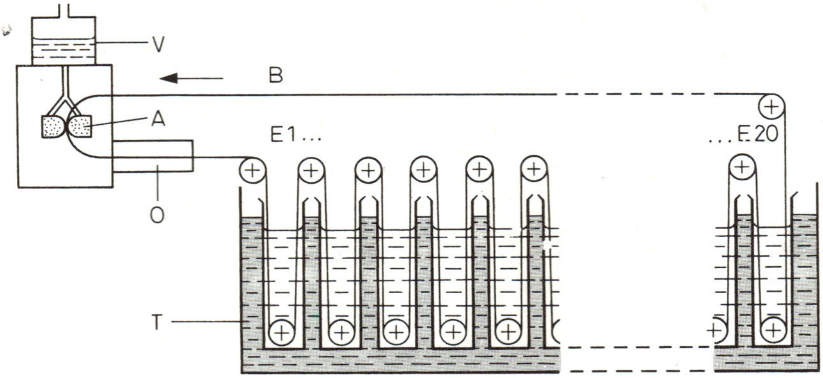

Fig. 9-4
Schematic representation of an automatic fractionation apparatus with 20 extraction stages (according to BLAIR, 1970)

A coating device; B conveying belt, 15.24 m long, 70 mm broad; E1 ... E20 extraction vessels containing the eluents of graduated solvency; O drying chamber for fixing the sample on the belt; T thermostat; V supply vessel containing the dissolved sample.

precipitation chromatography the gel phase is stationary, whereas in this case it is transported by the supporting belt. The liquid counter phase in the extraction vessels exhibits a gradient from one chamber to another. Within each glass vessel the solution is homogenized by the motion of the belt. The apparatus performs an automatic 20-stage gradient extraction.

9.5.2. Chromatographic sol-gel fractionation with a temperature gradient

The chromatographic separation on the basis of sol-gel equilibria requires a stable gel layer with a large surface area, which is capable of exchanging substance with the sol phase flowing over it. In the technique discussed in the preceding section the gel layer is prepared before the start of the actual chromatographic process. Now a method will be discussed in which the gel layer is formed only during elution. The technique in question is the *precipitation chromatography* developed by BAKER and WILLIAMS (1956), which is carried out in packed columns with an antiparallel temperature gradient. This technique has been reviewed by COOPER (1978). The sample is applied as a thin layer to sand or glass beads in the uppermost, hottest part of the column. For elastomers, Kieselguhr is suitable as a support (HULME and MCLEOD, 1962; PANTON et al., 1964). The eluent, flowing slowly over the sample, takes up an amount corresponding to its solvency at the respective temperature. In the cooler zones below, the solutions are no longer stable. If the conditions are favourable, a uniform film is precipitated on the surface of the packing material and the column wall, being viscous enough to adhere to the supporting material. Adsorption is conducive to this process particularly from θ solutions (cf., Section 6.2.3.). However, if the gel phase remains suspended in the form of fine droplets or flows off the packing material, then a chromatographic process cannot take place, and the efficiency of the separation decreases (VAUGHAN and GREEN, 1963). To prevent the flowing-off of the gel phase, which occurs mainly with polymer samples in the lower range of molar mass PEPPER and RUTHERFORD (1959) used a very fine-grained packing material. SCHNEIDER (1961) recommended the use of a relatively poor solvent in order that the gel layer should not become too soft. The gel layer must not clog the column and should be qualitatively and quantitatively uniform at all points of the column cross section.

In this respect the *temperature profile* is of great importance. For the Baker-Williams chromatography to be operable, there must be a defined temperature gradient along the column, which remains constant over the whole diameter. As the temperature is usually controlled from the wall, there are limits to the column cross-section. BAKER and WILLIAMS used a glass tube 24 mm in outside diameter. A few authors have employed thicker columns with diameters of about 40 mm (SCHNEIDER and HOLMES, 1959), 60 mm (SCHNEIDER et al., 1959), or 70 mm (SCHULZ et al., 1962). In order to fractionate larger quantities, CANTOW et al. (1961, 1963, 1964) used an apparatus with six separating tubes each with an inside diameter of 25 mm, which were connected in parallel. Annular cross sections or cross sections subdivided by bridges where the distance from a temperature-regulated wall was nowhere greater than 13 mm, were also employed (see Section 17.9.2.).

Every irregularity in the temperature profile disturbs the temperature-dependent sol-gel equilibria. In precipitation chromatography a warm eluent flows continuously into cooler zones of the separation tube. The effect of the *temperature gradient* applied from outside can be as intended only if the transport of heat towards the wall is so rapid that every liquid volume flowed in can assume the temperature corresponding to its position. Of course this cannot occur completely, because a heat flow always presupposes a temperature difference. The more the temperature in the interior approaches that of the wall, the slower the heat exchange becomes. Even in the stationary state a slight temperature difference exists. SCHULZ et al. (1962) investigated this problem experimentally on a column 70 mm in inside diameter. The temperature gradient along the column was 40 K over 340 mm. The authors determined the temperature difference between the wall and the column centre as shown in Fig. 9-5 as

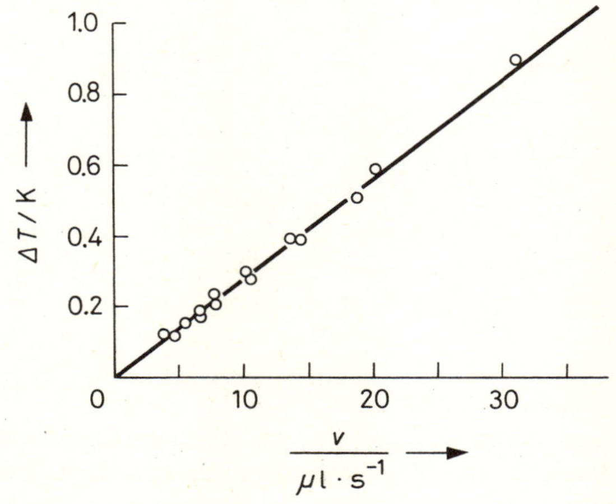

Fig. 9-5
Temperature difference between the wall and the centre of a Baker-Williams column 70 mm in diameter, which is filled with glass beads (0.1 mm), as a function of the rate of elution

Temperature gradient: 1.175 K · cm^{-1}; flowing medium: benzene-methanol (according to SCHULZ, SCHOLZ and FIGINI, 1962).

a function of the flow-rate. For a flow-rate of 40 ml/h the difference was 0.3 K, i.e., the surface of equal temperature sagged by about 3 mm.

In the chromatographic sense, a deep sagging means a great plate height. The sagging occurs especially for a high flow-rate, a steep temperature gradient and an impeded heat transfer. In quite extreme cases the break-through of the mobile phase, i.e., *channelling* in the centre of the column, is expected. The decrease in the viscosity of the mobile phase and the reduction of the gel proportion with increasing temperature both would maintain the channelling once the break-through has occurred. At the same time the rate of flow in the channel would be so high that the heat transfer as designed for normal operation would not suffice for a subsequent correction. From this it follows that the Baker-Williams fractionation should not be carried out with too steep a temperature gradient, that the column cross section should not be made arbitrarily large and that the elution should not on any account start before the column has been completely temperature-regulated.

Precipitation chromatography is operated as a *gradient elution*, starting with a mixture of low solvency. The antiparallel temperature gradient and the likewise antiparallel elution gradient promote the separation. However, the elution gradient also involves disadvantages. Thus, once the fractionation has been started it must be completed because the axial diffusion of the small molecules of the eluent would upset the sol-gel equilibria during an interrup-

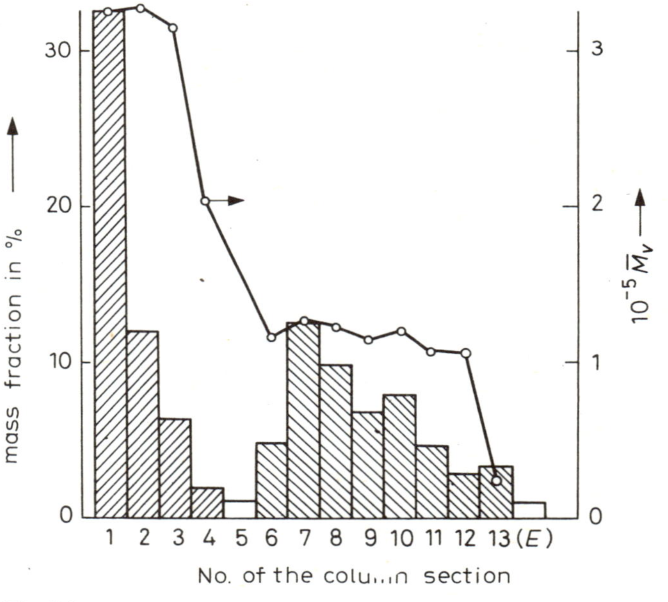

Fig. 9-6
Detection of the stationary phase in precipitation chromatography of a 1:1 mixture of two poly-α-methylstyrene samples with $\overline{M}_v$ = 337000 and 128000 g · mole

Column: L = 1.40 m; d_c = 36.7 mm; temperature at the top: 58 °C; bottom temperature: 10 °C; precipitant: n-hexane; solvent: benzene

The bar diagram shows the distribution of the polymer on the packing material of the column. The points connected by the combined curve indicate the molar masses of the components extracted. Only 1% of the polymer was contained in the eluate (E) (according to YAMAMOTO, NODA and NAGASAWA, 1970).

tion. Moreover, the gradient elution creates difficulties in the detection of the eluted sample components. If the solvents are transparent to UV light while the polymer absorbs UV, the elution may be monitored by means of a UV detector. SCHOLTAN and KWOLL (1972) employed a combustion detector. CAPLAN (1959) separated labelled polymers by a precipitation-chromatographic microtechnique, recording the activity in the eluate. H. J. CANTOW et al. (1966, 1968) showed that the Baker-Williams principle can also be realized with a single solvent (a θ solvent) by use of a temperature gradient and programmed increase of the temperature of the whole column.

For precipitation chromatography, the *stationary gel phase* is a conditio sine qua non. Even if the technique is operated most carefully, the gel phase does not always develop properly. In the Baker-Williams chromatography of polystyrene by methanol-methyl ethyl ketone mixtures in the low-molecular-weight range up to $P = 6000$, YAMAGUCHI and SAEDA (1969) obtained poorer results than expected. They also observed a turbidity in the eluates in some cases, which could not have developed outside of the column. Apparently, in the cooler parts of the column, the polymer had precipitated in droplets without forming a stationary phase.

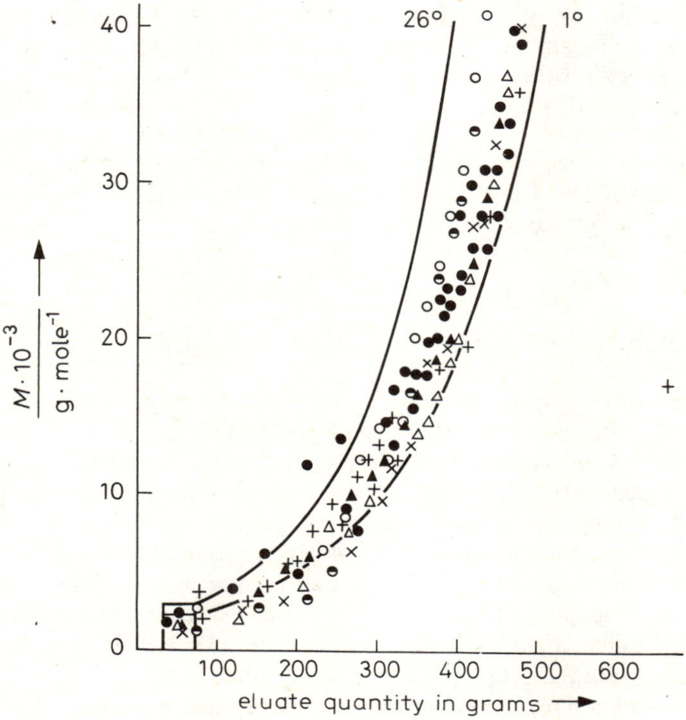

Fig. 9-7
Detection of retention in the precipitation chromatography of polycarbonates in a 32 cm column with a head temperature of 26 °C and a bottom temperature of 1 °C

Nearly all of the elution volumes of the fractions are greater than the values determined from the solubility curve at 26 °C. The points indicate the experimentally determined molar masses for the fractions from eleven independent separations of five different starting samples.

On the other hand, in a poly-α-methylstyrene/benzene/n-hexane system the adherent gel film required did develop. This has been shown directly by YAMAMOTO et al. (1970). They stopped the solvent supply at the very moment when the first polymer fractions had left the column in their preliminary test. The experiment was carried out using a 1:1 mixture of two polymer samples with narrow distributions. The mobile phase contained in the column was blown out by nitrogen. Then the column was cut into thirteen sections and the amount of polymer contained in each section was determined (Fig. 9-6). Section 1, the sample bed, still contained about $^1/_3$ of the substance. The remainder (except for 1 % in the eluate) was found on the glass beads in the other sections. Consequently the postulated reprecipitation due to the temperature field had occurred. (In a fractional extraction, the polymer, which had been dissolved from the sample bed, would have been removed from the column together with the liquid blown out.)

Another test is based on the retention of the polymer fractions: Fig. 9-7 shows the results of eleven polycarbonate fractionations in a column with a head temperature of 26 °C at the top of the column and a temperature of 1 °C at the bottom (GLÖCKNER, 1964, 1965a). The molar masses measured for the individual fractions are plotted as a function of the eluate quantity. The flanking curves are the calculated solubility curves for the outside temperatures. The data required had been determined by turbidimetric titration. If the Baker-Williams fractionation consisted only of the extraction of the material in the sample bed, then the experimental data points would have oscillated around the 26 °C solubility curve. They lie, however, with a few exceptions in the interval between the two curves. This seems to indicate a retention.

While an adequate adherence of the gel film on the supporting material is required on the one hand, there must not be any irreversible adsorption on the other (SCHULZ et al., 1965). The eluotropic strength (cf., Table 7-3) of the solvent must be high enough finally to displace the polymer from the surface of the column packing.

9.6. Resolution of partition chromatography

Partition chromatography with a given stationary phase can be judged, like other forms of chromatography, in the light of the resolution equation (3-25).

However, if the stationary phase is generated only during the chromatographic process, continuously changed during the separation and finally redissolved as in precipitation chromatography, then the phase ratio and the plate number cannot be determined in the usual way. Nevertheless the resolving power of precipitation chromatography is, in fact, a much-discussed subject, because in some studies unexpected disadvantages were found to occur due to the temperature gradient. Of special interest was whether isothermal elution would give better results, as found theoretically by MCLEAN and WHITE (1972) who took account of the change in the gel proportion in the column.

The following criteria have been used for an experimental evaluation of the separation performance:

— Are the fractions discharged in the expected order, i.e., with regularly increasing molar mass, or are there any inversions? (PEPPER and RUTHERFORD, 1959; SCHNEIDER et al., 1959; JUNGNICKEL and WEISS, 1961; HULME and MCLEOD, 1962; COOPER et al., 1962; FLOWERS et al., 1964)

- Which technique yields the broadest distribution curve which might be the most probable one? (Böhm et al., 1974)
- Which technique yields the greatest $\bar{M}_w/\bar{M}_n$ quotient for the polymer fractionated? (Cooper et al., 1962)
- Which method yields fractions with particularly high values of molar mass? (Cooper et al., 1962)
- Which method enables known mixtures to be completely separated? (Yamamoto et al., 1970; Kato et al., 1973; Böhm et al., 1974)
- Are the elution volumes of the fractions in accordance with the values to be expected from the theory of precipitation chromatography? (Glöckner, 1964; Yamaguchi and Saeda, 1969)
- Which method yields the narrowest fractions? (Moore et al., 1962; Schulz et al., 1965; Glöckner, 1966; Johnson et al., 1969; Spatorico and Coulter, 1973; Böhm et al., 1974)

As an illustration of the unfavourable effect of the temperature gradient, the careful investigation by Flowers et al. (1964) on the fractionation of Ziegler copolymers of dodec-1-ene and octadec-1-ene is frequently cited. Gradient elution at 23 °C yielded for the upmost fraction $[\eta] = 4.70$ ml · g^{-1} ($M = 13.5 \cdot 10^6$ g · mole^{-1}; proportion: 3.9%), whereas the Baker-Williams fractionation separated in the correct order of succession only up to $[\eta] = 2.6$ ml · g^{-1} ($M = 5.1 \cdot 10^6$ g · mole^{-1}; proportion: 1.6%) (see Fig. 9-8). The authors commented on this finding with reserve, saying that "a temperature gradient does not necessarily improve the separation".

In that study the results compared were obtained with different *elution gradients*. The isothermal elution was started with $\varphi_A = 0.50$ and carried out with a total volume of 7880 ml of mixture. The critical value $\varphi_B = 0.225$ (i.e., $\varphi_A = 0.775$), at which the solubility curve of the products investigated changes its slope drastically, was reached after more than 6000 ml had flowed through the column. On the other hand, the Baker-Williams fractionation was started with $\varphi_A = 0$ and had already finished with 5245 ml. The critical value was reached with about 3000 ml (see Fig. 9-9). The isothermal elution was carried out with a more gentle gradient. Thus it is not proven whether the *inversions* observed (i.e., the reversal of elution order) were solely due to the temperature gradient.

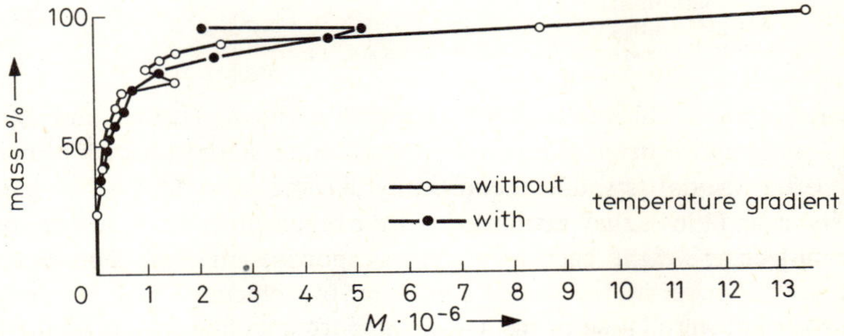

Fig. 9-8
Fractionation of a copolymer of octadec-1-ene and dodec-1-ene, $\bar{M}_w = 710000$ g · mole^{-1}, $\bar{M}_w/\bar{M}_n = 152$ (according to Flowers, Hewett and Mullineaux, 1964)

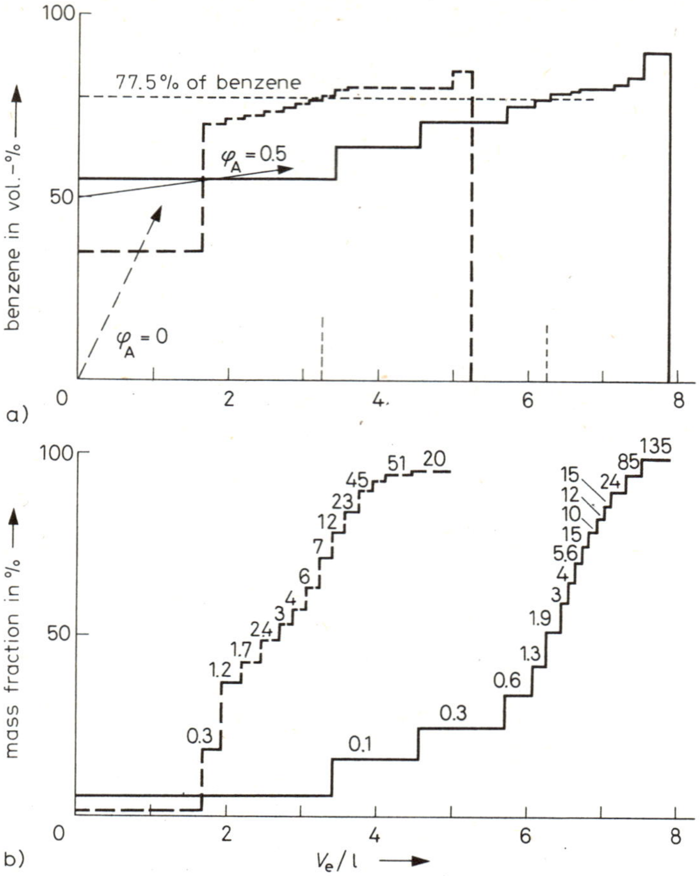

Fig. 9-9
Elution gradients and eluted quantity in the column fractionation of the copolymer of octadec-1-ene and dodec-1-enc (see Fig. 9-8)

a) Variation of the benzene content in the mixture with ethanol as a function of the eluent volume flowing through the column

b) Polymer content of the eluate fractions (the numbers indicate M in 10^5 g · mole^{-1})

——— isothermal elution (23 °C)

– – – elution with a temperature gradient

(according to FLOWERS, HEWETT and MULLINEAUX, 1964).

In the precipitation chromatography of polybutadiene, HULME and MCLEOD (1962) found that inversions occurred at high loadings. While a smooth distribution curve was obtained with 0.3 g, they observed irregularities with samples of 0.6 g or 1.1 g (see Fig. 9-10). Like PANTON et al. (1964), they attributed these irregularities to an adsorption of the first-eluted low-molecular-weight fractions on the supporting material. The separation process takes place on this covering layer until a polymer-free eluent of high solvency finally takes up the adsorption layer. These concepts are supported by the results of the fractionations shown in Fig. 9-11, where the temperature gradient, covering an interval of 50 K in each case, was adjusted to different levels (between +90 and —4 °C). With a sample amount of 0.3 g, inversions were also found to occur when the lowest temperature was —4 °C.

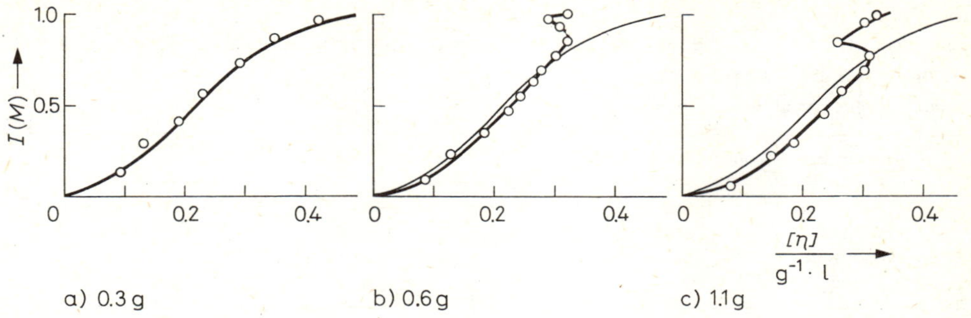

Fig. 9-10
Precipitation chromatography of cis-polybuta-1,4-diene with isooctane-diisobutene mixtures: effect of the loading

Temperature at the top: 90 °C, bottom temperature: 40 °C
a) A sample of 0.3 g gives a normal fractionation curve
b) Samples of 0.6 g and
c) 1.1 g give an inversion
(according to HULME and MCLEOD, 1962).

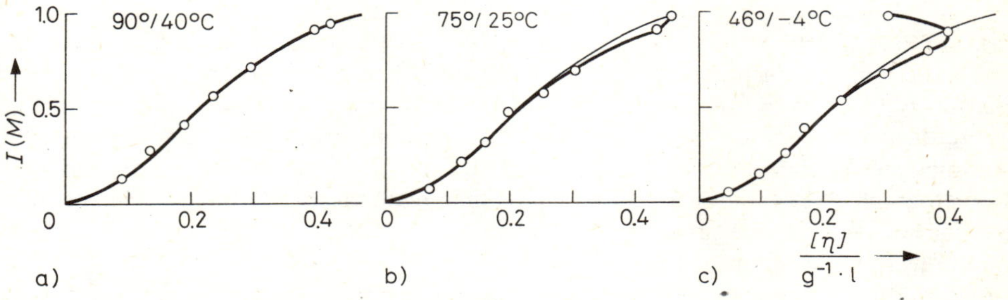

Fig. 9-11
Precipitation chromatography of cis-polybuta-1,4-diene (see the legend to Fig. 9-10): effect of the temperature for a constant head-bottom difference of 50 K

Inversions occur at the low column temperature (according to HULME and MCLEOD, 1962).
(For orientation, the normal curves are plotted as thin solid lines in Figs. 9—10b, c and 9—11b, c.)

In all of the papers published, inversions were mainly observed for unsaturated, polar and crystallizable polymers (polycarbonate, polyolefins), rarely for polystyrene and not at all for polyisobutylene. This supports the assumption that adsorption effects are involved.

It is often assumed that the temperature gradient is conducive to the occurrence of inversions, as is observed at first sight from the above results of FLOWERS et al. However, in the column fractionation of polybutadiene, COOPER et al. (1962) obtained the best resolution without any inversion by means of a smooth elution gradient in combination with a temperature gradient (cf., Table 9-2, No. 3).

Inversions were also observed in the isothermal elution of polyethylene (FRANCIS et al., 1958; KENYON and SALYER, 1960) and polypropylene (SHYLUK, 1962). If no inversions occur, the efficiency of the separating technique can hardly be evaluated from the experimentally determined distribution curves. With a temperature gradient in most cases the resolution as judged by the distribution curve is somewhat better only in the high-molecular range

Table 9-2
Separation efficiency in the column fractionation of polybutadiene (according to COOPER, VAUGHAN and YARDLEY, 1962)
Solvent: benzene; precipitant: ethanol

No.	Experimental conditions			Results of fractionation		
	Elution gradient (vol. fract. of ethanol)	Elution volume ml	Temperature gradient	$\bar{M}_w/\bar{M}_n$	$[\eta]_{max}$	Inversion in %
1	$\varphi_A = 0.81$ $\varphi_E = 0.15$	105	60—15°	1.36	2.4	20
2	$\varphi_A = 0.81$ $\varphi_E = 0.15$	105	nil	1.37	3.0	16
3	$\varphi_A = 0.45$ $\varphi_E = 0.16$	200	60—15°	1.43	3.2	0
4	$\varphi_A = 0.45$ $\varphi_E = 0.16$	200	nil	1.12	3.0	7

(GUILLET et al., 1960; YAMAGUCHI and SAEDA, 1969; BÖHM et al., 1974). For samples with narrow distributions the differences are more distinct than for broad ones.

GUILLET et al. (1960) fractionated polyethylene isothermally (133) and with a temperature gradient (temperature of column head 152, bottom temperature 100 °C) providing the same layer thickness in both cases and obtained a better separation by precipitation chromatography (see Fig. 9-12). Moreover the time required for the Baker-Williams fractionation was shorter; the process was completed within 24 hours, whereas in isothermal elution this time was required for each fraction.

YAMAGUCHI and SAEDA (1969) examined the column fractionation of polystyrenes by methanol-methyl ethyl ketone mixtures in several variants and observed an improvement

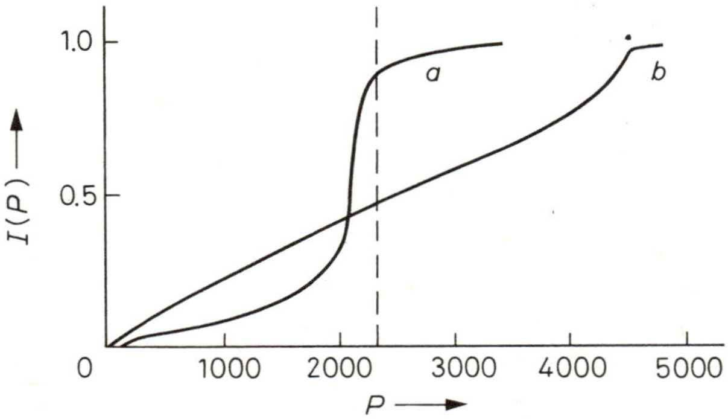

Fig. 9-12
Column fractionation of polyethylene (P = 2300)
a isothermal; b with a temperature gradient (52 K over 1.15 m) (according to GUILLET, COMBS, SLONAKER and COOVER, 1960).

due to the application of a temperature gradient in the high-molecular-weight range (above 600000 g · mole^{-1}). With the parameters applied this finding would be understood by the theory of SCHULZ et al. (1963). The theoretical transport curves showed that fractions with degrees of polymerization up to $P = 1000$ are retarded by the temperature gradient to nearly the same degree as in isothermal elution. Consequently a better separation cannot be expected in this range. On the other hand, the fractions with higher degrees of polymerization are more strongly retained, i.e., their separation is improved.

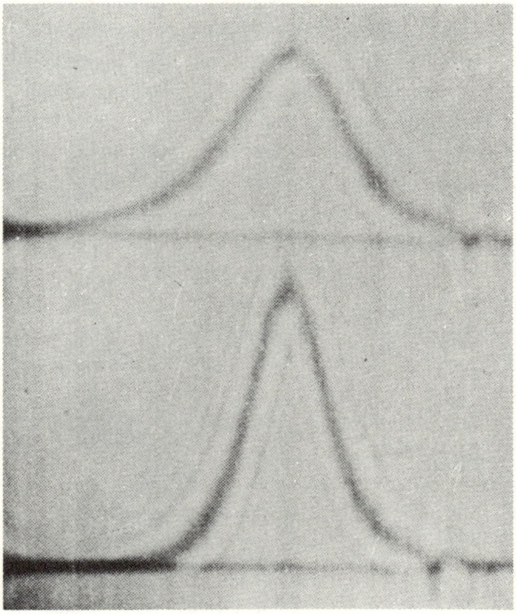

Fig. 9-13
Reproduction of the schlieren photographs of two polyethylene fractions of equal solution viscosity during sedimentation in the ultra-centrifuge
a) isothermally eluted fraction (upper curve)
b) fraction from precipitation chromatography with a temperature gradient of 52 K over 1.15 m (according to MOORE, GREEAR and SHARP, 1962).

The deepest insight into the separation efficiency can be expected from an analysis of the distribution within the fractions. This can be done by methods exhibiting at least as good a resolution as the technique to be tested. Moreover they should not be too time-consuming and should work with as small an amount of substance as possible. Thus the ultra-centrifuge, SEC and the cloud point titration are suitable.

MOORE et al. (1962) investigated the polyethylene fractions obtained by GUILLET et al. (1960) by means of the ultra-centrifuge, observing a markedly narrower distribution in the Baker-Williams fractions than in those obtained isothermally. They concluded that precipitation chromatography is superior to isothermal elution (see Fig. 9-13).

JOHNSON et al. (1969) investigated fractions of polyisobutylene, which had been prepared by gradient elution of 6.2 g of material, either with a temperature gradient or isothermally at 30° and 60 °C. The column had an inside diameter of 37 mm and a length of 1.20 m. The fractions were characterized by SEC. Their $\bar{M}_w/\bar{M}_n$ *ratio* had a minimum at about $\bar{M}_w = 50000$ g · mole^{-1}. Here the fractions eluted at 30 °C exhibited the lowest heterogeneity. However, below $\bar{M}_w = 25000$ g · mole^{-1} (about 40% of the total sample) and above $\bar{M}_w = 135000$ g · mole^{-1} the Baker-Williams fractions were narrower. Moreover, if a temperature gradient was used, a fraction of about 5% with $\bar{M}_w = 166700$ g · mole^{-1}, which had not been found in the isothermal elution, could be separated from the upper end of the distribution.

Spatorico and Coulter (1973) investigated the influence of the column packing mode. The usual technique is to precipitate the sample on a small amount of support material and transfer the latter into the uppermost column section. This technique was contrasted with some variants, in which the coating of the supports was performed by a fractional precipitation of the polymer in the presence of subsequent portions of the support. The column was packed in such a way that the preloaded batches were placed on the top, whereas the lower parts of the column were filled with the blank packing material. In the case of prefractionation the molar masses of the subsequent parts of the sample always decreased from top to bottom. The total loading was 3 g in methods A and C, respectively, and 5 g in method B (cf., Fig. 9-14). Polystyrene, polymethyl methacrylate and a polyester were fractionated in a gold-plated steel column 1.50 m long, with a difference of 25 and 20 K, respectively, between the head and the bottom temperature. The fractions were investigated by SEC with a correction for band widening. The ratio $\bar{M}_w/\bar{M}_n$ determined in this way was lowest for the fractions with medium molar masses ($\bar{M}_w \approx 10^5$ g · mole^{-1}). Here the results obtained by different packing techniques showed relatively small differences (see Fig. 9-15). For lower and higher fractions the quotients were greater, the inferiority of the packing technique C becoming more serious. This confirms that a chromatographically active gel film develops in a correctly operating Baker-Williams fractionation if the parameters are chosen properly. If the method really were only a fractional elution (as it has repeatedly been expressed after

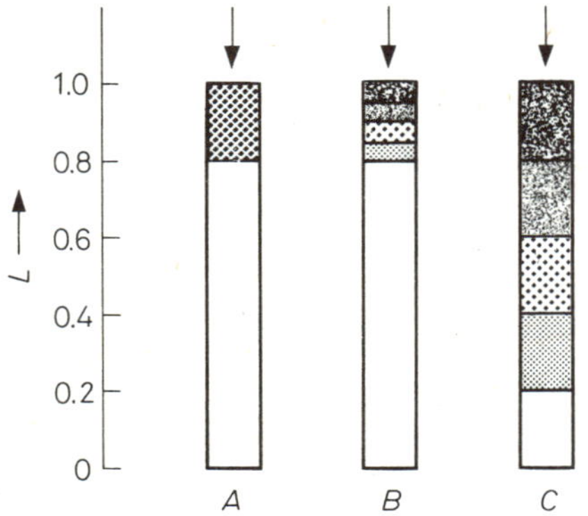

Fig. 9-14
Column packings for elution with a temperature gradient (according to Spatorico and Coulter, 1973)

A standard packing; sample bed 0.2 *L*; quantity supplied: 3.0 g of the sample; B sample bed 0.2 *L*; sample (5 g) prefractionated. Above the bed of blank silica gel, the components with the lowest molar mass are arranged first, followed by those with increasingly higher molar masses. C sample bed 0.8 *L*; sample (3 g) prefractionated; arrangement as in B, i.e., the highest molar mass is arranged at the top of the column.

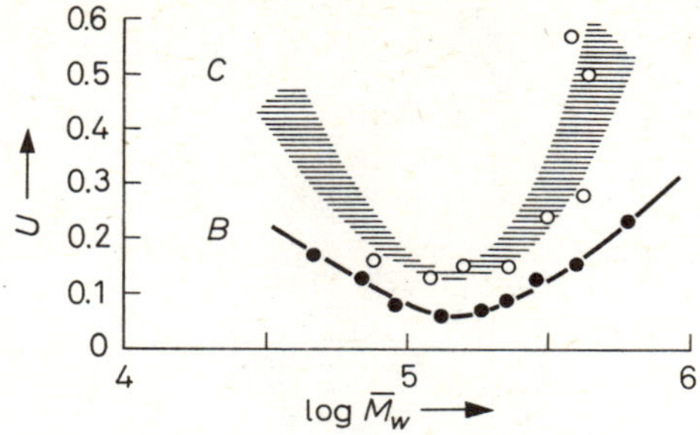

Fig. 9-15
Non-uniformity of polystyrene fractions as a function of their molar masses
Comparison of the packing techniques B and C of Fig. Fig. 9—14. The results obtained by the packing technique A in the fractionation of a polyester were nearly as good as those of technique B. (according to SPATORICO and COULTER, 1973).

failures), then the extraction of a prefractionated and much thinner polymer film as in version C should be more efficient.

The *non-uniformity* of Baker-Williams fractions was expressed by SCHULZ et al. (1965)

$$U = \frac{1}{12}\left(\frac{\Delta P}{P}\right)^2 \left[1 + 2.16\left(\frac{2\sigma}{V_{\mathrm{Fr}}}\right)^2\right] \tag{9-6}$$

where V_{Fr} is the volume of the fraction, ΔP is the width of the fraction (as calculated from V_{Fr} and $\mathrm{d}P/\mathrm{d}V$ at the point P), P is the average degree of polymerization of the fraction and 2σ is the half-width of the elution curve of a non-retained sample. Using experimental data, the authors obtained a mean value of $U = 5.5 \cdot 10^{-4}$ and a worst-case limit of $U = 2 \cdot 10^{-3}$. The non-uniformities determined experimentally in a later investigation were of the same order of magnitude (BÖHM et al., 1974). Consequently the Baker-Williams fractions are extremely narrow.

The same result was derived from cloud point titrations, which were initially used in order to enable the molar masses of very small fractions to be determined (GLÖCKNER, 1965a, 1965b). The titration of normal fractions yielded S-shaped turbidity curves, whereas in most cases Baker-Williams fractions led to curves rising abruptly from the abscissa. This seemed to suggest that the cause lies in a different width of the distribution. The smoother rise of the curves of normal fractions was attributed to parts of the sample precipitating earlier than the bulk. It appears that good Baker-Williams fractions do not contain these components. In a column where the temperature was controlled by means of a liquid jacket, polystyrene was fractionated with and without a temperature gradient, taking care that all other conditions were the same. Abruptly rising, steep curves resulted for the Baker-Williams fractions, and curves with a transitional section, for the fractions prepared without a temperature gradient (GAHNER, 1967).

Fig. 9-16 shows analogous investigations in an azeotropic copolymer of styrene and acrylonitrile, which also included an isothermal separation at 5 °C, the temperature of column bottom in precipitation chromatography (WEBER, 1976). Illustrated are the titration curves of the first fractions, the cloud points of which ranged between 53 and 70 vol.% of methanol. The solvent was methylene chloride. The isothermal fractionation at 25 °C yielded

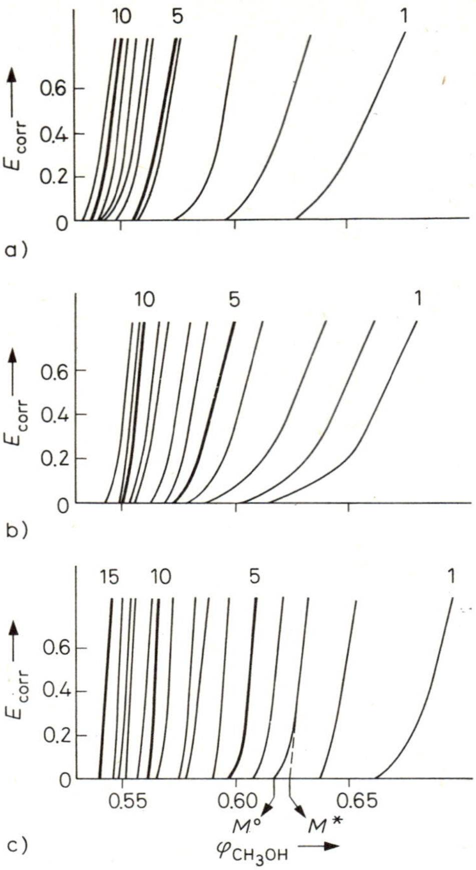

Fig. 9-16
Turbidity curves of fractions of an azeotropic styrene-acrylonitrile (38.5 mole %) copolymer

The fractions were obtained
a) isothermally at 5 °C
b) at 25 °C, or
c) with a gradient (25°/5 °C), under conditions which were otherwise exactly equal. Column: L = 1.00 m; d_C = 2.4 cm; sample quantity supplied: 0.3 g on 30 g of glass beads. Elution by methanol in mixtures with methylene chloride (φ_A = 0.57; φ_E = 0.35). The numbers written above the curves indicate the fraction numbers.

a total of 27 fractions, the two others, 30 fractions each. The following conclusions can be drawn from Figure 9-16:

- Only with the temperature gradient did the first fraction (No. 1) need more than 65% of methanol. A fraction with such a high precipitation point could not be isolated under isothermal elutions
- The turbidity curves of the fractions obtained by precipitation chromatography are steeper and, from No. 11 upwards, do not show any transitional section to the abscissa. Both results indicate narrower fractions. (At curve No. 3 in Figure 9-16c it is indicated how M^* and M^0 are derived from the inflectional tangent and the rising point, respectively.)

Further investigations using polystyrene and azeotropic styrene-acrylonitrile copolymers showed that for the highest and the lowest fractions the effect of the temperature gradient is greater than in the medium range around $M \approx 100000\ \mathrm{g \cdot mole^{-1}}$ (see Fig. 9-17). This is in accordance with results obtained by SPATORICO and COULTER (1973).

Under favourable conditions the temperature gradient improves the efficiency (GLÖCKNER and KAUFMANN, 1977). The fact that gradient elution of polystyrene yielded excellent separations even without a temperature gradient may indicate that the polymer was temporarily adsorbed on the supporting material, thus likewise forming a chromatographically active layer (SCHNEIDER et al., 1961).

Generally, isothermal stepwise elution yields good separations if the sample is deposited as a very thin layer on the surface of the total packing material, being already prefractionated, e.g., by gradual cooling of solutions in the column. True, the analysis time is higher than in precipitation chromatography, but the elution can be stopped at any time and is less susceptible to disturbances.

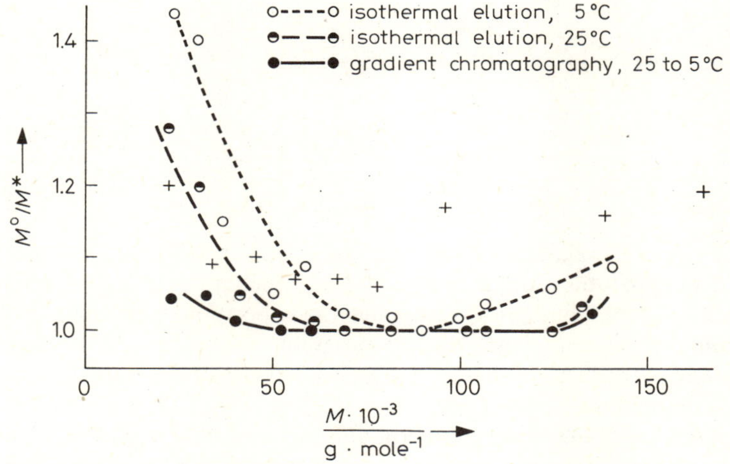

Fig. 9-17
Molar mass ratio determined from turbidity curves according to Fig. 9-16 for styrene-acrylonitrile copolymer fractions

M^0 is the value calculated from the lowest point of the turbidity curve, M^* has been calculated from the point of intersection of the inflectional tangent with the abscissa. For "monodisperse" fractions, $M^0/M^* = 1$. (The crosses indicate $\bar{M}_w/\bar{M}_n$ values obtained by SPATORICO and COULTER for polyester fractions by means of the packing technique A; see Fig. 9—14).

9.7. Supercritical fluid chromatography (SFC)

It has been known for a long time that supercritical media can dissolve liquids and solid substances. Because of this property, they were at first of interest as mobile phases for separation problems which could not be treated by gas chromatography, because the samples could not be evaporated without decomposition. In comparison with HPLC, on the other hand, chromatography in supercritical fluids deserves attention because the viscosities of these media are nearly as low as those of gases and their diffusion coefficients exceed

Table 9-3
Potential mobile phases for supercritical fluid chromatography (according to KLESPER, 1978)

Substance	Critical data			
	Boiling point / °C	ϑ_{cr} / °C	p_{cr} / MPa	ϱ_{cr} / $g \cdot cm^{-3}$
CF_3Cl	−81.4	28.8	3.83	0.58
CO_2	−78.5	31.3	7,15	0.448
CF_2Cl_2	−29.8	111.7	3.87	0.558
n-C_4H_{10}	0.5	152.0	3.68	0.228
$CF_2Cl \cdot CF_2Cl$	3.5	146.1	3.48	0.582
$CHFCl_2$	8.9	178.5	5.00	0.522
$CFCl_3$	23.7	196.6	4.09	0.554
n-C_5H_{12}	36.3	196.6	3.27	0.232
$(CH_3)_2CH \cdot CH(CH_3)_2$	58.0	226.8	3.04	0.241
CH_3OH	64.7	240.5	7.74	0.272
n-C_6H_{14}	69.0	234.2	2.90	0.234
C_2H_5OH	78.4	243.4	6.18	0.276
C_6H_6	80.1	288.9	4.74	0.302
H_2O	100.0	374.4	22.25	0.344

those of liquids by more than two orders of magnitude. To explore fully the possibilities of the method an apparatus is required by means of which the essential parameters of separation (pressure, temperature, flow-rate, eluent composition) can be kept constant or varied according to mutually independent programs (KLESPER and HARTMANN, 1978). A review was given by KLESPER (1978). Table 9-3 lists data for compounds which can be considered as supercritical mobile phases.

JENTOFT and GOUW (1969) used this method to separate styrene oligomers up to $P = 16$ in n-pentane containing 5% methanol at a temperature of 205 °C. The pressure required was 4.14 MPa (41.4 bar): 8 minutes after the sample had been introduced the pressure was increased at a rate of 690 $Pa \cdot s^{-1}$ to 6.21 MPa (62.1 bar). The separation in a 4 m steel column with an outside diameter of 3.2 mm took 1 h, and gave a good resolution of individual peaks (see Fig. 9-18). NIEMAN and ROGERS (1975) separated oligosiloxanes in supercritical fluids. In 1977, KLESPER and HARTMANN (a) reported on preliminary results obtained from an apparatus by means of which 49 individual peaks could be obtained from a polystyrene standard with a nominal molar mass of $\bar{M}_w = 2200 \ g \cdot mole^{-1}$ (see Fig. 9-19). In subsequent studies they investigated the effects of the various parameters (1977b, 1978). The separation is not due to steric exclusion, for the higher the molar mass of the homologous sample components the later they are eluted. The solubility of the oligomers rapidly increases with increasing pressure. The pressure is the most important quantity influencing the chromatography in supercritical fluids; it has to be increased continuously during the elution. At a constant pressure the higher fractions are eluted at increasingly longer intervals, i.e., the sample is practically arrested in the column. If a higher but likewise constant pressure is used, then the first part of the substance leaves the column almost unresolved. Repeated stepwise increases of the pressure restart the elution, which, however, breaks down at each step after a short

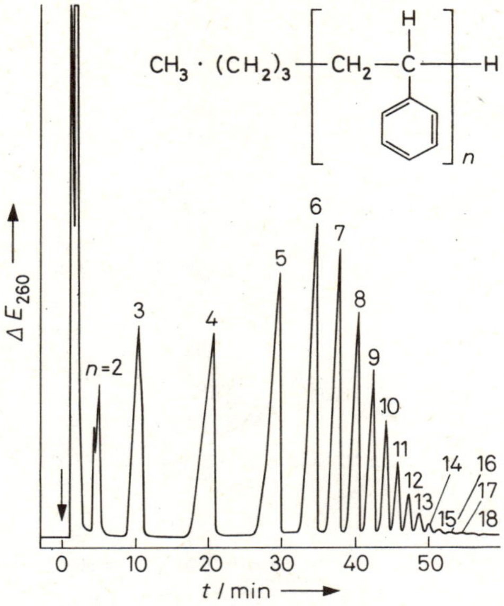

Fig. 9-18
Separation of styrene oligomers in n-pentene with 5% methanol on a chemically bonded n-octyl phase at 205 °C (according to JENTOFT and GOUW, 1969)

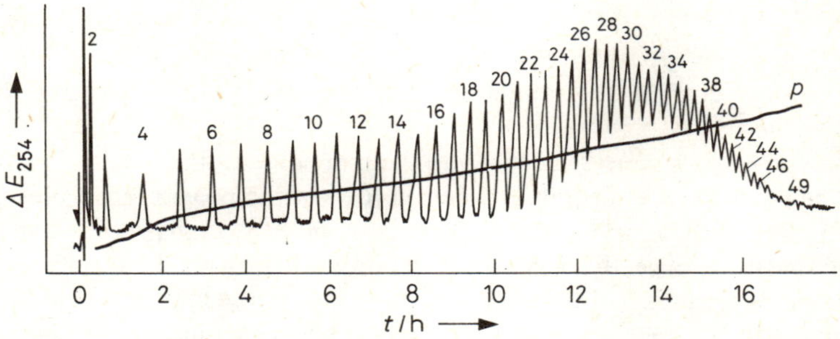

Fig. 9-19
Chromatogram of polystyrene ($\bar{M}$ = 2200) in n-pentane/methanol (90:10) at 235 °C
During the elution at 1.1 ml · min^{-1}, the pressure is increased from 2 MPa (20 bar) to 13 MPa in the manner shown. Column: L = 3 m; d_C = 2 mm; packing: Porasil® A, d_p = 37—75 µm, d_0 < 10 nm. Sample: 20 mg in 100 µl of cyclohexane. The numbers indicate the degree of polymerization (according to KLESPER and HARTMANN, 1977a).

time. Good separations are achieved by pressure programming with an adequately gradual increase. Fig. 9-20 shows three elugrams in which the increase in pressure was varied in the ratio 1:2:3. The pressure values written beside the curves for peaks 10, 20 and 30 indicate that, to a first approximation, the elution of a certain component always occurs at the same pressure, with only a slight dependence on the elution volume.

Good separations are achieved only above the critical temperature, which is 202 °C for pentane-methanol (95:5% v/v). At 190 °C, and even more obviously at 180 °C, i.e., with

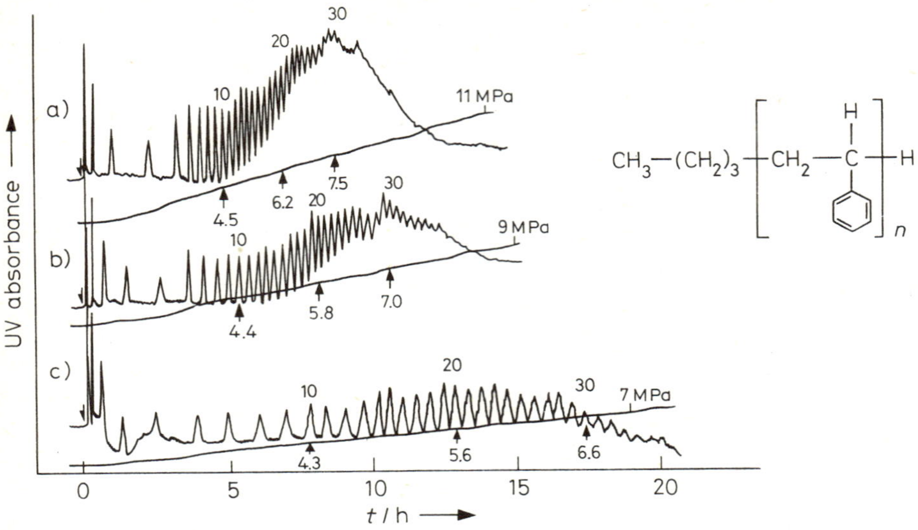

Fig. 9-20
Separation of polystyrene in supercritical pentane with 5% methanol at 220 °C
Column: L = 6.00 m; d_c = 3 mm; packed with Porasil A®; Pressure drop along the column: 2.5 MPa;
Sample: 40 mg of PS ($\bar{M}_w$ = 2200 g · mole^{-1}) in 100 µl of solution
a) pressure increase 660 kPa/h; v = 2.6 ml/min
b) pressure increase 440 kPa/h; v = 3.0 ml/min
c) pressure increase 220 kPa/h; v = 3.2 ml/min
(according to KLESPER and HARTMANN, 1977b)

the eluent having liquefied, the sample was eluted much more rapidly and in a plug-like manner.

As the pressure drops considerably along the 6 m column, the mechanism of separation is likely to be based on a multistep dissolution (due to the programmed pressure rise) and reprecipitation of each component (due to the advance into ranges of lower pressure). In this context, the observation that a good separation could be gained only using an eluent with a small amount of methanol added (5%) is of interest. The use of pure n-pentane yielded also chromatograms with a poor resolution, but increasing the methanol content from 5 to 10% did scarcely improve the resolution. This was observed with silica gel used as a column packing material, which naturally adsorbs methanol to a much higher degree than pentane. Consequently the surface of the supporting material will have a layer with a higher methanol content, which is obviously of importance for a good separation of the homologous oligostyrenes.

The capacity of the method is amazingly high. The injected amount of 1 mg per cm^3 of empty column volume was about 1000 times the value considered to be the maximum load in SEC (for higher molar masses, admittedly) (cf., Section 19.1.). In 1978 HARTMANN and KLESPER carried out preparative separations with injected amounts of 100 mg, using a 6 m column with an inside diameter of 5 mm. With 16 repetitions they obtained, for the first fifteen peaks, up to 20 mg of substance each, permitting further investigations. By refractionating, gas chromatography and spectrometry, the authors showed that each peak contains only a single oligomer and that the peaks succeeding one another in the elugram in fact differ from each other by exactly one monostyrene unit (M_0 = 104). Thereafter the

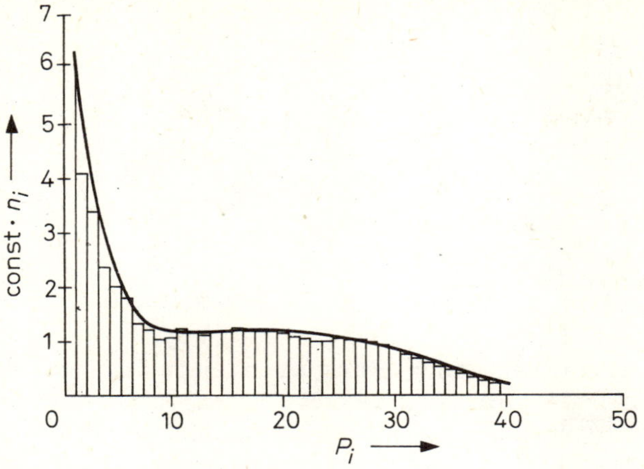

Fig. 9-21
Histogram of a commercial polystyrene standard (M = 2200 g · mole^{-1}. nominal)

Established on the basis of a preparative separation in supercritical pentane with 10% of methanol and an analysis of the fractions by refractionation, gas chromatography and mass spectrometry. The fractions plotted along the ordinate for the components with the degrees of polymerization P_i were obtained by cutting the chromatogram into parts, weighing the peak areas and dividing them by P_i (const · $n_i = m_i/P_i$) (according to KLESPER and HARTMANN, 1978).

authors established the histogram shown in Fig. 9-21 for the frequency of the individual oligomers in a commercial polystyrene standard prepared with butyl lithium as the starting compound. Using the algorithm of eqns. (4-3a) and (4-5a) this gives $\bar{M}_n$ = 1480 g · mole^{-1} and $\bar{M}_w$ = 2370 g · mole^{-1}, respectively, the latter being in good agreement with the declared value (2200). On the other hand, the heterogeneity, $H = 2370/1480 = 1.60$, differs considerably from the nominal value ($H \leqq 1.06$).

In 1981, SCHMITZ and KLESPER reported the use of a modified commercial HPLC apparatus (HP 1084 B) for the SFC separation of styrene oligomers. They used a temperature, pressure or elution gradient. The eluent was pentane with additions of cyclohexane.

10. Support materials

Although the mechanism of the separation techniques treated in Chapters 7-9 are different, a high resolution (see eqn. 3-25) in each case requires a high capacity for the separation path. In adsorption chromatography this means a large number of adsorption sites, in partition chromatography as large an interface as possible between the stationary film and the mobile phase and in exclusion chromatography the greatest possible number of pores selectively accessible. These requirements lead to the condition that the support materials should have a large surface area. Consequently, porous column packings are almost always employed in chromatography. (The glass beads in precipitation chromatography belong to the few exceptions.) Some supports are suitable for use in all the variants of liquid chromatography. This group includes silica gel, which can be used for liquid-solid adsorption chromatography (LSC) because of the activity of its surface hydroxyl groups, for liquid-liquid partition chromatography (LLC) because of its large surface area and for size exclusion chromatography because of its pore structure.

This chapter deals with general aspects. Inorganic support materials will be dealt with in detail in Chapter 11, organic ones in Chapter 12. Table 10-1 classifies the support materials according to their chromatographic behaviour. A survey of HPLC columns and packing materials was given by MAJORS (1980).

Table 10-1
Classification of the support materials according to their chromatographic effects (according to PORATH, 1979)

Class	Designation	Examples
A	without adsorption sites but with cavities	ideal molecular sieves
B	with unsupported adsorption sites	inorganic supports, primary adsorbents
C	sorbents of class B, mixed with (or deposited on) supports of class A or other inert solids	
D	sorbents of class A with covalently bonded substituents (adsorption sites)	polymer ion exchangers, bio-affinity adsorbents, secondary adsorbents
E	sorbents of class D, loaded with bi- or multifunctional adsorbates, which in their turn act as specific adsorption sites	"sandwich type" immunosorbents, indirect adsorbents
F	without adsorption sites, with solvency for the substance	bonded polymer layers

10.1. Chemical aspects

The chemical structure of the support materials is decisive for the thermal and chemical stability as well as for their swelling property in the eluent, but also affects the mechanical parameters and the chromatographic behaviour.

The number of support materials which are in principle suitable for use in adsorption chromatography is virtually unlimited. More than 100 substances were tested by CVET in his pioneering work in the first decade of this century. Inorganic materials like silica gel or alumina as well as organic products like cellulose powder, polyamide or polysaccharides can be considered. For instance, CVET had employed inuline, a starch-like substance, in the separation of chlorophyll components. White supports, on which the substance spots can be clearly observed, offer advantages in development chromatography.

In partition chromatography the stationary phases are applied to primary adsorbents (class B in Table 10-1). Silica gel is the most suitable material, but alumina is also used. Polar liquids such as β, β'-oxydipropionitrile or oligo- and polyglycols can be fixed to these adsorbents by means of their natural activity. A pure partition mechanism can only develop if the surface of the solid is completely covered; this requires about 100—200 mg of liquid phase per gram of adsorbent if the surface area is about 200 $m^2 \cdot g^{-1}$. To enable non-polar liquids also to be used as a stationary phase, the support material must be hydrophobized.

The impregnation by the stationary liquid can be carried out with bulk material if the size and the condition of the particles permit the application of a dry packing technique (cf., Section 17.1.2.). If slurry packing is required, then the packing material must be impregnated in the column: an eluent, saturated with the stationary liquid, is pumped through the column until the latter does not take up any more, and a stable coating has been achieved [D 5]. Polymer layers on glass beads (d_P = 44—50 μm) as a stationary phase for partition chromatography were also precipitated from solutions (SAGE et al., 1976). The extremely high-molecular-weight polystyrene layers on which the *phase distribution chromatography* of polymers was realized were also precipitated from solutions (see Section 9.4.2.).

The separating materials for size exclusion chromatography are organic or inorganic substances with a three-dimensional network structure and open pores. The *organic gels* are prepared either by cross-linking copolymerization (Section 12.1.) or by the cross-linking of macromolecular substances (Section 12.2.).

Besides the standard gels there are cross-linked polymers which are efficient in separation under suitable experimental conditions. For example, it was possible to separate oligomers up to $M = 10^4$ on cross-linked, chlorous butyl rubber in a simple, open column with a remarkably good resolution; however, the elutions too up to 18 hours (BREWER, 1965).

The soft organic gels are weakly cross-linked and porous due to swelling. The swelling volume is considerable and depends on the solvent (cf., Table 12-1). Therefore soft gels cannot usually be used together with changing solvents. The swelling property increases with decreasing cross-linking density, which on the other hand determines the pore size. This leads to a relationship between the swelling property and the separating range, which may be observed from Fig. 12-1.

The rigid and semi-rigid organic gels are *permanently porous*. They can be prepared by a cross-linking copolymerization in the presence of inert diluents (MOORE, 1964).

In Table 10-2 the terms *permanent porosity and porosity due to swelling* are compared with other definitions, which are also used for the characterization of porous materials and gels. The term *macroreticular* is applied to polymers which (in a dry condition) appear cloudy,

Table 10-2
Survey of porous packing materials for liquid chromatography

Type of porosity	Permanent porosity ⟵			⟶ Porosity due to swelling		
Classification	macroreticular			microreticular		
Structure	inorganic porous materials		polymer networks			
			rigid organic gels	supercross-linked isoporous networks	natural reversible gels	soft gels
Examples	silica gel	controlled porosity glass (CPG)	polystyrene gels for GPC	isoporous cross-linked polystyrene	agarose gel	polyacrylamide gel
Formation of the pore system	by precipitation	by selective dissolution	by cross-linking copolymerization in the presence of inert substances	by cross-linking of chain molecules in solution via chemical reactions	physical structuring processes	by homogeneous copolymerization or cross-linking in bulk
BET surface (in $m^2 \cdot g^{-1}$)	up to 900		$10 \leqq {}^1A \leqq 700$	$\triangleq 10^3$	$0 \leqq {}^1A \leqq 5$	0
Condition of the material as supplied	Aerogel		Xerogel[1])		Hydrogel	Xerogel[1])
Behaviour towards solvents	non-swelling		slightly swelling	highly swelling in all fluids	in good solvents	
Structure two-dimensional pattern	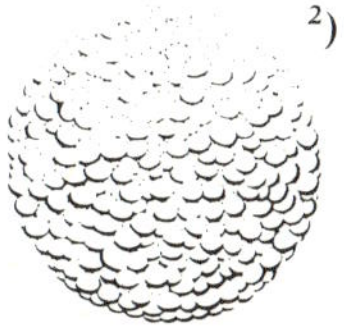 [2])		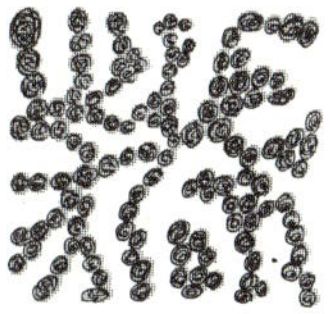	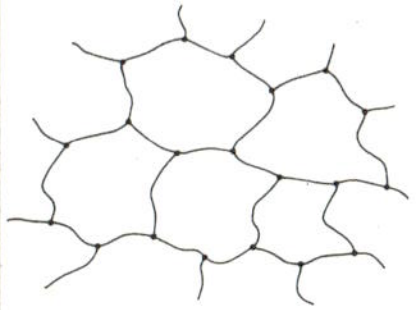	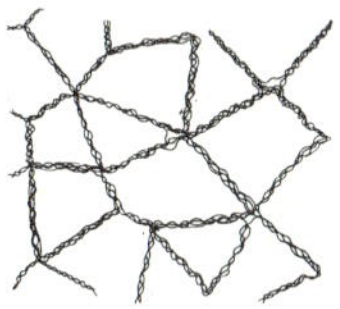	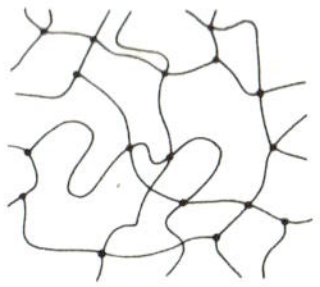

[1]) Efficient in separation as a lyogel or a hydrogel.
[2]) according to KIRKLAND (1973)

having pore sizes greater than 10 nm, an internal surface area of at least 5 $m^2 \cdot g^{-1}$ and an apparent density which is smaller than the true density by at least 0.05 $g \cdot cm^{-3}$ (Frölich et al., 1979).

Isoporous polystyrene gels were prepared by Davankov et al. (1973) by cross-linking of polystyrene in solution with bifunctional compounds, for example by Friedel-Crafts reactions with xylylene dichloride or dichloromethane (Cilipotkina et al., 1976; cf., Fig. 10-1).

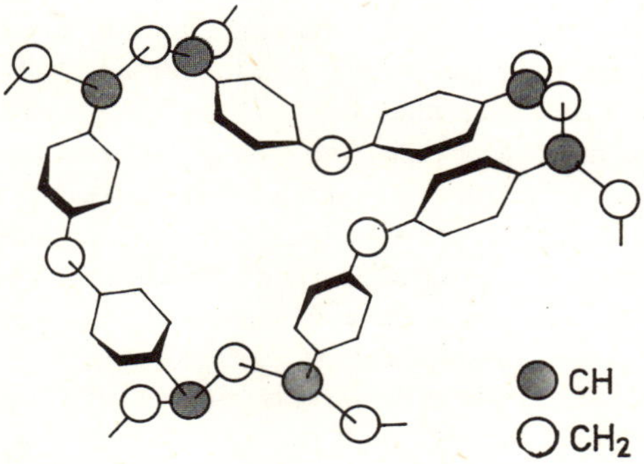

Fig. 10-1
Isoporous polystyrene gel
Schematic representation of the network structure obtained with dichloromethane as a cross-linking agent (according to Davankov and Cjurupa, 1980).

They were called isoporous because the homogeneous reaction may lead to gels with a homogeneous cross-linking density, and hence a narrower pore size distribution. (A cross-linking copolymerization, even if it takes place without any phase separation, usually leads to gels with spatially different degrees of cross-linking.) In a dry condition, the isoporous gels have a low density, and hence a rather large pore volume, and they have a very large specific surface area (Davankov et al., 1974a). They exhibit a remarkable swelling property, which in polystyrene solvents not only exceeds that of other gels of equal degrees of cross-linking but also occurs in nonsolvents such as alcohols and even in water (Davankov, 1974b). This is explained by a shrinking of the chains due to drying: in the solution where the cross-linking takes place, the chains have a relatively loose conformation.

In the dried-up gel these chain segments between the cross-linking points show a greater shrinking than that corresponding to the equilibrium conformation. The gels are subject to an internal pressure resulting from the chain deformation, which is just compensated by interactions occurring at a few chain contact points. Even in media having only a low solvation power the gels yield to the conformation-based pressure, and swell (Davankov et al., 1976).

Inorganic support materials are either porous natural products (such as diatomaceous earth) or are prepared by the etching of microheterogeneous starting materials (porous glasses), or they are given the desired porous structure by a carefully controlled precipitation process (silica gel, alumina).

The chemical structure of the porous support materials also determines whether they are mainly suitable for separations in aqueous or in non-aqueous media. Polar gels, especially those having free hydroxyl groups, are more or less *hydrophilic*, and hence suitable by nature for chromatography in aqueous solutions. Nonpolar gels such as cross-linked polystyrene are suitable for the separation of slightly polar substances in polar solvents, but not for investigations in aqueous systems, low-molecular-weight alcohols or acetone. On the other hand, the dextran, agarose and acrylamide gels, which are porous due to swelling by water, are above all usable for aqueous systems. Modification of the gel skeleton may change its suitability for certain solvents. Thus blocking the OH groups may convert a hydrophilic gel into a *lipophilic* support substance. Structurally analogous reactions carried out with the aim of extending the range of available separating materials had already gained importance in the early sixties. Semi-rigid and rigid skeletons have wider applicability than the soft ones in different solvents. Porous glasses and silica gels are not only usable for aqueous solutions but also for organic ones, in spite of their free hydroxyls (KOHLSCHÜTTER et al., 1966). As the inorganic packings exhibit advantages which are of great importance for an up-to-date elution chromatography, e.g., their pressure resistance and the constancy of their volumes during a change of the solvent, they are frequently modified by surface reactions in order to avoid unfavourable adsorption phenomena in exclusion chromatography. For example, the surface hydroxyls of glasses or silica can be masked by methyl trichlorosilane (LANGHAMMER and SEIDE, 1957) or hexamethyl disilazane (COOPER and JOHNSON, 1969), cf., also Section 11.10.

To estimate the possibilities and limits of application of a certain support material, it is useful to know its chemical structure.

10.2. Shape and constitution of porous supports

For a packing to be as homogeneous as possible with uniform flow channels, beads of equal size are most suitable. They enable columns to be prepared with the most favourable ratio of efficiency to flow resistance. The cross-section is filled by the closest spherical packing in the equatorial plane, with $\pi/(2\sqrt{3}) = 90.69\,\%$. The percentage bulk factor is $[\pi/(2\sqrt{3})]\sqrt{2/3} = 74.05\,\%$. If the bead diameters have a broad distribution, then the small particles can fit into the interspaces of the larger-sized ones. The packing density may even exceed the ideal value mentioned above, but at the expense of the permeability of the column (cf., Section 17.2.). For most support materials it was possible to find preparation techniques which yield beads with narrow size distributions.

Ground packing materials have irregularly shaped particles. Sieving is required to obtain batches with narrow size limits. Geometrically diverse support material has a poorer packing quality and makes it more difficult to prepare columns of high resolution and low flow resistance, especially if the particles are small.

Totally porous packing materials are commonly used, but some supports contain glass beads 30 μm in diameter as a core, on which a thin porous layer is fixed (porous layer beads, pellicular supports). These relatively large-sized and uniform particles are rather easily packed and yield geometrically perfect packings. With respect to the pore depth, i.e., the length of the dead channels, porous layer bead particles of 30 μm are equivalent to totally porous ones of 5 μm. As unnecessarily deep pores lead to a band widening, porous layer bead packing materials enable a higher resolution to be achieved than fully porous ones of equal size. For example, the efficiency of a slurry-packed column with 20 µm silica gel particles was also achieved by a column packed in the dry state with porous layer beads, 44 µm in size (SNYDER, 1971 b). However, highest resolutions were always achieved with extremely fine, totally porous beads.

Besides the pore depth, other effects contribute to peak broadening: the capacity of the column decreases if the chromatographic processes are restricted to the outer layer of the particles. This diminishes the resolution. Typical porous supports have an interior surface area ranging from 50 to 400 $m^2 \cdot g^{-1}$, whereas that of analogous pellicular supports is only between 1 and 14 $m^2 \cdot g^{-1}$. Accordingly the sample load has to be smaller. In LLC there are no problems if 3 mg per 100 cm^3 of column volume are injected into a column with a good porous packing material. (Some authors even used 15 mg/100 cm^3.) For pellicular supports the loading must be reduced by a factor of 10 at least. For size exclusion chromatography of polymers, their capacity does not suffice, but due to their good mass transfer characteristic they are of some interest for the high-speed chromatography of colloidal particles, which exhibit very low rates of diffusion (KIRKLAND, 1979; cf., Section 19.10.).

As regards the dimensional stability and the pressure resistance, inorganic supports satisfy all requirements of HPLC (pressure-resistant up to 30 MPa and more), provided that their porosity does not exceed 90%. However, care should be taken in handling irregularly shaped materials, in order to avoid the formation of fines. Organic supports are less brittle, but they exhibit high pressure resistance only at high degrees of cross-linking. Swollen gels are rather easily deformable.

10.3. Classification by sizes

Spherical particles can be described by their particle diameter, d_P. Usually this is a mean value, so that it is advisable to indicate the algorithm for calculating this mean value (cf., Section 4.2.1., where this problem is described for the mean values of the molar mass). For particles above 5 µm numerous methods are available for the determination of d_P, whereas in the range between 1 and 5 µm only a few methods are of use in practice (RUMPF et al., 1967). UNGER et al. (1978) investigated packing materials in this range of sizes. Spherical particles were measured under a microscope. The size of irregularly shaped particles was determined by means of a wide-angle scanning photosedimentometer. The mean values determined by this apparatus were in accordance with the Stokes diameters, but a packing material with 400 nm macropores was an exception (UNGER and GIMPEL, 1979).

GIDDINGS et al. (1979 a) showed that the field-flow fractionation (cf., Section 13.1.) is suitable for the characterization of particles of ca. 5 µm, providing information about the distribution in addition to the average particle size. Fines were clearly observable in the FFF chromatograms.

In a chromatographic laboratory, among the methods mentioned usually the microscope is employed for obtaining information about the particle size and the size distribution. If a stage micrometer is not available, then — according to a proposal by VERZELE et al. (1979) — particles of ca. 10 µm can simply be evaluated by comparison with a preparation of red blood cells which exhibit a narrow size distribution between 7 and 10 µm. This reference preparation can be obtained by smearing a drop of blood on a specimen holder. In the microscopic evaluation, irregularly shaped separating materials are overestimated by about 15%, because their particles are flatly arranged and always the largest cross-section is observed.

Frequently the particle size is simply derived from the mesh size of the sieves employed in the classification. The lower *mesh number* indicates the number of openings per inch (= 2.54 cm) of the sieve which just allows the particles to pass through. Accordingly the higher mesh number, e.g. 200—400 mesh, indicates the number of openings per inch of that sieve which retains all the particles of the material. As a rule of thumb, $d_P = 14800/MN$ can be used to estimate the particle size (in µm) from the mesh number. Table 10-3 lists some information about the nominal mesh number, the actual number of openings per inch, the clear width of these openings and the wire size for sieves of the U.S. series.

Table 10-3
Correlation of mesh numbers and particle sizes

Mesh number (openings per inch)	Clear width of the openings[1]) µm	Wire gauge of the U.S. Sieve Series (ASTM Specifications E-11-61) µm
40 (38.02)	420	250
50 (52.36)	297	188
60 (61.93)	250	162
70 (72.46)	210	140
80 (85.47)	177	119
100 (101.01)	149	102
120 (120.48)	125	86
140 (142.86)	105	74
170 (166.67)	88	63
200 (200.00)	74	53
230 (238.10)	62	46
270 (270.26)	53	41
325 (323)	44	36
400	37	(26)

[1]) particle size (upper limit)

Characterizing the packing material by the mean value $\bar{d}_{P,w}$ (or $\bar{d}_{P,n}$) and the standard deviation, which, for Gaussian distributions, can easily be determined by plotting them on a cumulative probability paper (UNGER, 1974), is better than the usual indication of the limits. The standard deviation should not exceed 20% of the mean value; thus for instance $d_P = 10 \pm 2$ µm should hold for microspheres.

The particle size affects several essential factors; the smaller the particles, the better is the resolution, but this will be at the expense of a more difficult packing of the columns, a higher flow resistance and a higher risk of clogging. Because of the latter effect, soft gels are marketed only as relatively coarse-grained grades: e.g., acrylamide gels are available in the size classes from 28 to 74 µm (unswollen) for "high-resolution GPC" and from 74 to 147 µm as a "general-purpose type". For large columns and higher flow rates even a grain size between 147 and 297 μm is recommended. The data for agarose and dextran gels are quite analogous. For soft gels, particle sizes of less than 28 μm are provided only for TLC. With soft gels, high numbers of theoretical plates can better be achieved by recycling (cf., Section 17.8.) than by means of long columns with a high pressure drop.

For *rigid support materials* there is no risk that the particles agglutinate and clog the column. Therefore the particles should be as small as possible in order to achieve a maximum resolution. This conflicts with the practical problems of flow resistance and packing technique: The packing problems increase with the decreasing size of the particles (cf., Section 17.1.). Nevertheless, microspheres 3 μm in diameter have been used with success, but are probably the lower limit for application in HPLC (MAJORS, 1980). For a sufficient separation efficiency the particles should not be larger than 50 μm. At present the normal range is between 5 and 25 μm. In TLC particles in the range between 0.1 and 30 μm were successfully used. Here the theory predicts an optimum at about 10 μm.

10.4. Characterization of the pore system

For the effect required, only the *open pores* accessible from outside are of importance. *Closed pores* which are fully enclosed by solid walls, and hence inaccessible, should not be present at all. For supports which are porous due to swelling, the spaces between the polymer chains are called pores.

Depending on the size, the cavities can be arranged in several classes (DUBININ, 1961):

— *Macropores:* pores of this type contribute to the specific surface area by no more than 0.1 $m^2 \cdot g^{-1}$. In principle, macropores with diameters of about 1 μm are only the entrance to the actually effective pore system.

— *Mesopores:* these *medium-sized pores* have diameters of about 0.1 μm, producing a specific surface area of 1—10 $m^2 \cdot g^{-1}$. When tested with nearly saturated vapours, they become filled by capillary condensation.

— *Micropores:* pores of this type have diameters of a few nm, yielding specific surface areas between 250 and 900 $m^2 \cdot g^{-1}$.

Some authors only distinguish between micro- and macropores, drawing the line at 2 nm.

10.4.1. Specific surface area

The surface area generated by a certain mass of a substance increases with the decreasing size of its particles, but this geometric *outer surface* is no more than 0.27 $m^2 \cdot g^{-1}$ even for microspheres 10 μm in diameter (for SiO_2, with a density of $\varrho = 2.2\ g \cdot cm^{-3}$). Consequently the large specific surface areas of the chromatographic support materials are mainly caused by the *interior surface* of the pore system.

The specific surface area of porous substances can be calculated from the nitrogen adsorption at 77.4 K, the boiling temperature of liquid nitrogen (BET method, BRUNAUER et al., 1938). VERZELE et al. (1979) described a BET apparatus suitable for duplicating, which proved successful in practice, and its utilization in the chromatographic laboratory. The specific surface area can also be estimated by the heat of wetting with hydrocarbons, alcohols, water, etc. (KISELEV, 1961).

Typical adsorbents exhibit specific surface areas of some hundred $m^2 \cdot g^{-1}$. Support materials for LLC should have no more than 50 $m^2 \cdot g^{-1}$, whereas for SEC permanently porous products with surface areas ranging from 5 to 300 $m^2 \cdot g^{-1}$ or corresponding gels are suitable. The optimum for supports used for bonded phases lies at about 300 $m^2 \cdot g^{-1}$ (KARCH et al., 1976).

10.4.2. Pore volume

The pore volume (also called *pore capacity* in some papers) can be determined from the maximum quantity of liquid taken up via the vapour phase. This presupposes that the molecules are small enough to penetrate into all of the pores. If this is the case, one obtains consistent results even with different liquids. However, there must not be any macropores ($d_P \sim 1$ µm), because they are difficult to fill from the gas phase (ŽDANOV, 1961). For some adsorbents, usable results can also be obtained by the approximation method of FISHER and MOTTLAU (1962), which is rather simple and, moreover, illustrates the concept of the pore volume: As a rule, dry adsorbents are flowable powders which, when a small quantity of liquid, e.g., water, is added, initially retain this property. However, as soon as the pores do not take up any more of the liquid added, the particles adhere to one another. The volume added until this sudden change occurs is set equal to the pore volume. The specific pore volume, ${}^1V_P$, normally ranges between 0.5 and 1 $cm \cdot g^{-1}$. For hydrophilic adsorbents such as silica gel it is determined by adding water, whereas for materials with bonded phases an organic solvent (preferably methanol) is used (VERZELE et al., 1979). In this case one has to consider that errors due to evaporation may occur.

10.4.3. Pore geometry

It is rare that information is available about the pore shape. However, investigations by electron microscopy have shown that cylindrical or funnel-shaped pores are ideal shapes. from which real systems differ rather widely. This is one of the reasons why the determination of the pore size by different methods frequently leads to different results.

The classical method is based on the fact that liquids contained in capillaries exhibit a vapour pressure which decreases with the decreasing width of the capillaries. The radius of the capillaries is calculated from the lowering of vapour pressure.

For non-wetting liquids, e.g., mercury, an external pressure is necessary to force them into the pores; this pressure must be increased as the pore size decreases. *Mercury porosimetry* (RITTER and DRAKE, 1945) can be carried out today using commercial instruments, which also enable an automatic data collection and evaluation. Naturally an adequate strength of the particle skeleton is a prerequisite to mercury porosimetry, because the pressures range up to 200 MPa in the determination of small-sized pores. The range of application of the method extends over three decimal powers, thus practically covering the whole range of pore sizes which is of interest. The pore-size distribution can be determined from the relationship between the pressure and the quantity of liquid forced into the pores. In mercury porosimetry, wide pores having narrower entrance openings are evaluated by the size of the entrance, whereas in capillary condensation they are evaluated by the size of their cavities. This "ink-bottle" effect yields a marked hysteresis loop in the diagram of the amount of mercury intruded vs. pressure: the mercury does not re-emerge from the pores until the pressure has dropped to about 20% of the value required for forcing it into the pores (LePage et al., 1968).

Another possibility of measuring the pore entrance sizes lies in the application of "molecular probes" (Dubinin, 1961). These are samples with known molecular dimensions in the range of the pore sizes. Samples with particles larger than the openings of the pore system are excluded, whereas smaller ones can enter the cavities and become adsorbed. This method was also utilized in the macromolecular range for the characterization of SEC separation gels (Halász and Martin, 1975, 1978). As compared with the above methods it has the advantage that it is also applicable to networks which are porous due to swelling (Freeman and Poinescu, 1977; Halász and Vogtel, 1980; Cjurupa and Davankov, 1980; Kuga, 1981), and to silica gels coated *in situ* (Nikolov et al., 1980). Freeman and Poinescu compared this *inverse SEC* with other methods, finding a rather good agreement of the results for porous glass. However, narrow pore size distributions can hardly be measured using this method, because very narrow distributions have only a small effect on the relationship between the elution volume and the logarithm of molecular size (Kubin and Vozka, 1978a). This is clear from the fact that even with pores of uniform size a monodisperse polymer sample is eluted within an interval of a certain extension; cf., Section 8.5. and Fig. 8-6. This effect is also reflected by the experimental results obtained by Nikolov et al. (1980). For silica, the authors found that the pore size distribution determined by exclusion chromatography appears broader than that determined by capillary condensation or mercury porosimetry.

As for the particle size (see Section 10.3.), in stating average pore sizes it should be indicated whether the value concerned is a number average or a volume average, the latter corresponding to the weight average of particle sizes.

For cylindrical pores, a simplifying geometrical consideration leads to the approximation

$$d_0 = 4000\,{}^1V_P/{}^1A \tag{10-1}$$

which establishes a relationship between the pore size, d_0 (in nm), the pore volume, ${}^1V_P$ (in $cm^3 \cdot g^{-1}$), and the specific surface area, ${}^1A$ (in $m^2 \cdot g^{-1}$). According to this relationship, within a type series of supports where the pore volume is approximately constant, the specific surface area should increase with decreasing pore size. This is reflected in the tables listing properties of commercially available adsorbents.

The distance over which the pores would extend if they were equal in width and arranged in a row can be calculated from the pore volume and diameter. For one gram of a silica gel with mesopores of $d_0 = 100$ nm and a pore volume of 0.65 $cm^3 \cdot g^{-1}$, this fictive length amounts to almost 83000 km (Halpaap, 1973). This incredible value may give an impression of the complex processes occurring in the porous support materials.

Support materials for LLC should have pore sizes greater than 100 nm. Adsorbents used for AC investigation of low-molecular-weight substances must have pore sizes of 4 nm at least, so that the establishment of equilibrium is not retarded by a hindered diffusion. In SEC the pore size required depends on the separation range aimed at. For the application of bonded phases, 10 nm pores are most suitable (Karch et al., 1976).

10.4.4. Porosity

The *internal porosity*, ε_P, indicates the contribution of the pore volume of an individual particle to the geometric total volume of the particle as the mean value for a packing material. The value can be calculated from the specific pore volume, ${}^1V_P$, and the density of the wall material, ϱ_w:

$$\varepsilon_P = \frac{{}^1V_P}{{}^1V_P + (1/\varrho_w)} \qquad (10\text{-}2)$$

The internal porosity of porous particles ranges between 0.5 and 0.9. It is shown in Fig. 10-2 as a ratio of the hatched sector to the area of the whole circle. For porous layer beads it reaches values of only a few per cent. UNGER et al. (1973) obtained silica with $\varepsilon_P = 0.9$ by a technique described in Section 11.1., using solutions in cyclohexane instead of pure polyethoxysiloxane.

Depending on the chemical nature of the separating materials, the internal porosity is a genuine capillary porosity or occurs only as a porosity due to swelling or gel porosity.

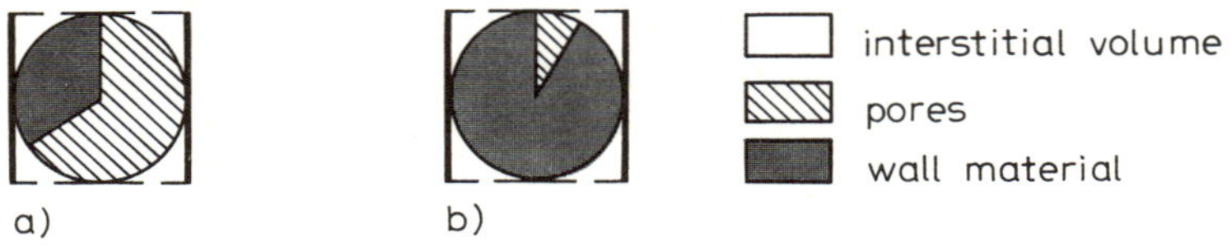

Fig. 10-2
Schematic representation of the porosity
a) of packings with fully porous particles
b) of columns with porous layer beads.

The *interstitial porosity*, ε_I, is the ratio of the interstitial spaces to the total volume (empty volume) of the column, being represented in Fig. 10-2 as the ratio of the sum of the four unshaded corner areas to the total square. The numerical value of the interstitial porosity is identical with the mean value of the free cross section of the column, q. For the closest sphere packing with a percentage bulk factor of 74%, the interstitial porosity would be 0.26. Usually it is greater, ranging between 0.35 and 0.40 for well packed columns. Values between 0.20 and 0.25 have been observed for packings of soft particles, which flatten at their contact areas under pressure.

The total porosity, ε_t, is the contribution of the total space accessible for small-sized particles to the empty volume of the column. In Figure 10-2 this is the ratio of the sum of the hatched and unshaded areas to the total square. Thus we can write the following equation:

$$\varepsilon_t = \varepsilon_I + \varepsilon_P (1 - \varepsilon_I) \qquad (10\text{-}3)$$

For columns with fully porous packings, ε_t ranges from 0.70 to 0.94. The difference from 1 gives the contribution of the wall material.

The interstitial volume is accessible for all sorts of the solutes. It is a prerequisite to the transport of the mobile phase. The chromatographic process itself, however, takes place in the pores and at their surface. Therefore ε_t should be as small as possible, but ε_P as large as possible. This requirement is of special importance in SEC, because in this case it is only the distribution of the sample molecules between the pore volume and the interstitial volume which effects the separation. A simple numerical example may illustrate this requirement: from a column with $V_C = 100$ ml and $\varepsilon_I = 0.4$, molecules larger than the exclusion limit are eluted at $V_e = 40$ ml. If $\varepsilon_P = 0.5$, then the maximum elution volume in which the molecules being fully capable of permeation are discharged is 70 ml. Assuming as in Section 8.1. that with $\Delta V_e = 70 - 40 = 30$ ml molar masses in a 1:100-range can be separated, then a volume of 15 ml corresponds to a molar-mass interval of one decimal power. However, if $\varepsilon_P = 0.9$, then one decimal power of the molar mass is distributed over 27 ml. The constant C_2 in eqn. (8-2) is proportional to the product $\varepsilon_P (1 - \varepsilon_I)$. This, in addition, means that in SEC high resolution requires columns with the largest possible internal porosity, ε_P, and the smallest possible interstitial porosity.

If a linear calibration relation is described, then the differences between the internal porosities of the gels combined with each other must not exceed 10%. (YAU et al., 1978a).

10.5. Selection and characterization of the chromatographic activity

In AC, the relative rate of migration, *R*, depends significantly on the chromatographic activity of the adsorbent. To obtain reproducible *R* values, attempts are made to standardize the adsorbent by certain pretreatments.

The AC or LLC of macromolecular substances is frequently carried out by gradient development. In this case the influence of the eluent composition should be so high that the activity of the adsorbent has almost no effect. However, if only slight deviations in the structure of the macromolecules should be detected, then under carefully balanced development conditions the activity of the adsorbent is of importance.

In a drying cabinet, surface water can be removed from the adsorbents in order to activate the latter. TLC plates are usually activated over 1 h at 110 °C. Investigations using carefully dried TLC plates have shown that at room temperature highly activated adsorbents rapidly take up atmospheric moisture: alumina layers 0.2 mm thick reach the moisture content determined by the respective relative air humidity within a period as short as 4 minutes (GEISS et al., 1965a). Silica gel layers adsorb about one half of the equilibrium moisture content within 3 minutes (DALLAS, 1965). There are few adsorbents for which this process takes a much longer time.

Fig. 10-3 shows how the moisture content of thoroughly pre-dried adsorbents increases with the air humidity. Silica gel takes up about twice as much water as alumina. As the loading increases, both the activity factor, α_A, and the surface volume, V_a, available for the adsorption decrease. The extent of re-moistening (and hence of deactivation) depends on the chemical nature and the surface structure of the adsorbent (Fig. 10-4).

From the relatively rapid water adsorption it follows that "activation at 110 °C" usually will not yield the standard activity required in TLC development. The exchange of moisture with the laboratory atmosphere during the application of the substance, during the pre-treatment of the plates as well as during the development itself can be excluded only at a remark-

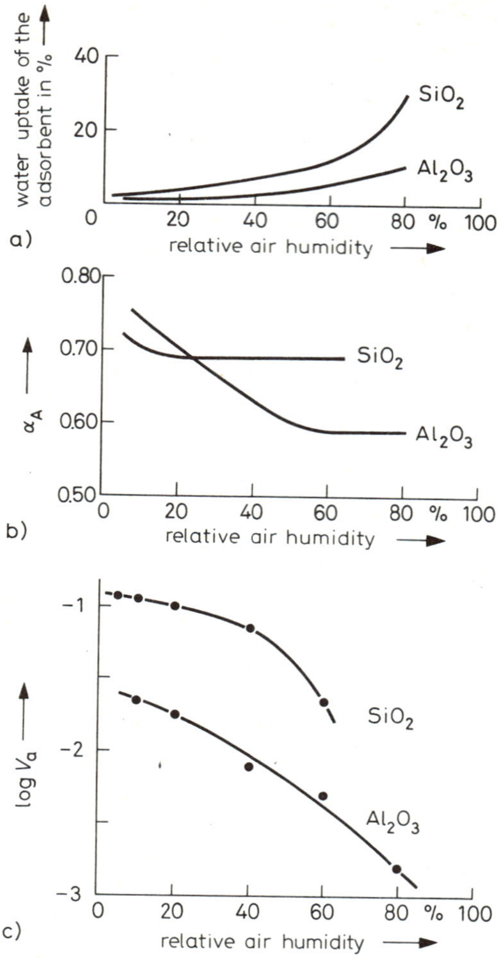

Fig. 10-3
Water content and activity of dried adsorbents after the establishment of equilibrium with air of different relative humidities (according to SNYDER [A 4])

a) Water content in % by wt.
b) Activity parameter α_A of eqn. (7-11)
c) Chromatographically utilizable surface volume of the adsorbents (cf. eqn. (7-8))

While the activity of silica gel in the normal operating range (20 to 60% relative humidity) is rather constant, the surface volume varies considerably even for this adsorbent. A standardization of the activity is indispensible.

ably high expense. Therefore, after drying at 110 °C, the plates should still be conditioned in a normal atmosphere (65% relative air humidity) — particularly if the development cannot be carried out in chambers at a well defined level of atmospheric moisture.

Plates kept in desiccators adsorb moisture during the development. At the commencement of migration the development still takes place on a dry layer, but parts reached at a later time have a lower chromatographic activity. This means a parallel gradient which effects a spot spreading in the direction of travel (and hence a smaller number of theoretical plates).

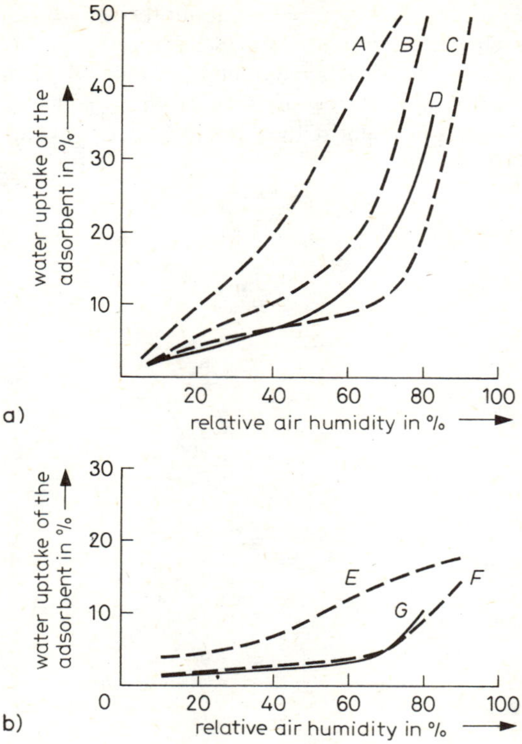

Fig. 10-4
Effect of the pore structure on the water content of dried adsorbents after the establishment of equilibrium with air of different relative humidities (according to SNYDER [A 4] and GEISS [E 5])

a) Silica gels
specific surface: A 650; B 500; C 400; D 400 $m^2 \cdot g^{-1}$; pore volume: A 0.65; B 0.75; C 1.00 $cm^3 \cdot g^{-1}$
average pore sizes: A 4.0; B 6.0; C 10.0 nm

b) Aluminas
specific surface: E 150; F 75; G 100 $m^2 \cdot g^{-1}$
(Curves D and G taken from Fig. 10-3a)

A large specific surface under otherwise equal conditions leads to a higher water absorption.

For thick layers there is the additional risk that the activity varies with the layer depth, i.e., the adsorbent deposited immediately at the plate has a higher chromatographic activity than that at the layer surface. In order to avoid the annoying problem of water adsorption during the development HALPAAP and REICH (1968) suggested the use of Al_2O_3 annealed at a higher temperature.

For the development, the adsorbent must be in an activity equilibrium and the real activity should be known. The determination according to BROCKMANN and SCHODDER (1941) is performed by column chromatography using dyes and a benzene-benzine mixture, which leads to classification of the adsorbents according to the "Brockmann grades" I—V. A higher grade index indicates a higher water loading of the adsorbent, and hence a lower activity.

For TLC, methods are needed which yield information about the activity of a layer ready to use. A simple test has been suggested by GEISS and SCHLITT (1972): the R_f value of Michrome No. 539 is determined by means of carbon tetrachloride and, after a multiplication by 100, indicated as an *activity index*, which likewise increases with the decreasing activity of the layer. Such additional information might reduce the difficulties existing with respect to the comparability of TLC data.

11. Inorganic supports

Initially, inorganic supports were used in adsorption chromatography. Then, after the production of adsorbents with wide pores, these were used in liquid-liquid partition chromatography and finally also in exclusion chromatography. In the last-mentioned technique the adsorption interactions with the naked inorganic surface led in some cases to disturbances. Thus methods for a chemical masking of these groups were developed, which led to an extensive range of inorganic adsorbents with a chemically modified surface, which play a decisive rôle in modern high-performance liquid chromatography.

Inorganic supports are pressure-resistant and, being truly permanently porous materials, do not show any variation in their volume when the eluent is changed.

11.1. Silica gel

Silica gel, which is described in detail in the monograph [A 25] by UNGER, is a porous, amorphous, hydrous silica with accessible hydroxyl groups. Silica gels can be prepared by precipitation from aqueous solutions of sodium silicate by the addition of an acid or by the decomposition of silicon tetrachloride with water. In the former case, the monomeric silicic acid produced initially condenses to polysilicic acid with an ever-increasing molar mass, which finally precipitates, passing from soft gels into hard aerogels by a progressive dehydration. The pore structure is affected by the conditions of preparation. At pH 10 the coagulate yields an aerogel with an interior surface area of about 200 $m^2 \cdot g^{-1}$ and a pore size of 10 nm, while at pH < 4 very small pore sizes of 2.5 nm and products with an interior surface area of about 800 $m^2 \cdot g^{-1}$ are produced. By a suitable control of the coagulation process it is also possible to prepare supports with wide pores for liquid-liquid partition chromatography and exclusion chromatography. Thus a precipitation with ammonia, carried out at 500 °C under pressure, leads to products with pore sizes of several hundred nanometres (LEPAGE et al., 1967, 1968). Fractosil® 25000, by means of which SINGH and HAMIELEC (1978) investigated latex particles with diameters of up to more than 1 µm, exhibits pore sizes of 3000 nm.

Microspheres of very uniform sizes can be prepared by the agglutination of colloidal silica particles. The process can be initiated and controlled by a urea/formaldehyde polycondensation carried out in the collodial solution. The resin produced is embedded in the microspheres and must be burnt out. What is left is a skeleton consisting of a corresponding number of the original colloidal particles, which adhere firmly to one another, the whole constituting a microsphere with a diameter of 5—10 µm. The pore size increases with the size of the colloidal primary particles. Values ranging between 6 and 350 nm can be achieved

(KIRKLAND, 1976). The internal porosity of supports prepared in this way is circ. 0.5; the remaining space is occupied by the spherical packing of the colloidal particles. The uniformity of the primary particles ensures that the pore sizes are also largely uniform. The microspheres have a high mechanical stability and can be packed under a pressure of up to 34.4 MPa (345 bar).

Following the technique developed by UNGER et al. (1973), uniform silica microspheres can be prepared from tetraethoxysilane, which is first converted into viscous polyethoxysiloxane by hydrolytic polycondensation. In the second stage the latter is emulsified in an ethanol-water mixture and condensed to silica gel microspheres by adding a catalyst. This initially yields hydrogel particles, which are separated from the liquor and dehydrated to porous aerogel particles. The sizes of the microspheres produced decrease as the speed of stirring is increased. The pore size mainly depends on the catalyst and can be increased to 30 nm by increasing the concentration of the latter. Even larger pores of up to 3000 nm can be obtained by treatment of the gel particles with NaCl solution and calcination of the salt-loaded microspheres at high temperatures. Because of their uniformity and high pore volume ($0.7 \leqq {}^1V_P \leqq 1$ ml · g^{-1}) the products are suitable for size exclusion, partition and adsorption chromatography.

The *accessible hydroxyl groups* are located on the surface of the silica gel at irregular distances. Some are positioned so close to each other that they can form hydrogen bonds or even undergo condensation reactions. These "reactive hydroxyl groups" also affect the chromatographic behaviour of the silica gel in a way different from that of the isolated "free hydroxyl groups". Analytically, one may distinguish between the different types, e.g., by a conversion with trimethylchlorosilane, which at a temperature of 195 °C attacks only the isolated hydroxyls (SNYDER and WARD, 1966; SNYDER, 1966a). The total number of accessible hydroxyl groups can be determined by means of a reaction with hexamethyldisilazane or lithium aluminium hydride. It is up to 5 groups per nm^2. A direct pulse chromatographic determination of surface hydroxyls using the complex $Zn(CH_3)_2 \cdot 2$ THF was performed by NONDEK and VYSKOČIL (1981). This rapid and sensitive method (about 10 mg sample material and 50 µl reagent) was successfully applied to RP silica (cf., Section 11.10.) and allowed the differentiation between isolated and non-isolated hydroxyl groups.

BATHER and GRAY (1976) stated 4.6 OH per nm^2 for the total number, 1.2 OH/nm^2 for the isolated hydroxyl groups and 3.4 OH/nm^2 for the hydroxyl groups forming hydrogen bonds with each other. These data are in accordance with those given by SNYDER [A 4], ARMISTEAD et al. (1969) as well as HAIR and HERTL (1969).

Silica gel is activated by heating, whereby the water retained at the surface escapes. Moreover water is also formed by a chemical reaction of adjacent (reactive) hydroxyl groups at temperatures between 200° and 400 °C. When this condensation to siloxane bridges has been completed, then only the free hydroxyls are left, being finally removed in the temperature interval 500—1100 °C. A silica gel annealed at very high temperatures and over a long period no longer has any hydroxyl groups. It has become hydrophobic and has lost its capability of selective adsorption. Moreover its interior surface area is reduced.

For adsorption chromatography, the accessible hydroxyls are decisive (BASILA, 1961; KISELEV, 1967). They can fix polar and unsaturated substances which act as electron donors. The higher the adsorption energies, the more clearly the interactions can be detected by spectroscopy. A silica gel whose hydroxyls have been completely exchanged by methoxy groups or a halogen has lost its normal adsorption power. Like the previously mentioned decrease in activity due to excessive pretreatment temperatures, this also shows that the hydroxyl groups are decisive. The usual activation is intended to make these hydroxyl groups accessible for the adsorbate by removing adherent water. Silica gel reaches its maximum activity at 150—200 °C within 4—6 hours.

If activated silica gel is re-contacted with water, then the latter is preferentially added to the reactive hydroxyl groups (HAMBLETON and HOCKEY, 1966), which are thus deactivated first. During the deactivation the silica gel warms up, because the adsorption energy of the water is released. The reactive hydroxyls are also more active than the isolated one towards certain adsorbate molecules, the functionality and geometry of which allow a simultaneous sorption to several hydroxyls. However, the free hydroxyls show a higher activity towards molecules which have only one possibility each for interaction (SNYDER, 1966b). BATHER and GRAY (1976) investigated the effect of annealing in a column packed with HPLC silica gel. The dry column was heated to a certain temperature over 2 hours in each experiment; after cooling it was tested with different samples in hexane as an eluent, blown dry at 120 °C, and then annealed at the next highest temperature. Even after the separation of the reactive hydroxyls, i.e., after annealing at 450 °C and above, the silica gel still exhibited a remarkable retention ability towards polar samples. After an in-situ treatment of the packing by a 5% solution of trimethylchlorosilane, the capacity factors of these samples were much smaller, showing about the same reduction with increasing annealing temperature as the concentration of the surface hydroxyls (see Fig. 11-1). This demonstrates the importance of the isolated OH groups for the adsorption capacity of the silica gel. Since at temperatures above 500 °C the capacity factors on naked silica gel even increased with the annealing temperature, an adsorption obviously also occurred on the Si—O—Si groups formed from the isolated hydroxyls by dehydration. Such groups exhibit internal stresses due to

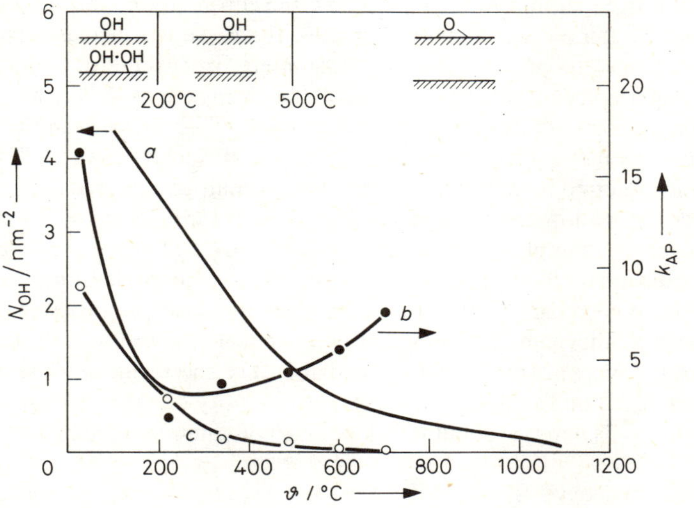

Fig. 11-1
Effect of the pretreatment temperature on the properties of silica

N_{OH}: Number of hydroxyl groups per 1 nm² of the surface (calculated from TGA data), curve a
k_{AP}: Capacity factor for acetophenone in n-hexane,
curve b — on naked silica
curve c — after capping of the free OH by $(CH_3)_3SiCl$
(according to BATHER and GRAY, 1976).

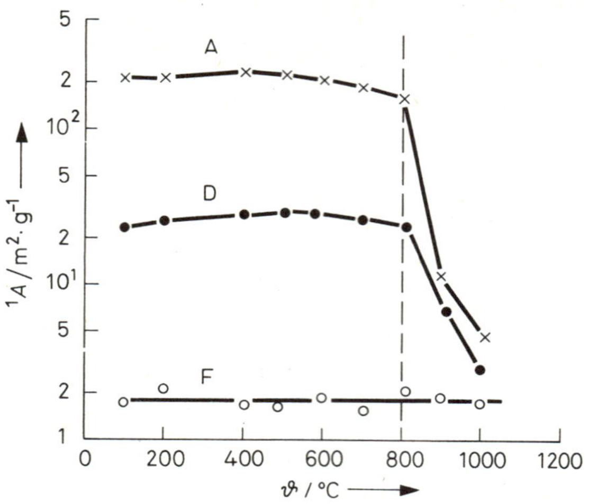

Fig. 11-2
Specific surface area of silica gel as a function of the pretreatment temperature
Annealing for 4 hours each. A Porasil® A, D Porasil D, F Porasil F (according to FELTL, LUTOVSKÝ, SOSNOVÁ and SMOLKOVÁ, 1974).

their complex geometric conditions. Heating to temperatures above 800 °C diminished the interior surface area (see Fig. 11-2).

In the normal range of application of silica gels, the chromatographic acitivity is adjusted by the addition of water. As the activity decreases, the R_f values increase. The smaller the ε^0 value of the eluent, the more distinctly is this effect. In chromatography using liquids of a very high displacement effect, e.g., methanol, the given activity differences play a minor rôle. Such strong eluents, thoroughly dried, may gradually remove even the deactivation water, forming a strongly adherent primary layer (SCOTT and KUČERA, 1979c).

Silica gel is the preferred material for adsorption chromatography. It can be easily activated, yields a good separation and is relatively inert even towards sensitive samples. For thin-layer chromatography, samples with surface areas of 300—600 $m^2 \cdot g^{-1}$ and pore sizes between 10 and 25 nm are used. The types with wide pores are used in partition chromatography and steric exclusion chromatography. If, in this case, the adsorption power of the surface hydroxyls interferes, they can be eliminated by chemical conversion (cf., Section 11.10.). About 100 different types are commercially available. The specification for several commercial products are listed in Tables 11-1 to 11-4. The quantity of spherical particles, which is required for column packing, is about 0.62 g per cm^3 of column volume (value for Spherosil®).

With its surface hydroxyls, silica gel behaves like a weak polyacid, the pH value of which ranges between 3 and 5. Therefore it adsorbs basic substances ($pK_B < 5$) most strongly. Eluents with a high pH attack the silica gel. The working range is between pH 2 and pH 8. Cadoxene (cadmium ethylene diamine hydroxide), an important solvent for native cellulose, dissolves silica gel (BEREK et al., 1977). Even water attacks it: from a 25 cm column freshly packed with 10 µm silica gel, 38 µg/ml were washed out at room temperature, and 100 µg/ml at 60 °C. The dissolved silica may interfere with the subsequent investigation

Table 11-1
Properties of irregularly shapéd porous silica particles

Products	Particle size / μm	d_0 / nm	^1A / $m^2 \cdot g^{-1}$	^1V_P / $ml \cdot g^{-1}$	Supplier[1])
LiChrosorb® Si 60	5, 7, 10, 30	6	500	0.75	EM
LiChrosorb® Si 100	5, 10, 30	10	300	1.0	EM
Polygosil® 60	5, 7.5, 10, 15, 20, 30, 40—63 63—100	6	500	0.75	MN
Vydac® 101 IR	10	10	350		SG
Silica gel 40[1])	15	4	650	0.65	EM
Silica gel 60[1])	15	6	500	0.76	EM
Silica gel 100[1])	15	10	400	1.00	EM
Partisil® 10	10	4—5	400	0.66	WTM
Fractosil® 5000	63—125	490			EM
Fractosil 10000	ditto	1400			
Fractosil 25000	ditto	3000			

[1]) TLC adsorbents with 10—15% gypsum are marked by "G"
EM E. Merck; SG The Separations Group; MN Macherey-Nagel; WTM Whatman

Table 11-2
Properties of spherical porous silica adsorbents

Products	Bead diameter / μm	d_0 / nm	^1A / $m^2 \cdot g^{-1}$	^1V_P / $ml \cdot g^{-1}$	Supplier[1])
Nucleosil® 50	5, 7.5, 10	5	500	0.8	MN
Nucleosil® 100	5, 7.5, 10, 30	10	300	1.0	MN
Nucleosil® 100 V		10		1.5	MN
Spherisorb® S—W	3, 5, 10	8	220		PS
Hypersil®	6	10	200	0.7	SS
Vydac® 101 TP	10	33	100		SG
Chromosorb® LC-6	5, 10	13	300		JM
Zorbax® SIL	7	7.5	250—300		DP
Pragosil®	5, 10, 20, 40	8	500		LA
LiChrospher®	cf. Table 11-3				
Porasil®	cf. Table 11-4				

[1]) DP Du Pont; MN Macherey-Nagel; PS Phase Separations Ltd.; SS Shandon Southern Products Ltd.; SG The Separations Group; JM Johns-Manville; LA Lachema

Table 11-3
Properties of LiChrospher® porous silica beads/MERCK

Products	Bead diameter µm	Exclusion limit[1]) 10^3 g · mole^{-1}	Linear fractionation range[2])	d_0 nm	m^2 · g^{-1}	^1V_P ml · g^{-1}
Si 100	5, 10, 20	50—70	3—50	10	250	1.2
Si 500	10	400—500	15—150	50	50	0.8
Si 1000	10	1000—2000	30—2000	100	20	0.8
Si 4000	10	>3500	100—>7000	400	6	0.8

[1]) for polystyrene in trichloromethane
[2]) according to KIRKLAND (1976)

Table 11-4
Properties of Porasil® porous silica beads/WATERS

Products	Particle size[1])	Separation range 10^3 g · mole^{-1}	d_0 nm	^1A m^2 · g^{-1}
A(60)	C, M	≦ 40	< 10	300—500
B(250)		≦ 200	15	140—230
C(400)		≦ 400	30	75—125
D(1000)		≦1000	60	38—62
E(1500)		≦1500	120	20—30
F(2000)		≦4000	>150	8—12
µ-Porasil	10 µm	0.1—10	6	500

[1]) C (coarse) 75—125 µm; M (medium) 37—75 µm

of preparative fractions, for instance with the evaluation of their $\bar{M}_n$ values (VAN DIJK et al., 1980).

In 1979, ATWOOD et al. succeeded in eliminating the disturbing effect of silica dissolution by means of a guard column, in which the eluent was saturated with silicic acid. In this way it was possible to use a column containing a 5 µm silica packing for 400 analyses at 65 °C and pH 10.74, but the precolumn had to be replaced three times. HANSEN (1981) also reported the helpful effect of a guard column. He used a LiChrosorb® Di 60 column (L = 0.15 m; d_C = 4.6 mm; d_P = 5 µm) which lasted for 6 months in daily use when equipped with an identical precolumn. Mixtures of methanol-water-0.2 M potassium phosphate (pH 7.7—8.0) were used as eluents.

BARKER et al. (1981 a) investigated the dissolution of silica in various solvents. They recommended that silica columns should be flushed with and stored in acetone after use in order to prolong their service life.

VERZELE et al. (1979) observed that an acid treatment increased the average pore size and at the same time diminished the specific surface area. Boiling a silica gel with $^{1}A = 432$ m^2 $\times$ g^{-1} and $d_0 = 8.8$ nm in hydrochloric acid, they obtained a product with $^{1}A = 284$ m^2 $\times$ g^{-1} and $d_0 = 13.0$ nm. According to their experience, a wet-treated silica gel with an original pore size of 6 nm in fact has pore sizes between 8 and 10 nm.

11.2. Highly disperse silicic acid, Aerosil®

Highly disperse silicic acid is produced in the combustion of silane or chlorosilanes, and also represents a silica with a large surface area. Unlike silica gel it is not porous. The high values of the specific surface area (about 150 to 400 m^2 · g^{-1}) result from the extraordinary fineness of the powder. The adsorption power, like that of silica gel, is due to free and bonded hydroxyl groups. The concentration of the surface hydroxyls is listed in Table 11-5. It should be mentioned that, at 200 °C, silica gel has a somewhat higher value of about 3.6 OH/nm^2.

Using non-porous silicas having large surface areas, many fundamental results concerning the adsorption of polymers have been found.

Table 11-5
Concentration of the SiOH groups on Aerosil 200, determined by conversion with $LiAlH_4$ from the amount of hydrogen liberated (according to DIETZ, 1976)

Pretreatment temperature in °C	BET surface m^2 · g^{-1}	Concentration of SiOH groups mval · g^{-1}	No. of SiOH groups per nm^2 of the surface	No. of SiOH groups per nm^2 (from IR measurements)	
				free	vicinal
200[1])	195	0.89	2.74	1.52—1.63	1.11—1.22
400[1])	196	0.68	2.09	1.40	0.69
800[2])	191	0.37	1.17	1.17	0

[1]) 24 h; 10^{-2} Pa (10^{-4} mbar)
[2]) 24 h; 0.1 MPa (1 bar); thereafter 400 °C; 10^{-2} Pa; 24 h

11.3. Alumina

A large number of crystalline forms of alumina are known. For chromatographic purposes, a material containing mainly the γ-modification is used. Additions of other modifications, defects in the crystal lattice, surface hydration and the formation of chromatographically active surfaces extending through different lattice planes make it difficult to give a straightforward description of the relationship between the structure of the adsorbent and its efficiency. The surface area ranges between 100 and 200 m^2 · g^{-1}, consisting mainly of cylindrical micropores (about 2.7 nm in diameter) and larger, irregular pores (BOWEN et al., 1967; JOHNSON and MOOI, 1967).

In most cases alumina is activated at temperatures as low as 150—200 °C. Supports annealed at a higher temperature could harm sensitive samples. Heating results in a liberation of water. After a vacuum treatment at 400 °C, the alumina still contains about 6 hydroxyl groups per nm^2, which can be detected spectroscopically and differentiated with respect to their bond types. However, for the chromatographic efficiency they are not as important as the hydroxyls of silica gel. This can be observed from the variation of the chromatographic activity with the annealing temperature. It also followed from the investigations by Bather and Gray (1978): these authors heated columns blown dry with nitrogen at 120 °C to a certain temperature (up to a maximum of 950 °C) for 2 hours each and then re-tested their chromatographic efficiency by means of xylene isomers and nitrobenzene in n-hexane. Under these conditions the column capacity factors increased to a maximum value occurring between 650 and 750 °C, although the hydroxyl groups disappeared more and more in the interval between 300 and 600 °C. The results suggest that sorption on alumina is predominantly due to electrostatic interaction. The steep drop of the capacity factors beyond the maximum at about 700 °C is related to damage of the surface due to the annealing (see Fig. 11-3).

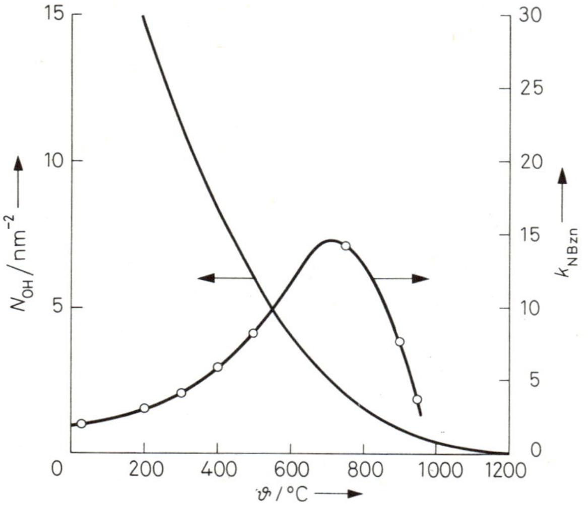

Fig. 11-3
Effect of the pretreatment temperature on the properties of alumina
N_{OH}: Number of hydroxyl groups per 1 nm^2 of the surface (calculated from TGA data)
k_{NBzn}: Capacity factor for nitrobenzene in n-hexane
(according to Bather and Gray, 1978).

According to King and Benson (1966), at the surface of the adsorbent a layer of aluminium ions is coated with a layer of oxygen ions. In this cation layer, the valence ratio implies that only 2/3 of the geometrically available sites are occupied. This affects the electron density, which at some points of the surface differs widely from the average value. Sites with a high positive field strength behave like acid groups, adsorbing basic or easily polarizable molecules. These acidic sites are decisive for the adsorption power of alumina towards most substances. Moreover, there are sites having a high negative field strength which basically act as proton acceptors. This results from the ability of alumina to separate weak acids (pK_S $\leqq$ 13), especially in a development by basic solvents, and from its tendency irreversibly to bond strongly acidic substances (pK_S $\leqq$ 5) by chemisorption. Finally, alumina obviously

Table 11-6
Properties of porous alumina particles

Products	Particle shape[1])	Particle size (μm)	d_0 (nm)	${}^1A$ ($m^2\ g^{-1}$)	${}^1V_P$ ($ml\ g^{-1}$)	Supplier[2])
LiChrosorb® AloxT	i	5, 10, 30	15	70	0.3	EM
Alox 60-D	i	5, 10, 20	6	60		MN
Spherisorb A—Y	s	5, 10, 20	13	93		PS
Spherisorb alumina A10Y	s	10	13.4[3])	95[3])	0.36[3])	
Alusorb 200	i	5—30	—	200	—	LA
Chromosorb LC-3	i	40				JM
Woelm Alumina	i	18—30	10—20	200		W
Aluminiumoxid 60 (Type E)	i	5—40 40—150	6	180—200	0.3	EM
Aluminiumoxid 90	i	5—40 40—150	9	100—130	0.25	EM
Aluminiumoxid 150 (Type T)	i	5—40 40—150	15	70	0.2	EM
Aluminiumoxid 60 G (with about 10% gypsum)						EM

[1]) i: irregular; s: spherical
[2]) EM E. Merck; MN Macherey-Nagel; PS Phase Separations Ltd.; JM Johns-Manville; W Waters; LA Lachema
[3]) according to Bather and Gray (1978)

also has surface sites which act as electron acceptors and may form charge transfer complexes. Above 1100 °C, α-alumina is formed, which can hardly be used for chromatography because of its low adsorption capacity, although it may exhibit a quite remarkable efficiency in difficult separation problems, e.g., in the separation of testosterone and 17-epitestosterone (Halpaap and Reich, 1968; cf., Section 10.5.).

The order of elution on γ-alumina is similar to that on silica gel, but substances with double bonds are more strongly adsorbed. Above all the separation is good for aromatic hydrocarbons.

Freshly precipitated alumina shows an alkaline reaction. It can be neutralized by washing and even slightly acidified by acid loading. Table 11-6 lists the specifications of several commercially available types. For the packing of columns, 0.94 g of adsorbent are required per cm^3 of column volume (value for Spherisorb® A-Y).

11.4. Magnesia

Magnesia is a basic adsorbent, on which weakly acidic substances can be separated with approximately the same results as on alumina. Unsaturated and aromatic compounds are adsorbed preferentially. Polycyclic aromatics may possibly be retained irreversibly. Substances differing from one another only by double bonds can be separated better on magnesia than on any other adsorbent.

In one respect magnesia resembles silica gel: as the temperature of activation increases, their activity passes through a maximum which is reached between 100 and 500 °C. Materials annealed at 1000 °C are no longer capable of a selective adsorption. Perhaps also for magnesia, the adsorption power is essentially due to hydroxyl groups. However, water is bonded rather loosely. The activity adjusted by the addition of water can easily change if dry eluents are used in a chromatographic process.

11.5. Magnesium silicate (Florisil®, Magnesol®)

Florisil is a coprecipitate of magnesia and silica with acid surface groups, which preferably adsorbs alkaline substances. The adsorbent, which contains a small quantity of water, separates in a manner intermediate between those of silica gel and alumina (SNYDER, 1963). It exhibits a much lower catalytic activity than alumina, and is therefore very suitable for the separation of highly sensitive substances.

11.6. Kieselguhr (diatomaceous earth)

Kieselguhr is a very weak adsorbent, the bonding power of which is much lower than those of other adsorbents. Its specific surface area only reaches values of up to 7 $m^2 \cdot g^{-1}$. One product marketed has ${}^1A = 0.5 \ldots 1.0\ m^2 \cdot g^{-1}$ and a particle size between 37 and 44 µm, which can be used, say, in liquid-liquid partition chromatography. In adsorption chromatography, Kieselguhr mainly acts as a "diluent": in mixtures with other adsorbents it reduces the activity by decreasing the V_a value.

11.7. Carbon materials

Charcoal and carbon made from animal wastes, e.g., blood charcoal, were used in early chromatographic investigations as the most popular materials until 1955. Activated charcoal exhibits a broad pore size distribution. The interior surface area (300—1000 $m^2 \cdot g^{-1}$) is highly differentiated. Adsorption takes place on carbon atoms as well as on carbonyl, carboxyl and hydroxyl groups, which are also present. Carbon adsorbents made from animal wastes still contain residual salts which also affect the adsorption.

KISELEV (1967) has shown that the impurities can be removed by annealing at a temperature above 1000 °C, which effects the conversion into graphite. The graphitized charcoal has a defined, non-polar surface and is most suitable for fundamental investigations (KISELEV, 1976).

Graphitized carbon black is used as a support material in gas chromatography. For high-pressure LC most products are too fragile and unsuitable with respect to their particle sizes. COLIN et al. (1976) crushed commercial material with large-sized particles on sieves, thus preparing particles with sizes ranging from 15 to 30 µm. Using the technique described by BARMAKOVA et al. (1974), these particles were then hardened by pyrocarbon obtained from the decomposition of benzene. In many respects the products behaved like reversed phases (cf., Section 11.10.). In 1980 (a, b), UNGER et al. reported liquid chromatography using porous carbon packings. One of the essential advantages of these supports is that they can also be used in alkaline liquids with pH values above 9. The maximum possible loading, however, is relatively low, being smaller than that of C 18 reversed phases by a factor of 25.

CICCIOLI et al. (1981) ground commercially available graphitized carbon black (80 to 100 mesh) of very high mechanical strength, and then further reduced it in size by means of a set of metal screens and acetone. Dried fractions (25—33, 33—45, 75—88 μm) were used to pack columns (d_C = 1.6 mm) which exhibited rather good chromatographic properties. Care had to be taken to avoid using eluents with a viscosity greater than 0.7 mPa · s.

Carbon adsorbents with a very regular surface geometry for gas chromatography can be prepared by thermal decomposition of polyvinylidine chloride (KAISER, 1970). The interior surface area exceeds 1000 $m^2 \cdot g^{-1}$. Adsorbents of this type were characterized in detail by LI-RU (1979).

A carbon adsorbent which has well defined and reproducible surface properties and is very suitable for packing high-efficiency HPLC columns was obtained by the reduction of PTFE with lithium amalgam (SMOLKOVA et al., 1980).

11.8. Porous glass

Glasses, in a wider sense, are amorphously frozen melts of any chemical composition whatsoever. Their characteristic properties include gradual softening over a broad temperature range. In a narrower sense, glasses are the vitreously frozen melts of alkali- and alkaline-earth metal silicates. If they are heated to sufficiently high temperatures for a long time, then a segregation occurs due to the crystallization of the SiO_2. This process is called devitrification, a process feared in glass blowing. While water usually dissolves the alkaline components only from a layer near the surface, devitrified melts can be extracted after cooling. This process leaves a skeleton consisting mainly of SiO_2. Alkali-metal borosilicates (96% SiO_2; 3 ... 4% B_2O_3; 0.5 ... 1% Na_2O) are most suitable for the preparation of such porous glasses (DEMENT'EVA et al., 1962; Bresler et al., 1963), which at first were used as support materials in gas chromatography. The size, number and shape of the pores can be influenced by the B_2O_3/Na_2O ratio (ŽDANOV et al., 1962; DOBYČIN et al., 1962). In 1965 (a) HALLER investigated the kinetics of segregation and found relationships between the annealing conditions and the morphology of the heterogeneous zones. The pore size of materials leached with aqueous hydrochloric acid increased with the duration of the preceding heat treatment (at a fixed temperature). Thus a way of preparing such materials with a predetermined average pore size for steric exclusion chromatography had also been found. The particles are produced by the mechanical crushing of the frozen melts and are irregularly shaped. Table 11-7 shows specifications of commercial types.

During the thermal treatment, a small quantity of the skeletal SiO_2 dissolves in the Na_2O/B_2O_3 phase. This highly disperse silica forms the *colloidal deposits* (HALLER, 1965a) inside the cavities after the extraction by mineral acids. Its quantity increases with the duration of the annealing process, i.e., it follows the same tendency as the dimensions of the cavities produced. The actual porous glass with macropores is obtained by a final treatment with alkaline solutions, which attack highly disperse SiO_2 more rapidly than the silicic acid skeleton of the pore walls (HEYER, 1980). In some cases the highly disperse silica is so closely packed

Table 11-7
Properties of Controlled Pore Glass (CPG-10)/ELECTRO-NUCLEONICS, Inc.

Products	Particle size[1])	Fractionation range 10^3 g · mole^{-1}	d_0 nm	${}^1V_P$ ml · g^{-1}
CPG-10-75		2.2—10	7.5	0.7
-120	C, M	4—33	12	0.9
-170		6—70	17	0.8
-240		9—170	24	1.1
-350		12—300	35	0.9
-700		40—1200	70	0.9
-1400		70—6000	140	
-2000		1000—12000	200	
-3000			300	

[1]) C (coarse): 75—125 μm; M (medium): 37—75 μm

that the dissolution is kinetically hindered (HEYER et al., 1977). Thus it cannot be excluded that small portions of the colloidal deposits are left in the pores, imparting some degree of microporosity to the macroporous glasses (HEYER, 1981). This would be a plausible explanation of the curious fact that in some cases good separation efficiencies were observed in the macromolecular range, whereas the values of the heights equivalent to a theoretical plate, as measured by benzene or o-dichlorobenzene, were rather low.

HALLER demonstrated the separation efficiency of porous glasses by an investigation of viruses (1965b), while CANTOW and JOHNSON (1967b) showed that porous glass is also suitable for synthetic polymers. Meanwhile it has been successfully employed in many SEC investigations.

The silanol groups of the silica matrix may cause difficulties by the adsorption of polar macromolecules. In aqueous media, even in the neutral pH range, the silanols form negatively charged sites to which positively charged groups are bonded ionogenically. Molecules containing amino groups or unshared electron pairs may be retained by chemisorption on B_2O_3 residues of the matrix, which form Lewis acids. These effects are influenced by the ionic strength of the solutions (COOPER and MATZINGER, 1979). To suppress the undesired interactions with the surface, the latter can be chemically modified (LANGHAMMER and SEIDE, 1967; COOPER and JOHNSON, 1969; CHANG et al., 1976; TALLEY and BOWMAN, 1979), or deactivated by coating with polyethylene glycol (HIATT et al., 1971) or silicone oil (MIZUTANI, 1980).

As regards the pore size distribution, there are rather large differences between products of different origins. HALLER (1965b) observed uniform and channels with a narrow size distribution in the products which he had prepared very carefully. He reconfirmed this finding in 1977. In 1971, YAU et al. investigated commercial products, also observing narrow distributions (see Fig. 11-4). Analogous results were obtained by ŽDANOV et al. (1977) and WAKSMUNDSKI et al. (1979) always with their own products. In contrast to this, the commercial materials investigated by CANTOW and JOHNSON (1967b), BARRALL and CAIN (1968) or COOPER et al. (1971) exhibited relatively broad pore size distributions.

It is interesting that the values of the height equivalent to a theoretical plate, as determined by different authors on columns containing porous glasses, differed widely and in part were

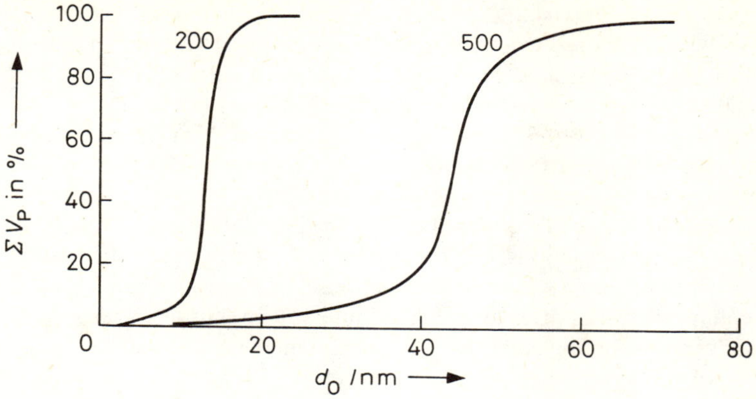

Fig. 11-4
Integral pore size distribution (volume distribution) for porous glasses (Bioglas® 200 and Bioglas® 500)
(according to YAU, MALONE and SUCHAN, 1971).

rather high. In 1977, using tobacco mosaic viruses as a sample. HALLER et al. found a reduced height equivalent to a theoretical plate of $h^* = 3.6$ and 8.9, respectively. The value of $h^* = 4.8$ determined by BASEDOW et al. (1976) with glucose as a sample was similarly favourable. OTOCKA (1973) packed columns with narrow sieve fractions ($d_p = 36$—44 µm), obtaining values between 10 and 14. SPATORICO (1975), using dry-packed columns ($d_p = 75$—125 µm) found values ranging between 30 and 50 with benzene as a sample. The results determined by COOPER et al. in 1971 with the use of o-dichlorobenzene for particles of the same size class were similarly high. Their h^* values ranged between 18 and 46, increasing with the average pore size.

The relatively poor values of the height equivalent to a theoretical plate are inconsistent with the high separation efficiency of the porous glasses as observed in many investigations (e.g., by CANTOW and JOHNSON, 1967b; OTOCKA, 1973; SPATORICO, 1975; SPATORICO and BEYER, 1975; BASEDOW et al., 1976). This discrepancy has been mentioned already in conjunction with the existence of colloidal deposites, vide supra.

11.9. Materials for precipitation chromatography

Precipitation chromatography requires inert solid surfaces, on which the gel can precipitate as a stationary phase. CANTOW et al. (1968) simply used the interior surface of a long capillary. Generally, however, filled columns are used. In this connection the requirement for an inert support material means that its surface should not influence the separation by either steric exclusion or adsorption. In precipitation chromatography, $dR/dM > 0$, whereas in steric exclusion chromatography $dR/dM < 0$ holds. A superposition of these two tendencies impairs the separation according to the chain-length. In adsorption chromatography, $dR/dM > 0$ is indeed also valid, but with this mechanism any difference in the polarity of the components usually has a much higher effect than their mass values.

In most cases, following the example given by BAKER and WILLIAMS (1956), glass beads with a diameter between 0.04 and 0.3 mm, preferably of 0.1 mm ("ballotini"), are used for the column packing. Glass wool has also been employed (Lovrič, 1969). The beads are cleaned with hot, concentrated hydrochloric acid until fresh acid no longer turns yellow. This procedure is followed by a treatment with hot, concentrated nitric acid and a thorough washing-out by distilled water. Flushing by acetone completes the treatment (Jungnickel and Weiss, 1961). The use of chromatosulphuric acid is not recommended, because it leads to a contamination of the surface with chromium ions.

A specific disadvantage of the glass beads is the gradual release of alkali, which catalyzes certain decomposition reactions, e.g. the splitting of ester bridges. The leaching of the glass can be clearly detected in the first eluate fractions of a Baker-Williams fractionation, above all if the column is re-run after an extended interval. In elutions with relatively non-polar solvents under fractionating conditions, polar polymers may adsorb on the glass. By means of labelled polymethyl methacrylate, Schulz et al. (1965) have shown that the extent of adsorption on the beads is much higher from benzene than from acetone. Glass surfaces treated by dichlorodimethylsilane showed much less adsorption.

To avoid disturbances due to alkali, sand was used as a support material in several investigations (Krigbaum and Kurz, 1959; Glöckner, 1965a). The cleaning was carried out in the same way as for the glass beads.

To improve the heat conduction, Pepper and Rutherford (1959) used copper grits as a column packing. They fractionated polystyrene, obtaining results which were, for a relatively high rate of elution, just as good as those obtained on glass beads at a lower rate. Finally, however, they nevertheless preferred the uniformly shaped glass beads to the copper powder in order to exclude any catalytic damage to the sample. Slonaker et al. (1966) used steel beads and graphite to achieve a better heat conduction.

Success was achieved in isothermal elution with diatomaceous earth (Celite, Chromosorb), especially in fractionations using sample sizes of 50 g (Henry, 1959) and 500 g (Kenyon et al., 1965). Spatorico and Coulter (1973) carried out their Baker-Williams fractionation on silica gel packings.

The support material employed for the sample bed is usually the same as that for the column packing. In most cases, following the example given by Baker and Williams, 0.3 g of polymer are applied to 30 g of support material, i.e., a ratio of 1:100 is used. Hall (1959) stated that overloading occurred for a proportion smaller than 1:40. Alvariño et al. (1978), using a bimodal polystyrene sample, obtained the best resolution at 1:50.

In the fractionation of polyethylene, Kenyon and Salyer (1960) found that in the column a surface area of the support material of 50 m^2 per gram of polymer is required for a successful separation.

The application of polydienes to glass beads is diffcult, because the mixture agglomerates. On diatomaceous earth, the adsorption tendency of which was saturated by a preloading with a highly macromolecular polydiene, Hulme and McLeod (1962) successfully introduced polybutadiene samples into the sample bed. The column was packed with glass beads.

11.10. Supports with a chemically modified surface (bonded phases)

The designation "(chemically) bonded phases" contrasts these supports with the adsorbents having physically fixed liquid films, which are used in liquid-liquid partition chromatography. The conditions to be fulfilled in order that films of this kind may not be removed by the mobile phase are stated in Section 9.4. The *chemical anchoring* of the chromatographically active molecular layer on the support eleminates the stability problem. In 1969, HALÁSZ and SEBESTIAN introduced silica supports with chemically bonded organic residues. Reviews of the rapidly growing literature in this field are given by LOCKE (1973), PRYDE (1974), MAJORS (1976), GRUSHKA [A 24], GRUSHKA and KIKTA (1977), and COLIN and GUICHON (1977).

11.10.1. Preparation of chemically fixed coatings

Bonded phases can be prepared according to the principles summarized in Table 11-8. Because of the extraordinary stability of the Si—C linkage, silanization with chloro- or alkoxysilanes as indicated under III is preferable (AUE and HASTINGS, 1969; LOCKE et al., 1972; PRYDE, 1974).

Procedures for the bonding of octadecyl phases are described by KIRKLAND (1975) and by HEMETSBERGER et al. (1977). UNGER (1969, 1976) reported on silanization without any solvent. The reactivity of the chlorosilanes is higher than that of the alkoxy derivatives (ENGELHARDT and MATHES, 1977). For steric reasons, monofunctional silanes can only react with isolated hydroxyl groups. However, as the complete coverage of the silica gel surface is decisive for an optimum chromatographic behaviour of the bonded phases, in most cases di- or trifunctional silanes are used. These compounds can also fix themselves to the so-called reactive hydroxyls, e.g., by the reaction of two chlorine atoms with the two adjacent surface groups. For steric reasons the third chlorine cannot find a co-reactant at

Table 11-8
Bonded phases on surfaces with Si—OH groups: Principle of preparation and properties

	Type and course of the reaction	Stability
I		hydrolyzable
II	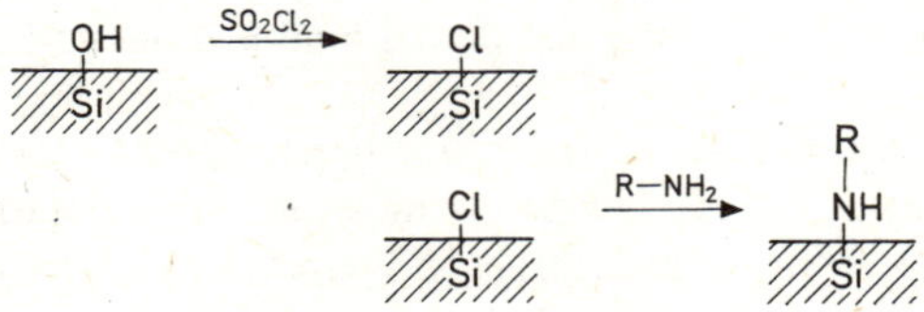	stable in the interval $4 \leqq pH \leqq 7$
III	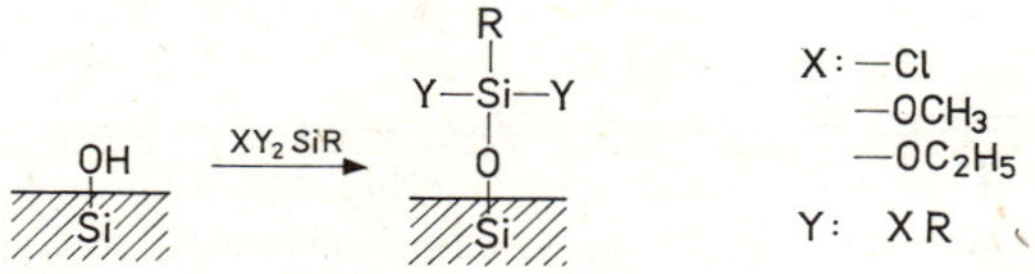	stable in the interval $1 \leqq pH \leqq 8$

the surface (UNGER et al., 1976). The silanization by either methyl dichloro octadecylsilane or trichloro octadecyl silane yield fully consistent results if the reaction is carried out under absolutely anhydrous conditions and followed by a TMCS treatment (KARCH et al., 1976). If the reaction of the di- or trifunctional silanes is not carried out with a complete exclusion of moisture, siloxane polymers may also be produced. On the other hand, excessive functional groups, which are preserved during the build-up of the bonded phases, may later be hydrolyzed to SiOH groups. The latter, like possibly unreacted silanol groups on the surface, develop their own chromatographic activity, which is superimposed on the intended effect of the bonded phase, and usually interferes. This leads to tailing peaks in the chromatogram. For this reason, and because of the risk involving the formation of polymer layers (mentioned in the following section), at present *monofunctional* silanes are preferred in the preparation of RP 18 phases, and the unreacted siloxane groups are removed by capping (see below) (MAJORS, 1980).

In the *methyl red test* according to SHAPIRO and KOLTHOFF (1950), silica gels with unreacted or uncovered silanol groups can be identified by their red-violet colouring, caused by a solution of this dye in benzene. The colour cannot be removed by washing with benzene. However, VERZELE et al. (1979) found that this test indicates the general acidity of the separating material rather than the presence of free silanol groups, depending, among other things, on whether the supports have been treated with alkali. UNGER et al. (1976) checked the quality of reversed-phase material by the chromatogram of a polar sample in a non-polar eluent: if a symmetric peak without any retention was obtained, then the complete conversion of the SiOH groups was assumed.

For this test it is necessary to flush the reversed-phase column very carefully in order to remove any traces of polar eluents. TANAKA et al. (1977) flushed their columns ($L = 0.15$ m; $d_C = 4.6$ mm) with at least 100 ml of methanol, 150 ml of THF and 200 ml of n-heptane. Then the polar solute was injected in as small a quantity as possible (<0.1 µg), and this injection was repeated after each following volume of 100 ml n-heptane until there was no further increase in retention. The solutes employed were anisole, methyl benzoate or acetophenone.

If the packing material contains entirely unreacted particles, they can be separated from the hydrophobized ones by floating with water (PASTUSKA and JUST, 1979). Residues of silanol groups can be capped by trimethylmonochlorosilane (TMCS) according to the pattern shown in Fig. 11-5 (KNOX and PRYDE, 1975). This reagent also allows an *in-situ silanization* or the resilanization of packed columns (BATHER and GRAY, 1976; KIRKLAND and ANTLE, 1977). OLIEMAN et al. (1981) reported the regeneration of deteriorated RP 18 columns by *in situ silylation* with chlorodimethyloctadecylsilane in toluene ($0.1 \text{ g} \cdot \text{l}^{-1}$) at temperatures of 30 and 48 °C.

For the conversion of silica gel with octadecyltrichlorosilane, LITTLE et al. (1979a, b) investigated the influence of the solvent and the suspension medium and found that the water-miscible ones yield markedly poorer results. Coatings prepared in carbon tetrachloride exhibited the highest chromatographic resolution. The reaction mixture (5 g Partisil®-10 in 50 ml of 1% silane solution) was allowed to stand at room temperature for over 24 hours, with occasional shaking. From a conversion vs. time curve given in this paper (Fig. 11-6a) it can be observed that even in toluene, which has a slightly lower efficiency, the conversion is already complete after 2 hours. In this suspension agent, refluxing is inefficient. Some effect can be observed in methanol, where the reaction proceeds much more slowly. In this case the coating actually takes place with octadecyltrimethoxysilane, which is produced in a pre-

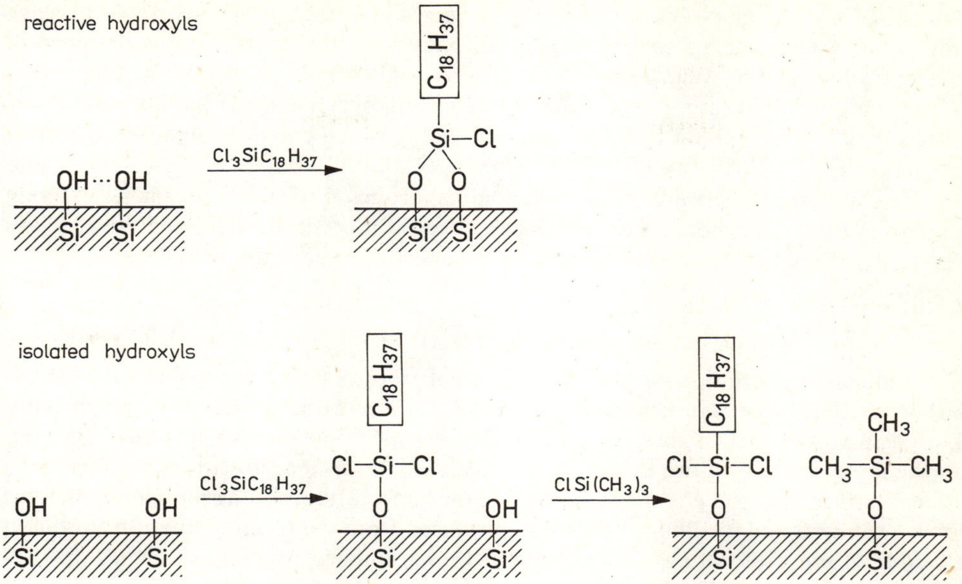

Fig. 11-5
Reaction of trichloralkylsilane with reactive or isolated hydroxyl groups and capping of unconverted isolated hydroxyls by trimethylmonochlorosilane (TMCS)

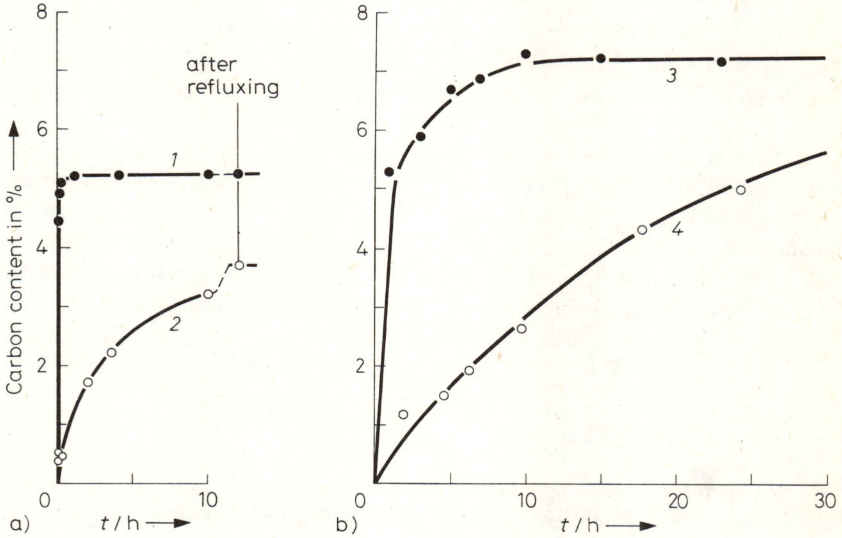

Fig. 11-6
Increase of the coverage as a function of time in the reaction of silica surfaces with silanes

a) Reaction of $Cl_3SiC_{18}H_{37}$ with Partisil-10 at room temperature: 1 in toluene; 2 in methanol (according to LITTLE, WHATLEY, DALE and EVANS, 1979b)

b) Reaction of $(C_2H_5O)_3Si(CH_2)_3NH{-}CO{-}CH_3$ with LiChrosorb Si 100 at 65 °C: 3 in benzene, 4 in tetrahydrofuran (according to ENGELHARDT and MATHES, 1977).

ceding reaction. The rate of reaction of alkoxysilanes with the surface hydroxyls of silica in the polar, water-miscible solvent tetrahydrofuran does not reach in value obtained in toluene (see Fig. 11-6b). WICKRAMANAYAKE and AUE (1980) observed that the conversion increases with the temperature; in a silylation carried out over 6 hours in boiling hexadecane (287 °C) they achieved about twice the carbon loading which could be obtained in boiling octane (126 °C).

The best way to synthesize layers with functional groups is to start with silanes carrying the appropriate groups, because a subsequent chemical modification does not always give good results. These silanes required can be prepared by the addition of functional olefins to trichlorosilane:

$$Cl_3SiH + CH_2{=}CH{-}CH_2{-}X \rightarrow Cl_3Si{-}(CH_2)_3{-}X$$

This reaction course presupposes that the functional group X does not react with the chlorosilane bonds. This is the case with nitrile; other functions can be masked by protective groups. Alkoxysilanes with their higher general stability are more suitable for this Si-C synthesis than the more reactive chloro derivatives. The difficulties in the subsequent anchoring on the silica surface can be overcome by increasing the temperature, adding p-toluenesulphonic acid as a catalyst, and distilling the alcohol produced from the reaction mixture, so that it is possible to prepare coatings just as good as those obtained from chlorosilanes (ENGELHARDT and MATHES, 1977).

Apart from the common techniques of surface modification, support materials of this kind can also be prepared by a cohydrolysis of tetraethoxysilane, polyethoxysilane and organoethoxysilane as a modifying agent (UNGER et al., 1976). This *bulk modification* yields porous, modified silica gels which exhibit higher capacity factors, and hence are of interest especially for the investigation of traces of organic substances in aqueous media; the selectivity values are approximately the same as those of a surface-modified silica gel. This is plausible, because in both cases the interface with the mobile phase has the same structure.

A different kind of surface modification is the deposition of carbon, e.g., by the pyrolysis of benzene vapour at 800 °C (BARMAKOVA et al., 1974). As the pyrocarbon coating increases, silica gel, which is hydrophilic by nature, is gradually converted into an extremely hydrophobic support material (KISELEV, 1976). COLIN et al. (1978a, b) investigated such products with respect to their behaviour in HPLC. As in the case of graphitized carbon black, it was found that these adsorbents exhibit only a low linear capacity, which is smaller than that of silica gel by nearly one order of magnitude.

11.10.2. Properties

For the chromatographic properties of the supports it is of importance that the bound layer does not seal the originally existing pore system. For small pore silica gels it is quite necessary to consider this possibility. An unbranched chain with eighteen carbon atoms, having a straight length of almost 2.3 nm, reaches the order of magnitude of the pore radii. In 1979 ENGELHARDT and MATHES, investigating the reaction of silica gels of different pore sizes with fourteen different silanes, found that the surface concentrations achievable in pores with diameters of more than 10 nm correspond to each other rather closely, whereas only about 90% of this value could be achieved in 6 nm pores. For silica gels with pore diameters ranging between 10 and 1800 nm, WICKRAMANAYAKE and AUE (1980) found that in the total range every doubling of the pore size increases the loading by 15—20%. In 1976, KARCH et al. determined the total porosity of columns packed with C_{18}-coated silica gel with an original pore size of $d_0 = 6$ nm. For methanol and water used as eluents, they found values of 0.43 and 0.49, respectively, which are typical of porous layer packing materials (cf., Section 10.4.4.). Shorter "bristles", on the other hand, effect a much smaller change in the existing porosity: Likewise for 6 nm silica gel as a support, but coated with $Si—(CH_2)_3—NH—CO—CH_3$, ENGELHARDT and MATHES (1979) determined an internal porosity of $\varepsilon_P = 0.25$ and a total porisity of $\varepsilon_t = 0.71$ for the column.

Table 11-9 shows a survey of the bonded phases currently in use. Besides alkyl phases, cyano or amino bonded phases find a widespread application, representing a useful alternative to silica gel in gradient elution, because the equilibrium with the mobile phase is attained more rapidly. The arrangement in Table 11-9 represents a scale of polarities which ranges from the hydrophilic adsorbents, including original silica and alumina, to the extremely hydrophobic supports with aliphatic coatings. The last-mentioned supports are usually called *reversed phases*, because they retard samples different polarities in an order reverse to that of the so-called *normal phases*, which include silica and alumina. These designations have no fundamental meaning, for there is such an abundance of support materials which can be arranged between the antipodes that an almost uninterrupted scale of polarity would result (HUBER, 1976). Moreover, it was more or less by chance that systematic investigations were initially (and for some time exclusively) carried out with polar adsorbents. If they had been started with graphite or the like, then the elutions orders observed in this case would surely have been designated as "normal". This must not be overlooked when the supports with the bound alkyl layers are briefly designated as RP 8 or RP 18.

The various bonded phases often show remarkable differences in selectivity, the causes of which are still rather obscure. According to experience there is little chance of finding a universal column by means of which all separation problems can be solved equally well. It is more advantageous to choose, from the abundance of possibilities offered by the scale of polarities, a stationary phase which enables a specific problem to be treated with optimum success. However, the essential factor in reversed-phase chromatography is the mobile phase; cf., Section 7.6.

From the kinetic behaviour it can be concluded that bonded phases with substituents arranged in a bristle-like manner ("*molecular fur*", HORVÁTH and MELANDER, 1977) become effective by interfacial processes. Bristles are produced by the reaction of monochlorosilanes with the surface hydroxyls, whereas di- or trifunctional silanes produce such a structure only if water is completely excluded. In the presence of water, hydrolytic condensation of the excessive functional groups produces polymer layers with a honeycomb structure. Upon an abrupt change of the mobile phase, the bulky polymer coatings reach the new state of equilibrium sooner than bonded phases with monomeric bristles, where an organic solvent such as methanol requires up to 1 hour to leave the fur by diffusion. This was observed by SCOTT and SIMPSON (1980) in experiments where they suddenly changed the solvent from 100% methanol to 100% water. This finding likewise indicates the influence of the mobile phase in RPC. Naturally, in water there were also dispersive interactions between the hydrophobic aliphatic chains, and corresponding conformational changes.

The chromatographic efficiency of the bonded phases depends on several factors:

— Structure and polarity of the functional groups

Table 11-9
List of commercial adsorbents in the order of decreasing hydrophilic property

For comparison, the last column shows the increment values which the polar end groups contribute to the adsorption behaviour or Al_2O_3 (cf., Table 7-1).

Designation of the chromatographically active phase	Structure	Relative retention of acetophenone (benzene = 1)	Polarity increment of the outer group(s)
a glycol glycerolpropyl	$\geq Si{-}(CH_2)_3{-}O{-}CH_2{-}CH(OH){-}CH_2(OH)$		13
b ether			
silica (naked)		10.5	
alumina (naked)			
c amino aminopropyl	$\geq Si{-}(CH_2)_3{-}NH_2$	6.3	6.24
d nitro	$\geq Si{-}(CH_2)_3{-}NO_2$		5.40
e oxynitrile	$\geq Si{-}(CH_2)_3{-}O{-}CH_2{-}CH_2{-}CN$		
f nitrile, cyano	$\geq Si{-}(CH_2)_3{-}CN$	8.1	5.00
g aminoaryl	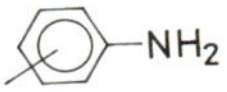		4.41
h nitroaryl	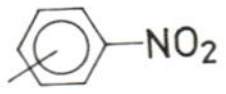		2.75
i dimethylamine	$\geq Si{-}(CH_2)_3{-}N(CH_3)_2$		
j allylphenyl	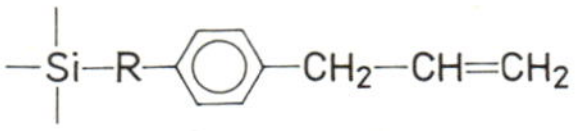	6.1	
k phenyl diphenyl	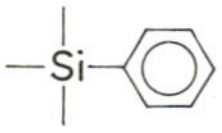	5.6	1.86
l fluoroether		4.9	
m methyl	$\geq Si{-}(CH_3)_2$		
n n-hexyl	$\geq Si{-}(CH_2)_5{-}CH_3$		−0.03
o n-octyl	$\geq Si{-}(CH_2)_7{-}CH_3$		−0.03
p n-octadecyl	$\geq Si{-}(CH_2)_{17}{-}CH_3$		0.03

— Structure and length of the chains by which the functional groups are fixed to the particle surface
— Pore size, porosity and accessible surface of the supports after (and before) the application of the stationary phase
— Coverage of the surface

The structure and the polarity of the functional groups represent the criterion by which the bonded phases are ordered, leading to the scale of polarity mentioned above. However, conclusions should be drawn with care because in most cases the other factors are also of influence.

The analysis of bonded organic residues by a hydrolysis of the silica gel with 2N KOH, a conversion of the hydrolysate with trimethylsilylimidazole and a separation by capillary gas chromatography was described by VERZELE et al. in 1980. HANSSON and TROJER (1981) performed column GC after pyrolysis at 800 °C for 5 s. They succeeded in characterizing the hydrocarbon chain-length as well as the functionality of the silanizing agent and in deciding whether a post-silanization had been performed or not. Their paper also contains a list of earlier work on the characterization of organic material bonded to silica particles.

With the aliphatic bonded phases the effect of the chain-length can be studied quite directly. In 1976, KARCH et al. found an increase in the capacity factors and the chromatographic resolution with increasing length of the bristles. The samples employed were aliphatic hydrocarbons, alcohols and phenols. SCOTT and KUČERA (1977) obtained an analogous result with aromatic solutes. LOCHMÜLLER and WILDER (1979) found a continuous increase of the capacity factors with the length of the bonded alkyl chain only in cases where solutes with large molecules, e.g., polyaromatics, were used, whereas they observed a limit when using small-sized solutes. BERENDSEN (1980) likewise found a limit and determined the *critical chain-length* from the course of the log k vs. RP chain-length curves. This critical chain-length is the point of intersection defined by the extension of the steep starting end of the curve with the extension of the horizontal (or even declining) branch of the curve at long RP chain-lengths. The use of solute molecules which were larger by one methyl or phenyl group increased the value of the critical chain-length by an increment of 0.5 and 1.3, respectively.

LITTLE et al. (1978a) carried out a column screen with eight test mixtures. Their report lists eighteen different bonded phases, which were eventually arranged in the order of the resolutions achieved for each test mixture. In six cases the RP 18 columns took first place. However, for the arabinose/xylose/fructose/sucrose test mixture and in the separation of two azo dyes in methanol-water (90:10) they came off the worst. The linear increase of the separating efficiency of hydrophobized supports with the increasing length of the aliphatic chain, which had been observed for the other test mixtures, could be shown to occur beyond the usual threshold of C 18 up to C 22 (LITTLE et al., 1978b).

As regards the influence exerted by the pore system, two aspects at least to be discussed: if bonded phase chromatography (BPC) is based on interfacial effects, then naturally the capacity of the packing, and hence its separation efficiency, increases with the surface area. Hence from this aspect small pore starting materials should be preferred to (too) large-pore ones. If, however, the bristle length is comparable with the pore radii, then the coating decreases the interior porosity. These effects, counteracting one another, lead to an optimum value of the primary pore diameter, which depends on the length of the bristles and is about 10 nm. If supports with bonded phases are to be used in size exclusion chromatography, then of course the pore system, which is required for the separation problem in question, must still be available after the silanization.

For size exclusion chromatography, inorganic supports with hydrophilic coatings are of greatest interest. Most of the water-soluble polymers, especially the biopolymers, can successfully be separated by gel filtration, but the organic gels used for this purpose are porous due to swelling and are not sufficiently pressure-resistant for HPLC. Although the most important inorganic adsorbents such as silica gel, alumina or glass are hydrophilic by nature,

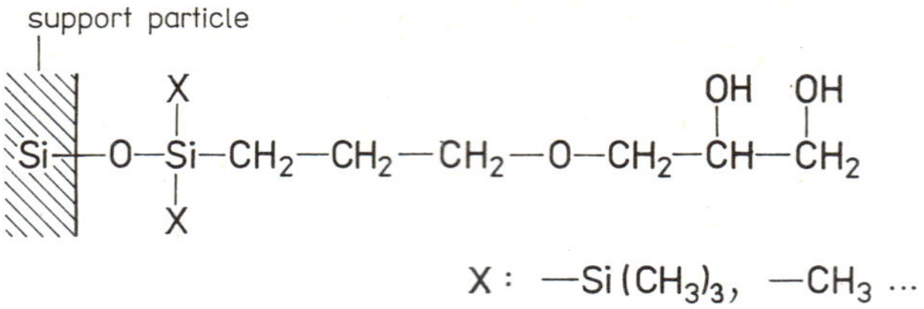

Fig. 11-7
Glycolic ether as a hydrophilic bonded phase on silica gel or porous glass ("glycophases")

they can hardly be used for the steric exclusion chromatography of biopolymers, because the samples are adsorbed or catalytically altered. ELTEKOV et al. (1973) modified a macroporous silica gel with an average pore diameter of 100 nm by reaction with γ-aminopropyltriethoxysilane, so that proteins were no longer irreversibly adsorbed by the gel. However, it remained hydrophilic enough to still be suitable for use in the SEC of lysozyme and cytochrome. The break-through to pressure-resistant, hydrophilic materials for steric exclusion chromatography was initiated by the *glycophases* developed by REGNIER and NOEL in 1976. These contain the structures shown in Fig. 11-7 on porous glasses or silica gels. The surface of the *TSK SW gels*, for which little structural information has been given till now, is likewise covered by hydroxyl groups. The basic skeleton is formed by silica gel (FUKANO et al., 1978). The TSK SW gels (cf., Table 11-10) are suitable for use in the range pH 3—8, and they are employed in the prefabricated columns made by many firms due to their excellent separation efficiency ($h^* = 6$ for $d_p = 10$ µm), their inertness, which allows an almost quantitative protein recovery, and their resistance to pressures of up to more than 10 MPa (WEHR and ABBOTT, 1979). Their separating power covers a wide range of molar masses (WICHMAN and PAVLU, 1980).

The importance of pressure-resistant supports for steric exclusion chromatography in aqueous media is also reflected by the systematic investigations of ENGELHARDT and MATHES (1977, 1979). In addition to the diols mentioned above, the authors fixed twelve other hydrophilic residues to silica gel by silanization and characterized the supports with respect to the coverage achieved and to their chromatographic efficiency. The best results were obtained by means of the coatings listed in Table 11-11.

Table 11-10
Types of TSK GEL SW columns/TOYO SODA MANUFACTURING CO.

Type	Particle size	Separation range[1])			
		Globular protein	Dextran	Polyethylene glycol	Plate height
	µm	10^3 g · mole^{-1}	10^3 g · mole^{-1}	10^3 g · mole^{-1}	µm
G 2000 SW	10	5— 100	1— 30	0.5— 15	≦60
G 3000 SW	10	10— 500	2— 70	1 — 35	≦60
G 4000 SW	13	20—7000	4—500	2 —250	≦60

[1]) according to Y. KATO et al. (1980)

Table 11-11
Hydrophilic bonded phases on LiChrosorb® SI 100 silica gel, which are suitable for steric exclusion chromatography in aqueous media (according to ENGELHARDT and MATHES, 1979)

Designation	Structure	Surface concentration µmole · m^{-2}	Mean surface area covered by one group nm^2
Diamine	$\geq Si-(CH_2)_3-NH-(CH_2)_2-NH_2$	2.5	0.66
Amide	$\geq Si-(CH_2)_3-NH-CO-CH_3$	4.4	0.37
Glycinamide	$\geq Si-(CH_2)_3-NH-CO-CH_2-NH-CO-CH_3$	3.2	0.52

The coverage of the inorganic support, has a significant influence on the properties of the bonded phases. This fact has been clearly stated by ENGELHARDT and MATHES (1979): "Reversed phases only show identical selectivity if their surface coverage is identical. Therefore, it is only possible to attribute a certain selectivity to a functional group of a bonded silane if the surface coverages of the phases to be compared are identical."

Incomplete coverage means that residual silanol groups may be accessible for solute molecule under certain conditions. Silanol groups are strongly polar and will therefore cause normal phase retention.

NAHUM and HORVÁTH (1981) as well as BIJ et al. (1981) discussed the *dual retention mechanism* which is due to the superposition of solvophobic and silanophilic interactions. Although the latter may sometimes favourably influence the chromatographic selectivity, BIJ et al. recommended the addition of a silanol-masking agent (such as amine, for example) to the eluent in order to obtain reproducible results, which can eventually facilitate the development of an elution strength system (cf., Section 7.6.).

In any case it should be ensured that the chromatographic process is not disturbed by free silanol groups, i.e., only those reversed-phase materials which do not exhibit a methyl red adsorption should be used. NICE and O'HARE (1978) as well as O'HARE et al. (1980) determined the latter for some commercial RP 8 materials. Among the parameters investigated (d_P, ^1A, N, carbon loading, wettability), the methyl red adsorption was the only quantity which correlated with the rather large differences of the various packings in the resolving power obtained in the separation of some steroids.

The effects which result from the interaction with the eluent must be added to the properties caused by the stationary phase itself. Comparisons between different bonded phases can be conclusive only if this influence is taken into account or eliminated by means of uniform mobile phases.

Bonded phases for the exclusion chromatography of water-soluble polymers must be wetted by water. This is the case with the structures mentioned in Table 11-11 and the glycophases (Fig. 11-7). These materials sink in water, whereas non-wettable ones float on the surface. Silica gel with a coating of $\geq Si-(CH_2)_3-NH-CO-CF_3$ floats on water.

It is wetted if at least 10% of methanol is added, whereas RP 18 requires a minimum of 60% methanol or 40% acetonitrile (ENGELHARDT and MATHES, 1977).

For the alkylsilane phases, the permissible sample amount increases with the length of the carbon chain, reaching values which are much higher than those of the naked silica gel. For example, an RP 18 column had ten times the capacity of a reference column containing the raw material.

Other advantages of the bonded phases lie in their ability rapidly to reach equilibrium with the eluent, and the fact that they are not dependent on traces of water in the eluent, which have a considerable influence especially on the activity of silica gel. Therefore bonded phases are most suitable for use in gradient chromatography. Because of their versatility and reliability, they are used in about 80% of all HPLC separations. In some cases care should be taken to avoid *damage by liquids* used as eluents or in the slurry technique (cf., Section 17.1.3.). Thus, for instance, bonded amino phases can react with acetone or other ketones and aldehydes, forming Schiff bases. Amino phases are susceptible to oxidation, so that for this reason, too, the mobile phase must not contain any dissolved oxygen nor, of course, any peroxides. In pure water, bonded amino phases may also become impaired, because they give rise to a pH shift into the alkaline region. Therefore, as a matter of principle, buffer solutions or an addition of acid are recommended for amino phases.

11.10.3. Polymer layers on inorganic support particles

Among the numerous stationary phases commercially available there are also porous layer beads with a coating of polycaprolactam (Polyamide 6), on which a separation takes place due to the ability to form hydrogen bonds (RABEL, 1973). Such supports are most suitable for the analysis of phenolic components. A technique for the conversion of silanol groups with vinyltrichlorosilane for a subsequent deposition of layers by polymerization, which can be carried out with different acrylate monomers, has been described by WHEALS (1975).

As polymer molecules are irreversibly adsorbed on solid surfaces (cf., Chapter 18), firmly adherent layers can also be prepared without a chemical bond (AUE et al., 1973; DANIEWSKI and AUE, 1978). SAGE (1976) also succeeded in depositing chromatographically active polymer layers from solutions on glass beads 44 to 50 μm in diameter. Systematic studies of polymer coatings on silica gel were carried out by LECOURTIER et al. (1978a, b, c, 1979). These investigations showed that a partition of the sample components between the mobile phase and the (swollen) polymer phase occurs if there is an adequate similarity. The rather low diffusion coefficients in the polymer matrix cause an undesired retardation of the substance (cf., Section 15.3.1.). In this context it is also of importance whether the polymer coats the interior surface of the pores of the silica gel (Fig. 11-8a) or seals the pores like a cover (Fig. 11-8b). The former structure was indicated by LECOURTIER et al. (1979) for polyethylene oxide on silica gel, the latter one for polystyrene on the same adsorbent. By grafting L-proline on linear polyacrylamide, the same team prepared a chiral polymer which, after adsorption on Spherosil® and complexing with copper ions, was suitable for use in the racemate separation of α amino acids (BOUÉ et al., 1980). The advantage of this variant of ligand-exchange adsorbents was the low swelling capacity, which allowed the use in different solvents.

Organic polymers (gels) are largely used for the preparation of the chromatographic bed in exclusion chromatography. They are dealt with in Chapter 12. Nevertheless, in connection with the coated supports discussed here, some remarks concerning the use of polymer

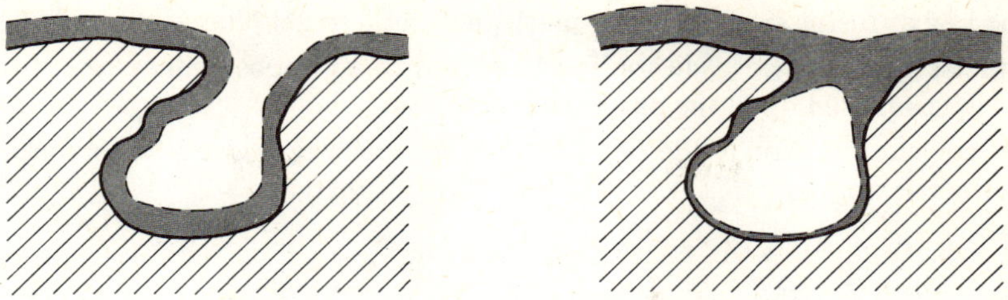

a) b)

Fig. 11-8
Models for the formation of polymer layers on porous supports
a) polyethylene oxide on silica gel
b) polystyrene on silica gel
(according to LECOURTIER, AUDEBERT and QUIVORON, 1979).

particles as a support in adsorption chromatography may be relevant here. These special materials include the polyamide powder which is available in the size ranges from 5 to 20 μm and from 20 to 32 μm. The cross-linked cellulose acetate with acetyl contents of 30 or 40% which is available with a particle size between 15 and 35 μm, is suitable for polycyclic aromatic hydrocarbons.

Relatively cheap but yet rather efficient columns for the preparative separation of bacteriochlorophylls were prepared by CHOW et al. (1978), using a commercial polyethylene powder with a particle size between 3 and 45 μm (mean value 9 μm) as a packing material. The separations were carried out in the usual way for alkane-covered bonded phases. According to the authors' information, the cost of such bonded-phase material would have been $ 50,000 for a 2.75 m column with an internal diameter of 9 cm.

Some types of cross-linked polystyrene have been developed especially for reversed-phase chromatography, a particular advantage being their stability to alkaline eluents (cf., Section 12.1.1.).

Mention should also be made of the chiral synthetic polymers for the separation of optical antipodes, which were reviewed by DAVANKOV and SEMECHKIN in 1977. Enantiomeric separations can also be achieved on reversed phases by adding metal chelates to the mobile phase (KARGER et al., 1980; GRUSHKA et al., 1980), or on chiral reversed phases by addition of copper(II) acetate to the eluent (DAVANKOV et al., 1980). In this case, the chiral behaviour was obtained by the adsorption of N-alkyl-L-hydroxypyroline onto the surface of a conventional RP 18 packing material. The alkyl chain (C_7 ... C_{16}) sticks fast in the "molecular fur" and serves as an anchor for the chiral group. A similar modification of plain silica simply by adsorption was observed by GHAEMI and WALL (1979) and by HANSEN (1981): in eluents containing long-chain quaternary ammonium salts, naked silica behaved like RP materials with chemically bonded chains.

The use of polymers as a support or a stationary phase for the separation of polymers was described by DAVE in 1975. This principle, which has so far only been employed in individual cases (NATTA et al., 1958; PINO et al., 1962, 1966; LANGHAMMER and QUITZSCH, 1961; VAUGHAN and GREEN, 1963), possibly promises solutions for difficult separation problems, but certain restrictions have to be taken into account:

— the incompatibility, which will prohibit the separation of all kinds of substances using one and the same stationary phase

- the low diffusion coefficients of macromolecules in gel phases
- the risk that a steric exclusion mechanism with an opposite direction of the separation is superimposed upon the partition mechanism
- the solvent retention in the stationary phase, which may also lead to interfering changes in the packing volume (VAUGHAN and GREEN, 1963).

12. Organic supports

The organic support materials for liquid chromatography are macromolecular particles which are insoluble in the respective eluents. In most cases this is achieved by covalent cross-linking, but there are also non-cross-linked support materials whose macromolecules in the ordered zones are linked so tightly by intermolecular forces that insolubility is ensured, e.g., agarose, cellulose. Covalently cross-linked polymers are prepared either by copolymerization with divinyl compounds or by the cross-linking of natural macromolecules. In order to achieve beads the processes are mostly performed in a suspension. There are numerous possibilities of preparing organic support materials, which for a good part have already been tested. The statements in this chapter will be restricted to commercial materials, which are used to a greater extent.

12.1. Cross-linked copolymers

Under suitable conditions, the polymerization of a mixture of monomers yields a homogeneous copolymer with a statistical distribution of the various structural elements. If one monomer species is multifunctional, e.g., divinylbenzene (cf., Fig. 12-5), then cross-linked copolymers are obtained because the two vinyl groups are usually inserted into different chains. HILD and REMPP (1980) investigated the mechanism of cross-linking and found that the second vinyl group reacts rather late, being present as an unsaturated side group over a longer time. The reactivity of the second double bond is about $^1/_{10}$ of that of the first one, even in symmetric divinyl compounds (HORIE, 1980). In kinetic considerations of cross-linking copolymerization this is frequently neglected.

To achieve a high cross-linking density for a given concentration of the cross-linking agent, the first copolymerization parameter of the cross-linking agent should be equal to or greater than that of the chain monomer. HEITZ et al. (1977) estimated such values of several potential cross-linking agents for vinyl acetate by the investigation of the corresponding monofunctional model compounds.

In *homogeneously cross-linked copolymers* the cross-links are statistically distributed. The mean network mesh size is determined by the proportion of the cross-linking agent in the polymerization mixture. When in contact with solvents, these networks swell, and the degree of swelling increases with the increasing mesh size of the network and the improving thermodynamic quality of the solvent (see Table 12-1). The larger the mesh sizes of these *gels porous due to swelling*, the higher is their exclusion limit (see Figure 12-1), but the lower is the mechanical strength of the gel granules. Therefore, for reasons of

Table 12-1
Gel bed volumes of several organic gels in various solvents, in $cm^3 \cdot g^{-1}$

Solvent	Fractogel® PGM 2000	Fractogel® PVA					Sephadex® LH 60	Spheron® 300
	Exclusion limit in 10^3 g · mole^{-1}							
	2[1])	0.5[2])	2[2])	6[2])	80[2])	1000[2])	10[3])	700[3])
Tetrahydrofuran	6.8	3.2	4.7	7.3	6.1	5.3	9.8	4.5
Ethyl acetate	—	3.6	4.9	9.7	7.8	13.2	3.1	4.3
Dimethyl sulphoxide	—	—	—	—	—	—	13.6	4.9
Dimethylformamide	4.3	3.2	4.7	7.4	5.5	6.1	13.1	4.9
Acetonitrile	—	—	—	—	—	—	3.2	4.5
Benzene	9.5	3.2	4.7	7.8	7.6	5.0	2.5	3.7
Trichlorobenzene	—	3.0	4.0	6.0	5.1	7.0	—	—
Acetone	—	3.5	4.7	5.4	4.8	5.4	5.7	4.4
Methanol	4.3	2.6	3.2	3.3	3.6	3.7	12.1	4.7
Ethanol	—	—	—	—	—	—	12.2	4.9
Chloroform	6.0	4.0	5.8	9.6	7.2	7.0	12.5	—
Water	4.5	—	—	—	—	—	12.6	4.3
Toluene	7.5	3.0	3.9	6.0	4.8	4.6	2.0	3.6
Dichloromethane	—	3.8	5.4	9.0	6.4	6.0	11.2	—
Carbon tetrachloride	—	3.0	3.6	4.0	2.1	2.4	2.0	3.3
Cyclohexane	—	1.8	2.2	1.7	1.6	2.0	—	—

[1]) Exclusion limit determined with polyethylene glycol samples; eluent: water.
[2]) ditto, but PS in THF
[3]) ditto, but dextran in water

stability homogeneously cross-linked gels must be built up with a proportion of the cross-linking agent of at least 1%. Their pores are accessible for molecules up to about $M = 5000$ g · mole^{-1}.

The exclusion limit of homogeneously cross-linked gels is proportional to their specific gel bed volume, ${}^1V_G$. For homogeneous copolymers of styrene-divinylbenzene, of methyl methacrylate-ethylene glycol dimethacrylate, of vinyl acetate-butanediol divinyl ether and of vinyl acetate-divinyl adipate, one has consistently

$$M_{\lim} = 500\,{}^1V_G - 850 \tag{12-1}$$

in the range $1.7 = {}^1V_G \leqq 10.0$ (Heitz, 1970).

For the fractionation of larger-sized molecules by the principle of steric exclusion it is possible to use *heterogeneously cross-linked gels*. The latter are produced by cross-linking copolymerization in the presence of inert diluents. These have to be miscible with the starting monomers and must not dissolve in the aqueous phase if the gels are to be prepared as beads by suspension polymerization. While for homogeneous copolymers the exclusion limit generally decreases with the increasing content of the cross-linking agent, in a heterogeneous cross-linking process it may even increase with the content of the cross-linking agent (see Fig. 12-2). This is because in heterogeneous gels the pores which are effective in the separation are not contained in the network matrix, but are represented by cavities between agglomerates of microbeads of the rather highly cross-linked skeleton substance. These cavities are largely free of the gel-forming polymer,

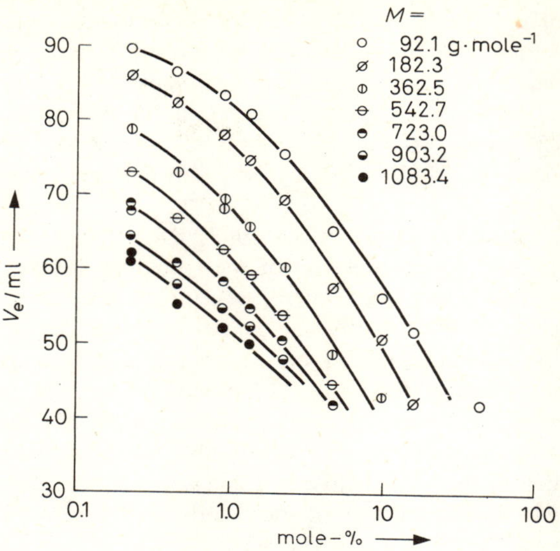

Fig. 12-1
Influence of the content of cross-linking agent (divinyl adipate) in the preparation of vinyl acetate gels on their separation effect on methylphenyl oligomers

Eluent: tetrahydrofuran
(according to HEITZ and PLATT, 1969).

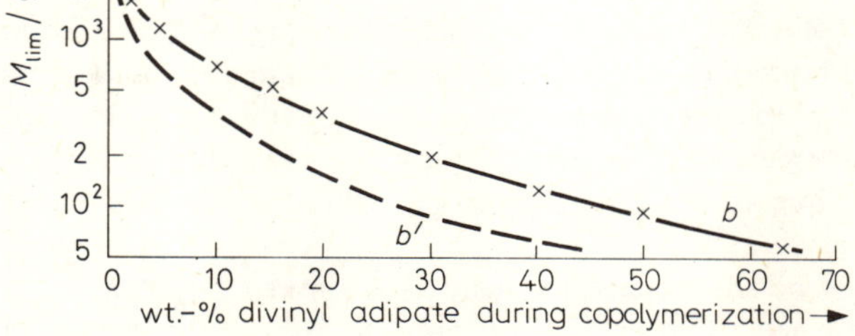

Fig. 12-2
Exclusion limit, M_{lim}, of polyvinyl acetate gels prepared with different concentrations of cross-linking agents

a : Gels prepared in the presence of 67 vol. % n-butyl acetate
b : Gels prepared without any inert diluent
b′: curve b plotted vs. mole-% divinyl adipate
(according to HEITZ, 1970).

whereas the polymer material in the pore walls hardly contains any solvent because of its high degree of cross-linking. It is these *intentionally* introduced local differences in the polymer density which are meant when heterogeneous cross-linking is spoken of in this chapter.

The composition of a gel by the skeleton substance and the cavities can be clearly observed on an electron micrograph (MOORE, 1964; ALTGELT and MOORE, 1967), but the holes on the gel surface cannot directly be related to the chromatographically detectable porosity (CRIGHTON and PEAKER, 1978).

These structures develop as follows (SEIDL et al., 1967; KUN, KUNIN, 1968): a cross-linking polymerization performed in the presence of inert mixture components initially produces networks in submicroscopic ranges (microgels), into which the diluent can be embedded to a certain degree. If the starting mixture contains more of the diluent than the amount corresponding to the swelling capacity of the microgels, phase separation will occur. In a certain composition range (II in Fig. 12-3b) close to the boundary between homogeneous and heterogeneous gel formation, inert liquid with a low monomer content is segregated from the gel matrix in microscopic droplets after the gel point is exceeded. However, the pores preformed in this way reclose in the usual post-treatment procedures (HÄUPKE and PIENTKA, 1974). If the content of inert substance or cross-linking agent is slightly higher, the segregation will lead to the separation of polymer-rich microgel particles with a reduced content of diluent, while a correspondingly more dilute solution is left as a continuous phase. The interfacial tension condenses the soft gel particles into microspheres which subsequently agglomerate while the medium in the interspaces becomes gradually poorer in monomer. The time when the phase separation occurs depends on the content of cross-linking agent, the quantity of the inert component and very much on the solvating power of the latter for the polymer produced. If a very large amount of diluent is present, the phase separation occurs with rather loosely reticulated gels, i.e., in this case the amount of the cross-linking agent may be relatively small. On the other hand, if the polymerizing system contains only a small quantity of inert medium, then narrower meshes and, consequently, a larger quantity of cross-linking agent are required. Fig. 12-3a shows this relationship, at the same time illustrating the influence of the solvent quality of the inert component: if, for instance, the precipitant heptane is used, the phase separation is achieved sooner than with toluene as solvent.

Very high additions of an inert component lead to flocculent polymers with particle diameters between 1 and 100 µm, which do not exhibit any measurable capillary surface (HÄUPKE and HOFFMANN, 1968). Consequently the range of macroporous copolymers also exhibits an upper boundary. Fig. 12-3d shows that, in the application of different inert media, the ranges of existence keep their shape but shift to different locations.

The heterogeneous structure of gels polymerized in the precipitation range causes the gel granules to appear opaque under the microscope, while gels which are porous due to swelling are transparent.

The permanent pores of the heterogeneously cross-linked copolymers also cause the apparent density of these gels to be as low as about 30—50% of the density of the polymer matrix. In good solvents the cross-linked wall material swells slightly, but the total increase in the volume of the gel granules is smaller than that of homogeneously cross-linked gels of the same cross-linking density, because the walls expand partially at the cost of the macropores.

In exceptional cases heterogeneously cross-linked copolymers acquire the porosity decisive for their efficiency in separation only by the absorption of a swelling agent, as it was

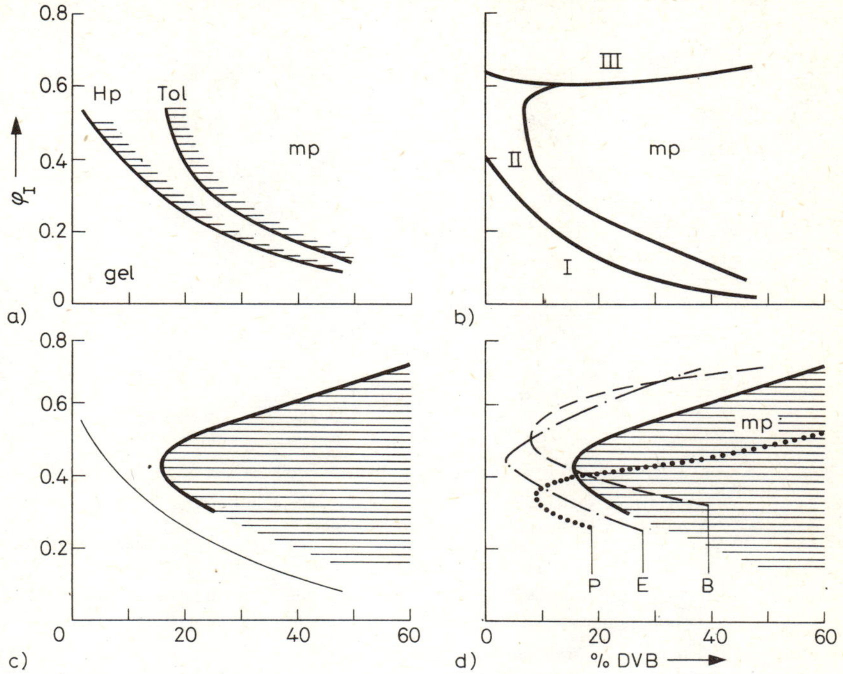

Fig. 12-3
Styrene-divinylbenzene gels: the influence of the content of cross-linking agent and inert solvent on copolymer quality

a) Boundary lines between homogeneous and heterogeneous cross-linking according to MILLAR and KRESSMAN (1965), Hp: n-heptane, Tol: toluene as the inert component I, mp macroporous copolymer
b) Phase diagram according to HÄUPKE and HOFFMANN (1968) for a mixture of aliphatics as an inert component (boiling range 150—200 °C)
c) Upper (and lower) limit of the macroporous range for n-heptane used as an inert component (according to JACOBELLI, BARTHOLIN and GUYOT, 1979). For comparison, the curve Hp from (a) is also indicated.
d) Location of the macroporous ranges for different inert components (suspension polymerization at 88 °C):
I: transparent gel
II: opaque polymer with sealed pores
III: flocculent aggregates with isolated cavities of sizes between 10^3 and 10^4 nm
P pentanol
E ethylhexanoic acid
B benzyl alcohol
as well as n-heptane (cf. (c)).
(according to JACOBELLI, BARTHOLIN and GUYOT, 1979)

The concentration data indicated on the abscissa refer to the monomer mixture, but the data on the ordinate to the total system.

found by HEITZ et al. (1977) with copolymers of vinyl acetate and bismaleimides. However, a heterogeneous cross-linking does not always lead to usable separating materials with the expected high exclusion limits. Thus diluents which would dissolve the wall material, if non-cross-linked, do not in all cases yield usable separating gels for SEC. But even with a precipitating diluent the final result very much depends on the mutual interactions of the components in the polymerizing system.

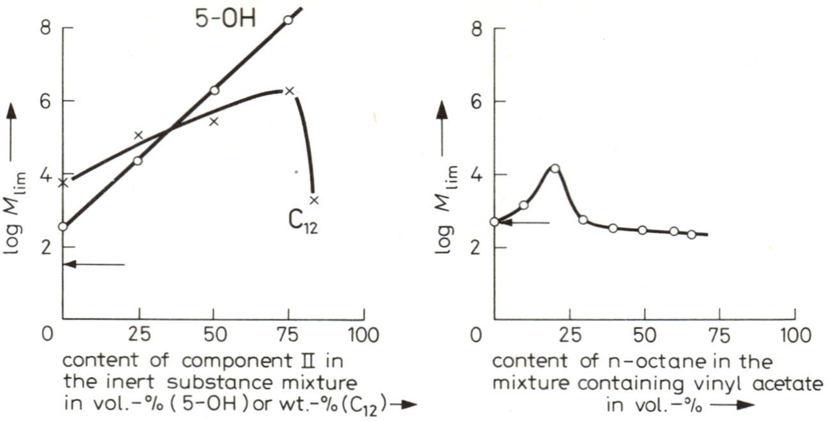

Fig. 12-4

a) Exclusion limit of styrene divinylbenzene gels prepared in the presence of inert solvents

C_{12}: Styrene divinylbenzene (75/25 wt./wt.) with toluene (I)/n-dodecane (II) as an inert medium (I + II: 60% of the total formulation) (according to MOORE, 1964)

5-OH: Styrene-divinylbenzene (45/55 wt./wt.) with toluene (I)/isopentanol (II) as an inert medium (I + II: 66.7% of the total formulation) (according to HEITZ, 1970).

b) Exclusion limit of vinyl acetate gels prepared in the presence of the indicated octane content by copolymerization with 20 wt.-% divinyl adipate (according to HEITZ, 1970)

(The arrows indicate the values of the exclusion limit of the gels prepared without the addition of an inert medium; in Fig. 12-4a: with a divinylbenzene content of 55%; according to HEITZ, 1970).

Fig. 12-4a shows that the exclusion limit of styrene divinylbenzene gels polymerized in the presence of toluene exceeds that of a homogeneous one cross-linked in a solvent-free system. A still higher effect is achieved if the solvent toluene (I) is replaced stepwise by a precipitant (II). For isopentanol as a precipitant the curve rises continuously, whereas for n-dodecane, having passed through a maximum, it slopes down steeply as the dodecane concentration is further increased. A maximum value for the steric exclusion limit also occurs in the cross-linking copolymerization of vinyl acetate as the amount of n-octane added as a diluent is increased (Fig. 12-4b). In all such cases, gels with a thin impermeable skin are produced on the right-hand of the maximum. A possible interpretation is that if the precipitant is not sufficiently polar, the primarily produced gel microspheres accumulate at the phase boundary between the monomer droplets and the aqueous dispersion agent. There they coalesce, finally forming a skin around the gel granule. Polar precipitants such as isopentanol do not exhibit such a displacement effect and allow the formation of structures with accessible pores, provided that the suspension polymerization is a pure one. A mixed mechanism including an emulsion polymerization may also lead to skin formation (HÄUPKE, 1980).

A review of the general aspects of polymer networks is given by HEITZ (1977, 1979).

12.1.1. Cross-linked polystyrene

Cross-linked polystyrene with a structure shown schematically in Fig. 12-5 results from the copolymerization of styrene with divinylbenzene. Technical divinylbenzene is a mixture.

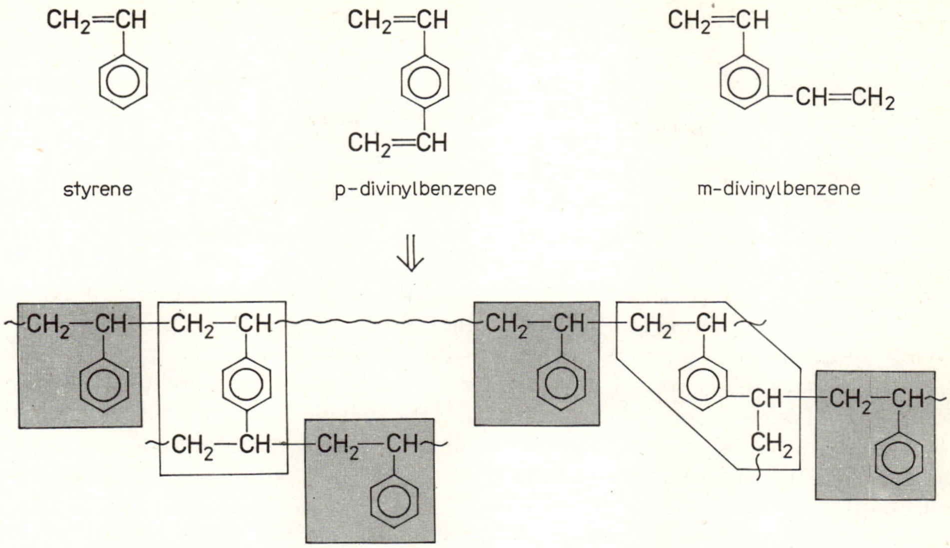

Fig. 12-5
Structure of a cross-linked copolymer of styrene and divinylbenzene (e.g., Styragel®) (schematic)

MOORE (1964) stated a divinylbenzene content of 55% (sum of m- and p-isomers) and an ethyl vinyl benzene content of 45%. By means of gas chromatography, WULFF and VESPER (1978) found in a distilled technical grade material 18.1% p-divinylbenzene, 35.8% m-divinylbenzene, 11.8% 4-ethylstyrene, 29.3% 3-ethylstyrene and 4.6% diethylbenzene and other saturated compounds. After a systematic enrichment, the p-divinyl content was 73.2% and that of p-divinylbenzene 14.8%. The pure isomers can be isolated by preparative gas chromatography.

The p-compound is slightly more reactive than the m-compound and, in copolymerization with styrene, leads to the gel point within a shorter time.

Polystyrenes cross-linked homogeneously or heterogeneously and their separation efficiency in SEC have been described by MOORE (1964). The gel prepared with 1% cross-linking agent but without any diluent was just usable; it exhibited a rubber-like quality and a steric exclusion limit of about 3500 g · mole^{-1}. On the other hand, a mixture of 24.8% styrene, 2.5% divinylbenzene and 72.7% toluene yielded a very soft and unsuitable gel. However, CRIGHTON and PEAKER (1978) successfully used the latter formulation for the preparation of a *composite polystyrene gel.* They allowed the soft gel to deposit by polymerization into a dimensionally stable, permanently porous polystyrene gel prepared in advance, thus obtaining a composite gel which could easily be packed and exhibited a separation efficiency in the range 400—40000 g · mole^{-1}.

The cross-linking of polystyrene in the presence of special template molecules can be utilized for the preparation of gels with geometrically well defined cavities, on which racemates can be separated (WULFF and VESPER, 1978).

The dependence of the chromatographic resolution on the particle size distribution was investigated by DAWKINS et al. (1977, 1979).

Heterogeneously cross-linked polystyrene gels are commercially available, for instance as Styragel® in the size classes $d_P < 37$ µm and 35—75 µm and as µStyragel® with an average particle size of 10 µm (see Table 12-2). The additional parameter in Ångström units (1 Å = 0.1 nm) indicates the "*nominal pore size*". This is about twice the value of the straight length of polystyrene molecules with $M = M_{lim}$. The straight length can be calculated from the molar mass by means of the Q factor (cf., Section 8.3.).

Bio-Beads S® are homogeneously cross-linked styrene divinylbenzene copolymers, the separation range of which is adjusted by the content of cross-linking agent. Bio-Beads SM-2® are heterogeneously cross-linked polystyrene gels. LI-RU (1979) described cross-linked copolymers of ethylstyrene and divinylbenzene and their classification into narrow particle size ranges by sieving in alcohol with the help of pressure and ultrasonics.

Styrene divinylbenzene copolymers are non-polar. Under suitable conditions these gels exhibit the efficacy of separation according to the principle of steric exclusion without superposition of adsorption interactions. The gels are suitable for use in the fractionation of non-polar and weakly polar polymers in almost all of the common solvents which effect a swelling or wetting of the cross-linked polystyrene (exceptions: AC, MEK, DMSO, W, alcohols) [F 31].

Poragel® adsorbents are styrene gels with a third comonomer which imparts certain adsorption properties to the products. This third monomer is ethylene glycol dimethacrylate, vinylpyrrolidone or vinylpyridine.

The Shodex® S gels are hydrophilic substances with a sulphonated polystyrene backbone which is cross-linked with divinylbenzene. The gels, although to be reckoned among the

Table 12-2
Properties of Styragel®, rigid, porous gel particles of cross-linked copolymers of styrene and divinylbenzene (prepacked columns)/WATERS Associates

Products	Bead diameter d_P/µm	Fractionation range 10^3 g · mole^{-1}
Styragel 60 Å	37—75	—0.5
100 Å		0.1—0.7
200 Å		0.2—4
500 Å		0.5—10
10^3 Å		1—20
3×10^3 Å		3—60
10^4 Å		10—200
3×10^4 Å		30—600
10^5 Å		100—2000
3×10^5 Å		300—6000
10^6 Å		1000—20000
10^7 Å		>20000
µStyragel (prepacked columns)		
µStyragel 100 Å	10	—0.7
500 Å		0.5—10
10^3 Å		1—20
10^4 Å		10—200
10^5 Å		100—2000
10^6 Å		10000—>20000

cation exchangers, have also been used in the size exclusion chromatography of polyvinyl alcohol and dextran (MILLER and VANDEMARK, 1980).

Non-modified polystyrene gels are sometimes used to isolate aromatic components from aqueous solutions. In such cases a decrease in gel activity with time is frequently observed; if the elution is carried out with 20% ethanol instead of purely aqueous solutions, then this decrease can be avoided almost completely (JAHANGIR and SAMUELSON, 1980).

Cross-linked styrene copolymers can also be used to solve some problems in reversed phase chromatography. RAMSDELL and BUHLER (1981), using the commercial product PRP-1, separated alkaloids in AcN/0.1 M NH_4OH mixtures, thus utilizing the stability of this packing material to alkalis.

12.1.2. Cross-linked polyvinyl acetate

Polyvinyl acetate gels for size exclusion chromatography are produced by copolymerization with butanediol(1,4) divinyl ether or adipic divinyl ester as cross-linking agents with the structures shown schematically in Fig. 12-6. These gels were first described by HEITZ and PLATT (1969). They are known by their trade name Fractogel® PVA (previously: Merckogel®OR, cf., Table 12-3).

With the use of homogeneously cross-linked polyvinyl acetate gels it is possible to achieve exclusion limits up to $M_{lim} = 4000$ g · mole^{-1}. A cross-linking copolymerization in the presence of inert diluents produces heterogeneous structures with steric exclusion limits above 10^6 g · mole^{-1}. What is remarkable is the width of the molar mass range covered by a single gel type. HEITZ et al. (1970b) demonstrated the simultaneous separation of polystyrene standards up to $M = 830000$ on the one hand and toluene on the other ($M = 92$ g · mole^{-1}) using Fractogel PVA 1000000.

The gels are prepared as spherical beads in the particle-size classes <30 μm and 32—63 μm by suspension polymerization. In organic solvents they swell to about six times their original volume. Nevertheless they have an adequate dimensional stability. Organic solvents with solubility parameters between 9 and 13 can be used as eluents (HEITZ, 1977), but not the higher alcohols and glycols.

Table 12-3
Properties of Fractogel® PVA, cross-linked copolymers of polyvinyl acetate/MERCK

Products	Bead diameter	Fractionation range	Bed volume of swollen gel in ml · g^{-1}		
	d_p/μm	10^3 g · mole^{-1}	in tetra-hydrofuran	in benzene	in ethyl acetate
PVA 500	32—63	—0.3	3.2	3.2	3.6
PVA 2000		—1	3.7	4.7	4.9
PVA 6000		—4	7.3	7.8	9.7
PVA 20000		—14	7.2	7.6	—
PVA 80000		—50	6.1	7.6	7.8
PVA 300000		—200	7.2	5.6	5.4
PVA 1000000		—1000	5.3	5.0	13.2

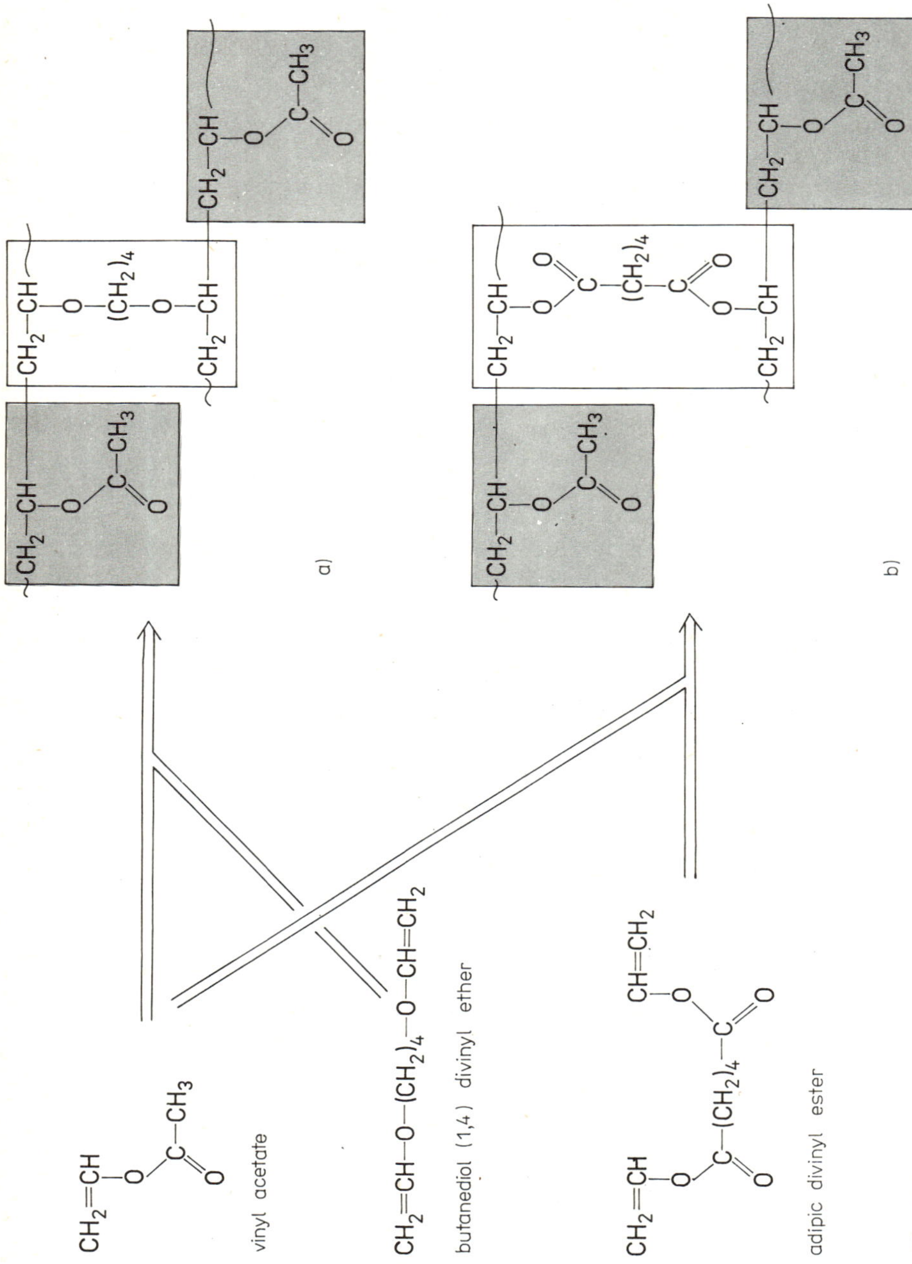

Fig. 12-6
Structure of polyvinyl acetate gels (e.g., Fractogel® PVA) with butanediol(1,4)divinyl ether (a) and adipic divinyl ester (b) as cross-linking agents (schematic)

From the copolymerization parameters of monofunctional model substances for potential cross-linking agents, HEITZ et al. (1977) concluded that a high cross-linking density can be achieved by the use of bismaleimides. In fact gels with exclusion limits of up to M_{lim} = 900000 g · mole^{-1} could be prepared with ethyl acetate as an inert diluent. However, no macroporosity was detectable in the unswollen condition and in tetrahydrofuran the gel-bed volume of these products with an obviously heterogeneous cross-linking was about five times the bulk volume. The gels could be saponified to corresponding polyvinyl alcohol gels without a destruction of their network structure; but these gels only showed a normal steric exclusion behaviour if the original gels were homogeneously cross-linked. From the heterogeneously cross-linked products the saponification produced porous supports, which retained oligoethylene oxides up to a molar mass of about 4800 g · mole^{-1}. This was obviously caused by adsorption (cf., Chapter 16).

12.1.3. Methacrylate gels

Copolymerization of ethylene glycol dimethacrylate with hydroxyethyl methacrylate (I) or polyethylene glycol dimethacrylate (II) yields support materials for steric exclusion chromatography with the structure shown schematically in Fig. 12-7.

Copolymers of type I were described by ČOUPEK et al. (1973), and are commercially available under the trade name Spheron®. They are prepared by suspension polymerization in the presence of inert diluents and are heterogeneously cross-linked. Their macroporous structure is determined by the type and the quantity of the inert component, while the micropores are determined by the proportion of hydroxyethyl methacrylate to the cross-linking agent ethylene glycol dimethacrylate (BORÁK et al., 1978). Table 12-4 gives a review of the various Spheron types. The gels can withstand pressures of up to 20 MPa (200 atm); they can be used in water as well as in numerous organic solvents and can be sterilized by heat. The primary hydroxyl groups can be converted in various ways, so that chemically modified gels for ion-exchange or affinity chromatography can be obtained.

Type II is described by HEITZ and WINAU (1970). Suspension polymerization in the presence of toluene as a diluent yields permanently porous, spherical gel particles. A

Table 12-4
Properties of Spheron® gels, cross-linked copolymers of hydroxyethyl methacrylate/LACHEMA

Products	Bead diameter d_P[1])	Fractionation range[2])	Bed volume of swollen gel in ml · g^{-1}	
		10^3 g · mole^{-1}	in water	in tetrahydrofuran
Spheron 40	C, M, F. S	20—60	4.3	4.5
100		70—250	4.3	4.5
300		260—700	4.3	4.5
1000		800—5000	4.4	4.7
100000		>10000	6.0	7.0

[1]) grades: C (coarse) 63—100 µm; M (medium) 40—63 µm; F (fine) 25—40 µm; S (superfine) ≦25 µm
[2]) for dextrans in water

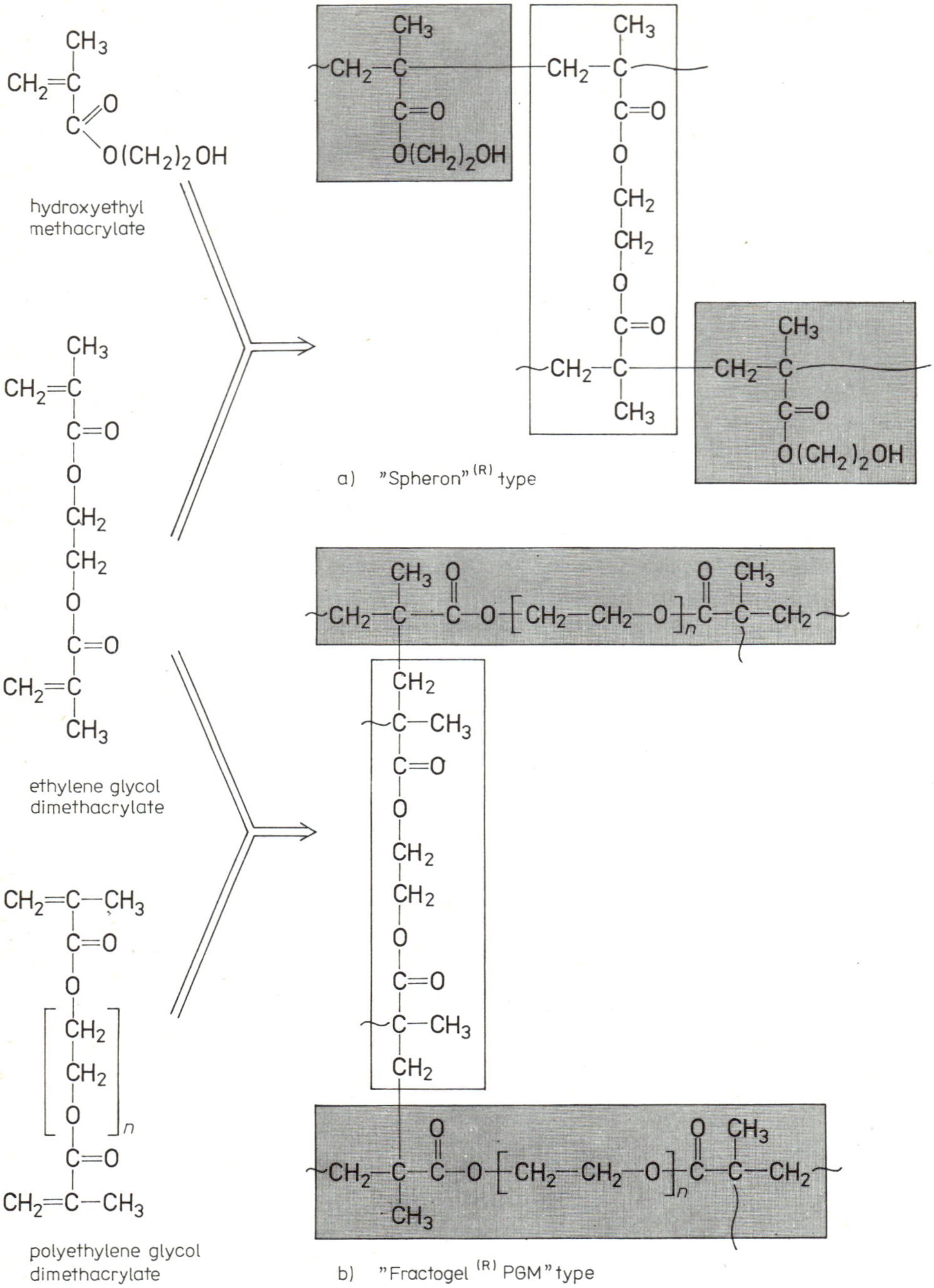

Fig. 12-7
Schematic structure of hydroxyethyl methacrylate gel (a) and polyethylene glycol dimethacrylate gel (b)
Cross-linking agent: ethylene glycol dimethacrylate (in both cases)
Examples: (a) Spheron®; (b) Fractogel® PGM.

commercial product is Fractogel® PGM 2000 with an exclusion limit of 2000 g · mole^{-1} for polyethylene glycol in water. The gel exhibits a high chemical stability and is compatible with all of the common eluents. It is preferentially used in water. Gel-bed volumes in some solvents are listed in Table 12-1. Mechanically, the material withstands pressures up to 10 MPa (100 atm).

12.1.4. Cross-linked polyacrylamide

Copolymerization of acrylamide with N,N′-methylenebisacrylamide yields gels with the structure shown schematically in Fig. 12-8. They were described by HJERTEN and MOSBACH (1962) and LEA and SEHON (1962), who used these gels for the separation of proteins. Polyacrylamide gels are commercially available under the trade names Bio-Gel® P and Acrylex®. Table 12-5 lists the various types of Bio-Gel P and their properties. The Acrylex types have the same specifications. Water and aqueous buffer solutions can be used as eluents. The gels are supplied in a dry condition and must be swollen in the buffer solution to be used as a eluent over 4 hours before they are packed into a separating column. The hydration yields the gel-bed volumes listed in Table 12-5. The particles of Bio-Gel P are spherical. Packings in the size classes 40 µm, 40—80 µm, 80—150 µm and 150—300 µm (size of the hydrated particles) are commercially available. An addition of 0.02% sodium azide can prevent a microbial contamination. The polyacrylamide gel described by DAWKINS and GABBOTT (1981) consists of beads with diameters between 7 and 12 µm, and separates within the range 10^2—10^4 g · mole^{-1}.

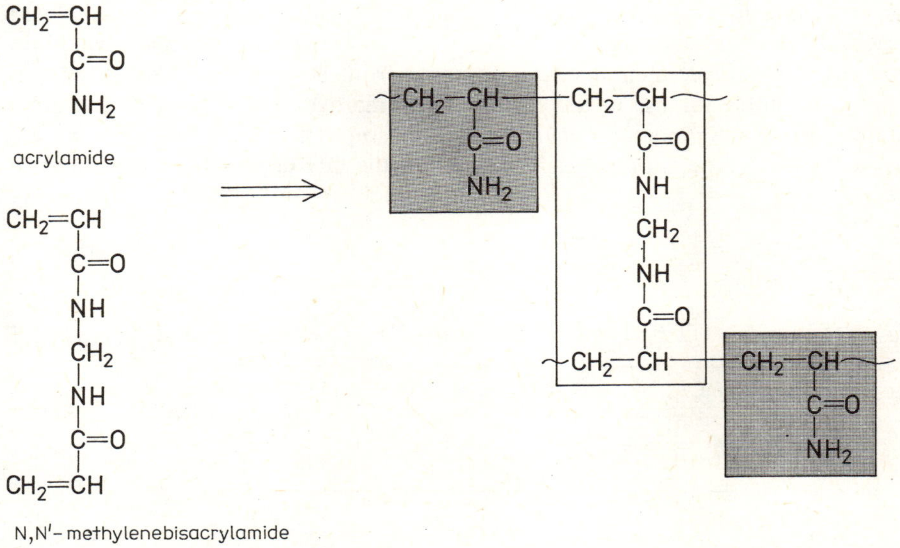

Fig. 12-8
Structure of a cross-linked copolymer of acrylamide and N,N′-methylenebisacrylamide (schematic)
Examples: Bio-Gel® P, Acrylex®.

Table 12-5
Physical characteristics of Bio-Gel® P, cross-linked copolymers of acrylamide/Bio-Rad

Products	Bead diameter d_P[1])	Fractionation range 10^3 g · mole^{-1}	Bed volume of swollen gel ml · g^{-1}	Water regain g · g^{-1}
P2	C, M, F, S	0.2— 1.8	3.5	1.5
P4		0.8— 4	5	2.4
P6		1 — 6	8	3.7
P10	C, M, F	1.5— 20	9	4.5
P30		2.5— 40	11	5.7
P60		3 — 60	14	7.2
P100		5 —100	15	7.5
P150		15 —150	18	9.2
P200		30 —200	25	14.7
P300		60 —400	30	18.0

[1]) grades: C (coarse) 150—300 µm; F (fine) 37—75 µm; M (medium) 75—150 µm; S (superfine) 10—37 µm

A *composite gel* of polyacrylamide and agarose (cf., Section 12.2.) has been described by Uriel et al. (1966, 1971). It has a skeleton of cross-linked polyacrylamide which is surrounded by agarose gel (Monsigny et al., 1978). The pore size, and hence the steric exclusion limits, are determined by the polyacrylamide network. These combination gels have been commercially available under the trade name Ultrogel® since 1975. Table 12-6 gives a review of the types and their properties. The gels are supplied in damp condition. The particles are spherical, their diameters ranging from 60 to 140 µm when swollen. The material is suitable for use in the exclusion chromatography of proteins and other water-soluble polymers. Supports for affinity chromatography can be prepared by coupling reactions on the NH_2 groups of polyacrylamide and the OH groups of agarose, respectively.

Table 12-6
Properties of the various Ultrogels®, composite gels of poly(acrylamide) and agarose/LKB

Products	Composition		Bead diameter d_P/µm	Fractionation range 10^3 g · mole^{-1}
	Acrylamide (%)	Agarose (%)		
ACA-22	2	2	60—140[1])	100—1200
ACA-34	3	4		20— 350
ACA-44	4	4		10— 130
ACA-54	5	4		5— 70

[1]) swollen in water

12.1.5. Cross-linked polyacryloylmorpholine

Copolymerization of acryloylmorpholine with N,N′ methylenebisacrylamide yields polymers with cross-linked structures as shown schematically in Fig. 12-9. Suspension polymerization of the monomers dissolved in water, carried out in paraffin oil as a dispersing agent, yields spherical particles. The gels were described by EPTON et al. (1974a, b, c, 1975). They are commercially available under the trade name Enzacryl® gel. Table 12-7 gives a review of the types and their properties. BROUGH et al. (1978) reported the experimental conditions of a successful suspension polymerization and stated interesting relationships between gel properties and preparation conditions (with regard to the design of the stirrer, addition of surfactants etc.; see Table 12-8).

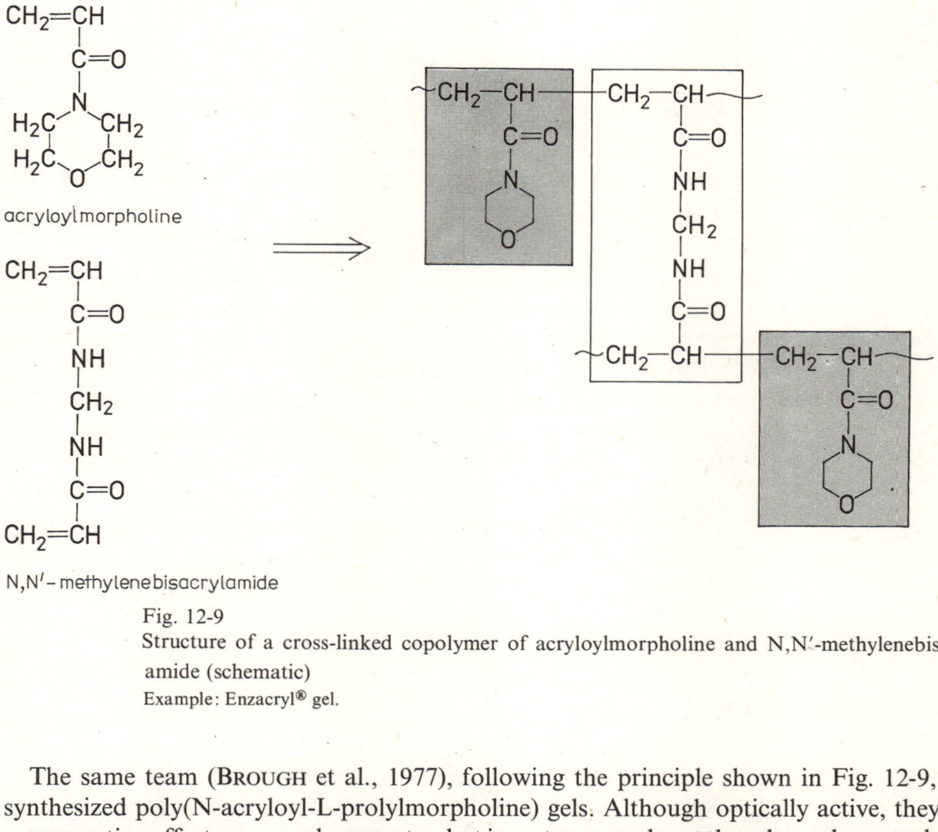

Fig. 12-9
Structure of a cross-linked copolymer of acryloylmorpholine and N,N′-methylenebisacrylamide (schematic)
Example: Enzacryl® gel.

The same team (BROUGH et al., 1977), following the principle shown in Fig. 12-9, also synthesized poly(N-acryloyl-L-prolylmorpholine) gels. Although optically active, they had no separating effect on several racemates, but in water as an eluent they showed a remarkable, reversible temperature effect in the range between 0 and 60 °C: the higher the column temperature, the sooner the ethylene oxide oligomers separated by exclusion effects emerged from the outlet of the 1 m column. At 20, 40 and 60 °C the partition coefficient of a sample with $M = 332$ g · mole^{-1} was 121%, 139% and 176%, respectively, of the value obtained at 0 °C. At the same time the swelling value in water dropped from 1.80 cm^3 · g^{-1} (20 °C) to 1.50 cm^3 · g^{-1} (60 °C). (An analogous effect for Sephadex® LH 20 was discussed by DETERMANN and LAMBERT (1971).)

Table 12-7
Properties of Enzacryl® gels, cross-linked copolymers of acryloylmorpholine/KOCH-LIGHT

Products	Bead diameter $d_P/\mu m$	Fractionation range in 10^3 g · mole^{-1} [1])	Fractionation range in 10^3 g · mole^{-1} [2])	Bed volume of swollen gel in ml · g^{-1} TCM	W	THF	M	Bzn
K 0	40—80	— 1		1.52	1.50	0.17	0.52	0.00
K 1		—10		—	—	—	—	—
K 2		—50	— 100	2.6	2.4	1.8	1.3	0.4
K 4			— 800	—	—	—	—	—
K 10			—2000	—	—	—	—	—

[1]) for dextrans in water
[2]) for proteins in water

Table 12-8
Properties of heterogeneously cross-linked polyacryloylmorpholine as functions of the polymerization conditions

Water content in monomers (in mg W per 1 g acryloylmorpholine)	Monomer composition (in moles acryloylmorpholine per 1 mole N,N'-methylenebisacrylamide) 5	10	20	40	80
	Swelling of the gel (in ml of aqueous buffer solution per 1 g of Xerogel)				
2.8	—	2.9 (K 2)[1])	—	—	—
5.7	—	5.1 (K 3)	—	—	—
11.4	9.7 (J 4)	12.1 (K 4)	12.4 (L 4)	17.4 (M 4)	20.6 (N 4)
14.9	—	14.9 (K 5)	—	—	—

[1]) The characters in parentheses are the gel notations chosen by the authors (BROUGH, EPTON, MARR, SHACKLEY and SNIEZKO-BLOCKI, 1978).

12.1.6. TSK Gel PW

Under this designation, packed columns for steric exclusion chromatography in aqueous media were offered by the Toyo Soda Manufacturing Co. without any detailed information about the gels employed. The various types are listed in Table 12-9. According to the manufacturers' information the particles are spherical and totally porous.

The PW type is a semi-rigid gel with a high porosity and a high hydrophilic surface with OH groups in the structural unit $—CH_2CH(OH)CH_2O—$ (HASHIMOTO et al., 1978). In aqueous media the gel can be used in the range pH 2—12. HASHIMOTO et al. (1978) also used it in tetrahydrofuran, but for the types G 3000 PW and G 4000 PW they observed a shrinking of the gel when water was replaced by THF. Polysaccharides, nucleic acids and water-soluble synthetic polymers such as polyvinyl alcohol, polyvinyl

Table 12-9
Properties of TSK gel PW/Toyo Soda Manufacturing Co. (prepacked columns)

Products	Bead diameter $d_P/\mu m$	Fractionation range in 10^3 g · mole^{-1}	
		for poly(ethylene glycols) in water	for dextrans in water
G 1000 PW	10	—1	
G 2000 PW	10	—5	
G 3000 PW	13	—20	—60
G 4000 PW	13	0.5—300	1—700
G 5000 PW	17		50—7000
G 6000 PW	17		500—30000

pyrrolidone and polycrylamide were fractionated on these materials with a good resolution, using pressures of up to 9 MPa (90 atm).

The Toyopearl® materials are likewise polymer gels with surface hydroxyls. They are softer, have larger sizes (20—40, 30—60, 50—100 μm), and can be used for pressures of up to 1 MPa at most. For these gels the relationship between the achievable efficiency and the packing technique has been described in several variants by Kato et al. (1981).

The TSK gel SW columns also offered by the same manufacturer, for steric exclusion chromatography in aqueous media, are packed with a modified silica gel and are discussed in Section 11.10.2.

12.2. Separating materials based on natural macromolecules

12.2.1. Cross-linked dextran

Dextran is a polysaccharide with D-glucopyranose units, which is built up from suitable disaccharides in an aqueous solution by bacteria of the lactobacilleae family. Thus *Leuconostoc mesenteroides* produces a branched dextran from sucrose with separation of fructose (see Fig. 12-10). The water-soluble polymer can be used as a synthetic blood plasma if the molar mass is within a certain range. Moreover, with its numerous hydroxyl groups it offers possibilities for covalent cross-linking. Thus epichlorohydrin reacts with hydroxyl groups on the carbon atom 3 and links neighbouring chains through ether bridges (see Fig. 12-11).

Porath and Flodin (1959) showed that the fractionation of water-soluble polymers by steric exclusion is possible on cross-linked dextran gels. The gels are porous due to swelling. The pore size, and hence the range of separation, depend on the cross-linking density. Water and aqueous solutions are suitable as swelling agents and working media in the range pH 2—10. With an adequate control of the reaction spherical gel beads which are suitable for packing chromatographic columns can be prepared. The products are commercially available under the trade names Sephadex® and Molselect®. Table 12-10

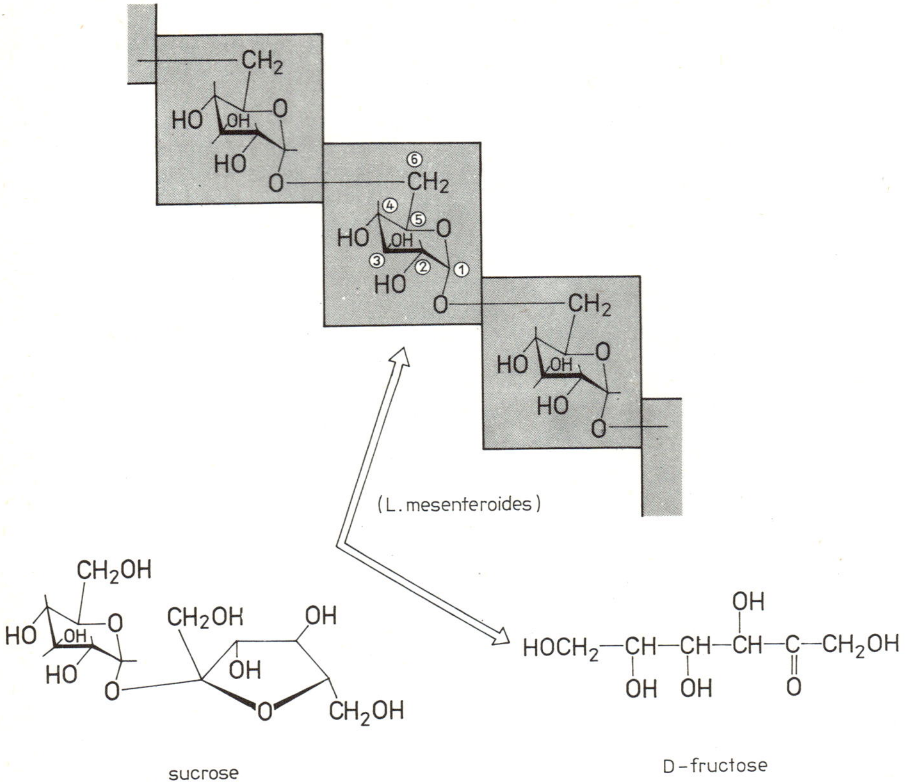

Fig. 12-10
Structure of a dextran chain (part) and schematic illustration of its biosynthesis from sucrose

gives a review of the Sephadex types. The Molselect gels are designated and classified analogously. The gels can be sterilized by autoclaving. Strong oxidants and mineral acids should be avoided. The relatively highly cross-linked types G 10 and G 15 can also be used in dimethylformamide.

Hydroxypropylation of cross-linked dextran produces gels with the structure shown schematically in Fig. 12-12. The modification does not change the number of hydroxyl groups, but it imparts a partially lipophilic character to the gel by the introduction of alkyl chains. Corresponding commercial products are Sephadex LH 20 and LH 60, which can also be used in polar organic solvents.

The main field of application of cross-linked dextrans are separations based on the principle of steric exclusion. However, as early as 1960 GELOTTE reported that adsorption phenomena may also occur, especially in investigations carried out with aromatic and heterocyclic compounds. In this so-called "aromatic adsorption" on hydrophilic gels, in part charge transfer interactions are involved (PORATH, 1979; SADA et al., 1979). The phenomena are influenced by the adsorption properties of the solvents (cf., Table 7-3), and therefore are manifested much more distinctly in organic media than in aqueous solutions. Logically, the hydroxypropylated dextrans (Sephadex LH 20 and LH 60 and others) are,

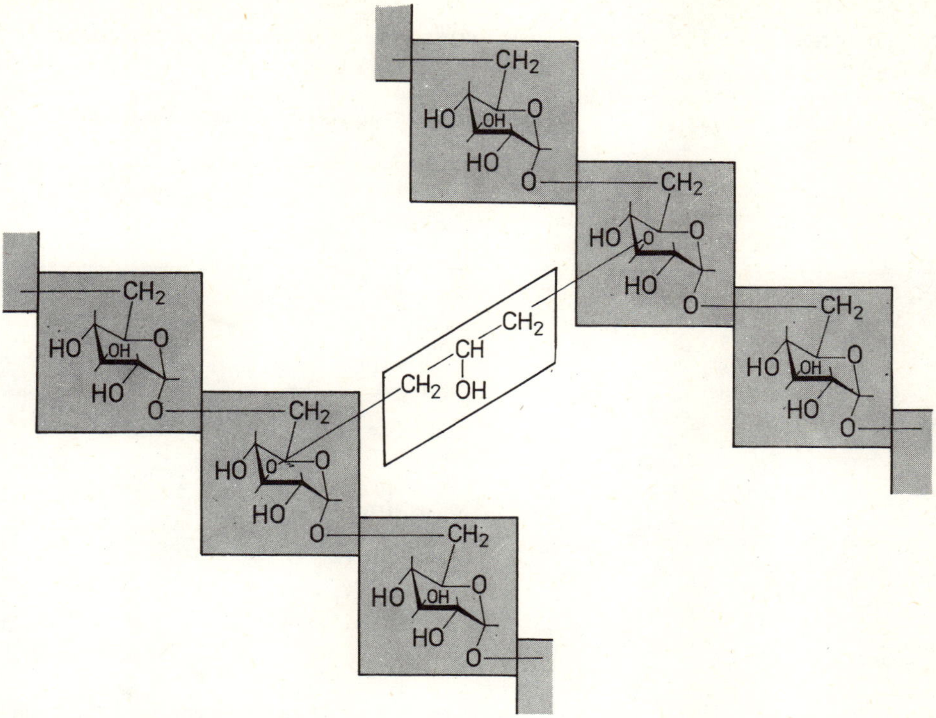

Fig. 12-11
Structure of dextran cross-linked with epichlorohydrin (e.g., Sephadex®) (schematic).

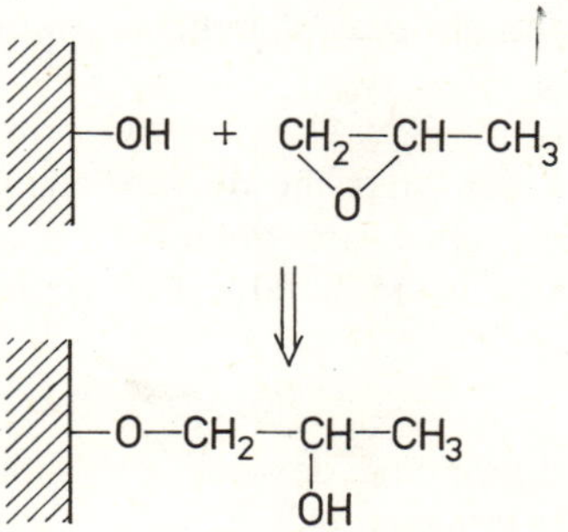

Fig. 12-12
Hydroxypropylation of a hydroxyl group

in addition to steric exclusion chromatography, especially used as supports for adsorption- and partition-chromatographic separations. The hydroxypropyl dextran gel Sephasorb® HP Ultrafine, which has a rather dense matrix, its structure being shown schematically in Fig. 12-13, has been developed especially for adsorption chromatography. It withstands pressures of up to 10 MPa (100 atm.) and is most suitable for use in the isocratic elution of compounds which either cannot be separated on inorganic adsorbents or

Table 12-10
Physical characteristics of Sephadex® gels, cross-linked dextrans, and Sephasorb®/ Pharmacia

Products		Bead diameter $d_P/\mu m$	Fractionation range in 10^3 g · mole^{-1}		Bed volume of swollen gel in ml · g^{-1}		
			for peptides and globular proteins	for dextrans	in water	in benzene	in THF
Sephadex	G-10	40—120	—0.7	—0.7	2.3	—	—
	G-15	40—120	—1.5	—1.5	2.5—3.5	—	—
	G-25	C, M, F, S[1])	1—5	0.1—5	4—6	—	—
	G-50	C, M, F, S[1])	1.5—30	0.5—10	9—11	—	—
	G-75	40—120, S	3—80	1—50	12—15	—	—
		10—40					
	G-100	40—120	4—150				
		10—40	4—100				
	G-150	40—120	5—300	1—100	15—20	—	—
		10—40	5—150	1—150	20—30	—	—
	G-200	40—120	5—600	1—200	30—40	—	—
		10—40	5—250		20—25	—	—
	LH 20	25—100	—	—~4	4.2	1.8	3.5
	LH 60	40—120	—	—~10	12.6	2.5	9.8
Sephasorb Ultrafine	HP	10—23	—	—	2.1—2.3	1.5—1.7	—

[1]) grades: C (coarse) 100—300 μm; M (medium 50—150 μm; F (fine) 20—80 μm; S (superfine) 10—40 μm

undergo undesirable chemical changes. Its capability of forming hydrogen bonds plays an important rôle in this respect. Sephasorb is stable against aqueous solutions in the range pH 2—12.

Dextran having allyl groups inserted can be cross-linked by N,N′-methylenebisacrylamide (see Fig. 12-14). Besides the copolymerization, which links the dextran chains together, blocks of bisacrylamide with a cyclopolymer structure are obtained (cf., Fig. 4-2c). The products are gels porous due to swelling, which are suitable for the separation of

Table 12-11
Physical characteristics of Sephacryl® gels, allyl-modified dextrans cross-linked by copolymerization with N,N′-methylenebisacrylamide/Pharmacia

Products	Bead diameter $d_p/\mu m$	Fractionation range in 10^3 g · mole^{-1}		Bed volume of swollen gel relative to volume in water		
		for proteins	for dextrans	DMF	THF	Tol
S-200 Superfine	40—105 (swollen in water)	5— 250	—80	1.00	0.65	0.60
S-300 Superfine		10—1500				

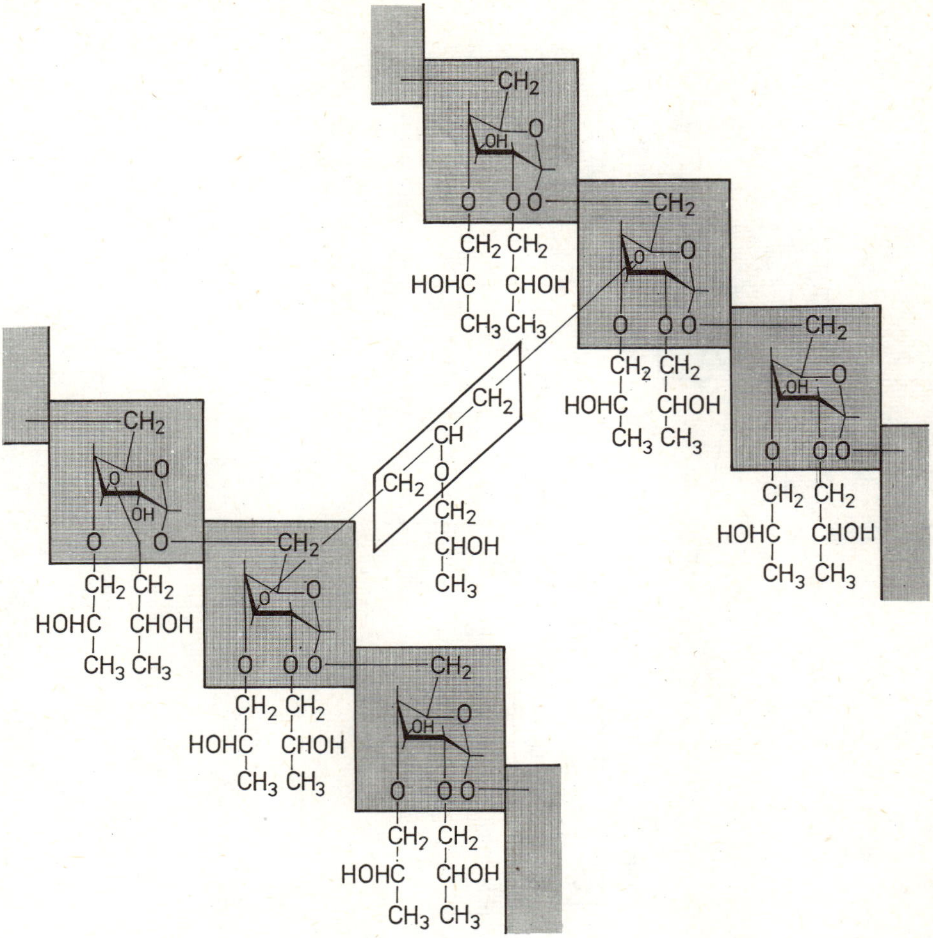

Fig. 12-13
Structure of hydroxypropylated dextran (e.g., Sephadex® LH 20) (schematic)

water-soluble proteins with molar masses of up to $1.5 \cdot 10^6$ g · mole^{-1}, and yet have sufficient mechanical strength that flow rates up to $U = 0.5$ cm · min^{-1} in 0.40 m columns can be used. The gels are available under the trade name Sephacryl®. Table 12-11 gives a review of the properties of the types. The gels are stable in the range pH 3—11. Exchanging the water by organic solvents has a much smaller effect on the gel volume, and hence on the pore size, than for other cross-linked dextrans. Outside of the column the gel can be adapted to DMF, DMSO or acetone with two intermediate stages (70% water and 30% water).

12.2.2. Agarose gels

Agarose represents the main component of agar, which is obtained from certain oceanic algae by extraction. It is a polysaccharide, namely a polygalactan, in which β-D-galactose

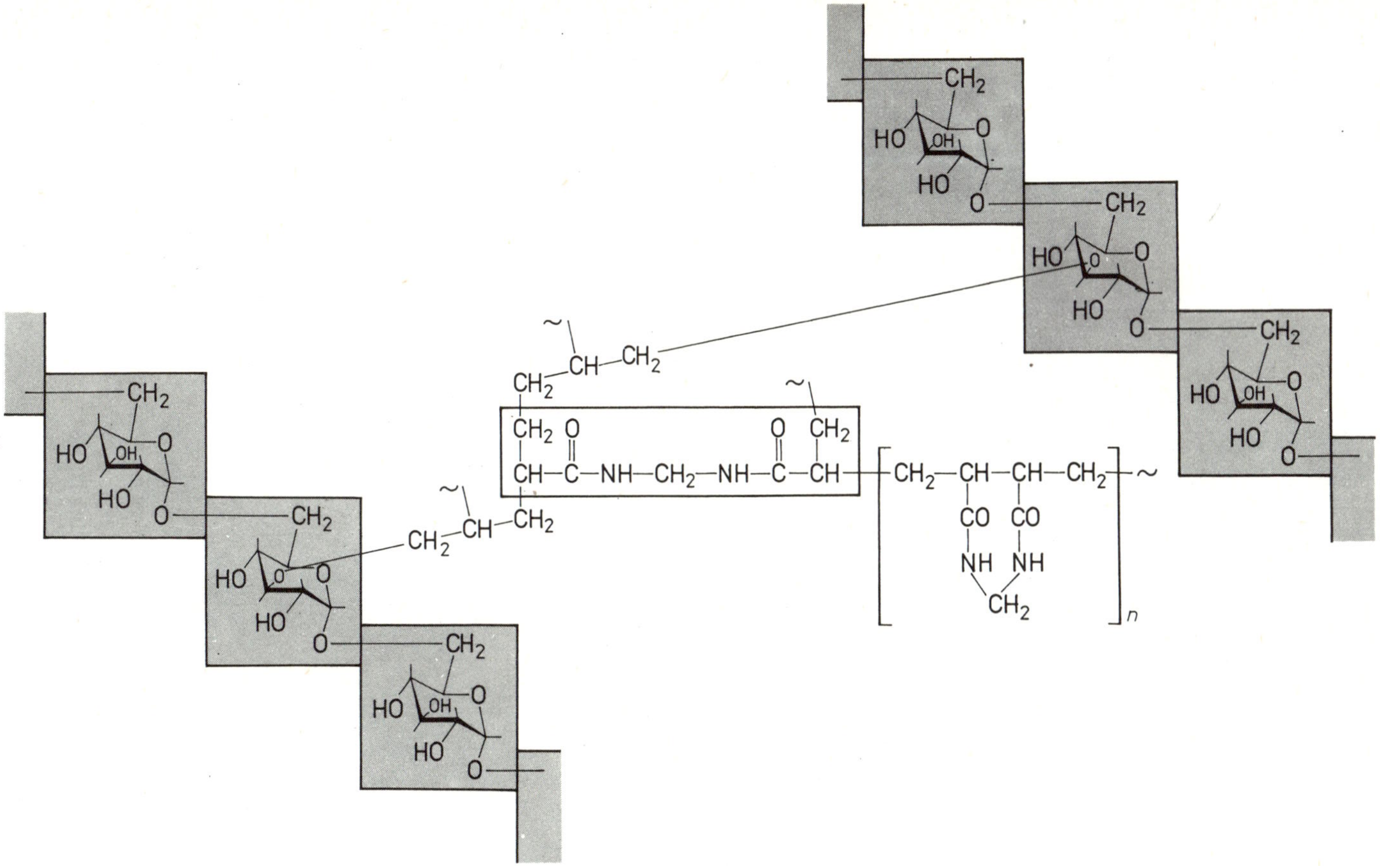

Fig. 12-14
Schematic structure of allyl-modified dextran after a cross-linking copolymerization with N,N′-methylenebisacrylamide (e.g., Sephacryl®).

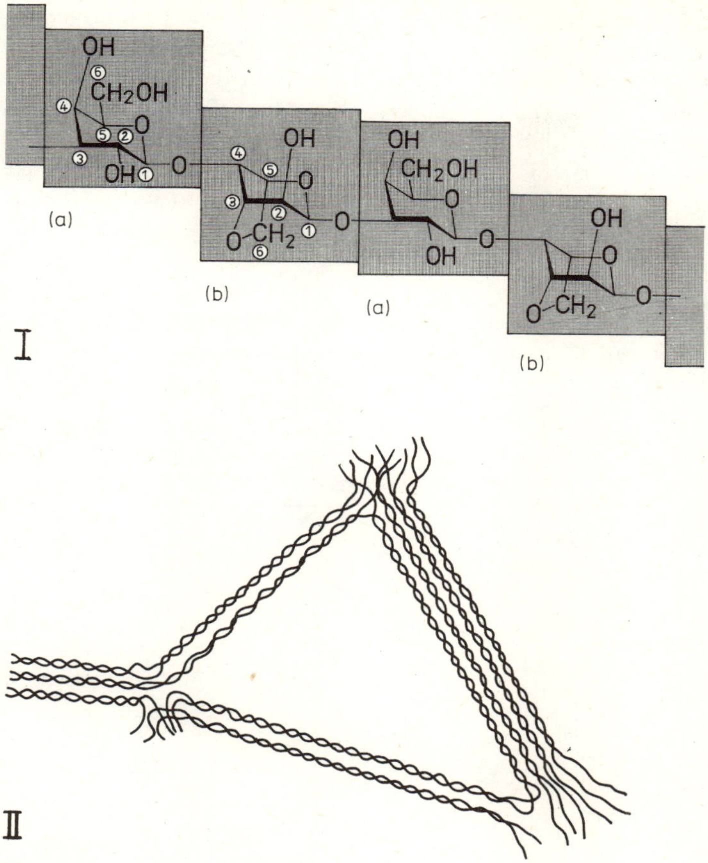

Fig. 12-15
I Structure of an agarose chain (part)
II Schematic representation of an agarose gel

Each filament corresponds to an agarose chain with the fine structure shown in Fig. 12—15 I
Examples: Bio-Gel A®, Sepharose®, Sagavac®.

(a) and 3,6-anhydro-α-L-galactose units (b) alternate (see Fig. 12-15 I). Agarose is easily soluble in hot water. Even with a content of 1 % solid matter, these solutions yield well shaped gels when cooling to ambient temperature. They represent three-dimensional networks in which the agarose chains form dense clusters with large cavities left between them (see Fig. 12-15 II). These cavities are very easily accessible. The size of these pores decreases with increasing agarose concentration in the gel. Gel beads can be prepared by dispersing agarose solutions in a hot inert organic solvent (HJERTEN, 1964).

Corresponding commercial products are Sepharose®, Bio-Gel A® and Sagavac® (see Table 12-12). Agarose can be used in the range of pH 4—9 and at temperatures between 2 and 30 °C. At higher temperatures the gels melt. Freezing or drying the gels impairs the pore structure. As the stability depends on hydrogen bridges, contact with substances such as urea or the like must be avoided.

A covalent cross-linking of the performed gels, e.g., by 2,3-dibromopropanol, yields materials with unchanged exclusion limits but improved stability (PORATH et al., 1971). Obviously the cross-linking takes place entirely within the rigid molecule clusters of the skeleton (see Fig. 12-16). The cross-linked agarose gel (Sepharose CL) can be sterilized at 120 °C, and is compatible with non-aqueous media. For the CL 6B gel it was even possible to use 0.5 N NaOH as an eluent (BAO et al., 1980). Cross-linked agarose is primarily used in exclusion chromatography, but owing to its stability it can also be

Table 12-12
Properties and types of Bio-Gel® A/BIO-RAD, or Sepharose®/PHARMACIA (agarose gels) as well as of Sepharose CL (cross-linked agarose)/PHARMACIA

Products	Bead diameter¹) $d_P/\mu m$	Fractionation range in 10^5 g · mole^{-1}	
		for polysaccharides	for proteins
Bio-Gel A			
—0.5m	C, M, F		0.1—5
—1.5m	C, M, F		0.1—15
—5m	C, M, F		0.1—50
—15m	C, M, F		0.4—150
—50m	C, M		1—500
—150m	C, M		10—>1500
Sepharose			
2B	60—200	1—200	0.7—400
4B	60—140	0.3—50	0.6—200
6B	45—165	0.1—10	0.1—40
Sepharose CL			
2B	60—250	—200	—400
4B	40—190	—50	—200
6B	40—120	—10	—40

¹) grades: C (coarse) 150—300 μm; M (medium) 80—150 μm; F (fine) 40—80 μm

modified in various ways and developed into support materials for affinity chromatographic separations (AFC). Due to the above-mentioned good accessibility of the pores and their size it is most suitable for AFC (PORATH, 1978). However, this interesting technique, which is so important for the isolation of enzymes and the like, is beyond the scope of this book, and we shall therefore refrain from a detailed description of these supports. Only the cross-linked agarose gels modified by phenyl or octyl groups for hydrophobic chromatography will be mentioned because of their similarity to reverse phases. They contain one hydrophobic group per about five galactose units, to which proteins having hydrophobic amino acids in the molecular surface may be attached. This occurs if the interaction with water as a solvent is reduced by salt addition (salting-out effect). The detachment may be caused by a reduction in the ionic strength (HOFSTEE, 1979).

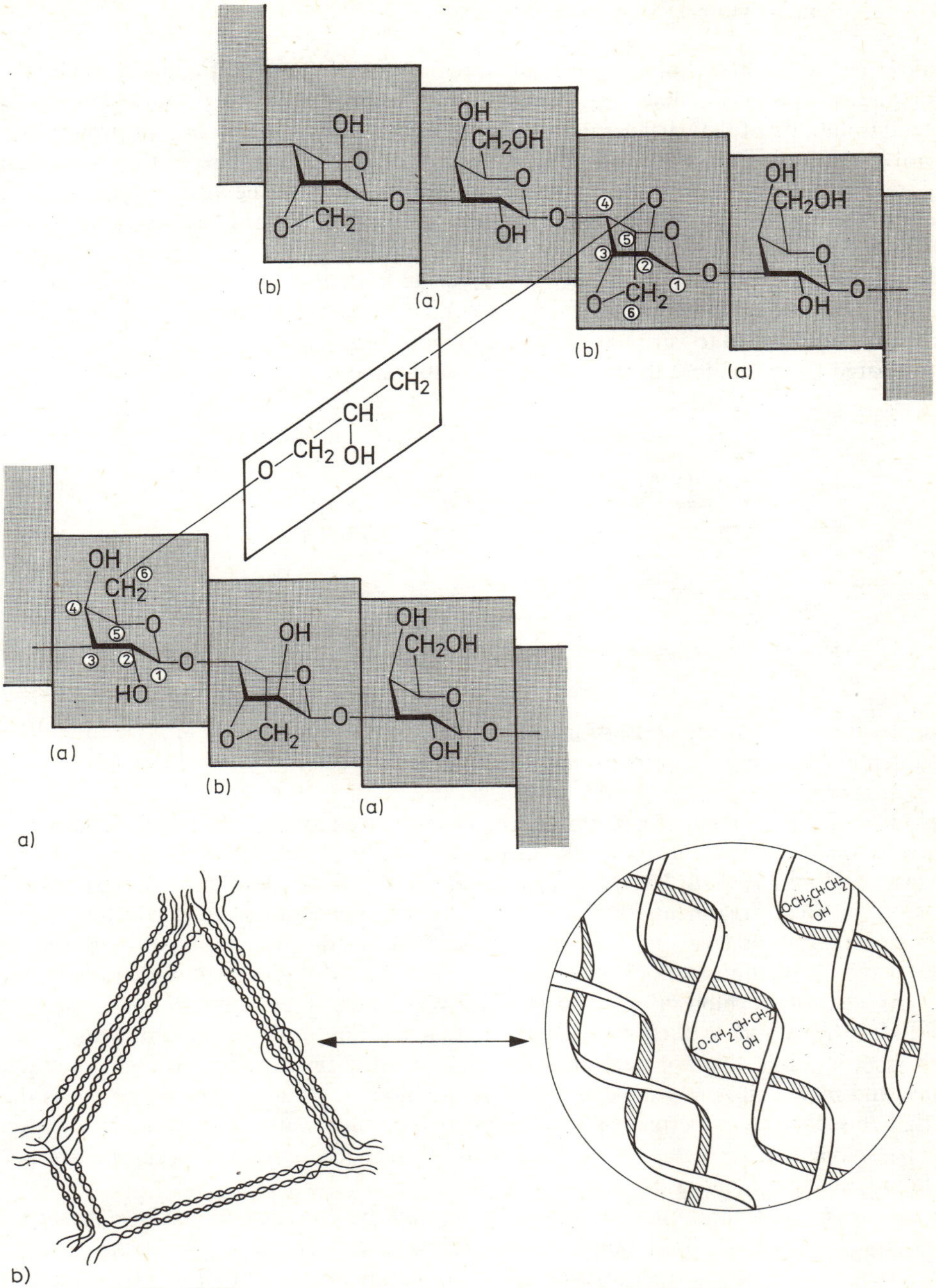

Fig. 12-16
a) Part of a structure of cross-linked agarose molecules (schematic)
b) Schematic representation of a cross-linked agarose gel, indicating the bridges between two agarose molecules

Example: Sepharose CL®.

12.2.3. Support materials based on cellulose

Cellulose in the form of fibres or ground fibres is one of the oldest support materials. Crystalline or microcrystalline zones affect the cohesion of the macromolecular chains and the insolubility of the particles in most of the solvents, while pores and the amorphous zones allow various interactions with the components of the mobile phase. In this respect the cellulose used for chromatographic purposes is comparable with the agarose gels, i.e., the principle of a network formation by intermolecular forces in the ordered zones, as shown in Fig. 12-15II, can analogously be transferred to cellulose.

Chemically, cellulose is poly-*β*-glucose anhydride with the chain structure shown in Fig. 12-17. Each glucose unit contains two secondary and one primary hydroxyl groups, which may be subjected to various polymer-analogous reactions. For example, ion exchangers prepared from cellulose in this way are of importance.

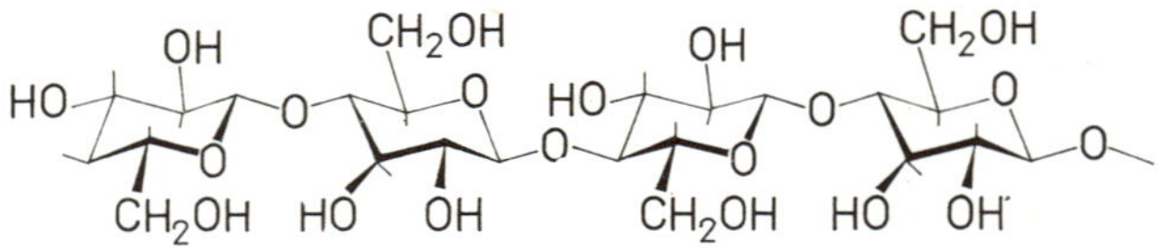

(1) cellulose = poly-β-glucose anhydride

Fig. 12-17
Structure of a cellulose chain (partial)

The disadvantages of the classical use of cellulose in chromatography are the difficulties encountered in packing the columns and the high flow resistance. These disadvantages are due to the fibre structure.

Cellulose can be obtained in a bead form by dispersing viscose, in which cellulose is dissolved as xanthogenate, in an organic solvent (Determann et al., 1968; Peška et al., 1976; Štamberg and Peška, 1978; Baldrian et al., 1978). The primary droplets undergo a sol-gel transition by heat treatment, which is followed by regeneration of the cellulose by splitting off the xanthogenate groups. This technique allows the preparation of beads with a diameter of 20—60 µm (or more) and different porosities. Column packings consisting of this type of bead exhibit a good permeability, because they are free of powdered fines and rather stable in their dimensions. It was possible to prepare beads with pores which were so large that for dextran standards the SEC relationship between retention volume and molar mass was satisfied up to 10^5 g · mole^{-1}. What is very important is the fact that this porosity is permanent, and is preserved when water is replaced by organic solvents. The network structure is similar to that of the macroreticular adsorbents. The gel bed volume is about 8 ml · g^{-1}

The process of preparation of cellulose beads and the situation of the raw materials are so advantageous that in a production designed to an appropriate scale the product may not be much more expensive than other things made from regenerated cellulose, e.g., cellophane or rayon. This reveals further possibilities for the industrial utilization of chromatographic techniques.

13. Other mechanisms of separation

Adsorption (cf., Chapter 7), steric exclusion (cf., Chapter 8), and the partition between two phases (cf., Chapter 9) are the mechanisms upon which most of the chromatographic methods for small-molecule and macromolecular samples are based. Ion exchange and chemically selective mechanisms are of importance especially for biopolymers.

In this chapter the description of the principles of the chromatographic separation of polymers is rounded off. Only those principles where the substance is transported together with the mobile phase will be considered. Isotachophoresis is beyond this scope, as is the characterization of polymers by their use as the stationary phase. Pyrolytic gas chromatography and thermofractography cannot be dealt with, nor can the other important methods which start with the disintegration of the polymers and analyse the low-molecular-weight decomposition products. Fractionation techniques which are not based on the fundamental principle of chromatography, i.e., the coupling of transport and distribution, must also be omitted.

13.1. Field-flow fractionation

The "3-F mechanism" has been proposed by GIDDINGS (1966, 1973) as a general principle for the separation of macromolecules (GRUSHKA et al., 1974). The separation path is a channel in which a field or a gradient acts in a direction normal to that of the flow (see Fig. 13-1). The field can be an electric field (CALDWELL et al., 1972), a gravity field (GIDDINGS et al., 1974), a temperature field (THOMPSON et al., 1967, 1969; HOVINGH et al., 1970) or a flow field caused by a transverse pressure gradient (GIDDINGS et al., 1976, 1977a).

The field disturbs the homogeneous distribution of the macromolecules, concentrating them at one of the walls of the flow channel. From this wall towards the interior the particle density decreases exponentially. The characteristic distance l of this distribution can approximately be identified as the mean thickness of the solute layer. Different components have different l values.

The longitudinal flow is laminar and has a hyperbolic velocity profile. Sample components having very low l values are contained in the slowly flowing layers close to the wall. They require more time to pass through the channel than other components having higher l values. The quotient $\lambda = l/w$ (w — channel width) determines the retention.

$$R = 6\lambda[\coth(1/2\lambda) - 2\lambda] \qquad (13\text{-}1)$$

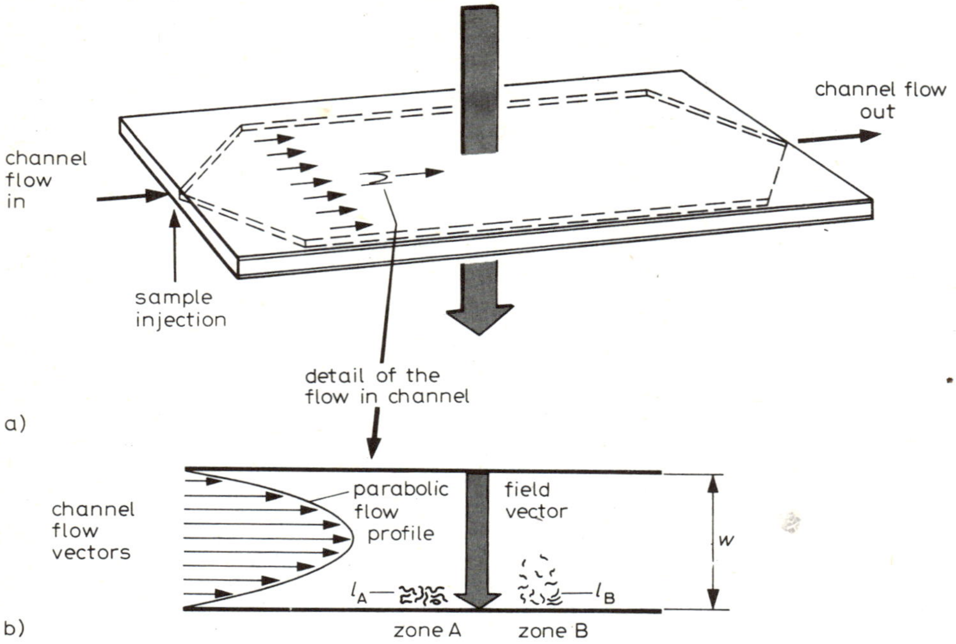

Fig. 13-1
Principle of field-flow fractionation

a) Shape of the channel and directions of the flow and the field
b) Flow profile in the channel and distribution of the components A and B under the action of the field; $l_A < l_B$
(according to GIDDINGS, 1979).

Up to $\lambda = 0.1$ one can approximate (GIDDINGS, 1973).

$$R = 6\lambda \tag{13-2}$$

The measured retention values agree excellently with the calculated ones, whereas the height equivalent to a theoretical plate (HETP) is given too low values by the theory (GIDDINGS et al., 1974, 1976; HOVINGH et al., 1970). In thermal field-flow fractionation (TFFF) of PMMA in dimethyl formamide, MARTIN and HES (1980) used a light-scattering detector (cf., Section 19.8.3.4.) connected to a differential refractometer, thus obtaining absolute information about the values of the molar mass in the eluate.

TFFF can be considered in connection with the fractionation of polymers by thermal diffusion (DEBYE and BUECHE, 1948; LANGHAMMER et al., 1954, 1955, 1958; GUZMAN and FATOU, 1958; KÖSSLER and KREJSA, 1959, 1962), cf., the summary by EMERY (1967). In thermal diffusion, the separation tube is vertical and the temperature gradient is horizontal. Density differences occur at the differently heated walls, causing a convection. This allowed a very good separation of small-molecule two-component mixtures (CLUSIUS and DICKEL, 1938), whereas problems occurred in the case of multicomponent mixtures. LANGHAMMER (1961) as well as KÖSSLER and KREJSA (1962) improved the method by superimposing a longitudinal flow.

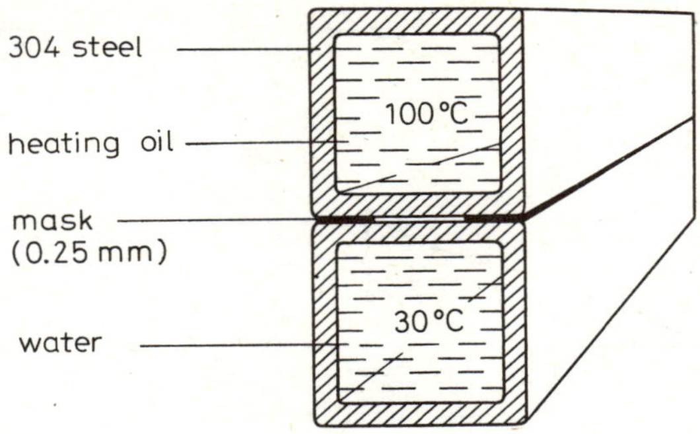

Fig. 13-2
Separation column for thermal field-flow fractionation (TFFF) (according to THOMPSON, MYERS and GIDDINGS, 1969)

The actual separation channel is the gap 0.25 mm high between the two steel tubes. The temperature-regulating liquids flow through these tubes. The sides of the channel are formed by polytetrafluoroethylene inserts. The steel walls of the separating path are polished.

In TFFF, convection is entirely eliminated. The separation channel is horizontal. A gap, 0.25 mm high and 12 mm wide, kept open by a mask made of PTFE or PETP and inserted between the cooled bottom plate and the heated top plate made of stainless steel (see Fig. 13-2) has proven suitable. In the apparatus described in 1969, the channel was 3.05 m long. With a temperature difference of about 70 K, and a channel only 36 cm long, a separation power was achieved exceeding that of the earlier GPC (MOORE, 1964) referred to for comparison (GIDDINGS, 1975). The measure employed was the so-called "*fractionating power*":

$$1/\theta = 0.25N^{1/2}\,\frac{\mathrm{d}\log(V/V_0)}{\mathrm{d}\log M} \tag{13-3}$$

The fractionating power, $1/\theta$, corresponds to the quotient $M/\Delta M$, where ΔM denotes the molar mass difference which can be separated with the resolution $R_S = 1$.

Figure 13-3 compares the fractionating power of TFFF and GPC. TFFF is also superior with respect to the *peak capacity*, n_p, i.e., the maximum number of distinguishable peaks in a chromatogram. The peak capacity can be expressed as

$$n_p = 1 + (N/16)^{1/2}\ln(V_{max}/V_{min})\ , \tag{13-4}$$

where V_{max} and V_{min} denote the limits of the elution range. In SEC, $V_{min} = V'$ and $V_{max} = V' + V''$ (cf., Section 8.1.), whereas in TFFF V_{max} might be arbitrarily large. However, the dilution of the sample and the duration of elution set a limit which gives about $V_{max}/V_{min} = 25$. For TFFF this yields

$$n_p = 1 + 0.8N^{1/2}\ , \tag{13-5}$$

whereas a value of

$$n_p = 1 + 0.2N^{1/2} \tag{13-6}$$

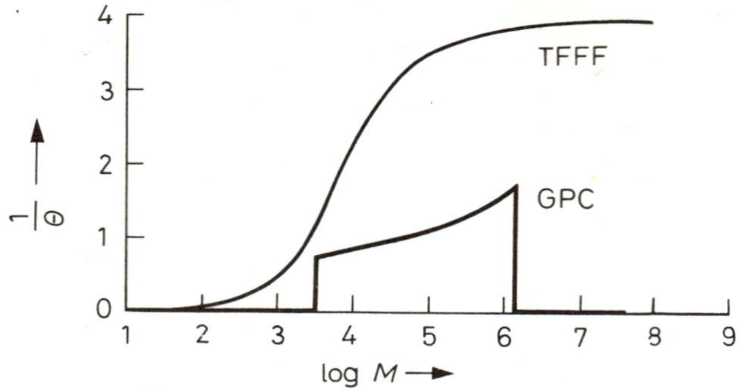

Fig. 13-3
Fractionating power, $1/\theta$, according to eqn. (13-3) for TFFF and a typical GPC column (according to GIDDINGS, YOON and MYERS, 1975)
Plate number in both cases: $N = 1000$.

can at most be obtained in SEC (where $V_{max}/V_{min} = 2.3$) (GIDDINGS, 1967). Consequently, for an equal number of theoretical plates, TFFF may reveal four times the number of peaks shown by SEC (see Fig. 13-4). This requires, however, a longer time.

Quick analyses can be carried out by means of very thin channels. Using a channel 0.051 cm high and 44.6 cm long it was possible to separate a mixture of three polystyrene standards (5000, 51000 and 160000 g · mole^{-1}) within only 4 minutes (GIDDINGS et al., 1978b).

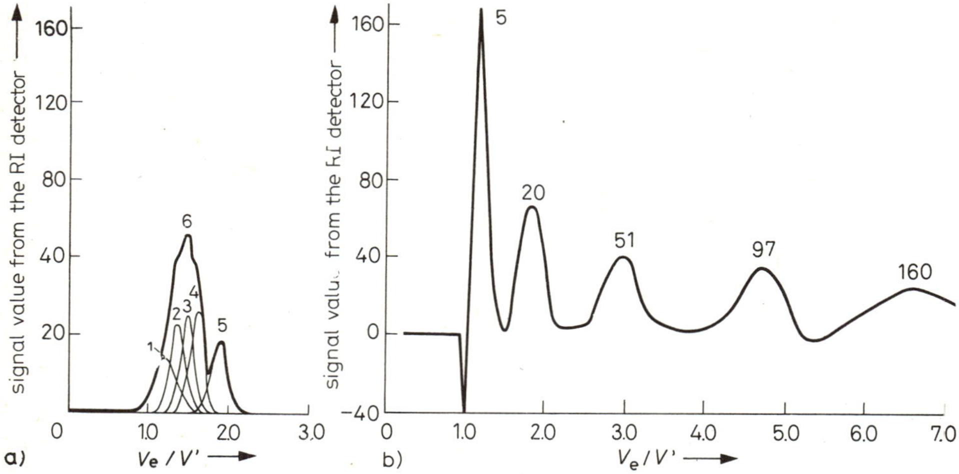

Fig. 13-4
Gel chromatogram (13-4a) and TFFF elugram (13-4b) of columns having approximately equal numbers of plates (polystyrene in toluene)

a) GPC column; $L = 3.66$ m; $d_c = 7.75$ mm; $V' \cong 60$ ml; $v = 1$ ml · min^{-1}; *1* 570000 g · mole^{-1}; *2* 267000 g · mole^{-1}; *3* 154000 g · mole^{-1}; *4* 82000 g · mole^{-1}; *5* 13850 g · mole^{-1}; *6* mixture of 1 ... 5

b) TFFF column; $L = 0.36$ m; channel volume $V' = 2.5$ ml; $v = 0.042$ ml · min^{-1}; $\Delta T = 65$ K. The indices at the peaks give the molar mass of the mixture component concerned, in 10^3 g · mole^{-1}

(according to GIDDINGS, YOON and MYERS, 1975).

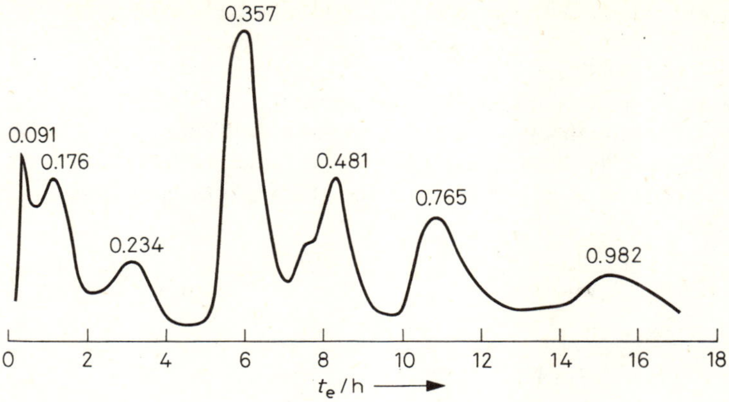

Fig. 13-5
Separation of polystyrene latex beads by programmed sedimentation FFF
The reduction of the speed from 28.6 Hz (1714 rpm) to 6.1 Hz (365 rpm) during the fractionation made it possible to separate particles with a diameter graduation of 1:10 (corresponding to a particle mass range of $1:10^3$) in a single analysis. Flow-rate: 14.5 ml · h^{-1}. The numbers written beside the peaks indicate the particle sizes in µm.
(according to GIDDINGS, MYERS and MOELLMER, 1978a).

The method was developed using PS samples, but GIDDINGS et al. (1979) have since demonstrated that it can also be applied to other polymers. In addition to polystyrene, samples of polyisoprene, polytetrahydrofuran and PMMA in a molar mass range from 7600 to 270000 g · $mole^{-1}$ were investigated in ethyl acetate and tetrahydrofuran as solvents.

The FFF methods exhibit a good separation effect even for very high particle masses (up to 10^{12} g · $mole^{-1}$), and hence can be used for the analysis of colloidal mixtures. For that purpose, the sedimentation FFF technique with a suitable gravity field as well as the method with a superimposed transverse flow are suitable (GIDDINGS et al., 1978a; 1979a).

Figure 13-5 shows the analysis of a mixture of polystyrene latex beads by means of sedimentation FFF. The separation channel used for that purpose was 90 cm long, 2.32 cm wide and 0.025 cm high, and was bent into a circular ring and placed in the rotor of a centrifuge. Introducing and discharging the liquid with the rotor running presents a difficult technical problem, for the solution of which BERG and PURCELL (1967) independently reported an ingenious device.

In 1980, KIRKLAND et al. described a sedimentation FFF apparatus which creates force fields of 15000 gravities in the separating channel (58 × 2.5 × 0.025 or 0.0125 cm) at 200 Hz (12000 rpm). The high acceleration made it possible to extend the separating range to 0.01—1 µm. YAU and KIRKLAND (1980) even reported the separation of water-based titania dispersions, polychloroprene and PMMA latices in a range from 0.001 to 2 µm by *programmed force field techniques*.

The flow-FFF method is realized in channels having semi-permeable walls, which permit a transverse flow of the dispersion medium but retain the dispersed particles. Fig. 13-6 shows a result obtained in this way in a mixture of four sorts of silica beads. The peak shape of the component with a particle size of 130 nm suggests the presence of aggregates.

Since the FFF methods are hardly subject to restrictions with respect to the particle shape, size, and condition, they are also suitable for the separation of biological particles such as viruses and the like.

Field-flow fractionations are liquid-liquid distribution techniques within a single phase. However, the latter is not homogeneous but, under the influence of the field, assumes such a property profile that the distribution processes become possible. For TFFF it has been shown that no separation effect occurs in the absence of a lateral temperature gradient (THOMPSON et al., 1969).

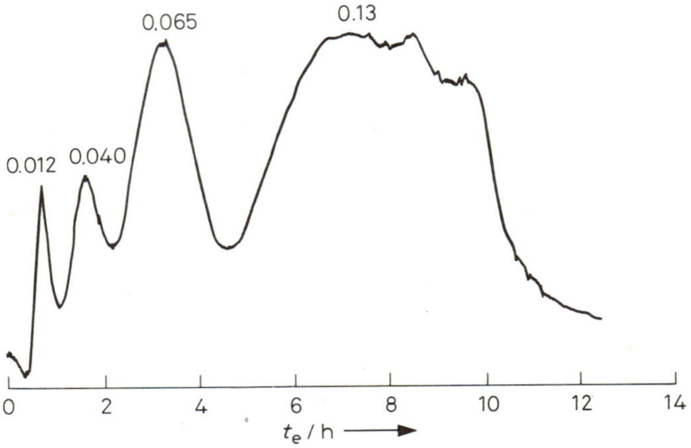

Fig. 13-6
Separation of colloidal silica beads by flow FFF
Flow rate: 3.1 ml · h^{-1} in a channel with semipermeable walls, through which a lateral flow of 10.8 ml · h^{-1} forms the flow field. The numbers written beside the peaks indicate the particle sizes in μm. (according to GIDDINGS, MYERS and MOELLMER, 1978a).

13.2. Hydrodynamic chromatography

This method is carried out using packed columns. SMALL et al. (1974) used beads of cross-linked polystyrene or cation exchangers made from the latter by sulphonation. The mean particle sizes were 18 μm, 40 μm or 58 μm, SILEBI and MCHUGH (1979) used columns filled with beads of a styrene-divinylbenzene copolymer. The packing material was not porous and did not exhibit a particular adsorption power.

Hydrodynamic chromatography is used for the separation of colloidal particles in a size range from about 1 to 1000 nm. The separation effect is caused by the velocity profile in the interstitial volumes. In capillaries with a laminar flow, the profile would have the well known parabolic shape, but also in the geometrically irregular channels of the packing the flow velocity rapidly decreases towards the boundary surfaces. Through Brownian motions, the particles carried along by the flow reach all zones of the cross-section, and are transported altogether with a velocity representing a mean value of very different velocities. The larger the particles the less they penetrate into the liquid layers near the wall and the sooner they appear at the column exit, i.e., the larger particles are eluted before the smaller ones, as is the case in SEC. On the basis of investigations using polystyrene latices with particle diameters, $\bar{d}$, ranging between 88 and 357 nm, SMALL (1974, 1977) stated the following relationship:

$$V' - V_e = m\bar{d} - n\bar{d}^2 \qquad (13\text{-}7)$$

For colloidal particles, the elution volume, V_e, is smaller than that of a low-molecular-weight marker used to determine the interstitial volume, V'. Consequently, the retention ratio as given by eqn. (3-1) is greater than unity. This unusual finding is, however, in accordance with the above model: the low-molecule-weight marker can penetrate most deeply into the slowly flowing boundary layers, and therefore is transported most slowly.

An increasing ionic strength of the eluent delays the transport of the colloidal particles. This indicates that the repulsion between particles and the surface of the packing material is promoted by electrostatic interactions. At very high ionic strengths the separation effect disappears.

Hydrodynamic chromatography has been used for process control in emulsion polymerization. At a properly chosen ionic strength, variations in the chemical nature of the colloidal particles were of only little influence. Recording the turbidity in the adsorption range is more suitable than refractive index measurement for monitoring the colloidal eluates (SILEBI and MCHUGH, 1979). The signal wavelength (possibly less than 254 nm) is decisive for an optimum detection, especially if one is dealing with a broad size distribution (NAGY et al., 1981).

13.3. Membrane chromatography

This method is very closely related to gel chromatography. Because of the peak broadening due to flow in packed columns, MEYERHOFF and SHIMOTSUMA (1967) attempted a separation

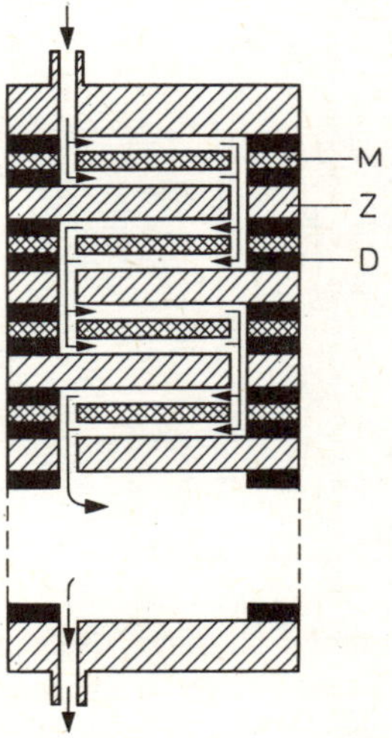

Fig. 13-7
Device for membrane chromatography, with an unidirectional flow on both sides of each membrane

M membrane; D sealing; Z baffle (metal plate)
(according to MEYERHOFF and SHIMOTSUMA, 1967).

according to the steric exclusion principle in channels with porous walls. They constructed a device with slit-like channels 8 mm wide and 0.125 mm high, which were formed by the sandwich-packing of cellophane membranes, Teflon® inserts and metal plates (see Fig. 13-7). In such a column containing 100 membranes each 1.4 cm in length, they investigated the behaviour of polystyrene and oligomeric propylene glycols and found a relationship between log M and V_e, which was analogous to eqn. (8-2). However, the arrangement had only a low pore capacity, V'', compared to normal GPC columns. Polystyrene ($\bar{M}_w = 867$ 10^3 g · mole^{-1}) was eluted with 8.16 ml, and benzene with 9.5 ml.

PŘISTOUPIL (1970) used the term "membrane chromatography" for a method alternative to paper chromatography or thin-layer chromatography.

13.4. Foam fractionation

Partition processes, which can be designed as multistage processes, are also possible between a foam as the upper phase and the foaming solution as the lower phase. In this way IMAI and MATSUMOTO (1963) separated polyvinyl alcohol in an aqueous solution according to its stereoregularity. SCHRÖDER (1977) investigated the suitability of the method for polymers in organic solvents (polymethyl methacrylate in benzene, polydimethylsiloxane).

C Chromatography under real conditions

14. Gradient technique

14.1. Definitions and systematics

Chromatographic techniques in which the migration conditions along the separating path or across the separating plane are locally different are called gradient techniques. In vector analysis, grad φ denotes the direction in which the scalar position function, φ, increases most steeply. Correspondingly, in chromatography a gradient is represented by an arrow pointing from the position of the lowest rate of migration to that of the highest one.

14.1.1. Orientation of the gradient

The direction of the gradient can be stated unambigously by referring it to the flow direction (see Fig. 14-1). In the case of a parallel gradient the rate of migration increases in the direction of travel. Gradients with opposite directions are designated by the prefix "anti" (*antiparallel, antidiagonal*).

In *planar chromatography* virtually all gradients can be realized, whereas *column chromatography* of course permits only parallel or antiparallel ones. Cross-sectional gradients, caused by an inhomogeneous packing for instance, usually lead to disturbances. The methods of field-flow fractionation (cf., Chapter 13) are based on the effect of orthogonal gradients.

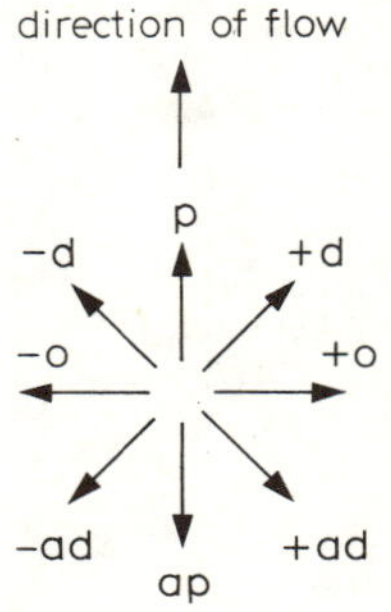

Fig. 14-1
Notation of the gradients according to NIEDERWIESER (1969a)
p parallel; d diagonal; o orthogonal; ad antidiagonal; ap antiparallel.

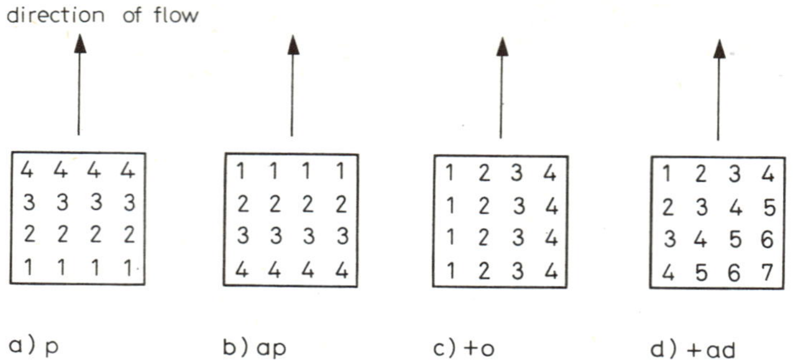

Fig. 14-2
Schematic representation of gradients in thin-layer chromatography

The higher the figure at a certain point of the plate, the greater is the mobility of the substance at this point. Fig. 14-2d shows that an antidiagonal gradient acts like a series of staggered ap gradients lying side by side.

In *thin-layer chromatography* an orthogonal gradient helps to determine the optimum development conditions quickly (see Fig. 14-2).

14.1.2. Form of the gradient

Along the gradient arrow towards the point of the highest migration rate the chromatographic migration conditions may vary linearly or in some other way (see Fig. 14-3). This also depends on which quantities are plotted versus one another. For example, Fig. 14-4

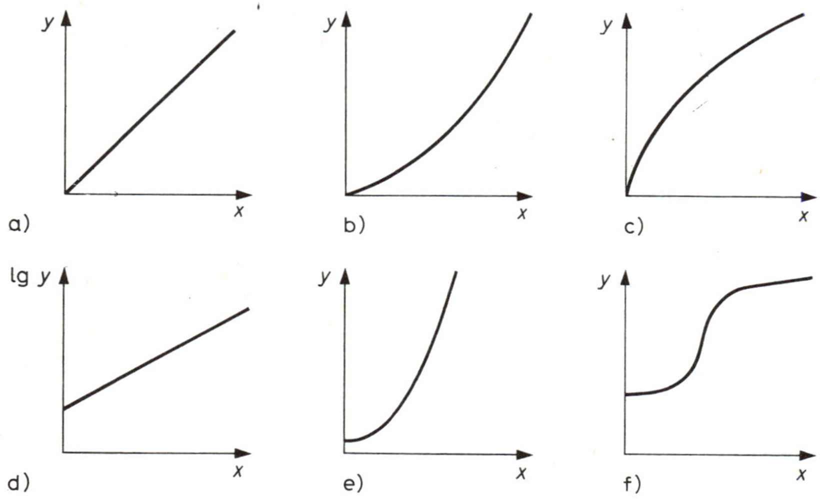

Fig. 14-3
Different shapes of gradients

a) linear; b) concave; c) convex; d, e) logarithmic; f) compound
y is the gradient-generating variable, e.g., the temperature, an activity quantity or the concentration of an eluent component.

shows a linear gradient of the eluotropic strength, which is derived from concave concentration gradients and yields a logarithmic variation of the distribution coefficient. This form produces elugrams with peaks of a constant width (SNYDER and SAUNDERS, 1969a).

Composite gradients consist of steep and less steep sections. The discontinuous gradients resulting from a stepwise variation of the chromatographic conditions form the borderline case. For a sufficiently large number of steps which are not too high, the discontinuous gradients are practically equivalent to the continuous ones.

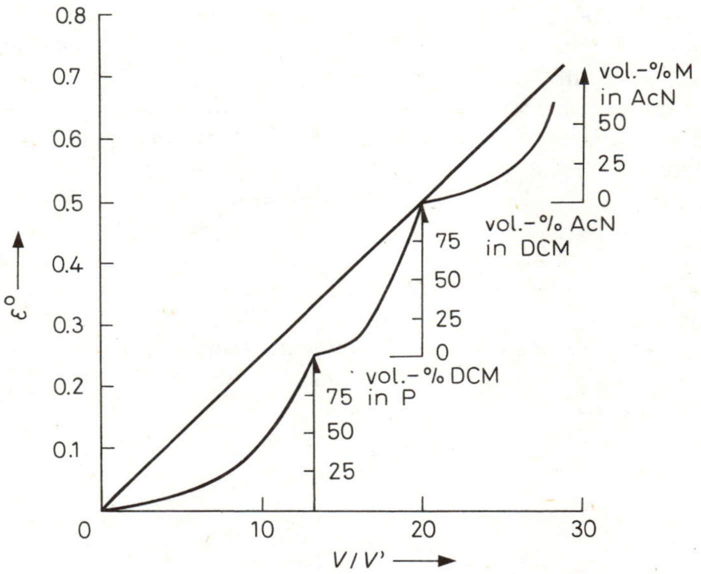

Fig. 14-4
Generation of a linear gradient of the eluotropic strength ε^0 (on silica gel) by addition of dichloromethane to pentane ($0 \leqq \varepsilon^0 \leqq 0.30$) of acetonitrile to dichloromethane ($0.30 \leqq \varepsilon^0 \leqq 0.50$) and of methanol to acetonitrile ($0.50 \leqq \varepsilon^0 \leqq 0.70$) (according to SNYDER, 1974b).

Composite gradients can satisfy the demands in the development of multicomponent mixtures: they can be flat where many components have to be separated, and steep where no bands occur. Thus the empty sections of the chromatogram are passed rapidly, whereas sufficient resolution is achieved in the densely occupied sections.

14.1.3. Gradient-analogous variations

Certain variations of the migration conditions, which influence all parts of a chromatographic bed simultaneously, have effects very similar to the gradient techniques. This, for instance, holds true for the variation of the rate of elution or the temperature.

14.2. Objectives of gradient chromatography

Initially the gradient techniques were used to overcome disturbances due to non-linear isotherms (HAGDAHL et al., 1952; TISELIUS, 1952; ALM et al., 1952). For low-molecular-weight-adsorbates, as early as 1965, SNYDER found that adsorbents with an adequate linear capacity were available. Here curved isotherms no longer call for gradient elution, but flexible macromolecules which are adsorbed with a change in conformation and cannot be desorbed without a change of solvent require gradients for that reason.

The other argument for the use of gradients is the fundamental problem of separation (cf., Section 3.5.). In chromatographic techniques the distribution coefficients may differ by orders of magnitude. If a sample contains such components, then either one has an extremely long wait for the last components or the first ones are not well resolved (see Fig. 3-4). Antiparallel gradients overcome this fundamental separation problem by compressing the chromatogram, so to speak, from behind. This gives the third argument in favour of the gradient methods: the last bands of a multicomponent chromatogram are frequently so flat that they are almost lost in the noise of the detector. The mentioned *compression* by antiparallel gradients affects the bands themselves: they not only appear sooner, but are also higher, and hence more easily detectable (Fig. 14-5). A review by ENGELHARDT (1975) gives examples of this phenomenon.

Parallel gradients accelerate the travelling substances along the separating path. This is also true for the individual components of the bands, which thereby altogether become wider. Parallel gradients do not improve the resolution; at best they are useful in order to separate quickly a rapidly travelling concomitant from the components of interest (GEISS et al., 1969).

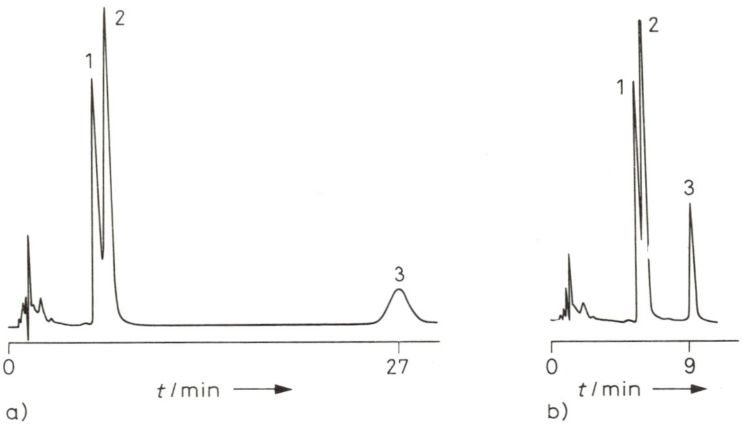

Fig. 14-5
Effect of an elution gradient

a) Isocratic elution with 20 vol.-% methanol in water, flow rate 2 ml · min^{-1}
b) Gradient elution of the same sample (agent mixture of a tablet, peak 1 is caffeine) on the same column with the same flow rate; concave gradient of 20 vol.-% methanol to 45% methanol in water (according to MCILWRICK, in: EISENBEISS, 1976).

14.3. Survey of gradient types

The chromatographic rate of migration of a substance depends on the quality of the mobile and the stationary phase as well as on the phase ratio, the temperature and the flow-rate. There is a correspondingly large number of possibilities of gradient generation.

Elution gradient. Antiparallel elution gradients are caused by solvents with a higher eluotropic strength which enter the separating path at a later time. They are primarily used in adsorption, precipitation and ion-exchange chromatography.

Elution gradients enable samples with $K_\alpha : K_\omega = 1:1000$ to be represented in one chromatogram. (K_α and K_ω are the distribution coefficients for the first and the last component, respectively, in the sample.) In chromatography on polar bonded phases or on the classical adsorbents dealt with in Chapter 11, gradient elution is carried out with increasing ε^0 values, whereas on the hydrophobic reversed phases one starts with as polar a solvent as possible, to which increasing quantities of a less polar liquid are added. Precipitation chromatography requires gradients for which the difference in the solubility parameters of the polymer and the elution mixture decreases, and hence the solvency increases.

In *elution chromatography* with simple apparatus, the gradients are generated in mixing devices. In some cases, the relationship between the geometry of the device and the form of the gradient can be exactly stated (reviews: Bock and Ling, 1954; Snyder, 1965; Mikeš and Vespalec, 1975). High-performance liquid chromatography equipment with gradient generators transforms every desired function into a corresponding gradient by electronic means. Even compound gradients can be generated in this way.

In *planar chromatography* an exchange of matter with the adjacent vapour phase may take place, therefore, in addition to intentionally generated gradients one also has to consider concealed ones, which develop spontaneously via the vapour phase. The preferential adsorption of an eluent component may also generate a spontaneous gradient.

Gradient elution is especially suitable for the survey chromatograms of unknown samples. Its disadvantages lie in the influence on the detector baseline, in the expenditure required for a reproducible generation, and in the necessity to bring the column back to the initial condition by means of a return programme after each gradient development. For silica gel this requires such high expenditure that gradient elutions are seldom performed on this material. Also the comparatively rugged reversed phases need some time for conditioning in the most common gradient technique with mixtures of water and an organic modifier. It is recommended to use a return programme lasting 10—15 minutes, followed by a 10 minute run under the (isocratic) initial conditions (Dolan et al., 1979).

Great care must be taken regarding the *chromatographic purity* of the solvents in order to prevent the occurrence of ghost peaks or other artefacts. The requirements are much stricter than in isocratic elutions: liquids still giving valid results in those techniques are not in each case pure enough for gradient elution. To check this, a blank experiment should be carried out.

If the purpose of the gradient elution is only rapidly to find the best eluent mixture that enables an isocratic separation of the sample, then, in view of the difficulties mentioned above, it may be better to perform a "sequential isocratic step" LC (Berry, 1980). Advanced instrument engineering makes it possible to carry out such search programmes unattended overnight.

Refractive index detectors and other devices responding to the composition of a mixture are unsuitable for gradient elution, except when isorefractive mixtures and an apparatus with dual columns are used (BOMBAUGH et al., 1969b). Ultra-violet detectors can be used for a great number of solvents; combustion detectors and crystal detectors can be employed universally.

Flow gradients. A gradient, i.e., a different rate of migration at different points of the separating path, usually requires different values of the distribution coefficients of the substance. The head of the gradient arrow points in the direction of decreasing K values. However, the migration rate can also vary due to the fact that the flow rate of the mobile phase differs from one point to another. If for instance an eluent evaporates from a thin-layer plate, then the forward flow of the mobile phase near the front will be smaller than at the start of the migration path, which additionally is flowed through by an eluent portion that evaporates thereafter.

On linear chromatographic beds, flow gradients are only possible in open devices which permit a lateral outflow of the eluent. In column chromatography a corresponding effect can be achieved by means of sections with semipermeable walls (cf., Section 17.8., POLSON and RUSSELL, 1966). In the circular development technique of planar chromatography [E 6, E 7] the flow gradient resulting from the geometry is of great importance.

Temperature gradient. Precipitation chromatography is a column technique with an antiparallel temperature gradient. Generally elution methods are not operated with a temperature drop along the column, as in this case, where it is desirable to re-precipitate the components dissolved immediately before and the temperature gradient is required for the formation of the stationary phase. Generally a constant temperature drop does not lead to substantially different separation conditions for the individual components, all of which indeed have to travel all the way through one and the same temperature field. This is different in development techniques, which are stopped before all of the components have arrived at the end of the separating path. Here it is possible to achieve marked effects, especially by means of coupled processes involving the eluent (cf., Section 21.6.).

Activity gradient. This term means the different quality of the stationary phase at different points of the separating path. In thin-layer chromatography the retardation required for an antiparallel gradient can be effected by layers whose activity increases in the direction of travel. For example, this effect can be produced by mixtures of silica gel or alumina with kieselguhr, which are poor in the active component at the start and become increasingly richer along the separating path. The minimum content of the strong adsorbent is about 1%, because otherwise the capacity at the start will be too low and overloading would occur (SNYDER and SAUNDERS, 1969b). For preparative TLC, however, *thin-layer plates* are commercially available which have concentration zones containing silica of an extremely small internal surface area (less than $1\ m^2 \cdot g^{-1}$) and a correspondingly low chromatographic activity (HALPAAP and KREBS, 1977).

Continuous material gradients are steeper in layers which, at the discharge end, contain only the active component, i.e., the undiluted alumina or silica gel. Naturally the layer-spreading technique is not entirely simple for such plates. It is much easier to generate activity gradients on homogeneously coated plates by a locally differentiated *vapour preconditioning* with a deactivating substance.

Activity gradients can occur *spontaneously* in an unsaturated standard chamber if the development is carried out with mixtures of solvents having different polarities. The activated layer preferentially absorbs the polar component, so that, as the vapour mixture rises in the chamber, it becomes poorer and poorer in this polar component. Consequently the upper parts of the plate are surrounded by a vapour which only contains a much diminished portion of the polar components. The quantity which can be absorbed from this vapour is correspondingly small, so that the deactivation is reduced here. In this way a spontaneous antiparallel activity gradient is generated on an initially homogeneous plate during the development (GEISS et al., 1969). Of course, convection processes in the vapour phase, which are hardly reproducible, likewise affect the corresponding gradient, and hence the separating process. For these and other reasons, GEISS called the N chamber "the least reproducible of all common development systems" [E 5]. To counteract the uncontrolled gradient generation, it has been proposed to make the vapour phase homogeneous by stirring (BUNGENBERG DE JONG and HOOGEVEEN, 1960, 1961).

In elution chromatography, the net effect of an activity gradient can be rated like that of a temperature gradient: if all of the components must fully pass through the activity field, the net effect is small, but if the chromatographic path is subdivided in such a way that the separating conditions for parts of the sample can be modified by a *column switching technique* (see Fig. 14-6), then good effects can be achieved (SNYDER, 1970, 1971b, 1974, 1974b).

Of course the subdivision cannot be carried so far that the optimum distribution coefficients result for all of the components; therefore the column switching technique cannot achieve such a high resolution as gradient elution does (see Fig. 14-7). Moreover the chromatogram should approximately be known in advance so as to ensure that switching is done at the right points. Thus the technique is less suitable for exploratory work than for series analyses.

Column switching has four essential advantages:

a) The whole elution is carried out with a uniform eluent and a constant flow rate, so the detector baseline is not influenced

b) The sample range can be as large as in gradient elution ($K_\alpha : K_\omega = 1:1000$)

c) In the various partial columns different separating mechanisms may operate. For example, the precolumn may separate on the principle of size exclusion, while the components pre-assorted by sizes are further resolved by adsorption or partition chromatography in the following columns (*multidimensional chromatography*, HUBER, 1976)

d) In column switching, the column is ready for the next analysis immediately after the elution of the last component. The troublesome resetting to the initial condition of gradient elution is eliminated. Even staggered injections are possible.

The column switching technique is a promising variant of high-performance liquid chromatography. It was carried out by JOHNSON et al. (1978) as a combination of steric exclusion chromatography and reversed-phase chromatography with C 18 adsorbents for the determination of antioxidants and vulcanizing accelerators in rubber stocks. A preseparation was carried out by the principle of size exclusion with tetrahydrofuran as an eluent. A 10 μl volume of the eluted fraction of the most interesting peaks was transferred to the RP 18 column and further separated there by means of a water-acetonitrile gradient.

In 1980a, SNYDER presented *boxcar chromatography* as a variant in which only that portion of the eluent which contains the components of interest is passed from a relatively

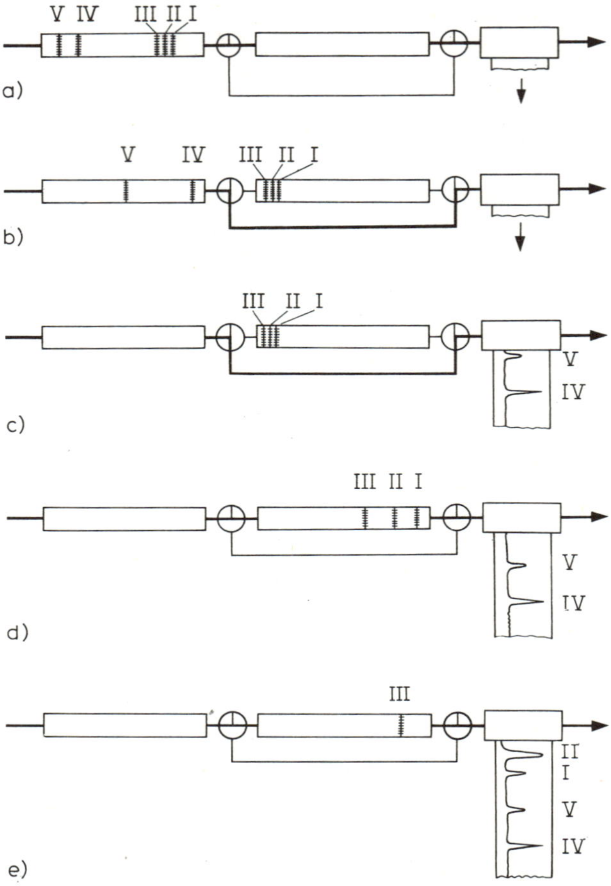

Fig. 14-6
Separation with coupled columns

The total column consists of a precolumn with a lower activity and a main column with a higher activity. The precolumn does not suffice for separating the components I—III with low *K* values, but it gives a good resolution of the components IV and V with high distribution constants. As soon as III has entered the main column, the latter is bypassed by means of a six-way cock, so that IV and V pass from the precolumn directly into the detector (14-6b). When they are eluted (14-6c), the components I—III stored in the main column are developed. This column switching technique offers great advantages for difficult routine separations and can be carried out with all detector types.

short precolumn into the much longer main column, while the unimportant constituents are discharged to the waste. Column switching is a promising technique especially for copolymer analysis (cf., Section 19.7.3.5.).

In principle, the enrichment of components in cartridges filled with separating material [F 24] is also an application of the column switching technique.

Elution programming. In the elution techniques, elution programming bears such a close relation to gradient elution (Chapter 14, p. 245) that a separate discussion is not necessary; programmed elution can be considered a development with a coarsely stepwise gradient.

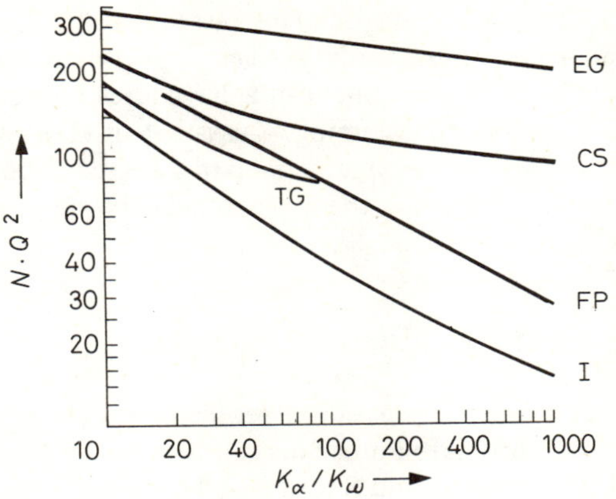

Fig. 14-7
Maximum values of the effective plate number, NQ^2, of the band travelling most rapidly as a measure of the resolution achievable in elution chromatography, plotted vs. the sample range

The range is indicated by the ratio, K_α/K_ω, of the distribution coefficients for the limiting components. The curves have been calculated with the assumption that the elution time (10 min) and the operating pressure are constant (pressure/time normalization). On the other hand, the flow-rate and the column length are not constant.
EG elution gradient; FP flow programming; I isocratic elution; CS column switching; TG temperature programming (according to SNYDER, 1970).

In the development methods there are differences between the two techniques if each part of the program is executed as a dry-bed development. The multiple development in thin-layer chromatography is an example of this technique.

Flow programming. It is possible to shorten an overlong chromatogram by increasing the flow-rate. This can be done in elution chromatography if the separation range is not larger than $K_\alpha/K_\omega = {}^1/_{20}$. This technique is most suitable for eluting a *much delayed component*. Just as in the column switching technique, the column need not be regenerated. With respect to detection there are also fewer limitations than in gradient elution. For most detectors, though, the baseline drifts with the rate of flow, because the measured quantity is pressure-dependent or the temperature deviates from the stationary value.

While flow programming reduces the duration of a chromatogram, it cannot lift the bands flattening out more and more towards the end. This is because the effect is due to an external mechanical cause, not to an action upon the capacity factors.

Temperature programming. If the column temperature is increased in the course of elution chromatography, then the adsorption coefficients decrease. Consequently adequate temperature programming might also solve the fundamental separation problem. However, an investigation carried out by SNYDER (1970) showed that only a very wide temperature range, which covers about 100 K, yields the desired effect. This calls for the use of high-boiling solvents, which on the other hand have such a high viscosity at the initial temperature that only a low efficiency can be achieved. Thus temperature programming only slightly improves the resolution compared to the ordinary elution.

Pressure programming in fluid chromatography. Fluid systems above the critical temperature offer various advantages as chromatographic eluents. Besides their solvency and their low viscosity (cf., Section 15.5.), the point of special interest here is that the distribution coefficients can be varied over wide ranges only by varying the working pressure (Bartmann, 1972). Figs. 9-18 to 9-20 show what fine results have already been achieved in this way. On a 3 m column with pressure programming, Klesper and Hartmann (1977a) were able to obtain individual peaks for styrene oligomers up to a degree of polymerization of 49 (Fig. 9-19). Without pressure programming there was practically no separation.

14.4. Resolving power of the gradient technique

Gradients should overcome the fundamental separation problem and compensate the increase in bandwidth at the end of the chromatogram. This can be done by means of elution gradients having a logarithmic form (Snyder and Saunders, 1969a)

$$\log k_t = \log k_0 - b'(V_t/V') = \log k_0 - b'(t/t') \tag{14-1}$$

where V_t is the volume passed through the column until the time t, V' is the volume of the mobile phase, t' is the mobile phase hold-up time and k is the capacity factor.

The variation of the capacity factor, which is expressed in this equation, requires that for a logarithmic gradient the *strength* of the eluent increases *linearly* with the volume of the eluate. For this reason, such *logarithmic solvent programs* were alternatively called *linear solvent strength gradients* (Snyder and Saunders, 1969a), and this latter designation is now preferred ("LSS", Snyder et al., 1979).

The column capacity factor indicates the total amount of substance in the stationary phase, referred to the amount contained in the mobile phase; it can be calculated from the distribution constant and the phase ratio (cf., Section 3.2.):

$$k = K \cdot q = \frac{n''_S/n''_E}{n'_S/n'_E} \cdot \frac{n''_E}{n'_E} = \frac{n''_S}{n'_S} = \frac{m''_S}{m'_S} \tag{14-2}$$

If the solvent added has a higher eluotropic strength than the preceding one, then $b' > 0$, and k_t at time t is smaller than k_0, which is the capacity factor at the time before the gradient becomes effective. Then eqn. (14-1) describes an antiparallel, logarithmic gradient.

The analogous condition for a gradient of the stationary phase is:

$$\log k_s = \log k_0 + b''(s/L) \tag{14-3}$$

where k_0 is the capacity factor at the start of the separating path, k_s is the capacity factor at the point s of this path and L is the total length of the path.

Again, an antiparallel gradient is given by $b'' > 0$, i.e., the migration is delayed more and more due to the increase of k_s with increasing distance s.

The effect of a gradient can again be discussed in the light of the general resolution equation (3-25), which by substituting the capacity factor takes the form:

$$R_s = \underbrace{\frac{1}{4} \cdot \sqrt{N}}_{(a)} \cdot \underbrace{\frac{k_{II} - k_I}{k_I}}_{(b)} \cdot \underbrace{\frac{k_I}{1 + k_I}}_{(c)} \tag{14-4}$$

Gradients primarily affect the efficiency factor, a, and the retention factor, c. It is useful to combine the two terms and to investigate how the *effective number of plates*

$$N_{\mathrm{eff}} = N\left(\frac{k_1}{1 + k_1}\right)^2 = N \cdot Q^2 \tag{14-5}$$

depends on the working conditions. The higher NQ^2, the better is the resolution.

To find the desired relationship, the continuous variation is subdivided into small steps 1, 2, 3, ... , *j*. The partial volume V_1 carries the neighbouring bands which are to be separated along the separating path by the distance L_1. Using the total volume, V', the total length, L, and the value, k_1, of the capacity factor in the volume fraction V_1, one obtains:

$$V_1/V' = k_1(L_1/L) \tag{14-6}$$

The expressions for the fractions V_2, V_3, etc., are derived analogously.
From the sum of all path sections, L_i, for two successive bands and the sum of variances, σ_i^2, obtained from the bands on these path sections, it is possible to calculate the final resolution. One obtains an expression for R_s^2, which, to a first approximation, differs from the squared form of eqn. (14-4) only by the retention factor, Q. With $\bar{L}_i = L_i/L$ this term reads:

$$Q = \frac{\sum^{j}\left(\prod^{i \leq m \leq j} G_m\right) \bar{L}_i k_i/(1 + k_i)}{\left[\sum^{j}\left(\prod^{i \leq m \leq j} G_m^2\right) \bar{L}_i\right]^{0.5}} \tag{14-7}$$

G_m is the *band compression factor*. For an elution gradient this factor is

$$G_m = \frac{k_{m+1}(1 + k_m)}{k_m(1 + k_{m+1})} \tag{14-8}$$

and indicates the fraction to which the width of a band is reduced in gradient elution as compared to the value obtained in isocratic elution. For a linear solvent-strength gradient, G_m depends only on the parameter b' (see Fig. 14-11).

The corresponding expression for a gradient of the layer quality in TLC is:

$$G_m = \frac{1 + k_m}{1 + k_{m+1}} \tag{14-9}$$

The value, k_m, for the capacity factor can be calculated using eqns. (14-1) and (14-3), respectively. In both cases one obtains $G_m < 1$ for *antiparallel gradients*.

In the derivation of equation (14-7) the following relationship is used instead of equation (3-11):

$$\sigma_{1+2}^2 = G_1^2 \cdot \sigma_1^2 + \sigma_2^2$$

From eqn. (14-7), the effect of gradients can be derived quantitatively. Fig. 14-7 shows curves calculated for column chromatography, which allow a comparison between normal development, temperature and flow programming, the column switching technique and gradient elution (SNYDER, 1970). As one would expect, the resolution achievable within a given time decreases with increasing separation range. However, for $K_\alpha : K_\omega = 1:1000$ the first bands can be separated with $NQ^2 = 200$ by gradient elution within 10 minutes, because the subsequent chromatogram is adequately compressed. If there is no gradient, for this time limitation the working conditions have to be modified in such a way that the effective number of plates is as small as about 15 for the starting bands.

The *optimum slope* of an elution gradient depends on the parameters specified. If the pressure is to remain constant, as in the present example, the most favourable value is $b' = 0.4$. For constant column length, an optimum elution gradient has the slope $b' = 0.2$. A certain component of the sample is always eluted at the same composition of the eluent — irrespective of the slope of the gradient.

Fig. 14-8 shows corresponding curves for TLC as NQ^2 values for components with the initial capacity factor k_0. The slope parameters b' and b'' refer to eqns. (14-1) and (14-3), respectively. The curve $b = 0$ for gradient-free chromatography shows a maximum at $\log k_0 = 0.3$, i.e., the best resolution is achieved *isocratically* for $k = 2$. On both sides of the maximum the curve slopes down sharply. Consequently a good separation is achieved only for components whose capacity factor, k_0, lies within a rather narrow interval. The permissible *sample range* increases with increasing b. This advantage of the gradient technique is gained, however, at the expense of a decrease of the maximum, and hence with a loss of resolution. With elution gradients this is even more pronounced than with gradients in the stationary phase.

The advantages shown by the broad curve for activity gradients with $b'' = 6$ or $b'' = 10$, i.e., a very large separating range with a resolution which is still rather good, are, however,

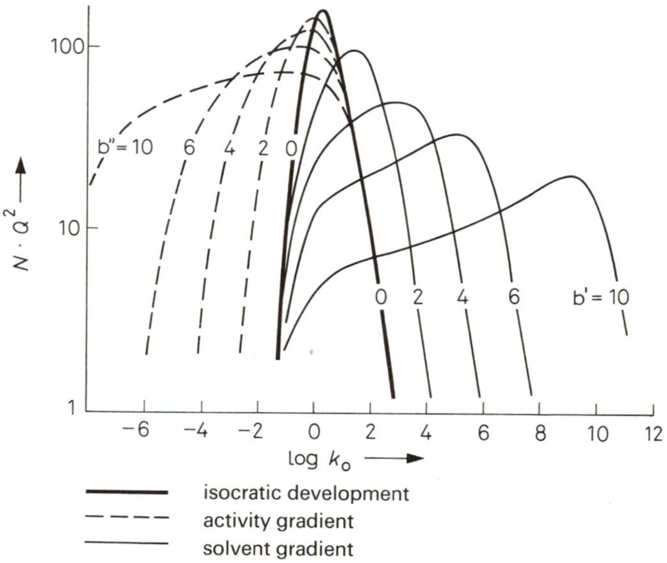

Fig. 14-8
Effective plate number, NQ^2, for logarithmic antiparallel gradients in TLC, as a function of the initial capacity factor, k_0

Parameter: slope of the gradient
For activity gradients the k values increase with increasing distance from the start. Therefore the starting values, k_0, must be very small in order to achieve an optimum development.
For elution gradients the strength of the eluent increases during the development; thus also components with high k_0 values are finally transported (according to SNYDER and SAUNDERS, 1969 b).

unrealistic insofar as gradients of the stationary phase which are so steep can hardly be generated — at least not by mixing different materials.

With the progress made in instrument engineering and with the development of bonded phases, gradient elution has found widespread application. In gradient elution on reversed phases, which today forms the basis of the bulk of all chromatographic investigations, the distribution coefficients increase

- with increasing polarity differences between the sample and the polar mobile phase
- with decreasing difference in polarity between the sample and the non-polar stationary phase and
- with increasing molecular size of the sample

The optimum value of b' is about 0.2 if 10 μm particles are used and the total elution time is about 15 minutes. For 5 μm particles and shorter times the optimum is approximately $b' = 0.1$, while larger particles and a slower elution require that $b' = 0.3$ (SNYDER et al., 1979).

In most cases the gradients are generated by adding an organic modifier (II) to water (I). In reversed-phase chromatography (RPC) the organic component II is the stronger solvent (cf., Table 7-5), so that the capacity factors decrease with increasing volume fraction φ_{II}:

$$\log k_t = \log k_W - S \cdot \varphi_{II} \tag{14-10}$$

The value k_W is usually determined by extrapolation, corresponding to the capacity factor which would be associated with the substance in pure water. (Direct measurements are often impracticable, because some reversed phases are not wetted by water and solubility problems are encountered.)

The higher the value of S, the greater is the variation of k with a given amount of the medium II. Thus S is a measure of the strength of the eluent II in RPC. Like the ε^0(RP) values discussed in Section 7.6., it depends on various experimental parameters, e.g., on the sample.

As already mentioned in Section 7.6., investigations by SCHOENMAKERS et al. (1979) into the dependence of the capacity factor on the volume fraction of the organic modifiers led to a quadratic equation. The linear set-up (14-10) only represents an approximation, which, however, deviates only slightly from the exact course of the curves, especially in the range $1 \leqq k \leqq 10$ which is used practically for the analyses (see Fig. 14-9). Apart from this statement which was already given by SCHOENMAKERS et al., one should not overlook their information that (at least for methanol and tetrahydrofuran as modifiers) S increases with $\log k_W$ (see Fig. 14-10). Using the experimentally determined dependence

$$S = S_1 + q \cdot \log k_W \tag{14-11}$$

one can derive the value of S which should occur for a sample having a capacity factor chosen as a standard. The form of eqn. (14-11) suggests choosing $k_W = 1$. If one assumes that such a reduction to S_1 is generally better than a comparison of the S values of samples with different k_W factors, then the data given by SCHOENMAKERS et al. yield the values listed in Table 14-1. The deviations from the methanol value, which are calculated from these data, correspond rather well with the ε^0(RP) values measured by COLIN et al. (1976) on graphitized carbon black, which are likewise referred to methanol with ε^0_M(RP) = 0.

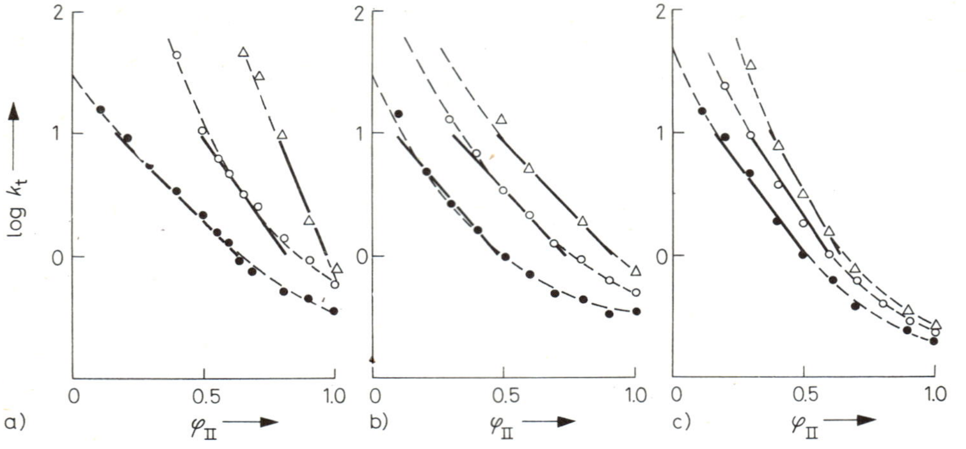

Fig. 14-9
Dependence of the capacity factor, k_t, on the volume fraction, φ_{II}, of the organic modifier in mixtures with water for
phenol (●), anisole (○) and naphthalene (△)
a) water-methanol; b) water-acetonitrile; c) water-tetrahydrofuran
Linear approximation according to eqn. (14-10) in the interval $0 \leqq \log k_t \leqq 1$
(according to SCHOENMAKERS, BILLIET and DEGALAN, 1979).

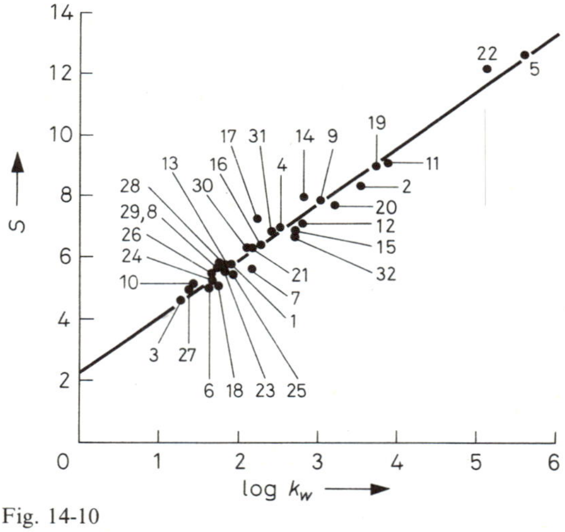

Fig. 14-10
Slope factor, S, of eqn. (14-10) (plotted vs. the ordinate section log k_w
for methanol-water mixtures at 25 °C and RP 18 as a stationary phase (L = 0.30 m, d_C = 4.6 mm, v = 1.5 ml/min)
The points determine the straight line $S = 2.23 + 1.83 \log k_w$ ($r^2 = 0.969$).
For tetrahydrofuran-water mixtures, the analogous relationship is $S = 4.33 + 1.80 \log k_w$
Samples: 1 acetophenone; 2 anethole; 3 aniline; 4 anisole; 5 anthracene; 6 benzaldehyde; 7 benzene; 8 benzonitrile; 9 benzophenone; 10 benzyl alcohol; 11 biphenyl; 12 chlorobenzene; 13 o-cresol; 14 diethyl o-phthalate; 15 N,N-dimethylaniline; 16 2,4-dimethylphenol; 17 dimethyl o-phthalate; 18 m-dinitrobenzene; 19 diphenyl ether; 20 ethylbenzene; 21 N-methylaniline; 22 naphthalene; 23 p-nitroacetophenone; 24 o-nitroaniline; 25 nitrobenzene; 26 m-nitrophenol; 27 phenol; 28 1-phenylethanol; 29 2-phenylethanol; 30 3-phenylpropanol; 31 quinolone; 32 toluene
(according to SCHOENMAKERS, BILLIET and DEGALAN, 1979).

Table 14-1
Strength of several organic modifiers for gradient elution on C 18-reversed-phases

	Methanol	Acetonitrile	Tetrahydrofuran
S_1 (Eq. 14-11; data according to SCHOENMAKERS et al., 1979)	2.23	2.75	4.33
$S_I - S_{I(M)}$	0	0.52 (1:	2.10 4.03)
ε^0 (RP) (according to COLIN et al., 1976; cf., Table 7-5)	0	0.039 (1:	0.139 3.56)

In practice, the relationship expressed by eqn. (14-11) implies that samples with a very broad range (expressed as K_α/K_ω or $k_{W,\alpha}/k_{W,\omega}$) exhibit a larger deviation from the linear approximation (14-10) than samples of narrower range. However, as regards the derivation of favourable working conditions for the gradient technique, the effects are so small that SNYDER et al. (1979), DOLAN et al. (1979) as well as JANDERA et al. (1979) preferred the simpler linear set-up (14-10). This procedure is followed here.

In reversed-phase gradient elution the resolution also increases with the increasing values of the effective plate number, NQ^2. Q can be calculated by eqn. (14-7). However, rather good approximate values are obtained from the empirical relationship

$$Q = \bar{k}/(1 + \bar{k}) \tag{14-12}$$

with a mean column capacity factor $\bar{k} = 1/(1.3 \cdot b')$; (SNYDER et al., 1979).

For gradient elution, the plate number, N, can be calculated by:

$$N = \left[\frac{(2.3b' + 1) \cdot G \cdot t'}{2.3b'\sigma_t}\right]^2 \tag{14-13}$$

(Eqn. (3-18) for isocratic elution yields too high values.)

Linear solvent strength gradients can easily be achieved by adding the stronger solvent with a constantly increasing volume flow-rate $\dot{\varphi}$. Then the following relationship holds for the volume fraction, φ_{II}, of this component:

$$\varphi_{II} = \varphi_{II,0} + \dot{\varphi} \cdot t \tag{14-14}$$

The percentage variation of the eluent composition per minute amounts to $100\dot{\varphi}$. Insertion of eqn. (14-14) into eqn. (14-10) gives:

$$\log k_t = (\log k_W - S\varphi_{II,0}) - S\dot{\varphi}t \tag{14-15}$$

From a comparison with eqn. (14-1) it follows that:

$$b' = S\dot{\varphi} \cdot t' \tag{14-16}$$

On the one hand this relationship enables the rate of variation, $\dot{\varphi}$, to be preselected on the basis of the recommended value $b' = 0.2$, the value S for the organic solvent chosen and the mobile phase hold-up time, t'. Here the strong influence of the molecular size on the actual value of S should be taken into consideration (LEA, 1980). On the other hand, using eqn. (14-16), the exact value of b' required for calculations with eqns. (14-12) and (14-13) can be determined from an experimental $\dot{\varphi}$ obtained by fine tuning.

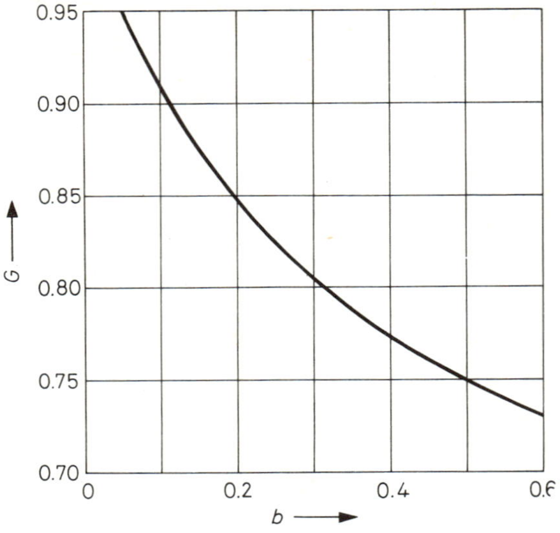

Fig. 14-11
Band compression factor, G, as a function of the slope, b, of the gradient (according to SNYDER, 1980a).

The band-compression factor, G, must likewise be known for the calculation of the plate number with the help of eqn. (14-13). For LSS gradients it can be taken from the diagram 14-11.

With the behaviour $\dot{\varphi}$ calculated from eqn. (14-16) for the optimum value of b', a blank is run in order to evaluate the suitability of the solvents and to detect possible effects on the baseline. Then a known mixture of standard samples is investigated, and finally the first analysis of the unknown sample is carried out.

If, after a survey chromatogram has been obtained by gradient elution, an isocratic fine resolution of a narrow range of the sample is desired, then the most suitable composition is that which appeared at the column head $2.5t'$ seconds before the fraction of interest emerged from the column in the gradient experiment.

SCHOENMAKERS et al. (1981) used gradient elution for a rapid selection of isocratic conditions in RP-HPLC. Water-methanol mixtures were used to determine the volume fraction, φ_M, at which the components of the sample exhibit capacity factors lying in a suitable range. If the isocratic elution by means of this eluent, W—Mφ_M, is not yet satisfactory because of an insufficient selectivity, the transfer rules

$$\varphi_{AcN} = 0.32\varphi_M^2 + 0.57\varphi_M \tag{14-17}$$

or

$$\varphi_{THF} = 0.66\varphi_M \tag{14-18}$$

can be used to find that W—AcN and W—THF mixture, respectively, in which the retention remains approximately constant while a good separation is achieved because of the possibly better selectivity (cf., Section 7.6.).

Finally let us return to the *parallel gradients* mentioned in Section 14.2., for which the distribution coefficients decrease with increasing distance. This acceleration may cause already separated bands to move still further apart, thus simulating an improved resolution. SNYDER and SAUNDERS (1969b) applied eqn. (14-7) to parallel layer gradients (i.e., $b'' < 0$ in eqn. (14-3)), and also confirmed by calculation that parallel gradients reduce the separating range and decrease the resolution. The steeper a parallel gradient, the more pronounced its adverse effect will be.

According to GIDDINGS (1967b), the number of peaks separable by gradient technique or programmed chromatography is

$$n_p = 0.1 N^{1/2} t/t' \tag{14-19}$$

where t' is the time required for a non-retained component to pass through the column (mobile phase hold-up time, cf., Section 3.1.); t is the *transition time* required for the change of the chromatographic conditions from the initial state to the final one at a certain point of the separating path. (The steeper the gradient, the shorter is t and the less peaks are detected.)

15. The influence of kinetic factors

In an optimum chromatogram the bands are completely separated but without any large interspaces. In Chapter 14 it has been shown how the succession of peaks can be accelerated; now we are concerned with the even more important question of how to proceed from overlapping bands to a baseline separation.

In Figure 15-1 it is indicated that thermodynamic and kinetic measures can be considered. The effect of kinetic factors on the course of the chromatographic process has been investigated by LAPIDUS and AMUNDSON (1955), VAN DEEMTER et al. (1956), HUBER et al. (1967, 1969). HORVÁTH and LIN (1976, 1978) and especially by GIDDINGS, whose monograph [A 9] gives a comprehensive description of the relationships. The peak width in GPC was treated in detail by GIDDINGS et al. (1966, 1977b) as well as by KELLEY and BILLMEYER (1971). Liquid chromatography using open tubes was investigated theoretically by HALÁSZ (1979).

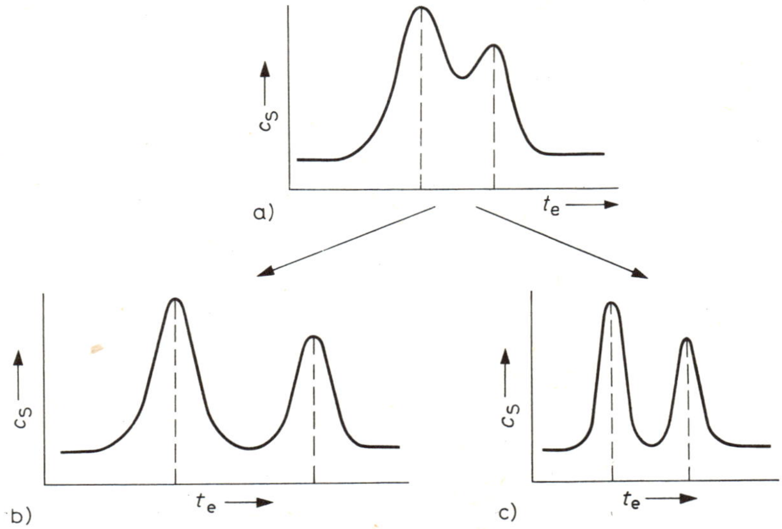

Fig. 15-1
Improvement of resolution

a) Initial chromatogram with overlapping bands
b) Improved resolution due to varied thermodynamic conditions (higher selectivity)
c) Improved resolution due to varied kinetic conditions (higher efficiency gives better separation with the peak distance remaining unchanged).

In the general resolution equation (3-25), it is above all the efficiency term, $\sqrt{N}$, which is influenced by the kinetic factors. So far we have been satisfied with the finding that the number of theoretical plates increases with the column length and that long columns yield better separations than (equivalent) short ones, but now it has to be investigated how as many plates as possible can be produced in one column of a given length. In view of the relationship $h = L/N$ (eqn. (3-20)), this is equivalent to the question of how the *minimum plate height* (height equivalent to a theoretical plate (HETP)) can be obtained. In addition to the influence dealt with in Section 3.3. and indicated in Fig. 15-2, a number of effects combine to produce the difference between the real and the ideal behaviour (see Fig. 15-3).

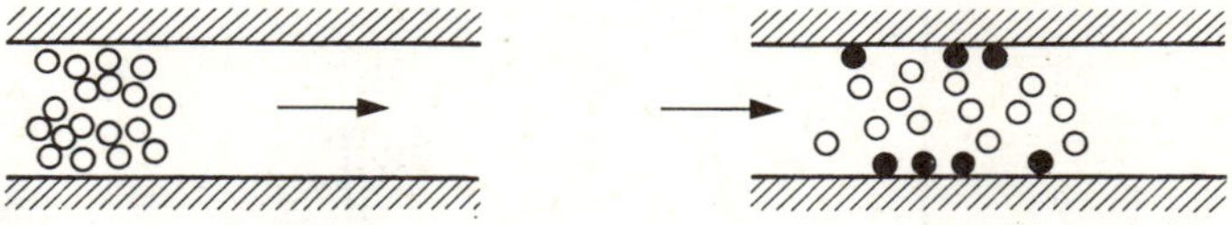

Fig. 15-2
Mechanism of the thermodynamically induced band formation
At the feed end of the separation path the molecules of the sample are still contained in the injection volume. The partition between the stationary and the mobile phase leads to the formation of a band which becomes broader and broader as the migration distance passed increases.

15.1. Band broadening due to axial diffusion

Let us say in advance: the diffusion coefficients in liquids are so small that in the normal operation of high-performance instruments the axial diffusion is of much less influence than the effects to be described in Sections 15.2. and 15.3.

The longitudinal diffusion is caused by the concentration gradient between the substance band and the adjacent liquid. The peak broadening is directly proportional to the diffusion coefficient, D, but inversely proportional to the linear velocity, u. According to GIDDINGS [A 9], for axial diffusion in the mobile phase

$$h'_{\mathrm{a}} = 2\gamma' D'/u \qquad (15\text{-}1)$$

and in the stationary phase

$$h''_{\mathrm{a}} = 2\gamma'' k \cdot D''/u \qquad (15\text{-}2)$$

where γ', γ'' denote resistance parameters of the packed column and the stationary phase, respectively, and k is the capacity factor (see eqn. (3-9)).

While in the mobile phase a longitudinal diffusion is always possible, it can only take place in the stationary phase if the latter has an adequate extension and does not too strongly resist a change of site. This is most likely to occur in partition chromatography. The total contribution of longitudinal diffusion to HETP is:

$$h_{\mathrm{a}} = h'_{\mathrm{a}} + h''_{\mathrm{a}} = \left(\gamma' + k\gamma'' \frac{D''}{D'}\right) \frac{2D'}{u} \approx \frac{2\gamma' D'}{u} \qquad (15\text{-}3)$$

In Section 15.4. it will be verified that in HPLC this contribution can in most cases be neglected. The situation is different in *planar chromatography*, where diffusion causes a

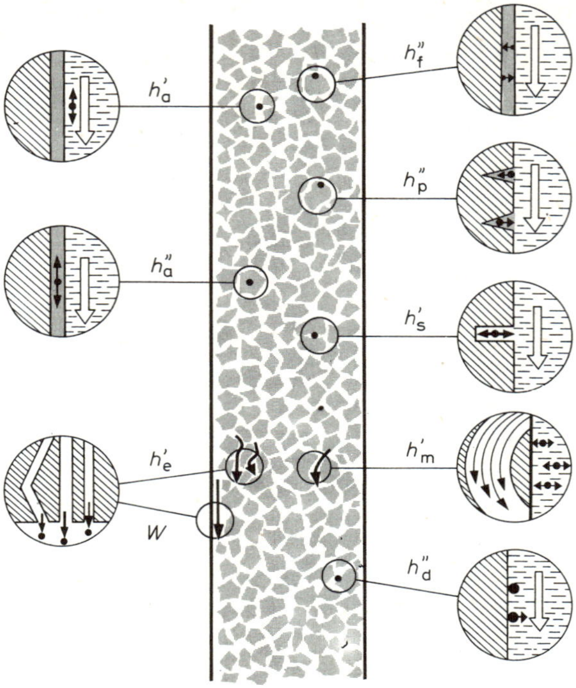

Fig. 15-3
Mechanisms of kinetic band broadening

h'_a longitudinal diffusion in the mobile phase; h''_a longitudinal diffusion in the stationary phase; h'_e eddy diffusion; W wall effect; h''_f diffusion into the interior of the stationary film; h''_p diffusion into pores filled with stationary phase; h'_s diffusion in a stagnant eluent; h'_m substance displacement in the flowing mobile phase and transverse diffusion; h''_d delayed establishment of equilibrium.

considerable disturbance mainly on too thick films. The coefficient γ' depends on the pore width and, in TLC on silica gel, passes through a minimum (PEKER et al., 1976).

15.2. Band broadening due to flow effects

15.2.1. Eddy diffusion

This means the spreading of a substance, initially contained in a narrow liquid band, over a larger volume as a result of flow effects. In a real packed column, which contains a labyrinth of interlocked channels of different widths and orientations, these effects cannot be avoided. According to GIDDINGS (1959), eddy diffusion is independent of the flow rate but increases with the particle diameter, d_P, with which the margin for the channel width increases as well:

$$h'_e = 2\lambda d_P \tag{15-4}$$

The shape and the regularity of the particles and above all the *packing quality* have a great effect. If the column is well packed, a broader trace of liquid soon encounters an obstacle which splits it up, while thinner traces merge together. In liquid-solid adsorption chromatography, values in the range 1—10 were found for 2λ.

15.2.2. Substance displacement in the flowing phase

The laminar flow in a capillary leads to a velocity profile which has the well known parabolic shape. This means that a solution volume introduced in the form of a plug is deformed because the central parts move faster than those flowing near the wall. A radial concentration gradient develops, which is reduced by *transverse diffusion*. Although the transport of the mobile phase through the packed column is only roughly comparable to a capillary flow, in principle the same phenomena occur. According to GIDDINGS [A 9] they are expressed by:

$$h'_m = \frac{\omega d_p^2 u}{D'} \tag{15-5}$$

The larger the particle size of the packing material, the wider the channels and the stronger the effect will be. The flow-rate increases the contribution, whereas the diffusion coefficient reduces it.

Eddy diffusion and *substance displacement* are closely interrelated. The effect of eddy diffusion decreases when the samples molecules frequently pass from fast-flowing liquid filaments into adjacent slow-flowing ones. The different substance displacement and the eddy diffusion interact in such a way that the band widening caused by the two effects in combination is smaller than their sum. The *flow contribution* to plate height of the two effects combined is [A 9]:

$$h'_t = \frac{1}{1/h'_e + 1/h'_m} = \frac{1}{1/(2\lambda d_p) + D'/(\omega d_p^2 u)} \tag{15-6}$$

This contribution vanishes as the flow rate tends to zero. This is plausible, but does not manifest itself in the isolated consideration of eddy diffusion. The dependence of the total flow contribution on the flow-rate, the particle size and the diffusion coefficient is difficult to determine because of the complex fraction, but in the range used practically it can be approximated by:

$$h'_t = A \cdot d_p(d_p u/D')^{1/3} \tag{15-7}$$

15.3. Band broadening due to resistance to mass transfer

In the idealized treatment of chromatography it is supposed that the *distribution equilibrium* is reached before the mobile phase flows on. It is assumed that at every moment and at each point any substance fraction in the mobile phase corresponds exactly to the substance fraction in the stationary phase as required by the distribution coefficient. In reality, however, the establishment of the equilibrium and the transport do not succeed one another, but the mobile phase already moves on during the mass transfer.

The liquid stagnating in pores and channels, the active points of which can be reached by the substance only by diffusion, has the same effect as well. Thus the band is widened by contributions being directly proportional to the flow-rate and inversely proportional to the mobility of the molecules.

15.3.1. Diffusion into the interior of the stationary phase

The chromatographically effective transition from the mobile to the stationary phase and vice versa takes place at the boundary between the phases. It can only be achieved by molecules which are near that boundary. Particles which have moved into the interior of the stationary phase must diffuse back to the phase boundary before they can cross it. This requires time, and hence contributes to band widening. If the stationary phase is a film (thickness: d_f) then the contribution to HETP is (GIDDINGS, 1959)

$$h''_f = \frac{q'' \cdot k \cdot d_f^2 \cdot u}{(1+k)^2 D''} = \frac{q'' R(1-R) \cdot d_f^2 \cdot u}{D''}, \tag{15-8}$$

whereas for the stationary phase in *saturated porous supports* one has

$$h''_p = \frac{k \cdot d_p^2 \cdot u}{(1+k)^2\, 30\psi'' \cdot D''} = \frac{R(1-R) \cdot d_p^2 \cdot u}{30\psi'' \cdot D''} \tag{15-9}$$

where q'' is the geometry factor for the stationary film ($q'' = {}^2/_3$ for a uniform liquid film), ψ'' is the resistance parameter and R is the retention ratio (cf., eqn. (3-2)).

These effects are of importance for partition chromatography (h''_f and h''_p) and steric exclusion chromatography (h''_p).

HERMANS (1968) investigated the effect experimentally in SEC. For a rapid permeation of the solute he found the essential conclusions drawn from eqn. (15-9) to be correct. For a particular gel with a constant pore depth, h''_p was directly proportional to the rate of elution and inversely proportional to the diffusion coefficient, D''. Naturally, particles which do not penetrate into the pores ($k = 0$) are also not subject to a band broadening in the stationary phase. Fig. 15-4 shows the increase in HETP for polystyrene samples with increasing flow-rate. Up to $M = 119000\ \text{g} \cdot \text{mole}^{-1}$ the molar mass mainly influences D'', so that these curves increase more steeply. On the other hand, in the neighbourhood of the phase boundary, the effect of the distribution coefficient is dominant, so that for $M = 860000\ \text{g} \cdot \text{mole}^{-1}$ the plate height is again rather low. For completely excluded samples, the plate height is independent of the flow-rate (KELLEY and BILLMEYER, 1979; GIDDINGS et al., 1977b).

The high curvature and the maximum of the curve for $M = 247000\ \text{g} \cdot \text{mole}^{-1}$, which behaves normally only up to about $5\ \text{ml} \cdot \text{min}^{-1}$, are due to the fact that the rate of permeation is too low. If the elution is too rapid, the slowly diffusing molecules have not enough time to penetrate fully into the pores. In this case, h''_p becomes proportional to $(D'')^2 \cdot K^4/v^2$, i.e., h''_p decreases as the flow-rate increases. The pore structure of the gel granules has a strong effect on the contribution h''_p. High values should be expected for pore systems branched in a labyrinthine manner, with outlets of different widths; but low values for funnel-shaped pores.

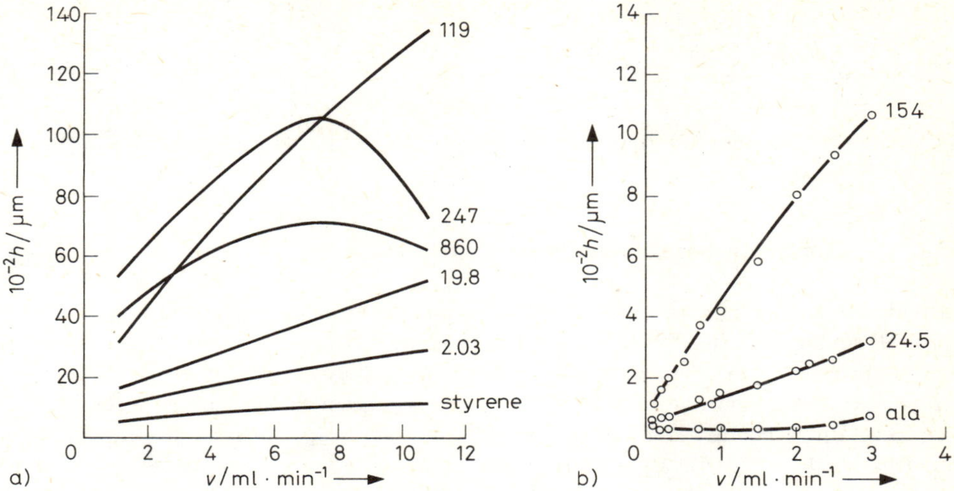

Fig. 15-4
Effect of the rate of elution on the height equivalent to a theoretical plate

(a) Polystyrene standards in chloroform on a Styragel® column (nominal pore size 10^4 Å); parameter: 10^{-3} M/g · mole^{-1}
(according to YAU, MALONE and SUCHAN, 1971)

(b) Proteins in 0.01 M phosphate buffer with 0.2 M Na_2SO_4 (pH 6.5) on TSK gel 3000 SW®; parameter: 10^{-3} M/g · mole^{-1} (154-aldolase; 24.5-chymotrypsinogen; ala-alanine, $M = 89$ g · mole^{-1})
(according to ROKUSHIKA, OHKAWA and HATANO, 1979).

The effect of the diffusion coefficient on the relationship between HETP and the rate of elution has also been investigated by VAN KREVELD and VAN DEN HOED (1978) and DAWKINS and YEADON (1979).

15.3.2. Diffusion in the stagnant mobile phase

Among the channels between the particles of the packing material, there are several which have no pressure differential acting upon them. Here, as well as in the pores of the packing particles, sample molecules move only by diffusion. The effect is significant if the sample molecules are small compared with the pore sizes. It is of great importance in *adsorption chromatography* with porous supports, where many active centres lie at the inner pore surfaces. The contribution to the plate height is (GIDDINGS, 1961)

$$h'_s = \frac{(1 - \varphi' \cdot R)^2}{30\psi'(1 - \varphi')} \cdot \frac{d_1^2}{D'} u \tag{15-10}$$

where φ' denotes the volume fraction of the stagnant mobile phase in the particles, d_1 is the pore depth and ψ' is a resistance parameter.

15.3.3. Retarded establishment of equilibrium at the phase boundary

This contribution may gain importance in adsorption chromatography if the rate constant for the adsorption step, k_{ads}, or for the desorption, k_{des}, is small. Then the contribution to the plate height is:

$$h''_{\mathrm{d}} = \frac{2(1-R)^2}{k_{\mathrm{ads}}} u = \frac{2R(1-R)}{k_{\mathrm{des}}} u \tag{15-11}$$

In classical TLC on silica gel, the spot spreading is largely due to this contribution (PEKER et al., 1972, 1976).

15.3.4. Combination of the retardation contributions

In a real chromatographic process the contributions formulated in eqns. (15-8) to (15-11) are not all equally effective. Depending on the specific separation conditions, one or other of the summands recedes. As a common characteristic they all exhibit proportionality to the *flow rate*. For a retardation due to *diffusion processes*, a further common feature is a dependence on a geometric quantity (d_{f}^2, d_{P}^2 and d_{l}^2, respectively) and the diffusion coefficient ($1/D'$ or $1/D''$). By means of suitable factors, the geometric data can be reduced to the *particle diameter*, and D'' can be expressed by D'. Thus the combined formula can be simplified as follows:

$$h''_{\mathrm{f}} + h''_{\mathrm{p}} + h'_{\mathrm{s}} + h''_{\mathrm{d}} = \left(C_1 \frac{d_{\mathrm{P}}^2}{D'} + C_2\right) u = \left(C_1 + \frac{C_2 D'}{d_{\mathrm{P}}^2}\right) \frac{u d_{\mathrm{P}}^2}{D'} \tag{15-12}$$

For high values of the distribution constant, as are possible especially in adsorption methods, and in high-efficiency columns packed with very small particles, the term denoted by C_2 is of importance (HORVÁTH and LIN, 1976, 1978; UNGER et al., 1978), whereas it can be neglected in ideal size exclusion chromatography.

15.4. Interaction of all kinetic factors

The total kinetically induced band widening (see Table 15-1) is composed of the diffusion contributions (eqn. (15-3)), the flow contributions (eqn. (15-7)) and the retardation contributions (eqn. (15-12)), and for partition and exclusion methods it is given by:

$$h = \frac{2\gamma' D'}{u} + \frac{A d_{\mathrm{P}}^{4/3}}{D'^{1/3}} u^{1/3} + C_1 \frac{d_{\mathrm{P}}^2}{D'} u \tag{15-13}$$

When written in this form, the combined expression almost coincides with the straightforward VAN DEEMTER equation (1956), in which the coupling of eddy diffusion and substance displacement as indicated by eqn. (15-6) is not taken into account:

$$h = \frac{2\gamma' D'}{u} + 2\lambda d_{\mathrm{P}} + \frac{\omega d_{\mathrm{P}}^2}{D'} u \tag{15-14}$$

Logically, the essential difference occurs in the second summand. In order to compare the efficiencies of different separation devices and techniques on the basis of (eqn. (15-13)), the *reduced plate height*,

$$h^* = h/d_{\mathrm{P}} \tag{15-15}$$

Table 15-1
Summary of the contributions to the kinetic band widening in the column

Effect	Symbol	Functional dependence	Adsorption chromatography	Exclusion chromatography	Partition chromatography
Axial diffusion					
in the mobile phase	h'_a 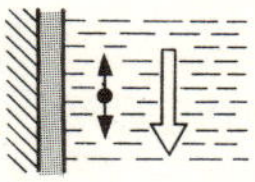	$\sim \frac{D'}{u}$	+	+	+
in the stationary phase	h''_a 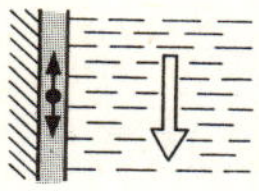	$\sim \frac{D''}{u}$	(+)		+
Flow effects					
Eddy diffusion	h'_e ↓ h'_t 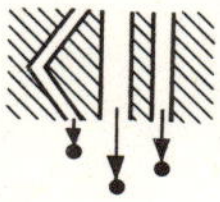	$\sim d_p$	+	+	+
Substance displacement in the mobile phase	h'_m ↓ h'_t 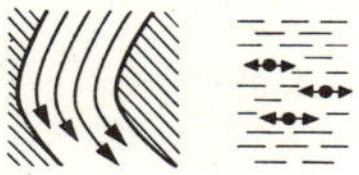	$\sim \frac{d_p^2}{D'} u$	+	+	+
Resistance to mass transfer					
Diffusion in the stationary film	h''_f 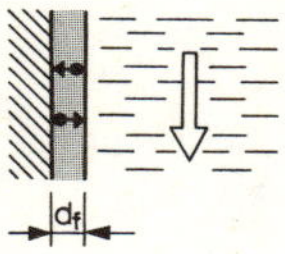	$\sim \frac{d_f^2}{D''} u$			+
Diffusion in pores filled with stationary phase	h''_p 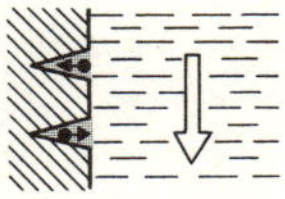	$\sim \frac{d_p^2}{D''} u$		+	+
Diffusion in a stagnant eluent	h'_s 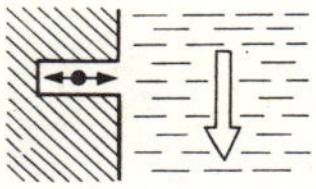	$\sim \frac{d_l^2}{D'} u$	+		(+)
Retarded establishment of equilibrium	h''_d 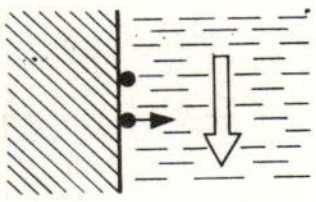	$\sim \frac{u}{k_{des}}$	+		(+)

and the *reduced velocity*

$$\nu = u \cdot d_p / D' \tag{15-16}$$

are introduced, and this finally gives:

$$\boxed{h^* = \frac{2\gamma'}{\nu} + A\nu^{1/3} + C_1\nu} \tag{15-17}$$

The coefficients of this equation can be experimentally determined for chromatographic columns by measuring HETP as a function of the flow rate. If sufficient data points are available over a wide range of the reduced velocity, ν, the parameters can be determined by curve fitting. Fig. 15-5 shows the behaviour of the curve according to eqn. (15-17). For

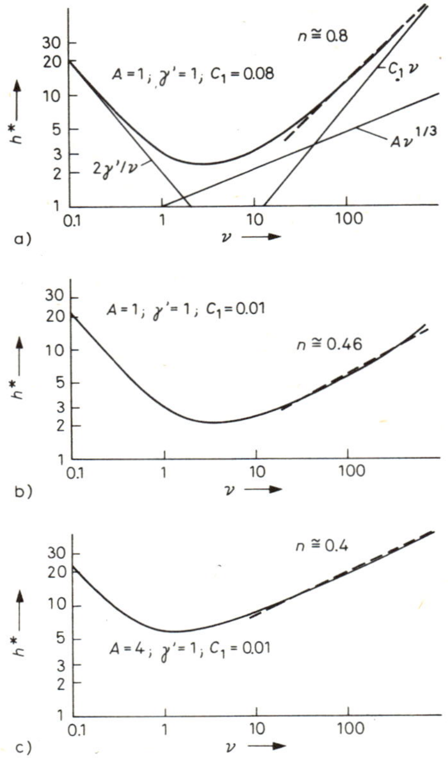

Fig. 15-5
Reduced plate height as a function of the reduced velocity according to eqn. (15-17)
a) $\gamma' = 1;\ A = 1;\ C_1 = 0.08$
b) $\gamma' = 1;\ A = 1;\ C_1 = 0.01$
c) $\gamma' = 1;\ A = 4;\ C_1 = 0.01$
The approximate evaluation of the right-hand branch of the curve by means of eqn. (15-18b) and Table 15-2 gives:
a) $n \approx 0.8;\ h^*_{100} = 12.7;\ A \approx 0.8;\ C_1 \approx 0.09$
b) $n \approx 0.46;\ h^*_{100} = 5.7;\ A \approx 1;\ C_1 \approx 0.01$
c) $n \approx 0.4;\ h^*_{100} = 19.6;\ A \approx 3.8;\ C_1 \approx 0.02$.

curve (15-5a), it is indicated how the three summands contribute to the overall behaviour. The *packing quality* is reflected in parameter A: a high value indicates an imperfect packing; for well packed columns, $A \approx 1$.

C_1 is small for a good column provided that in its design the optimum values have been chosen for the variables appearing in eqns. (15-8) to (15-10), i.e., for the pore structure, the particle shape and the thickness of the stationary film.

Then C_1 may be smaller than 0.05. This holds true for investigations using non-retained solutes. A sample with a substantial capacity factor, k, is present in the stationary phase to a correspondingly high degree. In this case, the resistance to mass transfer will increase as the effects discussed in connection with the eqns. (15-8), (15-9) and (15-11) become more and more pronounced. In these formulae this influence is expressed by the factor $(1 - R)$. Therefore plate-height investigations with probes showing a substantial retention lead to modified results, which manifest themselves especially in a steeper slope of the right hand branch of the h^*/v-curves. This means that C_2 in eqn. (15-12) can no longer be neglected. A quantitative approach gives an equation derived by GOLAY (1959).

The contributions from *longitudinal diffusion* are of little importance in comparison with the other terms in eqn. (15-17). For a reduced velocity greater than $v = 10$, the contribution $2\gamma'/v$ can be entirely neglected. Then h^* increases continuously but more and more slowly as v increases. The curve can be approximated by an exponential set-up.

$$h^* = Cv^n \tag{15-18a}$$

or

$$\log h^* = \log C + n \log v \tag{15-18b}$$

The value of n can be determined from the slope of the logarithmic plot according to eqn. (15-18b). The more the last term in eqn. (15-17) dominates over the second one, the higher the value of n will be, ranging from 0.33 to 1.00. Table 15-2 indicates the subdivision of the reduced plate height, h^*, among the second and the third summands, respectively, for different values of n. For example, let us consider the curve (15-5c). For the almost linear slope above $v = 10$, one obtains $n \approx 0.4$. For $v = 100$, $h^* \approx 19.6$. 10% of

Table 15-2
Contributions of $Av^{1/3}$ (band broadening due to flow) and $C_1 v$ (resistance to mass transfer) to the total value of the reduced plate height according to eqn. (15-17), determined from the slope, n, of the approximation $\log h^* = \text{const} + n \log v$ for $v > 10$ (according to DONE, KNOX and LOHEAC [D 6])

n	$Av^{1/3}/h^*$	$C_1 v/h^*$
0.33	1.00	0
0.4	0.90	0.10
0.5	0.75	0.25
0.6	0.60	0.40
0.7	0.45	0.55
0.8	0.30	0.70
0.9	0.15	0.85
1.0	0	1.00

this value belongs to the third summand, i.e., $C_1 v = 1.96$ or $C_1 \approx 0.02$; 90% gives $A v^{1/3} = A \sqrt[3]{100} = 17.6$, hence $A \approx 3.8$. (The starting values were $A = 4$ and $C_1 = 0.01$.) The minima of the h^* vs. v curves appear in most cases at a very small flow rate. With high polymers, it has not been observed as yet, cf., Figs. 15-4 and 15-6 as well as Section 19.5. Especially for polymers, axial diffusion in the column is of no influence.

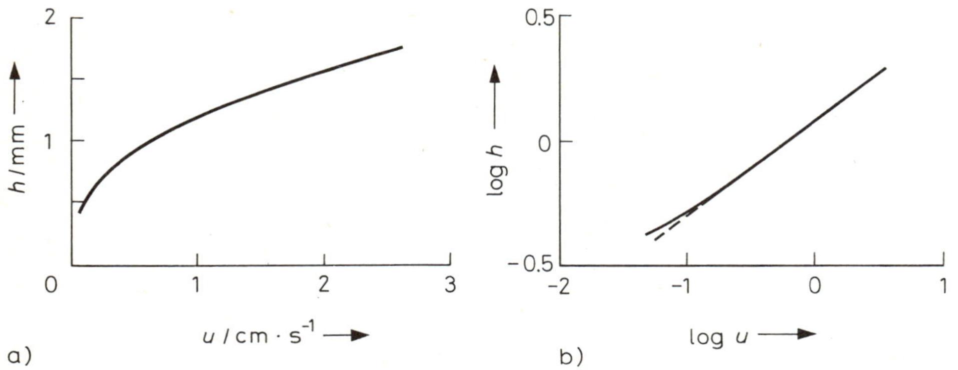

Fig. 15-6
Dependence of the plate height on the flow rate for a column packed with silica (d_P = 20 μm)
a) Cartesian representation (according to SNYDER, 1969)
b) Logarithmic representation.

An evaluation like that shown in Fig. 15-5 indicates on the one hand how well a column is packed, and on the other informs about the quality of the packing material and its contribution to the resistance to mass transfer. Improvements can be made if the respective contributions of the effects listed in Table 15-1 are known. GIDDINGS et al. (1977) carried out such investigations for exclusion chromatography of polystyrene on porous glass. If the plate height is not influenced by adsorption effects, then measurements using eluents of different viscosites should all form one and the same curve when h^* is plotted versus v. This is the case for 1,1- and 1,2-dichloroethane, whereas with 1,1-difluoroethane higher values were measured for the reduced plate height. This effect, together with the observation that the peak elution volume remained unchanged in the last mentioned eluent, led to the conclusion that the desorption required from 5 to 100 s, whereas the amount of the adsorbed polymer was very small. The resistance to mass transfer due to diffusion in the pores was determined by comparison with h^* vs. v curves measured under conditions where no permeation took place. For that purpose, on the one hand the standard sample (M = 4000 g · mole^{-1}) was measured in a geometrically analogous column containing compact glass beads, and on the other hand a sample having a higher molar mass was investigated in the standard column. The difference from the plate height as measured under permeation conditions shows that diffusion in the pores makes an essential contribution to the retardation term.

Similar investigations were carried out by HORVÁTH and LIN (1978) for adsorption chromatography, where the resistance to the establishment of equilibrium is of importance.

Fig. 15-7, which summarizes the results obtained under various conditions, shows that the minimum lies at very small values of ν. In most cases, too high rates are used. For reduced velocities of $\nu > 10^4$, the reduced plate height again decreases due to the occurrence of turbulence. CLUFF and HAWKES (1976) combined more than 750 pairs of values measured with bead packings by eighteen different teams under entirely different conditions into one h^* vs. ν diagram which extends up to $\nu = 10^7$. The common curve shows a minimum just below $\nu = 10$ and a maximum at $\nu = 10^4$.

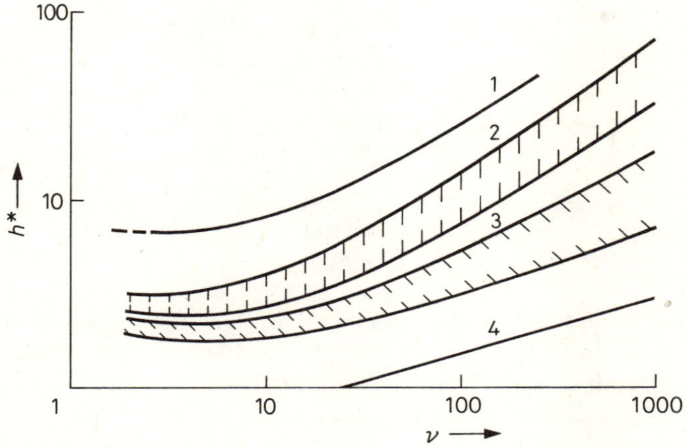

Fig. 15-7
Experimentally determined relationships between the reduced plate height ($h^* = h/d_P$) and the reduced velocity ($\nu = u \cdot d_p/D'$) for an unretained sample (according to DONE, KNOX and LOHEAC [D 6])

1 silica gel, 6—10 µm; KIRKLAND (1972)
2 Porasil, Corning porous glass; DONE et al. (1973)
3 Zipax (porous layer beads), 29—115 µm; DONE et al. (1973)
4 KNOX and PARCHER (1969); (d_p = 480 µm).

15.5. Conclusions drawn from the theory

The considerations summarized by eqn. (15-17) enable some conclusions to be drawn for effective separation paths with small h values: the columns must be packed as densely and uniformly as possible (HUBER, 1973). Low-quality packings give high values of A, and hence an earlier onset of the rise of h^* with increasing ν, i.e., high HETPs in the practically accessible range of operation. Fig. 15-8 shows the degree to which the efficiency depends on the packing technique: if the interstitial volume exceeds the minimum value by 3 cm^3, the efficiency is almost completely lost.

A most critical point is the packing near the wall. Here the flow conditions are different from those in the interior of the column, because along the wall there is an almost straight path (see Fig. 15-3: W). Large-sized particles must on no account be allowed to concentrate here due to vibrations essential for the packing procedure (cf., Section 17.1.2.). Also with a perfect packing technique, the influence of a rigid wall disturbs the regularity of the packing in a considerable part of the column cross-section. For rigid particles and narrow columns

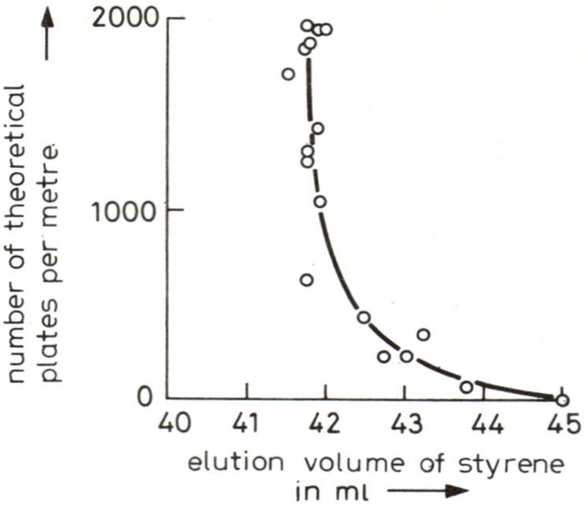

Fig. 15-8
Number of theoretical plates per metre as determined using monostyrene in chloroform for $v = 1$ ml · min^{-1} in a column with a length of 1.22 m and a diameter of 7.8 mm, which was differently packed with Bioglass® 500 (according to YAU, MALONE and SUCHAN, 1971)

this portion is very large and may amount to more than $^1/_3$ of the cross-section. It has been proposed that the column be lined with a soft film into which the outer particles of the packing can be pressed to a depth of $d_P/2$ (HIBY, 1968). An analogous effect can be achieved by radial compression of packings in tubing, e.g., made of polyethylene (FALLICK and RAUSCH, 1979). Preparative cartridges ($L = 0.30$ m; $d_C = 57$ mm) and analytical columns ($L = 0.10$ m; $d_C = 0.8$ mm) [F 19] are commercially available. In these the packing is subjected to a radially acting pressure of 1.8 MPa. Electron-microscope photographs show that part of the packing particles are pressed into the elastic wall. The chromatographic resolution of such columns is good (HANCOCK et al., 1981). Moreover the radial compression cartridges are quite inexpensive.

A generally suitable solution is presented by the *quasi-infinite* columns (KNOX and PARCHER, 1969), the diameters, d_C, and lengths, L, of which are mutually adjusted in such a way that a sample introduced in the column axis does not reach the wall on its way through the column. The geometric conditions depend on the particle size:

$$d_C^2 > 2.4L \cdot d_P \qquad (15\text{-}19)$$

For a column length of 30 cm with particles of 10 μm, an inside diameter of only 2.8 mm is sufficient to satisfy this condition. 5 mm columns with a length of 1 m and a particle size of the packing material ranging from 5 to 20 μm yielded an excellent resolution (KIRKLAND, 1972; MAJORS, 1973). The smaller and the more regular the particles of the packing material, the more plates can be achieved with a particular column. The epoch of *high-performance liquid chromatography* began with the use of particle sizes in the μm range (HUBER, 1969). 1000 or more of these particles can be arranged in the cross-section of a 3 mm column. Then the flow is very finely branched and the smoothing effect of the transverse diffusion is rather good because of the short paths. In the $0.5 \leqq u \leqq 5$ cm · s^{-1} interval of velocity, and with particles in the range $11 \leqq d_P \leqq 50$ μm, HETP increases with the particle size according to (HALÁSZ and NAEFE, 1972):

$$h = \text{const} \cdot d_P^{1.5} \qquad (15\text{-}20)$$

However, there is also a lower limit for the *optimum particle size*. HORVÁTH and LIN (1978) showed that in HPLC with reversed phases and aqueous eluents the particle size greatly influences the ratio of the contributions to HETP due to flow effects on the one hand and resistance to mass transfer on the other. While above $d_P = 10\ \mu m$ the flow contribution is dominant, below a limit ranging between 5 and 10 μm the resistance to mass transfer increases to a value many times that of the flow contribution. This effect is most pronounced for sample components exhibiting high distribution coefficients.

UNGER et al. (1978) also investigated the influence of particle size on HETP.

The smallest particles used in this case were angular alumina granules with a size of $1.8 \pm 1.1\ \mu m$. Fig. 15-9 shows an h^* vs. ν diagram obtained for spherical silica with a particle size of $3.8 \pm 1.4\ \mu m$ in a 0.1 m column with an inside diameter of 4.3 mm. All of the curves show a minimum which, for a (nearly) non-retained substance with a capacity factor of $k = 0.26$, lies at $h^* = 1.6$ and hence indicates an excellently packed column. As the capacity factor increases, h^*_{min} increases considerably, and the slope of the right-hand branch of the curve becomes steeper and steeper. The slope factor, n, in eqn. (15-18b) is close to unity, which means that the kinetics of the process is entirely governed by mass transfer. From this it follows that besides purely practical aspects, like the packing technique and the ratio of the extra-column effects to the dispersion within the column, there are also fundamental reasons suggesting an optimum of particle size which is about 5—10 μm. The specific value depends on the respective mechanism of separation.

In most cases, liquid chromatography is carried out at too high values of ν (GUIOCHON, 1976). The operating point lies on the right-hand side of the minimum of the curves shown in Fig. 15-5. There are two possibilities of getting closer to the h^* minimum: decreasing the *flow-rate* or accelerating the diffusion. A lower flow-rate extends the duration of the analysis. However, frequently for lower rates the column length can be reduced so as to

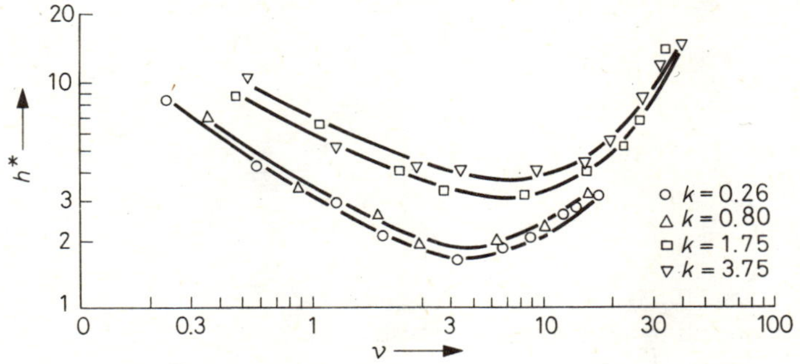

Fig. 15-9
Reduced plate height h^* vs. reduced velocity ν, as obtained from measurements with 4 samples of different retention

Column: $L = 0.10$ m; $d_c = 4.3$ mm, thermostated at 298 K, packed with silica gel microbeads, $d_p = 3.8 \pm 1.4\ \mu m$
Eluent: n-heptane, adjusted to a 33% relative water content
Samples: oligophenyls ($n \leq 5$)
The greater the capacity factor k of the sample, the higher is the h^* vs. ν curve obtained. The minimum shifts to higher values of the reduced velocity.
(according to UNGER, MESSER and KREBS, 1978).

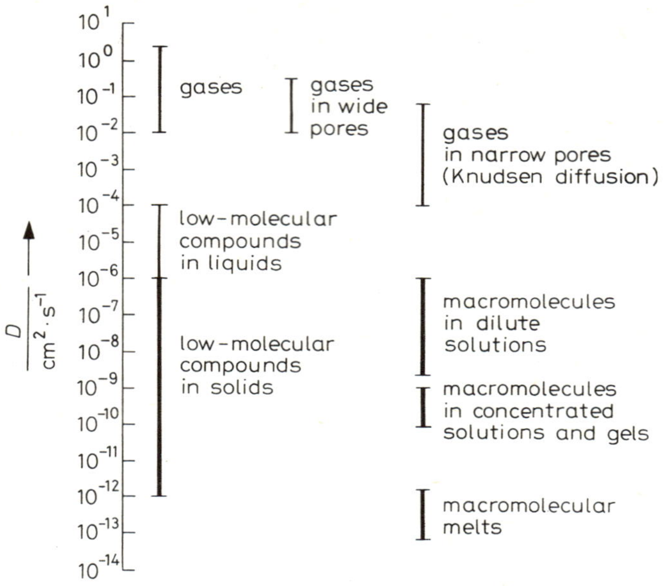

Fig. 15-10
Ranges of the diffusion coefficients for several substances

achieve higher performances even within the same time (GUIOCHON, 1976). As a desirable side effect, a much lower pressure is required.

The *diffusion coefficient* play an important rôle. In liquids they are smaller than in gases by orders of magnitude (see Fig. 15-10). Thus in liquid chromatography a resolution comparable with that of gas chromatography can be achieved only at a much higher expense. The diffusion coefficients are inversely proportional to the viscosity, η', of the solvent

$$D' = \frac{7.4 \cdot 10^{-8} (\psi M')^{1/2}\, T}{V^{0.6} \eta'} \tag{15-21 a}$$

(CHANG and WILKE, 1955), or

$$D' = \frac{1.4 \cdot 10^{-4}}{V^{0.6} \eta'} \tag{15-21 b}$$

(OTHMER and THAKER, 1953),
where ψ is the association coefficient (in most cases $\psi = 1$, but 2.6 for water, 1.9 for methanol, and 1.5 for ethanol as a solvent); M' is the molar mass of the solvent in $g \cdot mole^{-1}$; T is the absolute temperature in K and V is the molar volume of the solute in $cm^3 \cdot mole^{-1}$.

Therefore the viscosity of the solvent should be as low as possible ($\eta' \leqq 0.5$ mPa · s). The increase of the height equivalent to a theoretical plate with increasing viscosity was experimentally confirmed (SNYDER, 1967; HORVÁTH et al., 1967). For very high flow rates the effect of the viscosity is smaller than for low ones (KROL et al., 1972).

As the molecular size increases, the diffusion coefficients decrease. Thus in the macromolecular range a high resolution is even more difficult to attain than in normal liquid chromatography. Here, too, the thermodynamic quality of the solvent is of some importance because it promotes the expansion of the coils. To reduce the viscosity and to increase the diffusion coefficients, it is advisable to operate at an elevated temperature (LITTLE and PAUPLIS, 1971; COOPER, 1974). If the other conditions permit, this should in fact be done. Of course there are additional difficulties because the highest effect can be obtained by means of solvents having a low initial viscosity, but these solvents, as a rule, exhibit rather low boiling points. Therefore appropriate measures such as cooling or detection under pressure are required at the outlet of the apparatus. In this respect, fluid chromatography in supercritical media is of special interest (SIE and RIJNDERS, 1967; GIDDINGS et al., 1969; JENTOFT and GOUW, 1969; GOUW and JENTOFT, 1972; KLESPER and HARTMANN, 1977a) (cf., Section 9.7.).

At this point, some remarks on the optimization of liquid chromatographic separations may be useful (reviews: MORGAN and DEMING, 1976; KAISER and OELRICH [D 24]).

MARTIN et al. (1974, 1975a, b) investigated the question of the column dimensions which should enable a certain separation efficiency (e.g., $N = 5000$ for $k = 2$) with as little effort as possible. For this purpose they developed an *economical optimization model* which relates the DARCY equation for the pressure drop, (17-5), with the equations for the retention time and the plate height, (3-10) and (15-13), respectively. By means of this model it is possible to develop nomograms for the desired separation efficiency, which combine the column length, L, the operating pressure, Δp, and the particle diameter, d_P. GUILLEMIN et al. (1977) reported on the application of this model in the development of a routine technique with spherical silica gel. Using packing materials of $d_P = 5—7\ \mu m$, 800 plates/cm were achieved in short columns ($L \leqq 0.10$ m) at a moderate pressure (1—3 MPa). The analysis time was 5—15 min.

To find the optimum of the experimental conditions for a given separation problem, LAUB and PURNELL (1978) experimentally determined the ratio k_I/k_{II} (which governs the resolution, cf., eqn. (14-4)), for all components of the sample as a function of one main variable, e.g., the temperature. For experienced chromatographers, the graph of k_I/k_{II} vs. this single, essential quantity is frequently sufficient to determine the optimum working range. If necessary, the investigation is carried out in the same way for another variable, say, for the composition of the eluent or the plate number.

GANT et al. (1979) used an optimization method which also requires the experience of the chromatographer, but manages with fewer experiments and is quicker, simultaneously taking into account the dependence on both the composition of the eluent and the temperature. The starting point was the experimental determination of the capacity factors for the two most demanding components of the sample at two different temperatures and in two different eluents. From these 2×4 data, the dependence on the temperature and the composition was calculated by means of simple, linear approaches. Using these dependences and certain parameters of the column, the k values for other working conditions were calculated by means of a programmable pocket computer, thereafter calculating the diffusion coefficients (by eqn. (15-21a)), the reduced plate height by eqns. (15-15) to (15-17), the pressure drop across, the column by eqn. (17-3b), and finally the resolution, R_S, by eqn. (14-4).

The optimization of the eluent composition in reversed-phase chromatography using water and three other components was investigated by GLAJCH et al. (1980). Here, as a yardstick, the *chromatographic optimization function* (COF) was initially used, in which for all the components the logarithms of the R_S values (cf., eqn. (3-25)), as referred to the desired value (e.g., $R_S = 1.5$), are included as summands. Outside the optimum, the COF has a negative value, being zero in the optimum. After ten test chromatograms had been obtained with a systematically varied eluent composition, it was possible to calculate the course of the COF and to determine the position of the optimum. Since in the general case it is possible that a reversal in the order of peaks occurs upon a change in the eluent composition, the authors finally used the method of *overlapping resolution maps*. This starts from a graphical representation of the resolution, R_S, for a pair of substances each as a function of the modifier composition. In each of these triangular graphs the range of insufficient resolution was indicated by shading. Then the combination of all these *resolution maps* in a common diagram (similar to the *window diagram* method of LAUB and PURNELL, 1975) shows the optimum for the total sample: the optimum modifier composition lies in the remaining part of the area of the triangle — this is the part which is left between the partially overlapping shaded areas.

LINDBERG et al. (1981) carried out the optimization of a reversed-phase ion-pair chromatographic system for the separation of four alkaloids. Following a full factorial design, they investigated the influence of the variables (i) eluent strength (i.e., the methanol-water ratio), (ii) pH, (iii) the concentration of the phosphate buffer and (iv) of the ion-pair reagent.

In order to get a better picture of the probable optimum, the authors expressed the influence of the two most important variables, the concentration of the ion-pair reagent, x, and the eluent strength, y, on the capacity factor by an equation of degree 2. Using this mathematical model, response surfaces (for each of the four alkaloids) were generated which were represented in an x/y plane as isoresponse contour lines, i.e., by curves connecting points of equal k values. The optimization made it possible to reduce the time required for an analysis by about 30%, without loss of resolution.

A different aspect of optimization was discussed by GUIOCHON (1979), i.e., the question of how as high a peak capacity as possible can be achieved within a given analysis time, which is assumed to be

$$t_e = t' + t''$$

for the sample retained over the longest time. Using this relationship, eqn. (13-4) can be rearranged as follows:

$$n_p = 1 + (N/16)^{1/2} \ln (t_e/t') \qquad (15\text{-}22)$$

The desired maximum value is obtained by setting the first derivative equal to zero:

$$\ln (t_e/t') = 2 \qquad (15\text{-}23)$$

This gives

$$n_{p,\,max} = 1 + \frac{\sqrt{N}}{2} \qquad (15\text{-}24)$$

and using eqn. (3-10) one finally obtains:

$$1 + k = e^2 = 7.4 \qquad (15\text{-}25)$$

Thus within a given analysis time the maximum peak capacity can be achieved by eluting all of the components in the range $0 \leqq k \leqq 6.4$. If a component is retained in the column too long, then this can be remedied by using a stronger eluent and a longer column. Conversely, for too rapid an elution ($t_e < 7.4t'$) a weaker eluent used in a shorter column offers a possibility of separating more peaks within the given time.

16. Special problems

The statements made in the preceding chapters were selected and presented with a view to the specific problems of polymer chromatography, but they also referred to general problems of liquid chromatography. This chapter presents relationships encountered in work on polymers, which are mainly related to SEC.

16.1. Determination of the molar mass distribution from a chromatogram

The band broadening discussed in Chapter 15 also occurs with polymer samples. However, while for normal mixtures there is a chance of obtaining a peak for each individual component by means of suitable measures, for polymers one has to be satisfied with distribution curves. Although in the range of oligomers there are already pretty good examples of *single-peak separations* of polymer homologues, in the macromolecular range this cannot be accomplished by exclusion chromatography alone. Owing to the limits for the distribution coefficient, $0 \leqq K \leqq 1$, a gel chromatogram can only contain a limited number of peaks. GIDDINGS (1967a) stated a number of 20 for orientation. HEITZ (1975) even succeeded in obtaining individual peaks for the first 26 styrene and 29 tetrahydrofuran oligomers (see Fig. 8-8). Nevertheless, the peak capacity of a normal gel chromatogram is not sufficient for an indication of the individual polymer homologues. What can be obtained is a curve which in all of its points corresponds to the mass proportions of the differently sized macromolecules in the sample. This pessimistic statement might worry an expert in chromatography more than a polymer scientist, for whom the quasi-continuous distribution curves (cf., Section 4.2.) are part of his professional knowledge.

Thermodynamic and kinetic effects cause an individual component to be contained in a rather large volume at the end of the separating path, even though it might have been injected in an infinitesimally small volume. Frequently the concentration of each component follows a Gaussian distribution. A special problem of polymer chromatography lies in the fact that the polymer homologues succeed one another so closely. This causes a high degree of overlapping of their concentration curves (see Fig. 16-1). The larger the band broadening due to the apparatus and the broader the *dispersion function*, the higher the degree of this overlapping will be. As a consequence of band broadening due to the apparatus, an elugram is always broader than the underlying chain-length distribution. As the elution volume in exclusion chromatography represents the measure of the degree of polymerization, the raw, uncorrected elugram at its ends seems to reflect components which

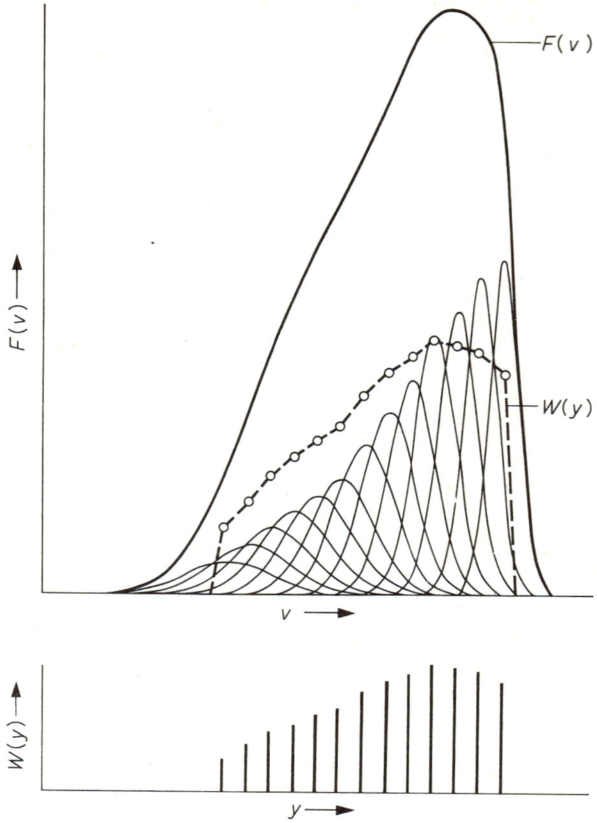

Fig. 16-1
Calculated "gel chromatogram" of a polydisperse sample with thirteen components
For every value of v, the height, $F(v)$, of the chromatographic curves is the sum of the contributions of all components whose elution curves extend over this abscissa value. The desired quantity is the mass fraction, $W(y)$, of that component which has its peak maximum at v and would alone determine the value of $F(v)$ if there were no spreading.

in reality are not present in the sample. The curve has spread as compared to the molar mass distribution. Narrow, bimodal or multimodal distributions may yield levelled elution curves lacking any profile, especially if the resolution power of the apparatus is low. It is the task of the correction methods to restore the original, narrow curve (see Fig. 16-2).

Even with a good resolution, each point of the elution curve is always determined by contributions from several components:

$$F(v) = \Sigma A_i G(v - y_i) \tag{16-1}$$

Here $F(v)$ is the ordinate value of the elution curve for the elution volume, v, and A_i are the proportions of the components which, with their elution curves, appreciably contribute to the ordinate value $F(v)$. Constituents whose elution maximum, y, is far enough from v can be ignored. $G(v - y_i)$ is the density distribution of the components to be taken into

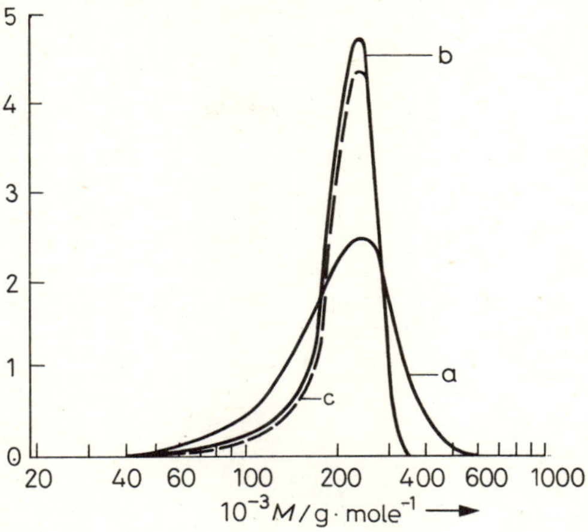

Fig. 16-2
Molar mass distribution of a polystyrene sample

a taken from the uncorrected gel permeation chromatogram; *b* derived from curve *a* by a correction method; *c* distribution curve derived from sedimentation investigations in the ultracentrifuge. (according to TUNG, 1971 b).

account in the elugram. Frequently the normal distribution

$$f(x) = \frac{1}{\sigma\sqrt{2\pi}} \mathrm{e}^{-\frac{(x-x_0)^2}{2\sigma^2}}$$

is set for $G(v - y_i)$.

This gives $x = v$, and $x_0 = y_i$ corresponds to the abscissa of the mode for the respective component considered. σ denotes the *standard deviation*, σ_a, *due to the apparatus*. If the number of components is large and their graduation is infinitesimally fine, then the summation can be replaced by integration (TUNG, 1966a):

$$\boxed{F(v) = \int W(y)\, G(v - y)\, \mathrm{d}y} \tag{16-2}$$

$W(y)$ represents the proportions of the components in the sample as a function of the peak position, y, which depends on the molar mass of the sample. Consequently, in eqn. (16-2) it is also possible to replace $W(y)$ by one of the distribution functions $H(M)$ or $H(P)$ discussed in Chapter 4. However, as it is practical to apply the corrections directly to the elugram, the formulation (16-2) chosen by TUNG will be retained. The integration has to be extended from the starting point to that value of v and y, respectively, where the curve returns to the baseline. Outside these limits there are no other contributions. Therefore one may formally extend the integration from $-\infty$ to $+\infty$. This holds for eqn. (16-2) as well as for all other integrals in this chapter.

The equations for the individual proportions $W(y)$ can be linearly combined if the *spreading functions*, $G(v - y)$, are independent of the concentration. Initially this is an open question, because the calibration curves determined from samples with narrow distributions depend on the concentration (see Fig. 16-3). However, investigations using ^{14}C-labelled

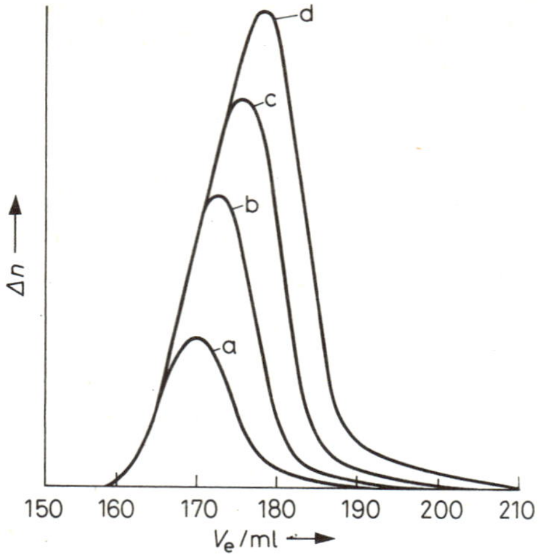

Fig. 16-3
Elution curves of a high-molecular-weight polystyrene ($\bar{M}_v = 1.5 \cdot 10^6$ g · mole^{-1}) in tetrahydrofuran as a function of the sample quantity

a 0.6; b 1.2; c 1.8; d 2.4 mg; injected volume: 0.3 ml (according to BERGER, 1975).

monodisperse polystyrene in mixtures with inert polymers showed that this effect disappeared when the viscosity gradient was small enough. Consequently the linear combination according to eqn. (16-2) is permissible for samples with broad distributions (BERGER, 1975). The individual elution curve of a sample with $M = 51\,300$ g · mole^{-1} remains unchanged even with an excess of other components (Fig. 16-4). The situation is different in the extremely macromolecular range, as will be discussed in Section 19.1. (p. 384).

To derive the molar mass distribution from the elugram, one has to determine $W(y)$. $F(v)$ is experimentally accessible and can be recorded directly as a detector signal in elution chromatography. $W(y)$ is the desired distribution function, which can be transformed into the true molar mass distribution by means of calibration relationships. Determining $W(y)$ from $F(v)$ requires a knowledge of the dispersion function, $G(v - y)$, as well as mathematical methods for the solution of the integral equation (16-2).

In principle, the spreading makes the profile of a mixture composition appear more or less levelled in the chromatogram. In the inverse transformation of $F(v)$ into $W(y)$, this "attenuation effect" has a corresponding "amplification effect", which increases with the increasing deviation of $G(v - y)$ from unity. This amplification effect may cause *artefacts*, which appear as oscillations of the $W(y)$ curve. These disturbances, which had already been found by TUNG (1966a, b), were investigated more closely by DUERKSEN and HAMIELEC (1968a/b, 1971). Countermeasures to be considered are the smoothing of the $F(v)$ curve, but above all the improvement of the precision achieved in evaluating the chromatographic data (TUNG, 1971a). Problems especially arise in the correction of the chromatograms of samples with narrow distributions. On the other hand, the latter are more affected by the broadening due to the apparatus. A resolution power characterized by $\sigma_a^2 = 12.5$ ml^2 effects a considerable change of the narrow curve of a sample with a logarithmic normal distribution with $H_t = 1.02$ (Fig. 16-5a), but hardly changes the curve of a sample with $H_t = 2.0$ (Fig. 16-5b). The molar mass distribution of normal laboratory

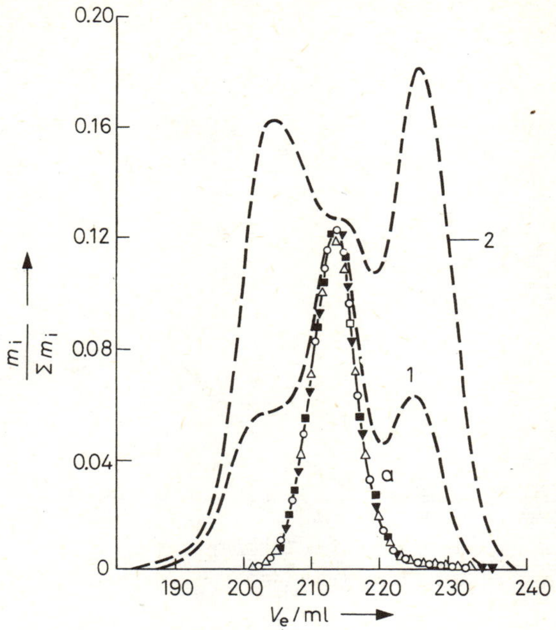

Fig. 16-4
Elution curve of a ^{14}C-labelled polystyrene (A, $M = 51\,300$ g · mole^{-1})

represented as an averaged normalized activity curve (a) in SEC in THF without and with additions of unlabelled polystyrene samples (B, $M = 30\,000$ and C, $M = 80\,000$ g · mole^{-1}); $V_0 = 0.3$ ml. Curve I: refractometer curve for a mixture of 0.9 mg A and 0.6 mg of B and C each. The associated activity curve is defined by the square symbols on curve a. Curve 2: ditto for a mixture of 0.9 mg A and 1.8 mg of B and C each. The associated activity curve is defined by the solid triangles on curve a. The other symbols refer to the elution of A without additions (circles: 0.6 mg; open triangles: 2.4 mg) (according to BERGER, 1975).

samples and technical grade polymers can thus be obtained from the *uncorrected chromatogram* if the resolving power of the apparatus is not too low. The large errors occur for fractions and samples with a very narrow distribution.

Precise information about the spreading function and its determination will be given in Section 16.3. For the time being we shall restrict ourselves to the simple case where the Gaussian distribution with a constant variance, σ^2, can be set for $G(v - y)$.

16.1.1. Solution of eqn. (16-2) by minimization methods

For N points of the chromatogram with a constant abscissa interval, $(\Delta y)_i$, a set of linear algebraic equations

$$F(v_k) = \sum_{i=1}^{n} W(y_1)\, G(v_k - y_i)\, (\Delta y)_i \tag{16-3}$$

where $k = 1, 2, \ldots, N$, is formulated. In principle such an equation can be established for every point, $F(v_k)$, of the curve. If the number of these equations equals that of the

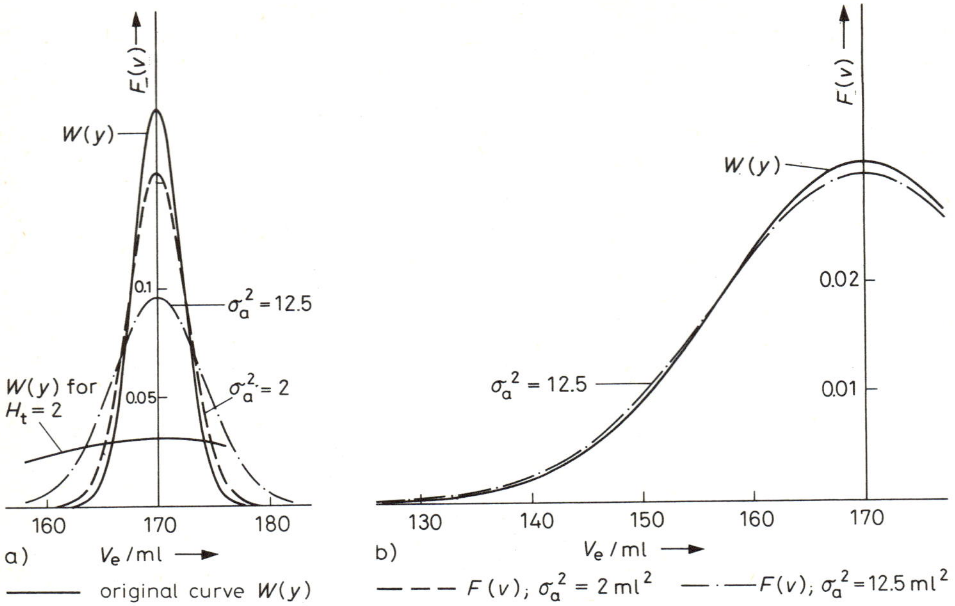

Fig. 16-5
Logarithmic normal distributions before and after spreading by a Gaussian function according to eqn. (16-2)

a) $H_t = \bar{M}_w/\bar{M}_n = 1.02$
b) $H_t = 2.0$
(This initial curve $W(y)$ is also indicated in Fig. 16-5a.)
For the sample it was assumed that $M_{max} = (\bar{M}_n \cdot \bar{M}_w)^{0.5} = 2 \cdot 10^5$ g · mole^{-1}; for the apparatus a linear calibration relationship was assumed: $V_e = 358.5 - 35.55 \cdot \log M_{max}$.

desired components ($N = n$), one has a linear set of equations, which can be solved in a relatively simple manner. The result is, however, unsatisfactory, because oscillations frequently occur at the boundaries of the functional range.

If more equations are set up ($N > n$), i.e., if an overdetermined set is established, then an approximate solution can be calculated by minimization methods such as the least-squares method. The evaluation methods given by Tung (1966a), Hess and Kratz (1966), Smith (1967) and Pickett et al. (1968) are based on this conception. A tendency to oscillations, which decreases as the number of equations is increased, can be suppressed by a programme step rejecting any negative values of $W(y_i)$ (Tung, 1966a). Timm and Rachow (1975) applied the method to systems with non-linear calibration relationships by subdividing the latter into individual sections, which could be considered linear to an adequate approximation.

16.1.2. Solution of eqn. (16-2) by iteration

In the correction methods given by Chang and Huang (1969, 1972) as well as Ishige et al. (1971), eqn. (16-2) is solved by iteration. Ishige et al. start from a form rearranged to equal zero

$$0 = F(v) - \int W(y)\, G(v - y)\, \mathrm{d}y \tag{16-2a}$$

into which they substitute $F(v)$ as a first approximation for the desired function $W(y)$. Integration yields a remainder, $\Delta F_1(v)$, for the difference on the right-hand side:

$$\Delta F_1(v) = F(v) - \int F(y)\, G(v - y)\, \mathrm{d}y \tag{16-2b}$$

This is used in calculating the second approximation. The difference left in this step is inserted in the next one, and so on:

$$\Delta F_2(v) = \Delta F_1(v) - \int \Delta F_1(y)\, G(v - y)\, \mathrm{d}y \tag{16-2c}$$

$$\vdots$$

$$\Delta F_i(v) = \Delta F_{i-1}(v) - \int \Delta F_{i-1}(y)\, G(v - y)\, \mathrm{d}y \tag{16-2d}$$

The iteration is continued until ΔF_N is close enough to zero. Then the sum of all these equations is taken, in which most of the summands cancel each other pairwise. Finally this gives

$$\Delta F_N = F(v) - \int \sum_{i=0}^{N-1} \Delta F_i(y)\, G(v - y)\, \mathrm{d}y \tag{16-2e}$$

where $\Delta F_0(y) = F(y)$.

A comparison of this final equation with eqn. (16-2a) shows that, for $\Delta F_N \to 0$:

$$W(y) = \sum_{i=0}^{N-1} \Delta F_i(y) \tag{16-4}$$

As the values $\Delta F_i(y)$ represent differences, $W(y)$ may also become negative. In this case oscillations occur in the tails of the corrected curves.

Another correction method, which was proposed in the same paper (Ishige et al., 1971), helps to avoid this error. The method again starts from the assumption that $W_1(y) = F(v)$, i.e., to a first approximation the experimentally determined elution curve is considered to be the sought-after molar mass distribution curve. Of course this is not true, because $F(v)$ is, owing to the instrumental spreading, always broader than $W(y)$. Hence the elution curve $F_1(v)$ calculated from $F(v)$ by eqn. (16-2b) is likewise broader than the measured one. The calculated curve is normalized to the area of the measured curve and compared with the latter point by point, i.e., the ratio $q_1 = F(v)/F_1(v)$ is determined for a great many abscissa values, v. This ratio is greater than 1 in the neighbourhood of the curve maximum, whereas $q < 1$ at the slopes of the curve. Point-by-point multiplication of the first approximation $W_1(y)$ by the factor q_1 gives $W_2(y)$. The curve of this function is narrower, because its ordinate values are higher than those of $W_1(y)$ in the centre, but lower on the slopes. From the second approximation $W_2(y)$, an elution curve $F_2(v)$ is calculated, which is again normalized and compared with the experimental elution curve $F(v)$. As was done with q_1 before, the ratio $q_2 = F(v)/F_2(v)$ is used to calculate a third approximation, $W_3(y)$, for the distribution. This procedure is continued until satisfactory agreement of the calculated elution curve, $F_i(v)$, with the measured one is achieved. The function used to calculate this curve, $W_i(v)$, is the true distribution function. In most cases ten iterations will suffice.

The methods developed by Ishige et al. also allow a calculation with a variable σ_a^2 and skew dispersion functions. In comparative investigations they proved very efficient (Danielewicz et al., 1977; Vozka and Kubin, 1977).

PARK and GRAESSLEY (1977a) corrected the chromatograms iteratively until the shape of the corrected curve no longer showed a visible change from one step to another. Six iterations were sufficient for that purpose even with narrow distributions.

KISLOV et al. (1978) developed an iteration method in which the slopes of the chromatographic curves are expressed by exponential functions, whereas a polynomial set-up is used for the central part. By this method, the correct $\bar{M}_w/\bar{M}_n$ values for the polystyrene standards investigated by experiment could even be calculated from chromatograms exhibiting an unusually large broadening; however, this required 100—200 iterations.

The iterative correction methods can be adapted to different dispersion functions and yield non-oscillating solutions.

16.1.3. Solution of eqn. (16-2) after approximating it by a polynomial

The continuous distribution function $W(y)$ approaches zero as $y \to \pm\infty$. In this method it is replaced by an expression which also satisfies this condition, representing the true course of $W(y)$ as accurately as possible in the required range and being integrable in combination with the dispersion function $G(v - y)$. The function

$$\tilde{W}(y) = \exp\left[-p^2(y - y_0)^2\right] \sum_m a_m(y - y_0)^m \tag{16-5}$$

(where p, y_0 are adjustable parameters) has such properties if the coefficients a_m are suitably chosen. TUNG (1966a) used Hermitian polynomials, the generating function of which,

$$\exp\left(-\frac{t^2}{2} + x \cdot t\right) = \sum_{m=0}^{\infty} \frac{H_m(x)\, t^m}{m!} \tag{16-6}$$

yields

$$H_0(x) = 1$$

$$H_1(x) = x$$

$$H_2(x) = x^2 - 1$$

$$H_3(x) = x^3 - 3x$$

$$H_4(x) = x^4 - 6x^2 + 3$$

$$H_5(x) = x^5 - 10x^3 + 15x$$

How many terms of the series (16-5) have to be included depends on the complexity of the function and on how well the parameters p and y_0 are fitted. For the Gaussian curve, $m = 0$. However, even complicated bimodal distributions can be represented very well by a relatively small number of terms. Fig. 16-6 shows an example.

If in eqn. (16-2) the Gaussian distribution with a constant σ_a^2 is substituted for $G(v - y)$, then a term-by-term integration is possible. If the curve of the chromatogram, $F(v)$, is likewise approximated by an expression according to eqn. (16-5), then a comparison of coefficients can be carried out, giving the desired series $\tilde{W}(y)$ (TUNG, 1966a, 1969; ALDHOUSE and STANFORD, 1968). In practice it has proved useful to fit a polynomial of degree 4 to the chromatogram section by section, with nine interpolation points each, in such a way

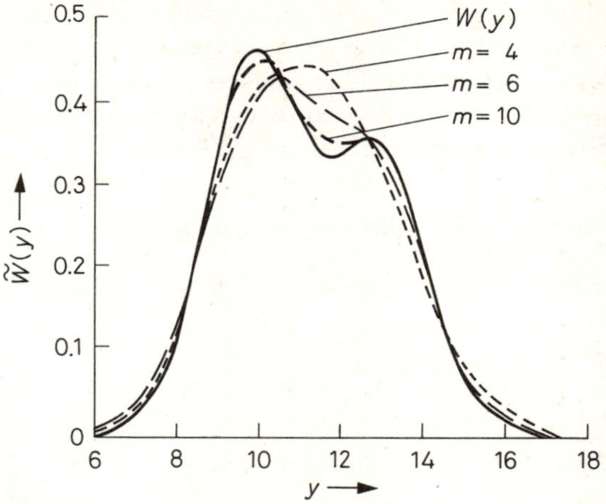

Fig. 16-6
Representation of a chromatographic curve according to eqn. (16-5)

A good approximation to the bimodal curve is already achieved with ten terms ($n = 10$) of a Hermitian polynomial. For $m \geqq 16$ there is no visible difference between $W(y)$ and $\tilde{W}(y)$.
(according to TUNG, 1971 a).

that each significant point is once used as the centre. The sectionwise procedure allows one to take into account the variation of σ_a^2 with the elution volume. The tendency to oscillations is suppressed by the type of fitting, which corresponds to a smoothing (TUNG, 1971 a).

16.1.4. Solution of eqn. (16-2) by Fourier transformation

Fourier analysis has been used by PIERCE and ARMONAS (1968), TUNG (1969), VLADIMIROFF (1970, 1971) as well as ROSEN and PROVDER (1971) to derive the corrected distribution $W(y)$ from the elugram $F(v)$. The Fourier transforms of the three functions $F(v)$, $W(y)$ and $G(v - y)$ of eqn. (16-2) are:

$$f(k) = \frac{1}{\sqrt{2\pi}} \int_{-\infty}^{+\infty} F(v)\, \mathrm{e}^{ikv}\, \mathrm{d}v \tag{16-7}$$

$$w(k) = \frac{1}{\sqrt{2\pi}} \int_{-\infty}^{+\infty} W(y)\, \mathrm{e}^{iky}\, \mathrm{d}y \tag{16-8}$$

$$g(k) \frac{1}{\sqrt{2\pi}} \int_{-\infty}^{+\infty} G(v - y)\, \mathrm{e}^{ik(v-y)}\, \mathrm{d}(v - y) \tag{16-9}$$

Use of the convolution theorem gives:

$$w(k) = \frac{1}{\sqrt{2\pi}} \frac{f(k)}{g(k)} \tag{16-10}$$

The transformed functions on the right-hand side of eqn. (16-10) can be calculated from the basic function, so that $w(k)$ is accessible. The inverse transformation yields the desired function $W(y)$. The method presupposes that $G(v - y)$ is independent of the elution volume, i.e., σ_a = const. If this is not the case, the method can be carried out section by section with the G value being applicable in each interval.

16.1.5. Solution of an equivalent partial differential equation instead of eqn. (16-2)

If the dispersion can be described by a Gaussian function, then eqn. (16-2) reads as follows:

$$F(v) = \frac{1}{\sqrt{2\pi\sigma^2}} \int W(y) \exp\left[-\frac{(v-y)^2}{2\sigma^2}\right] dy \tag{16-11}$$

VOZKA and KUBIN (1977) showed that the solution of a boundary-value problem for a certain diffusion problem has quite an analogous form. The equation to be considered is

$$D\frac{\partial^2 f(x,t)}{\partial x^2} = \frac{\partial f(x,t)}{\partial t} \tag{16-12}$$

with the boundary conditions

$$f(x, 0) = W(x) \tag{16-13}$$

$$f(\infty, t) = 0 \tag{16-14}$$

$$f(-\infty, t) = 0 \tag{16-15}$$

for the one-dimensional diffusion of a substance which at the time $t = 0$ is distributed over the distance co-ordinate x according to the function $W(x)$. The distribution after a diffusion time, t_d, for a constant diffusion coefficient, D, is obtained as a solution of this boundary-value problem:

$$f(x, t_d) = \frac{1}{\sqrt{4\pi D t_d}} \int_{-\infty}^{+\infty} W(x) \exp\left[-\frac{(x_0 - x)^2}{4Dt_d}\right] dx \tag{16-16}$$

This solution is equivalent to eqn. (16-11), where $2Dt_d = \sigma^2$, $x = y$ and $f(x, t_d)$ corresponds to the elution curve $F(v)$ with its broadening due to the dispersion. Thus the solution of Tung's integral equation can be reduced to the determination of the boundary condition $W(x)$ of the problem formulated by eqns. (16-12) to (16-15). The function $f(x, t_d)$ is obtained by experiment as an elution curve, and σ^2 is determined by one of the methods discussed in Section 16.3.2., or estimated utilizing the dependence shown in Fig. 16-14. For the numerical solution by means of a Taylor series, the functional values of the elugram are required at equidistant interpolation points. This procedure requires comparatively little computing work, which can, if necessary, even be done using a manual calculator; nevertheless the results achievable are pretty good.

16.1.6. Correction by the subtraction of ideal distributions

The subtraction method according to FRANK et al. (1968) can also be realized without a computer. Moreover it is most efficient for samples with a *narrow distribution*, for which the methods based on the solution of the integral equation (16-2) did not yield good results. For this subtraction method, a Gaussian curve which, at the mode and the sides, conforms well to the experimental curve is matched to the latter. This Gaussian curve is corrected for instrumental spreading; thereafter the segments cut off are re-added to the corrected bell-shaped curve.

The semiempirical method of SERVOTTE and DE BRUILLE (1975), which applies different corrections depending on the slope, is likewise a suitable compromise.

16.2. Determination of the mean values of the molar masses

The mean values can be calculated according to eqns. (4-2b) to (4-7b) by integration of the corrected, normalized chromatograms. This requires the abscissa transformation $V_e \rightarrow M$ (PICKETT et al., 1966). If the calibration relationship is non-linear, it should be available in an analytical form rather than only as a curve, in order to solve this problem in an economical way (cf., Section 8.4.).

If the integration is carried out for uncorrected elugrams, then the weight average, $\bar{M}_{w(u)}$, calculated in this way may by chance coincide with the true value, $\bar{M}_{w(t)}$, but $\bar{M}_{n(u)}$ is always smaller than $\bar{M}_{n(t)}$. In general there will be large deviations. This will be discussed in greater detail in Section 16.3.2.

If only the mean values are of interest and otherwise the uncorrected curves suffice, then Tung's equation (16-2) can be solved analytically by a *Laplace transformation* (HAMIELEC and RAY, 1969; HAMIELEC, 1970). This is possible if the dispersion follows a Gaussian function, and a linear relationship between V_e and $\log M$ exists. The result

$$M_{k(t)}/M_{k(u)} = \exp\left[(3 - 2k)\, D_2^2 \sigma_a^2/2\right] \tag{16-17a}$$

where k is the order of the mean values ($k = 1$), number average; $k = 2$, weight average; $k = 3$, z-average) and D_2 is a constant taken from eqn. (8-4) ($M = D_1\, e^{-D_2 V_e}$), in the form given in the 1970 paper

$$\frac{\bar{M}_{k(t)}}{\bar{M}_{k(u,\,z)}} = \exp\left[(3 - 2k)\, D_2^2 \sigma_a^2/2z\right] \tag{16-17b}$$

where z is the number of cycles, can be used directly for the determination of actual mean values. Taking the logarithm one obtains, for $k = 1, 2, 3, \ldots$:

$$\ln \bar{M}_{n(u,\,z)} = \ln \bar{M}_{n(t)} - (D_2^2 \sigma_a^2/2z) \tag{16-18a}$$

$$\ln \bar{M}_{w(u,\,z)} = \ln \bar{M}_{w(t)} + (D_2^2 \sigma_a^2/2z) \tag{16-18b}$$

$$\ln \bar{M}_{z(u,\,z)} = \ln \bar{M}_{z(t)} + (3D_2^2 \sigma_a^2/2z) \tag{16-18c}$$

With the recycling technique (cf., Section 17.8), $M_{k(u,\,z)}$ can be determined after $z = 1, 2, 3, \ldots$ cycles. Plotting the logarithms of the values against $1/z$, one obtains $\ln \bar{M}_{k(t)}$ as the ordinate intercept, and the variance σ_a^2, the characteristic value of the apparatus for a single cycle, can be calculated from the slope (see Fig. 16-7).

From Tung's equation (16-2), FIGINI (1979) derived equations for calculating the mean values. These equations are no longer based on the assumption of a linear calibration curve.

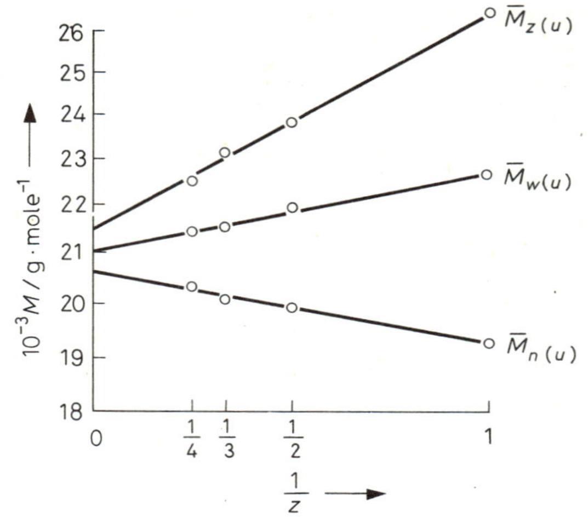

Fig. 16-7
Recycling evaluation as recommended by HAMIELEC (1970). Logarithmic representation of the calculated molar masses vs. $1/z$ (z: number of cycles). Results for polystyrene in tetrahydrofuran

$\bar{M}_{n(t)} = 20610$; $\bar{M}_{w(t)} = 21040$; $\bar{M}_{z(t)} = 21480$ g · mole^{-1} (in good agreement with the values of 1970, 20280 and 21020 g· mole^{-1}, respectively, as calculated from a corrected chromatogram). The standard deviation is $\sigma = 6.12 \pm 0.08$ ml (according to PROBST, UNGER and CANTOW, 1974).

16.3. The dispersion function $G(v - y)$

16.3.1. Symmetric and asymmetric distributions

It has already been indicated in Section 16.1. that the function $G(v - y)$ is not in each case represented by the Gaussian normal distribution. The latter can be applied if the band spreading is caused only by the thermodynamic and kinetic factors mentioned in Chapters 3 and 15, respectively, and if the isotherms are linear. The injected volume as well as mixing effects occurring outside the column contribute to the peak broadening without fundamentally disturbing the symmetry.

Concave isotherms imply a decrease of the distribution coefficients with increasing concentration, and hence a faster migration of the peak maximum as compared to the base. The consequences are obvious: beyond the maximum the substance curve tails off gradually, whereas the front edge slopes down sharply (like a wave surging against the beach).

For polymer samples, the skewing is primarily due to the *viscosity*. The higher the molar mass and the concentration, the more viscous the solutions will be. They are introduced as plugs, but in the interstitial spaces of the separation path, in the connecting pipes and the detector they are deformed by the flow profile in such a way that the portions of the substance flowing close to the walls lag behind. This *viscous fingering* yields bands with long tails and a marked skewing. The effect is independent of the separating mechanism and was also found to occur in columns with inert packings (YAU et al., 1971) and in

Table 16-1

Density function of the normal distribution $f(x) = \frac{1}{\sigma\sqrt{2\pi}} e^{-\frac{(x-x_0)^2}{2\sigma^2}}$
(Gaussian distribution)

Parameter	Symbol	Author
Variance	σ^2	
Standard deviation	σ	
Second statistical central moment	$\mu_2 = \sigma^2$	
Resolution factor	$h_{\text{Tung}} = \frac{1}{2\sigma^2}$	TUNG (1966a); NAKANO and GOTO (1975)
	$h_{\text{Kato}} = \sqrt{\frac{1}{2\sigma^2}}$	KATO and HASHIMOTO (1974) TUNG and RUNYON (1969)
	$h_{\text{Provder}} = \frac{1}{\sigma^2}$	PROVDER and ROSEN (1971)
Correction factor for M	$P = \exp(-D_2^2/4h_{\text{Tung}})$	
Skew	$\gamma = \frac{\mu_3}{\mu_2^{3/2}}$	
Excess	$\varepsilon = \frac{\mu_4}{\mu_2^2} - 3$	

capillaries (BILLMEYER and KELLY, 1968). The second statistical moment, μ_2, indicates the width of the distribution. The *skew*, γ, depends on μ_2 and on the third moment, μ_3. The width of the curve as compared to that of the normal distribution, irrespective of the symmetry, is called the *excess*, ε. This function includes the fourth moment, μ_4 (cf., Table 16-1 and Fig. 16-8).

μ_2 denotes the same as the symbol σ_a^2, i.e., the width of the dispersion due to the apparatus. The expression "resolution factor" with the symbol h used in various chromatographic literature will be avoided here since it is defined in different ways (see Table 16-1) and its symbol could be confused with the same one employed for the height equivalent to a theoretical plate. The units of measurement for the *dispersion parameters* are not consistently defined in the various papers, depending on how the retention and the peak width are measured (in ml, s, cm or counts). It is recommended to state σ_a^2 in ml^2.

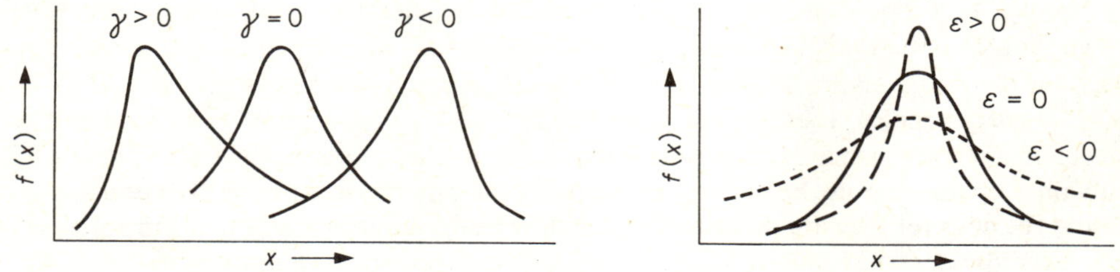

Fig. 16-8
Exponential distribution of the Gaussian type with skew γ and excess ε (x represents t_e or V_e)

16.3.2. Determination of the parameters μ_2, μ_3 and μ_4

For the correction of the experimental curves, $\mu_2 = \sigma_a^2$ must be known all over the elution interval. This important parameter can be determined in different ways.

Homogeneous test substances. Like low-molecular substances, chromatographically monodisperse polymers can be used to obtain chromatograms from which μ_2 and the higher moments can be determined directly. Such an investigation has been carried out by BERGER (1975) with the use of *chromatographically homogeneous* polystyrenes. The substances were prepared from anionically polymerized samples with a low non-uniformity by means of combined purification operations. The latter also included a repeated GPC fractionation in a column 7.32 m long. To obtain sufficiently large fractions, the starting substance was repeatedly injected. Corresponding eluate fractions of 5 ml each were combined, concentrated in vacuo and reinjected. As long as the substance was still separated according to molecular size in the preceding preparative run, V_e values increasing with the count number of the injected fraction were measured. On the other hand, if the starting substance was already chromatographically monodisperse, so that its elugram only reflected the instrumental spreading, then refractionation yielded equal elution volumes and identical curves for all of the eluate fractions. In this case the non-uniformity coefficient was only just 0.001—0.0001. This was achieved for a number of polystyrenes with molar masses ranging between 2100 and 750000 g · mole^{-1}. After passage through the 7.32 m column with a 70 µl detector cuvette, these chromatographically monodisperse samples in each case yielded skewed curves. The latter were represented as superpositions of four Gaussian distributions with the standard deviations $\sigma_1 = 8.755 - 0.0285 V_e$ (in ml), $\sigma_2 = 4\sigma_1$, $\sigma_3 = 8\sigma_1$ and $\sigma_4 = 16\sigma_1$. The maxima of the auxiliary curves 2—4 were shifted against that of the main curve by about $2\sigma_1$, $6\sigma_1$ and $8\sigma_1$, respectively.

This method has the advantage that it does not depend on restrictive assumptions. However, it can only be considered for special cases, because the preparation of chromatographically homogeneous samples requires a large amount of additional work.

Reverse flow. TUNG et al. (1966b) determined the variance due to the apparatus by means of a reverse flow technique using standard calibration samples: after a sample has travelled about half of the separation path, the direction of flow is reversed. The sample components which lag behind in the forward flow now have a lead, which is made up by the faster travelling components during the reverse flow. When the sample arrives at the column entrance, the chromatographic separation, i.e., the peak width caused by the sample, has decreased to zero, whereas the broadening due to the flow has doubled during the reverse flow. According to TUNG, the broadening, σ_a, in the whole column is calculated from contributions $2\sigma_v^2$ obtained in this way for the first half of the separation path and the contributions $2\sigma_h^2$ for the other half:

$$\sigma_a^2 = \frac{1}{2}(2\sigma_v^2 + 2\sigma_h^2) \tag{16-19}$$

Both values σ_v and σ_h must be measured, because generally the separation path consists of different columns (cf., Section 8.1.). This method requires additional effort for the reversal of the flow direction. Moreover it does not yield any information about the skewness; when the bands return to the column entrance, they are again symmetrical (TUNG and RUNYON, 1969).

Test samples with a known distribution. The instrumental spreading can be determined without great effort by means of samples whose molar mass distribution, or at least the mean values $\bar{M}_n$ and $\bar{M}_w$, are known exactly.

HENDRICKSON (1968) used the additivity rule, which is applicable for independent normal distributions (cf., Section 3.3.), in order to split up the measured peak width, $W = 4\sigma$, into a contribution due to the apparatus, σ_a, and a contribution due to the polydisperse sample itself, σ_p:

$$\sigma^2 = \sigma_a^2 + \sigma_p^2 \qquad (16\text{-}20)$$

The method is based upon the fact that the sample has a *molar mass distribution* which can be represented in the co-ordinates of the chromatogram by a Gaussian curve with the standard deviation σ_p. This is exactly true for samples with logarithmic normal distributions and for linear calibration relationships, but approximately also in cases of slight deviations. What is more important is the assumption that the *raw elugram* also represents a normal distribution. This is rarely the case. As a rule the experimental curves are skewed. The rear part of the peak is too flat, forming a long tail. Therefore the total peak width yields too high values for σ and σ_a. As a countermeasure, TUNG and RUNYON (1969) evaluated only the leading slopes (σ_f in Fig. 16-9). This method yielded good results in practical work (MAY and KNIGHT, 1971). However, highly skewed bands lead to systematic errors because their leading slopes are too steep. The quotient of the half-widths behind and before the maximum, raised to the second power, can even be used as an empirical measure of the skewing (BOMBAUGH et al., 1969a).

VOZKA et al. (1980b) have shown that the spreading function can be calculated using statistical moments for the normalized original chromatogram and for the corrected one. The latter can be obtained, for instance, by running the same sample at such a low speed that a high resolution is gained.

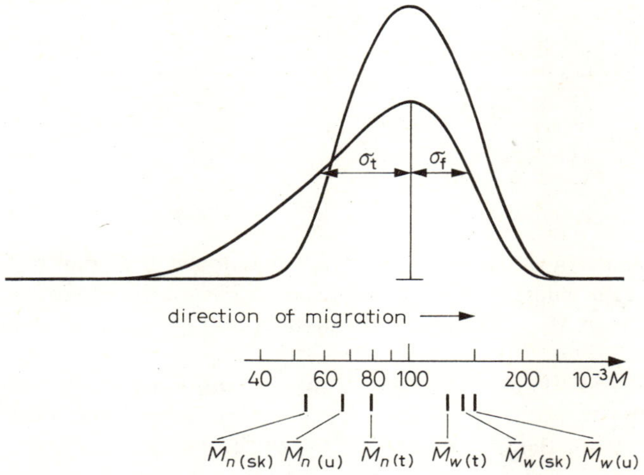

Fig. 16-9
Skewed distribution curve and symmetric Gaussian distribution with a coincident position of the peak maximum

Associated to the curves is a molar mass scale with the true mean values $\bar{M}_{w(t)}$ and $\bar{M}_{n(t)}$ as well as the mean values derived from the two gel chromatograms in the upper part of the figure (schematic representation).

Mean value method. If the mean values are known, the previously mentioned eqn. (16-17a) given by HAMIELEC and RAY (1969) offers the possibility of determining σ_a by means of samples with any distribution whatever. $\bar{M}_{n(u)}$ and $\bar{M}_{w(u)}$ are determined by integration of the experimental curves in the same way as in Section 16.2. For the heterogeneity one obtains:

$$H_t/H_u = \left(\frac{\bar{M}_w}{\bar{M}_n}\right)_t \bigg/ \left(\frac{\bar{M}_w}{\bar{M}_n}\right)_u = e^{-D_2^2\sigma_a^2} \qquad (16\text{-}21)$$

BALKE and HAMIELEC (1969) used this method to determine the instrumental spreading with the help of various samples, and compared the results with those of the reverse flow technique (see Fig. 16-12, "BH").

The following relationships are likewise derived from eqn. (16-17a):

$$\bar{M}_{n(u)} = \bar{M}_{n(t)}\, e^{-D_2^2\sigma_a^2/2} \qquad (16\text{-}22a)$$

$$\bar{M}_{w(u)} = \bar{M}_{w(t)}\, e^{D_2^2\sigma_a^2/2} \qquad (16\text{-}23a)$$

The number average, $\bar{M}_{n(u)}$, derived from the elugram is smaller than the true value, whereas the weight average is greater. This is easily understood: the elugram is broader than the molar mass distribution on both sides. The excess values included in the integration reduce the number average, because the latter is determined mainly by the lower contributions, whereas the weight average, depending mainly on the high contributions, is increased. These considerations do not apply to elution curves with skewed dispersion functions. The calculation of the mean values for a skewed curve yields lower values than for a symmetrical curve defining an equal area, even if the maximum lies at the same point (see Fig. 16-9). BALKE and HAMIELEC (1969) took this into account by introducing a skewing factor, sk:

$$\bar{M}_{n(u,\,sk)} = \frac{\bar{M}_{n(t)}\, e^{-D_2^2\sigma_a^2/2}}{1 + 0.5\ \mathrm{sk}} \qquad (16\text{-}22b)$$

$$\bar{M}_{w(u,\,sk)} = \frac{\bar{M}_{w(t)}\, e^{D_2^2\sigma_a^2/2}}{1 + 0.5\ \mathrm{sk}} \qquad (16\text{-}23b)$$

This decreases the two mean values. Now, as already mentioned in Section 16.2., the weight average derived from the elugram may by chance coincide with the true value, or $\bar{M}_{w(u,\,sk)}$ may even be smaller than $\bar{M}_{w(t)}$. The calculated number average is always smaller than the true one.

PROVDER and ROSEN (1971) likewise used a *Laplace transformation* and the assumption of a linear calibration relationship to solve Tung's integral equation. They substituted Hermitian polynomials for the integrand $G(v - y)$ of the solution integral, as did TUNG for the elution curve $F(v)$. In this way they derived the following eqns. (16-24) to (16-26), which contain all three parameters μ_2, μ_3 and μ_4

$$\frac{\bar{M}_{n(t)}}{\bar{M}_{n(u)}} = \left[1 + \frac{D_2^3\mu_3}{6} + \frac{D_2^4}{24}(\mu_4 - 3\mu_2^2) + \frac{1}{72} D_2^6\mu_3^2\right] \exp\left(\frac{D_2^2\mu_2}{2}\right) \qquad (16\text{-}24)$$

$$\frac{\bar{M}_{w(t)}}{\bar{M}_{w(u)}} = \left[1 - \frac{D_2^3\mu_3}{6} + \frac{D_2^4}{24}(\mu_4 - 3\mu_2^2) + \frac{1}{72}D_2^6\mu_3^2\right]\exp\left(\frac{-D_2^2\mu_2}{2}\right) \quad (16\text{-}25)$$

$$\frac{[\eta]_t}{[\eta]_u} = \left[1 - \frac{a^3D_2^3\mu_3}{6} + \frac{a^4D_2^4}{24}(\mu_4 - 3\mu_2^2) + \frac{1}{72}a^6D_2^6\mu_3^2\right]\exp\left(-\frac{a^2D_2^2\mu_2}{2}\right) \quad (16\text{-}26)$$

where a is the exponent of the $[\eta]$ vs. M relationship (5-8). On their left-hand sides, these three equations each contain the quotient of a directly measurable parameter of the test substance ($\bar{M}_{n(t)}$, $\bar{M}_{w(t)}$ and $[\eta]_t$, respectively) and the corresponding value calculated from the chromatogram. The latter are calculated assuming a perfect resolution of the chromatogram. As this is not the case, these values ($\bar{M}_{n(u)}$, $\bar{M}_{w(u)}$ and $[\eta]_u$), slightly deviate from the true values. The amount of this deviation depends on μ_2, μ_3 and μ_4.

If the coefficients D_2 and a are known from measurements of a number of standard polymers, then eqns. (16-24) to (16-26) can be used to calculate the three unknowns μ_2, μ_3 and μ_4. For that purpose, standard samples whose $\bar{M}_{n(t)}$, $\bar{M}_{w(t)}$ and $[\eta]_t$ values have been determined very accurately by osmometry, light scattering and viscosimetry are chromatographically separated, and the $\bar{M}_{n(u)}$, $\bar{M}_{w(u)}$ and $[\eta]_u$ values are calculated from the elugrams by integration. For $\mu_2 \leqq 0.5$ the exponential expressions containing $D_2^2\mu_2/2$ approximate unity, so that the equations can be simplified.

If not all of the starting quantities are available, then at least the two most important parameters, μ_2 and μ_3, can be determined according to this principle. If for instance the exponent a is unknown, then, neglecting $\frac{D_2^4}{24}(\mu_4 - 3\mu_2^2) + \frac{1}{72}D_2^6\mu_3^2$, the following simple, explicit expressions can be derived from eqns. (16-24) and (16-25)

$$\mu_2 = \frac{2}{D_2^2}\ln\frac{R_{n(t/u)} + R_{w(u/t)}}{2} \quad (16\text{-}27)$$

$$\mu_3 = \frac{6}{D_2^3}\frac{R_{n(t/u)} - R_{w(u/t)}}{R_{n(t/u)} + R_{w(u/t)}} \quad (16\text{-}28)$$

where $R_{n(t/u)} = \bar{M}_{n(t)}/\bar{M}_{n(u)}$ and $R_{w(u/t)} = \bar{M}_{w(u)}/\bar{M}_{w(t)}$.

On the other hand, if $\bar{M}_{w(t)}$ is unknown, then, neglecting the same terms as above and setting $R_{v(u/t)} = [\eta]_u/[\eta]_t$, the following relationships are derived from eqns. (16-24) and (16-26)

$$R_{n(t/u)}\exp\left(-\frac{D_2^2\mu_2}{2}\right) - R_{v(u/t)}\exp\left(-\frac{a^2D_2^2\mu_2}{2}\right) = \frac{D_2^3\mu_3}{6}(1 + a^3) \quad (16\text{-}29)$$

$$R_{n(t/u)}\exp\left(-\frac{D_2^2\mu_2}{2}\right) + R_{v(u/t)}\exp\left(-\frac{a^2D_2^2\mu_2}{2}\right) - 2 = \frac{D_2^3\mu_3}{6}(1 - a^3) \quad (16\text{-}30)$$

For μ_2 this gives:

$$\mu_2 = \frac{2}{D_2^2}\ln\frac{a^3R_{n(t/u)}}{(1 + a^3) + (1 - a^2D_2^2\mu_2/2)\,R_{v(u/t)}} \quad (16\text{-}31)$$

This equation can be solved by iteration, neglecting $a^2D_2^2\mu_2/2$ for the moment. From μ_2 determined in this way, μ_3 is then calculated by eqn. (16-29) or (16-30).

Eqn. (16-27) is the sum of eqn. (16-24) and the reciprocal of eqn. (16-25). On the other hand, division of eqn. (16-24) by (16-25) yields:

$$\frac{R_{n(t/u)}}{R_{w(t/u)}} = \frac{1 + D_2\mu_3/6}{1 - D_2\mu_3/6}\,e^{D_2^2\mu_2} \tag{16-32}$$

For symmetrical curves, $\mu_3 = 0$. Since

$$\frac{R_{n(t/u)}}{R_{w(t/u)}} = \frac{\bar{M}_{n(t)}}{\bar{M}_{n(u)}} : \frac{\bar{M}_{w(t)}}{\bar{M}_{w(u)}} = H_u/H_t \tag{16-33}$$

this again leads to the eqn. (16-21) given by Balke and Hamielec.

Proceeding analogously, one may observe that the empirical skewing factor, sk, is given by:

$$\text{sk} = D_2^3\mu_3/3 \tag{16-34}$$

A calculation method also using the mean values, which simultaneously yields the calibration and the dispersion function, has been described by Andreetta and Figini (1981).

Recycling technique. The instrumental spreading parameter $\mu_2 = \sigma_a^2$ can also be determined by means of the recycling technique. Such a method has already been referred to in Section 16.2. (see Fig. 16-7).

For samples with very narrow distributions, the following technique is applicable, which at the same time also yields the *heterogeneity*, H_t, of the sample (Grubišić-Gallot et al., 1976). Using the calibration relationship (8-4), from eqn. (4-34) it follows that

$$\ln(\bar{M}_w/\bar{M}_n) = \ln H_t = \sigma_M^2 = D_2^2\sigma_p^2 \tag{16-35}$$

where σ_M^2 is the variance of the ($\ln M$) distribution and σ_p^2 is the variance of the V_e distribution.

This equation is exact for samples with *logarithmic normal distributions*, but to a good approximation it is also valid for samples with a molar mass distribution following the generalized Schulz function as given by eqn. (4-21) (Berger and Schulz, 1970) as well as for samples with narrow distributions.

The total variance of the elution curve consists of σ_p^2, the contribution of the column, σ_c^2, and those of the other parts of the apparatus. The *peak broadening in the pumps*, which of course does not occur at all in normal elution chromatography, plays a rather important rôle in the recycling technique. It is useful to demonstrate this contribution, σ_P^2, separately. Finally, σ_i^2 reflects the contribution of injection to the peak broadening. If all of the contributions are independent of each other, and are represented by Gaussian functions, then the additivity rule can be applied. Then the following relationship holds after one cycle in the apparatus:

$$\sigma^2 = \sigma_p^2 + \sigma_c^2 + \sigma_i^2 = \frac{\ln H_t}{D_2^2} \tag{16-36}$$

For a repeated recirculation, $\sigma_p = [(\ln H_t)/D_2^2]^{1/2}$ increases with the number of cycles, z, whereas σ_c increases only proportionally to $\sqrt{z}$. Analogously, the last-mentioned relationship is also valid for a contribution due to the pumps, whereas σ_i of course is free of a factor. Thus the total variance of the elugram after z cycles is:

$$\sigma_z^2 = z^2\frac{\ln H_t}{D_2^2} + z\sigma_c^2 + (z-1)\,\sigma_P^2 + \sigma_i^2 \tag{16-37}$$

In order to determine the parameters of eqn. (16-37), McCrackin and Wagner (1980), in a second experiment, connected the pump directly with the detector, allowing the sample to complete z cycles in this loop. As the separating column had been omitted, in this case not only the broadening due to the column, σ_c, but also the contribution σ_p were eliminated, because without the column naturally the sample cannot be separated according to molar mass. For this shortened circuit, after z cycles:

$$\sigma_z^2 = (z - 1)\,\sigma_P^2 + \sigma_i^2 \tag{16-38}$$

Plotting this variance, σ_z^2 vs. $(z - 1)$ yields a straight line with the slope σ_P^2 and the ordinate intercept σ_i^2. To determine the remaining unknowns, eqn. (16-37) is rearranged as follows:

$$\frac{\sigma_z^2 - \sigma_i^2 - (z - 1)\,\sigma_P^2}{z} = \sigma_c^2 + z\,\frac{\ln H_t}{D_2^2} \tag{16-39}$$

σ_z^2 is determined from the elugram. As σ_i^2 and σ_P^2 have the same values as in the shortened circuit, which are known from the above experiments, the value of the expression on the left-hand side can be calculated and plotted vs. z. This gives a straight line, the ordinate intercept of which represents the dispersion of the column, while the slope yields the heterogeneity of the sample.

In similar, earlier investigations, the contributions σ_P^2 and σ_i^2 were not isolated, but combined with the contribution of the column to the dispersion due to the apparatus. Fig. 16-10 shows that such a summary treatment of the data yields non-linear relationships. Naturally, the smaller the contribution of the column to the total instrumental spreading, the more distinctly this non-linearity stands out. Today the columns employed are generally much better than a few years ago; so it was for objective reasons that in the past the error was not manifested so clearly as in Fig. 16-10.

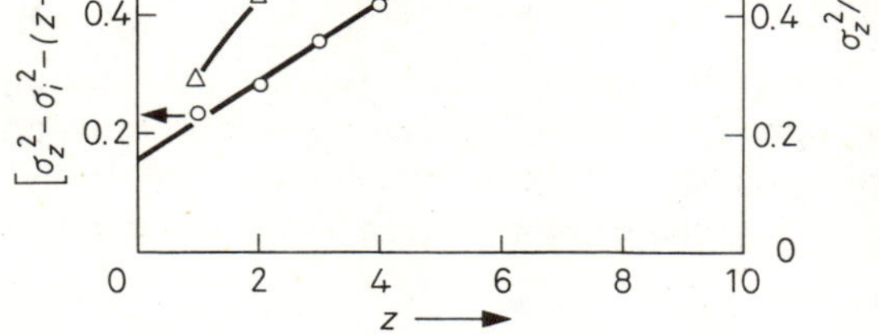

Fig. 16-10
Evaluation of recycling measurements on a polystyrene standard in toluene
Plotted according to eqn. (16-39) (○, left ordinate) and in a less adequate manner (△, right ordinate); z: number of cycles.
Column: 4 × 0.30 m μ-Styragel® (10^3, 10^4, 10^5 and 10^6 Å)
Sample: SRM 1478 polystyrene, $M = 36000$ g · mole^{-1}, $H = 1.0099$, $V_0 = 0.5$ ml
Results: $\sigma_i^2 = 0.056$ ml^2; $\sigma_P^2 = 0.246$ ml^2; $\sigma_c^2 = 0.16$ ml^2
(according to McCrackin and Wagner, 1980).

By the method described above, the heterogeneity of reference samples can be determined with a much higher accuracy than, say, by osmosis or light scattering. Samples defined in this way are then suitable for use in determining the dispersion due to the apparatus for other chromatographic equipment by a single cycle.

SEC with two detectors. If an SEC apparatus is equipped not only with the normal detector, the reading of which is proportional to the concentration of polymer, but also with a detector the response of which depends on the molar mass, then the dispersion function can be determined by evaluating the two different elution curves, and the true distribution function can be derived without any assumptions about the dispersion function (BERGER, 1978; HAMIELEC and OMORODIN, 1979; MCCONNELL, 1979). Detectors whose response depends on the molar mass use a low-angle light-scattering detector (cf., Section 19.8.3.4.) or a viscosity detector. As there is the relationship, $[\eta] = K_\eta M^a$, between the viscosity and the molar mass, a viscosity detector yields the curve

$$E(v) = K_1 \int M^a(y) \cdot W(y) \cdot G(v - y)\, \mathrm{d}y \tag{16-40}$$

whereas an R.I. detector or a UV detector yields the normal elution curve (eqn. 16-2).

BERGER has shown that the simultaneous solution of these equations by Fourier transformation for dispersion functions which are independent of the molar mass yields the true molar mass distribution, $W(y)$, with a good accuracy. To obtain more precise information also about the dispersion function $G(v - y)$, he suggested that one should assume the calibration function to be linear. Naturally this narrows the range of application, and moreover it complicates tracing the propagation of error.

SEC refractionation. A monodisperse component with the individual elution value y is broadened to the curve $G(v - y)$ due to instrumental spreading. Accordingly, a polydisperse sample with the molar mass distribution $W(y)$ is distributed over $W(y)\,G(v - y)$. The infinitesimal volume element $\mathrm{d}y$ contains the infinitesimal mass element $\mathrm{d}m$ (originating from different components i):

$$\mathrm{d}m = W(y) \cdot G(v - y)\, \mathrm{d}y \tag{16-41}$$

In the experimental realization of refractionation, naturally it is always a volume of finite width, Δy, which must be collected, with the distribution function $\omega(y)$ for the polymer fraction Δm contained in this volume:

$$\omega(y) = W(y) \cdot G(v - y)\, \Delta y \tag{16-42}$$

If this fraction is refractionated on the same apparatus, then $\omega(y)$ is substituted for $W(y)$ in eqn. (16-2). In this way one again obtains two different equations of the type of Tung's integral equation: for the total substance the elution curve following the usual relationship (16-2) is obtained, whereas the elution curve of the fraction is described by the function

$$f(v) = \int \omega(y) \cdot G(v - y)\, \mathrm{d}y \tag{16-43}$$

Similarly as in the technique using two detectors, there are two equations which allow the calculation of both the desired molar mass distribution and the dispersion function.

BERGER (1979b) proposed this method and indicated a solution by Fourier transformation, followed by iteration. Fig. 16-11 illustrates how in this way the elution curve can be split into the curve of the molar mass distribution and that of the dispersion function. Compared to the technique using two detectors, there are no special demands to be made upon the appa-

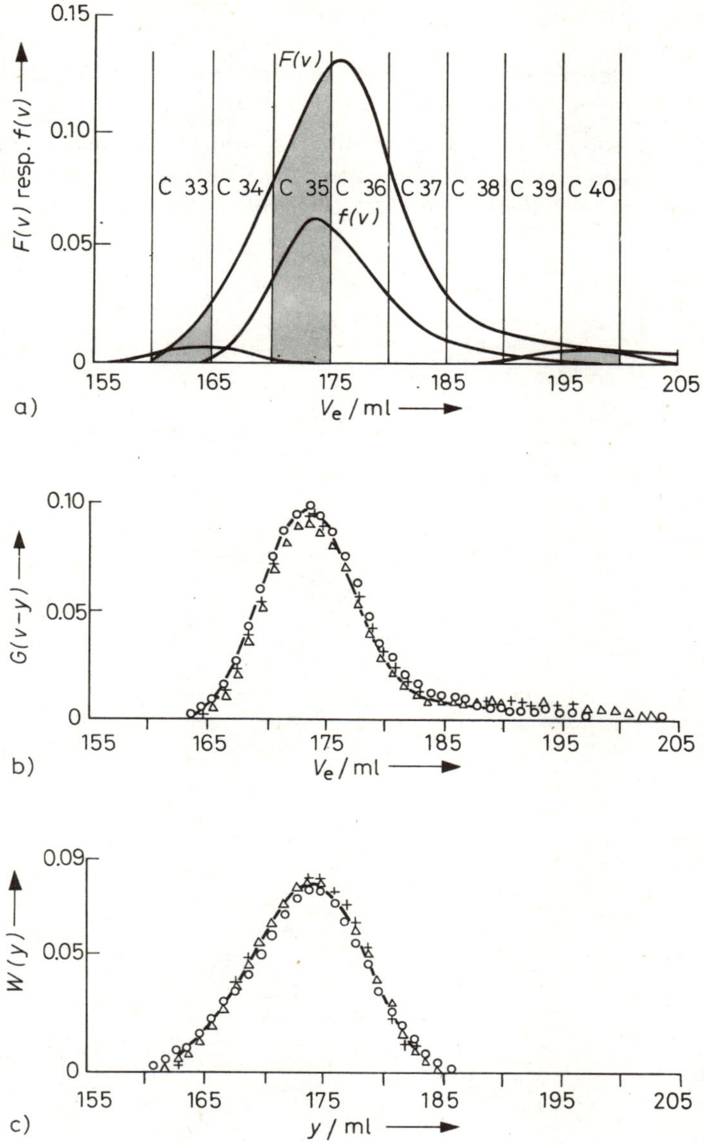

Fig. 16-11
Chromatographic refractionation

a) Elution curve, $F(v)$, for a polystyrene ($M = 670\,000$ g · mole^{-1}) in THF on a 7.32 m column packed with polystyrene gel; the fraction cuts and the elution curves, $f(v)$, are indicated for the three fractions C 33, C 35 and C 40.

b) Dispersion function calculated from the experimental curves of part a, obtained with C 33 (○), C 35 (+) and C 40 (△).

c) Corrected molar mass distribution derived from the elution curve $F(v)$.

(according to Berger, 1979c).

ratus. The dispersion function as well as the molar mass distribution can be determined without the assumption of a linear calibration relationship.

The effect of the molar mass on the dispersion can be investigated and taken into account (BERGER, 1979c). However, the amount of computing work required is rather extensive.

Results. Fig. 16-12 shows experimental results illustrating the relationship between the instrumental spreading and the elution volume. The various definitions of the resolution factor, used by the individual authors, are taken into account by converting all of the

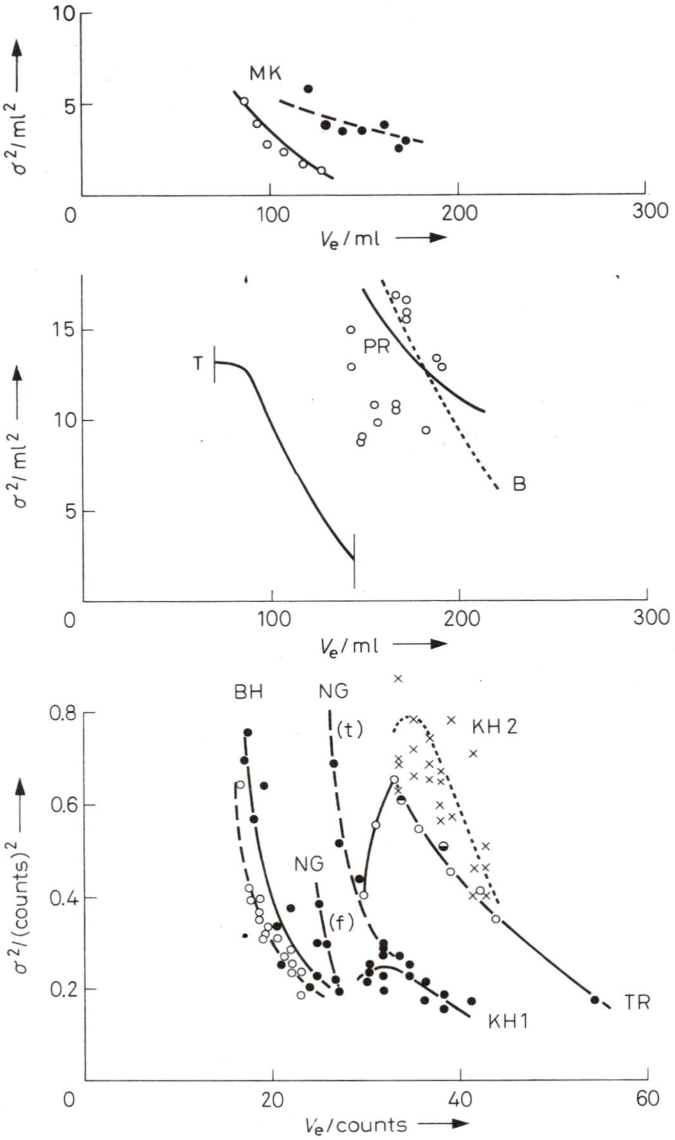

data into σ_a^2. The representation shows how widely this quantity varies in the respective range investigated.

Provder and Rosen (1971) as well as Nakano and Goto (1975) provided information about the relationship between μ_3 and the skewing on the one hand and the elution volume on the other. The last-mentioned authors approximated skewed chromatograms by two half-curves of normal distributions with the same height but different variances.

However important the knowledge of $\mu_2 = f(V_e)$ may be for the optimum correction of the chromatograms, it is only of limited value for the comparison of separation efficiencies, because the dead volume is implicitly included. It is mainly for this reason that curves from the different papers deviate so widely from one another in Fig. 16-12.

16.4. Effect of dispersion on the calibration curve

In this section we again return to the problem of calibration, which was discussed in Chapter 8 on the idealized assumption of a dispersion-free separating apparatus.

In the calibration using samples with a narrow distribution, the elution curves of the individual reference standards are indeed broadened as compared with the actual molar mass distribution, but the V_e value of the peak maximum remains unchanged. Therefore the "classical" *peak position calibration* generally leads to a correct calibration relationship. However, the dispersion has an effect in the calculation of the mean values, $\bar{M}_n$ and $\bar{M}_w$, for the reference standards from their elution curves and the experimental calibration relationship. This effect, as well as appropriate counter-measures, will be dealt with in Section 19.6.2. in the discussion of a practical example.

What shall be discussed here is the effect of dispersion on the calibration using samples with broad distributions. The advantages of this technique have already been mentioned in Section 8.3.3. Its disadvantage is that in the mean-value method an *effective calibration curve* is obtained. However, if the distributions of the samples to be investigated do not differ too widely from that of the reference samples, even this effective calibration gives

◀ Fig. 16-12 (on page 296)
Variance $\mu_2 = \sigma_a^2$ of instrumental spreading in the SEC of polystyrene samples

Where the siphon volume had been stated in the original papers, the data were converted into ml and ml². The lower diagram shows the curves in which the count data had to be retained. As far as 5 ml per count is applicable, these curves have the same scale as the other ones.

MK May and Knight (1971)
● "SEC 3" apparatus; ○ "SEC 4" apparatus

BH Balke and Hamielec (1969)
● mean value method; ○ reverse flow according to Tung (1966b)

KH 1,2 Kato and Hashimoto (1974c)
● 1 ml · min⁻¹; × 4 ml · min⁻¹

T Tung (1966b), reverse flow

PR Provder and Rosen (1971)

B Berger (1974), plot of the relationship $\sigma_a = f(V_e)$ given as an equation

NG Nakano and Goto (1975)
f leading edge; t rear part of the peak

TR Tung and Runyon (1969), measurement of σ^2 by reverse flow

The filled circle at about 55 counts was obtained with styrene dimer; the circle with the filled upper half ◓ was a polybutadiene sample; the circle with the filled lower part ◒ a PVC sample.

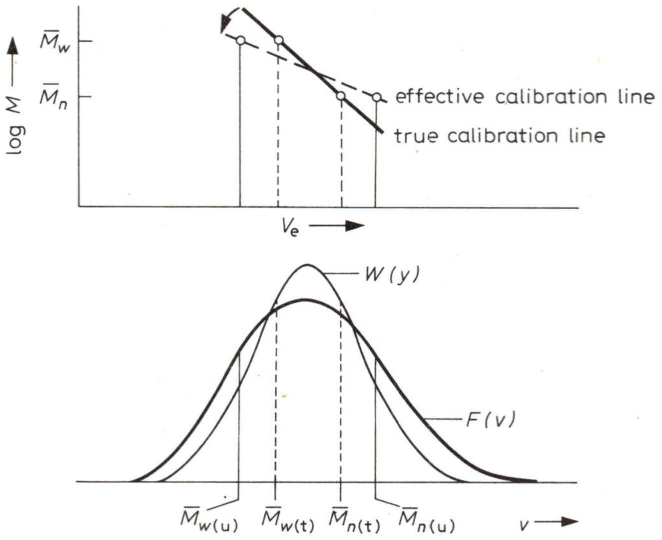

Fig. 16-13
Schematic representation of the deviation of the effective calibration line from the true one
The peak broadening will cause the mean values $\bar{M}_{w(u)}$ and $\bar{M}_{n(u)}$ calculated from $F(v)$ to differ too widely from each other. A calibration line derived from these data is anti-clockwise rotated from the true calibration line.

reasonable results. YAU et al. (1977) stated that the effective linear calibration curve obtained if eqn. (8-4) is applicable is rotated anti-clockwise relatively to the peak position calibration line. The centre of the rotation lies close to the mean retention volume of the reference sample, and the angle of rotation increases with increasing dispersion due to the apparatus and decreasing width of the molar mass distribution in the sample used for the calibration. Fig. 16-13 illustrates this effect: the lower part of the figure shows the curve of the true molar mass distribution, $W(y)$, together with the elution curve broadened due to dispersion, $F(v)$, plotted vs. v. As shown in Fig. 16-9, the dispersion leads to $\bar{M}_{n(u)} < \bar{M}_{n(t)}$ and nearly always to $\bar{M}_{w(u)} > \bar{M}_{w(t)}$. This means that on the V_e scale the uncorrected mean values $\bar{M}_{n(u)}$ and $\bar{M}_{w(u)}$ are further apart from each other than the true mean values. If the calculation described in Section 8.3.3. is carried out using the uncorrected mean values, then the slope of the calibration line determined in this way is too gradual.

YAU et al. (1977) have shown how this influence of dispersion can be counteracted: for an ideal resolution, the very small elution volume, V_i, contains only molecules of the same kind with the mass fraction m_i and the uniform value of the molar mass M_i. In this case the summation algorithms (8-28) and (8-29) yield the true mean values. However, if the equations, written in the form

$$\bar{M}_{n(u)} = \frac{D_1}{\Sigma\, F(V)\, e^{D_2 V}} \tag{16-44}$$

and

$$\bar{M}_{w(u)} = D_1\, \Sigma\, F(V)\, e^{-D_2 V} \tag{16-45}$$

are applied to elution curves reflecting dispersion effects, then they yield the uncorrected mean values, $\bar{M}_{n(u)}$ and $\bar{M}_{w(u)}$, since $F(V)$ deviates from the true value $W(V) = m_i$ due to the band broadening. A correction by means of eqns. (16-22a) and (16-23a), respectively, is required, so that for the true mean values one obtains:

$$\bar{M}_{n(t)} = \frac{D_1 \, e^{D_2^2 \sigma_a^2/2}}{\Sigma F(V) \, e^{D_2 V}} \tag{16-46}$$

$$\bar{M}_{w(t)} = D_1 \, e^{-D_2^2 \sigma_a^2/2} \, \Sigma F(V) \, e^{-D_2 V} \tag{16-47}$$

In these equations, dispersion is taken into account by the standard deviation of the instrumental spreading. To a good approximation, this broadening can be estimated from the experimental peak width of polystyrene standards with very narrow distributions. YAU et al. (1977) recommended investigating several reference standards and using the lowest value in order to exclude with certainty any artefacts due to overcorrection. Using this value for σ_a, the equations can be solved by iteration in the way indicated in Section 8.3.3. The authors call this technique GPCV2. Another variant, called GPCV3, also takes into account the skewing of the elution curve (YAU et al., 1978b).

16.5. Characterization of the separation efficiency in the chromatography of polymers

16.5.1. Reproducibility

As in every experimental investigation the first question in polymer chromatography refers to the reproducibility of the results. This reproducibility is especially important in SEC, because the elution curves are usually only evaluated by means of the previously recorded calibration curve, i.e. without an internal standard and, moreover, the elution volume bears a logarithmic relationship with the molar mass. An error in the elution volume causes a much larger one on the molar mass scale.

Systematic investigations have shown that the chromatographic process itself is reproducible. Differences in replicate measurements are largely due to errors in the measurement of the elution volume (SPATORICO, 1975).

In sol-gel partition chromatography, the reproducibility can be evaluated from the distribution curves obtained in repeated fractionations. As the results were quite favourable, these curves were frequently chosen for reference in the comparison of different separating techniques (SCHOLTAN and KRANZ, 1967; LANGHAMMER and SEIDE, 1967; ALTGELT, 1971).

SAMAY and FÜZES (1980) determined the standard deviation of the peak elution volume and of the $\bar{M}_n$ and $\bar{M}_w$ values calculated from the elution curves. They stated a sufficient short-time reproducibility, which was independent of the molar mass of the sample used, by making repeated measurements over a period of two weeks.

The desire to detect slight differences between similar chromatograms in a definite and objective way has a close relation to the question of reproducibility. This may be of importance in the quality control of plastics or in investigations of aging. For that purpose, HASSEL et al. (1979) programmed a PDP-1103 minicomputer in such a way that the primary data (elution volume and detector signal) are collected and stored as elution curves normalized according to eqn. (8-14). The comparison of any two curves is carried out by the substraction of the ordinate values corresponding to equal abscissa values. The method also has the advantage that in these calculated *difference chromatograms* the influence of dispersion is reduced.

16.5.2. Characterization by the height equivalent to a theoretical plate

It is customary to characterize the efficiency of chromatographic apparatus by the *number of theoretical plates*. This term has been taken from the distillation technique via the countercurrent distribution. Apart from the peculiarities discussed in Section 3.3., which result from the fact that the plate number is derived from the behaviour of one isolated component instead of a mixture, there are no fundamental problems in small-molecule chromatography. In *development techniques*, eqn. (3-14) is employed. The initial spots should be ideally small, and substance zones in the neighbourhood of the eluent front should be ignored. In *elution chromatography* the elution volume, V_e, is considered instead of the distance travelled, s, and eqn. (3-19) is used in the calculation. In contrast to development chromatography, most of the elution techniques use a chromatographic bed filled with the eluent from the beginning of the process (wet-bed chromatography). The *dead volume*, V', must have discharged before any component of the sample emerges. This means that eqn. (3-19) yields a non-zero number of theoretical plates even for components which are not retained at all. In spite of this obvious inconsistency, the evaluation is almost generally done in this way. The reasons for this may be elucidated by the following consideration: the number of theoretical plates indicates the quality of a separating path. The measured quantities V_e and W required for that purpose are determined by means of a detector. All investigations are carried out with a combination of a column and a detector, to which also the data obtained are referred. Consequently, in addition to the column contribution the bandwidth, W, also includes contributions due to the detector and other influences acting outside the column. The *effective number of theoretical plates*

$$N_{\text{eff}} = 16\left(\frac{V_e - V'}{W}\right)^2 \tag{16-48}$$

takes into account the dead volume, thus better describing the efficiency of the column for components subject to little or no retention. For $K = 0$, $N_{\text{eff}} = 0$ (see eqn. (14-5)), which is a reasonable result. However, at some distance from this limiting case it is of importance that only the elution volume is corrected and not the bandwidth, W. Due to this one-sided correction, the parameter resulting from eqn. (16-48) for the column-detector combination shows a larger deviation from the true column parameter than the conventional plate number according to eqn. (3-19). Therefore the latter is preferred.

Strictly speaking, these considerations do not involve a specific problem of polymer chromatography, but in exclusion chromatography they are of more importance than in separations with $V_e \gg V'$.

The plate number is proportional to the length of the column, if the packing is homogeneous. The plate number per unit length enables the comparison of different columns. The reciprocal $h = L/N$ defines the plate height. Starting from eqn. (16-48), one obtains the *effective plate height* (height equivalent to an effective theoretical plate HEETP):

$$h_{\text{eff}} = L/N_{\text{eff}} \tag{16-49}$$

h or h_{eff} may likewise serve to characterize columns. Compared to the plate number, the plate height has the additional advantage of being directly connected with the dynamic theory of the chromatographic process (see Chapter 14), and that the total height can be split up into the contributions of the different influences.

The characterization of columns by the plate number may suggest that this is a constant depending only on the chromatographic bed. The elugrams of low-molecular-weight substances would seem to support this. They exhibit broad bands for the components with high elution volumes and narrow bands for those with low ones, as expected from eqn. (3-19) for constant N. However, closer investigations show that the number of theoretical plates is different for the individual peaks of the chromatograms (HALÁSZ, 1966). This follows from the fact that the number of theoretical plates is eventually determined by the distribution coefficient. Nevertheless the number of theoretical plates is a valuable aid in evaluating the column quality.

Difficulties arise in the chromatography of macromolecular substances. In most cases the homogeneous substances which are required for the determination of the plate height are not available. A low-molecular-weight substance, which is readily available in a homogeneous form, on a separating path for polymers produces a peak close to the separating threshold. This peak is well-shaped and allows the calculation of a plate number by means of eqn. (3-19). To determine and indicate this quantity is one of the demands made upon SEC investigations by BLY et al. (1971). The plate number measured with the help of a low-molecular-weight substance provides useful information, e.g., about the quality of packing, but it does not agree with the plate number which is effective in the actual separation of polymers. This number is much smaller.

Fig. 16-12 shows how strongly σ_{a}^2 depends on the elution volume. The larger the molecules, the smaller is V_{e} and the higher is σ_{a}. Inspection of eqn. (3-18) shows that consequently the plate number decreases as the molar mass increases. Thus the plate height determined by means of a low-molecular-weight probe only provides information about the point from which the plate height increases if polymers are investigated.

In the macromolecular range the accessible data on the instrumental spreading converted into plate heights yield, on a logarithmic scale, straight lines starting from the low-molecular-weight value. For the data of TUNG and RUNYON (1969), MAY and KNIGHT (1971) and KATO et al. (1974), these lines are almost parallel (see Fig. 16-14). This gives the approximation

$$h_M = h_{\text{Bzn}}(M/M_{\text{Bzn}})^{0.3} \tag{16-50}$$

where h_M is the plate height for a sample with molar mass M and h_{Bzn} is the plate height for an injection of benzene or the like.

Results obtained with 5 µm polystyrene gel particles (KATO et al., 1974) or with porous (BASEDOW et al., 1976) yield also straight lines.

The values measured on a single column with Styragel with the nominal pore size of 10^5 Å gave a straight line if $1/h$ was plotted vs. log M (COOPER et al., 1973). As the value of h_{Bzn} was not reported, the log h/log M plot cannot be applied in this case.

BLY (1968b) plotted $\sqrt{N}$ vs. V_{e}. In the macromolecular range this gives a straight line; however, the value determined for acetone differs widely from this line.

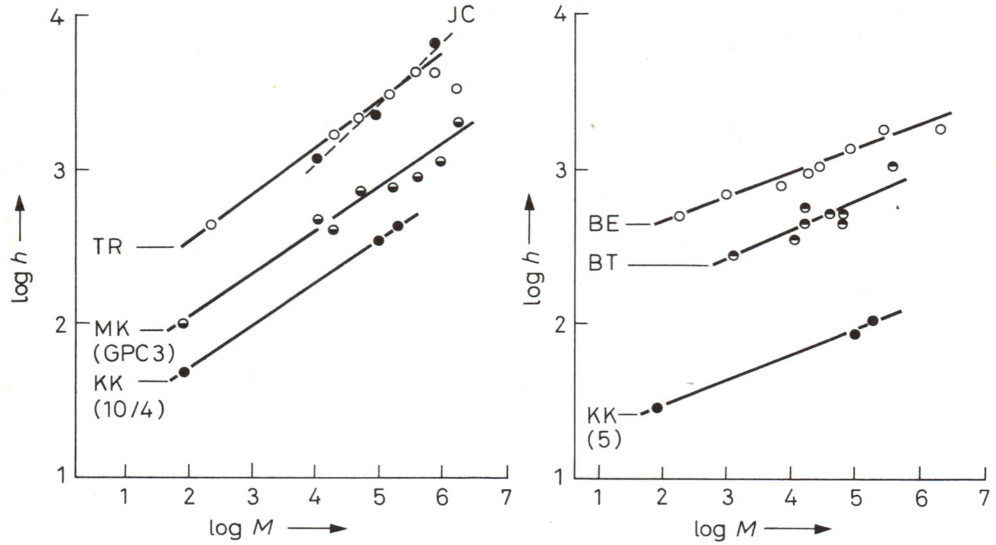

Fig. 16-14
Plate height (h, in µm) in SEC as a function of the molar mass; flow-rate 1 ml · min^{-1}

TR TUNG and RUNYON (1969)
(PS standards on cross-linked polystyrene)
MK MAY and KNIGHT (1971)
(PS standards on cross-linked polystyrene)
BE BASEDOW, EBERT, EDERER and HUNGER (1976)
(Dextran fractions on porous glasses)
JC JOHNSON, COOPER and PORTER (1973)
(PS standards on cross-linked polystyrene)
KK KATO, KIDO, YAMAMOTO and HASHIMOTO (1974)
(PS standards on cross-linked polystyrene, KK 10/4 $d_P = 10$ µm; KK 5 $d_P = 5$ µm)
BT BOSCHETTI, TIXIER and GARELLE (1974)
(Proteins on polyacrylamide-agarose gels).

16.5.3. Resolution, specific resolution, resolution index and separation power

Two successive substances are sufficiently separated if their peak maxima in the chromatogram differ by a distance of at least one base width (cf., Section 3.4.). As the resolution power in the polymer range must usually be tested by substances which themselves have a more or less broad distribution, a corresponding correction of the measured base width is necessary. A rather easy way of doing this is simply to refer the peak width to the heterogeneity. H, of the sample (BLY, 1968a, b). Then instead of eqn. (3-21) one obtains:

$$R_{s,\,corr} = \frac{2(V_{e,\,II} - V_{e,\,I})}{W_I/H_I + W_{II}/H_{II}} \tag{16-51}$$

However, this *corrected resolution* still strongly depends on the distance between samples I and II. For low-molecular-weight substances the resolution equation is applied to neighbouring components which are difficult to separate; for polymers there are no substance pairs of such significance.

The polymer samples are the usual reference standards. Using two successive standards, the resolution can be determined by eqn. (16-51); Z standards yield $Z - 1$ values. The effect of the distance of the samples is easily ascertained by, for instance, combining the standards with their respective next-but-one neighbours. For a constant efficiency of the separating equipment the values determined by eqn. (16-51) increase with increasing distances between the test samples. Thus $R_{s,\,corr}$ respectitively determined by means of the same substances, is only suitable for comparing different columns (COOPER, 1974), not for an absolute characterization of the separating power. The dependence on the sample distance can be counteracted by introducing $\log(M_I/M_{II})$ (BLY, 1968a). $V_{e,\,II}$ is the elution volume for the fraction with M_{II}; the requirement that $V_{e,\,II} - V_{e,\,I} > 0$ leads to $M_I > M_{II}$. This gives the *specific resolution*:

$$R_{sp} = \frac{2(V_{e,\,II} - V_{e,\,I})}{(W_I/H_I + W_{II}/H_{II})\log(M_I/M_{II})} \tag{16-52}$$

This equation implicitly contains the *selectivity factor*, S, of exclusion chromatography (cf., eqn. (8-3)), which indicates the slope of the calibration curve. The limit

$$\lim_{\substack{H_I,\, H_{II} \to 1 \\ W_I = W_{II} = W}} R_{sp} = S/W \tag{16-53}$$

was called the *resolution index* by BLY et al. (1971). As W, like σ^2, varies within the separating range (see Fig. 16-12), also this expression is not yet an optimum criterion for the separation efficiency of a column. In 1976, YAU et al. transformed eqn. (16-53) by substituting $S = C_2 = 2.303/D_2$ (cf., eqn. 8-5b) and $W = 4\sigma$ (cf., Fig. 3-2, caption):

$$R_{sp} = \frac{0.576}{\sigma D_2} \tag{16-54}$$

Since D_2 is inversely proportional to the column length, L, and σ varies with $\sqrt{L}$, the packing resolution factor

$$R_{sp}^* = \frac{0.576}{\sigma D_2 \sqrt{100\, L}} \tag{16-55}$$

(KIRKLAND, 1976) expresses R_{sp} for a 1 cm column (L given in m). The expression

$$RI = (M_{II}/M_I)^{1/Rs}, \tag{16-56}$$

which likewise follows from eqn. (16-52) for $H_I = H_{II} = 1$, was also called a resolution index by COOPER and KISS (1973). Eqn. (16-56) yields numerical values ranging between 0 and 1. The higher the RI, the better is the separation efficiency (CHUANG et al., 1973).

To eliminate the limitation imposed by $H \to 1$, R_s can be replaced by $R_{s,\,corr}$. Taking the reciprocal, one obtains an index

$$T = (M_I/M_{II})^{1/R_{s,\,korr}}, \tag{16-57}$$

which is no longer determined by the distance between the samples, does not require monodisperse substances and, moreover, provides clear information: it states by what factor M_I must be greater than M_{II} in order to achieve unity resolution with the apparatus concerned. $T = 10$ means a 4σ separation for $M_I/M_{II} = 10$. This index T is called the *separating power*. It can be determined in the linear range of the calibration curve by means of two samples which need not be monodisperse. If their distributions do not differ too widely from one another, it does not matter whether the number averages or the weight averages are inserted for M_{I} and M_{II} (GLÖCKNER, 1980).

16.5.4. Accuracy of molar mass values calculated from SEC curves

In 1976, YAU et al. recommended a performance criterion which includes the molar mass values, $\bar{M}_n$ and $\bar{M}_w$. If these values are calculated from the normalized chromatograms by means of eqns. (16-44) or (16-45) the results are $\bar{M}_{n(\mathrm{u})}$ and $\bar{M}_{w(\mathrm{u})}$, respectively, whereas eqns. (16-46) to (16-47) yield $\bar{M}_{n(\mathrm{t})}$ and $\bar{M}_{w(\mathrm{t})}$. These data are interrelated by eqns. (16-22a) to (16-23a), from which the definitions used by YAU et al.

$$\bar{M}_n^* = \frac{\bar{M}_{n(\mathrm{u})} - \bar{M}_{n(\mathrm{t})}}{\bar{M}_{n(\mathrm{t})}} = \mathrm{e}^{-D_2^2\sigma_\mathrm{a}^2/2} - 1 \qquad (16\text{-}58)$$

$$\bar{M}_w^* = \frac{\bar{M}_{w(\mathrm{u})} - \bar{M}_{w(\mathrm{t})}}{\bar{M}_{w(\mathrm{t})}} = \mathrm{e}^{D_2^2\sigma_\mathrm{a}^2/2} - 1 \qquad (16\text{-}59)$$

can easily be derived. The true molar mass values can also be obtained by independent measurements such as osmometry for $\bar{M}_{n(\mathrm{t})}$ or light scattering for $\bar{M}_{w(\mathrm{t})}$. The better the chromatographic resolution, the closer the molar mass criterion, $\bar{M}^*$, approaches its limiting value of zero. For a certain sample, the data obtained at different flow rates correspond well to the curves of $\bar{M}^*$ vs. σ_a as calculated from the right-hand part of eqns. (16-58; 16-59).

16.6. Real GPC

Under real conditions, the interactions between the sample and the stationary phase, or the porous packing material, are frequently not characterized by only one of the mechanisms discussed separately in Chapters 7—9. It is only ideally that GPC follows a pure size exclusion mechanism. In reality the gel matrix does not usually act as an inert pore system, but additionally has attractive or repulsive effects on the dissolved substance, which have been reviewed by AUDEBERT (1979) and DAWKINS (1979). The disappearance of samples in the column due to adsorption on the packing material, as occurs, for instance, with SAN copolymers in dichloromethane on silica gel, gives a clear indication of such interactions. However, substance-specific sample-gel interactions in GPC were already observed before the introduction of inorganic separating materials, on which such drastic effects may occur. For example, let us consider the investigations by HEITZ and KERN (1967), from which Fig. 16-15 has been taken. The figure shows the elution behaviour of polar and non-polar oligomers in tetrahydrofuran on various gels. This behaviour is naturally influenced by the dimensions of the columns used and the pore structure of the gels, so it is no wonder that the V_e values are different in the two parts of the figure. However, the fact that, on the polar polyvinyl acetate gel, the less polar oligophenyls have shorter elution times than the polar samples, whereas on polystyrene gel the opposite is the case, indicates chemical interactions between the samples and the separating gels.

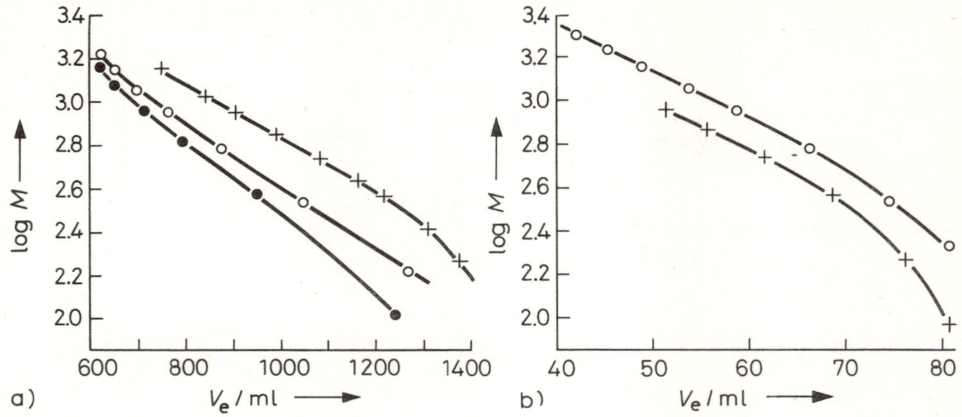

× methyl-substituted p-oligophenylene
○ linear oligourethane from diethyleneglycol and hexamethylene diisocyanate with phenyl end groups
● ditto, with glycol end groups

Fig. 16-15
GPC of oligomers on two different gels in tetrahydrofuran at 25 °C (according to HEITZ and KERN, 1967)

a) Polystyrene gel; b) Polyvinyl acetate gel.

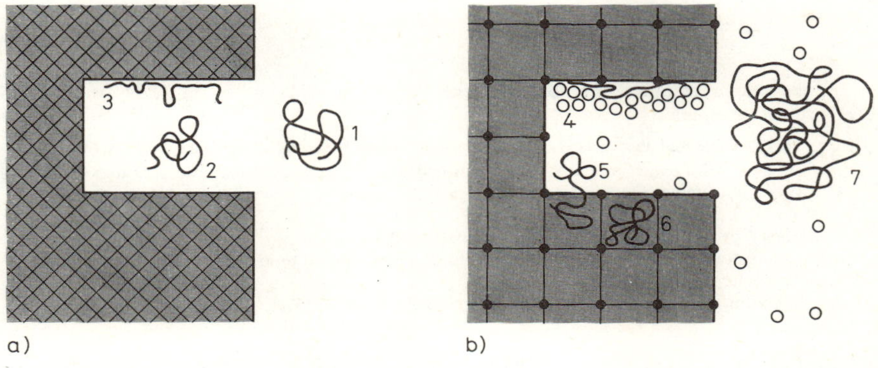

Fig. 16-16
Schematic representation of the various interactions between coiled macromolecules and porous materials

a) Silica; b) Polystyrene gel

1 molecule capable of permeation; 2 movable in the pore; 3 adsorbed; 4 forced against the wall of a pore by solvophobic interactions; 5, 6 network-limited partition in the wall material; 7 excluded.

A superposition of several separating mechanisms is not necessarily a disadvantage. Many a difficult separation problem in the range of small molecules has been solved in just this way. For macromolecular substances, however, in most cases the problem is the separation of

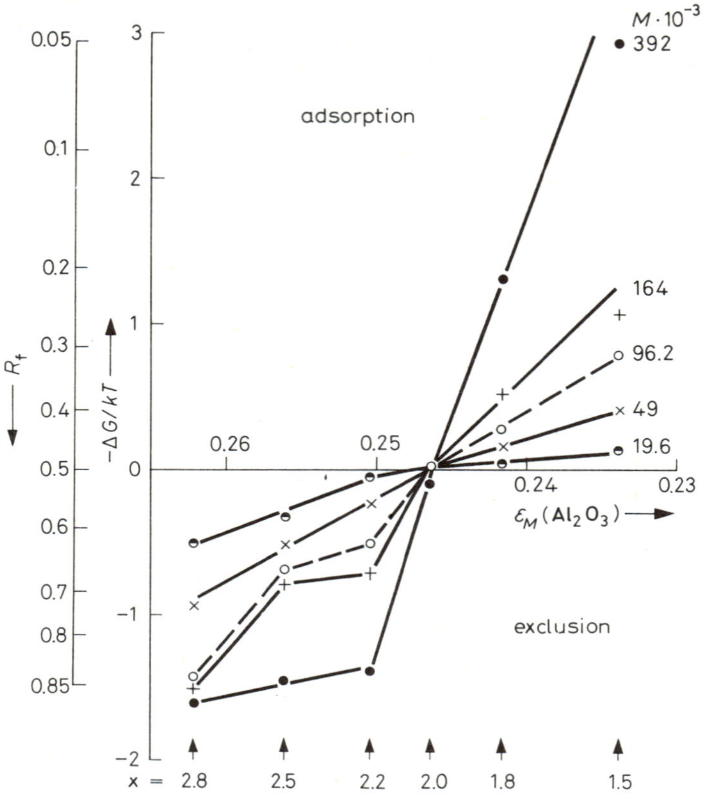

Fig. 16-17
Behaviour of polystyrene samples with narrow distributions in TLC, as a function of the eluent strength

The R_f value and the variation of the thermodynamic potential determined from R_f by eqn. (3-7) and $\Delta G = -kT \ln K$ are plotted vs. the eluotropic strength, ε_M, calculated according to SNYDER [A 4] (cf., curves and amounts to $\varepsilon_c(Al_2O_3) = 0.246$ or $\varepsilon_c(SiO_2) = 0.77 \times 0.246 = 0.189$

Eluent: cyclohexane-benzene-acetone = 40:16:x ($2.8 \geqq x \geqq 1.5$)
Chromatographic bed: KSK silica (d_0 = 10 nm), pre-exposed to the eluent vapour for 2 hours
Parameter: molar mass (number average)
(according to BELENKIJ, GANKINA, TENNIKOV and VILENČIK, 1976).

polymer homologues. The ideal exclusion chromatography is based on entropy effects (cf., Section 8.5.), for which

$$dK/dP < 0 \qquad (16\text{-}60)$$

whereas the mechanisms which are (mainly) due to enthalpy effects, e.g., adsorption, solvophobic interaction or interaction with a swollen phase, lead to an increase of the chromatographic retardation with increasing degree of polymerization, P, and consequently to:

$$dK/dP > 0 \qquad (16\text{-}61)$$

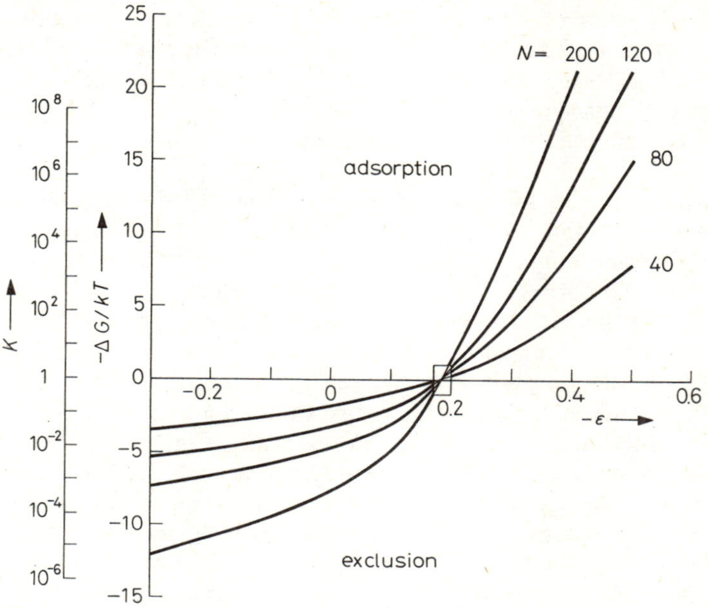

Fig. 16-18
Variation of the thermodynamic potential upon the penetration of macromolecules into pores, as a function of the adsorption energy, ε, of a polymer

Results of model calculations for chains with a different number of segments, Z ($200 \geqq Z \geqq 40$) and a slot with two flat parallel walls. The critical value of the adsorption energy according to this calculation is $\varepsilon_c(SiO_2) = 0.182$. The rectangle indicates the range of Fig. 16-17.
(according to SKVORZOV, BELENKIJ, GANKINA and TENNIKOV, 1978).

Therefore it usually means a disturbance when enthalpy-regulated mechanisms are superimposed upon an SEC separation. Some of the additional effects occur preferentially with organic porous substances, which are generally capable of swelling in organic solvents, while other types of interaction occur preferentially with inorganic separating materials, which are non-swellable but exhibit a higher adsorption activity. Fig. 16-16 gives a schematic survey of the various effects in question.

16.6.1. Adsorption and exclusion

In chromatography on porous separating materials, enthalpy- and entropy-controlled effects may concur, so that which dominates only depends upon strength of the interaction. BELENKIJ et al. (1976, 1978) found that the migration of polystyrene samples on thin layers of silica gel in the ternary eluent cyclohexane-benzene-acetone (40:16:x) is so strongly influenced by slight variations of the acetone concentration that the separation by adsorption changes into a separation by exclusion (see Fig. 16-17). If the acetone concentration is high enough to suppress enthalpy interactions between the polymer sample and the silica gel surface, then the exclusion mechanism dominates. In mixtures containing less than 2 parts by volume of acetone the polystyrene is adsorbed, so that, according to eqn. (16-61), the large-sized molecules are retained longer than the smaller ones.

A computer simulation of the process on the basis of the theory developed by DiMarzio and Rubin (1971) for the adsorption of macromolecules between two flat plates led to the curves shown in Fig. 16-18 (Skvorzov et al., 1978), the course of which is quite analogous to that of those determined experimentally. This agreement shows that the combined effects of enthalpy and entropy, as assumed in the calculation, are real. Moreover, two other points have to be stressed:

— For a *critical value*, ε_c, *of the energy of adsorption*, the enthalpy- and entropy-controlled effects cancel each other. While $dK/dP > 0$ in the adsorption range and $dK/dP < 0$ in the size exclusion range, at the critical value, ε_c, molecules of all sizes travel at the same speed and do not respond at all to the pore system.

— By means of enthalpy interactions, macromolecules may also penetrate into pores which are *smaller* than the coil dimensions. This can easily be observed from the curves shown in Fig. 16-18 for the range of adsorption energy immediately below the critical value, ε_c: ideal size exclusion chromatography takes place on inert pore systems, i.e., at the value $\varepsilon = 0$. Increasing enthalpy interactions lead to higher ordinate values of the curves. With an increase in $-\Delta G/kT$, the distribution coefficient, K, increases by orders of magnitude, as can be observed from the second ordinate scale. However, the increase of K means that an ever increasing fraction of the total pore volume of macromolecules of a certain size can be utilized (cf., eqn. 8-1). The gain in energy associated with adsorption has, so to speak, a suction effect on the macromolecules, compensating the loss of conformation entropy which accompanies the incorporation into too narrow pores (Belenkij, 1979).

In the investigation of the plate height for polymers above the exclusion limit, Knox and McLennan (1979) found that under certain conditions macromolecules obviously "squeeze themselves" into narrow pores. Above the steric exclusion limit, the matter transfer term in eqn. (15-17) is eliminated, so that the latter reduces to:

$$h^* = \frac{2\gamma}{\nu} + A\nu^{1/3} \tag{16-62}$$

The probes used by the authors were polystyrene standards with molar masses of 200000, 470000 and 2700000 g · mole^{-1}, whereas Hypersil® with an exclusion limit of 100000 g · mole^{-1} was used as a packing material. The eluent was dichloromethane. While in a very short column ($L = 55$ mm) the plate height for polymers (as usual) showed no minimum, in 101 mm and 257 mm columns it increased sharply with a decreasing rate of elution, and slightly, with an increasing rate (see Fig. 16-19). At first sight this seems to be the behaviour required by eqn. (16-62). However, the minimum lies at much too high a value of the reduced rate of elution, shifting more and more to the right as the column length and the molar mass of the samples increase. Moreover, at very low rates of elution the peaks were highly deformed, exhibiting considerable tailing. The authors interpreted these phenomena as resulting from a *partial penetration of excluded macromolecules* into outer pores of the packing material particles. This process takes a rather long time because a decoiling is necessary. The rate-determining factor is not the diffusion of the macromolecules, but the capture probability of the coils. As the shift of a peak as a whole towards higher elution volumes was never observed, and only tailing ocurred, it was concluded that only a relatively small number of the molecules present had an opportunity to penetrate into pores. Tailing is caused by the slow desorption.

Using very long columns (L > 10 m) and correspondingly long hold-up times, Ambler et al. (1977) observed an increase in V_e for PS and poly-α-methylstyrene samples as the flow-

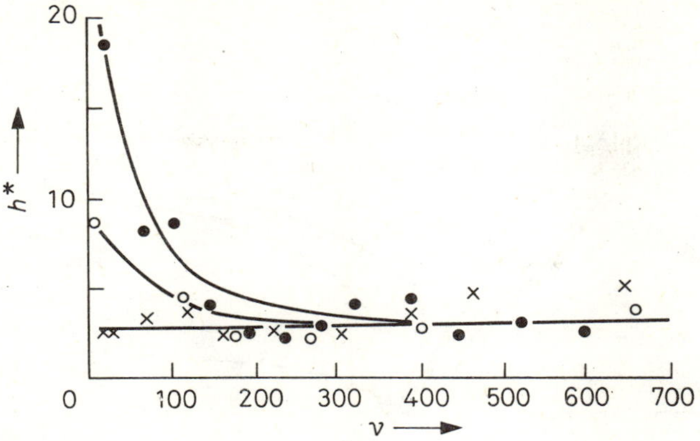

Fig. 16-19
Dependence of the reduced plate height, h^*, on the reduced velocity, ν, for polystyrene ($\bar{M}_w$ = 200 000 g · mole^{-1}) excluded from silica

Packing material: Hypersil®; column lengths: 0.055 m (×), 0.101 m (○), 0.257 m (●)
Eluent: dichloromethane
The increase of the plate height in the longer columns for a very low rate of elution is accompanied by considerable tailing (according to Knox and McLennan, 1979).

rate was decreased from 1 to 0.25 ml/min. In the latter case the samples were in contact with the polystyrene gel in the column for a period of about 10 hours. The additional delay was longest for high molar masses (about 10^6 g · mole^{-1}), whereas it was no longer observed below 10^4 g · mole^{-1}. As measures had been taken to prevent possible evaporation errors (cf., Section 17.7.), the observed effect is probably due to the fact that in the course of time the macromolecules may penetrate into smaller pores.

Too high values of the adsorption energy are irrelevant for the chromatography of polymers, because they lead to such high distribution coefficients that migration becomes practically impossible. In this range the sample is irreversibly adsorbed.

Tennikov et al. (1977) studied the transition from steric exclusion to adsorption by means of column elution chromatography. They used KSK silica gel (d_0 = 10 nm; ${}^1A$ = 350 m^2/g; ${}^1V_P$ = 0.9 cm^3/g) with grain sizes of 63—90 μm in a thermostated 0.60 m column with an inside diameter of 4 mm. Polystyrene standards with narrow distributions in the molar mass range from 730 to 50 100 g · mole^{-1} (corresponding to degrees of polymerization from 7 to 481) as well as oligomers were investigated. Chloroform, tetrachloromethane and mixtures of these solvents were used. Fig. 16-20 shows that at 30 °C, in media with at least 5.9 vol.-% chloroform, the elution curves are similar to the usual log M/V_e calibration curves of GPC. As the chloroform content decreases from 100% to the value mentioned, in effect only a shifting of the elution characteristic towards higher volume values can be observed. This, however, is largest for the samples with medium molar masses, and hence also leads to an increasing bending of the curves. The shift indicates an increase of the distribution coefficient, and consequently an increase of the accessible fraction of the total pore volume.

The shift of the exclusion limit indicated by the broken line in Fig. 16—20 directly shows the penetration of macromolecules into smaller pores. This may be due to the suction effect of the enthalpy interactions or to a coil-size reduction in the poorer solvent. If the

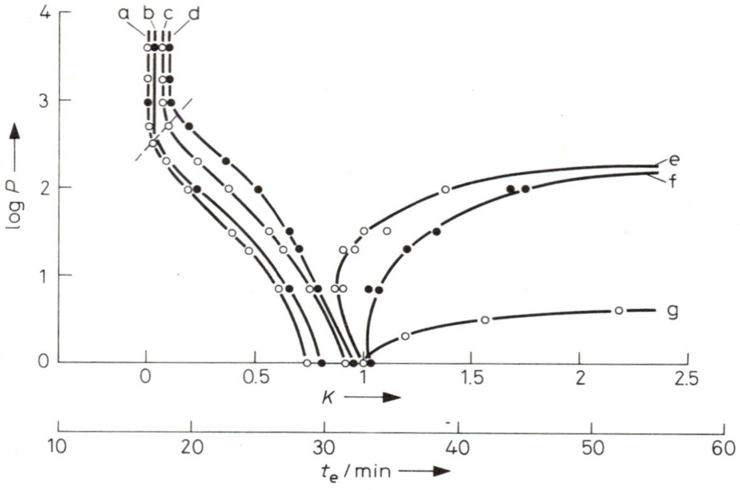

Fig. 16-20
Elution characteristic of polystyrene on silica gel in chloroform-carbon tetrachloride mixtures at 30 °C

Vol.-% of chloroform in the mixture: a 100; b 20; c 7.5; d 5.9; e 5.5; f 5.0; g 0. Column: $L = 0.6$ m; $d_c = 4$ mm; packed with KSK silica gel, $d_0 = 10$ nm.
(according to TENNIKOV, NEFEDOV, LAGAREVA and FRENKEL, 1977).

chloroform content of 5.9% is further reduced to 5.5%, then the curves bend to the right. As a result of this small step of 0.4%, the critical value of the adsorption energy is exceeded. The adsorption, increasing with the molecular size, leads to a very great retardation of the samples with degrees of polymerization above $P = 96$. Using eqn. (7-18) and the ε^0 values for chloroform and tetrachloromethane on silica gel with activity $\alpha_A = 1$, from these experiments one obtains $\varepsilon_c = 0.177$. This value is in satisfactory agreement with $\varepsilon_c = 0.189$ derived from the TLC experiments (see Fig. 16–17). In pure tetrachloromethane, only oligomers with seven monomer units at most can pass through the column, whereas all the higher polymer homologues get stuck.

Fig. 16-21 shows the *effect of the temperature*. At 12 °C the elution behaviour in the critical mixture containing 5.5% chloroform still fully corresponds to an ordinary GPC calibration curve; the bend observable in Fig. 16-20 cannot yet be detected. This occurs only at higher elution temperatures, being fully developed at 40 °C. Here adsorption effects extend the elution time, e.g., from 21.5 to more than 80 min. for the sample with a degree of polymerization of 186. This retention, increased by a factor of 4, occurs in a non-varied elution system, and is due only to the increase in temperature from 12 to 40 °C. (An increase of adsorption with increasing temperature has frequently been observed from polymers; cf., Section 6.2.2.). At 20 °C and 30 °C the log P vs. V_e curves split up into two branches. Thus obviously part of the sample is eluted in a normal way whereas the other part undergoes an additional, substantial retardation due to adsorption.

In a way very similar to that just described, a sudden change in the elution characteristic can also be caused by a modification of the packing material (Fig. 16-22).

For polymers the thermodynamic quality of the eluent also has a strong influence on the elution behaviour. In θ *solutions* even relatively weak interactions between the polymer and

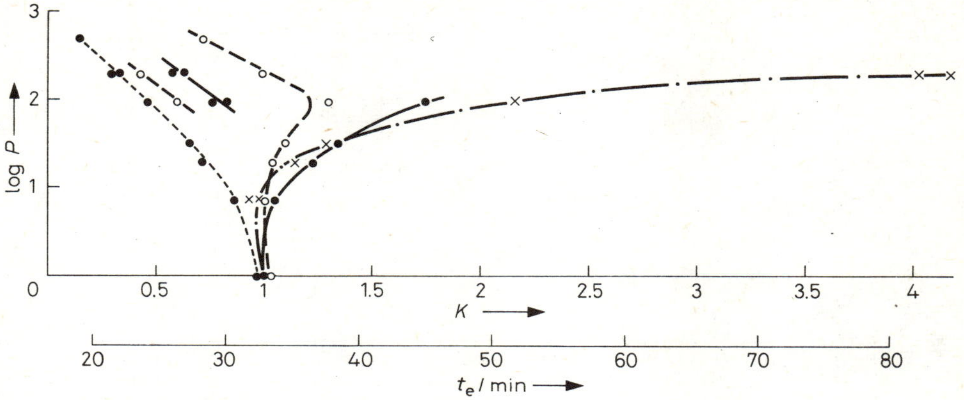

Fig. 16-21
Effect of temperature on the elution behaviour of polystyrene in carbon tetrachloride with 5.5 vol.-% chloroform on silica gel

-------- 12 °C; ——— 20 °C; ——— 30 °C; —.—.— 40 °C. Other conditions like those in Fig. 16-20. (according to TENNIKOV, NEFEDOV, LAGAREVA and FRENKEL, 1977).

the separating material may cause high retardations of the sample, because these solutions are at the margin of the stability range. In Chapter 6 the results of static measurements have been mentioned, which show the particularly high adsorption tendency of polymers in θ systems. Now corresponding observations in GPC will be discussed: Fig. 16-23 shows the universal calibration curves in chloroform and cyclohexane for polydimethylsiloxane, polyisoprene and polystyrene. For the first two polymers cyclohexane is a good solvent, which elutes them together from a polystyrene gel column. Compared with chloroform, the calibration curve is shifted to slightly higher V_e values. This effect can be explained on the basis of the swelling property of the polystyrene gel. For polystyrene, however, cyclohexane

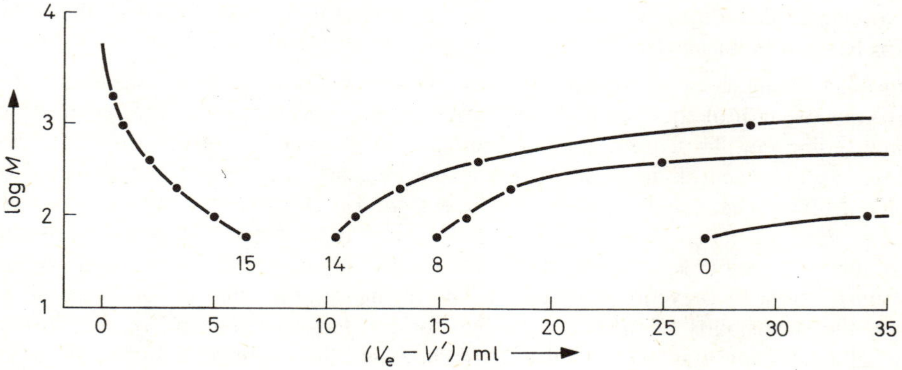

Fig. 16-22
Elution behaviour of polyethylene oxide samples in acetonitrile on surface-modified silica gels

0 — starting material (Si 60 silica gel); 8 — Si 60 with 8 wt.-% of heat-fixed polyethylene oxide (PEO; $M = 400$ g · mole^{-1}); 14 — ditto, with 14% PEO; 15 — ditto, with 15% PEO.
(according to LECOURTIER, AUDEBERT and QUIVORON, 1979).

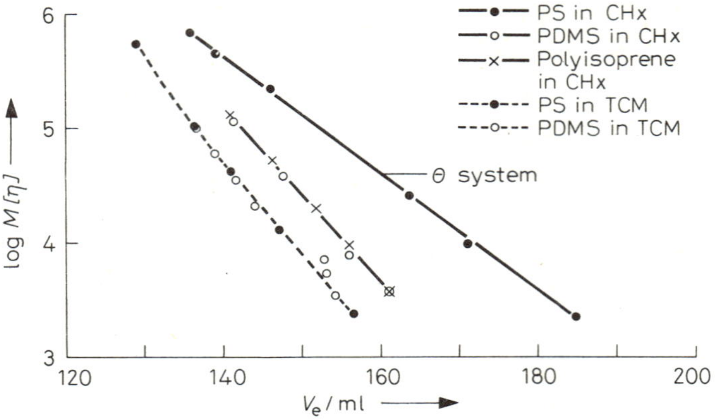

Fig. 16-23
Universal calibration curves in chloroform (TCM) and cyclohexane on polystyrene gel at 35 °C

Cyclohexane is a θ solvent for polystyrene. In this solution, the polymer is additionally retarded by adsorption.
(according to DAWKINS and HEMMING, 1975a).

at a temperature of 35 °C is a θ solvent, in which this polymer is retained for a much longer time than the two other ones. On the other hand, in chloroform, a good solvent, polystyrene also lies on the common calibration curve.

The question of practical importance is how a disturbance of a size exclusion mechanism due to adsorption can be avoided. The following possibilities are available:

— Choosing a separating gel with a low adsorption activity. For the size exclusion chromatography of weakly polar polymers, non-polar or weakly polar separating gels should be used, those with highly polar surface groups fail in most cases. The SiOH groups of silica gel and glass surfaces exhibit a strong adsorption effect. To enable an unrestricted utilization of the advantages of the inorganic separating materials, these groups have to be masked by reaction with suitable reagents (cf., Section 11.10.).

— Choosing a strong eluent. On polar supports, (weakly) polar samples are adsorbed if the eluotropic strength of the solvent is too low. For instance, the irreversible adsorption of polystyrene samples from tetrachloromethane on silica gel is a process of this type. (Normal exclusion chromatography is possible if benzene or tetrahydrofuran is used as an eluent.) The eluent must exhibit a sufficient elution effect, which can be observed from its ε^0 value (cf., Table 7-3), and at the same time it must be a good solvent in the thermodynamic sense. Although a strong self-adsorption of the solvent on the separating material prevents the adsorption of the sample, on the other hand it reduces the accessibility of the pores, so that for this reason certain differences in the universal calibration may occur. This will be discussed in Section 16.6.4.

— Additions to the eluent. If the polymer to be investigated is insoluble in eluents of a sufficient strength, solutions in good solvents with an insufficient polarity can be adjusted by adding a polar medium, so that SEC becomes practicable without complication by adsorption (ŽDANOV et al., (1973). The polar components of the eluent deactivate the adsorbent by blocking the adsorption sites. In this connection, the so-called *displacer effect* encountered in thin-layer chromatography should be mentioned (see Fig. 21-15).

The deactivation of the surface also occurs in competition with other effects, when mixed eluents containing components of different polarities are used: Fig. 16-24 shows calibration curves of polystyrene on silica gel in various pure solvents and mixtures. Although the mixtures are of the θ type, with the addition of methanol the retardation is lower than predicted by the universal calibration curve. This is in contrast to the higher retardation shown in Fig. 16-23. Apart from the reduction of the hydrodynamic volume due to the lower thermodynamic quality of the solvent mixtures, and besides the displacer effect of the methanol, a reduced accessibility of the pores may also contribute to this effect. FIGUERUELO et al. (1980) performed similar experiments using PS samples ($450 \leqq M \leqq 24700$ g · mole^{-1}) on Spherosil with a separating threshold of about 500 g · mole^{-1} in different solvents and mixtures. The elution volumes for some substances, the sizes of which were smaller than the separating threshold (monostyrene, ethylbenzene, dimer of α-methylstyrene), were about 10% greater than those of the oligostyrene samples, whose behaviour at around 600 g · mole^{-1} was almost independent of the molar mass. The different eluents had a great effect on this limit. The V_e values were lowest in B/M (75:25) and highest in B/Hep (92:8). To explain this result it was assumed that there were

Eluents with salt additions will be discussed in Sec. 19.3.2.

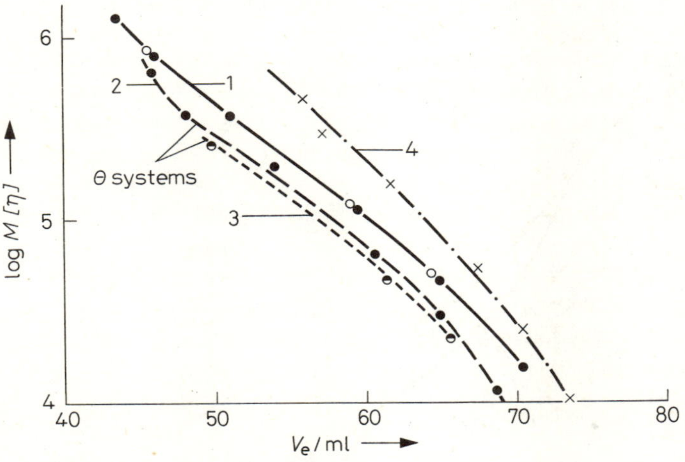

Fig. 16-24
Universal calibration curves for polystyrene standards in good solvents and in θ mixtures on silica gel

In non-polar, poor solvents (4) the polystyrene is retarded on the silica gel, whereas in θ mixtures with a methanol content (2, 3) it is eluted earlier than in good solvents (1)
1 universal calibration in pure benzene (●) and chloroform (○)
2 Bzn/M (77.8:22.2)
3 TCM/M (74.7:25.3)
4 MEK/Hp (50:50)
Column: $L = 2 \times 1.05$ m; $d_C = 9.5$ mm, packed with Porasil D and Porasil E
(according to BEREK, BAKOŠ, BLEHA and ŠOLTÉS, 1975 (curves 1, 2, 3) and BAKOŠ, BLEHA, OZIMÁ and BEREK, 1979 (curve 4)).

16.6.2. Solvophobic interactions in GPC

In columns with non-polar packing materials, an increased retention of non-polar substances occurs when polar eluents such as dimethylformamide are used. This effect manifests itself most clearly in the chromatography of low-molecular-weight substances, for which full accessibility of the total surface can be assumed and, to a first approximation, influences of the solvent on the size of the solute can be ignored. DUBIN et al. (1977) found that in dimethylformamide as an eluent the retention time of toluene on cross-linked polystyrene exceeds that of aniline by 24%. As compared with benzoic acid, the increase even amounted to 40% (see Fig. 16-25). In tetrahydrofuran there were only slight differences. Quite obviously the effect is caused by interactions between the substance and the substrate surface, which greatly depend on the solvent. This can also be observed from Fig. 16-26, which shows the elution behaviour of n-alkanes in nine different solvents. Naturally the differences in curves 1 to 8 are also influenced by the different swelling of the polystyrene gel in the various eluents, but the fact that in acetone there is a sudden increase in retention with increasing molecular size indicates solvophobic interactions (cf., Section 7.6.).

In addition to the data points for the low-molecular sample, Fig. 16-25 also shows the curves for some polymers of different polarities. In contrast to the adsorption phenomena observed on polar separating materials, as discussed in the preceding section, under the present conditions the polar samples are eluted before the non-polar ones. The characteristic for the polystyrene standards lies farthest to the right, approaching the elution value of toluene, while the curves for polyethylene oxide or polyacrylic acid point to much more polar small-molecule models. In dimethylformamide the calibration curves established by means

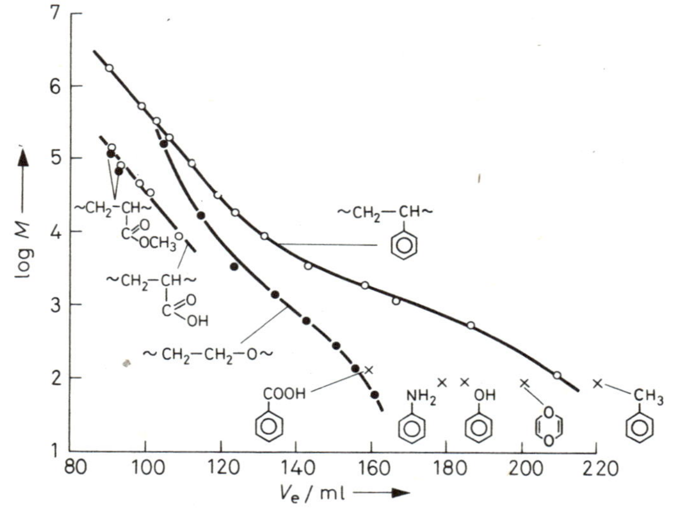

Fig. 16-25
Effect of the structure of the samples on their elution behaviour on cross-linked polystyrene in DMF

Calibration curves for polystyrene standards with narrow distributions and for unfractionated samples of polyethylene oxide, polyacrylic acid and polymethyl acrylate on a 4 m column with 4 Styragel® columns (10^5, 10^3, 10^2 and 60 Å), and elution values for five small-molecule substances of different polarities. (according to DUBIN, KOONTZ and WRIGHT, 1977)

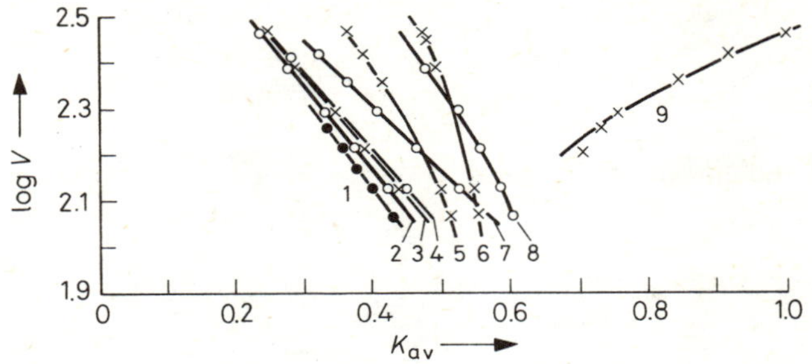

Fig. 16-26
Elution behaviour of n-alkanes (C_5—C_{16}) in different solvents on Styragel® 60 Å
Column: $L = 1$ m; $d_c = 5$ mm
1 tetrahydrofuran; 2 toluene, 3 benzene; 4 chlorobenzene; 5 1,2-dichlorobenzene; 6 butyl acetate; 7 ethyl acetate; 8 1,2-dichloroethane; 9 acetone
Logarithmic plot of the molar volume vs. the Laurent-Killander distribution coefficient K_{av} (according to OZAKI, SAITOH and SUZUKI, 1979).

of polystyrene standards lie at too high values of the elution volume. Consequently the molar masses found for polar polymers with weaker solvophobic influences appear too high. However, as DMF is indispensable for the investigation of polyacrylonitrile and other poorly soluble polymers and in most cases the calibration curves are determined by means of polystyrene standards, it is essential that the disturbances described are suppressed. The effect of the solvophobic distribution decreases if packing materials are used with a polarity approaching as closely as possible that of the eluent. Therefore silica gels and porous glasses are more suitable for polar eluents than silanized materials or entirely non-polar substrates. DUBIN et al. (1977) established a common universal calibration curve on an untreated porous glass using polystyrene, polyethylene oxide, polymethyl acrylate and poly-p-nitrostyrene as samples in dimethylformamide as an eluent, whereas considerable differences were found to occur on polystyrene gels (see Fig. 16-27).

Besides the polarity, the thermodynamic quality of the eluent has some influence. In poor solvents characterized by low values of the virial coefficient, A_2, the effect of solvophobic distribution equilibria is more likely to occur than in good solvents, which develop strong interactions with the dissolved polymer (DAWKINS and HEMMING, 1975c).

Dimethylformamide, which in any case is not a good solvent for polystyrene, is in certain cases used with salt additions (cf., Section 19.3.2.). This further reduces its solvating effect, as can be observed for example from the lower values of the intrinsic viscosity. In addition to the decrease in the coil dimensions, the stronger solvophobic interactions lead to an increase in the elution volume (see Fig. 16-28). This effect was also observed by CHA (1969) and by COPPOLA et al. (1972).

BOOTH et al. (1980) found that LiBr reduces the intrinsic viscosity of polystyrene in DMF by approximately 15%. However, while this effect was independent of the actual LiBr concentration within a relatively wide range, with increasing salt content the elution curves were shifted towards higher and higher values of the elution volume (see Fig. 16-29). A microgel with particle masses of more than 10^7 g · mole^{-1} even disappeared in the column if the LiBr content exceeded 0.1 mole · l^{-1}. The counterpart to these *salting-out effects*, which were repeatedly observed, was observed by SIEBOURG et al. (1980) in the exclusion chromatography of polystyrene samples on a column packed with polystyrene gel: by addition of 2 g/l $LiNO_3$ to a 10:1 (v/v) mixture of tetrahydrofuran and

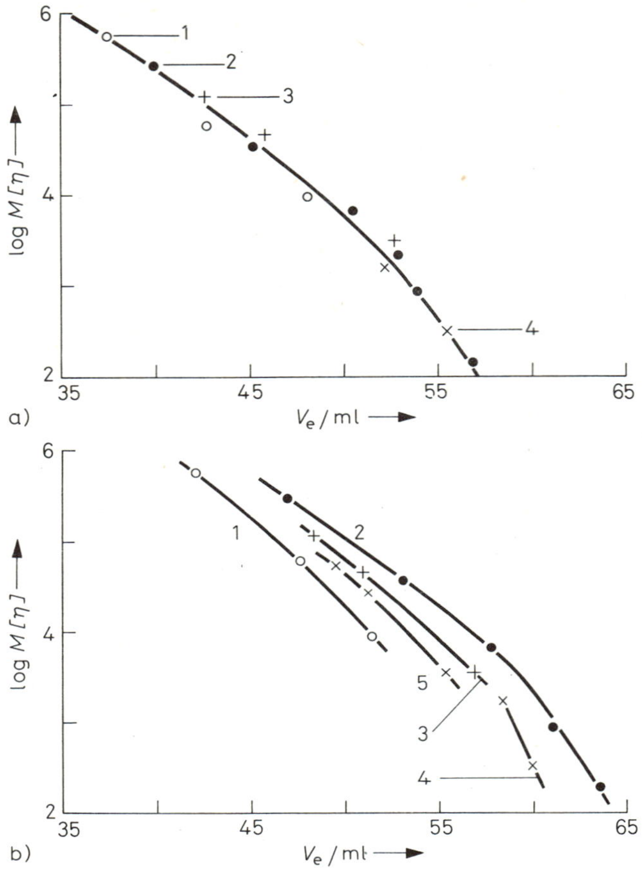

Fig. 16-27
Universal calibration curves in dimethylformamide on

a) CPG-10 porous glass; b) polystyrene gel
1 poly-p-nitrostyrene; 2 polystyrene; 3 poly(methacrylate); 4 poly(ethylene oxide); 5 poly(vinylpyrrolidone) (according to DUBIN, KOONTZ and WRIGHT, 1977).

dimethylformamide, it was possible to eliminate the increase in the elution volume of a polystyrene standard as compared to its elution in THF without DMF content. In the molar mass range from 2300 to 1400000 g · mole^{-1} (8 standards), the authors established a common straight line by plotting log M vs. V_e for the three eluents THF, THF + $LiNO_3$ (2 g/l) and THF + DMF (10:1) + $LiNO_3$ (2 g/l).

16.6.3. Partition in the wall material

On organic gels prepared without an inert solvent (cf., Chapter 12), the separation by the exclusion mechanism is based on the network-limited distribution of the sample between the mobile phase and the meshes of the network. In this case of course there are close

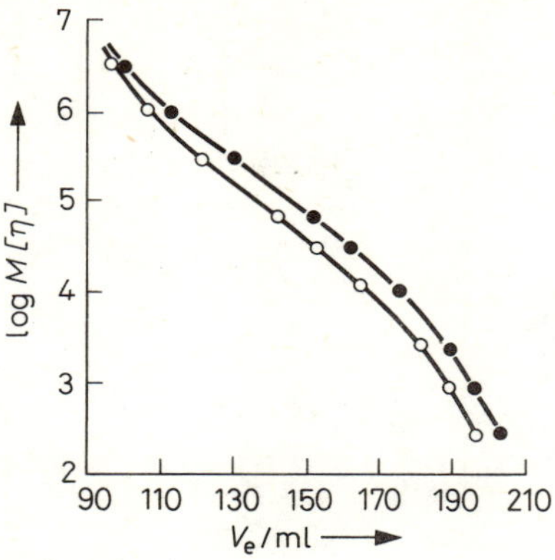

Fig. 16-28
Effect of LiBr on the calibration curve in DMF

Column: four columns packed with Styragel® with a nominal pore size of 2.5×10^4 A
Samples: polystyrene standards
Temperature: 80 °C
Flow rate: $v = 1$ ml/min
○ measurement in pure DMF; ● measurement in DMF containing 0.05 M LiBr (according to HANN, 1977).

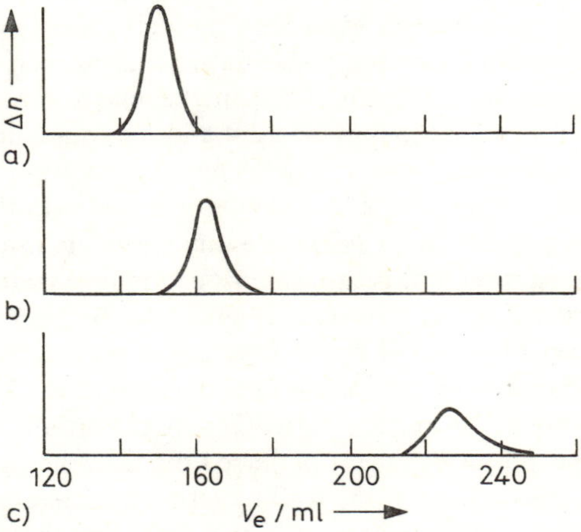

Fig. 16-29
Elution curve of a polystyrene standard

($M = 51\,000$ g · mole^{-1}) in DMF with different LiBr concentrations
a) 0.5; b) 86; c) 196 mmole LiBr/l; Styragel® column, the same as in Fig. 16-34 (according to BOOTH, FORGET, GEORGII and PRICE, 1980).

contacts between segments of the gel matrix and segments of the permeating polymer molecules, so that the permeation is regulated not only by the mesh size alone but also by the similarity or dissimilarity in the chemical structure. This was found in 1967 by HEITZ and KERN (Fig. 16-15), and can be clearly observed from Fig. 16-30. As the mechanical strength of such gels decreases with increasing mesh size, gels were developed which, due to cross-linking in the presence of inert solvents, contain macropores embedded in a relative-

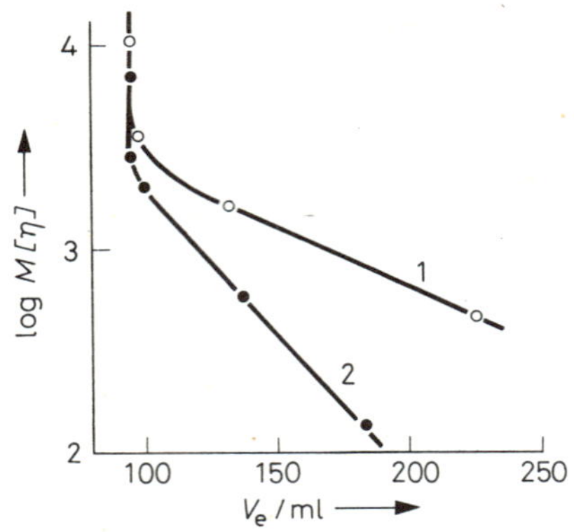

Fig. 16-30
Elution behaviour of dextran (1) and polyethylene glycol (2) on cross-linked dextran

Column: L = 0.96 m; d_c = 20 mm; packed with Sephadex® G-75, d_p = 40—120 μm
Eluent: 0.3 % NaCl in water
The similarity in the chemical structure of the sample and the gel leads to longer retention times of the dextran samples, and hence to a failure in the universal calibration.
(according to BELENKIJ, VILENČIK, NESTEROV and ŠAŠINA, 1973b; BELENKIJ et al., 1975b).

ly highly cross-linked polymer matrix. These gels are capable of separating samples with high molar masses (cf., Section 12.1.). However, even with its high cross-linking density, this polymer matrix represents a gel which is capable of swelling and allows sufficiently small molecules of the sample to penetrate by diffusion. Thus in the macropores a *network-limited partition* (HEITZ et al., 1967) between the pore content and the wall material may be superimposed upon the (intended) separation by steric exclusion, this partition being largely determined by enthalpy interactions. For that reason it should increase with the molecular size. However, as the smallest sample molecules can most easily penetrate into the narrow meshes of the wall material, the retardation in this case is most pronounced with these components. If the elution is performed with a mixed eluent (cf., Section 19.3.1.) the partitions of its ingredients between the gel and the mobile phase may vary. BLEHA and BEREK (1981) investigated the behaviour of a benzene-methanol mixture (77.8:22.2 v/v) and found that polar gels such as dextran and its derivatives preferentially took up the methanol from the mixture, whereas polystyrene gels preferred the benzene. In addition to the change in pore volume due to the swelling of the wall material, a preferential solvation may vary the thermodynamic interactions between sample molecules and the gel phase.

The inorganic porous separating materials are free of such partition effects. If the inorganic skeleton, however, is coated by a polymer layer, as is the case for certain modifications of the surface, then the sample may be distributed between the mobile phase and this polymer layer in a similar way as in gels. If the layers are not cross-linked, then the limitation by means of the network is eliminated, so that the large-sized molecules are also involved in this partition. In this case the increase of the enthalpy interaction with increasing degree of polymerization can develop freely.

The penetration of a macromolecule into the gel layer is accompanied by an entropy loss which, without a sufficient compensating enthalpy effect, will lead to a decrease in the partition coefficient, K, with increasing chain length. In this case the representation of $\log P$ vs. K has the same shape as the calibration curves in GPC. For sufficiently high interaction energies, the shape of the curve then undergoes similarly dramatic changes as at the onset of adsorption phenomena; the partition coefficients increase beyond 1 and rapidly reach very high values. On the basis of the Flory-Huggins lattice theory, LECOURTIER et al. (1979b) discussed the absorption of macromolecules by bonded polymer layers and calculated the dependence of the partition coefficient on the molar mass of the permeating molecules. A typical result is shown in Fig. 16-31.

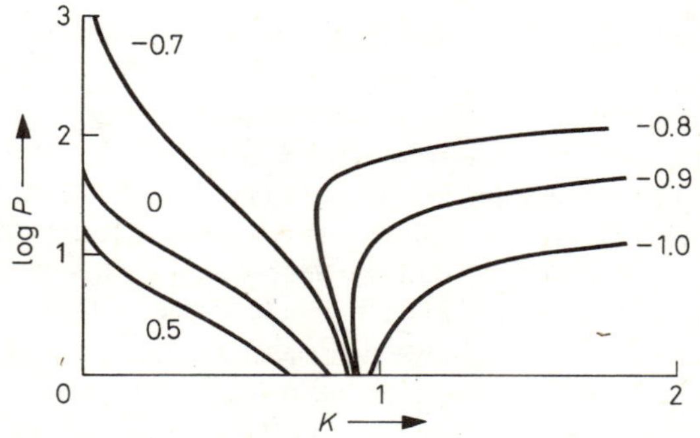

Fig. 16-31
Elution characteristics for different interactions between the sample and a polymer layer fixed to the support material

Calculated for $P_3 = 10$, $\chi_{12} = \chi_{13} = 0$, and the values of χ_{23} indicated as parameters.
χ-Huggins constant (cf., eqn. 5-9)
P degree of polymerization; K distribution coefficient
Indices: 1 eluent; 2 polymer sample (with the degree of polymerization indicated on the ordinate); 3 fixed polymer phase. The curve for $\chi_{23} = 0$ shows the distribution characteristic which is only due to entropy contributions. For high values of the energy of attraction ($\chi_{23} \leqq -0.8$) the samples are highly retarded.
(according to LECOURTIER, AUDEBERT and QUIVORON, 1979b).

The phase partition chromatography discussed in Section 9.4.2. is closely related to the phenomena discussed here. In gels, macromolecules have very low diffusion coefficients. The penetration into the depth of a swollen polymer phase therefore considerably impairs the kinetics of mass transfer. This leads to tailing and to poor resolution values, unless the elution is carried out very slowly.

So far we have discussed the effect of attractive interactions on the relationship between the partition coefficient and the molar mass. If the interactions are too weak or if the enthalpy increases, then macromolecules of a different chemical nature repel each other. Additional retardation of a sample or premature elution may therefore also be discussed from the aspect of *compatibility* and *incompatibility*, respectively, if a close interpenetration of gel and sample segments is possible or necessary in the chromatographic process. The fact that the elution of dextran samples on cross-linked dextran as a separating material, as shown in Fig. 16-30, is retarded as compared with polyethylene oxide samples results from differences in compatibility. In the discussion of irregularities observed with macroporous packings with densely cross-linked wall material, this interpretation should, however, be used with caution. The very different elution behaviour of polyvinyl acetate on the one hand and polystyrene on the other in GPC on macroporous polystyrene gels in tetrachloroethane (ALTGELT, 1971) cannot be finally judged until a better knowledge about the state of solution in this solvent is available. As PARK and GRAESSLEY (1977 a) as well as ATKINSON and DIETZ (1979) found an excellent agreement of the hydrodynamic volume calibration for PS and PVAC on macroporous polystyrene gels in tetrahydrofuran, the above deviations in tetrachloroethane are not necessarily due to incompatibility. For example, association phenomena would also lead to the shift of the PVAC elution volumes towards lower values, and increase with increasing concentration and increasing molar mass.

16.6.4. Reduction of the available pore volume by solvent adsorption

If the pore size and the molecular size are of the same order of magnitude, i.e., if the conditions are the same as in SEC, then solvent layers covering the pore walls may appreciably affect the accessibility of the pores. GROH and HALÁSZ (1980) investigated the behaviour of C_6 cyclic hydrocarbons on microporous silica gel and found that in "dry" dichloromethane (containing less than 5 ppm water) cyclohexane, cyclohexene and cyclohexadiene were eluted separately and before CD_2Cl_2 in the order stated. The accessible portion, θ, of the total pore volume, V'', available decreased for the three samples in the same order (V'' was determined as the difference between the elution volume, V', for an excluded polystyrene standard with molar mass $M = 110000\ \mathrm{g \cdot mole^{-1}}$ and the elution volume of CD_2Cl_2 in dry CH_2Cl_2). With increasing water content in CH_2Cl_2, θ varied in the way shown in Fig. 16-32, with a dramatic drop at 950 ppm. For this value the silica investigated had adsorbed 70 mg water per gram. One molecule covers an area of 0.3 nm^2, which corresponds to a localized adsorption. Above 950 ppm the pores fill with water, so that they become inaccessible. As expected, all of the samples used are not strong enough to displace the water adsorbed by the silica gel in the equilibrium with water contained in

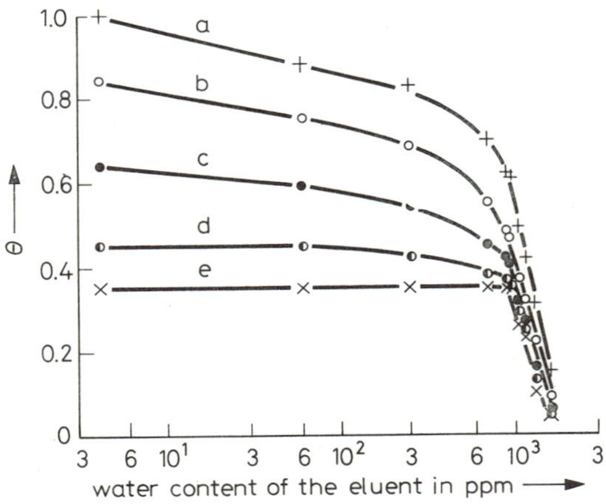

Fig. 16-32
Available portion, θ, of the total pore volume of a microporous silica gel as a function of the water content in dichloromethane as an eluent

Samples: a CD_2Cl_2; b cyclohexadiene; c cyclohexene; d cyclohexane; e Fluorinert (perfluorinated hydrocarbon mixture)
(according to GROH and HALÁSZ, 1980).

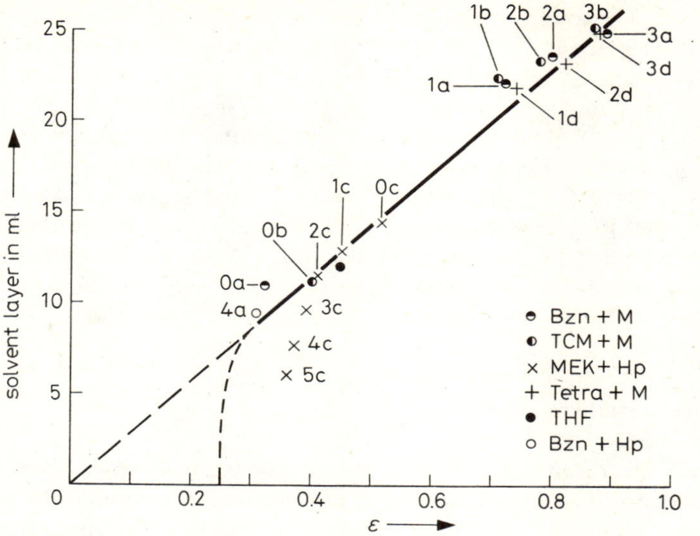

Fig. 16-33
Volume of the bonded solvent on silica gel as a function of the eluotropic strength

The layer volume shown on the ordinate was determined using a 0.45 m column with an inside diameter of 25 mm (SR 25, PHARMACIA), which was packed with 74 g Spherosil® XOA 200. For a specific pore volume of ${}^{1}V_{P} = 0.95$ cm³/g, the total pore volume was $74 \times 0.95 = 70.3$ ml. The available pore volume is equal to the difference between the steric exclusion volume, V', and the total permeation volume, V_{mobile}, determined by means of a polystyrene of $M = 2000$ g · mole^{-1}. V' and V_{mobile} depend on the solvent. $V_{mobile} - V' = V''$ is always smaller than 70.3 ml. As silica gel does not swell, the difference of 70.3 $- (V_{mobile} - V')$ is the volume of the adsorbed solvent layer.

Solvent mixtures used:
1a—3a: Bzn + M (*1* 10; *2* 16; *3* 25 vol.-% M)
4a: Bzn + Hp (8 vol.-%)
1b—3b: TCM + M (*1* 10; *2* 16; *3* 25% M)
1c—5c: MEK + Hp (*1* 25; *2* 40; *3* 50; *4* 55; *5* 60% Hp)
1d—3d: tetra + M (*1* 10; *2* 16; *3* 25% M)
(according to CAMPOS, SORIA and FIGUERUELO, 1979).

Pure solvents:
0a: Bzn
0b: TCM
0c: MEK

the eluent. For a lower water content in the eluent, the amount of water on the surface also decreases, and consequently the accessible portion of the total pore volume increases.

In dry CH_2Cl_2 the samples compete with adsorbed CH_2Cl_2. The higher the polarizability of the samples, which increases in the order cyclohexane < cyclohexene < cyclohexadiene, the more successful they are in this competition. As in these investigations there is no indication to assume differences in the molecular size and corresponding exclusion phenomena, the different elution of these "inert" compounds, which takes place before that of CD_2Cl_2, is due to the phenomenon of a *negative adsorption.*

If, as in SEC, the sample molecules and the pores are of about the same size, the accessible pore volume of polar separating materials may increase with the polarity or the polarizability of the sample, because then the latter can better displace the solvent adsorbed. In the previously described investigation, this was demonstrated by means of different samples in one and the same eluent.

In independent investigations, CAMPOS et al. (1979) have shown that the accessible pore volume also increases as the adsorption energy of the eluent decreases. They determined the exclusion volume of a certain column using 21 eluents and a polystyrene standard of

660 000 g · mole^{-1}. The total permeation volume, $V' + V''$, was evaluated by means of a polystyrene of 2000 g · mole^{-1}. The column was packed with Spherosil XOA 200 silica gel, so that volume variations due to swelling could be excluded. Fig. 16-33 shows that the volume of the adsorption layer depended linearly on the eluotropic strength $\varepsilon^0(Al_2O_3)$, see Table 7-3 and eqn. (7-18). However, the experimental points do not follow the straight line down to the origin, because the polystyrene used as a sample is itself adsorbed in weak eluents. In benzene-heptane (70:30) with $\varepsilon^0(Al_2O_3) = 0.275$ and in carbon tetrachloride with $\varepsilon^0 = 0.18$, the sample stuck in the column. Its displacing effect on the adsorbed solvent layer began at $\varepsilon^0 = 0.4$ in MEK-heptane mixtures, whereas the behaviour in benzene and benzene-heptane can be represented by a curve pointing towards the critical value, $\varepsilon_c = 0.246$, determined by BELENKIJ et al. (see Fig. 16-17).

16.6.5. Electrostatic repulsion

A unipolar static electrification of the column packing and the sample effects a repulsion, which may build up to a degree where the particles to be separated do not penetrate at all into the pore system of the packing. The electrostatically excluded parts of the sample are discharged in a liquid volume which, in an extreme case, may even be smaller than the interstitial volume of the packing. In any case, if interpreted with the help of the calibration curve, it would correspond to a very high value of the molar mass.

Electrostatic repulsion phenomena were observed by KIRKLAND (1979) in the chromatography of silica gel particles on silica packings (cf., Section 19.9.3.): while the 8 nm particles were normally eluted in water containing 0.02 M triethanolamine (pH = 8, adjusted by means of HNO_3), in 0.001 M NH_4OH a premature elution was found to occur, which suggests a steric exclusion from the 6 μm pores of the packing. From the data, the interstitial porosity of the column can be evaluated as $\varepsilon_I = 0.35$.

In GPC of phenol-formaldehyde condensates in DMF on silica gel packings, SCHULZ et al. (1981) observed a massive leading peak, which did not occur if the elution was carried out with an additon of 0.1 M LiCl, or if the novolak resin was investigated in the form of its acetyl derivative.

In SEC, the addition of salt is frequently used in order to suppress electrostatic disturbances; thus, an observation by BOOTH et al. (1980) in the chromatography of *polystyrene microgels* in dimethylformamide on Styragel® columns deserves attention (cf., Fig. 16-34): with increasing LiBr content there was an increase in retention, which was very pronounced mainly for small additions (up to 5 mmole/l). This additional retention was not at all perceptible for linear polystyrene within the same range. Starting from the hypothesis that the phenomenon is due to an adsorption of bromide ions on microgel and on the column packing material, and hence ultimately is caused by an overlapping of electrostatic double layers, the authors applied the theory by GOUY and CHAPMAN, according to which the thickness of the double layer should increase as $[LiBr]^{-1/2}$. As is shown in Fig. 16-34, part (b), the peak elution volume indeed depended linearly on $[LiBr]^{-1/2}$, so that the electrostatic nature of the effect was confirmed.

An electrostatic repulsion also possibly played a rôle in the passage of methacrylic acid oligomers in DMF through a Styragel® column (NEFEDOV et al., 1979). For a sample with $\bar{M}_v = 3200$ g · mole^{-1}, the molecules of which exhibit a straight chain-length of 10 nm, an elution volume corresponding to that of a polystyrene standard with a hydrodynamic

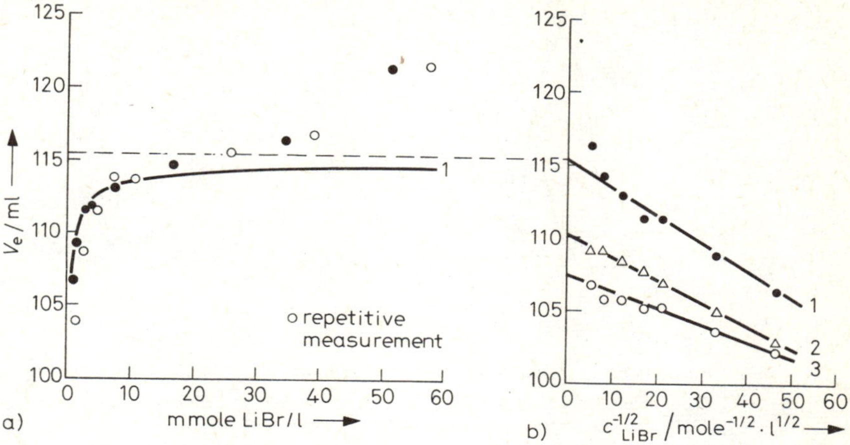

Fig. 16-34
Peak elution volume of a polystyrene microgel as a function of the LiBr quantity added

Column: four columns ($L = 4 \times 1.22$ m) with Styragel®
(nominal pore size 10^3; 5×10^3 ... 1.5×10^4; 10^5 and 10^6 Å)
Eluent: DMF at 80 °C with the indicated LiBr addition; flow rate $v = 1$ ml · min^{-1}
Samples: microgel particles, prepared by emulsion polymerization of styrene-divinylbenzene mixtures.
1 $M_w = 18 \times 10^6$ g · mole^{-1}; $\langle S^2 \rangle^{1/2} = 25.2$ nm; 2 64×10^6; 38.8; 3 85×10^6; 40.3 nm
The curve in the left part of the figure represents the relationship $V_e = 115.43 - 6.098/\sqrt{c}$, corresponding to the straight line 1 in the right part of the figure.
(according to Booth, Forget, Georgii and Price, 1980).

radius of 350 nm was measured. This phenomenon corresponds to the behaviour of sodium polystyrene sulphonate with a chain-length of 26 nm in contact with 500 nm pores, which is dealt with in Section 19.2.3.1.

Finally, in this connection the elugrams recorded by means of a conductometric detector should also be mentioned. They were obtained by Domard et al. (1979) on a silica gel column for $NaNO_3$ injections into DMF. Fig. 16-35 shows the curves observed after injecting a standard amount of $NaNO_3$ into DMF containing different initial concentrations of this salt. If the eluent initially contained less than $5 \cdot 10^{-2}$ mole $NaNO_3$, the additionally injected quantity was eluted in a volume smaller than the sum $V' + V''$, i.e., for an insufficient ionic strength the pores partially became inaccessible due to electrostatic repulsion (Domard: "The variation of the elution volume with the salt content of the eluent is attributed to the screening of the electrostatic repulsion between the ions and the polar gel."). This electrostatic exclusion of the low-molecular salt occurs in the same way if the latter is used as an electrolyte added in the chromatography of a polar polymer.

Naturally the electrostatic effects manifest themselves most distinctly in exclusion chromatography in aqueous eluents (cf., Section 19.2.3.).

16.6.6. Combination of adsorption, partition and exclusion

The additional influences occurring in real GPC alter the elution volume as compared with its value in pure SEC. As a rule the elution volume increases as an effect of the non-exclusion phenomena. For a quantitative determination of this effect we start from

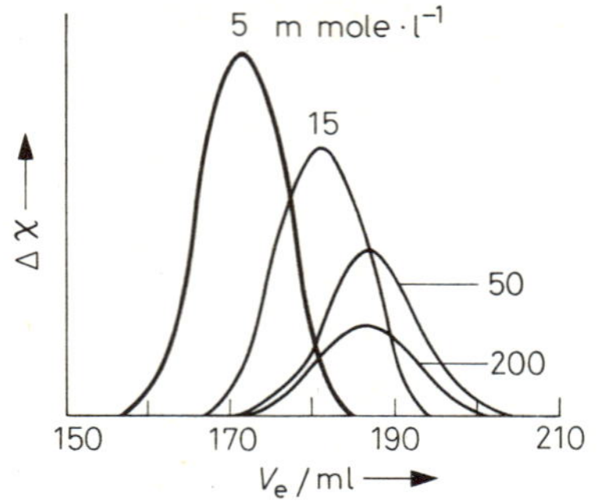

Fig. 16-35
Elution curves of $NaNO_3$ in dimethylformamide

The amount of the sample was $5 \cdot 10^{-3}$ mole. $NaNO_3$ in each case, dissolved in 0.5 ml of the eluent (DMF with the $NaNO_3$ content indicated as a parameter).
Column: $L = 1.47$ m; $d_c = 15$ mm;
Packing: mixture of five Spherosil® types
Detector: conductivity cell
(according to DOMARD, RINAUDO and ROCHAS, 1979).

eqn. (3-10), which relates the total elution time, t_e, with the total time spent in the mobile phase, t':

$$t_e = t'(1 + k) \tag{16-63}$$

Multiplication by the volume flow-rate and substitution of the relationship $k = K \cdot q = K(V''/V')$ gives:

$$V_e = V' + K \cdot V'' \tag{16-64}$$

This relationship is identical with eqn. (8-1) for ideal size exclusion chromatography, where V' is the interstitial volume and V'' the total pore volume. $K = K_{excl}$ denotes the Wheaton-Bauman distribution constant, cf., Section 8.3.4. For enthalpy-controlled interactions with the distribution constant K_{LC}, the following relationship can be derived from eqn. (16-63) in a analogous way:

$$V_e = V_{mobile} + K_{LC} \cdot V_{stat} \tag{16-65}$$

This equation also holds for ideal adsorption chromatography (AC) and ideal partition chromatography (LLC). However, the equation is rarely written in this form, because in most cases the volume of the stationary phase, V_{stat}, is unknown, so that the value of the distribution constant, K_{LC}, cannot be determined on the basis of eqn. (16-65) (cf., Section 16.7.). For that reason, the relationship with the capacity factor

$$V_e = V_{mobile}(1 + k_{LC}) \tag{16-66}$$

is preferred in AC and LLC. On the other hand, in SEC on non-swelling separating materials, the maximum pore volume available can be determined by experiment, so that in this case the formulation (16-64) using the Wheaton-Bauman distribution constant is indeed suitable. In GPC using swollen gels the Laurent-Killander distribution constant is preferable; see Fig. 16-36.

Starting from eqn. (16-65) it is possible to derive the relationship for GPC under real conditions. However, in this case it is worth noting that for AC and LLC V_{mobile} includes the interstitial volume, V', and the accessible pore volume. For a solute which, like the solvent, can freely penetrate into all of the pores, V_{mobile} is the sum of the interstitial volume and the total pore volume, V_P, diminished by the volume of the stationary phase, V'' (cf., Fig. 16-36). The quantity $V_{mobile} = V' = V_I + V_P - V''$ is the *eluent hold-up volume*. In AC and LLC, V_{mobile} is at the same time the elution volume of an inert component, and in SEC it is often called *total permeation volume*. Components with $K_{LC} > 0$ emerge *after* the solvent peak. On the other hand, in SEC the elution volume of a non-retained (excluded) sample is $V' = V_I$, and retained samples emerge *before* the solvent peak. For partially excluded samples with AC and LLC, $V_{mobile} = V_I + K_{excl}(V_P - V'')$ and hence V_{mobile} is almost identical to the elution volume of a component of a corresponding molecular size in ideal SEC, as given by eqn. (16-64).

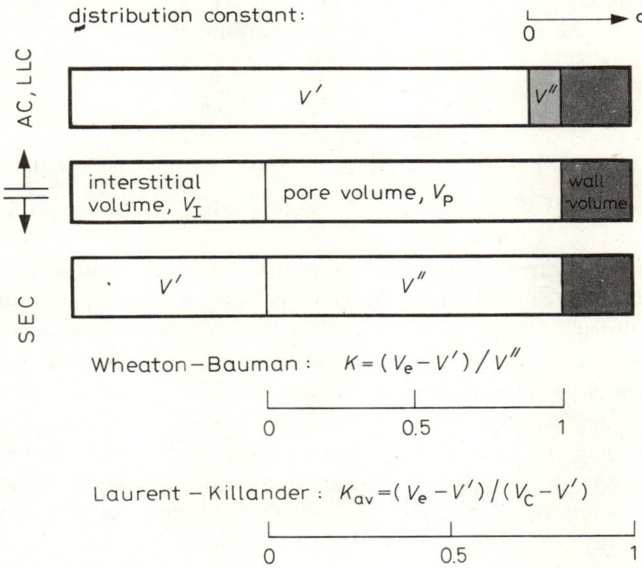

Fig. 16-36
Schematic representation of the phase ratios in a column with porous packing material
Use either in adsorption chromatography (AC) and partition chromatography (LLC) or in size exclusion chromatography (SEC) and in the definition of the SEC distribution constants.
Total volume: $V_C = V_I + V_P + V_W$; mobile phase volume: V'; stationary phase volume: V''.

In SEC, just as in the enthalpy-controlled retention mechanisms, the eluent hold-up volume is equal to the sum of the interstitial volume and total pore volume.

This statement assumes that the volume, V'', of the stationary phase in AC and LLC is negligible in comparison with V_P, which is correct in most cases. Consequently the peak of an inert component marks *the end* of the elution interval in ideal SEC but *the beginning* of the elution range in ideal AC or LLC. In Figs. 16-20 and 16-21, this peak would occur at $K = 1$.

Substitution of $V_{mobile} = V_I + K_{excl} \cdot (V_P - V'')$ into eqn. (16-65) yields the following equation for real GPC (with V_{stat} corrected for the excluded portion):

$$V_e = V' + K_{excl} \cdot V'' + K_{LC} \cdot V_{stat} \tag{16-67}$$

This relationship makes allowance for the fact that there are two processes which are closely interrelated but associated to different reference systems: the distribution constant, K_{excl}, is referred to the distribution between the volumes V_I and V_P, whereas the constant K_{LC} is referred to the distribution between the phases with the volumes $(V_I + K_{excl} \cdot V_P)$ and V_{stat}. Here V_{stat} is usually not known, but definitely not identical with V_P.

The relationship (16-63) was stated by Bakoš et al. (1979) and by Yau et al. [D 19]. As expected, for $K_{LC} = 0$ it includes the equation for ideal GPC as a limiting case.

If the fact that the reference systems for K_{excl} and K_{LC} are different is ignored, then for real GPC one obtains an expression in which the two distribution constants occur as a product. Unlike the usual definition $K = 0$ of the distribution constant, for an inert surface this algorithm would require $K_{LC} = 1$ as the limiting value in order that the equation for ideal SEC be derivable from the general expression as a limiting case.

16.7. Experimental determination of the volume portions in LC columns

In SEC, $V' = V_I$ if the contribution of the connection lines, which is very small in most cases, can be neglected. The interstitial volume (void volume) V_I can be determined by means of suitable colloidal particles or by means of a polymer sample with a molar mass greater than M_{lim}, i.e., whose molecules are too large to penetrate the pores of the packing. Electrostatic interactions or adsorption phenomena must not occur. In non-aqueous eluents V_I is usually determined by means of high-molecular-weight PS standards ($M \geqq 10^6$ g $\times$ mole^{-1}). Dextran blue is frequently used in aqueous gel filtration. Another interesting method proposed by Krejči et al. (1980) is based on the fact that, if a low-molecular-weight organic sample such as butanol is injected into a silica gel column, an electrokinetic detector responds to this injection by *two* signals, the first of which corresponds to the void volume. This first peak is due to an electrokinetic phenomenon and occurred neither in the differential refractometer nor in the UV detector. The effect was also demonstrated on columns with alumina or glass beads as a packing material. An essential advantage of this method is that it is even possible to characterize packings with a high adsorption activity.

In AC and LLC the greater part of the pore volume, V_P, contributes to the volume of the mobile phase. The total sum $V_I + V_P = V_C - V_W$ can be determined by means of pycnometry with a liquid which does not effect a swelling of the wall material of the packing. Slaats (1980), Slaats et al. (1981) investigated RP silica packings in this way, using acetonitrile or methanol. Another possible technique is helium pycnometry.

The volume of the mobile phase is a little bit smaller than the sum $V_I + V_P$, because the volume, V'', of the stationary phase must be subtracted (see Fig. 16-36). The problem of an exact determination of V' is therefore closely related to a precise evaluation of the small value of the stationary phase volume, V''. These topics have been discussed in detail by Slaats et al. (1981). In AC the following possibilities exist:

Breakthrough method: After the column has been equilibrated with an eluent of a constant composition, e.g., the pure component I, it is changed over to the desired mixture ratio. The component II contained in this mixture is incorporated into the stationary phase. The amount of component II taken up by the column can be calculated from the volume which has flowed through the column in the period between the solvent change and the step on the recorder.

Desorption method: This is the counterpart to the breakthrough method; the column is changed over from an eluent of a given composition back to the pure component I.

Minor disturbance method: A small quantity of component II (or of a mixture whose content of II differs from that of the eluent) is injected into an eluent flow of a given composition (I/II) which is in equilibrium with the column. This minor disturbance is indicated by the detector with a certain delay, which depends on the adsorption of the disturbing component. A single measurement yields one value of the derivative dc''_{II}/dc'_{II}. If these measurements are extended over the whole range of concentrations with different initial concentrations, c'_{II}, then the whole adsorption isotherm of component II can also be determined in this way.

Injection of an isotopically labelled solvent component: If a labelled component II* does not undergo any exchange reactions in the column, then the determination of the elution volume, V_{e, II^*}, after an appropriate injection offers an easy way to determine the mobile phase volume and the adsorption isotherm. However, in general the value V'_{II} obtained in this way differs from that determined by the method of minor disturbance. Therefore, if the injection of the labelled component also involved a disturbance of the total concentration of II, then even two peaks would occur. This phenomenon was sometimes misinterpreted as indicating a chromatographic separation of isotopes.

Injection of an unretained solute: Investigations of this kind have been carried out for instance on reversed phases with uracil or phenol (in methanol as an eluent), with cytosine or a UV-absorbing salt. What is measured is the retention volume of these samples, which are assumed to be unretained. BERENDSEN (1980) found that the volume measured with a low salt concentration on reversed phases is markedly smaller than the true mobile phase hold-up volume. Ion exclusion possibly occurs (cf., Section 19.3.1.).

Linearization of the net retention time for homologous series: This method is based on the assumption — which has frequently been verified — that there exists a linear relationship between the logarithm of the net retention time and the number of carbon atoms of the investigated compound in homologous series. If a plot of $\log t''$ vs. the carbon number is non-linear, this is considered to indicate that the value subtracted from the measured elution time, t_e, was not the true value, t', of the mobile phase hold-up time. Consequently the quantity subtracted is varied until the postulated linearity is achieved. This method was tested on RP 22 packings by BERENDSEN (1980). With M/W (90:10), the investigated homologous series yielded values which coincided within the limits of error. It is recommended to carry out this method using n-alcohols. In THF/W mixtures the postulated linearity was not achieved.

Peak maximum method: After an appropriate calibration, the height of the peak maximum can be taken as a measure of the substance concentration at the column outlet. By combining this value with the initial concentration, $c_{0, II}$, it is possible to determine a point on the adsorption isotherm (DE JONG et al., 1980).

Batch method: In this case the adsorption isotherm is determined outside the column by adding a well defined quantity of the mixture I + II to a certain amount of the dry packing material and measuring the change in concentration in the supernatant solution.

Recycling method: Here the change in concentration is observed from an enclosed volume of the eluent mixture, which is recycled through the column, whereby it reaches equilibrium with the adsorbent contained in the column as a packing. As in the batch method, the phase ratio is rather small, so the concentration must be determined very precisely.

16.8. Degradation by shear

High-pressure liquid chromatography with an input pressure of 10 MPa or above takes place on a high mechanical level. Generally the pressure drops across the column to an insignificant residual value. The generating of heat associated with the *mechanical work* causes problems in AC and LLC the distribution coefficients of which vary with the temperature. In polymer chromatography, *degradation by shear* may occur. SLAGOWSKI et al. (1974) have shown that in SEC under normal working conditions molecules with more than 10^7 g · mole^{-1} are degraded. After a single passage through a 6.10 m column, a polystyrene sample with $\bar{M}_w = 43.7 \cdot 10^6$ g · mole^{-1} had only an $\bar{M}_v$ of $19 \cdot 10^6$ g · mole^{-1}. The reduced specific viscosity decreased to 40% of the initial value. The degradation even occurred when the rate of elution was decreased from 1 ml · min^{-1} to 0.25 ml · min^{-1}. In the log M vs. V_e diagram, the point for the degraded sample lay on the extension of the calibration line, while the initial value was too high. Consequently, above $M = 10^7$ g · mole^{-1}, the curvature of the calibration relationship need not in each case correlate with the neighbourhood of the exclusion limit. Up to $5 \cdot 10^6$ g · mole^{-1} SEC can yield reliable results (APPELT and MEYERHOFF, 1980).

The degradation by shear greatly depends on the flow-rate. For a sample with $M = 7.1 \cdot 10^6$ g · mole^{-1} it proved feasible to avoid degradation when the elution was carried out with a linear velocity of less than 0.1 cm · s^{-1} (YAU et al. [D 19]), however, the degradation reached high values when a sample with only $M = 8 \cdot 10^5$ g · mole^{-1} was forced through a nozzle with $u = 11.85$ cm · s^{-1} (LEOPAIRAT and MERILL, 1978).

16.9. Energy aspects

For energy-induced transitions between the mobile and the stationary phase, the following relationship holds in the low-molecular-weight range (MARTIN, 1949).

$$\Delta\mu^0 = \Sigma\, \Delta\mu_i^0 \tag{16-68}$$

where $\Delta\mu^0$ is the variation of the standard chemical potential in the transition between the mobile and the stationary phases and $\Delta\mu_i^0$ are the contributions of the structural units of the molecule to this variation.

If $\Delta\mu_M^0$ represents the contribution of a monomeric unit, and the contributions of the end groups and other additional groups are described by $\Sigma\, \Delta\mu_E^0$, then the application of this additivity rule to a macromolecule with the degree of polymerization P gives:

$$\Delta\mu^0 = P\, \Delta\mu_M^0 + \Sigma\, \Delta\mu_E^0 \tag{16-69}$$

Even if the variation of the chemical potential for an individual monomeric unit is very small, the multiplication by P leads to a very large range of $\Delta\mu^0$ values for the members of a homologous series of polymers.

Chromatographic separations yield satisfactory results if the R *values* are neither too low nor too high. With $0.1 \leqq R \leqq 0.9$, the limits of the permissible range are already fixed rather generously.

As eqns. (3-2) and (3-5) yield, in view of $n''_E/n'_E = q$,

$$R = \frac{1}{1 + Kq} \tag{16-70}$$

and the thermodynamic partition coefficient K is related with $\Delta\mu^0$ by

$$K = \exp\left(-\Delta\mu^0/\boldsymbol{RT}\right) \tag{16-71}$$

the approximation $n''_E = n'_E$ gives for the lower limit ($R = 0.1$):

$$K = 9 = \exp\left(-\Delta\mu^0/\boldsymbol{RT}\right) = \exp -(-2.2) \tag{16-72}$$

At 25 °C, $\Delta\mu^0/\boldsymbol{RT} = 2.2$ gives $\Delta\mu^0 = 5.53$, and analogously for the upper limit ($R = 0.9$), $\Delta\mu^0 = -5.53$ kJ · mole^{-1}. Consequently, the relative rate of migration falls into an acceptable range only if $\Delta\mu^0$ lies within relatively narrow limits. Thus in the chromatography of macromolecules R values beyond the limits 0.1 and 0.9 are likely to occur. In fact it is almost a rule in AC that polymer samples either do not travel at all ($R = 0$) or travel together with the front ($R = 1$).

Because of the proportionality between $\Delta\mu^0$ and P as formulated by eqn. (16-69), the members of a homologous series of polymers can only travel at a reasonable rate if $\Delta\mu^0_M$ for the monomeric unit is only slightly different from zero. However, even if all of the factors such as the composition of the mobile phase, the activity of the stationary phase, the temperature, etc., are balanced so well that the requirement $\Delta\mu^0_M \approx 0$ is satisfied and the polymer in fact passes the separating path, then this does not necessarily mean that a separation by chain-length will be obtained. This would again require (in the optimum, inner range of R values) a difference of 0.4 kJ · mole^{-1} at least in the $\Delta\mu^0$ values of the components to be separated. Thus the condition for a reasonable migration inevitably excludes the other condition for the resolution. To overcome this contradiction, it is possible to use *gradient techniques*.

On the other hand, if the influence of molecular size is completely eliminated by carefully selected working conditions ($P\,\Delta\mu^0_M = 0$), then a separation depending only on $\Sigma\,\Delta\mu^0_E$ may be achievable. This term takes into account all of the features in which the macromolecules differ from each other, except for the degree of polymerization. It is just this kind of variation in the macromolecules — extending only over a few structural units and usually being undetectable by other methods — which may cause variations in $\Delta\mu^0$ of that order of magnitude which is required for the chromatographic separation.

17. Techniques in macromolecular elution chromatography

This chapter deals with important features of chromatographic techniques in greater detail, taking up suggestions from the preceding chapters for closer investigation. Some questions about the general column technique will be discussed with respect to polymerspecific problems.

17.1. Packing of HPLC columns

17.1.1. Preparation of the columns

The empty, open tubes are washed with a hot detergent solution. Contaminations sticking tightly to the walls may be removed mechanically by means of a cloth tampon, but great care should be taken to avoid any scratching of the internal surface. Only walls as smooth as a mirror enable homogeneous packings to be achieved. If necessary, 50% nitric acid can be used. Afterwards the tubes are flushed with water, acetone and chloroform. Finally they are rinsed out once more with acetone and blown dry with nitrogen or clean pressurized air.

17.1.2. Dry packing

Relatively large-sized particles can be packed in a dry state. Non-swelling, irregular particles larger than 40 μm or spherical particles more than 20 μm in diameter are suitable. The dry packing technique requires circumspection and patience, but no sophisticated apparatus. Mechanical energy must be applied to achieve as close a packing as possible. Violent shaking, however, should be avoided, because otherwise the larger particles accumulate at the wall.

The column to be packed is closed with a porous bottom plate, the upper end being extended by a tube of equal diameter, about 5 cm long, which is smoothly seated on top of the column and held in place by a tightly fitting, pressure-tight pipe clamp. The extension allows one to pack the column uniformly up to its upper edge and to compact the packing under pressure.

In each filling step, a quantity of packing material sufficient to produce a layer 5—10 mm thick is poured into the vertical column. This is about 300 mg for a 7.8 mm tube. After adding a portion, the vertically aligned column is slightly bounced on a wooden board 80—100 times. At the same time the column wall is tapped with the finger very gently at the respective filling level. With some practice, this takes less than one minute. Thereafter the bouncing is continued for another 15—20 s without tapping laterally, then the next portion is poured in.

After the column has been filled step by step up to the extension, it is again set down gently under a slow rotation about its axis for another five minutes. Now the packing can be compacted by means of a solvent. For this purpose the column is connected to a pump. For 10 minutes at least, a solvent is passed through the column at such a rate that a pressure of 8—10 MPa above the actual operating pressure is built up.

The packing process can be greatly facilitated by the use of a simple device (see Fig. 17-1). This device lifts the column about 100 times per minute by a very small distance and drops it down again. At the same time the column is rotated while material is fed continuously at a low rate. The vibrations should be such as to stir up a layer of only a few millimetres at the respective level of the packing being produced. The packing density achieved increases with the increasing duration of the vibration (COOPER and KISS, 1973).

In relatively short glass tubes, columns of a good and reproducible quality have successfully been packed even with irregularly shaped particles of sizes down to 20 μm, which was achieved by carefully tapping each portion with a ramrod (HUBER et al., 1972a). *Radial compression* is an interesting variant for achieving an optimum quality of packing (EON, 1978, [F 19]; cf., Section 15.5.).

Large columns can be dry-packed without any additional problems, whereas slurry packing (cf., Section 17.1.3.2.) becomes more and more complicated as the volume of the suspension increases. (Most workers use a maximum of 100 ml.) Therefore preparative columns with dimensionally stable packing material are generally filled according to the dry-packing technique. If only for economic reasons, the particles for preparative columns are in most cases chosen rather coarse, so that this technique can be applied. ROUMELIOTIS and UNGER (1979), however, used a 23.5 mm I.D. slurry-packed column with a length of 0.25 m, packed with 5 μm particles, for the separation of proteins.

17.1.3. Wet packing technique

17.1.3.1. Gels

Swelling gels must be allowed to reach the swelling equilibrium with the eluent before they can be flushed into the column. Great care should be taken with soft gels, because in most cases they can be pressurized only up to a few kPa.

Introducing the gel by sedimentation from a solvent flowing continuously through the column at a low rate has proved to be successful (FRITZSCHE, 1967; HEITZ and ULLNER, 1968). Fresh gel suspension is continuously fed into the column to make up for the solvent discharging through the porous bottom plate. Thus columns with a high resolving power were obtained.

Sufficiently pressure-resistant gels are stirred to give a homogeneous suspension with as high a gel content as possible, and forced so rapidly into the column that the size-dependent

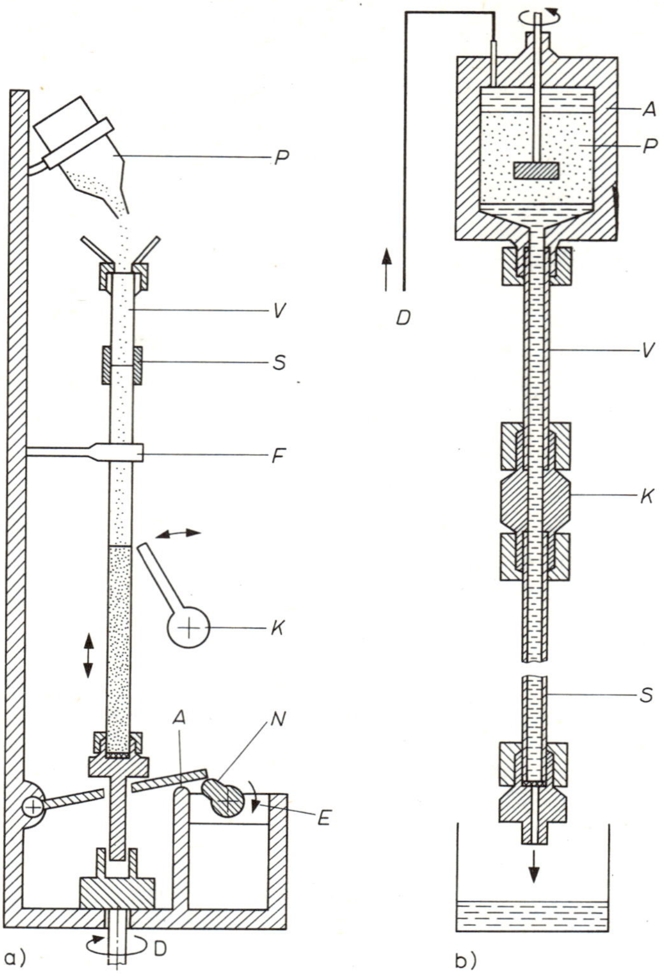

Fig. 17-1
Devices for column packing

a) Device for dry packing
P packing material; V extension tube; S connecting pipe clamp; F column guide; K tapping device; E eccentric cam; A stop; D drive for the rotary motion of the column

b) Slurry packing device
A stirring autoclave; P suspension of the packing material; D hydraulic fluid; V extension tube K tube connection S column

sedimentation is overridden. The eluent emerges through the porous bottom plate. The stirring autoclave is arranged on the top of the column.

A series of papers dealing with the packing of the semi-rigid gel Toyopearl® have been published by KATO et al. (1981).

17.1.3.2. Non-swelling packing materials

Rigid, non-swelling particles can in principle be suspended in any inert liquid desired, and packed under a high liquid pressure. Usually the slurry liquid is not identical with the

Table 17-1
Working conditions in the slurry packing technique
KG: silica gel, KGm: surface-modified KG; SG: polystyrene gel

Support material	Amount g	Column L / m	d_C / mm	V_C / cm³	Suspension medium (v:v)	Amount cm³	Reference
KG	9	0.50	4.5	7.95	TBE/TCE/M (45.6:53.9:0.5)	100	MAJORS (1972)
KG	2.8	0.30	4.0	3.77	TBE/Dx/tetra (40:30:30)	50	STRUBERT (1973)
KG					TBE/Dx (50:50)		[F 27]
KG					TBE/THF		
KG	4				methylene iodide/M (90:10)	20	KNOX in [D 14], p. 24
KG					Tri/Eol (40:60)		[D 13]
KG	2.5	0.25	4.0	3.14	tetra/Dx (50:50)		[F 27]
KGm(NH_2)	9	0.50	4.6		propanetriol/M (20:80)		
KG					0.001 m NH_4OH in water		KIRKLAND (1972)
SG					2-chloroethanol/Ac TCE/Ac		
KG	17.5	0.50	8.0	25.14	M[1])	750	BRISTOW (1978)
KG/KGm(NH_2)	5	0.25	4.6	4.99	M/W (90:10)[1])	60	[F 34]
KGm	5	0.25	4.6	4.99	Ac[1])	60	[F 34]

[1]) upward slurry method

eluent to be used. Columns containing particles smaller than 20 μm can be packed in this way. Packings of a reasonable quality can be obtained using either the *suspension method* or the *viscosity method*. Both methods have been realized in numerous variants. It must not be overlooked that a variant which yields good results with a polar, hydrophilic material, such as silica, is likely to fail with a hydrophobic material, e.g., RP 18.

In the *suspension method*, the particles are usually suspended in a liquid of the same density (SNYDER, 1969; MAJORS, 1972; DAWKINS and HEMMING, 1972b; KIRKLAND, 1972; STRUBERT, 1973). For that purpose, the media listed for silica suspensions in Table 17-1 contain a high-density solvent, such as tetrabromoethane (TBE, $\varrho = 2.967 \text{ g} \cdot \text{cm}^{-3}$), in such a concentration that there will be neither creaming nor sedimentation. Dioxane and methanol as polar additions prevent bunching of the particles. The suspensions with a balanced density may contain up to 25% of packing material. First they are homogenized by an ultrasonic treatment carried out over several minutes. If the density is exactly balanced, it is recommended to allow the suspension to stand for some hours; otherwise the filling process should be carried out rapidly so that it will be complete within 10 minutes. For the packing operation, the suspension is covered by a volume of heptane in the pressure-resistant vessel and pressed through the bottom opening into the connected column. The suspension medium emerges through the porous plate, while the suspended particles build up the packing. The pressure employed should be as high as the strength of the column, of the screw connections and all other parts of the apparatus permit. The more rapidly a column is filled, the higher the efficiency of the packing will be. Naturally this also makes high demands on the mechanical stability of the packing material particles, which must not break under the loads acting upon them.

The passage of the heptane can be detected by a pressure drop. Then heptane is pumped at a reduced rate until about ten times the column volume has been passed through the column.

The use of tetrabromoethane involves the risk of a separation of bromine or hydrogen bromide, which may lead to a chemical impairment of the packing materials. Therefore for those bonded phases of a lower stability the viscosity method or the *upward slurry technique* described below are recommended. In this variant of the suspension method, which was developed by BRISTOW (1977, 1978), a rather dilute suspension in methanol is pumped into the column, which is arranged above the filling device, pointing upwards with its discharge end. This variant has a number of advantages. For example, it allows the use of cheap and non-toxic solvents. A methanol-water mixture (90:10) is recommended for polar Spherisorb® materials (silica, nitrile or amino bonded phases), and acetone for non-polar materials [F 34]. These media prevent the agglomeration of particles, which represents one of the major difficulties in the slurry technique. Investigations on the stability of silica dispersions in different media were carried out by BROQUAIRE (1979).

Moreover, in this variant the supply vessel can easily be filled before the column is attached on top of it. Any air-bubbles can rise and escape before the packing process starts, while agglomerates settle on the bottom. The gradual sedimentation of the individual particles is compensated by the upward flow. If this flow in the supply vessel amounts to 20 times the settling velocity of the smallest particles, then stirring is not necessary. This method enabled silica gel particles of 5 and even 3 µm to be packed, yielding columns of excellent separation efficiency.

For alkyl silane modified silica, methanol is too polar to effect a good slurry. For particles of alumina or glass it was not possible either to achieve a satisfactory packing, presumably because of their higher density. In acetone, chloroform or n-hexane, however, the alkyl silane modified silica gels were successfully dispersed.

The settling velocity depends linearly on the density difference and quadratically on the particle size. During the packing procedure, a sedimentary fractionation of the particles must at all costs be avoided. Consequently, narrow particle size distributions are of advantage. The quadratic dependence on the particle size implies that for very small particles ($d_p \sim 5$ µm) the density of the suspension medium is no longer so decisive.

In the *viscosity method* the settling is counteracted by a rather high viscosity (40—60 mPa · s or 40—60 cP) of the slurry medium (ASSHAUER and HALÁSZ; 1974). Suitable liquids are, for example, paraffin oil or cyclohexanol. This technique enables columns of good quality to be packed without having great experience. A disadvantage is the increasingly higher pressure required to force the dispersant through the growing packing. Therefore it is expedient to heat the already packed part of the column to about 60—80 °C, so that in this part the viscosity decreases. Using this variant the time required in work with toluene-cyclohexanol (34:66) could be reduced from several hours at room temperature to 10—30 minutes [F 28].

The difficulties in wet packing increase with the increasing height of the packing. Therefore it is recommended to have the column length L (in m) not much longer than

$d_p^2/250$ (d_p in μm), especially with viscous suspension media. However, the upward slurry method even enabled 1 m columns to be successfully packed with 5 μm particles (Bristow, 1978).

The mechanical stability of the packing can frequently be improved by *slamming*. For this purpose the shut-off valve before the column is closed after the packing process has been completed and the column has been flushed with methanol; then the pressure in the pneumatic pump is increased to the maximum permissible value and instantaneously applied to the packing (Kirkland, 1975).

Generally the slurry liquid must exhibit a good wetting ability towards the particles of the packing. This is favoured by a close similarity in polarity. Problems were encountered for hydrophobic organic gels. These were successfully for styrene-divinylbenzene copolymers reduced by a ten-hour treatment with hot, dilute NaOH solution followed by neutralization and carefully rinsing of the particles (Cox and Anthony, 1975).

17.1.4. Final manipulations

After packing and compacting, the pre-column is removed. The column is prepared for use by carefully removing 1—2 mm of the packing, inserting the porous upper plate, and closing the column.

High separating efficiencies can be achieved with an injection into the centre of the packing. Naturally, in this case the column cannot be closed with the porous plate. On the other hand, pricking into the packing will soon destroy the upper zone, so that the efficiency decreases. Scott et al. (1967) recommended that 10—20 mm of the packing should be removed very carefully, an eluent volume equal to several times the column volume should be forced through the packing in order to restore a smooth surface and, finally, the packing should be refilled with glass beads of $d_p \approx 40$ μm. An injection into this layer can be carried out without producing disturbances. To prepare columns which are packed with microspheres for on-column syringe injection, it is also recommended to remove 5 mm of the packing material so carefully that again a smooth surface is obtained. A disk made of porous nickel, which is mechanically protected by a sieve plate of thin stainless-steel wire, is pressed upon this surface. The sample is injected into a 2 mm layer of silanized glass beads, which is arranged between the steel fabric and a closing plug of porous PTFE [F 34].

17.2. Flow resistance

The flow in a capillary of radius r and length L^*, to which the differential pressure Δp^* is applied, is governed by Hagen-Poiseuille's law

$$\Delta V = \frac{\Delta p^* r^4 \pi}{8\eta^* L^*} \Delta t \qquad (17\text{-}1\,a)$$

where η^* is the viscosity of the liquid. The volume, ΔV, flowing through the capillary in the time Δt can be expressed by the volume flow-rate, v^*, or the linear flow velocity, u:

$$\Delta V/\Delta t = v^* = ur^2\pi \qquad (17\text{-}2)$$

Thus eqn. (17-1a) can be rewritten as:

$$\Delta p^* = \frac{8\eta^* L^* u}{r^2} \tag{17-1b}$$

In a packed column, the liquid flows through a labyrinth of channels which are neither uniform nor circular. The only certain fact is that they increase in size as the sizes, d_P^*, of the particles of the packing material increase. The *pressure drop* across the column is:

$$\Delta p^* = \frac{\eta^* \cdot L^* \cdot u}{(d_P^*)^2} \cdot f^* \tag{17-3a}$$

The geometry factor, f^*, can be combined with $(d_P^*)^2$ into the *permeability*, $\varkappa^*$, of the column:

$$\varkappa^* \equiv (d_P^*)^2/f^* \tag{17-4}$$

For the linear flow velocity this gives:

$$u = \frac{\varkappa^* \cdot \Delta p^*}{L^* \cdot \eta^*} \tag{17-5}$$

The proportionality between u and $\Delta p^*/L^*$, which is expressed by this equation for the flow through a packed column, is known as Darcy's law.

The geometry factor, f^*, depends on the shape of the particles and on the packing density; its value is about 1000, if all the other quantities are given in cgs units. (This was to be indicated by the asterisk for d_P^*, η^*, v^*, etc.) The geometry factor is mainly determined by the interstitial porosity, ε_I (cf., Section 10.4.4.), which can be experimentally determined from the elution time t'_{ex} for an excluded solute, the column dimensions L^* and d_C^* and the flow rate, v^*, by means of the relationship:

$$\varepsilon_I = \frac{v^* t'_{ex}}{L^* (d_C^*/2)^2 \pi} \tag{17-6}$$

With this variable one obtains the *Kozeny-Carman relationship* for the permeability, $\varkappa^*$:

$$\varkappa^* = \frac{(d_P^*)^2 \varepsilon_I^3}{180(1 - \varepsilon_I)^2} \tag{17-7}$$

As ε_I is about 0.4 for a well packed column, one obtains (in agreement with the above approximation for f^*):

$$\varkappa^* \approx (d_P^*)^2/1000 \tag{17-8}$$

By means of this relationship, an expectation value for the permeability can be calculated, while the actual value can be determined by eqn. (17-5). Comparison of the two results allows an estimation of the quality of the column packing: if the actual value is much higher than that calculated by eqn. (17-8), then the column is packed too loosely and probably will not be stable enough. A packing which is too loose may change under the conditions of high-pressure liquid chromatography, allowing cavities and breakthrough channels to occur by a spontaneous after-compaction, which makes the column unserviceable. On the other hand, too low a permeability indicates a high content of fines. Such packings may be subject to clogging when operated. Therefore the fines must be removed before the packing process. This can be done by flotation in alcohol and decanting after a rest period of 30 minutes. For silica gel, 0.001 M ammonium hydroxide has also proved successful as a sedimentation medium. An ideal permability can most easily be achieved by means of particles as uniform in size as possible. Smaller particles occupy the interstitial spaces between the larger ones, thus drastically increasing the flow resistance. BRISTOW (1978) reported this effect, stating that the width of the particle-size distribution must not exceed a 5:1 ratio.

The linear flow velocity, u (in cm $\cdot$ s^{-1}), is equal to the quotient of the column length and the mobile phase hold-up time:

$$u = L^*/t' = 100L/t' \tag{17-9}$$

It is connected with the volume flow-rate, v (in ml/min), by the following relationship

$$u\left(\frac{d_C}{20}\right)^2 \cdot 60\pi\varepsilon_t v = v(= v^* \cdot 60) \tag{17-10}$$

where d_C is the inside diameter of the column (in mm) and ε_t is the total porosity of the column.

For the pressure required (in MPa) one obtains

$$\Delta p = \frac{\eta L v}{d_P^2 d_C^2} \cdot f \tag{17-3b}$$

where η is the viscosity of the mobile phase in mPa $\cdot$ s (1 mPa $\cdot$ s = 1 cP), L is the length of the column in m, d_C is the inside diameter of the column in mm, v is the rate of elution in ml $\cdot$ min^{-1} and d_P is the particle size of the packing material in µm. The factor $f = (4000/60\pi)(f^*/\varepsilon_t)$ depends on the internal geometry of the packing via f^* and ε_t. Its value is about $3 \cdot 10^4$.

Numerical example: Column length $L = 0.25$ m; inside diameter $d_C = 4.6$ mm; volume flow rate $v = 1.5$ ml/min; microspheres with $d_P = 10$ µm; pressure drop $\Delta p = 3$ MPa; elution volume for PS with $M_{\text{lim}} = 2 \cdot 10^6$ g $\cdot$ mole^{-1}: $V' = 1.83$ ml (exclusion volume); eluent THF, i.e., $\eta = 0.55$ mPa $\cdot$ s.

Interstitial porosity (17-6):

$$\varepsilon_I = \frac{(1.5/60) \cdot 73.2}{25 \cdot 0.23^2\pi} = \frac{1.830}{4.155} = 0.44$$

Expected value for the permeability:

$$\varkappa^* = \frac{0.0010^2}{1000} = 10^{-9}\ \text{cm}^2$$

Permeability according to eqn. (17-5)

$$\varkappa^* = \frac{u \cdot L^* \cdot \eta^*}{\Delta p^*} = \frac{0.342 \cdot 25 \cdot 0.0055}{3 \cdot 10^7} = 1.57 \cdot 10^{-9}\ \text{cm}^2$$

where $u = 1.5/(0.23^2\pi\ 60 \cdot 0.44) = 0.342$ cm $\cdot$ s^{-1}, $\Delta p = 3 \cdot 10^7$ dyne/cm² and $\eta^* = 0.0055$ g $\times$ $\text{cm}^{-1} \cdot s^{-1}$. The relatively high value obtained by eqn. (17-5) reflects the rather high interstitial porosity, $\varepsilon_0 = 0.44$, of the packing.

The pressure required for polymer solutions exceeds that of low-molecular-weight liquids of equal viscosity by a factor of 2—9 (LAUFER et al., 1976). The pressure rapidly increases with decreasing particle size. This is one of the reasons why preparative columns are not usually filled with superfine 5 or 10 μm particles. While for analytical steel columns the upper limit of the pressure range is defined mainly by the delivery head of the pump, for preparative columns with their much larger diameters the strength of the column wall plays a decisive rôle. In most cases the columns are filled with 37—75 μm particles in order to avoid the necessity of too thick walls.

The inlet pressure increases linearly with the column length, the viscosity and the rate of elution. For hard polystyrene gels the proportionality to the volume flow-rate has been confirmed by experiment (LITTLE et al., 1969). This shows that even under higher pressures the packing is not deformed. Soft gels are unstable and can only be subjected to small loads. In this case the flow-rate must remain low; a single elution may then possibly take several days (FRITZSCHE, 1967; HEITZ, 1973b). For soft gels it is expedient to use short columns, which require a low pressure, and to achieve the required plate number by recycling (Section 17.8.).

17.3. Exchange of columns

In SEC, the packing material, representing the stationary phase, is the decisive factor. Compared to this, all other variables are of only minor importance. Therefore, particularly in the development of a method, other column sets are required now and then, depending on the problem to be solved. If the single tubes are not equipped with changeover valves, the bank must be split up and rearranged. In this case an ingress of air into the column must be avoided. The separation efficiency of some gel packings is impaired by air-bubbles.

To dismember a column set the connecting line between the last column and the detector is disconnected. A syringe filled with a solvent is connected to the last column through a suitable plastic tubing. Now the last section can be fully removed from the assembly, because any ingress of air at the column inlet can be prevented by pressing the syringe in order to introduce more solvent. The disconnected section is sealed by a cap first at the inlet end, and thereafter at the outlet end (where the syringe had been connected). The other sections are removed in an analogous way.

For assembly of a column set, first the section next to the pump is installed. During this operation, the capillary to be connected is rinsed by solvent using a syringe at the far end of this section. Thus all the sections can be connected without ingress of air-bubbles.

17.4. The service life of a column

A trouble-free operation of elution chromatography requires that all sample components injected are also completely discharged. In the case of an isocratic operation, the column is then at once ready for the next analysis. The effort made in packing proves rewarding, because a high-performance separating column can be used for hundreds of analyses.

Of course errors in operation must be excluded. For instance, for highly swollen gels an exchange of the solvent may lead to a collapse. Rapid damage to the columns must also be feared if reactive substances have to be investigated, especially in adsorption chromatography, where the catalytic effect of the support surface favours a reaction.

Nevertheless, even with correct operation, changes of the chromatographic bed may occur in the course of time. Thus MEYERHOFF and JOVANOVIČ (1967), in a repetitive investigation of cellulose trinitrates in tetrahydrofuran, found that the calibration curve for an SEC column shifted towards smaller elution volumes. This indicated a decrease in the pore radii. Frequent monitoring of SEC instruments by means of test mixtures is therefore recommended.

In some cases oligomers were found to evolve from PS gels (GIAMMARSE et al., 1968). If such an effect occurs, then of course it will increase with the amount of the column packing material. It may interfere with preparative separations if the compositions of the fractions are to be investigated (NORRIS and PEAKER, 1973).

A correctly operated SEC column, which is kept free of any components which are irreversibly bound, can be used for years (HAZELL et al., 1968; MUKHERJI et al., 1978). Guard columns are recommended as a means of saving column lifetime.

SAMAY and FÜZES (1980) investigated the *long-time reproducibility* of Styragel® columns. There was no significant shift in the calibration curve, either in the standard deviation of the peak elution volume or in the peak widths, during two years of operation with THF at room temperature. A similar column operated at 130 °C with 1,2,4-trichlorobenzene as an eluent, however, exhibited a pronounced shift in its elution characteristic towards smaller elution volumes.

17.5. Sample introduction

Usually the sample is injected into the flowing eluent immediately before the separating bed. For the chromatographic process to take place in the linear range of the isotherms, the amount of substance should be as small as possible. In this respect, size exclusion LC is not as sensitive as AC, because the total pore volume of an SEC column is relatively large in comparison with the capacity of an AC column. Nevertheless, also in SEC analytical separations should be carried out using as small a sample amount as possible. The limited sensitivity of the detection methods, however, requires a certain minimum quantity for each component. Thus, in practice a compromise is always required.

The amount of substance required for detection can be injected in a smaller or in a larger solution volume. While in the chromatography of small molecules the injection volume is generally chosen to be as small as possible ($0.1\ \mu l \leqq V_0 \leqq 5\ \mu l$), because it yields the best resolution, the situation is not so simple for macromolecular solutions because of their high viscosity. If a larger sample volume is chosen the concentration can be reduced, which results in a lower viscosity; but then the peak broadening due to the apparatus occurs with a higher starting value. However, the injection volume is not critical as long as it is small in comparison with the peak volume.

The approximate equation

$$W/W_i \approx 1 + \frac{2}{3}(V_0/W)^2 \tag{17-11}$$

where W is the base width of the peak (4σ), $W_i = \lim_{V_0 \to 0} W$, and V_0 is the injection volume, shows that W exceeds the ideal value (injection volume infinitely small) only by

0.7% if the injection volume is 10% of the peak volume (WICKE, 1965). The validity of this relationship was experimentally verified by HEITZ and ULLNER (1968). However, for short high-resolution columns the size of the injection volume becomes critical, and the design of the sampling device is of great importance (COLIN et al., 1979). For high polymers more attention usually has to be paid to the concentration of the solutions, because the *viscous fingering* (cf., Section 16.3.1.) represents the higher risk. This distortion mainly occurs where the eluent and a highly viscous solution are immediately adjacent to each other. The higher the viscosity of the solution, the more distinct is the viscous fingering; consequently the distortion increases with the molar mass and the concentration of the sample, but also with an increasing slope of the viscosity gradient. In a common SEC separation according to molar mass, for a sample with a broad distribution the components concentrated in the injected volume are soon parted rather widely from each other. Due to this the concentration decreases, and the viscosity steps become flatter and flatter. The solvent and the solution with its initial concentration are no longer adjacent, but rather the solution component I and the solution component II, etc. Thus, for samples with broad distributions the viscosity problem is comparatively insignificant.

Samples with narrow distributions do not undergo this chromatographic dilution to the same extent. In this case an abrupt change in viscosity continues to exist, because only kinetic band broadening is of any importance. The same holds for chromatographic techniques which do not separate according to the chain-length but, for instance, according to a structural feature. The abrupt viscosity step is prone to viscous fingering and eluent breakthrough.

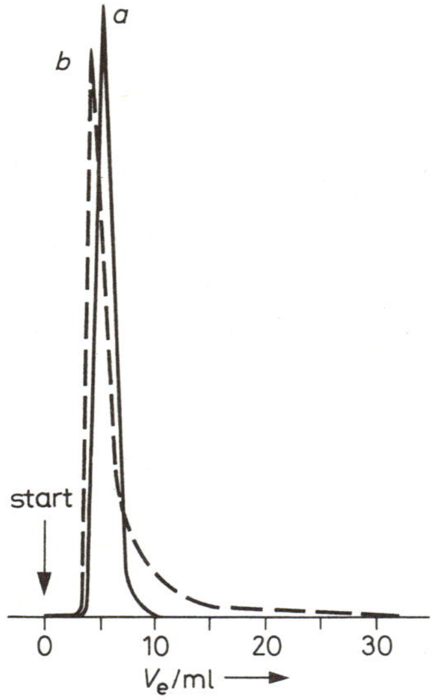

Fig. 17-2
Elugrams of samples which were only passed through the external parts of the apparatus (valves, capillary connections, detector).

Indication of the flow refractometer detector after an injection of 1 ml each of a solution containing 2 g · l^{-1} of the sample substance in tetrahydrofuran
a monostyrene; b polystyrene: $\bar{M}$ = 860 000 g · mole^{-1}
(according to OSTERHOUDT and RAY, 1967).

Recommendations for the permissible sample concentration are discussed in Section 19.1.

Fig. 17-2 shows two detector signals which indicate the effect of the viscosity gradient. Equal injections of monostyrene and polystyrene solutions led to different signals, although the solutions were only passed through the external parts of the apparatus, but not through the separating path (OSTERHOUDT and RAY, 1967).

The premature elution of the polystyrene peak, as can be observed from Fig. 17-2, was confirmed by OUANO and BIESENBERGER (1971) on the basis of systematic investigations on the diffusion phenomena in dilute polymer solutions flowing in capillaries. In a capillary 3.70 m long and 1.19 mm in diameter, 4—6 μl of polymer solution ($c_0 = 0.2\%$) even yielded multimodal peaks, which appeared distinctly earlier than the hexane peak. From this the authors concluded that molecular entanglement among the polymer coils causes radial concentration non-uniformities and virtual two-phase flow.

According to OUANO (1972b), the band broadening in capillaries depends on $(Q_t \cdot V_t \cdot v/D')^{1/2}$, where Q_t is the cross-section of the capillary and V_t its volume, v is the volume flow-rate and D' is the diffusion coefficient. ISHII et al. (1978) obtained an analogous result for the spreading effect of a glass-wool tampon. Through D', this phenomenon depends on the molar mass, and hence is related to hydrodynamic chromatography; cf., Section 13.2.

17.6. Stopped-flow technique

The effect of interruptions on the result of SEC has been investigated in detail by COOPER et al. (1969), using a commercial chromatograph equipped with either four polystyrene gel columns, each with a length of 1.22 m and a diameter of 7.8 mm, or with a long silica gel column 6.35 mm in diameter. Within the periods required for recording spectra and the like, no changes were observed in the record obtained from the differential refractometer. This was the case for interruptions of up to 90 min, and for samples with a molar mass down to 10300 g · mole^{-1}. When a sample with 160000 g · mole^{-1} was kept in the column for 62 hours, exactly the same record was eventually obtained as in a normal, non-interrupted analysis. A substance with $\bar{M} = 2030$ g · mole^{-1}, which was kept in the apparatus for 17 days (!), thereafter only yielded an $\bar{M}_w/\bar{M}_n$ ratio of 2.31 instead of 2.22 for a smooth passage, but the shape of the elugram was only slightly changed even in this extreme case.

The theoretical treatment of stopped-flow injection has been given by KUBIN and VOZKA (1978). If mechanically driven single-displacement pumps are used, the compressibility of the elution liquid may cause errors in this injection technique.

17.7. High-precision measurements of the elution volume

For apparatus with constant-flow pumps, the elution volume results, to a first approximation, from the chosen volume flow rate and the retention time. For a constant paper feed the length of the elugram can be taken as a measure of the retention time. For polymers, however, this approach is often not accurate enough. Because of the exponential relationship between M and V_e, every error in the elution volume has a great effect on the molar mass. To obtain exact values, the eluate can be collected and the quantity of each portion determined by means of volumetry or gravimetry. SCOTT and KUČERA (1976) collected the eluate in a 25 ml grade A burette, and took all the readings with the pump switched off in order to avoid errors due to the compressibility of the tetrahydrofuran used as an eluent. Most apparatus is equipped with a *siphon* as a standard accessory, which automatically discharges after it has been completely filled. This yields a *count* on the elugram. By weighing it was found that the successively discharged quantities of a liquid differ from each other by less than $\pm 1\%$ (YAU et al., 1968). They may, however, vary as a function of the solvent. Thus PROVDER et al. (1971) found a value of 5.024 ml per count for tetrahydrofuran, but 5.148 ml for tetrafluoroethylene. In both cases the flow-rate was $1\ \text{ml} \cdot \text{min}^{-1}$. The higher the flow-rate, the larger was the quantity of liquid per count (YAU et al., 1968; BALKE and HAMIELEC, 1969; PROVDER and ROSEN, 1970).

For very low rates the decrease is mainly caused by evaporation (YAU et al., 1968; LITTLE et al., 1969).

Systematic investigations with chloroform at room temperature have shown that the following relationship holds between the liquid volume per count, V_c, the flow-rate, v, and the rate of evaporation, r_v(YAU et al., 1968)

$$V_c = V_{c,0} + t_s v + \frac{V_{c,0} + t_s v}{v} \cdot r_v \qquad (17\text{-}12)$$

where $V_{c,0}$ is the volume per count at the flow-rate $v = 0$ and the rate of evaporation $r_v = 0$; t_s is the duration of a discharge in min. The authors determined V_c for flow-rates ranging between 0.11 and $10.5\ \text{ml} \cdot \text{min}^{-1}$ and obtained the values $V_{c,0} = 5.04$ ml, $t_s = 0.07$ min and $r_v = 0.0055\ \text{ml} \cdot \text{min}^{-1}$. Their results are shown in Fig. 17-3. LITTLE et al. (1969) obtained analogous results with toluene. The increase in V_c with increasing

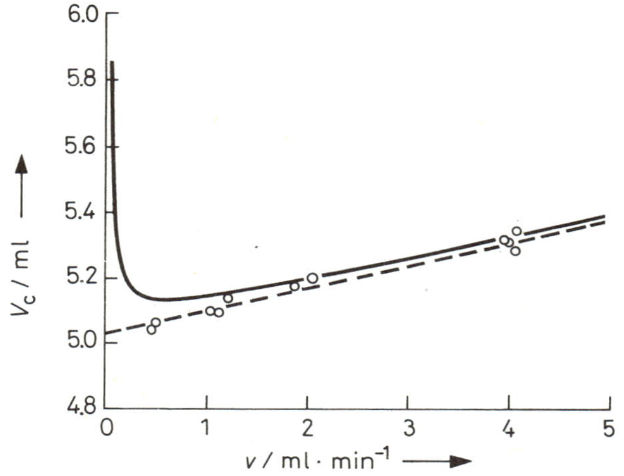

Fig. 17-3
Effect of the flow rate on the count volume discharged from the siphon per count

Measurements with chloroform at room temperature. The upper curve was calculated by eqn. (17-12), taking into account the evaporation loss, the effect of which increases with decreasing rate of elution (according to YAU, SUCHAN and MALONE, 1968).

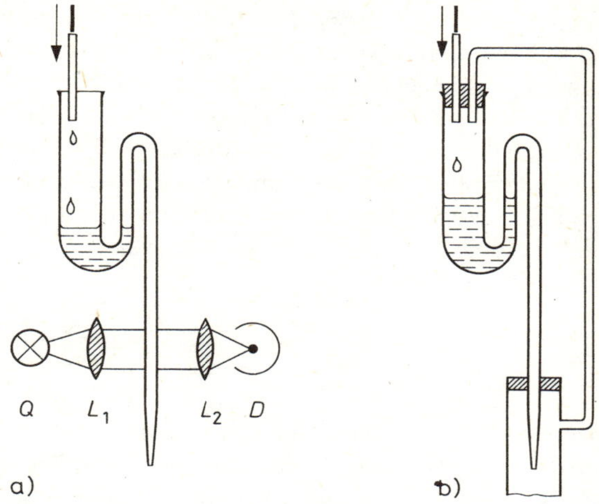

Fig. 17-4
Siphon measuring devices

a) Standard version
Q light source; L_1, L_2 combination of lenses of the light barrier; D photocell
b) Closed version with vapour feedback according to YAU, SUCHAN and MALONE (1968) (light barrier is not plotted).

flow-rate is due to the fact that liquid also flows in during the discharge of the siphon, being drawn off together with the liquid already contained in the siphon. The higher the flow-rate, the larger is the additional volume. The quantity evaporated per count is calculated as the product of the *rate of evaporation* and the *time* required to fill the siphon, $(V_{c,0} + t_s v)/v$. To eliminate the considerable error which may occur at low rates of elution, YAU et al. equipped the siphon with a vapour return line, which did not affect the discharge but prevented the evaporation (see Fig. 17-4). When the flow dependence of the siphon filling was taken into account, the apparent dependence of the elution volume on the flow-rate vanished in most cases (see for instance RÖSSLER et al., 1973; SPATORICO, 1975). As different flow-rates require different inlet pressures, for accurate measurements the compressibility of the liquids must be taken into account. Kinetic effects above all influence the position of the peak maximum; the proposal to use the *average elution volume* for the characterization, which will be discussed in Section 19.1., is therefore of interest in this connection. If great demands are made upon the accuracy of the elution values, the temperature dependence of the density has to be taken into account, which is rather high for organic liquids. BARLOW et al. (1977) found that a variation of 5 K in the siphon temperature leads to an error of 1% in the elution volume, and that at 135 °C a vertical temperature drop of 40 K can occur in a common air bath oven for the siphon. It was possible considerably to improve the precision of the determination of the elution volume by improving the control device of the oven and installing an additional fan.

The quantity of the eluate has also been measured by means of an automatic balance. While SPATORICO (1975) had still found that the precision of siphon measurements (±0.1 ml) is higher, VAN KREVELD and VAN DEN HOED (1978) preferred an electronic balance *because* of its accuracy. Also MOEBUS et al. (1979), using a self-made electronic device, found that the weighing technique was superior. Droppers have also been used.

MEYERHOFF (1971) employed a measuring device of this kind, where a stepping motor for the paper feed was controlled by the impulses generated by the individual drops. Problems may occur if the dissolved substance changes the surface tension, and hence the size of the drops. This has been observed in investigations with aqueous polymer solutions (HASHIMOTO et al., 1978). In SEC using microcolumns with an inside diameter of 0.5 or 1 mm (ISHII et al., 1978), the demands made upon the precision are so extreme that so far it has not been possible to evaluate the fine results of separation in only a 100 µl total eluent volume in the customary way by means of calibration curves.

17.8. Recycling

The plate number of a homogeneously packed column is proportional to the column length. However, the resolution varies only with the square root of the plate number (see eqn. (3-25)). Increasing the length of the column by a factor of 4 only doubles R_s, but the duration of the analysis and the pressure required multiply four-fold. Therefore increasing the column length as a means of improving the resolution is only resorted to if the possibilities lying in the *relative distribution factor* and the *retention factor* can no longer be utilized. Unfortunately, in SEC this margin is very narrow because of the limits of $0 \leqq K \leqq 1$ for the distribution coefficient. If the increase in efficiency achievable by improving the *packing quality* is also not sufficient to attain the resolution required, then indeed the only resort is to increase the length of the separating path, with the associated

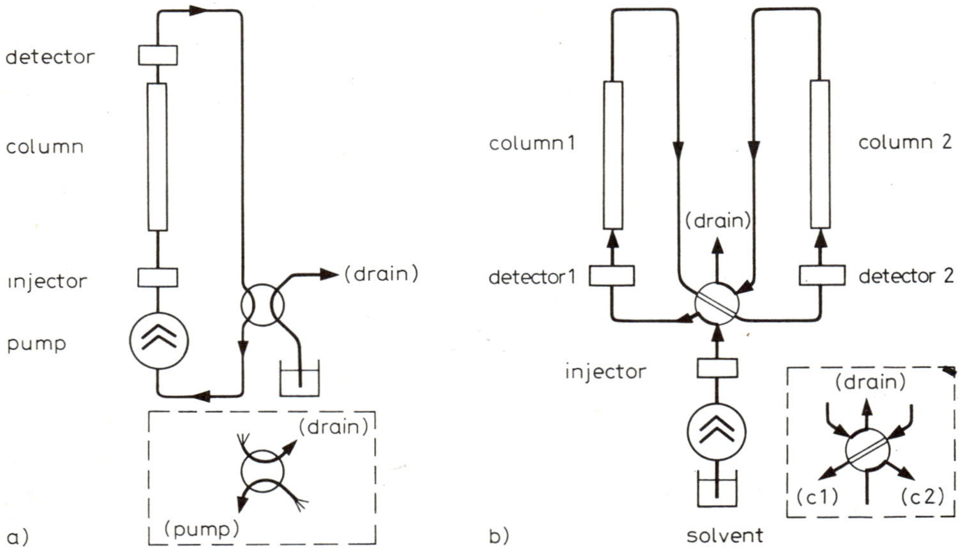

Fig. 17-5
Recycling
a) Closed loop recycling
b) Alternate pumping recycle method
The alternate positions of the switching valves are indicated in the box marked with a broken line.

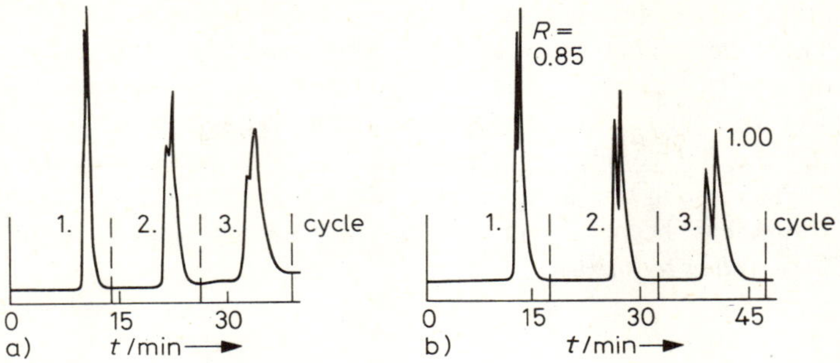

Fig. 17-6
Closed loope recycling

Effect of the pump construction on the resolution of two neighbouring components. The same column was used in both cases:
$L = 0.50$ m; $d_c = 10.7$ mm; with $N = 6200$ theoretical plates. The volume between the detector outlet and the column inlet, which is determined mainly by the stroke volume of the pump, was 6.6 cm³ in experiment (a) and 2.0 cm³ in experiment (b). (The flow rates were (a) 9.4 $cm^3 \cdot min^{-1}$ and (b) 7.6 $cm^3 \cdot min^{-1}$).
In the experimental arrangement with the smaller stroke volume, the resolution for the chosen anthracene/phenanthrene model system increased from 0.85 to 1.00 in three cycles, whereas R remained smaller than 1 for the arrangement (a).
(according to MARTIN, VERILLON, EON and GUIOCHON, 1976).

disadvantages mentioned above. Increasing the pressure is most critical. It can easily reach the order of magnitude where sensitive packings are damaged.

Recycle liquid chromatography enables long separating distances to be achieved without applying a higher pressure. For that purpose the sample is passed through the chromatographic bed several times in succession. Considerable investigations using this technique have been carried out, for instance, with adsorption columns. The method is of importance wherever components in close proximity to each other have to be separated and the total range of the distribution coefficients is narrow. Such cases occur with any retention mechanism, but most frequently in exclusion chromatography. Therefore recycling is of importance mainly for SEC.

The repetitive passage through the separating path can be realized by connecting the column outlet to the pump inlet; see Fig. 17-5(a) (PORATH and BENNICH, 1962; HEITZ and ULLNER, 1968; WATERS, 1970). Such a closed circuit naturally requires periodically operating pumps, i.e., where the uptake is continuous, having as low a stroke volume as possible. Nevertheless in each passage the sample undergoes undesirable mixing in the pump, which generally makes the greatest contribution to the band broadening due to the apparatus (cf., Fig. 16-10). Fig. 17-6 shows how two highly rated pumps yielded separation results of different quality in an apparatus which was otherwise identical.

The effect of the pump can be eliminated by using the technique specified by BIESENBERGER et al. (1971b, c) and adopted for HPLC by HENRY (1974). In this experimental arrangement shown in Fig. 17-5(b), the six-port valve is shifted from one position to another whenever the substance has completely passed from one column into the other. This variant allows the use of any pump desired.

The remarks made at the beginning of this section lead to the equation for the maximum resolution achievable after z cycles:

$$R_{s,z} = R_{s,1} \sqrt{z} \qquad (17\text{-}13)$$

Here $R_{s,1}$ denotes the resolution achieved after a single passage through the chromatographic bed. This relationship, which can also be derived from eqn. (3-25), represents an oversimplification in two respects: first, the resolution does not increase unlimitedly with the number of cycles, but reaches a maximum at the *optimum number of cycles* (KALASZ et al., 1975). Beyond this volume, component X_I overtakes component X_{II}. Secondly, not all of the factors increase in the same way with the number of cycles. A careful analysis of these relationships has been carried out by MCCRACKIN and WAGNER (1980) (cf., Section 16.3.2. *Recycling technique*).

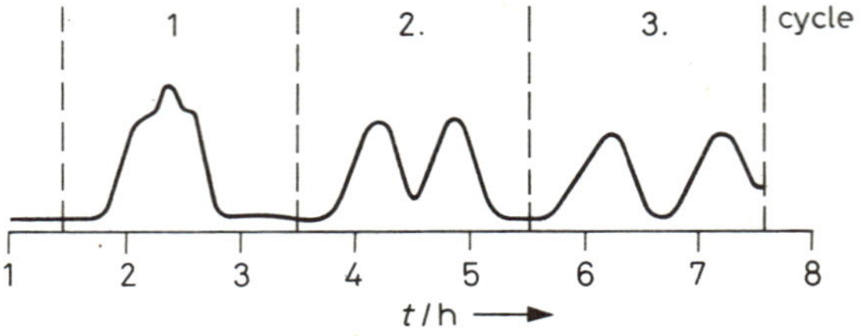

Fig. 17-7
Recycle SEC of two polystyrene samples in toluene on cross-linked polystyrene (nominal pore size $2.5 \cdot 10^4$ Å)

Preparative separation of 3.5 g (c_0 = 10 g · l^{-1}) in a 63.5 mm column 1.22 m long. Flow rate: 14.4 ml · min^{-1}. The resolution, R_s (cf., eqn. (3-21)) is 0.39 after the first cycle, R_s = 0.91 after the second cycle and R_s = 1.14 after the third cycle (according to BOMBAUGH and LEVANGIE, 1970 b).

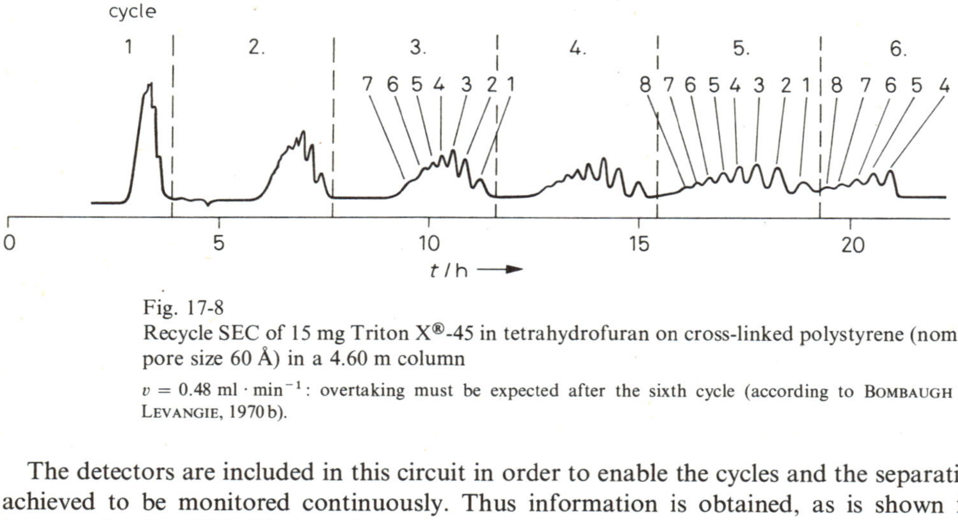

Fig. 17-8
Recycle SEC of 15 mg Triton X®-45 in tetrahydrofuran on cross-linked polystyrene (nominal pore size 60 Å) in a 4.60 m column

v = 0.48 ml · min^{-1}: overtaking must be expected after the sixth cycle (according to BOMBAUGH and LEVANGIE, 1970 b).

The detectors are included in this circuit in order to enable the cycles and the separation achieved to be monitored continuously. Thus information is obtained, as is shown for instance in Fig. 17-7. The continuous monitoring is also necessary in order to find out when a separation has to be interrupted in order to avoid overtaking. Although the separation shown in Fig. 17-8 has clearly been improved from the fourth to the fifth cycle and the same could be expected for the sixth cycle, the process had to be stopped because in the latter the band 8 reached the boundary of the preceding cycle. In such a case the resolution cannot be further increased for the whole band spectrum by means of the given apparatus. It is, however, possible further to differentiate a very important part of the

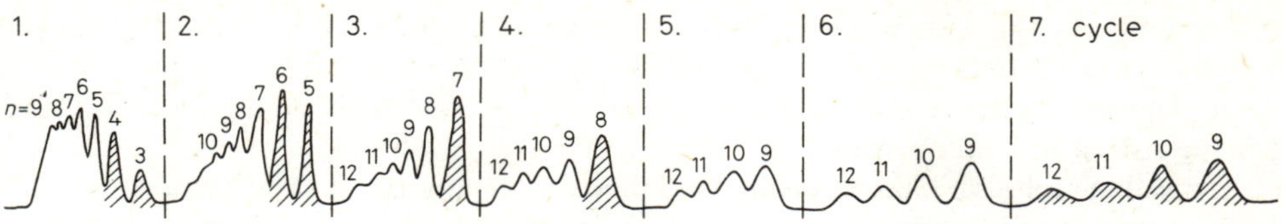

Fig. 17-9
Semi-preparative recycle SEC of a polystyrene standard with a molar mass of 600 g · mole^{-1} (nominal). The figures indicate the number of styrene units of the components.

Column: $L = 2 \times 1.20$ m, $d_c = 20$ mm; packed with polystyrene gel, cross-linked with 4% divinylbenzene; eluent: $CHCl_3$, $v = 3$ ml · min^{-1}. (The hatched peaks were cut after the respective cycles.) (according to NAKAMURA, ISHIGURO, YAMADA and MORIIZUMI, 1973).

spectrum by opening the circuit when less important components pass through the detector. If thereafter the closed circuit is restored, then the section of interest has empty eluent on both sides, so that the spreading can continue. Fig. 17-9 exemplifies this, showing how oligomers with 3—12 styrene units can be isolated from a polystyrene standard of the nominal molar mass of 600 g · mole^{-1}. Fig. 17-10 has been chosen intentionally to demonstrate of the overlap effect: the separation of the binary system becomes better and better up to the sixteenth cycle, but in the eighteenth cycle the faster component catches up the slower one with a lead of one cycle, and the formerly separated bands begin to overlap.

To keep all the bands separated but yet avoid an overlap of the cycles, the circulating liquid can be carefully concentrated (HEITZ and ULLNER, 1968). POLSON and RUSSEL (1966) recirculated the eluate through a semipermeable capillary tube. Part of the solvent diffused through the wall and evaporated from the outside. This caused the peaks to narrow, and the chromatogram as a whole was compressed (*antiparallel flow gradient*, cf., Section 14.3.).

On the one hand the examples show what separation efficiencies are achieved by recycle liquid chromatography, while on the other hand they demonstrate that the process has to be monitored very carefully. Therefore the blind mode of operation suggested by POILE (1980), where the number of cycles is preselected and the chromatographic surve is not

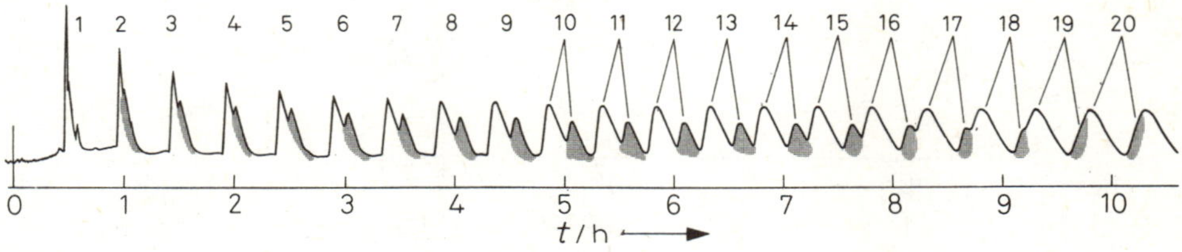

Fig. 17-10
Demonstration of the overtaking effect in the recycling of anthracene ($k = 3.58$) and phenanthrene ($k = 3.83$)

The position of the faster anthracene is indicated by a raster. The highest resolution is reached after the thirteenth cycle. Then the anthracene peak overlaps the phenanthrene peak. The experiment was carried out with the arrangement also used for Fig. 17-6 (a), but here the flow-rate was only 5.7 ml · min^{-1}. (according to MARTIN, VERILLON, EON and GUIOCHON, 1976).

recorded until the sample emerges from the apparatus, can only be used for routine investigations of mutually similar samples. The loss of resolution in a detector which is correctly dimensioned in proportion to the column is not high enough generally to justify the risk of a blind operation.

The recycling technique has pratical importance especially for separations on *soft gels* and for *preparative SEC*, because the permissible loading increases somewhat more rapidly than the number of cycles required (BOMBAUGH and LEVANGIE, 1970a, b). In a single column packed with polystyrene gel, the resolution achieved with an injection of 3.5 g and three cycles was equal to that obtained by an injection of 1 g and a single analysis. In both cases the test mixture consisted of two polystyrene standards.

The advantage of closed loop recycling lies in the fact that the separation efficiency can be improved without increasing the gel and solvent quantities.

17.9. Elution chromatography on a preparative scale

From a historical point of view, the preparative variant stands at the beginning of the development of elution chromatography. Today, however, with respect to the number of papers published and the number of applications, analytical elution chromatography occupies the premier position. However, one should not regard preparative chromatography as being of minor importance. Often it is the only possibility (and rather an efficient one in most cases) for the separation of components which are needed for further processing.

The term "preparative" does not refer to the size of the fractions, but rather to the fact that the substance components are isolated from the eluate. For instance, fractions for NMR measurements (DICKSON et al., 1971) were obtained by a preparative technique.

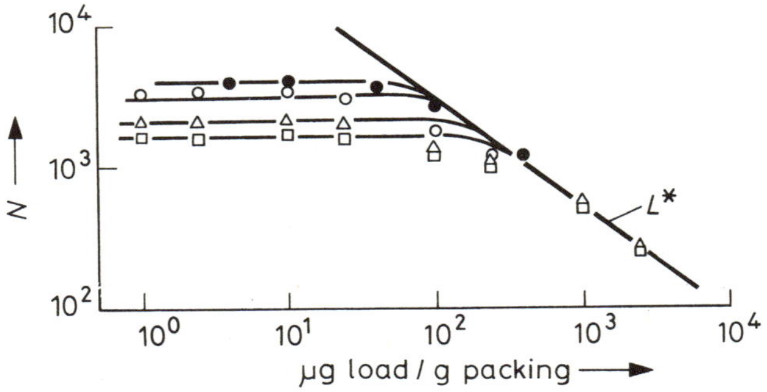

Fig. 17-11
Common loadability borderline (L) for different columns packed with Si 60 silica.

*Column plate number vs. specific load (solute: 2,4-dimethylphenol, $k \approx 3$). Eluent: dichloromethane
Column A (open symbols): $L = 1.00$ m; $d_C = 10$ mm; $d_p = 20$—25 µm; □ $u = 0.84$ cm/s; △ $u = 0.55$ cm/s; ○ $u = 0.18$ cm/s
Column B $L = 0.10$ m; $d_C = 10.8$ mm; $d_p = 5$—8 µm; ● $u = 0.21$ cm/s
(according to DEJONG, POPPE and KRAAK, 1981).

Even if the rules of thumb (a maximum of 1 g of the sample per 100 g of the column packing in AC, and 0.4 g at most in LLC [F 37]) are obeyed, preparative scale LC usually takes place in the non-linear range of the distribution isotherms (GAREIL et al., 1980; DE JONG et al., 1980). The higher the plate number of a column, the lower is its loadability per gram of packing material (DE JONG et al., 1978; BOMBAUGH and ALMQUIST, 1975). A small plate number means a high instrumental dispersion, i.e., a high dilution of the components on their way through the column. For their concentrations this improves the chance to reach the linear range of the distribution isotherms before the components arrive at the end of the column. For very small amounts of a sample the plate number is independent of the loading, i.e., the peak width is only determined by instrumental dispersion. After a certain concentration is exceeded, the plate number decreases as the loading is further increased. In this range the plate number is determined by the non-linearity of the isotherms; dispersion no longer plays the major rôle, and the curves for different flow-rates merge into a common *loadability borderline* (DE JONG et al., 1981; cf., Fig. 17-11).

An indication of the importance of preparative-scale LC can be derived from the quite abundant methodical and instrumental developments, reviewed by HAYWOOD and MUNRO (1980). Preparative GPC has been reviewed by COOPER et al. (1975b) as well as by VAUGHAN and DIETZ (1978).

Most of the devices used for preparative chromatography are rather demanding. For some problems, which can be solved by the principle of displacement development, the *chromatofuge* can be expected to yield an economical solution (FINLEY et al., 1978). In this apparatus developed by HEFTMANN et al. (1972), a centrifugal acceleration of 200 gravities forces the eluent through the chromatographic bed, which is packed into a drum, rotating at a speed of $16^2/_3$ Hz (1000 rpm).

17.9.1. Preparative SEC

In the oligomer range, for carefully selected working conditions it is possible to carry out preparative separations simply with a higher sample concentration. ALTGELT (1965), using 1 g of low-molecular-weight polybutene ($M = 1200$ g · mole^{-1}), introduced as a 10% solution in benzene-methanol (10:1), was able to achieve a separation by SEC which coincided with the result of precipitation chromatography. Also HEITZ and ULLNER (1968) worked successfully with 10% solutions in the low-molecular-weight range, but with styrene oligomers they in part observed disturbances such as bimodal peaks. From the great number of investigations on preparative SEC in the high-molecular-weight range, it was observed that the injection volume and the column cross-section had to be increased together with the sample amount. The flow-rate can be increased proportionally to the *column cross-section*; then the linear velocity, determining the kinetic band broadening, remains constant. The design of the column head must ensure that the sample and the eluent are uniformly distributed over the large column cross-section (COOPER et al., 1975a, COQ et al., 1979). The volume of the sample loop is increased to between 100 and 200 ml; thus the concentration of the sample solution need not exceed 1—3% (BOMBAUGH et al., 1968). Generally it is preferable to increase the load by increasing the injection volume instead of the concentration. BOMBAUGH and LEVANGIE (1970b) as well as Y. KATO et al. (1975) obtained chromatograms with a higher resolution if a certain mass of the polymer was loaded in a larger injection volume (BOMBAUGH: $V_0 = 350$ ml instead of 35 ml; KATO: 20 ml instead of 10 ml). In most of the studies on preparative SEC the injection volume V_0 was equal to, or less than, $^1/_{100}$ of the column volume, V_C. VAUGHAN and FRANCIS (1977), however, successfully performed preparative SEC of polypropylene and PVC with $V_0 = {}^1/_{20} V_S$.

Using a polystyrene column 1.22 m long and with an outside diameter of 63.5 mm, BOMBAUGH and LEVANGIE (1970a, b) achieved good separations of a maximum of 1 g of polymer in an injection volume of 100 ml. PEYROUSET and PANARIS (1972) found that the heterogeneity, H, of chromatographically obtained polystyrene fractions increased distinctly as the sample concentration was increased from 1 to 6%. The effect depended on the molar mass. For $M = 175000$ g · mole^{-1}, $H = \bar{M}_w/\bar{M}_n$ increased from 1.21 for $c_0 = 1\%$ to $H = 1.37$ for $c_0 = 6\%$. The best of all fractions obtained has a heterogeneity of 1.10. The apparatus employed was equipped with a series of up to five columns ($L \leqq 5 \cdot 1.22$ m, $d_C = 60$ mm) which were packed with silica. For PVC fractions prepared by HATTORI et al. (1978) on Styragel® columns with a diameter of 50.8 mm and a total length of 3.66 m, at a starting concentration of 0.5% the heterogeneity was even distinctly higher than for 0.2%, if samples with $\bar{M}_w = 380000$ or 174000 g · mole^{-1} were taken as the starting material. On the other hand, this effect did not occur in the SEC fractionation of a PVC with $\bar{M}_w = 35600$ g × mole^{-1}.

The adverse effect of a high starting concentration was also found by DARK et al. (1968) in the SEC of polyethylene on polystyrene gel in a column 1.22 m in length and with an outside diameter of 63.5 mm. The heterogeneity of the fractions obtained ranged between 1.23 and 1.44, without a systematic dependence on the molar mass. On the other hand, the heterogeneity of the GPC fractions of carboxy polybutadienes prepared on polystyrene gel in a 1.22 m column with an outside diameter of 50.8 mm increased from 1.15 to 2.66 with increasing molar mass (LAW, 1969). Nevertheless, MONTAGUE and PEAKER (1973), using polystyrene, found the narrowest distribution for the fraction with the *highest* molar mass. KATO et al. (1975) separated standard polystyrenes on a preparative scale in columns packed with polystyrene gels (particle size 10 μm) and achieved good resolution. Subsequently the fractions were characterized by analytical high-resolution SEC. The results are shown in Fig. 17-12, taking into account the mass proportions; this experimental result rather nicely confirms the assumptions upon which Fig. 16-1 is based.

COOPER et al. (1975a) carried out preparative separations of polystyrene in tetrahydrofuran, using a set of four columns. One column, 1.22 m in length and with an inside diameter of 57.2 mm (outside: 63.5 mm), was packed with polystyrene gel and another of the same size with CPG 10-350 porous glass. The other two columns, each with a length of 0.85 m and an inside diameter of 66.7 mm, were packed with CPG 10-2000 and CPG 10-120 porous glass, respectively. An equal interior volume was aimed at for all of the columns. From the chromatogram, the number of theoretical plates as well as the resolution

$$R_S = \frac{V_{e,\,II} - V_{e,\,I}}{2\sigma_{II} + 2\sigma_I}$$

and the resolution index

$$RI = (M_{II}/M_I)^{1/R_S}$$

were calculated. The plate number increased with decreasing molar mass. For a given pair of polymers, at $C_0 = 0.1\%$, R_S as well as RI were distinctly greater than at $C_0 = 1.0\%$. Moreover the authors chromatographed a commercial polystyrene with $\bar{M}_w \approx 2\bar{M}_n$ under different conditions, taking fractions from the eluate of the preparative separations for an

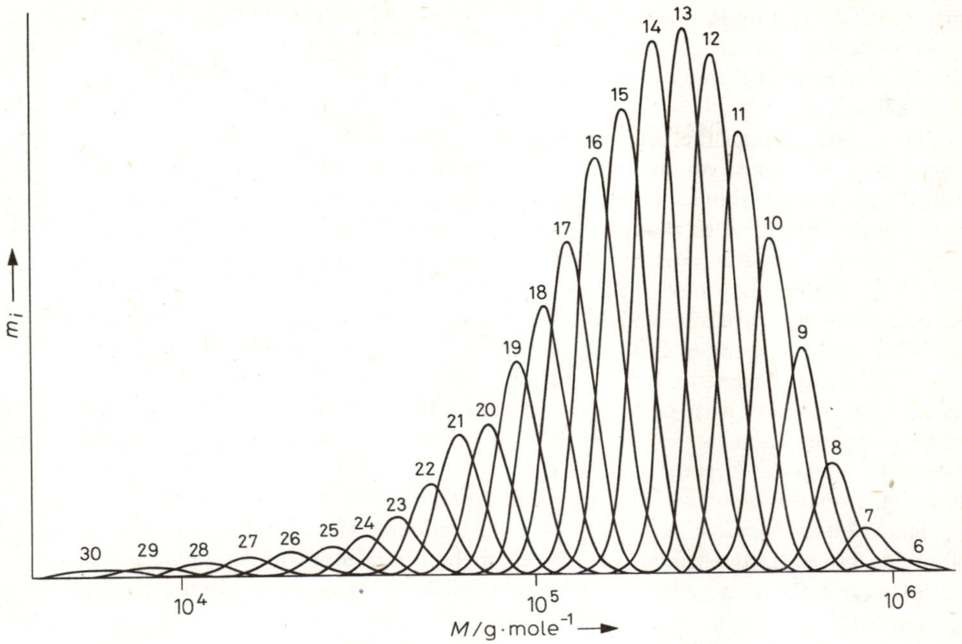

Fig. 17-12
Preparative SEC of a polystyrene sample with a broad distribution (NBS 706) in a θ-solvent consisting of methyl ethyl ketone and 11.3 vol. % methanol

(For the experimental conditions see Table 19-1, Nr. 10)
Representation of the distribution in the fractions on the basis of the analytical SEC in methyl ethyl ketone with $v = 0.5$ ml · min^{-1}
(according to Y. KATO, KAMETANI, FURUKAWA and HASHIMOTO, 1975).

analytical refractionation. For the separation of a 2% solution, the fractions with $\bar{M}_w < 10^5$ g · mole^{-1} exhibited a higher heterogeneity than for an injection of solutions with a lower concentration. Above $M = 10^5$ g · mole^{-1} the quality of fractionation was hardly dependent on the starting concentration in the range from $c_0 = 0.5$ to $c_0 = 2.0\%$. An increase in the linear velocity from 0.04 cm · s^{-1} (corresponding to a flow rate of 50 ml × min^{-1}, or 1 ml · min^{-1} in an analytical column with $d_C = 7.8$ mm) to 0.065 or 0.113 cm × s^{-1} did not impair the heterogeneity of the fractions obtained. On the other hand, for the preparative separation of two samples with molar masses of 10300 and 51000 g · mole^{-1}. BOMBAUGH and LEVANGIE (1970c) found that the resolution decreased almost linearly from 1.05 to 0.55 if the linear velocity was increased step by step from $u = 0.0085$ to 0.0700 cm · s^{-1}.

For columns with an inside diameter of about 5 cm and a sample volume of 100 ml, the *sample quantity* should not exceed 1 g per injection. The best fractions can be expected using starting substances with broad, uniform distributions.

As the separation is reproducible, larger fractions can be obtained by the combination of corresponding eluate fractions from repeated separations on the 1 g scale. PEYROUSET et al. (1975) in this way subdivided a high-density polyethylene with $\bar{M}_w/\bar{M}_n = 9.8$ into 38 fractions with a total mass of 227 g. An especially good separation could be achieved by at first subdividing it into three portions, which were then further fractionated on combinations of silica gel columns. The quotient $\bar{M}_w/\bar{M}_n$ of the fractions ranged between 1.18 and 1,05, decreasing as the molar mass decreased. The largest fraction contained 19.32 g ($\bar{M}_w$ = 8300 g · mole^{-1}, H = 1.10), the smallest one, 0.7 g ($\bar{M}_w$ = 909000 g · mole^{-1}, H = 1.18). Even though the smooth operation of the separation, which proceeds almost automatically, is impressive, the amount of work required with respect to the quantities of solvent to be handled should not be underestimated. A volume of 450 l of 1,2,4-trichlorobenzene probably had to be pumped through the steel columns with an inside diameter of 100 mm for the pre-fractionation and 60 mm for the main separation, from which the individual fractions finally had to be isolated BARLOW et al. (1971) equipped their apparatus for preparative SEC with a thin-film evaporator in order to obtain liquid volumes which were easier to handle. In this evaporator the eluate was concentrated to 25% of its volume before being passed into the fraction collector.

Preparative SEC is either controlled only by a time programme, or the eluate flow is split up in a proportion ranging from 1:50 to 1:100, the smaller quantity passing through the detector while the larger one flows directly to the fraction collector.

Initially, preparative columns did not reach the *numbers of theoretical plates* obtained with analytical columns, because in the dry packing technique there is obviously a tendency to accumulate larger-sized particles of the gel in the outer zones of the column. However, PEYROUSET and PANARIS (1972) succeeded in packing preparative columns which even exhibited twice the number of theoretical plates of analytical ones. The authors compacted the packing only by a slight tapping in the longitudinal direction. This observation was repeatedly confirmed: especially with fine-grained particles it has been possible to obtain preparative columns with high numbers of theoretical plates. Indeed the adverse effect of the outermost zone increases only linearly with the diameter, whereas the inner zone, where a particle segregation does not occur, increases with the square of the diameter (*quasi-infinite columns*, cf., Section 15.5.).

Preparative SEC is rather expensive: the quantity of the stationary phase required increases almost linearly with the sample amount. Together with the demanding packing technique this leads to high cost for preparative columns.

Nevertheless preparative SEC is carried out even *on an industrial scale*, if there is no other solution for a special problem (JANSON, 1971; WATERS, 1971; [D 18]). EK (1968) reported separations in columns with diameters up to 1.80 m, filled with coarse-grained, cross-linked dextran, which had a flow capacity of about 25 l · min^{-1}. Thus proteins and antibiotics could be purified, for example, ethanol removed from albumin solutions.

Economic considerations suggest that preparative SEC should be carried out with as high a loading as possible. This diminishes the resolution, but as *overloading* (cf., Section 19.1.) impairs the first-eluted, macromolecular component less than all other ones, it has been suggested to prepare fractions with narrow distributions and graduated molar-mass values from suitable starting samples in such a way that in each run only the first components are isolated (VAUGHAN and DIETZ, 1978). So preparative SEC with individual injections resembles the preparative counter-current chromatography mentioned in Section 17.9.3.

17.9.2. Preparative precipitation chromatography

Compared with the high cost of the columns needed for preparative SEC, the cost of the packing material for precipitation chromatography is quite negligible. The quality of the packing does not present any problems either: a bolting sample component is slowed down by the temperature gradient in the column until it is caught up by the substance band to which it belongs by its solubility. Consequently the broken-through component does not cause a band broadening. However, this presupposes that the temperature is constant over the column cross section (cf., Section 9.5.2.). For this reason, the wall-to-wall distance must not be arbitrarily increased as compared to the dimensions which proved suitable for the analytical work.

CANTOW et al. (1961) constructed a column with six tubes operated in parallel, each with a diameter of 25 mm and a length of 1 m, by means of which they fractionated polymer samples of 35—100 g (CANTOW et al., 1961, 1963, 1964). They obtained very good separations. The analytical refractionation of fractions yielded $\bar{M}_w/\bar{M}_n$ ratios of 1.020 and 1.013, respectively.

The apparatus described by HENDERSON and HULME (1967) uses an aluminium column with an outside diameter of 100 mm and a length of 1 m, and with a wall thickness of 12.7 mm. Six study protruding from the wall, each 25.4 mm long, formed an internal profile, such that the distance from the metallic heat conductor nowhere exceeded 13 mm. Using this apparatus, 45 g of polymer were fractionated with a total solvent quantity of 30 l within 15 hours, the heterogeneity of the fractions ranging between 1.1 and 1.3.

From the papers by CANTOW et al. and by HENDERSON and HULME, it is observed that very large quantities of polymer were successfully separated using relatively small volumes of solvent. In this respect precipitation chromatography is superior to preparative SEC. Another advantage lies in the fact that in precipitation chromatography the fractions are obtained in a single sweep. With a rotational-type evaporator at the column outlet they can continuously be collected as a solid. On the contrary, in preparative SEC the components obtained from many cycles, in each of which only 1 g is injected, must be combined in the right order.

17.9.3. Continuous preparative chromatography

Normally an elution chromatographic separation is initiated by an injection, and the separated components are collected at different times at the column outlet. In this case the chromatographic bed itself does not move. In *continuous elution chromatography* the chromatographic bed and the eluent flow are moved in relation to each other in such a way that the time sequence of the peaks is transformed into a spatial side-by-side arrangement. This is achieved by cross-current or counter-current processes.

In *counter-current chromatography* with a solid support phase the latter is moved in one direction while the mobile phase is moved in the opposite direction. The substance is introduced near the middle of the separating path. The solid phase carries along those components which have a low chromatographic rate of migration, whereas the mobile phase transports the other components to the opposite end of the separating path. Thus the substance is decomposed into only two fractions. This is sufficient for many preparative purposes and allows the use of loadings which are far above the usual limit (BARKER, 1978; BARKER et al., 1978a).

The motion of the solid support phase in the apparatus has been put into practice, but several difficulties, such as disturbances of the packing quality and a continuous abrasion of the support material, gave rise to further developments of the apparatus. By use of

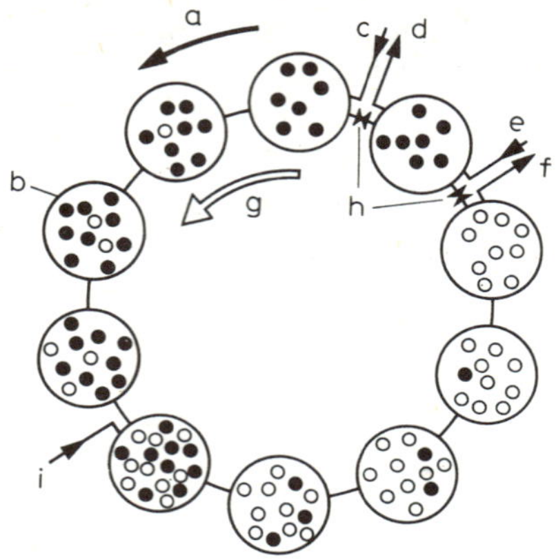

Fig. 17-13
Principle of operation of a sequential continuous chromatography unit

a direction of port rotation; b one of the ten columns interconnected by lines and valves; c eluent entry; d discharge of the slow-travelling component (●); e purge; f discharge of the fast-travelling component (○); g direction of flow of the mobile phase; h transfer valves closed; i introduction of substance
For the sake of clarity, all connection lines which are not in operation have been omitted (according to BARKER, ELLISON and HATT, 1978b).

close-packed annular columns, which were moved together with the external tube, this development led to stationary column batteries, in which inlet and outlet valves are operated according to a recirculation program (SZEPESY et al., 1975).

BARKER et al. (1978a, b) carried out a continuous SEC fractionation of dextran in an apparatus consisting of ten glass tubes connected in series, each 0.70 m long and with an inside diameter of 51 mm. Fig. 17-13 shows a schematic diagram of the corresponding circuit. In the fractionation, this circuit was interrupted at two points in each step (advancing in an anti-clockwise direction), so that it was possible to elute the components with the highest values of the distribution coefficient from the respective individual column isolated in each case, whereas the components with smaller distribution coefficients were obtained from the remaining nine columns. This apparatus, in which the motion of the stationary phase was replaced by a step-by-step change of the introduction and discharge points, successfully decomposed up to 50 g of substance per hour into two components of different molar masses, with about 15 l of solvent required per hour.

Counter-current chromatography with a *liquid* stationary phase without solid support particles has been investigated mainly by ITO et al. (cf., Section 9.3.). The preparative applications of this variant has been dealt with in a paper by ITO and BOWMAN (1978).

In *cross-flow chromatography* a bed of the solid support material is moved orthogonally to the direction of flow of the mobile phase. FOX et al. (1969a, b) as well as NICHOLAS and FOX (1969) developed and utilized an apparatus for continuous cross-current chromatography, where the packing was contained in an annular gap about 1 cm wide between two concentrically arranged PMMA cylinders. The eluent stood above the packing, which was about 15 cm high, flowing through the packing bed under its own hydrostatic pressure, and emerging from openings in the bottom. While the device was slowly rotating, substance was introduced continuously at some point at the upper edge of the packing. The components of this substance, carried along by the rotating bed over different distances depending on

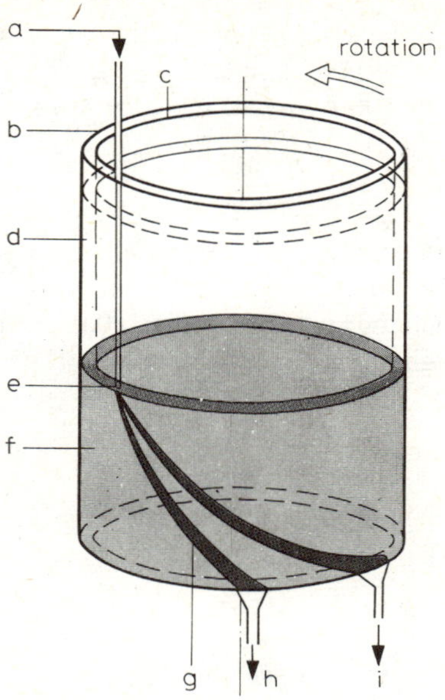

Fig. 17-14
Schematic diagram of an apparatus used for continuous cross-current chromatography
a sample feed; b outer cylinder; c inner cylinder; d eluent; e introduction of sample; f packing; g trace of component I; h discharge of component I; i discharge of component II
(according to FOX, CALHOUN and EGLINTON, 1969a).

their chromatographic rate of migration, emerged from the apparatus at spatially different points (see Fig. 17-14). In a stationary operation, each component discharged from the apparatus at a well defined position. This principle was used by SVENSSON et al. as early as 1955.

The operation of such an apparatus is not simple and it also requires a strictly homogeneous packing of the whole bed, but the method has the advantage that the separation is not restricted to two components only.

A review of continuous modes of chromatography, which also includes parametric pumping systems and electro-chromatographic techniques, has been published by SUSSMAN and RATHORE (1975).

17.9.4. Comparisons and conclusions

There are several possibilities for a preparative separation of polymer mixtures by the application of chromatographic principles. In a comparison of efficiencies, several points have to be taken into account:

— The quality of the fractions, characterized by as small a value as possible for the heterogeneity, $H = \bar{M}_w/\bar{M}_n$
— The purchase cost and the installation cost of the apparatus
— The direct and indirect running cost for carrying out a separation, which to a great extent depends on the quantities of solvent to be handled
— The safety of operation of the apparatus

As regards the last requirement, SEC can be considered highly satisfactory. For a comprehensive evaluation, MARTIN and JOHNSON (1973) suggested a *figure of merit*, FM which includes the separation quality, and the total cost, ΣE, to be spent per gram of substance obtained in a fraction:

$$FM = [(M_w/M_n - 1)^2 \cdot (\Sigma E)]^{-1} \qquad (17\text{-}14)$$

According to a classification given by COOPER (1978), in this valuation precipitation chromatography ranks far above SEC from an economic point of view. If one sets FM = 1 for precipitation chromatography, then the analogous value is 0.06 for normal SEC and 0.01 for high-temperature SEC. The higher the numerical value of FM, the more advantageous the method will be.

D Applications

18. Adsorption chromatography of polymers — experimental parameters and results

"The great success of the chromatographic method in the field of carotenes and other classes of pigments made it appear desirable to find out whether in the range of high polymer substances this method also allows a separation of mixtures which are otherwise difficult to separate." With this sentence MARK and SAITO, in their paper published in 1936, motivated experiments on cellulose acetate solutions with blood charcoal as an adsorbent, and preliminary tests, in which calcium carbonate, alumina and starch were included. They filtered solutions in acetone through a column, the packing of which, after being exhaustively washed with the solvent, was divided into three sections. From the material in each section the polymer was extracted by dioxane. Viscosity measurements showed that the largest molecules had travelled farthest or even passed through the column. This result, contradicting Traube's rule, was initially considered to be due to the rate of adsorption. However, soon it was realized that the frequently observed stronger retention of components with low molar masses is due to the pore structure of the adsorbents.

Whether or not adsorption chromatography works well depends on the relationship between the quantity adsorbed and the concentration, on the rate of adsorption and the desorption behaviour. All of the influencing factors should support each other.

18.1. Rate of adsorption

HELLER and TANAKA (1951) investigated the kinetics of the adsorption of polyethylene glycol on charcoal, finding a decrease in the rate of adsorption with increasing molecular size. While the monomeric ethylene glycol reached 90% of the maximum adsorption within less than 15 seconds, oligomers with 600 and 6000 $g \cdot mole^{-1}$ required as much as 2.5 and 9 minutes, respectively.

Other authors obtained similar results. Adsorption proceeds most slowly on porous adsorbents. In some cases it took as long as several weeks to reach the equilibrium (cf., FRISCH et al., 1959).

Even on smooth metallic surfaces the final value is not reached instantaneously; the time required increases with increasing molar masses and decreasing concentrations. In adsorptions on chromium from solutions of about 1 $g \cdot l^{-1}$ polystyrene ($M = 1.9 \cdot 10^6$ g $\times mole^{-1}$) in cyclohexane, a final value was reached after 24 hours. For $M = 537000$ g $\times mole^{-1}$ the adsorption on a gold surface had already become constant after 4 hours (STROMBERG et al., 1959). On the other hand, from a solution with only 0.1 $mg \cdot l^{-1}$

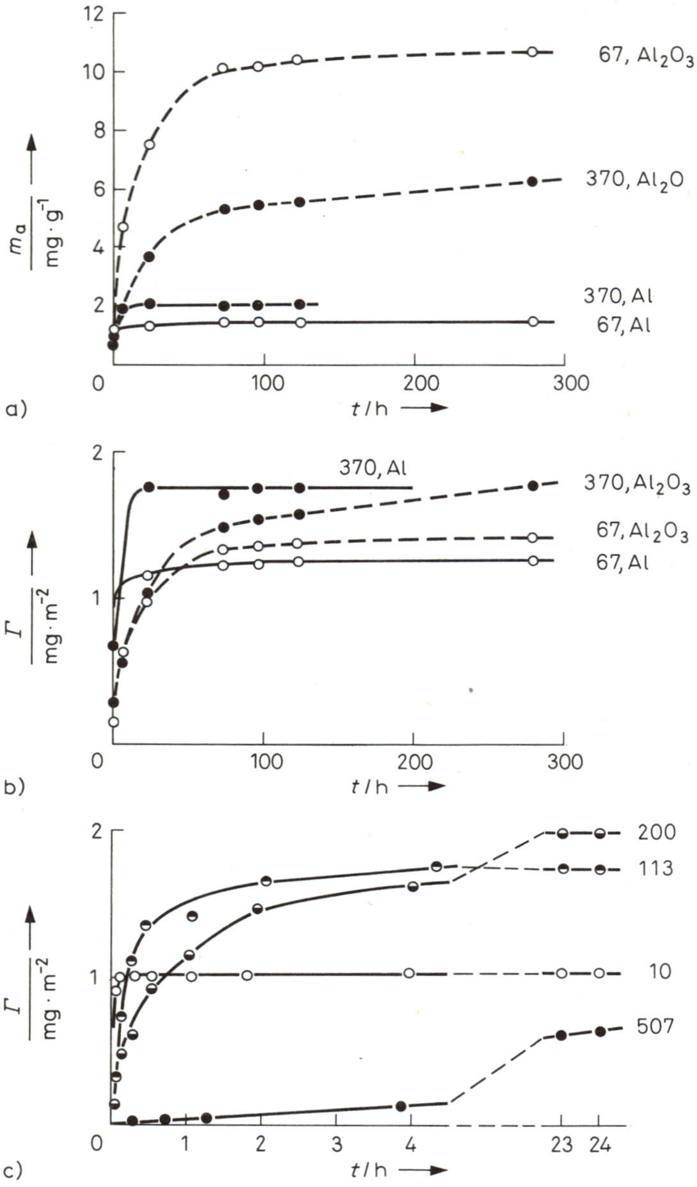

Fig. 18-1
Time dependence of the adsorption of polymers

a) Adsorption of polystyrene from cyclohexane on aluminium powder or porous alumina: increase of the amount of substance adsorbed per g of adsorbent as a function of time (according to Burns and Carpenter, 1968). The numbers beside the curves indicate the molar mass in $kg \cdot mole^{-1}$.

b) As 18-1a, but represented as the amount of substance adsorbed per unit area

c) Adsorption of polystyrene from cyclohexane on porous glass; increase of the amount of substance adsorbed per m^2 as a function of time (according to Furusawa, Nakanishi and Kotera, 1975).

polystyrene with a molar mass of 76000 g · mole^{-1} only reached the adsorption equilibrium on chromium after more than 30 hours.

The influence of the concentration as well as the high acceleration which can be achieved by stirring show that the adsorption of macromolecules is a diffusion-controlled process. As the coefficient of diffusion decreases with the increasing size of the diffusing particles (cf., eqns. (15-21 a, b)), it takes a long time for the quantity of polymer required for saturation, which is about 1 mg · m^{-2}, to arrive at the surface. For adsorbents with a large surface area, the amount of adsorbate to be transferred to this area is correspondingly large. Already this fact might contribute to the longer transition times on alumina, charcoal, silica gel and other porous materials. The effect of pore size adds to this. Very bulky molecules can only penetrate into pores which are wide enough, whereas many more pores are accessible for the smaller-sized molecules. Therefore the large molecules, which, moreover, are slower to diffuse, require a particularly long time to arrive at accessible surface sites (Fig. 18-1 a, b, c).

On flat surfaces the macromolecules are deformed by the adsorption. Numerous investigations have shown that the density of the adsorption layer exceeds that of the coils in solution by a factor ranging between 2 and 10. The first-adsorbed macromolecules cling to the surface very flatly. For those arriving subsequently, only part of the surface is left. This remaining part is more and more split up into residual areas of decreasing extension as the coverage increases. Consequently, macromolecules adsorbed at this stage of the process project more deeply into the solution than those attached earlier. This idea is supported by ellipsometric results, which, for polystyrene adsorbed from cyclohexane on chromium, showed a steep increase in the layer thickness as a function of time, but only a relatively slight increase in the quantity adsorbed. The density of the adsorbed layer dropped to half of its value within 90 minutes (Fig. 18-2, KILLMANN and v. KUZENKO, 1974).

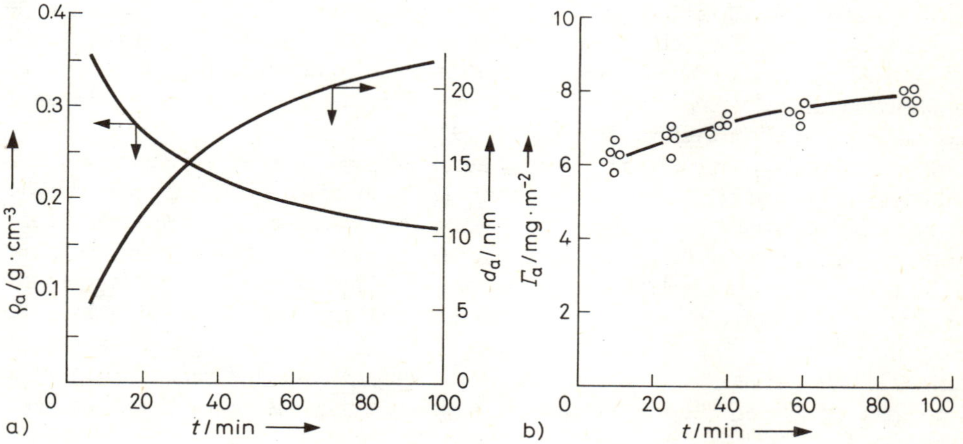

Fig. 18-2
Adsorption of polystyrene ($\bar{M}$ = 750000 g · mole^{-1}) from cyclohexane on chromium at 36 °C, c_0 = 1 g · l^{-1}

a) Increase of the layer thickness, d_a, measured by ellipsometry and decrease of the density, ϱ_a, in the adsorption layer as functions of time

b) Increase of the loading (according to KILLMANN and v. KUZENKO, 1974).

On the other hand, coils can occupy suitably sized pores without being compressed. The radii of the accessible pores as calculated from the pore size distribution and from the adsorption achieved after 25 days, corresponded to the radii of gyration of the macromolecules (BURNS and CARPENTER, 1968).

In a pore, a coil has more points of contact with the wall than on a flat surface. For PMMA THIES (1971) obtained values of $p_{CO} = 0.46$—0.49 in porous silica gel, whereas his results on non-porous alumina and silica were $p_{CO} = 0.23$—0.28 and $p_{CO} = 0.32$—0.39, respectively. (For the meaning of p_{CO} cf., Section 6.1.5.)

A most complex time dependence must be expected if the pores are a little smaller than the coil dimensions and strong surface forces result in a gradual deformation, so that the molecules slowly "squeeze themselves" into the cavities (cf., Section 16.6.1.).

18.2. Desorption behaviour

The mechanism of chromatographic migration requires that the transitions between the stationary and the mobile phase take place rapidly and reversibly and that the isotherms are linear (Figure 6-2a). However, even the experiments carried out by MARC as early as 1913 with gum arabic, albumen and starch on barium sulphate as well as the systematic investigations by JENCKEL and RUMBACH (1951) yielded non-linear isotherms, as shown schematically in Fig. 6-2c: the amount adsorbed increases steeply even at very low concentrations and soon reaches a saturation value. Such extremely non-linear isotherms are typical. This behaviour manifests itself the clearer (i) the more the S^0 value for the repeat unit exceeds the $A_S \varepsilon^0$ value of the solvent (cf., eqn. 7-11), (ii) the more the quality of the solvent decreases and (iii) the more the degree of polymerization increases. The high amounts adsorbed in equilibrium with almost infinitesimal solution concentrations suggest that the desorption can hardly be achieved without changing the solvent.

In fact a reversible adsorption of flexible macromolecules has been observed in only a few exceptional cases, which include the extremely dilute solutions referred to in Sections 6.1.1. and 6.1.2. (Here it should also be remembered that the adsorption energy in the polystyrene/toluene/glass system is certainly very low. Polystyrene dissolved in benzene is not adsorbed at all on Aerosil (HERD et al., 1971).)

Generally it is not possible to get a loaded adsorbent again free of polymer by either a dilution or even a treatment with the pure solvent (ROBB and SMITH, 1974). Neither can the desorption be achieved by a temperature increase (BURNS and CARPENTER, 1968). Frequently even a higher load was observed at a higher temperature. Only a stronger solvent with a high ε^0 value can displace the polymer from the surface. Thus PVAC is removed from iron by acetonitrile (KORAL et al., 1958), and PMMA from iron or glass by 25% acetonitrile in benzene (ELLERSTEIN and ULLMAN, 1961). An adsorption does not occur from solutions in such solvents. In multilayer adsorption, where polymer molecules are deposited on polymer molecules, desorption can be achieved along the usual lines. The last (lowermost) layer, though, again behaves as described above.

Fig. 18-3 shows that even glass retains PMMA. From the quantity (23 mg) and the total surface area of about 15—20 m^2, a loading of fully 1 mg · m^{-2} is derived, which is of the same order of magnitude as the statically determined Γ_a values (Table 6-1).

The adsorption-desorption behaviour of macromolecules can be considered a process which, for a single segment, takes a reversible course and largely corresponds with the

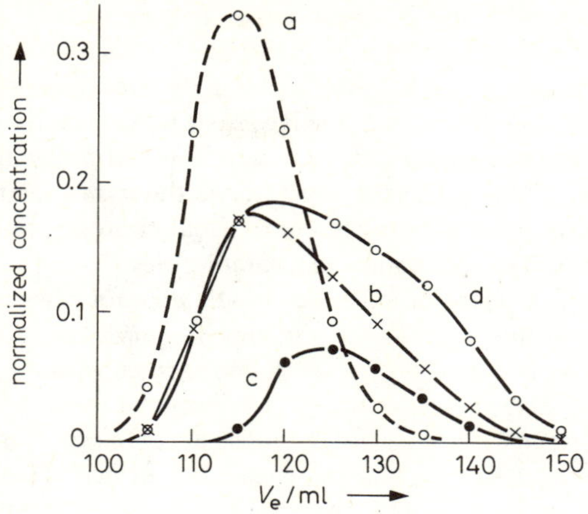

Fig. 18-3
Retention of PMMA (M = 680000 g · mole^{-1}) when filtered through a column packed with glass beads (d_P = 100 μm)

a solution in acetone on silanized glass (100% recovery); b solution in benzene on silanized glass (73.5% recovery); c solution in benzene on untreated glass (76% irreversibly adsorbed); d solution in benzene on untreated but pre-saturated glass (third sample, 100% recovery); (according to SCHULZ, BERGER and SCHOLZ, 1965).

adsorption-desorption transition of small molecules. A single segment can be readily adsorbed and desorbed. In contrast to free small molecules, however, it cannot leave the surface, since it is part of the chain adsorbed at many points. Therefore, after the desorption it has to stay so close to the surface, immediately in front of the adsorption site it has just left, that *re-adsorption* cannot be avoided. The segment can hardly stay until all segments of the train are also freed. However, if during the short period of desorption its site on the surface is occupied by a particle which has a higher adsorption energy, then a different situation arises: little by little, the statistical desorption of individual segments may lead to the displacement of the whole macromolecule.

Solvent molecules or segments of other macromolecules with a higher adsorption energy are such competing elements. PMMA displaces PS from silica gel in trichloroethylene completely within 2—4 hours (THIES, 1966), being displaced in its turn by polyvinylpyrrolidone in chloroform (ROBB and SMITH, 1977). Likewise, on silica gel and with trichloroethylene as a solvent, BOTHAM and THIES (1970) found that the displacing effect decreased in the following order: PVAC > ethyl cellulose > ethylene copolymer with 44.2 wt.-% vinyl acetate (EVA 44.2) > EVA 20.4 > PS. Ethyl cellulose displaces adsorbed PMMA only very slowly.

The probability of the desorption of a segment can be estimated from the distribution coefficient of the structurally analogous small molecules. This probability is less than 1 (since otherwise an adsorption would not occur), but normally it is distinctly above zero. The probability of a simultaneous and independent desorption of all adsorbed segments of a macromolecule is equal to the product of the individual probabilities, thus approaching zero at a high degree of polymerization. Very large-sized macromolecules are bonded more strongly than lower homologues, although the enthalpy of adsorption per segment has the same value. This may lead to a displacement of adsorbed macromolecules by larger ones of the same kind, which has been observed in some cases. The decrease of the bonded portion, p_{ESP}, from 0.9 to 0.6 by a polymer added subsequently, which was found in measurements carried out by Robb and Smith for polyvinylpyrrolidone on silica gel (cf., Sections 6.1.6. and 6.2.1.), likewise indicates a displacement by macromolecules of the same kind. In this case it applies only to individual adsorbed trains of molecule segments, which may be why it proceeds so rapidly. The driving force is the gain in entropy in the transition from the extreme starting structures to the more usual conformations, which is connected with the successive adsorption of further macromolecules.

To sum up, the adsorption for the macromolecules as a whole is not reversible. On the other hand, for the individual *segments* there is an *adsorption equilibrium*, which can be treated thermodynamically. As long as a coil only touches the surface with a few segments and the areas of contact do not yet extend over entire trains of segments, the molecule may again detach in the same way. However, if it does not get loose at this stage, it is irreversibly adsorbed by the subsequent deposition of the segments on the surface. There are many indications that adsorption chromatography with macromolecules is only possible in the first-mentioned stage.

The capture of a single macromolecule brings the P-fold number of segments to the surface. This multiplication of the number of molecules by the degree of polymerization is one of the causes for the high degrees of coverage achieved even at low solute concentrations. In addition the centre of the coil is still far from the surface when the capture begins. In the adsorbed conditions, however, the coils are compressed to a layer, the height of which is only a small fraction of this distance, and hence hardly affects the approach of other coils. The attachment only becomes difficult when the surface is largely covered. It stops when the remaining unoccupied surface groups are all buried under the cover of adsorbed macromolecules.

The *capturing mechanism* indicates that large-sized molecules are preferentially adsorbed from mixtures: as the coil volume increases proportionally to $M^{3/2}$, the large molecules are the first to come into contact with the surface. Moreover multiple contacts, which are required to prevent the re-detachment of a particle having arrived at the surface by diffusion, are more likely to occur with large molecules than with small ones. Even without a higher capturing probability, the macromolecular components would make a greater contribution to adsorption, because naturally the mass increase per capture step in the adsorption layer is proportional to the molar mass.

However, if the concentration in the solution is low with respect to the area to be covered, then the *diffusion towards the surface* becomes the rate-determining factor. Then the faster diffusing, smaller molecules are adsorbed before the large ones.

18.3. Conclusions for adsorption chromatography

The relatively slow adsorption of macromolecules from highly dilute solutions does not represent an absolute hindrance to chromatographic separation. Although of course it leads to a condition far from the ideal one assumed in Chapter 3, a chromatographic separation would nevertheless be possible.

What is much more important is the fact that fully adsorbed macromolecules cannot be desorbed by the same solvent. Thus it follows from the adsorption behaviour that the separation of macromolecules with repeat units of the same kind by isocratic elution only has a chance of success if the coils are adsorbed with only a few segments. Fully adsorbed macromolecules must be desorbed by a medium with a higher adsorptive power. Therefore *displacement development* and the *elution gradient technique* are of the greatest importance for the adsorption chromatography of polymers.

A separation according to the *molecular size* is most likely in the neighbourhood of the θ state.

Adsorption chromatography is of special importance for the detection and separation of molecules with slight *chemical* modifications. This problem can be rather easily solved if the extra groups are more strongly adsorbed than the unmodified segments. For example, PMMA was much more strongly adsorbed by iron powder if 1% of the ester groups were saponified (ELLERSTEIN and ULLMAN, 1961). The adsorption of PVAC from 1,2 dichloroethane on iron increased by almost a factor of three if 13% of the acetate groups were saponified. A further saponification up to 26% increased the adsorption only slightly beyond this value (KORAL et al., 1958). The separation of macromolecules carrying a small number of polar groups from the bulk of less polar chains requires such conditions that the ordinary molecules are not adsorbed. Only the extra groups are retained by the adsorbent. As the adsorption enthalpy is below the critical value for all other, normal segments, the coils are not deformed or extended over the surface. Thus there is the possibility of a reversible adsorption and a chromatography under isocratic conditions especially for problems which can hardly be solved otherwise.

Chromatographic separations, above all on planar beds, scarcely take place under conditions as ideal as those applied in the recording of adsorption isotherms (cf., Section 6.1.1.). To prevent unintended changes of the *activity* of the adsorbent during the separation, well defined conditions, e.g., with respect to the water loading, must be ensured. As some support materials, e.g., silica gel, bond water very strongly, in such a case the separation may take place on the bonded water layer rather than directly at the SiOH groups. This topic has been discussed very thoroughly by SCOTT (1980b).

For example, polyvinylpyrrolidone is fixed onto silica gel through a water film, as was shown by the measurement of the proton relaxation times by means of pulse nuclear magnetic resonance (ROBB and SMITH, 1974). Polyethylene terephthalate oligomers are likewise adsorbed on glass through a water film (STROMBERG and GRANT, 1963). Silica gel can also bond non-aqueous solvents by hydrogen bridges so strongly that the adsorption of moderately polar components may then take place on this primary layer. Only very strongly retained solutes with a polarity which can compete with that of the solvent are believed to interact directly with the silica gel surface (SCOTT and KUČERA, 1979c).

Table 18-1 summarizes some experimental investigations of the column adsorption chromatography of polymers. Papers on static adsorption measurements are also listed in the same order of succession (by the polymers investigated). The following statements can be made:

— Investigations on adsorption chromatography have been carried out since 1936, with some increase in activity in the fifties.

— The chromatographic experiments were preferentially carried out using polydienes, polyesters and polyethylene oxide as samples. As these substances are polymers which are difficult to fractionate by solubility methods, this illustrates the orientation of these investigations towards practical aspects.

Table 18-1
Bibliography of experimental work in the field of adsorption chromatography of polymers and the adsorption behaviour of macromolecules in solution (Separations yielding individual components are indicated by an asterisk*.)

Polymer	Chromatography		Adsorption measurements	
	Support materials	Authors	Adsorbent	Authors
Polyolefins, optically active	poly(S)-3-methyl-pent-1-ene	Pino et al. (1962) Pino et al. (1966)		
Polypropylene	polypropylene on SiO_2	Natta et al. (1958)		
Polyisobutylene	charcoal	Landler (1947)	carbon black	Krauss and Dugone (1955)
	charcoal, SiO_2	Losikov et al. (1956)	carbon black	Binford and Gessler (1959)
Natural rubber			carbon black	Krauss and Dugone (1955)
Butadiene styrene rubber	charcoal	Landler (1947)	carbon black	Goldfinger (1945)
	carbon black	Golub (1953)	carbon black	Amborski et al. (1950)
	silica gel	Tagata and Homma (1972)	carbon black	Kolthoff and Kahn (1950)
Butadiene acrylonitrile rubber	charcoal	Landler (1947)		
Polychloroprene	glass powder NaCl	Poláček and Matyska (1962)		
Polystyrene			activated carbon, glass, Al	Jenckel and Rumbach (1951)
			activated carbon	Treiber et al. (1953)
			activated carbon	Hobden and Jellinek (1953)
			sand	Sonntag and Jenckel (1954)
			charcoal	Jellinek and Northey (1954)
			glass	Fendler et al. (1955)
	charcoal	Yeh and Frisch (1958)	carbon black	Jurženko and Malejev (1958)
			carbon black	Frisch et al. (1959)
			foils of PE, PETP PA, cellophane	Patat and Estupinan (1961)
			Aerosil	Kiselev et al. (1963)
			chromium	Stromberg et al. (1964)
PS 600[1])	SI—C 18	*Waters Chromatography Notes 1974 [F 38]	chromium	Stromberg et al. (1965)

[1]) The numbers refer to the nominal value of the molar mass of the PS standards.

PS 600, 900	SI—C 18	*Kirkland (1975)	Aerosil	Thies (1966)
PS 800	SI—C 18	*Parris (1978)	glass	Rowland and Eirich (1966a, b)
PS 750	SI—C 18	*Lattimer et al. (1979)	Al, Al_2O_3	Burns and Carpenter (1968)
PS 1100	silica gel	*Beck and Halász (1978)	silica gel	Bogacheva et al. (1969)
PS 2100	silica gel	*Eisenbeiss et al. (1978)	silica gel	Howard and Woods (1972)
PS 2900	silica gel	*Knox and McLennan (1979)	chromium	Killmann and v. Kuzenko (1974)
$2 \cdot 10^3 \leqq M \leqq 2 \cdot 10^6$	silica gel	Klein and Leidigkeit (1979)	glass	Furusawa et al. (1975)
PS oligomers with polar terminal groups	silica gel	Månsson (1980)		
Polyvinyl acetate			activated carbon	Claesson and Claesson (1944/45)
			activated carbon	Claesson (1949a, b)
			glass	Tujnman and Hermans (1957)
			foils of Cu, Al, Pt	Patat and Schliebener (1961)
			iron and tin powder, Al_2O_3	Koral et al. (1958)
			chromium	Gottlieb (1960)
	activated carbon	Kangle and Pascu (1961)	cellulose	Luce and Robertson (1961)
	glass powder	Blazėjewicz (1968)	Aerosil	Kiselev et al. (1968)
Ethylene vinylacetate copolymers			Aerosil	Thies (1968)
			silica gel	Botham and Thies (1970)
			glass	Hara and Imoto (1970)
Polyvinyl alcohol			Pt, cellophane	Patat and Schliebener (1961)
			AgI-sol	Fleer et al. (1972)
			oxidized silicone	Fleer and Smith (1976)
Polymethyl methacrylate	activated carbon	Claesson (1949a)	activated carbon	Jenckel and Rumbach (1951)
			glass, sand, Al powder	
			silica sand	Sonntag and Jenckel (1954)
			foils of Cu, Al, Pt	Patat and Schliebener (1961)
	Al_2O_3, hydrophobized	Jurženko and Malejev (1958)	iron and glass powder	Ellerstein and Ullman (1961)
			Aerosil	Thies (1966, 1971)
			silica gel	Botham and Thies (1970)
			aluminium silicate	Patel et al. (1970, 1971)
			Aerosil	Miyamato et al. (1974)
Polymethacrylic acid and ester	activated carbon	Ulińska and Huppenthal (1960)	silica sand	Fontana and Thomas (1961)

Table 18-1 (continued)

Polymer	Chromatography		Adsorption measurements	
	Support materials	Authors	Adsorbent	Authors
Graft and random copolymers of styrene and methyl methacrylate or acrylate	silica gel	HORII and IKADA (1974)	silica gel	HERD et al. (1971)
	silica gel	BELENKIJ (1979)		
	silica gel	TERAMACHI et al. (1979)		
	silica gel	DANIELEWICZ and KUBIN (1981)		
Block copolymers styrene/ethylene oxide	cellulose powder	WESSLEN and MÅNSSON (1975)		
Polyethylene oxide	silica gel	BÜRGER (1967b)	charcoal	HELLER and TANAKA (1951)
		CALZOLARI et al. (1971)	foils of Pt, Al	PATAT and SCHLIEBENER (1961)
	silica gel	KISELEV et al. (1971)	activated carbon	JURŽENKO and MALEJEV (1958)
	silica gel	KUŽAEV et al. (1973)	glass, Al	KILLMANN and SCHNEIDER (1962)
	SI—C 8, SI—C 18	*MELANDER et al. (1979)	carbon black	HOWARD and MCCOUNELL (1967)
	SI—C 2	*NOZAWA and OHNUMA (1980)	carbon black	ELTEKOV et al. (1974)
	SI—C 18	*MURPHY et al. (1981)	Aerosil	KILLMANN (1976)
Polypropylene oxide	Al_2O_3	MORRIS and PERSINGER (1963)	terpene resins (opt. active)	FURUKAWA et al. (1966)
Polyester	urea	KERN et al. (1955)	silica gel	STROMBERG and KLINE (1961)
	charcoal	BRUCK (1958)	glass powder	STROMBERG et al. (1959)
	silica gel	HILT et al. (1966)	silica gel, Al_2O_3	
	SI—C 8	*COULOMBE et al. (1978)	chromium	PEYSER et al. (1967)
			Aerosil	LIPATOV et al. (1975)
			Aerosil	DIETZ (1976)
PETP oligomers	silica gel	*ZABORSKY (1977)		
	silica gel	*HUDGINS et al. (1978)		
Caprolactam oligomers	silica gel	*BRODILOVÁ et al. (1979)		
Polyvinylpyrrolidone	activated carbon	CLAESSON (1949a, b)	platinum	PATAT and SCHLIEBENER (1961)
			PE, cellophane	PATAT and ESTUPINAN (1961)
			chromium	KILLMANN and v. KUZENKO (1974)
			mercury	JEHRING (1974)
			silica gel	ROBB and SMITH (1974)

Styrene acrylonitrile copolymers	silica gel	Teramachi and Esaki (1975)		
Cellulose acetate	activated carbon	Mark and Saito (1936)	$CaSiO_3$	Trivedi et al. (1973)
		Levi and Giera (1939)		
Cellulose nitrate	activated carbon	Claesson (1949a, b)		
	starch	Brooks and Badger (1950)		
	activated carbon	Swenson (1955)		
Epoxy oligomers	SI—C 18	*Waters Chromatography Notes 1974 [F 38]		
	SI—C 18	*Shiono et al. (1980)		
	SI—C 18, silica gel	*Hagnauer and Setton (1978)		
	SI—C 18	*Hagnauer (1980)		
Dextran	activated carbon	Claesson (1949a, b)	activated carbon	Ingelman and Halling (1949)
			cellulose nitrate	Rothman et al. (1953, 1955)
Polyarabinose	charcoal	Goodban and Owers (1975)		
Polyvinyl chloride			sand, glass	Jenckel and Rumbach (1951)
			activated carbon, Al	Felter and Ray (1970)
			$CaCO_3$	Felter (1971)
Silicones	activated carbon	Bannister et al. (1954)	iron and glass powder	Perkel and Ullman (1961)
Proteins			$BaSO_4$	Marc (1913)
		Huisman et al. (1960)	polystyrene latex	Lyklema and Norde (1973)
		Cyperovic and Galic (1962)	polystyrene latex	Norde and Lyklema (1975)
		Hartley et al. (1962)		
		Finlayson and Wosesson (1963)		

— Successful separations were generally achieved by means of *elution gradients.*
— In many cases part of the polymer was irreversibly retained. Therefore several authors (KERN et al., 1955; POLÁČEK and MATYSKA, 1962) used adsorbents which could be dissolved after the separation (urea or sodium chloride).
— The results of the adsorption chromatography of polyethers are particularly instructive. As an isocratic elution can be successful for particles with a rigid structure rather than for flexible coils, the observations by MARON and FILISKO (1972), from which a helical structure for polyethylene oxide in certain solvents (e.g., in an aqueous solution) is inferred, are of special interest.
— Polystyrene was most frequently used in adsorption studies. This corresponds to the general preference given to this polymer in physico-chemically oriented investigations.
— For investigations with polydienes, carbon black was preferentially used as an adsorbent. This takes into account its special importance as a filler in rubber mixtures.

As a rule, static adsorption measurements are carried out in concentration ranges which are higher than the chromatographic ones. Moreover the chromatography of polymers is only possible in a very narrow energy band close to the critical value of the adsorption energy (cf., Sections 16.6.1. and 16.9.), whereas static measurements can also be carried out at higher values. For these reasons it is only with reserve that conclusions about the chromatographic behaviour can be drawn from the results of static adsorption experiments.

In the past few years, oligomer separations have been carried out by means of adsorption elution chromatography, where the resolution is so good that the individual polymer homologues are clearly separated from each other (see Fig. 18-4) and sometimes even structural isomers can be observed (see Fig. 18-5). Separations of this kind are indicated by an asterisk in Table 18-1. They correspond to the HPLC separations of small-

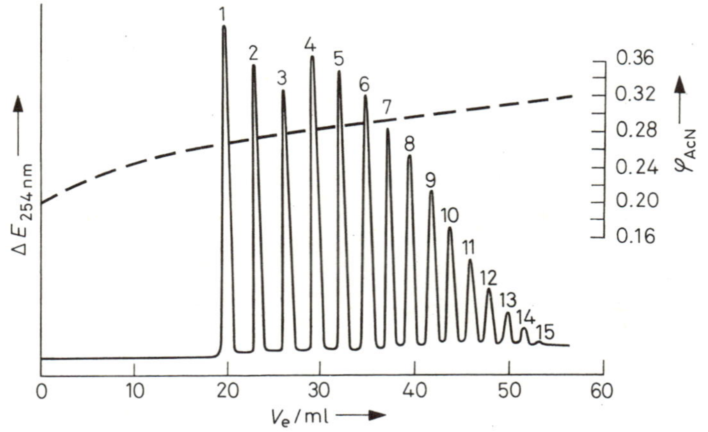

Fig. 18-4
Adsorption chromatography of phenyl ethylene oxide oligomers by gradient elution on C 8 silica gel (d_P = 7 μm)

Column: L = 0.25 m; d_c = 4.6 mm; ϑ = 25 °C
Eluent: water-acetonitrile (20 → 32 vol.-%), with a gradient as indicated; v = 2 ml/min
The numbers 1 ... 15 indicate the ethylene oxide residues in the individual components. Each chain has a phenyl end group.
(according to MELANDER, NAHUM and HORVÁTH, 1979).

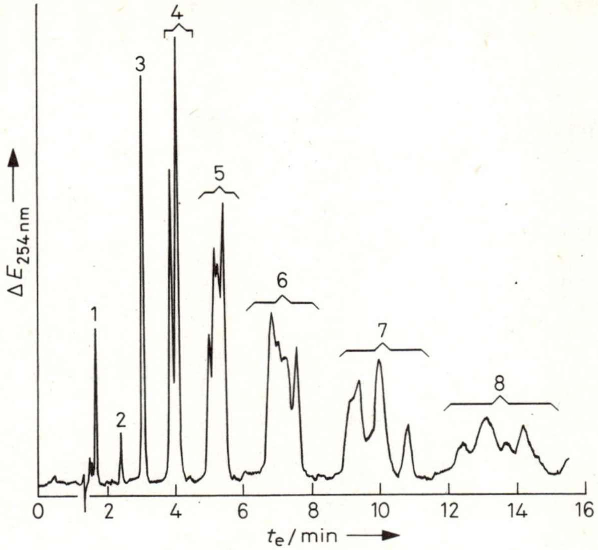

Fig. 18-5
Separation of the oligomer bands of polystyrene (M = 600 g · mole^{-1}, nominal) into four homologous series

Column: L = 0.25 m; d_C = 4 mm; with LiChrosorb® RP 18; d_P = 5 μm
Eluent: acetonitrile, v = 1.7 ml/min
The numbers indicate the monomeric units in the material of the individual substance bands.
(according to EISENBEISS, DUMONT and HENKE, 1978).

molecule compounds, frequently being obtained by a similar method, i.e., using C 18 or C 8 bonded layer silica and an elution gradient. In this way separations of low-molecular-weight PS (M = 600 g · mole^{-1}) as well as of epoxy resin intermediate products were realized as early as in 1974 [F 38]. There are remarkable separations into individual components under isocratic conditions, examples of which are shown by pictures (b) and (c) in Fig. 18-6. The samples investigated are PS standards which, among other things, are used for the calibration of SEC. The fact that these standards consist of a great number of polymer homologues can also be observed from the results obtained by exclusion chromatography using very long columns (see Fig. 8-8) and by chromatography in supercritical media (see Figs. 9-18 to 9-20). Compared to these methods, the isocratic adsorption chromatography used in this case can be considered to be relatively straightforward and quick. However, it must not be forgotten that, in isocratic adsorption chromatography, the success depends on a very accurate optimization of the eluent and of the stationary phase. The isocratic chromatograms shown in Fig. 18-6 (b) and (c) were both obtained with a high pentane content of the eluent. Pentane is a precipitant for macromolecular polystyrene. The picture (a) in Fig. 18-6 was obtained by gradient elution starting from pure methanol. Following a linear programme, tetrahydrofuran was added until, after an hour, pure THF emerged into the column. (Similar conditions have been applied by PARRIS (1978) with analogous results.)

Fig. 18-7 shows mass spectra obtained from components (C), (H) and (O) indicated in Fig. 18-6a. For (C) the mass value of a polystyrene trimer is obtained (with the terminal

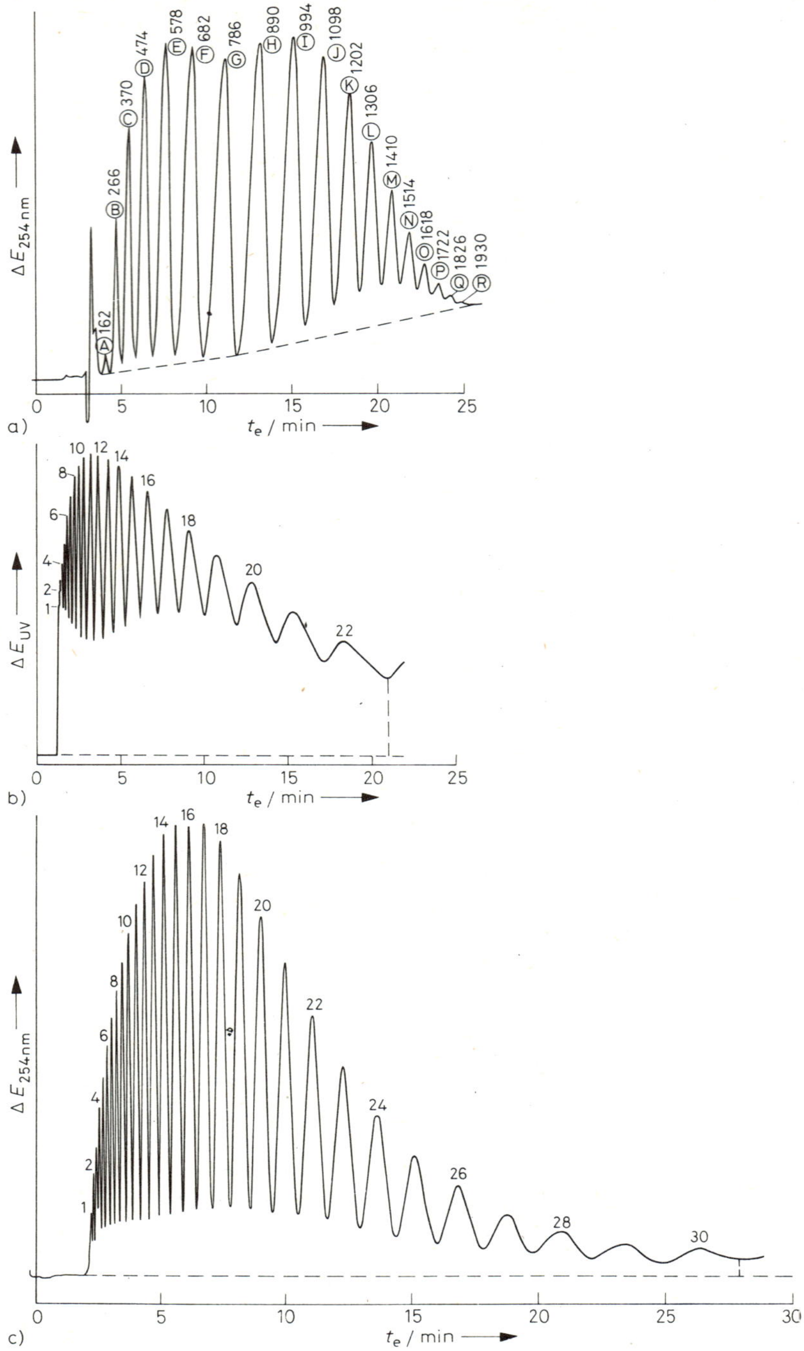

ΔE 254 nm
A 162
B 266
C 370
D 474
E 578
F 682
G 786
H 890
I 994
J 1098
K 1202
L 1306
M 1410
N 1514
O 1618
P 1722
Q 1826
R 1930
0 5 10 15 20 25
a)
t_e / min
ΔE UV
1 2 4 6 8 10 12 14 16 18 20 22
0 5 10 15 20 25
b)
t_e / min
ΔE 254nm
1 2 4 6 8 10 12 14 16 18 20 22 24 26 28 30
0 5 10 15 20 25 30
c)
t_e / min

group $-(CH_2)_3CH_3$ which stems from the butyllithium used as an initiator): 370 = 3 · 104 + 57 + 1. (H) is the eighth peak, its mass number is 890 = 8 · 104 + 57 + 1. According to the information derived from the mass spectrum, the fifteenth peak (O) contains a rather high amount of material with 1722 = 1618 + 104 in addition to the component with 1618 = 15 · 104 + 57 + 1. Such admixtures from the neighbouring bands are plausible, because in this range there are only slight indications of a chromatographic separation. However, peaks (C) and (H), which are almost baseline-separated from each other, only contain such admixtures to a minor degree. The authors attribute the trace components in the 400—600 mass range, which occur in the mass spectrum of peak (C), to the C 18 layer of the silica gel. ("This background is typical of many field desorption spectra obtained from reverse-phase LC separations, and is probably due to decomposition products from the column packing.")

With increasing molecular size, the difficulties of isocratic AC according to molar mass become more and more serious, because the degree of polymerization is included as a factor in the expression for the variation of the chemical potential (cf., Section 16.9.). In the actual macromolecular range, the conditions for a reasonable rate of migration and those for a resolution of individual components are in fact mutually exclusive. Here one has to use gradient elution. With an addition (5→50 vol.-%) of tetrahydrofuran to decahydronaphthalene, KLEIN and LEIDIGKEIT (1979) were able to separate a mixture of eight PS standards in the molar mass range from $2 \cdot 10^3$ to $2.3 \cdot 10^6$ on silica gel (see Fig. 18-8).

As SEC does not, or at least not directly, separate according to the chemical composition, adsorption methods are of special interest, e.g., for investigations of copolymers. TERAMACHI et al. (1979) separated copolymers of styrene and methyl acrylate by column chromatography on silica gel with a linear gradient of methyl acetate (7→35 vol.-%) in carbon tetrachloride. For that purpose, a silica gel was chosen with a mean pore size of d_0 = 5 nm. Therefore superposition of the AC by a size exclusion mechanism need not be

◄ Fig. 18-6
Separation of polystyrene oligomers by adsorption chromatography

a) Gradient elution of a commercial SEC standard
($\overline{M}_n$ = 750 g · mole^{-1}, nominal); V_0 = 25 µl; c_0 = 29 g · l^{-1}
Column: L = 0.30 m; d_C = 3.9 mm; packed with µ Bondapak® C (C 18 on Si)
Eluent: methanol/tetrahydrofuran (0 → 100% within 60 min, linear); v = 1 ml/min
The peaks are indicated by their mass numbers. A summation from the indicated baseline gives $\overline{M}_n$ = 673; $\overline{M}_w$ = 855
(according to LATTIMER, HARMON and WELCH, 1979).

b) Isocratic elution of a commercial SEC standard
(M = 2900 g · mole^{-1}, nominal)
Column: L = 0.101 m; d_C = 7 mm; packed with Hypersil® silica gel; d_p = 6 µm; d_0 = 12 nm
Eluent: pentane-dichloromethane (80:20)
The numbers indicate the degree of polymerization. Summation over the peak areas up to the limit plotted gives $\overline{M}_n$ = 1626; $\overline{M}_w$ = 1822
(according to KNOX and MCLENNAN, 1979).

c) Isocratic elution of a commercial SEC standard
(M = 2100 g · mole^{-1}, nominal); V_0 = 10 µl; c_0 = 30 g · l^{-1}
Column: L = 0.25 m; d_C = 4 mm; packed with LiChrosorb® Si 60 silica gel, d_p = 5 µm
Eluent: n-pentane-tetrahydrofuran (87:13); v = 1 ml/min
The numbers indicate the degree of polymerization. Summation over the peak areas up to the limit plotted gives $\overline{M}_n$ = 1723; $\overline{M}_w$ = 2041
(according to EISENBEISS, DUMONT and HENKE, 1978).

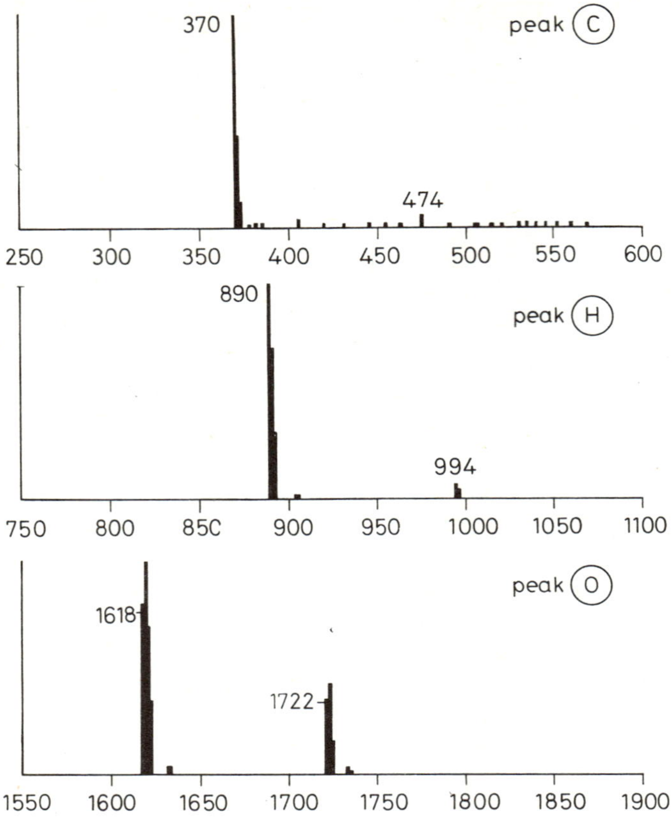

Fig. 18-7
Mass spectra of the polystyrene peaks (C), (H), and (O) of the chromatogram shown in Fig. 18-6a

The spectra were taken by a Varian 311A mass spectrometer; the temperatur of the ion source was circ. 425 K. The emitter wire was additionally heated by 5 mA (peak C), 10 mA (peak H) and 22 mA (peak O) (according to LATTIMER, HARMON and WELCH, 1979).

taken into account, as the coil size of the polymer molecular was about 20 nm. By means of internal standards, the authors were able to convert the elution curve into a curve of the chemical composition distribution. The preparatory work for this gradient technique included TLC and isocratic column elutions. Fig. 18-9 shows results obtained in this way, the arrow indicating the range in which the gradient elution was finally carried out. The authors point out that the activity of the adsorbent was not constant. In repeated experiments, all three copolymers were mostly eluted with a lower content of methyl acetate than in the preceding run. The cause of this phenomenon is certainly also responsible for the deviation of the two points 1* and 2* in Fig. 18-9.

On the other hand, DANIELEWICZ and KUBIN (1981) emphasized the good reproducibility of their experiments, which indicated constancy of the activity of the adsorbent under the conditions chosen. They also employed silica, using three types (A, B, C) whose exclusion limits were either below ($M_{lim} = 3 \cdot 10^4$ or $1 \cdot 10^5$) or above (10^7 g · mole^{-1}) the size of the

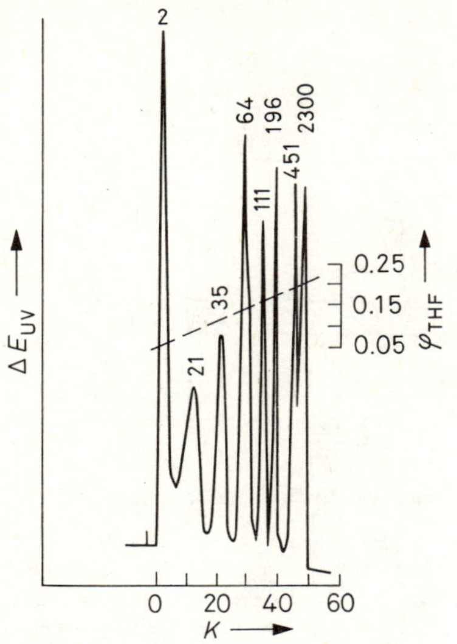

Fig. 18-8
Gradient elution of macromolecular polystyrene standards

Column: $L = 0.25$ m; $d_c = 2.6$ mm; packed with LiChrospher® Si 4000 silica gel
Eluent: decalin-tetrahydrofuran (5 → 50 vol.-%); $v = 0.5$ ml/min
Duration of the analysis: ca. 35 min
The numbers indicate the molar masses of the mixture components in 10^3 g · mole^{-1}. The distribution coefficient, $K = (V_e - V')/V''$, is marked on the abscissa (V'': total pore volume) see eqn. (8-31). (according to KLEIN and LEIDIGKEIT, 1979).

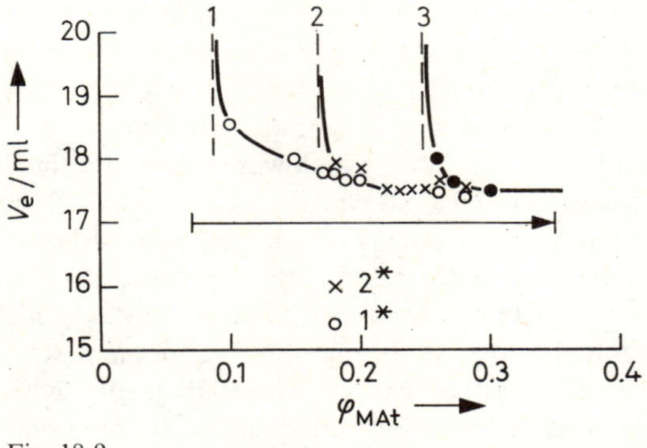

Fig. 18-9
Isocratic adsorption chromatography of styrene/methyl acrylate copolymers on microporous silica gel

Dependence of the peak elution volume in a 0.60 m column ($d_c = 7.5$ mm) on the amount of methyl acetate added to CCl_4 (7—35 vol.-% MAt).
Samples: 1 46.6 mole-% methyl acrylate ($\overline{M}_n = 261000$); 2 57.3% (276000); 3 77.9% (302000).
The data points marked by 1* and 2* were obtained with fractions of 1 ($140 \leq M \cdot 10^{-3} \leq 438$) and of 2 ($162 \leq M \cdot 10^{-3} \leq 636$), respectively, on the same column at a later time. (M had no effect on V_e.) (according to TERAMACHI, HASEGAWA, SHIMA, AKATSUKA and NAKAJIMA, 1979).

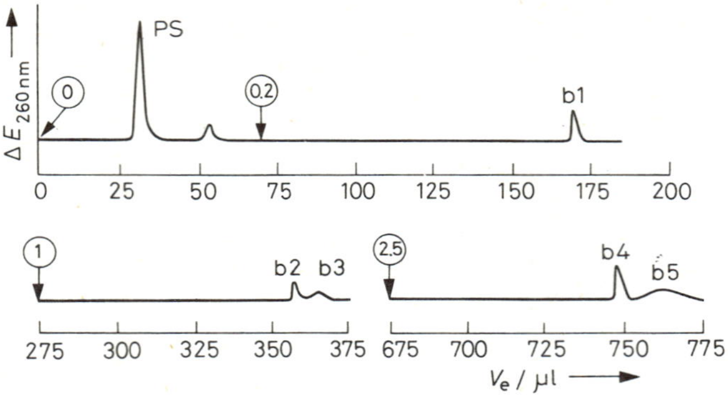

Fig. 18-10
Separation of block copolymers of styrene and methyl methacrylate by means of an eluent programme

Column: L = 0.22 m; d_C = 0.6 mm; packed with KSK silica gel, d_P = 9 ± 1 μm
Eluent: dichloromethane-methanol (0 → 2.5 %).
The methanol content was adjusted step by step to the values indicated at the times marked by arrows. v = 2.8 μl/min. PS polystyrene (precursor); b1—b5 block copolymers with an increasing content of MMA and an increasing molar mass.
(according to BELENKIJ, 1979).

poly(styrene-co-methacrylate) samples investigated. The separation according to the chemical composition was carried out by DCE/THF gradient elution (3→20 vol.-% THF). Before each run the used columns were flushed by at least ten column volumes of pure THF, so that there was no fear of any influence exerted by non-desorbed residues from the preceding run.

Block copolymers of styrene and methyl methacrylate were also investigated by adsorption chromatography. BELENKIJ (1979) reported a column elution method with silica gel as an adsorbent and dichloromethane as an eluent, to which methanol was added as a displacer for the copolymer components with a high content of methyl methacrylate (see Fig. 18-10). INAGAKI and DONKAI (1979) carried out a development chromatography in a glass cylinder with a diameter of 50 mm and a length of 0.15 m, using an isocratic mixture of 72.5 % ethyl acetate and 27.5 % benzene, which was sucked into the dry bed from below (see also TANAKA et al., 1980b). A silica gel packing was used, which could be split up into layers of a height of about 1 cm by means of inserted paper filters. The sample was applied in a block of filter-paper at the bottom end of the separating path. The lower the content of methyl methacrylate in a certain component of the sample, the higher was the level of the section to which it had travelled during the development (cf., Fig. 19-41, Section 19.7.3.5.). The method resembles the "cascade chromatography with glass frits" used by BÜRGER (1967b) in the investigation of polyethylene oxide compounds.

The complications in the adsorption behaviour of flexible macromolecules are caused by their variable shape. Rigid molecules raise fewer difficulties. This point is of such importance that it justifies taking a look at biological macromolecules: the intramolecular forces which form the ternary structure of polypeptides can even effect an adequate stabilization. At a pH of 4.9 (the isoelectric point), human serum albumin forms *spheroids* with axes 14, 4 and 4 nm long, respectively. Here, at the isoelectric point, the adsorption on negatively charged polystyrene latex reaches a maximum of fully 2 mg · m^{-2}. This finding coincides well with the assumption of a total coverage by compact, flatly laid spheroids (LYKLEMA and NORDE, 1973). At some distance from the isoelectric point, the adsorption behaviour is quite different. Here the findings indicate unfolded molecules (NORDE et al., 1973). In the unfolded structure, the albumin molecule can approach the surface with more adsorbable groups than in its compact conformation. Consequently the exothermic adsorption enthalpy increases with increasing distance from the isoelectric point. In its neighbourhood this enthalpy is smallest. For adsorption on latex particles carrying a high electric charge, even an endothermic heat effect occurs at the isoelectric point. Then the entropy yields the driving force for the adsorption. Overall, the decrease in entropy due to the fixation of a macromolecule in the adsorption layer is outweighed by the larger entropy gain due to the simultaneous displacement of many solvent molecules, which recover their degrees of freedom of motion. The contribution of the solvent molecules to the total entropy depends on the numerical proportion, but not on the shape in which the macromolecule is adsorbed. If a *larger* decrease in entropy is associated with the fixation of a flexible coil rather than with the adsorption of a molecular ellipsoid having no internal degrees of freedom, then for equal amounts of material displaced, the gain in entropy must be highest at the isoelectric point. As it is at this point that the temperature has no influence on the amount of substance adsorbed, it can be assumed that the increase with increasing temperature, as observed in other pH ranges, is due to kinetic effects.

Under comparable conditions, the more compact molecules of beef ribonuclease show a much weaker dependence of the adsorbed quantity on the pH value (NORDE and LYKLEMA, 1975).

Rigid, conformationally stable macromolecules are more suitable for adsorption chromatography than flexible coils. Thus polypeptides can be separated by adsorption chromatography. *Hydrophobic chromatography* utilizes the fact that many proteins also have hydrophobic areas on the *surface* of the compact molecules. From solutions of a high ionic strength, these molecules adsorb on octyl or phenyl residues incorporated in coarse-meshed hydrophilic agarose gels (salting-out effect). At a lower ionic strength they are desorbed again.

While the hydrophobic retention takes place in a similarly general and unspecific way to many another kinds of adsorption separation, *affinity chromatography* allows highly specific operations. It utilizes the key-to-lock-relation existing between proteins and their conformational partners. The conformational partner, an inhibitor or the substrate, is covalently bonded to a support of large surface area. A column packed with this especially prepared support material adsorbs exclusively the desired enzyme from a mixture consisting of any components whatever. After the column has been flushed, the enzyme is liberated by changing the pH value or increasing the ionic strength (LERMAN, 1953; REINER and WALCH, 1971).

Polysaccharides and glycoproteins can be isolated according to an analogous principle using bonded wheat-germ agglutinin as a stationary phase with a specific affinity to N-acetylglucosaminyl groups.

Adsorption chromatography deserves attention mainly because it can separate according to features to which other methods do not respond. Just as small-molecule stereoisomers can be separated into the optically active components by means of ligand chromatography (ROGOŽIN and DAVANKOV, 1970; DAVANKOV, 1980), so a *stereoselective adsorption* of macromolecules has been observed as well. Using adsorption, NATTA et al. (1958, 1960) frationated stereo block polymers of propylene by their crystallizability. PINO et al. (1962,

1966) separated higher poly-α-olefins into components of different optical activities. Isotactic and syndiotactic PMMA were separated by MIYAMOTO et al. (1974) using adsorption on Aerosil. This separation utilized the fact that the isotactic PMMA is preferentially adsorbed (MIYAMOTO and CANTOW, 1972). The separation of stereoisomers does not necessarily require a solid phase boundary, as shown by the foam fractionation, by means of which IMAI and MATSUMOTO (1963) separated polyvinyl alcohol into isotactic and syndiotactic components.

Very satisfactory separations according to structural features have also been realized by the methods of thin-layer chromatography (cf., Section 21.5.2.2.).

19. Experimental parameters and results of size exclusion chromatography

19.1. Influence of the sample size

It has already been shown in Fig. 16-3 that the peak elution volume shifts towards higher values as the sample size increases. This effect shows up more distinctly as the molar mass of the sample increases, its heterogeneity decreases and the resolution power of the apparatus employed becomes higher. All of the curves shown in Fig. 16-3 were recorded with the same detector sensitivity, so that the areas under the curves vary directly with the sample size. Figure 16-3 shows that the elution of the sample always starts at the same volume value. The increase in the concentration only affects the position of the maximum and the end of the curve. However, if the curves recorded at different concentrations are normalized to equal areas, then an increasing deformation can be observed. This has most serious consequences, because these curves do not then represent the molar mass distribution of the sample, but more or less artefacts.

The question of how to proceed in order to obtain correct results is of an immediate, practical importance. For samples with a broad distribution, for long columns with a correspondingly high dilution effect or for columns with a low separating power, the risk of overloading is fortunately not too high, but naturally this risk becomes more important as the performance of the separating apparatus increases. Columns containing 5 μm particles are most sensitive to overloading (KATO et al., 1973). Table 19-1 lists some conditions which have been used successfully in experimental work.

It has been suggested to extrapolate all of the results, including the calibration, to zero concentration. A smaller amount of work is required if the analyses themselves are carried out only at one concentration and the chromatograms obtained are evaluated point by point using the calibration curve which corresponds to the concentration at this point. (The respective value can be observed from the ordinate value; MORI, 1976; NAKANO and GOTO, 1975). The mean values $\bar{M}_n$, $\bar{M}_w$ and $\bar{M}_z$ determined from the elution curves also depend on the concentration which was used in the experimental determination of the curves. The higher the molar mass of the sample, the stronger is this dependence. By means of a nomogram of $\bar{M}_{(t)}/\bar{M}_{(u)}$ vs. $[\eta]\,c$, which indicates a linear relationship, it is easy to eliminate the influence of concentration on the mean values (PODOSENOVA and ROŽKOVA, 1980).

Usually the measurements are simply carried out at a sufficiently low concentration, at which the calibration is also carried out. But where are the limits? As a rule of thumb, the viscosity of the solution should not exceed twice the viscosity of the solvent. For concentrations of less than 1 g · l^{-1} this condition is fulfilled in most cases. KNOX and MCLENNAN (1979) stated that not more than 1 μg of sample substance should be injected per cm^3 of the column packing. In systematic investigations of the elution behaviour of standard polystyrenes on polystyrene gel columns, MOORE (1970) found that the product of the *sample volume*, V_0 (in ml), the *sample concentration*, c_0 (in g · l^{-1}), and the intrinsic viscosity, $[\eta]$ (in l · g^{-1}), must be smaller than 0.05—0.1 ml. When this condition was satisfied, the elution volume as well as the peak width corresponded to the values obtained at the lowest detectable concentrations. The existence of such a limit is of a great practical importance, as it is a prerequisite to the application of SEC as a high-performance routine technique. The fact that, beyond a certain limit, further dilution no longer affects the chromatographic results has also been found by BERGER and SCHULZ (1970). In θ *solutions* it was even found that an influence of the concentration only exists

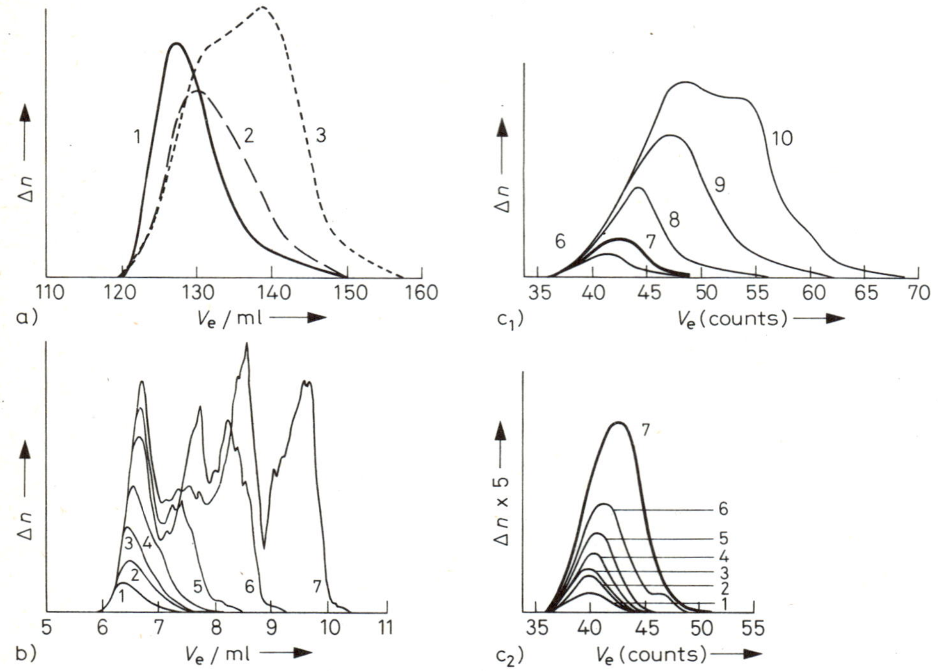

Fig. 19-1
Overloading effects in SEC

a) PS standard (867000 g · $mole^{-1}$) in trichlorobenzene at 130 °C on a Styragel® column, $L = 5 \times 1.22$ m, $d_C = 7.8$ mm
1 1; *2* 5; *3* 10 g · l^{-1}; $V_0 = 1.67$ ml
(according to SAMAY and KUBIN, 1979)

b) PS standard (2610000 g · $mole^{-1}$) in tetrahydrofuran (25 °C; $v = 0.334$ ml · min^{-1}) on silica gel ($d_p = 63$—71 μm; $M_{lim} = 200000$ g · $mole^{-1}$), $L = 0.30$ m, $d_c = 8$ mm.
1 0.111; *2* 0.222; *3* 0.444; *4* 0.888; *5* 1.775; *6* 3.55; *7* 7.1 g · l^{-1}; $V_0 = 0.08$ ml
(according to JANČA and POKORNÝ, 1978 a)

c) Polymethacrylic acid (16000 g · $mole^{-1}$) in dimethylformamide on Styragel®, $L = 5 \times 1.22$ m; $d_C = 7.8$ mm.
Concentration of the sample graduated from 0.15 g · l^{-1} (curve 1) to 20 g · l^{-1} (curve 10). Curves 6 and 7 in (c 1) and (c 2) are identical, reflecting the change of scale.
(according to NEFEDOV, LAZAREVA, BELENKIJ, FRENKEL and KOTON, 1979).

above 6 g · l^{-1} (KATO and HASHIMOTO, 1974a). An addition of precipitants likewise suppresses the dependence on the concentration (KATO and HASHIMOTO, 1974b; BEREK et al., 1975).

The lower limit of the utilizable range of concentrations is determined by the detector sensitivity and the stability of the baseline. In 1968 MOORE performed measurements using less than 0.1 g · l^{-1} for the solution of difficult problems in an extremely macromolecular range.

For some published examples, Fig. 19-1 shows to what degree an *overloading* deforms the peak shape. Although the curves, like those in the tolerance range, always start at the same volume value, they are not reproducible in their exact shape, not even on the same apparatus. For the comparison of different elugrams, the *peak elution volume* employed otherwise cannot be used. (In an extreme case, even several peaks may occur, all of which, however, represent artefacts; see Fig. 19-1 b.) On the other hand, the *average elution volume*, V_{av}, has proven suitable (JANČA and POKORNÝ, 1978a). It is calculated from the height, h_i, of the chromatogram at the elution volume, V_i, according to the relationship:

$$V_{av} = \Sigma\, V_i \cdot h_i / \Sigma\, h_i \qquad (19\text{-}1)$$

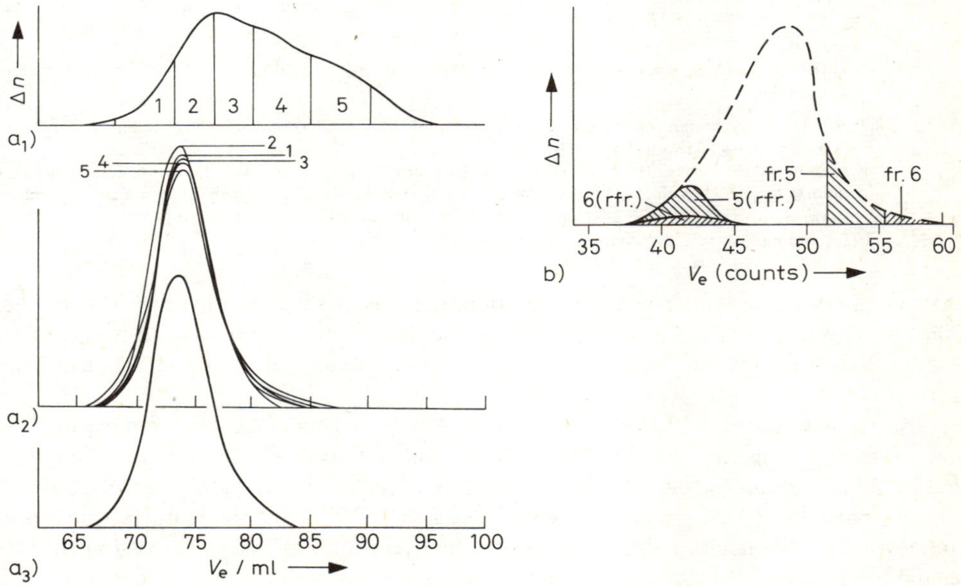

Fig. 19-2
Loss of separation efficiency by overloading

a) Elution of polystyrene with overloading
a 1) Elution curve of a PS standard ($\overline{M}_w$ = 867000 g · mole^{-1}) on a column (L = 3 × 1.22 m) packed with Styragel® with a nominal pore size of 10^5, 10^4 and 3 · 10^3 Å, respectively; eluent: THF, v = 1 ml × min^{-1}, c_0 = 10 g · l^{-1}, V_0 = 1.67 ml
a 2) Refractionation of the eluate slices indicated in (a 1) with c_0 = 1 g · l^{-1}
a 3) Elution curve of the starting sample with c_0 = 1 g · l^{-1}
(according to SAMAY and KUBIN, 1979)
b) Elution of polyamido acid with overloading
The envelope curve was obtained with c_0 = 7 g · l^{-1}. The eluate slices 5 and 6 in the tail are not true fractions; in the refractionation at a reduced concentration, they were eluted with 40—45 counts.
(according to NEFEDOV, LAZAREVA, BELENKIJ, FRENKEL and KOTON, 1979).

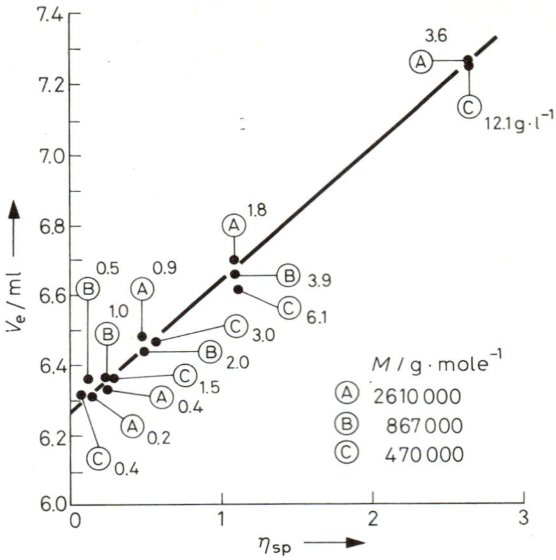

Fig. 19-3
Influence of the specific viscosity on the value of the elution volume of polystyrene standards

All three samples are above the exclusion limit of $M_{lim} = 50000$ g · mole^{-1}. Column: $L = 0.30$ m; $d_C = 8$ mm; packed with Porasil® B; $d_p = 63$—71 µm
In spite of the very different concentrations (rounded values are indicated in g · l^{-1} beside the data points), the values of all the samples lie on a common straight line if the peak shift is considered a function of the viscosity. Here as well as in Fig. 19-4, V_e is the mean value of the sum of the peak areas. V_{av}, eqn. (19-1). (according to JANČA and POKORNÝ, 1978 a).

ADAMS et al. (1966), when performing refractionations, noted that overloading impairs the resolution. This may reach an extent where all of the eluate components have an equal molar mass, so that there is no separation at all (MINDNER and BERGER, 1979), see also Fig. 19-2.

Several factors contribute to the whole concentration dependence. The extreme phenomena of overloading are mainly based on viscosity effects. This opinion, which was stated by ALTGELT and MOORE (1967), has recently been supported by JANČA (1977, 1979, 1980 a, b), partially together with POKORNÝ (1978 a, b, 1979). Thus the features which are characteristic of overloading also occur above the exclusion limit (see Fig. 19-1 b). Moreover the broadening of the elution curve towards higher values can be correlated with the specific viscosity, even if the latter has to be adjusted with quite different concentrations for samples of different molar masses (see Fig. 19-3). The dominating rôle of the specific viscosity also manifested itself in the differential elution of a polystyrene standard ($M = 670 \cdot 10^3$ g · mole^{-1}) in mobile phases, which were likewise tetrahydrofuran solutions of this standard (see Fig. 19-4). The left-hand part of this figure (with the negative values for η_{sp}) shows that this influence even extended into the region where the polymer concentration in the injected volume was lower than that in the eluent (cf., Section 19.9.1., Vacancy chromatography). Also in the normal working range, i.e., below the exclusion limit, up to 80% of the concentration effects in the injection of solutions with $\eta_{sp} > 1$ are due to the influence of viscosity (JANČA and POKORNÝ, 1978 b).

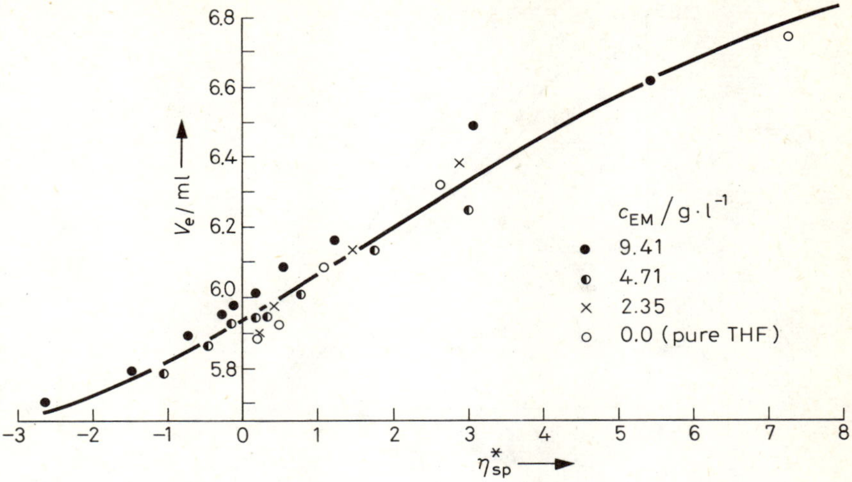

Fig. 19-4
Differential elution
Influence of the viscosity difference on the retention volume of a polystyrene standard (M = 670000 g · mole^{-1}) in THF solutions of different concentrations
Column: L = 0.30 m; d_C = 8 mm; packed with Porasil® B; d_p = 63—71 µm; exclusion limit appr. 10^5 g × mole^{-1}; eluent: THF or solutions of the sample with the indicated concentration, $\eta^*_{sp} = (\eta_{sample}/\eta_{EM}) - 1$. In the left-hand part ($\eta^*_{sp} < 0$), the concentrations are $c_{sample} < c_{EM}$. Here the elution was carried out with v = 0.038 ml · min^{-1} instead of v = 0.334 ml · min^{-1} otherwise (c_{EM}: PS concentration in the eluent). (according to Janča and Pokorný, 1979).

The slope shown in Fig. 19-3 divided by the elution volume, $V_{e,0}$, at $\eta_{sp} = 0$, i.e., the term $(dV_e/d\eta_{sp})/V_{e,0}$, enables overloading effects of different columns to be compared with each other. Janča et al. (1981) found that this term was almost constant for two columns with approximately equal plate height but very different L/d_C ratios, whereas for a third column with a higher efficiency they found twice the value.

Owing to the influence of the molar mass, the normal peak broadening (see Fig. 16-3) before the onset of actual overloading leads to a fan-like pattern of calibration curves recorded at different concentrations (see Fig. 19-5). This typical pattern also occurs with non-linear calibration relationships. For samples with molar masses of less than 10^4 g × mole^{-1}, the influence of the concentration can usually be neglected (Boni et al., 1968; Mori, 1977), so that for such samples one obtains lines parallel to the abscissa in the representation of the elution volume as a function of the starting concentration (see Fig. 19-6). The plots obtained for polymers with higher molar masses are straight lines, the slope of which increases with increasing M. In the separating range of the gel, this relationship between the peak elution volume and the molar mass can be related to the shrinking of the coil molecules with increasing concentration (Rudin, 1971; Vilenčik and Belenkij, 1971), but it must not be overlooked that, owing to the chromatographic dilution, the concentration of the sample in the column is lower than the starting concentration, c_0 (Janča, 1980b). In 1979, Mahabadi and Rudin showed that the chromatographic effects are in accordance with a theory developed by Yamakawa (1965), which contains general statements on the behaviour of macromolecules in solution. An essential argument in favour of this approach lies in the mentioned independence from

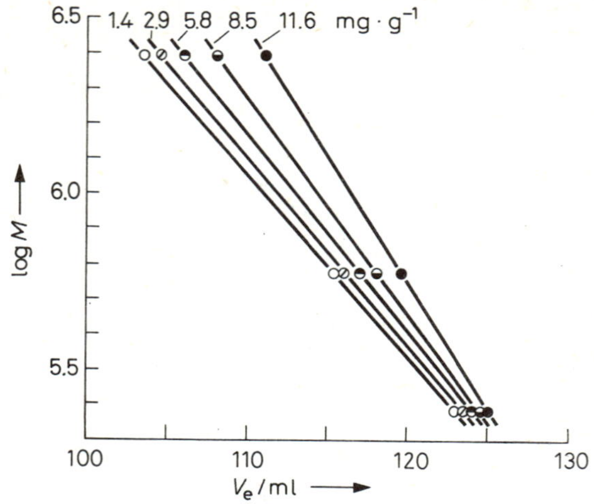

Fig. 19-5
Concentration dependence of the calibration relationship in the SEC of PMMA in THF on polystyrene gel

Rate of flow: 1 ml · min^{-1}; concentration stated in mg of substance per g of solvent (according to BERGER and SCHULZ, 1971).

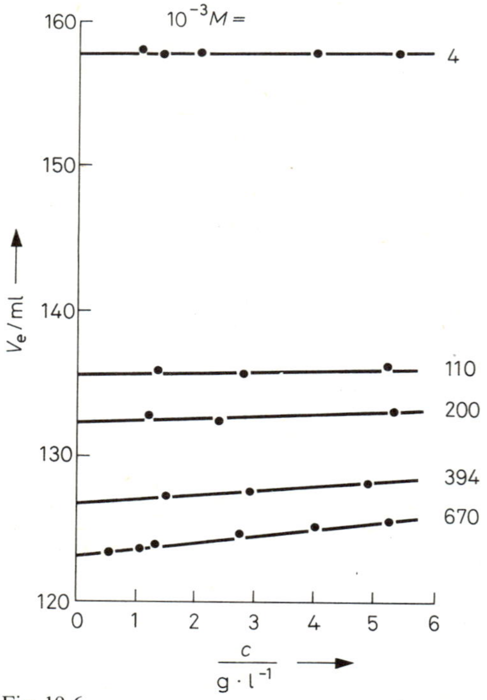

Fig. 19-6
Increase of the retention volume with increasing concentration

Polystyrene in 1,2,4-trichlorobenzene at 130 °C; v = 1 ml · min^{-1} (according to NAKANO and GOTO, 1975).

concentration in θ solutions (p. 378/79) and in the fact that in the latter the elution volumes are greater than in good solvents. The rising, straight lines of V_e vs. c obtained in good solvents, if extrapolated towards higher concentrations, intersects the corresponding extension of the horizontal line for the θ system at an abscissa value, c_x, which, in the light of the above considerations, is interpreted as that concentration at which the coils have shrunk to their θ dimensions. MAHABADI and RUDIN (1979) have shown that for a polystyrene with $M = 3.35 \cdot 10^6$ g $\cdot$ mole^{-1} in methyl ethyl ketone this value c_x is in fairly good agreement with that c_x which results from the formula established by YAMAKAWA (1965), but can also be calculated on the basis of the model developed by RUDIN and WAGNER (1976). BLEHA et al. (1980) likewise compared the increase in the elution volume with the results of classical theories about polymer solutions, taking into consideration YAMAKAWAS approach and the work published by EJZNER (1961). Both theories lead to the conclusion that the peak elution volume should increase with increasing concentration. However, from the ratio of the measured values to the theoretical ones, which is shown in Figs. 19-7 and 19-8, it appears that the change of the coil volume is not sufficient to explain the total concentration dependence of V_e.

During the passage over the separating bed the sample is diluted, so that an expansion of the coils can occur, which increases with the distance travelled. This should lead to an acceleration, which should improve the separation at the leading edge of the sample band but impair the separation in the intermediate elution range and at the rear edge of the band as a result of the catching-up effect of the components with the smaller molar-mass values. It appears that the importance of this effect is not very high, as generally the elution curves at higher concentrations do not start earlier than at lower ones (see

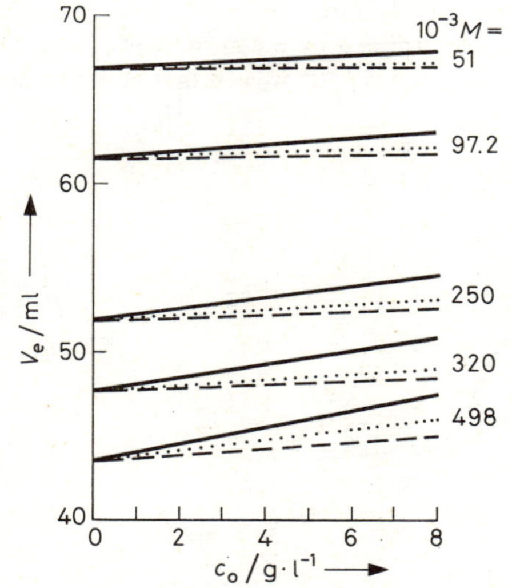

Fig. 19-7
Influence of the concentration on the peak elution volume of polystyrene standards in tetrahydrofuran

Experimental dependence (———) and curve calculated on the basis of the solution theories given by YAMAKAWA (1965, ...) and EJZNER (1961), — — —)
(according to BLEHA, MLÝNEK and BEREK, 1980).

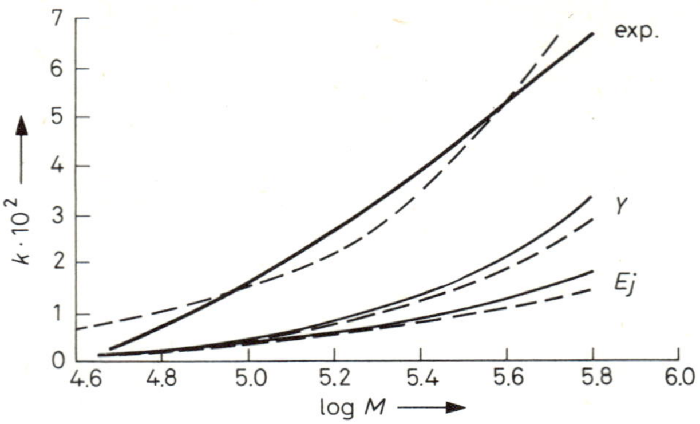

Fig. 19-8
Increase k of the peak elution volume with the concentration as a function of the molar mass, $k = \mathrm{d}V_e/\mathrm{d}c_0$

PS in tetrahydrofuran (———) and in toluene (— — —); exp.: experimental data, cf., Fig. 19-7 Y, Ej curves calculated for the above systems (according to BLEHA, MLÝNEK and BEREK, 1980).

Fig. 16-3), but the poorer separation in the medium and lower range of molar masses can clearly be observed from, say, the experimental results obtained by MINDNER and R. BERGER (1979). From their observations, the authors concluded that, as the concentration increased, the elution profile of each component of the sample was asymmetrically broadened towards higher elution volumes. Of course such a spreading of the components would have less serious consequences at the start of the elution than at its end. This asymmetric broadening of the elution profile of an individual component was observed by K. C. BERGER in 1975 (see Fig. 19-9): the elution curve determined by means of an activity detector for a ^{14}C-labelled polystyrene with a molar mass of $1.5 \cdot 10^6$ was in fact shifted towards higher

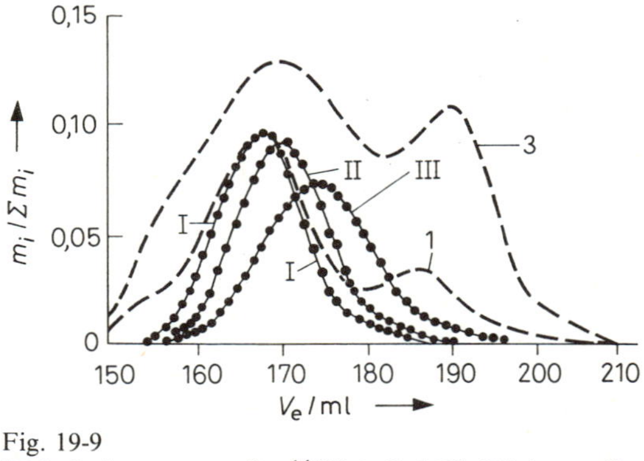

Fig. 19-9
SEC elution curves of a ^{14}C-labelled PS (A) (normalized activity curves) in mixtures with two inactive polystyrenes (B and C), eluent: THF

A $M = 1.5 \cdot 10^6$; B $M = 0.63 \cdot 10^6$; C $M = 2.67 \cdot 10^6$ g · mole^{-1}
Total sample volume: 0.3 ml each
I 0.9 mg A mixed with 0.3 mg B and 0.3 mg C (associated: curve 1 of the differential refractometer)
II 0.9 mg A mixed with 0.9 mg B and 0.9 mg C
III 0.9 mg A mixed with 1.2 mg B and 1.2 mg C (associated: curve 3 of the differential refractometer)
(according to BERGER, 1975).

elution volumes by increasing additions of other PS samples. The dependence of this effect on the molar mass can be observed from a comparison with the analogous Fig. 16-4, where investigations of the same kind with a labelled sample of only 51 300 g · mole^{-1} are shown, and do not exhibit any influence of concentration in the range investigated.

Beside the effect of viscosity and the coil shrinking, osmotic effects must also be taken into consideration (SCHWEIGER and LANGHAMMER, 1977; ANDERSON and BRANNON, 1981). A higher concentration of the sample necessarily leads to a higher concentration gradient between the sample solution and the adjacent pores, which are initially free of substance. This effects a penetration of the molecules to greater depths of the pores. After the passage of the substance band, it naturally takes a longer time until these molecules return into the mobile phase. This also contributes to the asymmetric peak broadening for an individual component.

As long as the pores can take in the increasing supply of substance, the normal concentration dependence with constant values of dV_e/dc can be expected. Overloading, i.e., non-reproducible and oddly shaped elution curves without a separation of the injected sample,

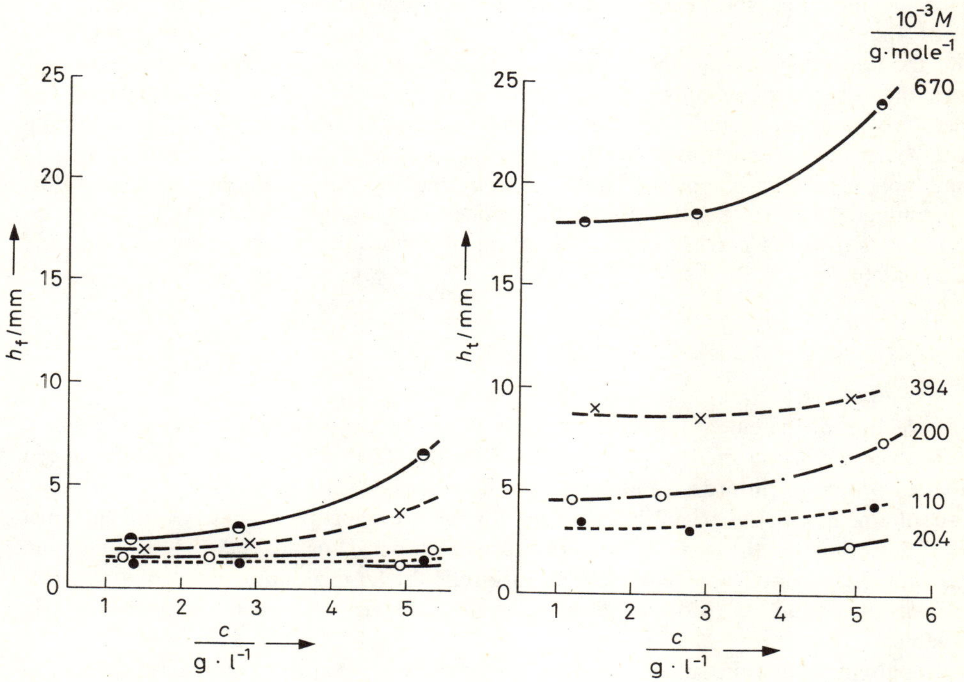

Fig. 19-10
Increase of the plate height with the concentration for polystyrene standards with molar masses ranging between 20400 and 670000 g · mole^{-1}
Rate of flow: 1 ml · min^{-1}; injected volume: V_0 = 2 ml;
eluent: 1,2,4-trichlorobenzene; 130 °C
Column set consisting of four columns (each 1.22 m long), packed with polystyrene gel, nominal pore sizes 10^7, 10^6, 10^5 and 10^3 Å.
h_f height of a theoretical plate, calculated from the front slopes of the peaks; h_t ditto, calculated from the rear slopes of the peaks
(according to NAKANO and GOTO, 1975).

will occur if the quantity of substance exceeds the capacity reserve connected with the depth of the pores. Thus columns packed with very small particles, which, at low concentrations, owing to the short diffusion distances (cf., Section 15.3. ff.) exhibit excellent separation efficiencies, are most sensitive to overloading. The fact that for samples with a high molar mass and a narrow distribution the effect occurs sooner is plausible, too, because of the smaller number of pores fitting these particles.

There is no sense in performing kinetic evaluations of the non-reproducible elugrams of overloaded columns. In the concentration range below the onset of overloading, an increase of the plate height with increasing concentration can be observed. Fig. 19-10 shows that this effect especially concerns the rear side of the peaks.

19.2. Working temperature

As the overloading effects are at least partially caused by too high a viscosity of the solutions, it is sometimes recommended to carry out investigations at an elevated temperature. As the temperature increases, the viscosity of both the polymer and the solvent decreases. The ratio of the viscosities is the main cause of the disturbances. If it decreases, then raising the temperature will indeed diminish the danger of overloading, but it may also happen that the ratio remains constant or even increases, so that the increase in the temperature does not eliminate the overloading effect at all. Then the remaining advantage lies in the reduction of the flow resistance and of the kinetic band broadening.

Some polymers, e.g., polyolefins, polyoxymethylene and polyamide, are only soluble at higher temperatures, so that necessarily an elevated temperature must be used. However, in most cases SEC is carried out at the ambient laboratory temperature without any special control.

19.3. Solvents

In the selection of the eluent for SEC, the criteria are not so well defined as in AC, where decisions can be made on the basis of the ε^0 scale. The solvent must exhibit quite a number of properties in order to allow successful exclusion chromatography.

A requirement which is trivial but decisive for the pre-selection is that the solvent must allow a detection by the type of detector provided for. Thus for UV or IR detection it must not itself exhibit any significant absorption; for an evaluation by means of a differential refractometer, the index of refraction must differ from that of the sample as widely as possible.

The requirement that the solvent should dissolve the sample appears similarly trivial. The higher the exponent a in eqn. (5-8), the more distinctly marked the dependence of the hydrodynamic volume on the molar mass (eqn. (8-17)) will be. Thus in good solvents one may expect higher values of C_2 (cf., eqn. (8-3)), and hence better separation efficiencies, but on the other hand the viscosity of the solutions is also higher, and the concentration dependence of V_e is more pronounced. The mean values $\bar{M}_w$ and $\bar{M}_n$ calculated from the chromatograms of polychloroprene in the θ solvent methyl ethyl ketone were independent of the concentration up to $6 \text{ g} \cdot \text{l}^{-1}$, whereas a drift was found to occur above $3 \text{ g} \cdot \text{l}^{-1}$ in the good solvent tetrahydrofuran (KATO and HASHIMOTO, 1974a).

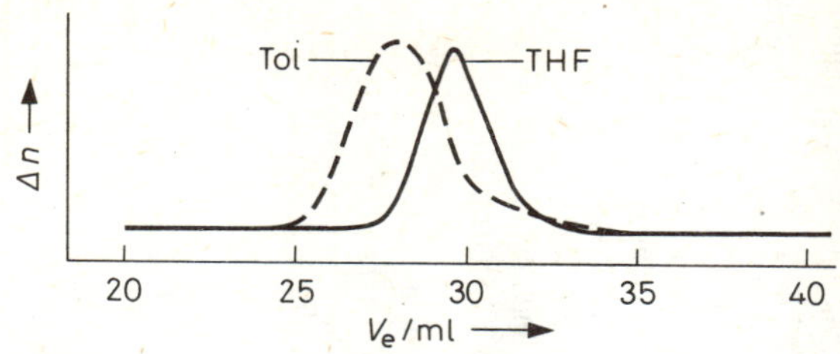

Fig. 19-11
Elution curves of a polystyrene standard ($\bar{M}_w$ = 860000 g · mole^{-1}) in the SEC on polystyrene gel of a nominal pore size of 10^5 Å in a 7.8 mm column with a length of 1.22 m

——— in tetrahydrofuran, with 0.0149 ml · min^{-1}
– – – – in toluene, with 0.023 ml · min^{-1}
(according to COOPER, JOHNSON and BRUZZONE, 1973).

The eluent should have a low viscosity in order to avoid working with too high a pressure and to keep the kinetic contributions to the peak width small, as summarized in Chapter 15. In this respect tetrahydrofuran is more favourable than toluene (see Fig. 19-11). In 1980, COOPER reported in detail the solvent influence in the SEC of polystyrene standards on controlled-porosity glasses. Here 1,2,4-trichlorobenzene was employed as a third eluent, whose viscosity of 1.42 mPa·s is even higher than that of toluene (0.59 mPa·s). With increasing viscosity, the elugrams broadened towards higher elution volumes.

In some cases the polarity, indicated by ε^0 (Table 7-3), must be taken into account. This is necessary if there is the risk that a polar sample is adsorbed on a polar separating material.

If the solvent used for the sample is identical with the eluent, then only the sample is indicated. However, if it contains additions such as traces of water or dissolved gases, then sensitive, non-specific detectors additionally record *ghost peaks* (ŠLAIS and KREJČI, 1974). The latter may also occur if the sample displaces a substance from the column or if in mixed solvents there are changes in the concentration, e.g., by evaporation, selective sorption (BEREK et al., 1976) or owing to Donnan-type equilibria (cf., Section 19.3.3.2.).

19.3.1. Exclusion chromatography with solvent mixtures

Generally SEC is carried out with pure eluents, but there are some cases where advantages may be achieved by the use of mixtures. In Section 19.1. reference was made to the possibility of decreasing the *concentration dependence* by the addition of a precipitant. In this case, however, one should be aware of the risk that such an addition can impair the precision of the measurements unless its quantity is kept constant. In this respect, hygroscopic eluents such as DMF or THF may cause considerable difficulties. SPYCHAI and BEREK (1979) investigated the influence of water (up to 8.9%) in THF on the exclusion chromatography of polystyrene in silica-based columns, and recommended using dry THF in order to avoid disturbances, and a guard column packed with activated silica gel, which is arranged before the injection point in the eluent flow.

The precipitation gel chromatography developed by PORATH (1962) in the isolation of proteins likewise uses solvent-precipitant mixtures. This interesting variant is achieved by

Table 19-1
Working conditions in size exclusion chromatography (selected examples)

SG polystyrene gel; PG porous glass; KG silica gel

No.	d_C / mm	L / m	Gel	d_P / µm	Polymer	Eluent (solvent)	V_0 / ml	c_0 / $g \cdot l^{-1}$	m_0/V_C / $mg \cdot l^{-1}$	v / $ml \cdot min^{-1}$	u[1]) / $cm \cdot s^{-1}$	h / µm	t'[2]) / min	Authors
1	7.75	3.66	SG	70	PS	Tol	1	13.4	77.5	0.44	0.039		170	MOORE (1964)
2	7.75	7.32	SG		PS	THF	1	1.6	4.6	2	0.177		80	TUNG and RUNYON (1969)
3	7.75	4.88	SG		PS	TCB	0.78	0.5 a) 1 b)	1.7 3.4	1	0.088	102	100	MAY and KNIGHT (1971) a) $M < 3 \cdot 10^5$ b) $M > 3 \cdot 10^5$
4	7.75	6.10	SG	5	PS	MEK	1	0.5	1.7	0.5	0.044	34.6	240	KATO et al. (1973)
5	7.75	6.10	SG		PS	THF	2	→0	→0	1	0.088	423	120	SLAGOWSKI et al. (1974)
6	7.75	8.53	SG		PS	THF	2	2	10.0	0.25	0.022	320	650	FETTERS (1976)
7	15	1.10	PG	75	dextr.	W	0.5	4...80	10.3...206	0.13	0.003	480	600	BASEDOW et al. (1976)
8	63.5	2.44	SG		PS	Tol	100	10	129.4	85	0.112	490	40	BOMBAUGH et al. (1968)
9	60	1.22	KG	100	PE	TCB	100	10	289.9	50	0.074	400	30	PEYROUSET et al. (1975)
10	21.1	4.88	SG	10	PS	MEK + 11.3% M (θ-LM)	20	6.5	76.2	8.5	0.101	43.5	90	Y. KATO et al. (1975)
11	7.75	2.44	SG	37	PS	Tol	1	0.5	4.3	12.5	1.10		5	LITTLE et al. (1969)
12	7.75	0.61	SG	5	PS	THF	0.05	1	1.7	2.5	0.221	26.6	5	KATO et al. (1974)

[1]) linear velocity (for an open cross section of 40% in the packing)
[2]) delay time, waiting time until the appearance of the chromatogram

a solvent gradient which is built up in the column before the start of the separation. The orientation of the gradient is antiparallel, i.e., the content of precipitant increases towards the column outlet. The sample is injected as soon as the eluent composition at the column inlet is such that precipitation will no longer take place there. Pure solvent is used as an eluent. As the macromolecular components on the gel column travel faster than the solubility boundary, which depends on the composition of the eluent mixture, they permanently run against this "wall" on their way through the column. The bands develop wherever the solubility threshold of the respective component is located. The continuing pressure acting upon this barrier compresses the bands.

The addition of a polar liquid such as methanol to a solvent with a low eluotropic strength, ε^0, for the suppression of adsorption phenomena (Ždanov et al., 1973) is of great importance if the sample is insoluble in a pure medium of a high eluotropic strength. Likewise, in the SEC of polyelectrolytes and highly polar polymers, electrolytes are added to the eluent in order to meet practical requirements. This set of problems will be discussed in the following section.

19.3.2. Addition of salts to organic eluents

In the GPC of PAN containing sulphonate groups, Cha (1969) observed multimodal peaks, finding that an addition of LiBr to the eluent DMF led to the simple curves which corresponded to expectation (see Fig. 19-12). In this case the main peak was shifted towards higher elution volumes. This effect, which was confirmed repeatedly, is even caused by very small additions; a further increase of the salt has only a minor influence on the course and the shape of the elution curves (see for instance Fig. 19-16(b)).

If the salt addition also reduces the viscosity of the polymer solution, then one may attempt to explain the phenomenon by the decrease in the hydrodynamic volume, which, in ideal exclusion chromatography, should lead to higher values of the elution volume. The intrinsic viscosity was found to be drastically reduced for PAN with sulphonate groups in DMF by 0.2 M $NaNO_3$ (—42%, Domard et al., 1979), for polyester urethanes in DMF by 0.05 M LiBr (—42%, Hann, 1977) and for polyacrylamide in formamide by 0.005 M

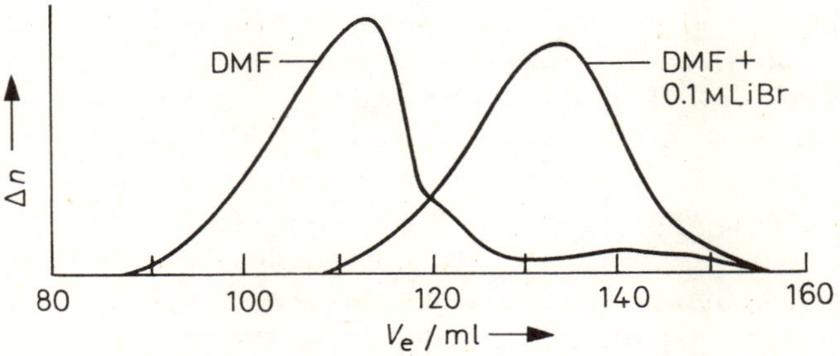

Fig. 19-12
Elution curves of a sample of polyacrylonitrile with four sulphonate groups per chain (M_n = 60 200 g · mole^{-1})
Column set: four columns packed with polystyrene gel (7 · 10^6; 3 · 10^6; 10^5; 10^3 Å)
Eluent: DMF or DMF with 0.1 M LiBr, at 72 °C, v = 1 ml · min^{-1}, V_0 = 2 ml
(according to Cha, 1969).

KCl (max. —22%, ONDA et al., 1979). In the last example the decrease in the hydrodynamic volume was obviously the only cause of the increase in the peak elution volumes. In this case ONDA et al. were able to show that the values for the solutions in pure formamide and with a salt addition lay on a common curve when log ([η] M) was plotted vs. V_e. On the other hand, for PS in DMF a salt addition shifts the universal calibration curve towards higher elution volumes. This was pointed out in Section 16.6.2. in the discussion about solvophobic interactions (see also Figs. 16-28 and 16-29). Consequently the coil shrinking is not in each case an adequate explanation for the change in the elution characteristics upon salt addition.

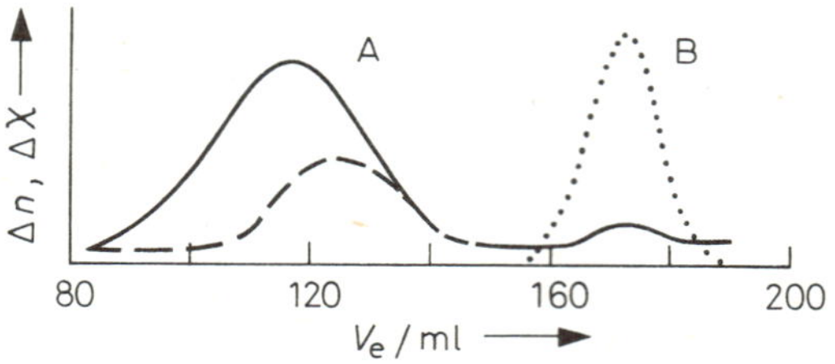

Fig. 19-13
Elution curve of a polyacrylonitrile sample with 1.2 · 10^{-3} mole/g sodium sulphonate groups

Simultaneous record of the refractive index Δn (———) and the conductivity $\Delta\chi$ (— — —: $\Delta\chi \times 10$; ....: $\Delta\chi \times 1$)
Column: L = 1.47 m; d_c = 15 mm; packed with a mixture of Spherosil® 75 Å, 150 Å, 500 Å, 1250 Å, 3000 Å
Eluent: DMF with 0.005 M $NaNO_3$
Rate of elution: v = 0.2 ml · min^{-1}
(according to DOMARD, RINAUDO and ROCHAŠ, 1979).

An essential contribution to this topic was made by the investigations performed by DOMARD et al. (1979) on PAN containing sulphonate groups in DMF, where the refractive index, the viscosity and the conductivity of the eluate were measured. Fig. 19-13 shows such a chromatogram. From the curves shown for the refractive index and the conductivity, the following conclusions can be drawn:

— Beyond the polymer peak (A) there was a second one (B) at higher elution volumes, which indicated a salt effect. This peak did not appear when PS samples were investigated in DMF or in DMF-$NaNO_3$ mixtures. The peak elution volume of B increased with increasing salt content in the eluent, approaching the total accessible volume, $V' + V''$. Up to a concentration of 0.05 M $NaNO_3$ in the eluent, the additional peak B behaved just like a peak which would appear upon a direct salt injection (see Fig. 16-35).

— The additional peak was caused by the salt injected together with the sample, although the concentration of this salt was equal to the concentration in the eluent (the sample was dissolved in a portion of the eluent). Consequently the electrolyte and the polymer substance travelled like independent sample components (see also Section 19.3.3.2.).

— As compared to the refractometric record, the conductivity peak of the polymer sample was shifted towards higher elution volumes. A quantitative evaluation of these two curves showed that the charge density per unit mass decreased with increasing chain-

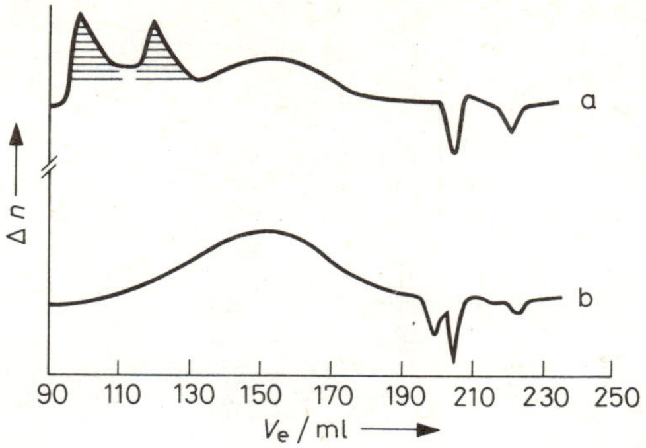

Fig. 19-14
Elution curves of a polyester urethane in DMF (curve a) and in DMF with an addition of 0.05 M LiBr (curve b)

Column set: four columns packed with Styragel® with a nominal pore size of $2.5 \cdot 10^4$ Å, calibration curve: see Fig. 16-28
Temperature: 80 °C
Rate of flow: $v = 1\ \text{ml} \cdot \text{min}^{-1}$
(according to HANN, 1977).

length. The investigated copolymers of acrylonitrile and sodium methallyl sulphonate thus exhibited a chemical heterogeneity which is coupled with that of the chain-length.

COPPOLA et al. (1972) attributed the premature elution of polyelectrolytes in salt-free eluents, which is generally reproducible to some extent only under very strict conditions, not only to the charge-induced coil expansion but also to *associations*. For polar macromolecules, association phenomena in solution are in fact possible. Certainly they were the cause of at least one of the extra peaks observed by HANN (1977) in the GPC of polyester urethanes on Styragel®. They disappeared upon addition of 0.05 M LiBr, without shifting the "standard" peak at 150 ml, which was also observable in pure DMF (see Fig. 19-14). As macromolecular associates of a certain structure cause a substantial increase in viscosity, the reported decrease by 42% upon salt addition may be connected with their destruction.

A conclusive answer to the question of whether premature peaks, which disappear upon salt addition, are indeed *associates* or reflect other artefacts (see for instance Section 19.3.3.1.) can be expected from the use of the LALLS detector, whose signal depends not only on the amount of substance but also on the size of the particles in the eluate.

The explanation for anomalous elution curves of polar samples and for the correcting effect of salt additions becomes difficult if the latter do not cause a corresponding reduction in viscosity. This applies to the investigations using poly(N-vinylacetamide) in DMF on silanized porous glass (DUBIN and MILLER, 1977, see Fig. 19-15). An addition of 0.01 M LiBr compressed the broad elution curves, shifted the maximum towards lower V_e values and eliminated the apparent bimodality of the macromolecular samples, which is exemplified by Fig. 20-15(b) and which had obviously been caused by the exclusion limit. As already mentioned, the reduction in viscosity by the LiBr addition was insignificant. The sample concentration likewise had only a minor effect on the elution curves, and the Huggins constant for the concentration dependence of η_{sp}/c had a normal value; thus the broad elution curves in salt-free DMF could not be explained by the occurrence of associates. As adsorption and partitioning effects could also be excluded, the broadening of the

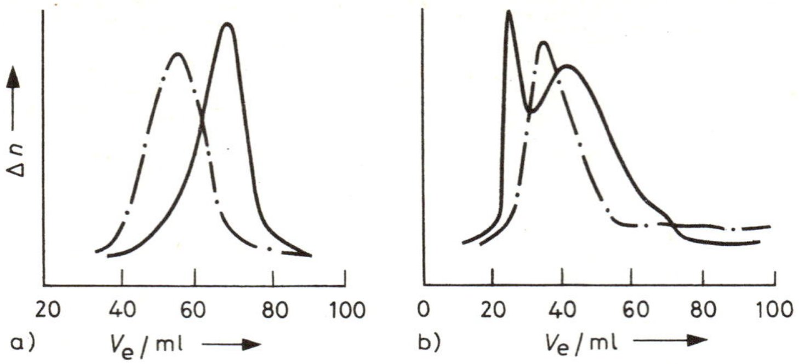

Fig. 19-15
Elution curves of a low-molecular-weight (a) and a macromolecular (b) sample of poly(N-vinylacetamide)

——— DMF; —·—·— DMF + 0.01 M LiBr
Column set: $L = 4 \times 0.45$ m; $d_C = 8.8$ mm
Packing: silanized porous glass, 493; 327; 172; 89 Å (one column unit each)
$V_0 = 0.5$ ml; $v = 2$ ml · min^{-1}
(according to DUBIN and MILLER, 1977).

curves must be interpreted as a broader distribution of the molecular dimensions in a salt-free solvent.

For polyelectrolytes, the qualitative effect of the salt addition can be understood fairly well: polyelectrolytes are polymers with cationic or anionic groups arranged at different points of the backbone chain. As long as the low-molecular-weight counter ions ensure a homogeneous distribution of the charge density, the coils can behave normally. In dilute solutions, however, the counter ions diffuse out of the coils so that electrostatic forces cause an additional and unusually large coil expansion. In the chromatographic migration of a polyelectrolyte sample, such conditions exist at both fronts between the sample and the eluent. In exclusion chromatography this means that the large molecules travelling in the lead undergo a further enlargement of their hydrodynamic volume, which causes them to migrate even faster. In the elugram this causes a flattening of the front slope. The small molecules travelling more slowly than the bulk are also expanded, but in this case the associated acceleration causes them to catch up with the bulk again. Consequently the band in which the sample is distributed is compressed from behind so that the back slope is unusually steep, whereas the leading part of the band is spread. This behaviour can clearly be observed from the elution curves of polyelectrolytes in Fig. 19-16. The magnitude of the effects depends on the distribution of the ionogenic groups along the polymer chain, on the charge and the mobility of the low-molecular-weight counter ions and naturally on their tendency to leave the interior of the coil. If a sufficient quantity of salt is added (see the right-hand side of Fig. 19-16), i.e., if the difference in the osmotic pressure between the interior of the coil and the surrounding medium is reduced, then the peak asymmetry disappears.

Chromatographic migration is accompanied by selective dilution. For polyelectrolytes this dilution has very specific consequences, because the tendency of the counter ions to leave the coils increases more and more as the latter are farther and farther separated from one another. At high dilutions, there are effects which may differ more widely from the normal behaviour in saliferous eluents than the viscosimetrically determined values of $[\eta]$

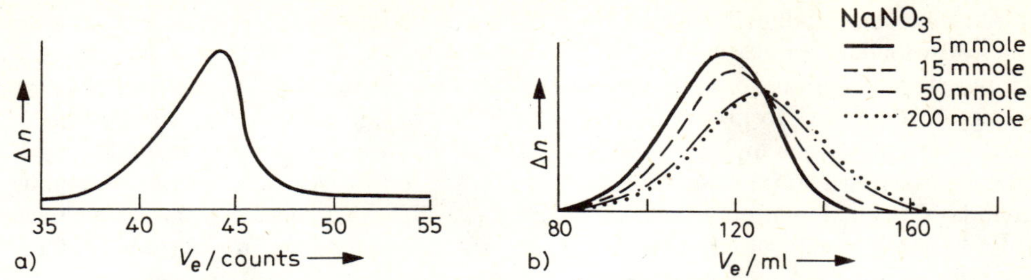

Fig. 19-16

a) Elution curve of a polyamido acid from pyromellitic dianhydride and 4,4′-diaminodiphenyl ether

Column set: $L = 5 \times 1.20$ m; packed with styrene-divinylbenzene gels (WATERS) with nominal pore sizes of 10^6; 10^5; $3 \cdot 10^4$; 10^4 and 10^3 Å

Eluent: DMF; $v = 0.83$ ml · min^{-1}

Sample: $[\eta] = 4.16$ dl · g^{-1}

(according to NEFEDOV, LAZAREVA, BELENKIJ, FRENKEL and KOTON, 1979)

b) Refractometer curves of the GPC of polyacrylonitrile with 1.2 meq/g sulphonate groups in DMF with $NaNO_3$

Column: $L = 1.47$ m; $d_C = 15$ mm; packed with a mixture of Spherosil® 75 Å, 150 Å, 500 Å, 1250 Å, 3000 Å

Rate of elution: $v = 0.2$ ml · min^{-1}

(according to DOMARD, RINAUDO and ROCHAŠ, 1979).

in solutions with and without salt additions. This is true especially if the $[\eta]$ values of the salt-free solutions are extrapolated from the rising ("normal") branch of the plot of η_{sp}/c vs. c, and hence found too small. In fact the *polyelectrolyte effect*, i.e., the enormous increase in the reduced viscosity, occurs only at high dilutions. Consequently, from the greater effects in exclusion chromatography as compared to the influence of the same salt addition on the viscosity one cannot simply infer an association of the coils. An association would require that the mutual repulsion of the like electric charges is eliminated.

Finally, it must be stressed that the qualitative description given above for the behaviour of polyelectrolytes is of course a simplification. The quantitative relationships between the volume required by macro-ions, their concentration and that of the low-molecular-weight electrolytes added are complex and not yet completely clarified. The theoretical approaches available at present, and their relation to the measurements of adequate accuracy have been reviewed by MANDEL (1981).

A general prediction of the effect of salt additions, which also includes the behaviour of macromolecules without ionogenic groups, is rather difficult. For example, adsorption effects have been found to be prevented as well as initiated by a salt addition: the adsorption on Styragel was *suppressed* by adding 0.05 M LiBr to solutions of quaternized polyurethanes in DMF (HANN, 1977), or by adding $\geqq 0.4 \cdot 10^{-3}$ mole/l LiBr to solutions of polystyrene microgel in DMF (BOOTH et al., 1980), but it was *initiated* on porous glass if partially saponified polyacrylamide was investigated with an addition of 0.005 M KCl rather than in pure formamide (ONDA et al., 1979).

Apart from the osmotic effect in polyelectrolyte solutions, the effect achieved by an addition of electrolyte mainly consists in the screening of charges which may occur on the polymer or the column packing material, and hence in the suppression of electrostatic interactions (cf., Section 16.6.5.). Moreover the frequently used lithium salts may form complexes with amines and amides (DUBIN and MILLER, 1977); e.g., LiBr forms a complex with four molecules of dimethylacetamide, LiCl with ε-caprolactam (1:4) and with a large number of secondary amines (1:1). Thus the conformation of the macromolecules existing in the system can be influenced in a variety of ways, because the mobility is possibly varied or ionogenic complexes formed. Complexing many also contribute to the *salting-in effect* of $LiNO_3$ on polystyrene in THF-DMF mixtures, which was observed by SIEBOURG et al. (1980) and referred to in Section 16.6.2. The effect of $LiNO_3$ on polystyrene containing 2—5 mole-% tributylammonium sulphonate groups, which was reported in the same paper, can also be considered from this aspect: this polymer is just soluble in pure tetrahydrofuran, but in a column packed with polystyrene gel it was eluted only slowly and with artefacts. When injected with a higher concentration, the sample did not emerge at all during the elution range of exclusion chromatography. An injection of a solution of $LiNO_3$ in THF caused the stuck sample to emerge from the column. As the salt travels more slowly than the polymer, a single injection was sufficient.

19.3.3. Size exclusion chromatography of aqueous solutions

Exclusion chromatography in aqueous media, also called gel filtration, was developed earlier than SEC in non-aqueous media. Later on the separation efficiencies achieved by this more recent technique again gave rise to a search for column packing materials which allow high process rates for separations in aqueous media. An ideal packing material for aqueous SEC should meet the following requirements:

- Hydrophilic behaviour
- No contact sites for hydrophobic interactions
- Compressive strength and dimensional stability for pressures up to more than 10 MPa (if possible, up to 50 MPa)
- Spherical shape, or at least the same shape for all particles
- Particle diameter 10 µm or less, with a narrow range of variation
- Chemical stability over a wide pH range
- A pore system independent of the ionic strength and the temperature
- Charge-free surface, which causes neither adsorption nor electrostatic repulsion
- Inert surface having no catalytic effect on unstable samples
- Resistance to chemicals used for the control of bacterial infestation

The materials available meet these requirements to a certain extent. Reviews of aqueous SEC have been given by DETERMANN [D 1], COOPER and MATZINGER (1977) and COOPER and VAN DERVEER (1978).

Under certain conditions the results obtained in water on the one hand and in organic liquids on the other are equivalent. Thus the value measured for the total pore volume in a column packed with a bare silica gel, with *toluene* as a sample and THF as an eluent, was equal to that obtained with *glucose* as a sample in 0.5 M sodium acetate (BUYTENHUYS and VAN DER MAEDEN, 1978). Consequently, for glucose in aqueous solutions it can also be assumed that it migrates only by size effects. On a column with porous glass as a packing material and 0.2 M Na_2SO_4 as an eluent, SPATORICO and BEYER (1975) found a *universal calibration curve* for dextrans, sodium polyacrylates, sodium polystyrenesulphonates and copolymers of acrylic acid and ethyl acrylate (mole ratio 1:1), which had been neutralized by NaOH. If the values were plotted vs. the Wheaton-Bauman distribution coefficient (cf., eqn. 8-31) rather than simply vs. $\dot{V}_e$, the data points for PS standards in three different non-aqueous eluents (THF, Bzn, TCM) also lay on the same universal calibration curve.

If larger differences between the results in aqueous and in non-aqueous media occur, then this usually means that some of the essential requirements are not satisfied. It is above all electrostatic interactions which cause deviations. The high dielectric constant of water favours the generation of charges; moreover, water-soluble substances are in most cases electrolytes.

In SEC, the charge interactions are more obvious than in other methods, since in this case the pore wall should be energetically inert and the separation should be based only on entropy effects. In enthalpy-governed separating techniques of course charge interactions occur as well, but there they are included in the whole spectrum of energy interactions, where they are not so conspicuous. Moreover in SEC the pore size and the size of the particles to be separated are of the same order of magnitude, which means that the distance between the pore wall and the permeating molecule will be rather small. If charges of the same sign build up, then the mutual repulsion will quickly neutralize the permeation ability of the sample under these special geometrical conditions. In other chromatographic techniques the pore systems are usually chosen so as to make the openings large compared with the dimensions of the molecule to be separated.

Column packing materials with a silicic acid skeleton bear SiOH groups, which after a dissociation build up a negative charge on the surfaces. Although these groups can be eliminated by a heat treatment, there is no sense in doing this for silica designed for application in aqueous media, because new SiOH groups will soon be produced by hydrolysis. Generally the organic gels hardly show such charging phenomena, but some observations indicate that in polyacrylamide gels and in cross-linked dextrans carboxyl groups are sometimes present, the dissociation of which at pH 4 results in a charge build-up.

19.3.3.1. Ion exclusion

Salts injected as a sample in pure water are eluted sooner than uncharged low-molecular-weight substances such as glucose. The elution curve of a salt is asymmetric; it rises gradually and slopes down steeply beyond the maximum (see Fig. 19-17). The shape indicates a convex isotherm (cf., Figs. 3-5 and 6-2), i.e., an increase in the distribution coefficient with increasing concentration. Consequently, at a high starting concentration the sample has a larger pore volume available, i.e., it penetrates more deeply into the pore system. For ions which are small compared with the pore size, this behaviour can be explained by the phenomena of diffusion: cations and anions diffuse independently of each other with a velocity depending on the difference of the osmotic pressure and on the respective rate of migration. For example, for NaCl the rate of migration of the anion exceeds that of the cation by about 50%. The faster diffusion of the chloride ion generates a potential difference which is largest where the concentration gradient is steepest, i.e., in the boundary layer between the sample and the pure water. The potential difference retards the penetration into the pore system. Therefore the fractions at the slopes of an (electrolyte) substance band moving over the chromatographic bed have a slightly higher rate of migration than the central fractions, in which the higher concentration has an additional shielding effect on the charges. The higher rate of migration occurring at the edges of the electrolyte band causes the leading fractions to travel farther and farther ahead

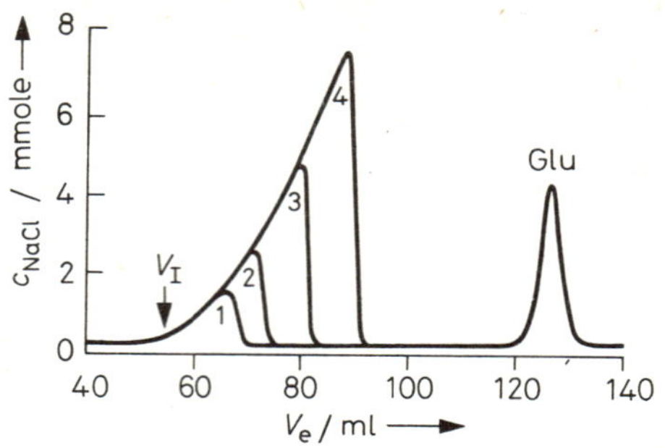

Fig. 19-17
Elution curves for sodium chloride at different sample concentrations
Column: Enzacryl-Gel® K 1
Eluent: water
Concentration of the sample: *1* 0.01 M; *2* 0.02 M; *3* 0.05 M; *4* 0.1 M
The arrow indicates the interstitial volume determined by means of Dextran Blue. The peak at 125 ml is due to a glucose injection.
(according to EPTON, HOLLOWAY and MCLAREN, 1976).

of the others, whereas at the backward slope it causes fractions lagging behind to catch up with the bulk again. This results in the asymmetric band shape, which is entirely different from the viscosity-induced tailing of the polymer bands, but resembles the polyelectrolyte behaviour mentioned in Section 19.3.2.

If the pores are sufficiently large compared to the ions, then, as the ionic strength increases, the peak maximum shifts so far ahead that the total pore volume is reached as a limit. The ionic strength, I, is calculated from the concentrations, c_i, of the individual ionic species with the respective valencies, z_i, according to the relationship:

$$I = \frac{1}{2} \Sigma\, c_i \cdot z_i^2 \qquad (19\text{-}2)$$

If the limit mentioned above is reached, the ions existing in the solution completely shield the electrostatic repulsion between the sample and the gel. RINAUDO and DESBRIERES (1980) investigated the salt exclusion effect with NaCl and sodium polystyrenesulphonate in two silica columns with very different pore sizes. The results shown in Fig. 19-18 were obtained for a silica with $d_0 = 500$ nm. As can be seen, the distribution coefficient can be represented as a linear function of $I^{-0.5}$, and hence of the screening parameter of the Debye-Hückel theory. This holds for NaCl, whose behaviour can be clearly supported by data points, and apparently also for the sodium polystyrenesulphonate, the molecules of which, having a straight chain-length of $P \cdot l_{eff} = 102 \cdot 0.252 = 26$ nm, are likewise small compared with the mean pore diameter. However, owing to the high viscosity it was not possible to investigate the concentrations required to approximate $K = 1$. The straight horizontal lines indicate the value, K_{ex}, of the distribution coefficient, which did not further diminish even with additional dilution. From this the relative exclusion volume, i.e., that portion of the total pore volume from which the sample is excluded by electrostatic repulsion, can be calculated for the respective electrolyte. For NaCl this amounts to 49.2 %, and for sodium polystyrenesulphonate ($P = 102$) 81.5 %. From the mentioned size ratio (26 nm : 500 nm) and from the fact that in both cases a 10^{-4} M NaCl solution used as an eluent reduces the

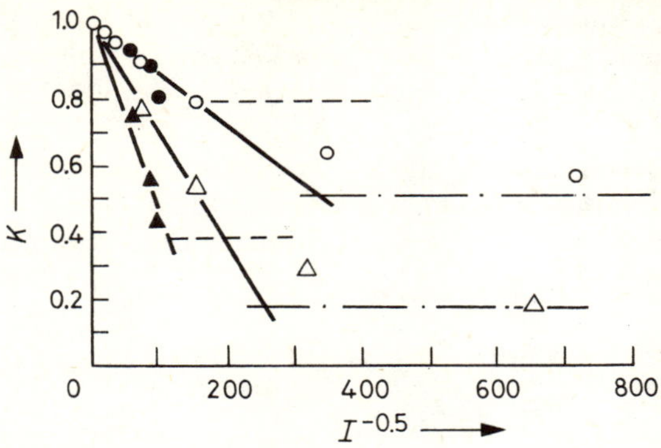

Fig. 19-18
Effect of the ionic strength, *I*, on the exclusion distribution coefficient, *K*, of electrolytes

Column: L = 1.47 m; d_c = 8 mm; packed with Spherosil®, d_0 = 500 mm; ○, △ — eluent: water; ●, ▲ — eluent: 10^{-4} M NaCl; ○, ● — sample NaCl; △, ▲ — Sample sodium-polystyrenesulphonate, $\bar{P}_w$ = 102
(according to RINAUDO and DESBRIERES, 1980).

inaccessible portion of the pores (49.2→22%; 81.5→61.7%), it can be concluded that in *both* cases the underlying process is electrostatic exclusion.

The narrower the pores, the higher is the salt concentration required for a complete screening of the electrostatic interaction. While $c_S \geqq 10^{-3}$ mole · l^{-1} was sufficient for 500 nm pores, a silica gel with d_0 = 30 nm required $c_S > 10^{-1}$ mole · l^{-1}.

Electrostatic exclusion can be utilized as a basis of a separating technique for small molecules which differ in their charges. Fig. 19-19 shows the behaviour of different amino

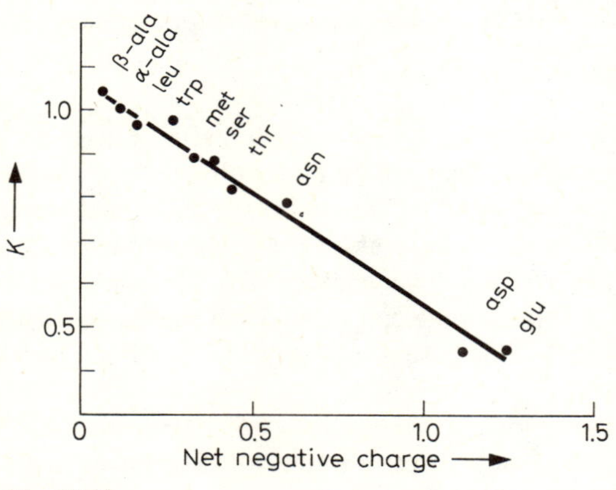

Fig. 19-19
Distribution coefficients of amino acids on CM-Sephadex® C-50. The elution in the 0.01 M sodium tetraborate buffer (+HCl, pH 9) is dependent on the net negative charge of the different amino acids under these conditions.

Column: L = 0.22 m; d_C = 21.2 mm; m_0 = 1 μmole per amino acid; V_0 = 0.5 ml; v = 0.53 ml · min^{-1}
ala- alanine; leu- leucine; trp- tryptophan; met- methionine; ser- serine; thr- threonine; asn- asparagine; asp- aspartic acid; glu- glutamic acid.
The K values for separate injections were determined using the ninhydrin method and calculated by eqn (8-31).
(according to CRONE, 1975).

acids in a column packed with an anionically modified dextran gel. As expected, the sample with the lowest charge (β-alanine) travelled most slowly. However, the fact that its K value was greater than 1 indicates attractive interactions (e.g., adsorption) in addition to the exclusion mechanism.

In most cases, however, exclusion chromatography is used to achieve a separation according to the molecular size, so that the electrostatic effects represent a disturbance. This can be suppressed by adding an electrolyte to the eluent. In many cases an ionic strength of 0.1 is sufficient.

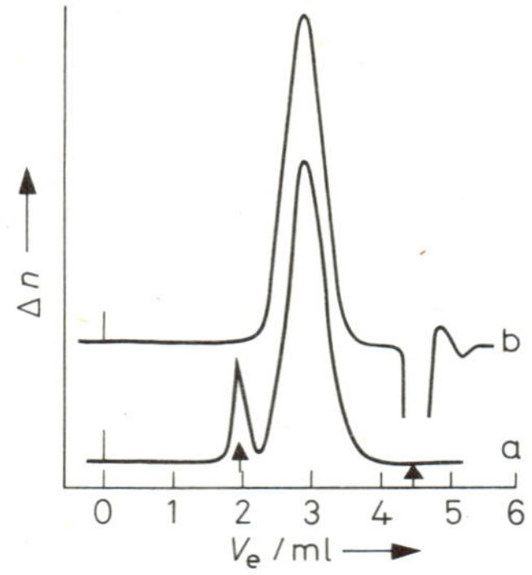

Fig. 19-20
Charge effect in the elution of dextran on silica gel

Column: L = 0.30 m; d_c = 4.6 mm; packed with LiChrospher® Si 100
Sample: dextran T 20, 20 μl of a 1% solution (w/v)
Rate of flow: 0.5 ml · min^{-1}
In pure water (curve a), a pre-peak occurs at the exclusion limit indicated by the first arrow; this artefact disappears in the elution in 0.5 M sodium acetate solution, pH 5 (curve b). The second arrow at 4.4 ml indicates the total pore volume
(according to BUYTENHUYS and VAN DER MAEDEN, 1978).

Finally, it should be mentioned that electrostatic exclusion effects need not be restricted to polyelectrolytes. Especially on SiOH surfaces, analogous effects have also been observed with non-electrolytes (see also Section 16.6.5.). Fig. 19-20 shows the behaviour of dextran on a silica gel column: in water there is a pre-peak at the exclusion limit. This completely excluded portion is likely to carry a negative charge, like the silica gel surface. It is remarkable, however, that most of the sample emerges within the normal elution volume, and is hence obviously uncharged. In this case a salt addition (0.5 M sodium acetate) also caused the artefacts to disappear.

The retention effects of non-ionic polyacrylamide on CPG-10 porous glass were investigated in detail by OMORODIN et al. (1980). In pure water, all samples were eluted with the interstitial volume, V_I, from a column packed with the glass type CPG-10-2000 with d_0 = 200 nm. The molar masses ranged between 55000 and 5 to 6 · 10^6, the fractionating range of the packing was 10^6—12 · 10^6, i.e., the exclusion cannot have been due to steric effects. In electrolyte solutions the samples exhibited larger elution volumes. Finally, normal elution characteristics were achieved by means of additions of polyethylene glycol or alkylphenoxy polyethoxyethylene (Tergitol®, a neutral surfactant from Union Carbide). A four column combination (4 × 1.22 m × $^3/_8$") packed with the CPG types 3000 (2 ×), 1000 and 370, respectively, yielded a linear calibration relationship for the whole range of molar masses if a solution of 0.0167 M Na_2SO_4 with 1% CH_3OH, 0.05% NaN_3 and 0.0021% Tergitol was used as the mobile phase. In water without any additions the non-ionic polyacrylamide is obviously repelled from the glass surface by ion exclusion.

Between the surface of silica gel and the polyacrylamide there are also energetic interactions, which manifested themselves, however, in a somewhat different way under the working conditions used by VAN DIJK et al. (1980). Upon repetitive injection of one and the same sample, the authors found an increase in the peak area and a decrease in the elution volume. Moreover the curves indicated more and more distinctly a bimodal profile. The authors attributed these effects to the saturation of adsorption sites without a deposition of voluminous layers. This was evidenced from the unchanged separating behaviour of the column for dextran samples, i.e., the pore structure of the packing did not vary.

No disturbances were observed in the elution of dextrans in desalted water on sulphonated, cross-linked polystyrene gels (MILLER and VANDEMARK, 1980). For polyvinyl alcohol samples an extra peak or a step appeared at the exclusion limit. The elution curves obtained by HEUBLEIN et al. (1980) for polyvinyl alcohol in water containing 0.1% ethanol also exhibited a step at the exclusion limit. For the Spheron® 100 employed, this exclusion limit lies at 250000 g · mole^{-1}. The phenomenon was interpreted as an associate formation, but it was independent of the age of the solutions. Incompletely saponified samples, i.e., PVAC/PVAL copolymers (containing up to 17% acetate groups), exhibited the extra step to a lesser extent, and the main peak occurred at higher elution volumes.

19.3.3.2. Ion inclusion

If the sample contains a polyelectrolyte and a low-molecular-weight electrolyte, and the pore system excludes only the macromolecular component, then the conditions are

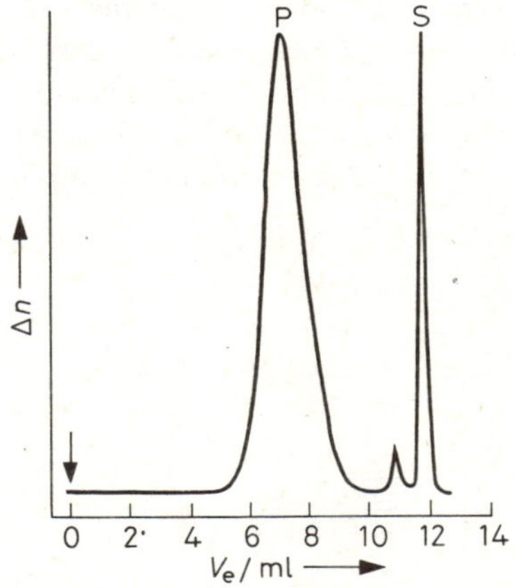

Fig. 19-21
Formation of a salt peak by Donnan-type equilibria (ion inclusion)

Column: $L = 0.90$ m; $d_C = 4.6$ mm; packed with LiChrospher® Si 100
Eluent: 0.5 M sodium acetate, pH 5
Rate of flow: $v = 1$ ml · min^{-1}
Sample: sodium-heparin, $V_0 = 40$ µl, $c_0 = 2\%$
After the elution of the heparin (peak P) a high and sharp sodium acetate peak appears, the elution volume of which approximately corresponds to the total pore volume. (The intermediate small peak is due to NaCl.)
(according to BUYTENHUYS and VAN DER MAEDEN, 1978).

similar to those at a semipermeable membrane which separates electrolyte solutions from one another: a *Donnan-type equilibrium* is established (LINDSTRÖM et al., 1977), in which the ions capable of permeation are distributed in such a way that their activity compensates for the fact that the macro-ions can only occur in one part of the system. Thus the low-molecular-weight electrolyte ensures the equality of the electrochemical potential in both phases. Fig. 19-21 shows an example which was observed by BUYTENHUYS and VAN DER MAEDEN (1978) after the injection of 0.8 mg sodium heparin in a 0.5 M sodium acetate buffer solution (pH 5) into a silica gel column with an exclusion limit of $50 \cdot 10^3$ g · mole^{-1}: the heparin anions were excluded from the silica gel, whereas the sodium ions, following the activity gradient, penetrated into the pore system. They were succeeded by additional acetate ions, so that under the heparin band the sodium acetate concentration was higher than in the other pores. The salt excess behaved like an independent sample. Naturally the latter travelled more slowly than the excluded heparin. Under certain conditions with respect to the concentration, the charge distribution and the column geometry, the included salt appeared as a separate peak. An analogous result was obtained by DOMARD et al. (1979) in a non-aqueous system (DMF with 0.005 M $NaNO_3$ and PAN as a sample) (cf., Fig. 19-13).

19.3.3.3. Polyelectrolyte swelling

If the ionic strength in the coil is greater than in the surrounding medium, then there is a driving force for the low-molecular-weight counter ions to diffuse off the coil. To maintain electric neutrality, and under the action of the remaining like charges, the polyelectrolyte chain expands beyond its equilibrium conformation. It occupies too large a hydrodynamic volume, appears too early in the eluate and consequently, on the basis of the calibration curve, is assigned an excessive value of the molar mass. Naturally, for polydisperse samples incorrect distributions are recorded under these conditions. Again this can be remedied by a sufficient ionic strength. Fig. 19-22 shows how the position of the peak maximum for sodium heparin shifted away from too low values by increasing the salt content of the eluent. Above an ionic strength of 1, the peak position remained constant.

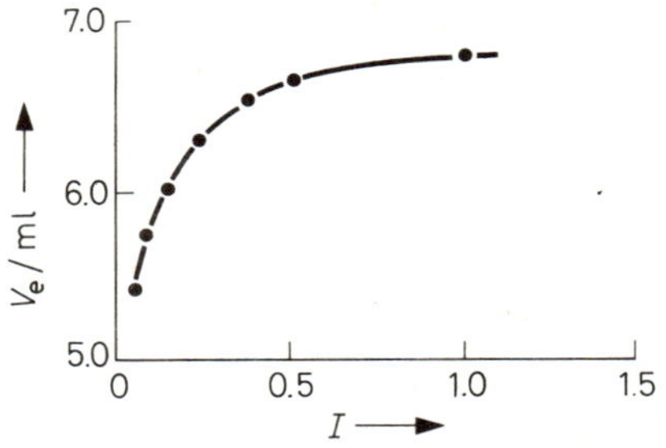

Fig. 19-22
Influence of the ionic strength on the peak elution volume of heparin

Column: $L = 0.60$ m; $d_c = 4.6$ mm; packed with LiChrospher® Si 100
Eluent: sodium acetate solutions with pH 5
Rate of flow: $v = 0.5$ ml · min^{-1}
Sample: sodium-heparin, $V_0 = 20$ µl, $c_0 = 2\%$
(according to BUYTENHUYS and VAN DER MAEDEN, 1978).

19.3.3.4. Adsorption and hydrophobic interactions

While in the case of *similar charges* of the surface and the polymer the electrostatic effects can be shielded and hence adequately controlled for many purposes by adding an electrolyte, *opposite charges* lead to an adsorption which is irreversible under chromatographic conditions.

So SPATORICO and BEYER (1975) reported on successful separations of uncharged hydrophilic polymers and polyanions, whereas they found considerable difficulties with polycations on porous glass.

Coating the surface with polyethylene oxides offers a possibility of suppressing the adsorption on glass or silica gel. For porous materials with d_0 = 25 nm, Carbowax® 20 M has proven suitable. Unfortunately such coatings are gradually washed off by the eluent and must be renewed from time to time. In this respect it is easier to add a certain amount of Carbowax (0.01—0.1%) to the eluent itself.

The chemical fixation of a hydrophilic layer on the inorganic support (REGNIER and NOEL, 1976) opened the way to the rapid exclusion chromatography of water-soluble

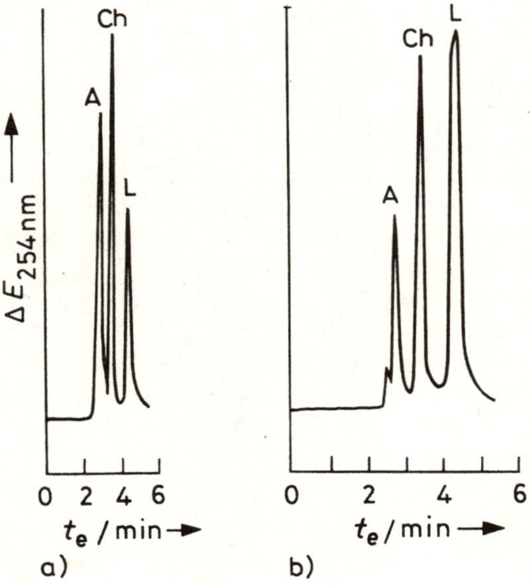

Fig. 19-23
Rapid size exclusion chromatography in an aqueous medium: Separation of proteins on LiChrosorb Diol®

a) Analytical separation. Column: L = 0.25 m; d_C = 6 mm
Rate of flow: v = 2.0 ml · min^{-1}
Sample: 0.1 mg A; 0.1 mg Ch; 0.05 mg L

b) Preparative separation. Column: L = 0.25 m; d_C = 23.4 mm
Rate of flow: 21.0 ml · min^{-1}. Sample: 25 mg each of A, Ch and L, in V_0 = 0.5 ml.

Column packing: LiChrosorb Diol, 5 μm particle diameter;
eluent: phosphate buffer (0.008 M K_2HPO_4 + 0.042 M Na_2HPO_4) with 0.1 M NaCl, pH 7.5 (in both cases).
A bovine albumin (65000)
Ch chymotrypsinogen A (25000)
L lysozyme (14300 g · mole^{-1})

polymers. By means of LiChrosorb® diol, which, like the glycophases, contains the structure shown in Fig. 11-7 as a bonded phase and is available in the form of 5 μm spheres, ROUMELIOTIS and UNGER (1979) realized high-resolution protein separations within only 5 minutes (see Fig. 19-23).

As glass and silica gel are hydrophilic by nature, for an understanding of the purposes of the modification it has to be taken into consideration that the difficulties with the bare support materials are mainly due to the *silicic acid anions.*

At pH 7 half of all the silanol groups have dissociated (UNGER, 1979), and to be certain that there are no anionic groups on the surface, the silica gels should only be used in media with pH $\leqq 3$. However, especially with biopolymers it is often necessary to work at pH values near the neutral point. Thus the modification must replace the hydrophilic SiOH groups by another hydrophilic structure which forms no ions in a certain pH range around the neutral point, but otherwise leaves all the good properties of the silica particles unchanged.

The TSK-gel SW-type columns (cf., Table 11-10) are widely used for aqueous SEC. Regarding the packing material it has only been said that it is a modified silica gel (FUKANO et al., 1978). With a particle size of 10 μm, the packings exhibit a high resolution (see Fig. 15-4b). As a minimum value of the reduced plate height a value of $h^* = 2.1$ was measured with alanine ($M = 89$ g · mole^{-1}) at a flow-rate of 0.3 mm · min^{-1} (ROKUSHIKA et al., 1979). KATO et al. (1980a) reported the separation ranges and the specific resolution of the SW columns; they compared (1980d) the SW types on a silica gel basis with the organic PW types and found the latter more suitable for the separation of synthetic oligomers and polymers with a broad distribution, whereas the SW types were superior in the exclusion chromatography of proteins. This also follows from the location of the separating ranges and from the steepness of the calibration curves. The high separation efficiency of the PW types for dextrans, polyethylene glycol, polyacrylamide, polyvinylpyrrolidone and polyvinyl alcohol had also been pointed out by HASHIMOTO et al. (1980).

The high resolution of the SW columns enables protein separations to be carried out within a few minutes (see Fig. 19-24). Eluents which have proven suitable for use include 0.01 M phosphate buffer (pH 6.5) containing 0.2 M Na_2SO_4 (ROKUSHIKA et al., 1979), 0.1 M phosphate buffer containing 0.3 M NaCl (pH 7; KATO et al., 1980b, d) as well as 0.067 M KH_2PO_4 containing 0.1 M KCl and $6 \cdot 10^{-4}$ M sodium azide (pH 6.8; WEHR and ABBOTT, 1979).

For proteins in a molar mass range from 6000 (insulin) to 480000 g · mole^{-1} (ferritin) ROKUSHIKA et al. (1979) found a linear relationship between log M and V_e, which could also be used to determine the molar mass of unknown samples. However, they found that lysozyme was eluted later than expected on the basis of its molar mass. In a salt-free buffer solution it even got stuck in the column. Additions of NaCl and Na_2SO_4 suppressed the adsorption, but even in eluents with more than 0.2 M Na_2SO_4 an adsorption was found to occur for this enzyme as well as for aromatic amino acids and the N-heterocycles of the nucleotides.

For numerous biologically important polymers the tendency to adsorption phenomena and to undesirable electrostatic effects can be suppressed by complexing. Sodium dodecyl sulphate (SDS), urea or 6 M guanidine hydrochloride have proven suitable as complexing agents. KATO et al. (1980c, e) also adopted these techniques for TSK SW columns (see Fig. 19-25).

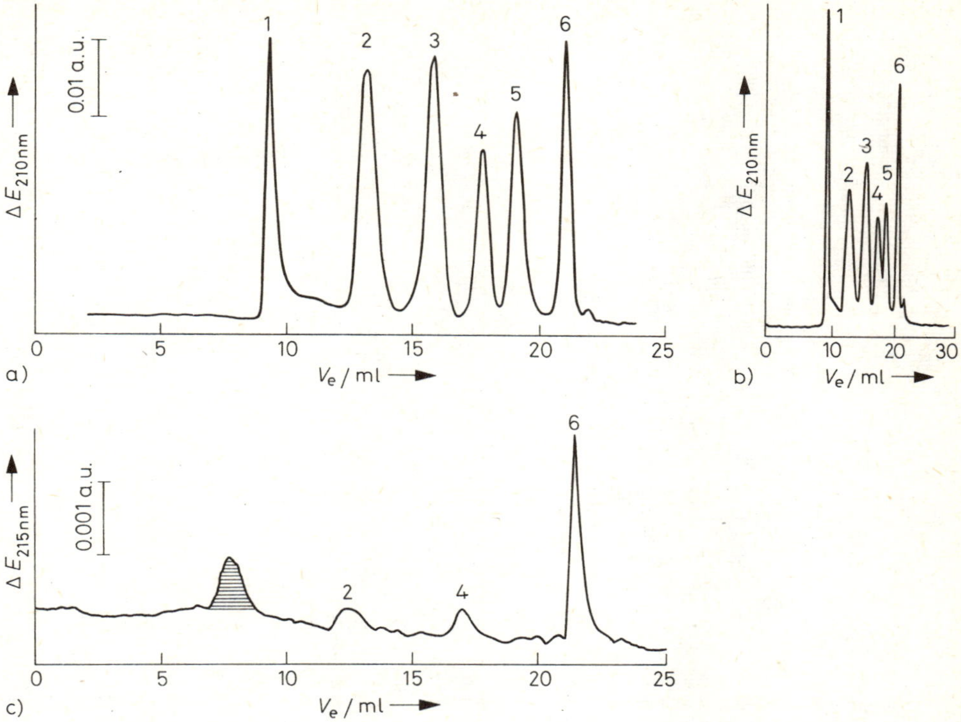

Fig. 19-24
Separation of protein mixtures on TSK-Gel SW 3000 in 0.01 M phosphate buffer with 0.2 M Na_2SO_4 (pH 6.5)

Column: $L = 0.60$ m; $d_C = 7.5$ mm; $d_P = 8$—12 µm
a) Rate of flow: $v = 0.5$ ml · min^{-1}
m_0: 2—4 µg each for components 1—5
40 µg for component 6
b) $v = 2.0$ ml · min^{-1}, i. e., separation within circ. 11 min.
m_0: as under a)
c) $v = 1$ ml · min^{-1}
m_0: 20 ng each for components 2 and 4
120 ng for component 6
1 ferritin (480000); 2 aldolase (145000); 3 ovalbumin (45000); 4 chymotrypsinogen (24500); 5 insulin (6000); 6 alanine (89 g · mole^{-1})
The shaded peak in chromatogram c) lies at the exclusion limit and is caused by an unknown substance. It occurs only at the highest detector sensitivity.
(according to ROKUSHIKA and HATAKANO, 1979).

Although generally hydrophilic packings are used in aqueous size exclusion chromatography, in special cases there may be phenomena corresponding to the solvophobic interactions discussed in Section 16.6.2.

In the chromatography of proteins on an ether-modified silica gel, VIVILECCHIA et al. (1977) found an additional retention which increased with increasing salt concentration. This behaviour is characteristic of solvophobic interactions. Obviously the bonded phase had a weakly hydrophobic character. This was unexpected because the material was wetted by water.

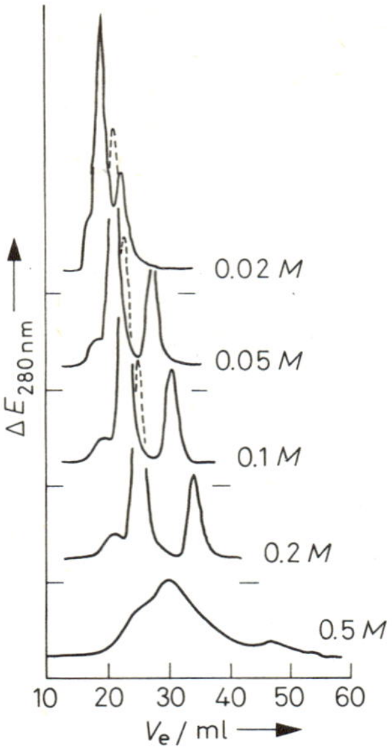

Fig. 19-25
Elution of γ-globulin (human serum) as an SDS complex at different sodium phosphate concentrations (indicated)

Column set: $L = 2 \times 0.60$ m; $d_C = 7.5$ mm; packed with TSK gel G3000SW; $\vartheta = 25$ °C
Eluent: 0.02—0.5 M sodium phosphate in water with 0.1% sodium dodecyl sulphate (SDS), pH = 7.0 flow rate $v = 1$ ml · min^{-1}
UV detection at 280 nm
(according to KATO, KOMIYA, SASAKI and HASHIMOTO, 1980c).

On special support materials, the hydrophobic interactions can be utilized for chromatographic techniques by means of which mainly proteins can be separated (hydrophobic chromatography, cf., Section 18.3.).

19.3.3.5. The calibration of aqueous exclusion chromatography

SEC calibration (cf., Sections 8.2. and 8.3.) requires either standards with a narrow distribution or samples having an accurately known broad distribution. Water-soluble standards with a narrow distribution can be prepared from the common PS standards by sulphonation. For this purpose CARROLL and EISENBERG (1966) stated a method which proceeds without any decomposition and up to the sulphonation degree 1. Moreover, to avoid a cross-linking by sulphone bridges, which will lead to a broadening of the molar mass distribution, BROWN and LOWRY (1979) carried out the sulphonation with SO_3 + triethyl phosphate (2:1) in dichloroethane. The reaction was started at —20 °C and finished at room temperature. The reaction products were used as sodium polystyrene sulphonates.

Defined samples of dextran and polyvinylpyrrolidone for calibration purposes are commercially available. A summary of their characteristics and the suppliers has been given by COOPER and VAN DERVEER (1978). These samples have indeed a broader distribution, but they are not polyelectrolytes like the polystyrene sulphonates. VRIJBERGEN et al. (1978) used dextrans of this kind for an iterative calibration. Polyacrylamides with exactly characterized distribution for calibration according to the method explained in Section 8.3.3. have been described by ABDEL-ALIM and HAMIELEC (1974).

The charge effects discussed in the preceding sections suggest recording the calibration curves under exactly the same conditions with respect to pH value, ionic strength etc. as employed in the succeeding analyses. Although SPATORICO and BEYER (1975) have shown that under favourable conditions a universal calibration curve can be obtained with very different substances, it is better to avoid any risk by using sodium polystyrenesulphonates in the calibration for the investigation of polyelectrolytes, and uncharged calibration standards in work with uncharged synthetic polymers. In this respect it is good practice to use standard proteins with an exactly known molar mass in the calibration for the SEC of proteins.

19.4. SEC investigations on band broadening

The peak width is caused by the column (σ_c), by the extra-column parts of the apparatus, such as the detector, the injection chamber and the connecting capillaries (σ_e) and by the heterogeneity of the sample (σ_p). If all the contributions lead to Gaussian distributions independently of each other, then it is possible to apply the addition rule of variances:

$$\sigma^2 = \sigma_p^2 + \sigma_c^2 + \sigma_e^2 \qquad (19\text{-}3)$$

On the basis of this relationship, LEPAGE et al. (1968) tested silica gel columns with benzene as a probe substance ($\sigma_p = 0$) by measuring the total variance, σ^2, on homogeneously packed columns of different lengths (50, 100 and 200 cm). As the additivity of variance also holds for the contributions made by paths which are travelled successively, the term σ_c^2 increases proportionally to the column length. This can be observed from Fig. 19-26 in which σ^2 is plotted vs. V_e^2/L (in accordance with the model of theoretical plates). The common ordinate intercept is the contribution σ_e^2. The band broadening in the column increases with increasing flow-rate.

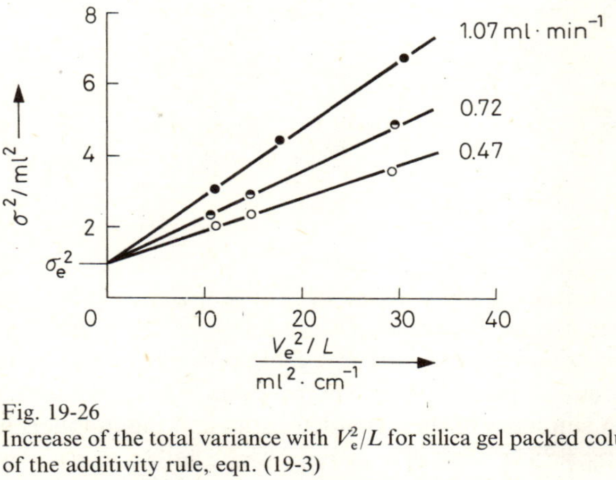

Fig. 19-26
Increase of the total variance with V_e^2/L for silica gel packed columns of different lengths. Test of the additivity rule, eqn. (19-3)
Benzene in tetrahydrofuran, v = 1.07, 0.72 and 0.47 ml · min^{-1}
(according to LEPAGE, BEAU and DEVRIES, 1968).

Table 19-2
Testing of polystyrene gel columns and column sets with low-molecular probes in tetrahydrofuran (experimental data; OSTERHOUDT and RAY, 1967)

A ... F column designation; R differential refractometer; A + B + 2R = 1.58 → $\sigma_e^2 = 0.28$; A + B + C + 3R = 2.18 → $\sigma_e^2 = 0.32$; 4(A + R) = 3.85 → $\sigma_e^2 = 0.43$

Column set	L/m	$\frac{V_{e,\,exp}}{ml}$ (styrene)	$\frac{V_{e,\,calc}}{ml}$	Plate number N_{exp} (benzene)	Plate number per m $\frac{}{N \cdot m^{-1}}$	Plate height h/μm	σ^2/ml^2	σ_c^2	$\sigma_e^2 + \Sigma\, \sigma_e^2$
1	2	3	4	5	6	7	8	9	10
1 A + R	1.22	51.80		2790	2290	437	0.96	0.66	(0.96)
2 B + R	1.22	51.45		4267	3500	286	0.62	0.32	(0.62)
3 C + R	1.22	44.90		3378	2771	361	0.60	0.30	(0.60)
4 A + B + R	2.44	97.70	97.60	7338	3010	332	1.30	—	1.28
5 A + B + C + R	3.66	136.75	136.85	12100	3310	302	1.55	—	1.58
6 D + E + F + A + R	4.88	193.05	(190.25)	14550	2980	335	2.56	—	2.95
7 R		5.65						0.30	

The additivity rule has also been used in establishing Table 19-2. The total variance (8) can be calculated for every combination (1) from the number of theoretical plates (5), as given by OSTERHOUDT and RAY (1967), and from the elution volume (3). The sum of rows 1 + 2 gives the combination (A + R) + (B + R) with $\sigma^2_{calc} = 0.96 + 0.62 = 1.58$. This corresponds to row 4, which shows the measured values. In $\sigma^2_{calc} = 1.58$, however, the term σ^2_e is contained twice. The difference $1.58 - 1.30 = \sigma^2_e = 0.28$ yields information about the *band broadening outside the column*. Rows 1 + 2 + 3 and 5 can be evaluated analogously. As pointed out in the original paper, columns D, E and F largely correspond to column A. Owing to the lack of precise information, D + E + F + A was treated approximately as A + A + A + A. It appears that this approximation is only a rough one, as shown by the larger deviations in the calculated elution volumes. Therefore the value $\sigma^2_e = 0.43$ calculated from this approximation was ignored in taking the mean value.

Both examples show that the band broadening in the detector and the other parts of the apparatus is of the same order of magnitude as the broadening in the column. For excellently packed, short columns, the detector contribution may even dominate.

19.5. High-speed SEC

Usually, columns with an interior diameter of 7.8 mm are used at a flow-rate of 1 ml · min^{-1}. If 60% of the cross-section is occupied by packing particles, this gives a linear velocity of 0.088 cm · s^{-1}. With column sets of a total length of ≧1 m, an analysis takes about 2 hours. Strictly speaking, however, this is the waiting time between the injection of the sample and its appearance at the column outlet. The next sample can be injected as soon as the preceding one has a sufficient lead, so the total number of analyses carried out per working day can be increased. On the other hand, the waiting time is in fact decisive in process control. Therefore there is a real interest in variants which yield the result more rapidly.

YAU et al. (1968) and many other chromatographers investigated the effect of higher flow-rates. Up to 35 ml · min^{-1} were forced through standard columns (LITTLE et al., 1970). In previous work with a maximum flow-rate of 12.5 ml · min^{-1}, it had been shown that the peak elution volume did not depend on the flow-rate (LITTLE et al., 1969). COOPER et al. (1973) summarized the literature on this topic, which also contains different findings. With their own measurements extended down to 0.052 ml · min^{-1}, they confirmed the independence of V_e of the flow-rate. Care must be taken to avoid the siphon error (cf., Section 17.7.). LITTLE et al. (1969, 1970) observed that the peak symmetry improved at high flow-rates.

Obviously the viscous fingering was reduced. This effect might be of importance especially at somewhat higher loads. On the other hand, the plate height increases with the flow-rate, and this may be of some consequence regarding the location of the peak area. The higher the flow-rate, the more the peak area will extend towards higher elution volumes. This effect will be especially pronounced with samples of a very narrow MMD. It may cause a flow-rate dependence of the mean values of molar mass ($\bar{M}_n$, $\bar{M}_w$, $\bar{M}_z$) as calculated from the peak area. These values would vary inversely with the flow-rate. Consequently, it is good practice to perform the analyses *and* the calibration at precisely the same flow-rate.

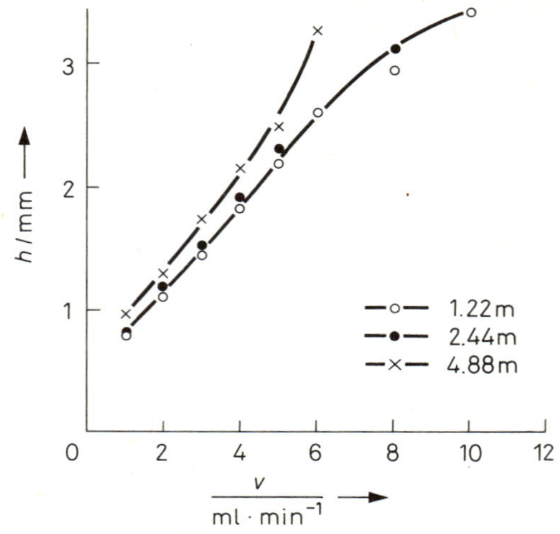

Fig. 19-27
Effect of the flow rate on plate height
Investigations with polystyrene ($M = 110000$ g · mole^{-1}) in tetrahydrofuran on 20 μm polystyrene gels in columns of different lengths: 1.22 m; 2.44 m; 4.88 m
(according to KATO, KIDO, YAMAMOTO and HASHIMOTO, 1974).

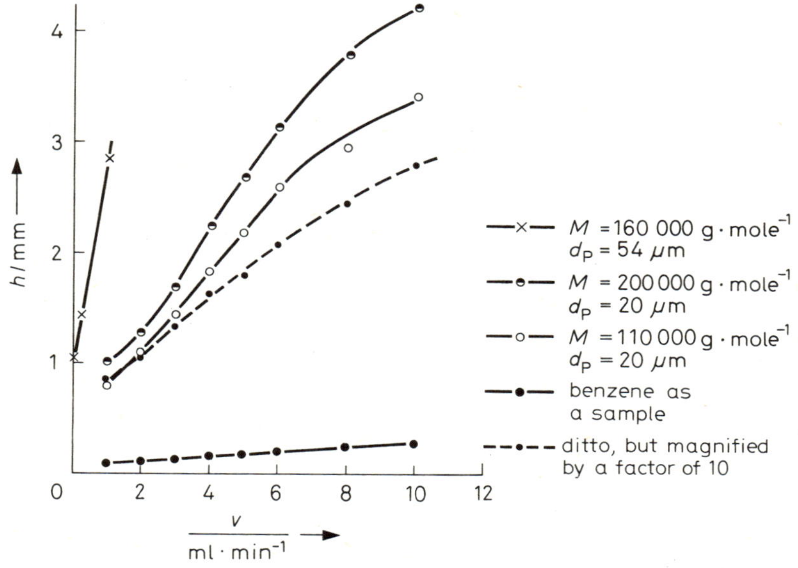

Fig. 19-28
Effect of the flow rate on plate height
Investigations with polystyrene standards in THF on polystyrene gel (7.8 mm columns $L = 1.22$ m) (KATO et al., 1974b). (×) according to COOPER; JOHNSON and BRUZZONE (1973).

Fig. 19-27 shows the increase in plate height for polystyrene in THF on 20 μm gel particles, indication that in longer columns the plate height is less favourable than in the shorter ones. Consequently the increase in efficiency with increasing length is partially cancelled. In any case a 1.22 m column with a flow-rate of 3 ml · min^{-1} is better than a 2.44 m column with 6 ml · min^{-1} — with respect to pressure and solvent quantity required as well as to its efficiency (KATO et al., 1974b). The influence of the molar mass is shown in Fig. 19-28.

For a benzene injection, the height of a theoretical plate is much lower at all flow rates, but it shows the same rate dependence. This is demonstrated by spreading the scale of the ordinate in Fig. 19-28. The values for the sample with 160000 g · mole^{-1} were measured on a column packed with 54 μm particles (COOPER et al., 1973). Relative to the particle size (Fig. 19-29), these values are in good agreement with those measured by KATO et al. (1974b). In a reduced representation according to eqn. (15-17), the correspondence is complete (Fig. 19-30). No minimum could be observed in the range investigated. The same result was obtained by ČOUPEK and HEITZ (1968) as well as by LITTLE et al. (1970), who plotted the logarithm of h vs. ud_p/D'^2. HEITZ and ČOUPEK (1967) found a minimum only with benzene as a probe substance for $v = 2$. For practical work with polymer solutions, this value is already beyond the useable range. Consequently *the highest resolution* is achieved with the *smallest possible values of* v. As in high-speed SEC the flow-rate cannot be reduced, the *particle sizes* have to be small and the *diffusion coefficients* should be as high as possible. The diffusion is rapid in low-viscosity solvents and at elevated temperatures. LITTLE et al. (1970) have demonstrated the influence of temperature by the resolution of a 1:1 mixture of two standard polystyrenes (cf., Table 19-3): in spite of a three-fold flow-rate at 80 °C it is possible to achieve a resolution which is practically equal to that at 23 °C (0.89→0.84).

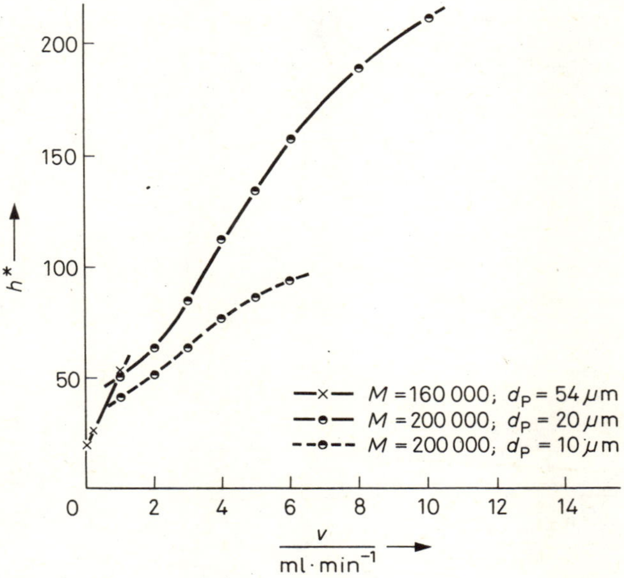

Fig. 19-29
Dependence of the reduced plate height on the flow-rate
Combination of the results obtained by COOPER et al. (1973), (×) and KATO et al. (1974b).

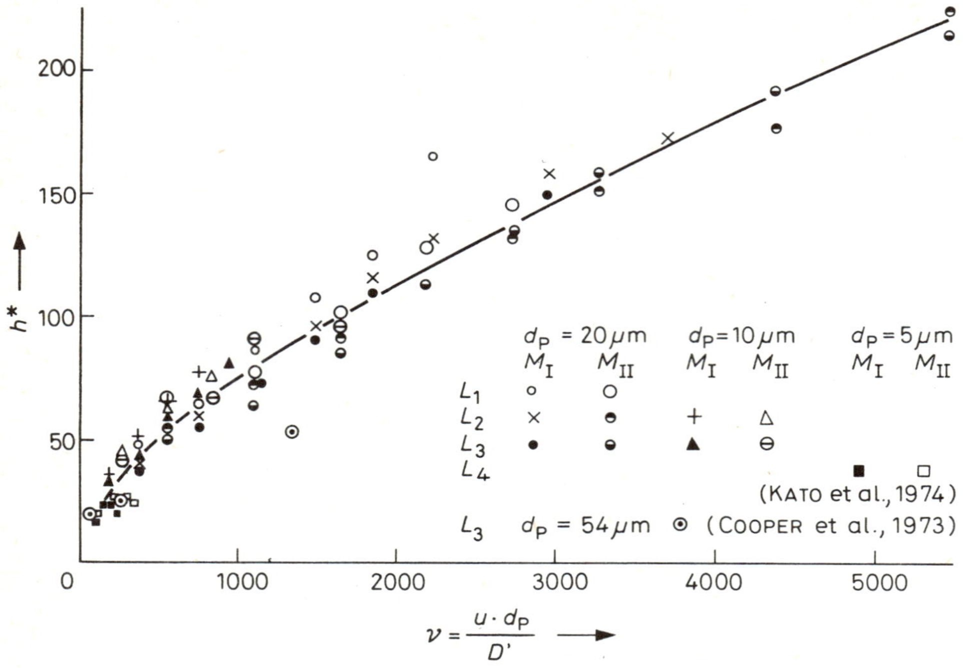

Fig. 19-30
Reduced representation according to GIDDINGS and MALLIK (1966) on the basis of eqn. (15-17)

— Measurements with polystyrene samples in tetrahydrofuran on polystyrene gels, with $v = 1$—10 ml × min^{-1} (according to KATO et al., 1974b); $d_p = 5, 10, 20$ μm; $L_1 = 4.88$ m; $L_2 = 2.44$ m; $L_3 = 1.22$ m; $L_4 = 0.61$ m; $M_I = 110000$ g · mole^{-1}; $M_{II} = 200000$ g · mole^{-1}
— ditto, but with $v = 0.054$—1.055 ml · min^{-1}; $M = 160000$ g · mole^{-1}; $d_p = 54$ μm, $L_3 = 1.22$ m (according to COOPER et al., 1973).

Table 19-3
Resolution, R_S, in SEC on a 2 × 1.22 m column set, interior diameter 7.8 mm, packed with polystyrene gel (37—42 μm), calculated by eqn. (3-21) (according to LITTLE, WATERS, BOMBAUGH and PAUPLIS, 1970)

Samples: polystyrene, 411 000 g · mole^{-1} and 19850 g · mole^{-1}, in toluene

Flow rate v / ml · min^{-1}	Resolution R_S at 23 °C, $c_0 = 2 \cdot 0.5$ mg	at 23 °C, $c_0 = 2 \cdot 5$ mg	at 80 °C, $c_0 = 2 \cdot 5$ mg
1.0	1.75	1.43	
5.0	1.33	1.19	
12.5	0.99	0.89	
35.0	0.66	0.63	0.84

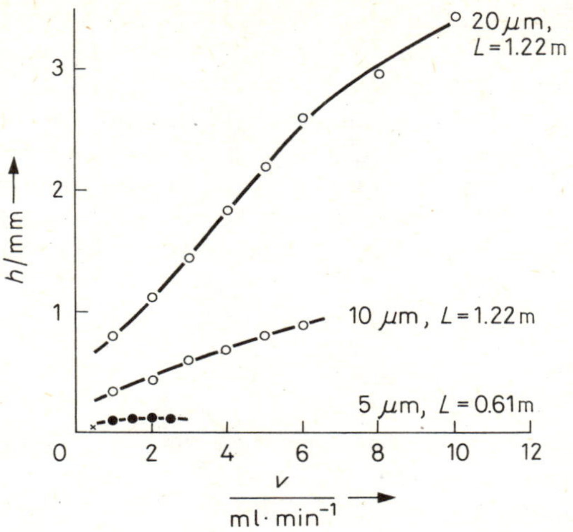

Fig. 19-31
Influence of the particle size on plate height, for different rates of flow. Measurements with polystyrene, $M = 110000$ g · mole^{-1}, in tetrahydrofuran on short columns (KATO et al., 1974b)

The value indicated by a cross was determined by benzene injection on a 6.10 m column packed with 5 μm particles (KATO et al., 1973).

The investigations carried out by KATO et al. (1974b) clearly show the interaction of the flow-rate, the particle size and the height of a theoretical plate in macromolecular chromatography (Fig. 19-31): the 5 μm packing yields a resolution which is so much better that it is possible to use shorter columns. This also reduces the waiting times. Very good results were obtained by means of a 0.61 m column with an interior diameter of 7.8 mm, packed with a mixture of three polystyrene gels with nominal porosities of 10^6, 10^4 and 10^3 Å and a uniform particle size of 5 μm. With an injected volume of 0.05 ml, a sample concentration of 1 g · l^{-1} and a flow-rate of 2.5 ml · min^{-1}, chromatograms with an excellent resolution were obtained within 10 minutes (KATO et al., 1974b).

DARK et al. (1975) likewise succeeded in reducing the waiting times to between 12 and 25 minutes. They used a column 1.52 m long, packed with 10 μm PS gel particles. The eluent was THF, the flow-rate 4 ml · min^{-1}.

A separation of three PS standards (M = 4, 40 and 400 kg/mole) within 90 sec was realized by YAU et al. (1976), who used a 0.40 m column with an interior diameter of 7.8 mm, packed with porous silica microspheres ($d_p = 7$ μm). Here the flow-rate was about 8 ml × min^{-1}.

For chromatography at a high flow-rate, the *number of effective plates per second* represents an important characteristic:

$$\frac{N_{\text{eff}}}{t_e} = \frac{u}{h} \frac{k^2}{(1 + k)^3} \tag{19-4}$$

This equation follows from

$$L = N \cdot h = u \cdot t'$$

using eqns. (3-1), (3-2), (3-9) and (14-5). For a peak with the retention time $t_e = t' + t''$, the number of effective plates per second varies directly with the linear velocity, u, and inversely with the height of a theoretical plate, h. Moreover eqn. (19-4) indicates the dependence on the capacity factor k. The maximum of this function lies at $k = 2$. As $k = K \cdot q$, and the SEC distribution coefficient, K, reaches at most 1, this demonstrates the importance of the pore volume, V''. Only for a high value of internal porosity, ε_P, of the packing material, the second factor, q, the phase ratio, will become large enough so that the product $K \cdot q$ will not differ too widely from the optimum.

19.6. Reliability of the results

Size exclusion permeation chromatography has rapidly become of widespread interest because it yields information about the molar mass distribution, more easily than other methods. However, there were doubts as to whether the curves obtained were real. As a contribution to this problem, CHRISTOPHER (1976) chromatographed 52 mixtures of known compositions, which were prepared from polystyrene standards with $\bar{M}_n$ values ranging between $1.78 \cdot 10^6$ and 104 g · mole^{-1}. He used several column combinations, each of which consisted of four or five columns with various PS gels. From the exclusion limits reported it can be observed that in some of the combinations the pore size values were graduated rather roughly. For example, combination I had columns with exclusion limits graduated in a ratio of 1250:125:3.75:1, with an upper value of $410 \cdot 10^6$ g · mole^{-1}; for combination II the ratio was 80:2.8:2.8:1, with an upper limit of $0.82 \cdot 10^6$ g · mole^{-1}. Both combinations operated rather poorly. What was evaluated was the difference between the mean values calculated from the chromatograms, $\bar{M}_{n(u)}$ and $\bar{M}_{w(u)}$, and the mean values resulting from the proportions of the mixtures. These deviations ranged up to ±175%; they were large for combinations I and II, but small for combinations III and IV, which consisted of better graduated columns. A good column combination must uniformly cover the required range of pore sizes, with gel types differing from one another by about one decimal power (cf., Section 8.1.).

19.6.1. Round robin testings

The ASTM Committee D-20.70.04 (1971) reported a round robin testing with PS standards. The results obtained in the contributing laboratories varied by more than 100% (ADAMS et al., 1973).

Butadiene oligomers with hydroxyl or carboxyl end groups were the samples in another test (ADAMS et al., 1973). The number averages of the six samples ranged between 3000 and 6200 g · mole^{-1}, and were calculated from the chromatograms with a coefficient of variation of 6.2% (overall mean value). The largest deviation of an individual $\bar{M}_n$ determination from the mean value of the measurements in all eight laboratories involved was +12.5 and −12.0%.

BLY (1980) reported another round robin testing in eight laboratories, the results of which will be included in the ASTM method D 3593-77. Seven different polymers were investigated (PC, PMMA, poly(octadecyl methacrylate), PS, PVAC, PVC and polychloroprene); the intrinsic viscosity was calculated from the SEC results by summation, with

Table 19-4
Results of an IUPAC round robin testing with polyethylene samples (STRAZIELLE, 1976)

Sample	High-density PE				Low-density PE		
Molar mass in 10^3 g · mole^{-1}	$\bar{M}_n$		$\bar{M}_w$		$\bar{M}_w$ (SEC)		
	Osm.	SEC	LSC	SEC			
	21	10.1	166	178	25 (fr. 16)	...	232 (fr. 6)
Coefficient of variation, %	—	18.0	6.1	18.7	mean: 11.7		
Maximum deviation of a measured value from the average	—	+22.9 —33.2	+7.3 —7.7	+18.6 —28.3	+25.2 —21.9		(+40.0)[1]) (—28.0)[1])

[1]) Such high values only for fraction No. 16, for which the coefficient of variation was 22.6%.

appropriate values of K_η and a (cf., eqn. 5-7), and compared with the $[\eta]$ value measured directly in the unfractionated sample.

Another test was carried out in Europe with a commercial sample of high-density PE and with fractions of a low-density PE (STRAZIELLE, 1976). In this case the coefficient of variation was 18.7 and 11.7%, respectively (see Table 19-4). A summary of the abundance of SEC papers is given in Table 19-5.

19.6.2. A working technique

MAY and KNIGHT (1971) used the following procedure: the chromatograms of twelve PS standards in THF were recorded, and the log M vs. V_e curve was plotted using the peak maxima. The log M values for every full cm^3, totalling about 70, were taken as interpolation points for the calculation of a *polynomial* of degree 5, by means of which the $\bar{M}_{n(u)}$ and $\bar{M}_{w(u)}$ values for the standards used were calculated from the chromatograms. Usually these values were too low, the deviations from the true values $\bar{M}_{n(t)}$ and $\bar{M}_{w(t)}$ ranging up to 10%. Therefore the log M vs. V_e curve was shifted, new interpolation points were taken, thus calculating a polynomial which eventually yielded the correct values for the standard samples.

Having established the calibration, the authors determined the *instrumental dispersion*. For that purpose samples with a known distribution were chromatographed and evaluated according to the method of TUNG and RUNYON (1969). The value σ_a^2 obtained for the instrumental dispersion in most cases yielded an overcorrection of the chromatograms for the calibration standards. After the fitting had also been completed in this respect, and the calibration curve and the final dispersion correction had been determined, for test purposes the $\bar{M}_w$, $\bar{M}_n$ and H values were calculated once again from the chromatograms of the calibration standards.

After this careful preparatory work, the experimental chromatograms were evaluated as follows: for a sufficiently great number of abscissa points, V_i, the curve was underlaid by Gaussian curves with a uniform standard deviation, σ_a. The area defined by each Gaussian curve was adjusted proportionally to the height, m_i, of the elution curve at the central

Table 19-5
Bibliography on gel chromatography

Polyethylene		Wagner and McCrackin (1979)
	in o-dichlorobenzene; 135 °C	Maley (1965); Nakajima (1968); Dawkins and Hemming (1975d); Ball and Scholte (1978)
	in trichlorobenzene	Maley (1965); Nakajima (1966, 1968, 1971a/b/c, 1972); Salovey and Hellman (1967); Wild and Guliana (1967); Bata et al. (1970); Holmström and Sörvik (1970); Cote and Shida (1971); Drott and Mendelson (1971); Hudson (1971); Miltz and Ram (1971); Otocka et al. (1971); Ward and Williams (1971); Wild et al. (1971, 1977); Ogawa et al. (1972a); Prechner et al. (1972); Ross and Frolen (1972); Williamson and Cervenka (1972); Cervenka and Williamson (1973, 1974); Westerman and Clark (1973); Servotte and De Bruille (1975); Peyrouset et al. (1975); Nakano and Goto (1976); Strazielle (1976); Barlow et al. (1977); Brauer et al. (1977a); Constantin (1977); Scheinert (1977); Scholte and Meijerink (1977); Ball and Scholte (1978); Ferguson et al. (1978); Eckhardt et al. (1980)
	in tetralin	Mindner and Berger (1979)
	in chloronaphthalene; 150 °C	Axelson and Knapp (1980)
Polypropylene	in o-dichlorobenzene	Ogawa et al. (1972); Atkinson and Dietz (1976); Vaughan and Francis (*prep.*) (1977)
Polybut-1-ene		Ukita and Abe (1972)
	in trichlorobenzene	Ring and Holtrup (1967)
Poly-1-olefins (C_4 ... C_{18})	in trichlorobenzene	Mori (1974)
Polyisobutylene	in trichlorobenzene	Cantow et al. (1967a, b); Chang et al. (1973)
	in tetrahydrofuran	Menin and Roux (1972); Mrkvičková et al. (1980)
	in carbon tetrachloride	Cooper and Johnson (1971)
	in heptane	Poznjak et al. (1977, 1980)
Polybutadiene		Smith and Thiruvengada (1970); Law (1971); Loo and Hsu (1971); Iwama (1972); Subramamian (1972); Fetters (1979); Janča et al. (1979a)
	in o-dichlorobenzene	Harmon (1965); Maley (1965); Hazell et al. (1968); Smith (1974)
	in dichloromethane	Perrault et al. (1971)
	in trichloromethane	Ambler (1980)
	in tetrahydrofuran	Grubišić (1967); Tung and Runyon (1969); Kraus and Stacy (1972); Perrault et al. (1971); Peyrouset and Panaris (1972), (*prep*), (1974); White et al. (1972); Adams et al. (1973); Smith (1974); Park and Graessley (1977a/b); Stojanov et al. (1978); Ambler (1980); Nentwig and Sinn (1980); Narasimhan et al. (*in mixture with PS*) (1981)
	in toluene	Ambler et al. (1974); Ambler (1980)
	in n-heptane	Poznjak et al. (1977)

Table 19-5 (continued)

Polybutadiene (branched)	in trichloromethane	MOTTWEILER and SCHEINERT (1981)
Polybutadiene (prepolymers, hydroxyl-terminated)	in tetrahydrofuran	ANDERSON et al. (1975); BACZEK et al. (1975)
Polyisoprene	in tetrahydrofuran	FETTERS and MORTON (1974); CAMPOS-LOPEZ and ANGULO-SANCHEZ (*Guayule rubber*) (1976); PASTUSKA et al. (*natural rubber*) (1979); NENTWIG and SINN (1980)
	in cyclohexane	DAWKINS and HEMMING (1975 a/b)
	in trichloromethane	DAWKINS and HEMMING (1975 a/b)
Polychloroprene	in methyl ethyl ketone	KATO and HASHIMOTO (1974 a); KATO et al. (1977)
	in tetrahydrofuran	KATO and HASHIMOTO (1974 a)
Polystyrene	in tetrahydrofuran	MEYERHOFF (1965 b, 1968); OSTERHOUDT and RAY (1967); GRUBIŠIĆ et al. (1967); ALLIET and PACCO (1968); LE PAGE et al. (1968); BEAU et al. (1969); HEITZ and PLATT (1969); TUNG and RUNYON (1969); HATTORI and HAMASHIMA (1970); HEITZ and WINAU (1970); MOORE (1970); HEITZ et al. (1971); PACCO (1971); MENIN (1972); MEUNIER and GALLOT (1972); MOHITE and MEYERHOFF (1972); PEYROUSET and PANARIS (1972), (*prep.*) (1974); ADAMS et al. (1973); COOPER et al. (1973), (*prep.*) (1975 a): JAMES and OUANO (1973); BERGER (1974, 1975); KATO et al. (1974) (*high velocity*), (1975); OTOCKA and HELLMAN (1974 a); PROBST et al. (1974); SLAGOWSKI et al. (1974) DAWKINS and TAYLOR (1975); HEITZ (1975); SPATORICO (1975); TIMM and RACHOW (1975); CHRISTOPHER (1976); GRUBIŠIĆ-GALLOT et al. (*recycling technique*) (1976); FETTERS (1976); DAWKINS et al. (1977, 1979, 1980); BRAKS and HUANG (1978); HASHIMOTO et al. (1978); ISHII et al. (1978 b); OUANO (*GPC Lalls*) (1978); HASSELL et al. (*UV exposed*) (1979); MCCRACKIN and WAGNER (1980); NARASIMHAN et al. (*in mixture with PBd*) (1981)
	in toluene	MOORE (1964); CANTOW and JOHNSON (1967 b); LITTLE et al. (1969, 1970); ZINBO and PARSONS (1971); PEYROUSET and PANARIS (1972), (*prep.*) (1974); COOPER and KISS (1973); COOPER et al. (1973); COOPER (1974); KATO, Ta. et al. (1975); MORI (1977); YAU et al. (1977); MCCRACKIN and WAGNER (1980)
	in trichloromethane	YAU et al. (1968); HEITZ et al. (1971); OTOCKA and HELLMAN (1974 a), (*vacancy chromatogr.*) (1974 b); BEREK et al. (1975); DAWKINS and HEMMING (1975 a)
	in dichloromethane	KNOX and MCLENNAN (1979); KEVER (1981 a/b)
	in methyl ethyl ketone	KATO et al. (1973, 1976); OTOCKA and HELLMAN (1974 a)
	in dimethylformamide	ZINBO and PARSONS (1971); UGLEA (1973); DAWKINS and HEMMING (1975 b)
	in benzene	ZINBO and PARSONS (1971); BEREK et al. (1975); MORI (1976)

Table 19-5 (continued)

	in cyclohexane	Dawkins and Hemming (1975a/b)
	in carbon tetrachloride	Cooper and Johnson (1971)
	in tetrachloroethylene	Mirabella et al. (1976)
	in o-dichlorobenzene	Hazell et al. (1968); Dawkins et al. (1971); Patel (1974)
	in trichlorobenzene	Determann and Michel (1966); Cantow et al. (1967b); Boni et al. (1968); Coll and Gilding (1970); May and Knight (1971); Cooper (1974)
	in trichloroethylene	Dawkins and Hemming (1975d)
	in dioxan	Zinbo and Parsons (1971); Otocka and Hellman (1974a)
	in trans-decalin	Dawkins and Hemming (1975c)
	in tetraphenylethylene	Dawkins and Taylor (1975)
	in mixtures	Berek et al. (1974): *Bzn/M* Berek et al. (1975): *Bzn/22.2% M*; *TCM/25.3% M*: *MEK/50% Hp* Kato and Hashimoto (1974b); Kato Y. et al. (1975): *MEK/11.3% M* Berek et al. (1976): *Bzn/22.2% M* Bakoš et al. (1976): *THF/PS* Šoltes et al. (1980): *MEK/10.7% M*
Polystyrene (isotactic)	in tetrahydrofuran	Guenet et al. (1977)
Polystyrene (branched)	in toluene	Berry (1971); Pannell (1972); Belenkij et al. (1973a); Ta. Kato et al. (1975, 1979); Budtov et al. (1976)
	in tetrahydrofuran	Park and Graessley (1977b); Popov et al. (1980)
	in methyl ethyl ketone	Berry (1971)
Polystyrene (microgel)	in dimethyl formamide	Price et al. (*0.1* M *LiBr*) (1977); Booth et al. (*0.0004 ... 0.03* M *LiBr*) (1980)
Polystyrene latex	in water/aerosol/KNO_3	Singh and Hamielec (1978); Coll and Fague (1980)
Polystyrene dicarboxylic acids (oligomers)		Månnson (1980)
Sodium polystyrene-sulphonate	0.2 M aqueous Na_2SO_4 solution	Spatorico and Beyer (1975)
	0.05 M $NaSO_4$ solution	Brown and Lowry (1979)
	phosphate buffer	Cooper and van Derveer (1978); Cooper and Matzinger (1979)
	10^{-4} M H_2O/NaCl	Rinaudo and Desbrieres (1980)
	THF-DMF/$LiNO_3$	Siebourg et al. (1980)
	dimethyl formamide	Dubin et al. (1977)
Tri-n-butylammonium salt	THF-DMF/$LiNO_3$	Siebourg et al. (1980)
Poly-α-methylstyrene		Leonard and Malhotra (1977); Malhotra and Leonard (1977)
	in tetrahydrofuran	Heitz et al. (1971); Kato et al. (1973); Probst et al. (1974); To. Kato et al. (1975); Stojanov et al. (1978)

Table 19-5 (continued)

	in trichlorobenzene	COLL and GILDING (1970)
	in toluene	Ta. KATO et al. (1979)
Polyphenyls	in benzene	SHULTZ et al. (1972)
Polymethyl methacrylate	in tetrahydrofuran	MEYERHOFF (1965b); SMITH et al. (1966); GRUBIŠIĆ et al. (1967); BALKE and HAMIELEC (1973); JANČA et al. (1975); SPATORICO (1975); SCHNABEL and SOTOBAYASHI (1976); JENKINS and PORTER (1980)
	in a tetrahydrofuran/PS mixture	BAKOŠ et al. (1976)
	in a chloroform/methanol mixture	KATIME (1979)
Polymethyl methacrylate (oligomers)	in dimethyl formamide	HEITZ (1973b)
	in tetrahydrofuran	SAMAY et al. (1978)
Poly(ethyl methacrylate)	in trichloroethylene	DAWKINS and HEMMING (1975d)
Poly(butyl methacrylate)	in tetrahydrofuran	SAMAY et al. (1978)
Poly(butyl methacrylate) (oligomers)	in dimethyl formamide	HEITZ (1973b)
Poly(octyl methacrylate)	in tetrahydrofuran	SAMAY et al. (1978)
Poly(lauryl methacrylate)	in tetrahydrofuran	MAHABADI and O'DRISCOLL (1977)
Poly(2-methoxy-ethyl methacrylate)	in tetrahydrofuran	JANČA et al. (1975)
Polyacrylic acid	in a buffer solution, pH = 4.95, 5.40, 2.00, 2.35	MIYATAKE et al. (1976)
	in a 0.2 M Na_2SO_4 solution	SPATORICO and BEYER (1975)
	in dimethyl formamide/(0.01 M LiBr)	DUBIN et al. (1977)
	in sodium acetate/(0.2, 0.5 M)	BUYTENHUYS and VAN DER MAEDEN (1978); FUKANO (1978)
Polyvinyl acetate	in tetrahydrofuran	GRUBIŠIĆ-GALLOT et al. (1972); PARK and GRAESSLEY (1977a, b); CANE and CAPACCIOLI (1978); ATKINSON and DIETZ (1979)
	in toluene	ATKINSON and DIETZ (*prep.*) (1979)
Polyvinyl acetate latex	in water/ 1 g/l aerosol; 1 g/l KNO_3	SINGH and HAMIELEC (1978)
Polyvinyl alcohol	in water	HASHIMOTO et al. (1978); MILLER and VANDEMARK (1980)
Polyester		VACHTINA et al. (1970)
	in tetrahydrofuran	BILLMAYER and KATZ (*aliphat. P.*) (1969); HELLWIG and SCHÖLLNER (1973); LEE (1978); WOJTANA (1979)
	in ethyl acetate	VACHTINA et al. (1976)

Table 19-5 (continued)

Polycarbonate		HOARE and HILLMAN (1971)
	in tetrahydrofuran	DUVDEVANI et al. (1971); CZARNECKA and DOBKOWSKI (1979); BRZEZIŃSKI and DOBKOWSKI (1980); DOBKOWSKI and BRZEZIŃSKI (1981)
	in trichloromethane	DAWKINS et al. (1973)
Polyethylene phthalate	in tetrahydrofuran	DÖRFFEL et al. (1976)
Polyethylene terephthalate	in o-tert-butylphenol, 110 °C	DÖRFFEL et al. (1976)
	in tetrachloroethane with 0.5% nitrobenzene, 25 °C	PASCHKE et al. (1977)
	in hexafluoro-2-propanol	DROTT (1977a)
	in chloroform	ISHIDA and KAWAI (1972)
	in tetrahydrofuran	MINARIK et al. (1977)
	in m-cresol	OVERTON et al. (1968)
	in an m-cresol/chlorobenzene mixture	ISHIDA and KAWAI (1972)
Polypropylene terephthalate	in trichloromethane	BIRLEY et al. (1980)
	in dioxan	BIRLEY et al. (1980)
	in tetrahydrofuran	BIRLEY et al. (1980)
Polybutylene terephthalate	in hexafluoroisopropanol	SLAGOWSKI et al. (1977)
Polyether		BEREK and NOVAK (1973)
	in methanol	BEREK and NOVAK (1972); BEREK et al. (1974)
	in tetrahydrofuran	BEREK et al. (1974)
	in dimethylformamide	BEREK et al. (1974); HEITZ (1979, 1981)
	in water	BEREK et al. (1974)
Polyethylene oxide		MORI (1978a)
	in tetrahydrofuran	KAZANSKIJ and SOLOVJANOV (1969); CROUZET and MARTENS (1975); Y. KATO et al. (1975); MORI and YAMAKAWA (1980); MURPHY et al. (1981)
	in trichloromethane	MORI and YAMAKAWA (1980); TALEB-BENDIAB and VERGNAUD (1980)
	in dimethylformamide	DUBIN et al. (1977)
	in water	HEITZ (1970); ENGELHARDT and MATHES (1977, 1979); FUKANO et al. (1978); HASHIMOTO et al. (1978); KATO et al. (1978, 1980a, 1980d)
Polypropylene glycol		VACHTINA et al. (1980)
	in tetrahydrofuran	SCHOLTAN and KRANZ (1967); CROUZET and MARTENS (1975)
	in toluene	SCHOLTAN and KRANZ (1967); AMBLER (1976)
	in dioxan/water	CONCIN et al. (1980)
Polymeric azo initiators	in dimethyl formamide	WALZ et al. (1977)
Polyvinyl ketones	in toluene	LYONS and CATTERALL (1971)

Table 19-5 (continued)

Cellulose	in cadoxene/water (1:1)	BEREK et al. (1977)
	in cadoxene/0.5 N NaOH	BAO et al. (1980)
Cellulose derivatives	in tetrahydrofuran	ALEXANDER and MULLER (1971)
Carboxymethyl cellulose	in 0.5 M sodium acetate	BUYTENHUYS and VAN DER MAEDEN (1978)
Cellulose nitrate	in tetrahydrofuran	MEYERHOFF (1965b); MEYERHOFF and JOVANOVIĆ (1967); SEGAL (1968); SEGAL et al. (1970); MULLER and ALEXANDER (1968); CHANG et al. (1973); OUANO et al. (1973)
	in ethyl acetate	HOLT et al. (1978)
Cellulose acetate butyrate	in tetrahydrofuran	VYAS et al. (1979)
Dextrans	in DMSO	VRIJBERGEN et al. (1978)
	in DMF	ENGELHARDT and MATHES (1977)
	in water	BOMBAUGH et al. (1968); BEAU et al. (1969); HEITZ and WINAU (1970); LANGHAMMER and SEIDE (1970); ENGELHARDT and MATHES (1977); BARKER et al. (*prep.*) (1978b), (1979); FUKANO et al. (1978); HASHIMOTO et al. (1978); SOETEMAN et al. (1978); KATO et al. (1980a/d); MILLER and VANDEMARK (1980)
	in water/0.2 M Na_2SO_4	GRANATH and KVIST (1967); SPATORICO and BEYER (1975)
	in water/NaCl	BASEDOW et al. (1976); HAGEL (1978)
	in water/1% diethylene glycol	LE PAGE et al. (1968)
	in a buffer solution, pH = 8.2	HALLER (1977)
	in water/0.2% KNO_3	BASEDOW and EBERT (*prep.*) (1979); BASEDOW et al. (1980)
	in water/1% NaN_3	ENGELHARDT and MATHES (1979)
	in 0.1 N HNO_3/0.1 N $NaNO_3$	TALLEY and BOWMAN (1979)
	in methanol/60% water	DREHER et al. (1979)
	in 0.1 M sodium acetate/acetic acid buffer	DREHER et al. (1979)
	in phosphate buffer	COOPER and MATZINGER (1979)
	in 0.02% potassium hydrogenphthalate aqueous solution	BARKER et al. (1981)
Dextran, grafted with polyacrylamides	in water/0.05 M potassium biphthalate	MCCORMICK and PARK (1981)
Oligosaccharides		NITSCH et al. (*sinistrin, prep.*) (1979)
	in water	JOHN and DELLWEG (1973)
	in 0.1 N acetic acid	MCCLENDON (1979)
Amylose	in dimethyl sulphoxide	VAN DIJK et al. (1976)
Polymers of β-lactam antibiotics	in water/0.1% NaCl	UENO et al. (1981)

Table 19-5 (continued)

Polyamides		Ede (1971)
	in m-cresol	Prince and Stapelfeldt (1968); Dudley (1972)
	in trifluoroethanol	Dark et al. (1968)
	in hexamethyl phosphorus triamide	Panaris and Pallas (1970)
	in o-chlorophenol	Walsh (1971)
	in benzyl alcohol	Pastuska and Just (1979)
	in hexafluoro-2-propanol	Drott (1977a)
Polyamides (trifluoroacetic derivatives)	in dichloromethane	Jacobi et al. (1981)
Poly(γ-benzyl L-glutamate)	in N,N-dimethylacetamide	Dawkins and Hemming (1975)
Polyamido acid	in dimethyl formamide	Nefedov et al. (1979)
	in dimethyl formamide/salt	Belenkij et al. (1977)
Aromatic polyamides	in DMF	Herold and Meyerhoff (1980)
	in H_2SO_4 (96%)	Arpin and Strazielle (1976)
Polyurethanes		Baker (1971)
(starting and intermediate products)	in ethyl acetate	Vachtina et al. (1976)
	in dimethyl formamide	Schulz (1976)
	in tetrahydrofuran	Maley (1965)
Polymeric isocyanates	in THF	Ambler et al. (1977)
Polyvinyl pyrrolidone	in water	Langhammer and Seide (1970)
	in dimethyl formamide	Dubin et al. (1977)
	in 1,2-dichloroethane/ethanol	Chaufer et al. (1977)
	in 0.1 M tris-HCl buffer	Engelhardt and Mathes (1979)
Poly(2-vinyl pyridine)	in N,N-dimethyl formamide	Ho-Duc et al. (1971); Gourdenne et al. (1972); Salomone et al. (1975); Mencer and Grubišić-Gallot (1979)
	in tetrahydrofuran	Mencer and Grubišić-Gallot (1979)
	in N,N-dimethyl acetamide	Dawkins and Hemming (1975)
	in 0.1 N HNO_3/0.1 N $NaNO_3$ mixture	Talley and Bowman (1979)
Poly(4-vinyl pyridine)	in 0.1 N HNO_3/0.1 N $NaNO_3$ mixture	Talley and Bowman (1979)
Polyacrylonitrile	in dimethyl formamide	Noack (1971); Albrecht (1972); Dawkins and Hemming (1975b)
	in dimethyl sulphoxide	Fritzsche (1967)
Polyacrylamide	in water	Abdel-Alim and Hamielec (1974); Hashimoto et al. (1978); El-Awady (1979)
	in formamide	Onda et al. (1979)
	in phosphate buffer	Cooper and Matzinger (1979)

Table 19-5 (continued)

	in 0.005 M KCl	ONDA et al. (1980)
	in 0.0167 M Na_2SO_4 (with 1% CH_3OH, 0.05% NaN_3, 0.002% surfactant)	OMORODION et al. (1980)
Poly(N-vinylacetamide)	in dimethyl formamide	DUBIN and MILLER (1977)
Polyvinyl chloride		BAISAL and KAUPPILA (1971); CHAN (1971); HSU (1971); LYNGAAE-JORGENSEN (1971); CHAN and WORMAN (1972); CRUSOS et al. (1972); NAKAO and KURAMOTO (1972)
	in tetrahydrofuran	MALEY (1965); GRUBIŠIĆ (1967); ROHN (1967); LE PAGE et al. (1968); BEAU et al. (1969); TUNG and RUNYON (1969); FELTER and RAY (1970); PACCO (1971); ABDEL-ALIM and HAMIELEC (1972); PEYROUSET and PANARIS (1972), (*prep.*) (1974); ANDERSSON (1973); SPATORICO (1975); JIŠOVÁ et al. (*prep.*) (1977); SÖRVIK (1977); HATTORI (*prep.*) (1978); LUKÁŠ (*chlorinated*) (1978); GILBERT et al. (*oligomers*) (1980)
Poly-1,4-dichloro-2,3-epoxy butane	in dimethyl formamide	PODDAR and FORSMAN (1976)
Fluoroether polymers	in 1,1,2-trichloro-1,2,2,-trifluoroethane (Freon 113)	KORUS and ROSSLER (1978)
Silicones		Larsen (1971)
	in cyclohexane	DAWKINS and HEMMING (1975c/d)
	in trichloromethane	DAWKINS and HEMMING (1975c/d)
	in toluene	SHAW and RODRIGUEZ (1967); KENDRICK (1969); ANDRIANOV et al. (1977); DODGSON et al. (*prep.*) (1978); MANDIK et al. (1979)
	in tetrahydrofuran	FRANCOIS et al. (1978)
	in trans-decalin	DAWKINS and HEMMING (1975b)
	in trichloroethylene	RODRIGUEZ et al. (1966); DAWKINS and HEMMING (1975d)
	in o-dichlorobenzene	DAWKINS et al. (1971)
	in isopropanol/TCM	SHIMONO et al. (1979, 1980)
Polyphenols	in methanol/4.4% Na_2SO_4	WILSON et al. (1979)
Lignins		LORA and WAYMAN (1980)
	in water	OBIAGA and WAYMAN (1974); CHUA and WAYMAN (1979)
	in a dioxan/water mixture	CONCIN et al. (1980)
Lignins (permanganate oxidation products)	in dimethyl formamide	MOROHOSHI and GLASSER (1979)
Lignosulfonates	in 0.1 M $NaNO_3$ (pH = 6.6)	[F 39, Chromatix]
	in water	FORSS and STEULUND (1973); BÖTTGER et al. (1976)

Table 19-5 (continued)

Phenolic resin intermediate products		DUVAL (1970); MONTAGUE et al. (1971)
	in tetrahydrofuran	QUINN et al. (1968); HEITZ et al. (1971); WAGNER, GREFF (1971, 1972); BRAUN et al. (1972); BRAUN and ARNDT (1978); MORI and YAMAKAWA (1980)
	in trichloromethane	MORI and YAMAKAWA (1980)
Aminoplastic intermediate products	in tetrahydrofuran	BRAUN and LEGRADIC (1972)
	in dimethyl formamide	BRAUN and PANDJOJO (1979a/b); BRAUN and BAYERSDORF (1980)
	in dimethyl sulphoxide	DUNKY et al. (*UF glues*) (1981)
	in dimethyl sulphoxide/water	KUMLIN and SIMONSON (1981)
	in ethyl acetate	BRAUN and BAYERSDORF (1979)
	in water	TAYLOR et al. (1980)
Epoxy resin intermediate products		BIESENBERGER et al. (1971a); HAMMOND et al. (1971)
	in tetrahydrofuran	MILES (1965); LARSEN (1968); EGGERS and HUMPHREY (1971); BATZER and ZAHIR (1975); BRAUN and LEE (1975, 1976); HAGNAUER (1980); MORI and YAMAKAWA (1980)
	in dimethyl formamide	HEITZ (1981)
	in trichloromethane	MORI and YAMAKAWA (1980)
Alkyd resin intermediate products		RYBICKA (1971); BELLENS (1973)
	in tetrahydrofuran	CHRISTENSEN and FINK-JENSEN (1973); WOJTANIA (1979)
Diallyl phthalate prepolymers	in tetrahydrofuran	BŁEDZKI et al. (1978)
	in trichloromethane	BŁEDZKI et al. (1978)
Phthalate ester	in tetrahydrofuran	MORI (1980a)
	in trichloromethane	MORI (1980a)
Paraffinic waxes C_{18-30}	in tetrahydrofuran	SOSA et al. (1978)
Copolymers:		
Ethylene-propylene		CEAUŞESCU et al. (1972)
	in tetrahydrofuran	SMITH (1974)
	in o-dichlorobenzene	SMITH (1974); OGAWA and INABA (1977)
	in trichlorobenzene	SCHOLTENS and WELZEN (*GPC/LALLS comb.*) (1981)
	in n-heptane	POZNJAK et al. (1977)
Ethylene-isobutylacrylate copolymer	in o-dichlorobenzene	FERLAUTO (1971)
Ethylene-vinylacetate copolymer	in o-dichlorobenzene	FERLAUTO (1971)
	in tetrahydrofuran	ECHARRI et al. (1979)
	in trichlorobenzene	BRAUER et al. (1977b)
	in xylene	BARLOW et al. (*prep.*) (1971)

Table 19-5 (continued)

Butadiene-acrylonitrile copolymer		Prokovskaja and Frolova (1969)
Butadiene-styrene copolymer	in tetrahydrofuran	Cantow et al. (1968); Runyon et al. (1969); Adams (1971); Morris (1971); White et al. (1972)
	in benzene	Barlow et al. (*prep.*) (1971)
	in toluene	Ambler et al. (1974)
α-methylstyrene-butadiene copolymer		Elgert and Wohlschiess (1977)
	in tetrahydrofuran	Stojanov et al. (1977)
But-1-ene — sulphone copolymer	in tetrahydrofuran	Bowden and Thompson (1975)
Styrene-acrylonitrile copolymer	in trichloromethane	Kranz et al. (1972); Mori (1980b)
	in dimethyl formamide	Kranz et al. (1972)
Styrene acrylonitrile copolymer, grafted on E-P copolymer	in tetrahydrofuran	De Chirico and Arrighetti (1979)
Styrene — methylmethacrylate copolymer		Terry and Rodriguez (1968)
	in tetrahydrofuran	Dondos et al. (1974); Bakoš (1976); Lovrič et al. (1976); Teramachi et al. (1978)
	in trichloromethane	Norris and Peaker (1973)
Styrene-ethylmethacrylate copolymer	in tetrahydrofuran	Samay et al. (1978)
Styrene-butylmethacrylate copolymer	in tetrahydrofuran	Samay et al. (1978)
Styrene-octylmethacrylate copolymer	in tetrahydrofuran	Samay et al. (1978)
Styrene-isoprene copolymer	in toluene	Ho-Duc and Prud'homme (1973)
Styrene — maleic-anhydride copolymer	in tetrahydrofuran	Chow (1976)
Styrene — vinylstearate copolymer	in tetrachloroethylene	Mirabella et al. (1975)
α-methylstyrene — methacrylonitrile copolymer	in tetrahydrofuran	Probst et al. (1974)
Alcyl acrylate — alcyl methacrylate copolymer	in tetrahydrofuran	Morimoto and Suzuki (1972)
	in dimethyl formamide	Varma and Patnaik (1979)
Methyl acrylate — vinylidene chloride copolymer	in tetrahydrofuran	Revillon et al. (1976)

Table 19-5 (continued)

Methyl methacrylate — vinylidene chloride copolymer	in tetrahydrofuran	REVILLON et al. (1976)
Methyl methacrylate — methacrylic acid copolymer	in tetrahydrofuran	OUANO (1978)
Methyl methacrylate — vinyl acetate copolymer	in trichloromethane	PROVDER and KUO (1976)
Methacrylate — methyl-N-vinylcarbamate	in tetrahydrofuran	RUNYON et al. (1969)
Acrylic acid — ethyl-acetate copolymer	in 0.2 M Na_2SO_4 solution	SPATORICO and BEYER (1975)
Acrylonitrile — Na-sulfonate copolymer	in dimethyl formamide/ $NaNO_3$	DOMARD et al. (1979)
Vinyl alcohol — vinyl acetate copolymer	in water/0.1% ethanol	HEUBLEIN et al. (1980)
Vinyl chloride — vinyl acetate copolymer	in tetrahydrofuran	CHEN and BLANCHARD (1972); JANČA and KOLINSKÍ (*prep.*) (1977); JANČA et al. (*prep.*) (1978, 1979c); MORI (*prep.*) (1978b)
Vinyl chloride — vinyl stearate copolymer	in tetrachloroethylene	MIRABELLA et al. (1975)
Vinyl chloride — vinylidene chloride copolymer	in tetrahydrofuran	REVILLON et al. (1976)
Block copolymers:		
Polystyrene — polyethylene oxide	in dimethyl formamide	WALZ and HEITZ (1978)
Polystyrene — polybutadiene block copolymer		OMOTO et al. (1979)
	in tetrahydrofuran	NESTEROV et al. (1978); TUNG (1979)
	in toluene	NESTEROV et al. (1978)
Polystyrene — polyisoprene block copolymer	in tetrahydrofuran	GRUBIŠIĆ-GALLOT et al. (1972)
Polystyrene — polymethylmethacrylate block copolymer	in tetrahydrofuran	GRUBIŠIĆ-GALLOT et al. (1972); DONDOS et al. (1974)
	in dimethylformamide	NESTEROV et al. (1978)
	in toluene	NESTEROV et al. (1978)
Polystyrene — polybutyl methacrylate block cop.	in trichloroethylene	DAWKINS and HEMMING (1975d)
Polystyrene — polyacrylonitrile block cop.	in dimethyl formamide	NESTEROV et al. (1978)
	in trichloromethane	MORI (1980b)
Polystyrene — poly-2-vinylpyridine block copolymer	in tetrahydrofuran	MENCER and GRUBIŠIĆ-GALLOT (1979)
	in dimethyl formamide	MENCER and GRUBIŠIĆ-GALLOT (1979)

Table 19-5 (continued)

Polyisoprene-poly-α-methylstyrene block copolymer	in toluene	NESTEROV et al. (1978)
Polymethyl methacrylate cellulose graft copolymer	in dimethyl formamide	SIMIONESCU et al. (1972)
Ethylene-co-propylene, grafted with styrene-co-acrylonitrile polymer	in tetrahydrofuran	DE CHIRICO et al. (1979, 1981)
Poly(1-pentenylene), grafted on butadiene-styrene copolymer	in tetrahydrofuran in dimethyl formamide	HOFFMANN and URBAN (1977) HOFFMANN and URBAN (1977)
Silicic acid	in 0.1 M NaCl solution (pH = 2)	TARUTANI (1970)
	in water/HCl (pH = < 2 — 9.5)	SHIMADA and TARUTANI (1979)
	in mineral acids (1—5 N)	IWASAKI et al. (1980)
Silicic acid (colloidal)	in 0.02 M triethanolamine	KIRKLAND (1979)
	in water/0.5 g/l aerosol OT/KNO_3	COLL and FAGUE (1980)
Vinyl silylate and its copolymer with VAC	in tetrahydrofuran	INOUE and KOHAMA (1981)
Aluminosilicate (colloidal)	in 0.001 M NH_4OH	KIRKLAND (1979)
Metal chelate -diketones	in ethyl acetate	NODA et al. (1979)
Polyphosphates		OHASHI et al. (1966); FELTER et al. (1968); NEDDERMEYER and ROGERS (1968, 1969)
	in 0.1 M NaCl	KOUCHIYAMA et al. (1978)

point, V_i. The sum of the ordinates of all the Gaussian curves at each value of V_i should agree with m_i. If this did not happen to be the case at the first attempt, then the evaluation was repeated with a modified proportionality factor or changed foot points, V_i, until the summation curve coincided with the elution curve within given limits. With the values $(\mathrm{d}m_i/\mathrm{d}V_i)$ determined in this way and the derivative of the calibration function, $(\mathrm{d}V_i/\mathrm{d}\log M_i)$, the logarithmic molar mass distribution

$$\frac{\mathrm{d}m_i}{\mathrm{d}\log M_i} = \frac{\mathrm{d}m_i}{\mathrm{d}V_i}\,\frac{\mathrm{d}V_i}{\mathrm{d}\log M_i} \qquad (19\text{-}5)$$

(YAU and FLEMING, 1968) was plotted.

MAY and KNIGHT used two chromatographic apparatus. One of them (GPC 3) consisted of four columns, each 1.22 m long and 7.8 mm in diameter, packed with PS gels with nominal pore sizes of 10^4, $2 \cdot 10^5$, $2 \cdot 10^5$ and 10^6 Å; the other one (GPC 4) had three analogous columns, but the respective pore sizes were 10^4, $2 \cdot 10^5$ and 10^6 Å. Parts (a) and (b) of

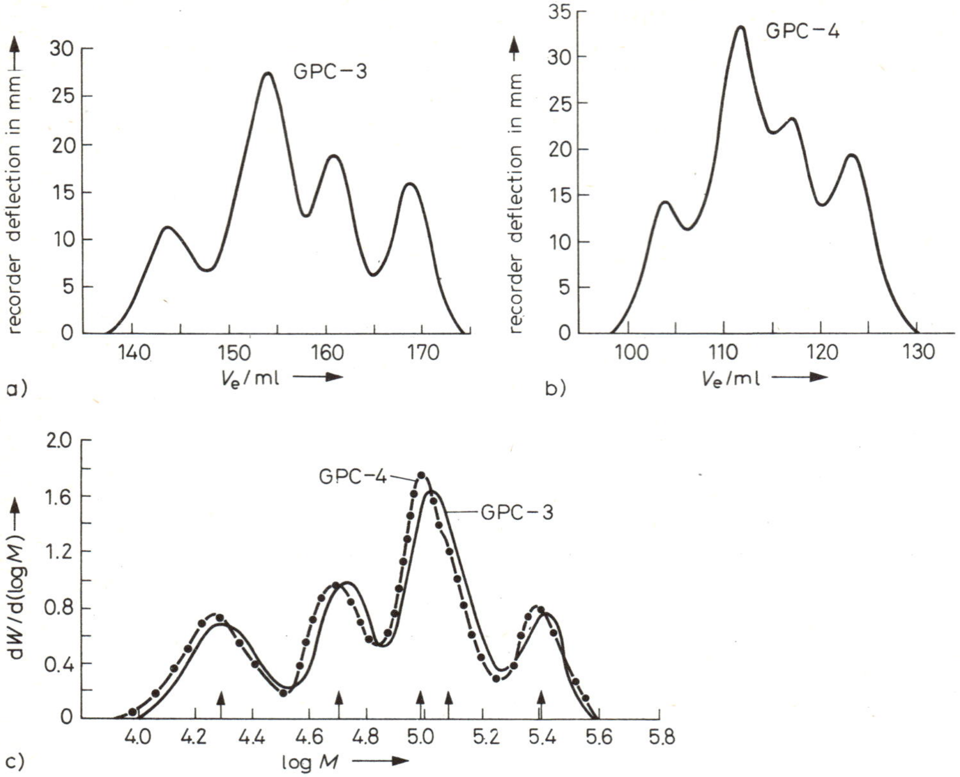

Fig. 19-32
Investigation of a mixture of five polystyrene samples with two different SEC apparatus

a) Raw elugram from a four-column set ("GPC 3")
b) Raw elugram from a three-column set ("GPC 4")
c) Molar mass distribution determined from a) and b). (The arrows pointing upwards from the abscissa indicate the components.)

(according to MAY and KNIGHT, 1971).

Fig. 19-32 show the chromatograms recorded using these column combinations for a test mixture of PS standards. From the raw curves it can hardly be observed that they were obtained from one and the same sample, but after the evaluation described above they yielded the corresponding molar mass distributions shown in Fig. 19-32 (c). Fig. 19-33 (a) shows the analogous curves for an ordinary polystyrene. The differences are not greater than the uncertainty in repetitive analyses on one and the same apparatus (see Fig. 19-33 (b)).

19.6.3. Size exclusion chromatography with long columns

Apart from the endeavour to achieve a high resolution by fine gel particles (KATO et al., 1973), efforts have also been made to achieve this aim by the use of long columns (FETTERS and MORTON, 1974). On such an apparatus with seven columns ($L = 7 \times 1.22$ m, $d_C = 7.8$ mm), packed with PS gels, FETTERS (1976) was able to establish the identity of polystyrene samples

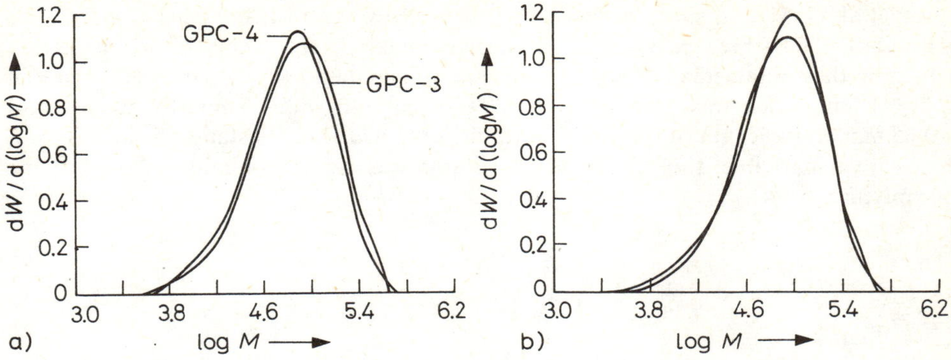

Fig. 19-33
Molar mass distribution of a broad polystyrene sample
a) Results of analyses on two different apparatus
b) Double analysis on one and the same apparatus ("GPC 3")
(according to MAY and KNIGHT, 1971).

and to detect a 4% concomitant of double the molar mass in the standard reference material NBS 705. In a paper published in 1977 (AMBLER et al.), even a twelve-column set with a total length of $12 \times 1.22 = 14.64$ m was used.

SCOTT (1980) pointed out that one of the advantages of *microbore columns* is that, by means of column concatenation, they yield plate numbers which really increase proportionally with the length of a separating path. A column set consisting of fourteen units, each 1 m long, $d_C = 1.0$ mm, packed with Spherisorb® 5 μm, was found to produce 750000 theoretical plates. Here, modified Swagelok® unions (KUČERA and MANIUS, 1981) ensured that the individual tubes directly butted against each other.

HEITZ (1975) used 10 m and 20 m columns, which simply consisted of a Teflon® tubing filled with the separating gel and wound up into a helical coil. As the gels employed could only resist a pressure of 0.8 MPa, the simple and inexpensive construction of the columns was of no disadvantage.

The purpose of such extra-long columns is the same as in recycling: a greater number of theoretical plates should be achieved with the separating material of a given quality. The higher prime cost of extra-long columns is counterbalanced by advantages in their operation: there is no additional peak broadening caused by valves, detectors or pumps, it is possible to carry out staggered injections and the operation requires almost no supervision.

19.6.4. Micro SEC

Using barely 1.5 μg of sample substance in an injection volume of 0.03 μl, ISHII et al. (1978) carried out the separation of five PS standards as shown in Fig. 19-34. For this purpose they used a PTFE column with an interior diameter of 0.5 mm, packed with 10 μm particles of a PS gel, which had an exclusion limit of $M_{\lim} = 10^5$. The detector employed was a UV spectrophotometer with a 0.13 μl flow-through type cell. (Another cell with a volume of 0.63 μl did not yield the required resolution.)

Kever et al. (1981a, b) successfully realized micro SEC with LiChrospher® Si-100 (two parts) and Si-1000 (three parts) as a separating material. By means of sedimentation classification they obtained a fraction of the separating material with a particle diameter of $d_p = 7 \pm 1$ μm, which was used to pack a PTFE column with an interior diameter of 0.6—0.65 mm and a length of 0.30 m. The results obtained with PS standards in dichloromethane are remarkable, too. The injected volume was 0.5 μl, containing 2.5 μg of the sample mixture.

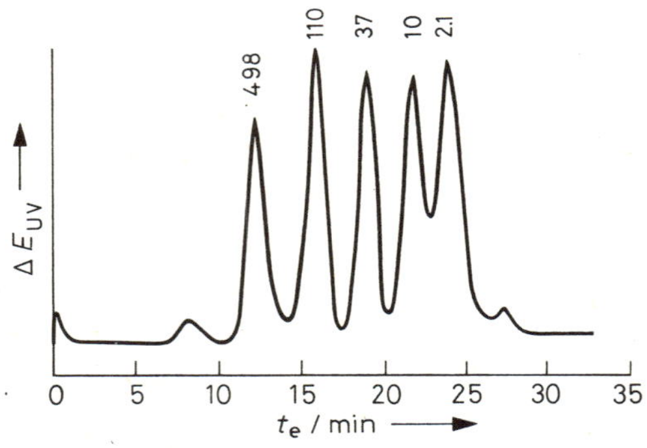

Fig. 19-34
Micro-SEC separation of five polystyrene standards, with a substance input of 1.41 μg
Column: $L = 0.34$ m; $d_C = 0.5$ mm; wall material: Teflon®; packing: Shodex® A-804 polystyrene gel, $d_p = 10$ μm; eluent: THF; $v = 2$ μl · min^{-1}; injection volume $V_0 = 0.03$ μl
The numbers beside the peaks indicate the molar mass in kg · mole^{-1}. The inputs were 0.27, 0.27, 0.24, 0.30, 0.33 μg, respectively (from 498 to 2.1).
(according to Ishii, Hibi, Asai and Jonokuchi, 1978).

19.7. Size exclusion chromatography of copolymers

Copolymers with no measurable chemical composition distribution (CCD) can be treated like homopolymers in exclusion chromatography. However, if there is a chemical heterogeneity, then considerable deviations from the normal behaviour may occur (Ogawa and Inaba, 1977; Ogawa, 1979). A general description of the SEC of complex polymers, which include copolymers with a CCD, was given by Hamielec (1981).

Let us consider the case of extreme heterogeneity in a binary system, i.e., a *mixture of two homopolymers* (Fig. 19-35): owing to differences in their hydrodynamic volume, chemically different polymer species will generally be eluted in different volume fractions, in spite of equal molar masses. In Fig. 19-35 (a) the vertical shading indicates the component which, owing to its chemical properties, has the smaller hydrodynamic volume, and hence exhibits the longer retention time. In a mixture where the smaller molecules belong to this component and the larger ones to a faster migrating component (part (a) of Fig. 19-35), a large additional broadening of the elution curve results (b). On the other hand, if the slower component has the larger molecules (c), then the elution curve is narrower than for a corresponding, chemically homogeneous polymer (d). This example shows that the elution curves of copolymers may differ widely even if their molar mass distribution (MMD) is the same. Elution curves always show the distribution of the hydrodynamic volume. If

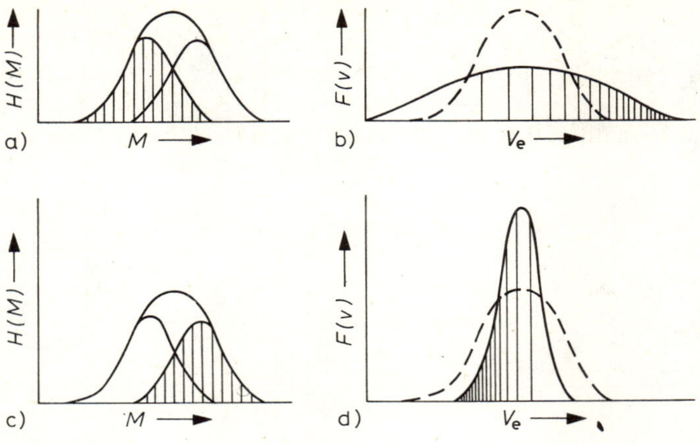

Fig. 19-35
Disturbance of the SEC elution curve by chemical heterogeneity: fictitious broadening (a → b) or narrowing (c → d)

Part (a) shows the molar mass distribution of a sample in which the short-chain components consist of the slower-travelling polymer (shaded). On their way through the column, these components are additionally delayed, so that the observed elution curve (part b) is broader than it would be if the molar mass distribution assumed in (a) consisted of polymer homologues having no deviations in their chemical structure. This elution curve is shown by the broken line in (b).
The counter example (c) → (d) was realized by polybutadiene, as the faster-component, and polystyrene. HOFFMANN and URBAN (1977) observed that in tetrahydrofuran PS with $M = 470\,000$ and PBd with 190000 were eluted together in a narrow band. The shading in parts (b) and (d) of the figure indicates where the slower polymer will accumulate.

all of the sample components have the same structure, this distribution is similar to the MMD, but for copolymers the hydrodynamic volume depends on the molar mass *and* the chemical composition.

19.7.1. Molar mass distribution (MMD) and chemical composition distribution (CCD)

As a matter of principle, copolymers have a *statistical* CCD, the width of which decreases with increasing chain-length (STOCKMAYER, 1945). In the macromolecular range it is below the experimental detection limit (STEJSKAL et al. 1981). The CCD due to conversion, which is caused by a depletion of one of the monomeric species in the polymerizing mixture, can be considerable for conversions above 10% (STEJSKAL and KRATOCHVIL, 1978).

The schematic diagram shown in Fig. 19-36 helps to explain the distribution as a function of the molar mass, M, and the molar composition, x: the curve $H(M)$ is a cross-section of the distribution, which shows the MMD at a certain value x, whereas the curve $H(x)$ shows the CCD at a certain value of M. The lowest points of the two cross-sections plotted lie (together with those of all other cross-sections, not shown) on a curve from which the *correlation between MMD and CCD* can be observed. The inserted diagrams a—d in Fig. 19-36 show such figures in the $x - M$ plane. In (a) and (b) there is no correlation. Fig. (a) applies to a polymer with a high chain-length heterogeneity, H, and a low

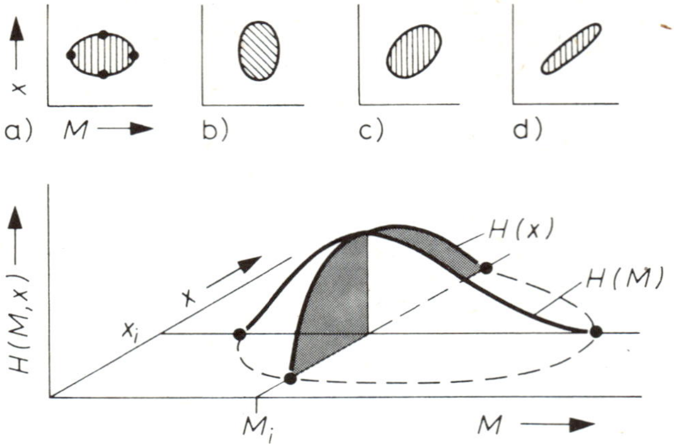

Fig. 19-36
Molar mass and composition distribution in copolymers (schematic)

The curve $H(x)$ shows the chemical composition distribution (CCD) for the molar mass value M_i, the curve $H(M)$ shows the molar mass distribution (MMD) for a selected value, x_i, of the composition. Such cross sections can be realized in an infinite number of ways. The lowest points of all the curves lie on a closed curve in the x-M plane.
In the parts (a)—(d) of the figure, the slices indicate possible ways of fractionation by SEC they are parallel to the x-axis if the hydrodynamic volume depends only on the molar mass, but not on the composition (Figs. a, c, and d). In these cases, from the chemical composition of the eluate components it can be concluded whether or not there is a correlation between MMD and CCD. In case (b) the hydrodynamic volume depends on M and on x. As indicated by the slices, in this case the eluate composition varies with the elution volume, although there is no correlation between MMD and CCD.

chemical heterogeneity, U, whereas in Fig. (b) U prevails. Fig. (c) indicates a low correlation, and Fig. (d), a strong correlation between M and x.

Information concerning CCD and MMD in copolymers can be obtained by *cross-fractionation* (Litmanovič and Štern, 1967), in which the polymer is first subdivided in such a way that the mole fraction, x, in the individual fractions increases with the molar mass. Then these pre-fractions are in their turn fractionated in such a way that x varies inversely with M (see Fig. 19-37). Only such a combined separation can yield information about the complex overall distribution. A single cross-section always yields only one profile of the distribution; this may give a useful piece of information if it is known how the sought-after distribution is projected on this profile. For SEC elution curves this is seldom the case. Among the reasons is the fact that these curves are functions of the hydrodynamic volume, V_h. If in a special case V_h is not influenced by the composition, then the successively eluted components represent cross-sections parallel to the x-axis, as indicated in the Figs. (a) and (c) of 19-36. In this case the components have a constant mean composition, $\bar{x}$, if there is no correlation between MMD and CCD (Fig. (a)), whereas they exhibit a drift of $\bar{x}$ with V_e if there is a correlation, (c). However, if V_h depends on the chemical composition, then the successively eluted components may be separated from each other, as shown by the oblique lines in Fig. (b). In this case $\bar{x}$ varies with V_e, although there is no correlation between MMD and CCD. Consequently the use of a second, composition-dependent detector is not in each case sufficient for a definite determination of the elution profile (see also Section 19.7.3.5.).

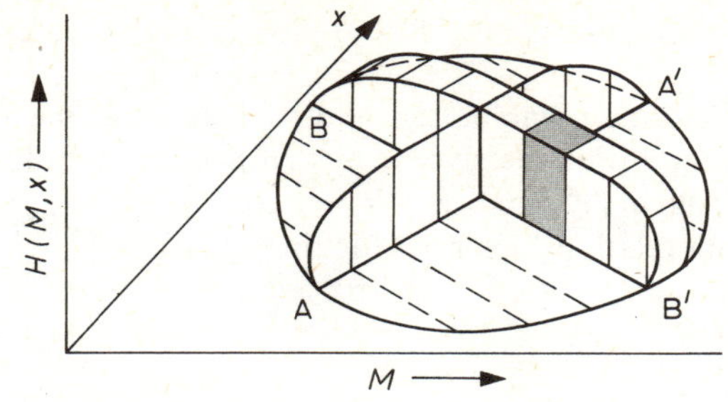

Fig. 19-37
Principle of the cross-fractionation of copolymers

The substance is first fractionated in such a way that x increases with increasing M (section A → A'). The prefractions are further separated in such a manner that x decreases with increasing M (section B → B'). The shaded figure indicates one of the final fractions.

19.7.2. Practical examples

One of the most important questions in the SEC of copolymers is to decide whether or not the hydrodynamic volume depends on the composition. For styrene-methyl acrylate copolymers with 46.6, 57.3 and 77.9 mole-% MA, this was determined by plotting [η] vs. $\bar{M}_n$, where the points for a total of 24 fractions determined a common straight line to a good approximation (see Fig. 19-38 (a)) (TERAMACHI et al., 1978). Thus, within the mentioned limits, for this system it was ensured that the composition had no significant influence on

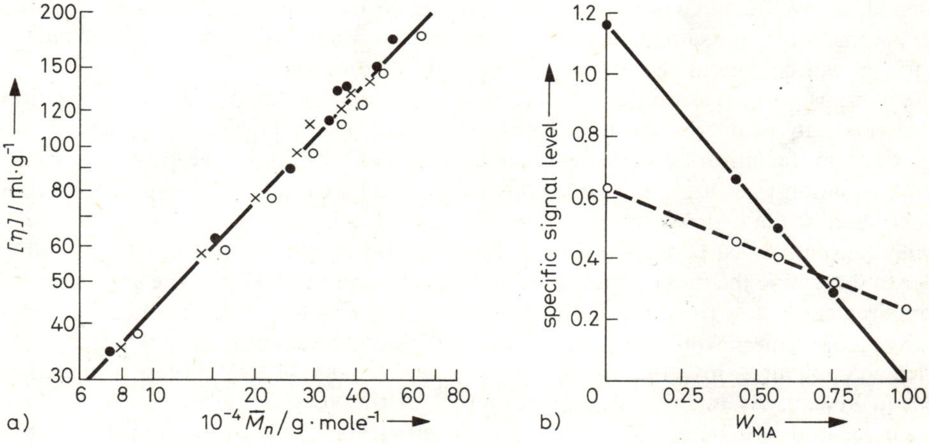

Fig. 19-38
Preliminary investigations for the SEC of copolymers of styrene and methyl acrylate in THF

a) Combined [η] vs. M plotting for copolymers with 46.6 (×) 57.3 (○) and 77.9 (●) mole-% MA
b) Response of the UV (●) and the RI detector (○) (cf., eqns. 19-6, 19-7) as a function of the MA content (according to TERAMACHI, HASEGAWA, AKATSUKA, YAMASHITA and TAKEMOTO, 1978).

the hydrodynamic volume, and consequently the SEC followed the scheme outlined in Fig. 19-36 (a) and (c).

The next task is to select a detector combination and to determine the response to composition, w_I, and concentration, c. For the styrene-methyl acrylate copolymers in the paper quoted, an R.I. detector (signal: r_{RI}) and a UV detector (signal: r_{UV}) were used. In their linear range, the following relationships hold for the signals:

$$r_{UV} = c(1 - w_I)\, K_{UV,\, II} \tag{19-6}$$

$$r_{RI} = c[w_I K_{RI,\, I} + (1 - w_I)\, K_{RI,\, II}] \tag{19-7}$$

The index I refers to the methyl acrylate, i.e., w_I is the mass fraction of MA in the copolymer and $K_{UV,\, II}$ is the contribution of the styrene content to the signal of the UV detector. The two $K_{RI,\, i}$ in eqn. (19-7) represent the specific contributions of styrene (II) and methyl acrylate (I) to the signal, r_{RI}, of the differential refractometer. These K_i values are determined from the plot shown in Fig. 19-38 (b).

The third problem is to show that, for the system under investigation, the universal calibration holds true, i.e., that the CCD and the deviations in the $[\eta]$ vs. M relationships of the different components do not disturb an evaluation by the methods discussed in Section 8.3.2. This can be proven as follows: the log $(M[\eta])$ vs. V_e curve recorded with reliable standards (usually PS) is used to convert the elution curves of copolymers with broad distributions point by point into $M_i[\eta]_i$ values. If, as assumed, the $M[\eta]$ calibration is in fact applicable to all the components of the eluted samples, then the summation of all the values $M_i[\eta]_i$ should give the $M[\eta]$ value, which results from direct measurements of the viscosity and light scattering of the whole sample. This test neither presupposes the knowledge of the $[\eta]$ vs. M relationships nor a chemical homogeneity of the SEC fractions.

After this preparatory work the elution curves of the samples under investigation can be recorded. Usually, for binary copolymers it suffices to use two detectors which respond to the composition of the sample in a different but known way, so that the average chemical composition can be determined for each component of the eluate.

If the composition has no influence on the hydrodynamic volume, then the evaluation does not raise any problems, because, just as for a homopolymer, the molar masses can be derived from the universal calibration curve (see Section 8.3.2.) by means of the Mark-Houwink equation (5-8) using the same constants K_η and a for all of the components. On the other hand, if the hydrodynamic volume depends on the composition, then the Mark-Houwink constants must be known for all chemically different components of the sample in order to determine the molar mass from the respective value of V_e and the composition. This requirement is not trivial insofar as it is not always easy to derive the dissolution behaviour of copolymers from the behaviour of the parent homopolymers. The ethylene-propylene copolymers investigated by Ogawa and Inaba (1977) represent a rather convenient system. As in this case the exponents of the viscosity relationship (5-8) are rather similar ($a_{PE} = 0.74$; $a_{PP} = 0.78$), and the coefficients K_η likewise do not differ from each other too greatly ($K_{\eta,\, PE} = 4.9 \cdot 10^{-4}$; $K_{\eta,\, PP} = 1.0 \cdot 10^{-4}$), it was possible to use the following approximations:

$$a_{Co-EP} = (a_{PE} \cdot a_{PP})^{1/2} \tag{19-8}$$

$$K_{Co-EP} = w_{PE} \cdot K_{PE} + w_{PP} \cdot K_{PP} - 2 w_{PE} : w_{PP} (K_{PE} \cdot K_{PP})^{1/2} \tag{19-9}$$

However, the determination of the Mark-Houwink constants for copolymers of different compositions usually requires measurements on an extensive sample material, which should include homogenous substances in a wide range of compositions and molecular sizes. This means a great amount of work.

It can, however, be reduced if the viscosity of the eluate can be continuously measured and converted into $[\eta]$ by means of the total concentration, which may be determined by a R.I. detector (GRUBIŠIČ-GALLOT et al., 1972). Then the determination of M from the hydrodynamic volume is again very easy, but the demands made upon the accuracy of the viscosity measurement and the reliability of the conversion of the viscosity measured at a single concentration into the limit $[\eta]$ are considerable. Moreover the influence of the composition on the refractive increment must be taken into account; cf., eqn. (19-7).

Thus reservations regarding this variant are plausible ("... it is not certain that the universal calibration method will necessarily bring about better results." TUNG, 1979). REVILLON et al. (1976) applied this method in the investigation of vinylidene chloride copolymers. However, the molar masses determined in different ways exhibit as a whole such a low correlation that arguments for or against the viscosity method could not be derived from these investigations.

In most cases largely simplified calculations are carried out to determine molar masses from the respective elution volumes and the universal calibration curve determined by means of PS standards.

For binary block copolymers, RUNYON et al. (1969) used log M vs. V_e curves of the relevant homopolymers, from which the values M_I for one of the parent polymers and M_{II} for the other would follow at a certain V_e. Using these data as well as the mass fractions w_I and w_{II}, the authors calculated the molar mass of the copolymer by linear interpolation on a log scale:

$$\log M_{Co,R} = w_I \cdot \log M_I + w_{II} \cdot \log M_{II} \tag{19-10}$$

Although problematic from a theoretical point of view (HO-DUC and PRUD'HOMME, 1973), for block copolymers this equation yields good results as compared to other methods (TUNG, 1979). Obviously the influence of hetero-contacts between the blocks is of minor importance for the polymer structures investigated. Deviations occur only if the values M_I and M_{II}, which would be calculated from a certain V_e value for the two parent polymers, differ from each other by more than 100% (CHANG, 1971).

In the characterization of statistical copolymers, MIRABELLA et al. (1975a) used the "*working molar mass*", $M_{Co,M}$. It was calculated from an averaged molar mass for the monomeric units, $\bar{M}_0$, the value M_{PS}, which can be derived from the PS calibration curve for the measured elution volume, and $M_{0,S}$ for styrene

$$M_{Co,M} = \frac{\bar{M}_0}{M_{0,S}} M_{PS} \tag{19-11}$$

$\bar{M}_0$ is obtained from the primary molar masses $M_{0,I}$ and $M_{0,II}$ of the chemical structural units of the copolymer:

$$\bar{M}_0 = w_I M_{0,I} + w_{II} M_{0,II} \tag{19-12}$$

As $M_{PS}:M_{0,S} = P$ is the degree of polymerization, the concept of the "working molar mass" amounts to the hypothesis that for copolymers, to a first approximation, log P vs.

V_e can be used like a universal calibration curve. This was explicity stated by Stojanov et al. (1977) and by Gilding et al. (1981).

For a copolymer of vinyl chloride and vinyl acetate, Mori (1978) calculated the molar mass by means of Q factors (cf., Section 8.3.1.), which were assumed to increase with the molar mass.

19.7.3. Determination of the chemical composition

In the SEC of chemically heterogeneous copolymers it is necessary to determine the chemical composition of successively eluted fractions. In this section, several possibilities of doing this will be discussed, which can mainly be used in an on-line operation. In principle, the use of a conductivity detector in the elution of PAN with sulphonate groups as mentioned in Section 19.3.2., also belongs to the set of problems discussed here. The same holds for the *scintillation analysis* of copolymers prepared from specially labelled monomers. Norris and Peaker (1973) reported such investigations of copolymers of ^{3}H methyl methacrylate and inactive styrene.

In what follows, methods will be discussed which in principle are generally applicable and do not require any specific preparations. It will largely depend on the properties of the copolymer system which of these methods is most suitable for a certain problem.

19.7.3.1. UV detection

The UV absorption at 254 nm, the main emission wavelength of mercury vapour lamps, can be used in determining the composition if at this wavelength the absorption of one of the components prevails sufficiently and the eluent itself is transparent. With the use of spectrophotometer detectors, such investigations can also be performed at any other wavelength. By use of combinations of an R.I. and a UV detector, elution curves of styrene-butadiene copolymers in THF (Runyon et al., 1969) (see Fig. 19-39), of mixtures of polystyrene and polybutadiene (Narasimhan et al., 1979), α-methylstyrene-methacrylonitrile copolymers in THF (Probst et al., 1974), copolymers of α-methylstyrene and butadiene (Elgert and Wohlschiess, 1977) and styrene-methyl acrylate copolymers in THF (Teramachi et al., 1978) were recorded. The examples are analogous to one another, because in each case the copolymers consisted of aromatic units (styrene or α-methylstyrene) and the units of a second monomer, which, like the solvent, would not absorb at the wavelength selected.

The number of the monomeric units with UV absorption which are immediately linked to each other, that is the sequence length, has an appreciable influence on the total absorption (Brüssau and Stein, 1970; Garcia-Rubio et al., 1979). As the sequence length is generally not known, variations may lead to errors of 10—20% with respect to the composition (Stützel, 1971). A spectrophotometer which repeatedly records and stores the whole spectrum during elution [F 35] is of great interest for copolymer analysis.

19.7.3.2. IR detection

The composition of the copolymer in the eluate can be analyzed by infrared measurements if the eluent does not itself absorb at the points where the polymer bands occur. Survey maps of the infrared windows of the various solvents facilitate the selection of a suitable eluent [F 14]. Methyl acrylate units were determined in THF by means of the carbonyl band

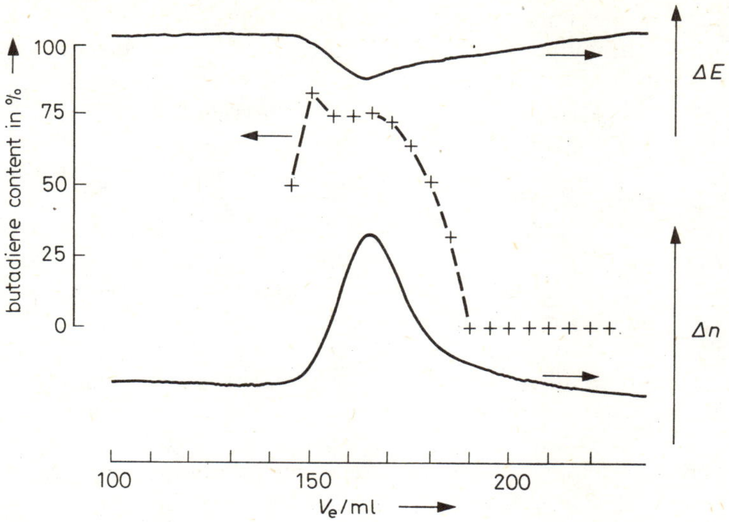

Fig. 19-39
Elugram of a butadiene-styrene block copolymer in tetrahydrofuran, recorded by a flow refractometer and a UV detector

The peak values differ slightly from one another, which indicates a chemical heterogeneity. The values of the butadiene content as calculated from the detector curves were inserted in the figure.
(according to RUNYON, BARNES, RUDD and TUNG, 1969).

at 1735 cm^{-1} (RUNYON et al., 1969), vinyl stearate in trichloroethylene at 1760 cm^{-1} (MIRABELLA, 1975a, b), t-butyl methacrylate in trichloroethylene at 1724 cm^{-1} (DAWKINS and HEMMING, 1975d) vinyl acetate and methyl methacrylate in chloroform at 1730 cm^{-1} (PROVDER and KUO, 1976). Styrene units were recorded in trichloroethylene at 697 cm^{-1} (DAWKINS), in chloroform at 1603 cm^{-1} (PROVDER) and in tetrachloroethylene at 1503 cm^{-1} (MIRABELLA et al., 1975b).

Ideally, the concentration of each monomer species is measured at a characteristic wavelength. Inspired by the sequence-length dependence of UV absorption, MIRABELLA (1980) investigated the effect of the molar mass on the IR absorption of polystyrene samples. In tetrachloroethylene, the absorption at 1493 cm^{-1} proved independent of the molar mass in the whole range investigated ($600—2 \cdot 10^6$ g · $mole^{-1}$), whereas another band at 1450 cm^{-1} decreased significantly with the transition from 2100 to 600 g · $mole^{-1}$, but was independent of the molar mass above 2100 g · $mole^{-1}$.

In some investigations, IR spectra were recorded in transmission cuvettes by the stopped-flow technique (e.g., by MIRABELLA et al., 1976). In other investigations the spectra of solvent-free eluate components were measured, e.g., of VC/VAC copolymers (JANČA and KOLÍNSKÝ, 1977; MORI, 1978). An apparatus specifically developed for the IR analysis of the small substance quantities which can be obtained from HPLC is commercially available [F 29].

19.7.3.3. Microchemical analysis

If an adequately efficient and sensitive method can be found, the composition of the eluate components can also be determined by a chemical analysis. For example, the determination of the chlorine content according to SCHÖNIGER in the SEC fractions of VC/VAC copolymers can be mentioned (JANČA and KOLÍNSKÝ, 1977), but also the use of pyrolytic gas chromatography for the analysis of the SEC eluates of block copolymers of P(S/MMA) (BELENKIJ et al., 1975), P(S/Bd), P(S/AN) and P(S/αMS) (NESTEROV et al., 1978). Using this technique, MORI (1980) investigated an industrial S/AN copolymer with an acrylonitrile content of 24.0 wt.-% (38.3 mole-%). About 70% of the polymer exhibited molar masses ranging between 50000 and 400000 g · mole^{-1} and a content of 24% AN. The components (circ. 15%) with higher molar masses had a *lower* AN content, which decreased with increasing M to a bare 22% (35.4 mole-%). At the other end it *increased* with decreasing M to 29.5% (45.1 mole-%). This heterogeneity is remarkable, for the system exhibits an azeotropic point at 38 mole-% of acrylonitrile.

An ozone *reaction detector* has been developed by POZNJAK et al. (1977). In this detector the eluate is continuously contacted with an ozon-containing gas flow, and the content of double bonds in the eluted polymer fraction is determined from the ozone consumption. The detector was used in the analysis of ethylene-propylene terpolymers. Although reaction detectors are not as simple as detectors measuring a physical quantity, their application as highly specific and sensitive detection instruments in the HPLC of small molecules provides an ever-increasing stock of experience, so that the reaction detectors will probably also find a more widespread application in the SEC of copolymers in future.

19.7.3.4. Automatic turbidimetric titration of SEC eluates

HOFFMANN and URBAN (1977) developed an apparatus for the automatic analysis of SEC eluates by turbidimetric titration, in which a precipitant is added to 5 ml eluate in a tit-

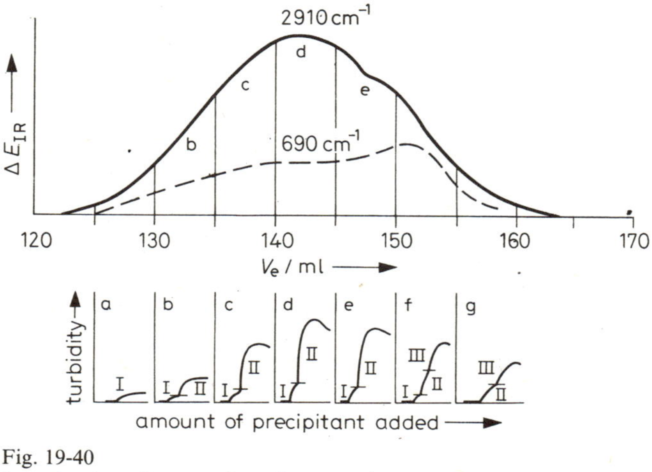

Fig. 19-40
Investigation of styrene-butadiene copolymers with grafted cyclopentene by SEC and turbidimetric titration

Elution curve after a separation in 4 × 1.20 m columns packed with Styragel®; IR — record: 2910 cm^{-1} — CH and CH_2; 690 cm^{-1} — phenyl.

Part of the highly homogeneous S-Bd copolymer (III) emerges ungrafted at $V_e \geqq 150$ ml. Apart from the graft product (II), the turbidity curves also indicate a small quantity of poly(1-pentenylene) homopolymer (I).

(according to HOFFMANN and URBAN, 1977).

ration vessel equipped with a stirrer, and the resulting turbidity is recorded as a function of the quantity of the precipitant. In each case the process is restarted as soon as another 5 ml of eluate have been produced. Fig. 19-40 shows results obtained in this way for a graft copolymer of cyclopentene on a styrene-butadiene copolymer.

This approach is of general importance insofar as the first separation according to the hydrodynamic volume is followed by a separation on the basis of solubility characteristics. Thus the principle of cross-fractionation is applied, which is the only way to obtain complete information about the heterogeneity of copolymers.

19.7.3.5. Combination of chromatographic techniques

Even if operated automatically, turbidimetric titration involves rather a large amount of work. It would be easier if this principle could be realized simply by a combination of chromatographic techniques. For that purpose, Balke and Patel (1980) coupled two SEC apparatus in such a way that the second apparatus (GPC 2) refractionated certain eluate slices from the first one (GPC 1). The GPC 1 unit included three μ-Styragel® columns and nine silica gel columns and was operated in the stopped-flow mode, GPC 2 included three columns packed with μ-Bondagel®, which is compatible with different solvents within a wide range. Copolymers of styrene and n-butyl methacrylate were investigated in GPC 1 with tetrahydrofuran as an eluent, and with THF/n-heptane (40:60), in GPC 2. In this mixture, polystyrene homopolymers were eluted later than copolymers. The separation was due to the fact that the components occurring in an eluate slice (0.1 ml) from GPC 1 were eluted with different elution volumes in GPC 2 if they exhibited different chemical compositions. The problems of this method, which was called "orthogonal chromatography" by the authors, lie in the fact that GPC 1 must have a high capacity in order to yield a sufficient quantity of the sample for GPC 2, and that very sensitive detectors must be used with the GPC 2 unit. The elugrams were recorded simultaneously in the UV range at 235 and 254 nm.

Naturally it is also possible to combine separating paths occurring via different mechanisms by column switching, but in the adsorption technique, which would be of special interest, the specific adsorption behaviour of polymers raises greater difficulties than those encountered in the low-molecular-weight range (cf., *sequential analysis* in Section 19.11.).

Inagaki and Tanaka (1981) combined a preparative AC as a first separating step with SEC as the second one in order to investigate the chemical composition distribution (CCD) and the molar mass distribution of block copolymers of styrene and methyl methacrylate. By means of Inagaki and Donkai's method (1979) as described in Section 18.3., a product containing 47.0 wt.-% styrene was separated into nine fractions whose styrene content ranged between 84.9 and 21.0%. The SEC curves of those fractions which are obtained in a sufficient quantity are shown in Fig. 19-41 in such a way that the rather broad CCD of the starting material is clearly demonstrated. However, if the same starting material was first separated into fractions by SEC, followed by determining the gross composition of these fractions, then the picture shown in Fig. 19-42 was obtained, from which only a narrow CCD can be observed. This results from the fact that for this polymer the hydrodynamic volume depends only slightly on the chemical composition, so that the broad CCD is rather uniformly transferred to the SEC fractions, the composition of which therefore differs only slightly. This is a good demonstration of the fact that SEC with detector combinations, which is much liked for use in the investigation of copolymers, only under

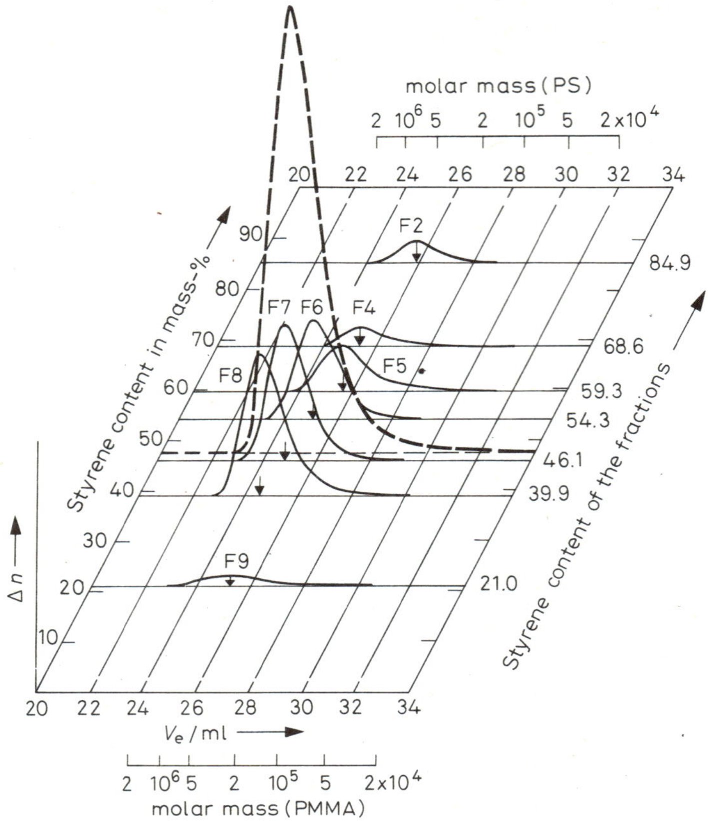

Fig. 19-41
SEC curves of the fractions of a styrene-methyl methacrylate block copolymer with a styrene content of 47 mass-%

The fractions were obtained from 300 mg of the starting material by preparative adsorption chromatography on silica gel. The broken curve shows the elugram of the unfractionated sample, the arrows indicate the position of the peak maximum on the V_e scale (baseline).
(according to INAGAKI and TANAKA, 1981).

most favourable circumstances allows reliable conclusions concerning CCD (see also Section 19.7.1.).

The SEC investigation of copolymers can already be considered successful if the problems discussed in Section 19.7.2. can be clarified. Then the elution curve can be represented as a molar mass distribution curve (see Fig. 19-43). The composition of the individual components is indicated by additional data. In the example shown in Fig. 19-43 the preparatory investigations had shown that the hydrodynamic volume was independent of the composition (see Fig. 19-38 (a)). Thus the increase of the methyl acrylate content with the molar mass definitely indicates a coupling between the MMD and the CCD. It can be assumed that towards the end of the copolymerization, during which the methyl acrylate accumulates in the starting mixture, a gel effect occurs, leading to the formation of very large macromolecules. That is why the high methyl acrylate content preferentially occurs in the molecules which have a high molar mass.

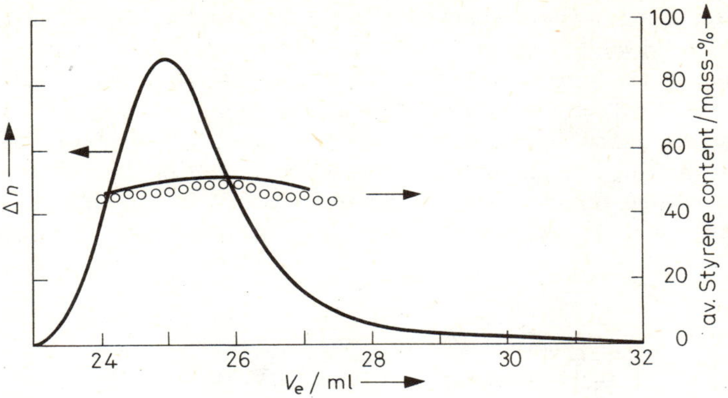

Fig. 19-42
SEC curve of the same styrene-methyl methacrylate block copolymer as in Fig. 19-41, and point-by-point composition curve (according to TANAKA, OMOTO, DONKAI and INAGAKI, 1980)

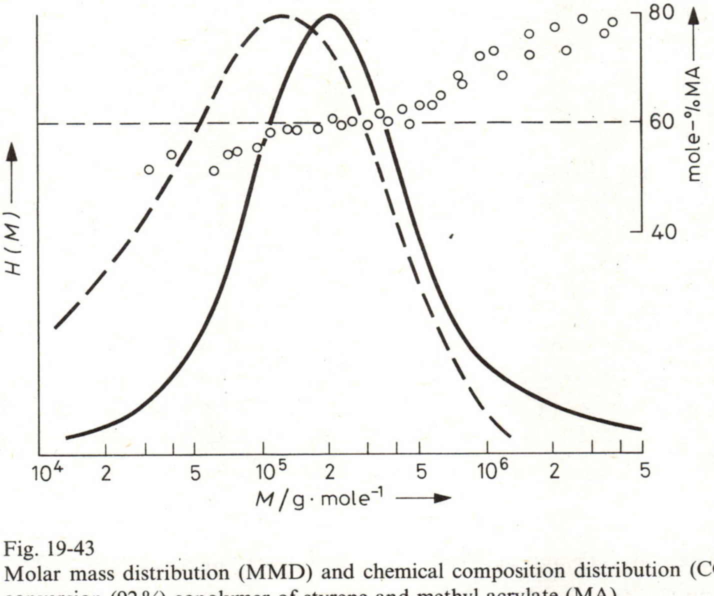

Fig. 19-43
Molar mass distribution (MMD) and chemical composition distribution (CCD) of a high-conversion (92%) copolymer of styrene and methyl acrylate (MA)

Sample: M_{osm} = 482000 g · mole^{-1}, 60.2% MA, 60% MA in the monomeric mixture.
SEC: 2.44 m column packed with polystyrene gels, eluent, THF; v = 1.2 ml · min^{-1}, c_0 = 0.1 g · ml^{-1}.
The MMD determined by SEC (———) extends to higher M-values than the curve calculated under the presumption of termination through disproportionation (– – –). As the azeotropic point of the system is at 23.4 mole-% MA, MA accumulates as the conversion increases, which finally leads to the formation of very large macromolecules with a high MA content. On the basis of the preliminary investigations (see Fig. 19-39), the position of the CCD points (○) suggests that there is a correlation between MMD and CCD.
(according to TERAMACHI, HASEGAWA, AKATSUKA, YAMASHITA and TAKEMOTO, 1978).

In the average chemical composition of the fractions obtained by SEC, differences which exceed the values derivable from the classical theory of copolymerization were also found to occur in copolymers of α-methylstyrene and methacrylonitrile (PROBST et al., 1974) or of vinyl chloride and vinyl stearate (MIRABELLA et al., 1975a). In an evaluation of this finding, it should also be taken into account that the quantity considered is the *average composition* of the eluate fractions. The *actual* CCD cannot be derived from this result. Presumably it is even broader. The further development of the effective technique of chromatographic analysis will certainly help to clarify some problems connected with copolymers.

19.8. SEC of polymers with long-chain branching

General branching problems have been reviewed by SMALL (1975). For the SEC of branched polymers, HARMON (1978) published a review, while DROTT (1977) compiled a bibliography of the relevant papers.

The coil volumes of macromolecules with long-chain branching are smaller than those of the linear isomers. On the other hand, short side chains have only a minor influence on the coil volume; their effect on the viscosity or the elution volume can be neglected.

The influence of long-chain branching on the coil volume can be described by the ratio of the radii of gyration (cf., Sec. 4.5.):

$$g = \left(\frac{\langle S^2\rangle_{0,\,\mathrm{br}}}{\langle S^2\rangle_{0,\,1}}\right)_{M_{\mathrm{br}}} \qquad (19\text{-}13)$$

The solution viscosity also varies with the coil volume; therefore the influence of branching can similarly be described by the viscosity ratio:

$$g' = \left(\frac{[\eta]_{\mathrm{br}}}{[\eta]_1}\right)_{M_{\mathrm{br}}} \qquad (19\text{-}14)$$

In these equations, the linear reference substance must have the same molar mass as the branched polymer. Strictly speaking, the viscosity values for the θ state would be required, but in view of all the other influences acting upon the final result it is permissible to use the approximation $g' = g'_\theta$ (HAMA et al., 1972; CERVENKA and WILLIAMSON, 1973; DIETZ and FRANCIS, 1979).

19.8.1. The relationship between g', g and the number, b, of branch points per molecule

The relationship between the viscosity ratio and the ratio of the radii of gyration

$$g' = g^{\varepsilon} \qquad (19\text{-}15)$$

was repeatedly made an object of theoretical investigations. ZIMM and KILB (1959) obtained $\varepsilon = 0.5$ for star-shaped branched products, and hypothesized that this value might also be applicable to other branched structures as an approximation. In many subsequent papers this was accepted without any further investigation, but there are many indications that for a statistical branching ε is greater (HAMA et al., 1972; MRKVIČKOVA-VACULOVA and KRATOCHVIL, 1972; AMBLER and MCINTYRE, 1975; SCHOLTE and MEIJERINK, 1977; PARK and GRAESSLEY, 1977b; see also Section 19.8.3.3.). An evaluation performed by BOHDANEĆKY (1977) confirmed that in most cases the value is about 0.8. The investigation of a great number of branched polyethylene samples by ^{13}C NMR spectroscopy and SEC yielded a value of $\varepsilon = 0.75$, whereas for PVAC a value of $\varepsilon = 1.0$ was found by comparing the SEC measurements with the conclusions drawn from a kinetic model (HAMIELEC, 1981).

An exact knowledge of ε is necessary to enable the number, b, of the *branch points per molecule* as a concrete, kinetically utilizable quantity to be calculated from the viscosity ratio. The true value of ε, however, must also be known if a theoretically founded expression $g = g(b)$ is needed for the evaluation of the experimental results, as is the case in the Drott-Mendelsohn method (cf., Sect. 19.8.3.1.). But the following statement made by OTOCKA et al. (1971) is still valid: "The unanswered question of whether any single approach is valid to branched molecules of arbitrary geometry is the least satisfying aspect of branched polymer analysis, representing the major obstacle to unambiguous results."

Equations describing the relationship between g and the number b were derived by ZIMM and STOCKMAYER (1949). For a *trifunctional* branching, the following relationships hold:

— for *monodisperse* polymers with a *uniform* number of branchings per molecule:

$$g_{3(\mathrm{u,\,u})} = 3(\pi/4b)^{1/2} - {}^5/_2 b \qquad \text{(19-16)}$$

— for *monodisperse* polymers with *statistically* distributed branch numbers, the number average of which is b_n:

$$g_{3(\mathrm{u,\,r})} = \left[\left(1 + \frac{b_n}{7}\right)^{1/2} + \frac{4b_n}{9\pi}\right]^{-1/2} \qquad \text{(19-17)}$$

— for *polydisperse* polymers with the weight average, b_w, of the number of branch points per molecule:

$$g_{3(\mathrm{r,\,r})} = \frac{6}{b_w}\left\{\frac{1}{2}\left(\frac{2 + b_w}{b_w}\right)^{1/2} \ln\left[\frac{(2 + b_w)^{1/2} + b_w^{1/2}}{(2 + b_w)^{1/2} - b_w^{1/2}}\right] - 1\right\} \qquad \text{(19-18)}$$

For a *tetrafunctional* branching, the following relationships hold:

— for *monodisperse* polymers with *statistical* branching:

$$g_{4(\mathrm{u,\,r})} = \left[\left(1 + \frac{b_n}{6}\right)^{1/2} + \frac{4b_n}{3\pi}\right]^{-1/2} \qquad \text{(19-19)}$$

— for *polydisperse* polymers with *statistical* branching:

$$g_{4(\mathrm{r,\,r})} = \frac{1}{b_w} \ln(1 + b_w) \qquad \text{(19-20)}$$

For $g(b)$ all these equations give continuously decreasing curves, which start at 1 and approach the value 0 asymptotically. The starting slope is different; for example, when $b = 10$ one obtains $g_{\mathrm{i,\,i}} = 0.591$, 0.580, 0.415, 0.412 and 0.240, respectively.

In most cases the branchings are generated by chain transfer, e.g., in low-density polyethylene (LDPE). Consequently the branch points are trifunctional and statistically distributed. Because even fractions are not monodisperse, eqn. (19-18) has the widest range of application.

From the number of branch points per molecule and the functionality, f, it is possible to calculate a mean value, $\bar{M}_b$, for the molar mass of the chain sections between two neighbouring branch points:

$$\bar{M}_b = M/[b(f-1)+1] \tag{19-21}$$

For $b \gg 1$, in view of the branching frequency

$$\lambda = b/M \tag{19-22}$$

eqn. (19-21) gives:

$$\bar{M}_b = 1/[\lambda(f-1)] \tag{19-23}$$

For a single branch point, the portion of the molar mass of the whole molecule is:

$$M^* = (\bar{M}_b/2)\cdot f \tag{19-24}$$

This is the *critical value* which demarcates the range of branched molecules from that range where branchings have no effect. At this threshold the plot of log $[\eta]$ vs. log M deflects from the straight line which is valid for linear products of the same type (see Fig. 19-44).

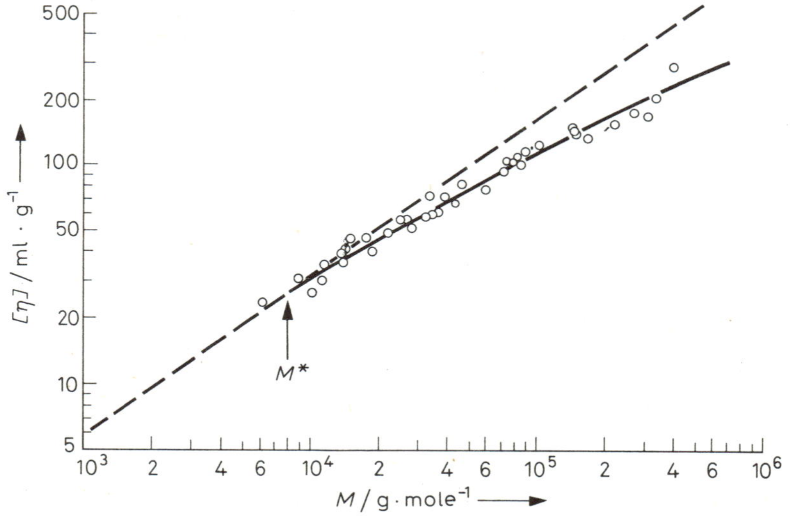

Fig. 19-44
Intrinsic viscosity of fractions of the SRM 1476 (NBS) polyethylene sample, plotted vs. the molar mass

Viscosity measurements in 1,2,4-trichlorobenzene at 130 °C, molar mass values obtained from light-scattering measurements, or SEC evaluations. For $M \geqq M^*$, the plotted curve represents the function $\log [\eta]_{br} = -1.4587 + 0.8658 \log M - 0.0326 (\log M)^2$.

The broken straight line is the viscosity function for linear polyethylene:

$\log [\eta]_l = -1.4067 + 0.725 \log M$.

(according to WAGNER and MCCRACKIN, 1977).

From an analogous diagram for the respective polymers to be investigated, their value M^* can be derived quite easily. AMBLER (1977) used this value to check the analytically obtained λ value by means of eqn. (19-23) and (19-24).

19.8.2. Universal calibration for branched polymers

If the universal calibration is applicable, then at a definite elution volume the following relationship holds for components having different structure:

$$M_1[\eta]_1 = M_{br}[\eta]_{br} \tag{19-25}$$

Branching reduces the viscosity, this has to be compensated for by a corresponding increase in the molar mass. Before investigating the relationship $(M_{br} > M_1)_{V_e}$ in greater detail, let us state that the mentioned compensation (decreasing $[\eta]$, increasing M) cannot always work well. In principle M can increase arbitrarily, but the coil volume determining the viscosity cannot assume arbitrarily small values. For a very dense branching and large-sized particles it must be expected that elution volumes occur which exceed the values of the universal calibration curve. For example, this was observed for styrene-divinylbenzene copolymers, which had been polymerized with 1% divinylbenzene up to a conversion of 15%, i.e., almost up to the gel point (AMBLER and MCINTYRE, 1975, 1977). In 1977, AMBLER once again explained the problem in detail, stating that a proposed empirical correction of the universal calibration is required only for an extremely high degree of branching.

For normally branched polymers the universal calibration relationship was confirmed many times. (In fact, in 1966 the differences between linear polymers and differently branched ones had given rise to the search for a universal calibration relationship, and this was found to be the plot of log $(M[\eta])$ vs. V_e, cf., Section 8.3.2.). Nevertheless it is necessary, at least for systems which have not yet been investigated, to prove the validity of eqn. (19-25) before drawing further conclusions. This can be done by measuring the universal curve with calibration standards and then using it to calculate, for any samples of the system to be analyzed, their $(M[\eta])$ values from elution curves by integration. These values should agree with the product of the intrinsic viscosity and the molar mass, which are directly measured for the respective total substance.

In this approach, which has already been referred to in Section 19.7., the integration is carried out with the product $(M[\eta])$. This is done without isolating the factors M_i and $[\eta]_i$, which is most difficult for branched polymers, because at this stage of the investigation the degree of branching in the individual components with different molar masses is generally not known. Even this knowledge would be of little use, because for branched polymers there are no viscosity vs. molar mass relationships which are as simple as for linear polymers (cf., eqn. (5-8) and Section 19.8.3.2.).

If the universal calibration is applicable, then, using eqns. (19-14) and (5-8), eqn. (19-25) can be rearranged as follows:

$$M_1[\eta]_1 = K_\eta M_1^{1+a} = M_{br}[\eta]_{br} = M_{br} g' K_\eta M_{br}^a \tag{19-26}$$

This gives:

$$g' = (M_1/M_{br})_{V_e}^{1+a} \tag{19-27}$$

This relationship, which is valid for components with equal V_e values, plays a decisive rôle in the SEC of branched polymers.

19.8.3. Evaluation of the elugrams of branched polymers

The first object of the evaluation is to determine the molar mass distribution, which in the macromolecular range extends farther than can be directly observed from the elugram (Fig. 19-45). Moreover the branching parameter, g, has to be determined. If possible, the relationship between the molar mass and the number, b, of the branch points per molecule should also be determined, because this allows one to answer the questions raised in connection with eqns. (19-21) to (19-24).

The known methods are based either on a mathematical evaluation of the elution curves alone (Sections 19.8.3.1. and 3.2.) or on the use of additional detectors, which allow a direct factorization of the hydrodynamic volume, $M[\eta]$, to be taken from the SEC curve. The first mentioned methods, 19.8.3.1. and 3.2., were also developed first.

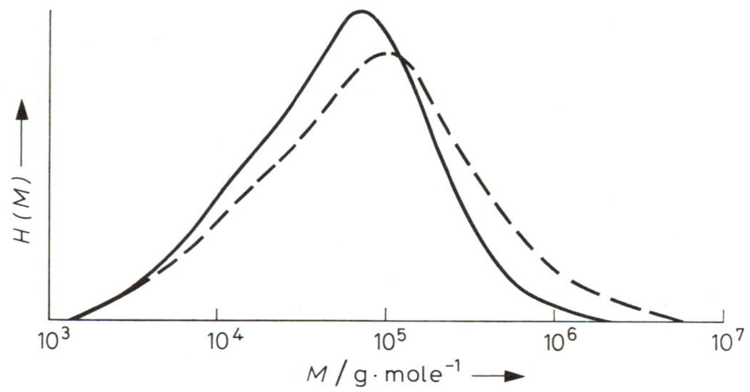

Fig. 19-45
Molar mass distribution of a polyethylene sample ($M_{w(LS)}$ = 195 000 g · mole^{-1}) with long-chain branchings
——— apparent distribution as derived from the SEC elugram
– – – – distribution after the branching correction
(according to CONSTANTIN, 1977).

19.8.3.1. The Drott-Mendelson method

It is assumed that the universal calibration relationship, eqn. (19-25), is valid and that the branching frequency λ defined by eqn. (19-22) is constant, i.e., that the molar mass and the number of branching points per molecule are proportional to each other. The second condition is required in order to enable eqns. (19-16) to (19-20), which have the general form $g = g(b)$, to be transformed into $g = g(\lambda \cdot M)$. For example, for eqn. (19-18) this gives:

$$g = \frac{6}{\lambda M}\left\{\frac{1}{2}\left(\frac{2+\lambda M}{M}\right)^{1/2} \ln\left[\frac{(2+\lambda M)^{1/2}+(\lambda M)^{1/2}}{(2+\lambda M)^{1/2}-(\lambda M)^{1/2}}\right]-1\right\} \qquad (19\text{-}28)$$

Here M denotes the weight average of the molar mass of the branched polymer. Finally, if it is known how the function g can be derived from the viscosity ratio g' (cf., Section 19.8.1.), then for each component i the molar mass, M_i, can be calculated from $V_{e,i}$ via the hydrodynamic volume:

$$(M_{\text{br}}[\eta]_{\text{br}})_i = M_i g' K_\eta M_i^a = K_\eta M_i^{1+a} g^\varepsilon(\lambda M) \qquad (19\text{-}29)$$

According to the summation rules discussed in Section 4.2.1., the M_i values should yield the mean values of the total substance. If so, this is considered to confirm the λ value used in the calculation.

The method stated by DROTT and MENDELSON (1970a, b) starts from the intrinsic viscosity, $[\eta]_{br}$, of the sample, using the following equations:

$$[\eta]_{br} = \Sigma\, w_i[\eta]_{i,\,br} \tag{19-30}$$

$$[\eta]_{br} = \Sigma\, w_i g_i'[\eta]_{i,\,l} = K_\eta\, \Sigma\, w_i g^{\varepsilon}(\lambda M_i)\, M_i^a \tag{19-31}$$

Iteration yields the value of λ, by means of which a value of $[\eta]_{br}$ which agrees with the measured one can be calculated from the elution curve. This λ value is used to construct the log M vs. V_e curve from the universal calibration curve and to calculate $\bar{M}_n$, $\bar{M}_w$ etc. of the starting material. In some cases, though, unreasonable values of λ are obtained by this method (JACKSON, 1973; NAKANO and GOTO, 1976).

19.8.3.2. The method by Ram and Miltz

From the representation of log $[\eta]$ vs. log M for unbranched and branched polymers, one can observe the lag of log $[\eta]_{br}$ behind log $[\eta]_l$, which begins at the value M^* of the molar mass and increases with increasing M. From Fig. 19-44 it is seen that $M^* = 8000$ g · mole^{-1}, but often $M^* = 5000$ or 6000 is used for low-density polyethylene (LDPE).

Up to $M = M^*$, eqn. (5-8) is valid:

$$\ln\,[\eta] = \ln K_\eta + a \ln M$$

For the non-linear relationships of branched polymers (for $M \geqq M^*$), RAM and MILTZ (1971) proposed the following formula:

$$\ln\,[\eta]_{br} = \ln K_\eta + a \ln M + b \ln^2 M + c \ln^3 M \tag{19-32}$$

The terms with the constants b and c take account of the lag of $[\eta]_{br}$ behind $[\eta]_l$. The values of b and c can be determined using the knowledge of the experimental relationship shown in Fig. 19-44 or, according to RAM and MILTZ, by iteration from the elution curve of a branched substance and its value $[\eta]_{br}$ measured directly. By means of the universal calibration and estimates for b and c, the elution curve is converted into a first approximation of the molar mass distribution. From the latter, the terms w_i and M_i are derived. The M_i values are used to calculate the terms $[\eta]_{i,\,br}$ by eqn. (19-32). Finally, the intrinsic viscosity of the whole sample is calculated by eqn. (19-30) and the data $[\eta]_{i,\,br}$ and w_i. Varying b and c, the calculation is repeated until a satisfactory agreement between $[\eta]_{exp}$ and $[\eta]_{calc}$ is achieved.

WILD et al. (1977) compared this method (RM) with the DROTT and MENDELSON method (DM) in an investigation of low-density PE samples. The DM method requires less computing time and, at first sight, has the advantage that it yields the λ value as an evident characteristic. However, the assumption of a constant λ is not generally justified. Nevertheless, for low-density PE this assumption obviously represents a good approximation. The RM method dispenses with the restrictive assumption of a constant value of λ, and, at higher degrees of branching, yields more realistic results than the DM method.

19.8.3.3. Branching analysis with a viscosity detector

A viscosity detector with a pressure transducer was described by OUANO (1972). The instrument measures the pressure drop across an exactly temperature-regulated capillary

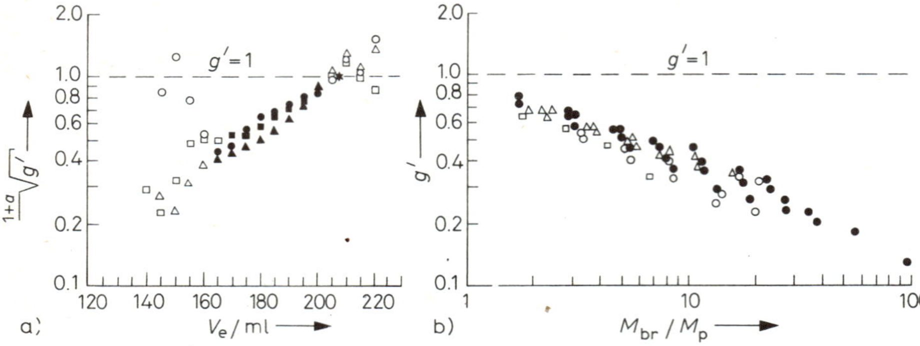

Fig. 19-46
Viscosity ratio, g', of branched polybutadiene samples

a) three repititive runs with a sample of $M_{w(LS)} = 353\,000$, which had been obtained from a precursor of $43\,300$ g · mole^{-1} by irradiation (86% of the gel dosage). The open symbols indicate the measured data for fractions containing less than 1.5% of the quantity injected (for $V_e \leqq 170$ ml), or less than 15% (for $V_e > 200$ ml). The asterisk indicates the precursor substance. The quantity plotted is the viscosity ratio at a certain V_e and thus related to g' as shown by eqn. (19-27). For the system investigated: a = 0.776.

b) Viscosity ratio as a function of the ratio of the molar masses before (M_p) and after irradiation (M_{br}): Combined plotting of all samples obtained from three different starting polymers (○, ●: $M_p = 43\,300$; △ — 77300; □ — 135000 g · mole^{-1}). The solid circles represent the reliable data from the left-hand diagram. The open circles are from additional experiments.
(according to PARK and GRAESSLEY, 1977a, b).

arranged at the end of the apparatus, this pressure drop being proportional to the viscosity for a constant rate of flow. Viscosimeters which automatically measure the time needed for the discharge of a certain liquid volume have been used in many other investigations. The device is filled which each successive content of the siphon, so that one measuring point per count is obtained (MEYERHOFF, 1971; GRUBIŠIĆ-GALLOT et al., 1972; SERVOTTE and DEBRUILLE, 1975; NAKANO and GOTO, 1976; SCHEINERT, 1977; CONSTANTIN, 1977).

PARK and GRAESSLEY (1977a) likewise used this method, starting with linear samples to investigate the effect of zone spreading in the column, the effect of substance delay and substance mixing in the siphon and in the connecting capillaries, as well as the precise course of the calibration curve in an extremely macromolecular range ($M > 10^6$ g · mole^{-1}). With branched polybutadiene, which was obtained from linear PBd of a very low heterogeneity by electron bombardment, they used a branched model substance which had great advantages for the development of the method. From the universal calibration curve and the $[\eta]$ vs. M relationship for linear PBd the viscosity ratio could be obtained by eqn. (19-14) for each siphon content, so that the value of the molar mass of the branched product in this eluate fraction, M_{br}, could be estimated by means of eqn. (19-27). The left-hand part of Fig. 19-46 shows the results of three replicate runs for the evaluation of reproducibility, while the right-hand part shows the results obtained with seven differently branched products prepared from three different precursor polybutadienes.

The advantage of the system chosen lies in the fact that the number of branch points per molecule can be calculated from M_{br} and M_p, the molar mass of the linear precursor:

$$M_{br}/M_p = b + 1 \tag{19-33}$$

Moreover the assumptions made for eqn. (19-19) are satisfied, so that the theoretical value, $g_{4(u,r)}$, can be calculated from the b values determined experimentally be means of eqn.

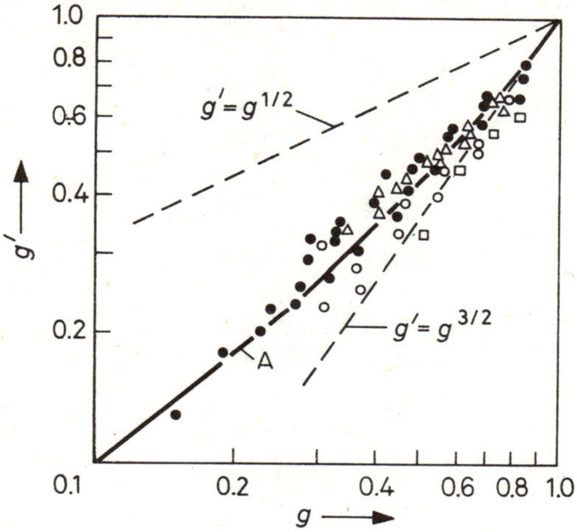

Fig. 19-47
Relationship between the viscosity ratio, g', and the ratio of the radii of gyration, g, for branched polybutadienes
Curve A follows the empirical relationship:
$\ln g = 0.735 \ln g' - 0.113 \ln^2 g'$.
The samples were prepared from linear low-heterogeneity polybutadiene by irradiation (starting data of M_p: ○, ● — 43300; △ — 77300; □ — 135000 g · mole^{-1}).
(according to Park and Graessley, 1977b).

(19-33). Fig. 19-47 shows a plot of these values vs. the independently measured viscosity ratio of the same components. It can clearly be seen that the much-used relationship for radially branched polymers, $g' = g^{1/2}$, is not a good representation of the results obtained here. For slightly branched products, $g' = g^{3/2}$ is a rather good approximation. In the whole range, the correlation is best reflected by the empirical formula:

$$\ln g = 0.735 \ln g' - 0.113 \ln^2 g' \qquad (19\text{-}34)$$

Using this relationship, Park and Graessley (1977b) investigated a divinylbenzene-styrene copolymer as well as PVAC (see Fig. 19-48). While the investigation of the former substance yielded a rather good support of the assumption that λ = const, on which the Drott-Mendelson evaluation is based, for PVAC the λ value increased steeply in the range from 300000 to 700000 g · mole^{-1}.

19.8.3.4. Branching analysis with a light-scattering detector

Light-scattering detectors yield a signal which directly depends on the molar mass of the sample component. If a second detector measures the concentration in the respective eluate volume, then the true molar mass can be determined without a prior calibration. Naturally, for branched polymers this is the molar mass M_{br} which is greater than the value M_l of linear molecules, which would emerge from the column in the same volume fraction V_e. As M_l can be taken from the SEC calibration curve for the linear polymer, the ratio g' can be calculated for each component by the relationship (19-27).

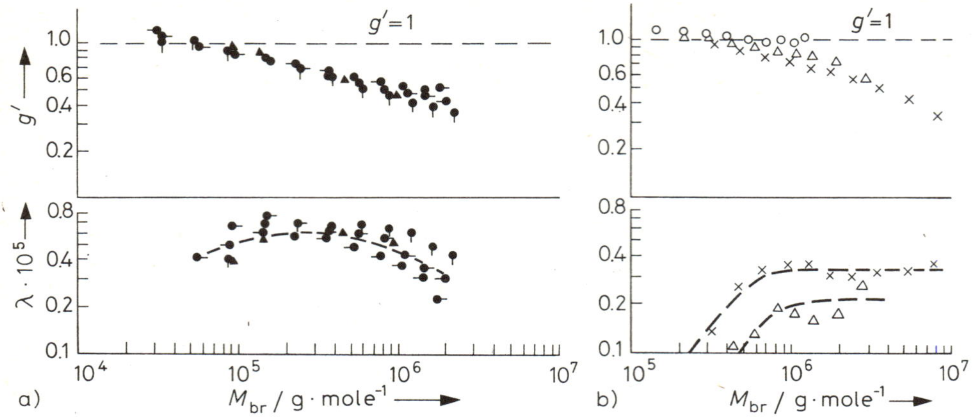

Fig. 19-48
Viscosity ratio, g', and branching frequency, λ, as functions of the molar mass, M_{br}, for a styrene-divinylbenzene copolymer (a) and for polyvinyl acetate samples (b)

a) $\bar{M}_{w(LS)} = 387000$ g · mole^{-1}, three repetitive SEC runs (circles) and four isolated fractions of this copolymer (triangles). To a first approximation, $\lambda = b/M$ (see eqn. (19-22), independently of the molar mass.

b) ○-PVAC-11, polymerized up to a conversion of 11%, $\bar{M}_{w(LS)} = 570000$ g · mole^{-1}, no long-chain branchings; △-PVAC-42, 42% conversion, $\bar{M}_{w(LS)} = 1020000$, long-chain branchings in the components $M \geqq 4 \cdot 10^5$; ×-PVAC-71, 71% conversion, $\bar{M}_{w(LS)} = 2370000$, long-chain branchings in the components $M > 3 \cdot 10^5$.

(according to Park and Graessley, 1977b).

The intensity of the scattered light emerging from a polymer solution mainly depends on the concentration and the molar mass of the solute as well as on the angle ϑ between the primary beam and the direction of observation. The concentration dependence implies the second virial coefficient A_2. The molar mass can be determined by measuring at different angles in solutions of graded concentrations and extrapolating to $c = 0$ and $\vartheta = 0$.

A new apparatus permits measurements down to $\vartheta = 2°$ (Kaye and Havlik, 1973; [F 21]), which eliminates the angular extrapolation. The determination of the scattered light in the immediate neighbourhood of the primary beam is achieved by means of a very precisely aligned laser primary beam 0.1 mm in diameter (Low Angle Laser Light Scattering, LALLS). The cuvette is formed by two thick quartz glass blocks with polished surfaces and a through-drilled PTFE spacer, against which the quartz windows are pressed. The remarkably thick windows ensure that light scattered from the air/glass interface does not influence the optical system which measures the light emerging from the cell.

Dust particles produce "spikes", which, although they do not disturb too much the usual evaluation of the detector record, cause considerable errors in the digital evaluation if they occur just when a measured value is recorded. Ouano (1976) and Brüssau (1977) reported this phenomenon and possible remedies. To avoid the dust problem, T. Kato et al. (1979) constructed a 90° light-scattering detector with a He-Ne laser, which could not, however, be used for branching analyses because of the difficult extrapolation to zero angle.

The high level of instrument engineering now enables the great variety of problems of long-chain branching to be dealt with on a broad experimental basis. A high-efficiency

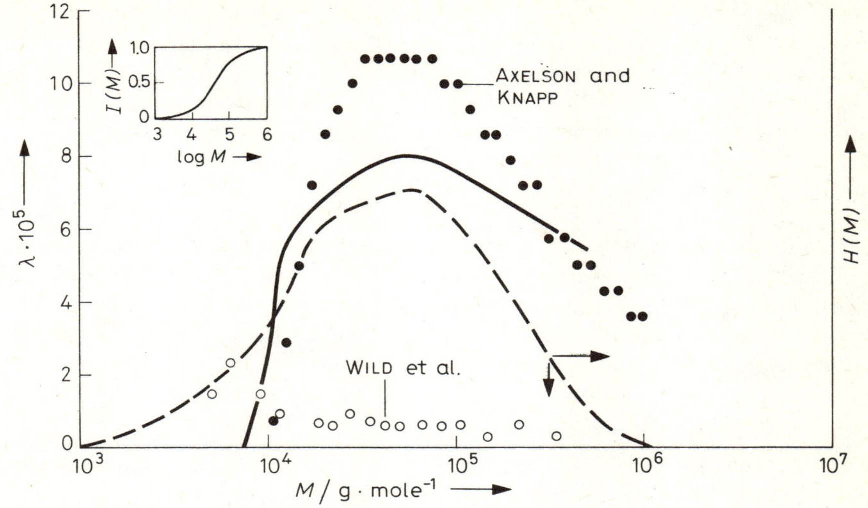

Fig. 19-49
Molar mass distribution (– – – –) and branching frequency (———) for LDPE SRM 1476 (according to WAGNER and MCCRACKIN, 1977).

The MMD is based on a SEC analysis of the total polymer, after filtration through a 0.45 μm Millipore® filter. The evaluation was carried out by eqns. (19-38) to (19-40). The curve of the branching frequency was obtained by fractionation and investigation of a total of 122 sub-fractions of 10 × 20 g SRM 1476.

The points ● likewise represent the branching frequency, they were stated by AXELSON and KNAPP (1980), who employed a LALLS detector.

The values of branching frequency indicated by ○ were determined by WILD et al. (1977) by means of SEC and viscosimetry according to the Drott-Mendelson method using fractions of the SRM 1476 sample. In the original paper the λ values are given as a function of $I(M)$ so that for the above representation the integral distribution curve (inserted diagram) had to be constructed from the differential distribution curve stated by WILD et al. WILD et al. used eqn. (19-15) with $\varepsilon = 0.5$, but this cannot be the cause for the low values of λ, because too low a value of ε will give too high λ values.

separation by SEC in combination with the LALLS detector, a viscosity detector and a third detector measuring the concentration of the substance would be ideal equipment. Then for all sample components both factors of the hydrodynamic volume, M_i and $[\eta]_i$, could be measured separately, and the ratio g' could be determined in two independent ways and indicated as a function of the molar mass. The work reported by AXELSON and KNAPP (1980) is a step in this direction. The authors investigated the standard reference material SRM 1476, a branched PE of the National Bureau of Standards, by SEC in α-chloronaphthalene at 150 °C on silica microspheres with a combination of the LALLS and the IR detector. The results coincide with those obtained by WAGNER and MCCRACKIN (1977) (see Fig. 19-49), if ε is assumed to be 0.65 which is rather low a value.

19.8.4. Branching analysis by a combined investigation by SEC and an ultracentrifuge

In size exlusion, a branched macromolecule behaves like a linear one which has a smaller molar mass, but with respect to its rate of sedimentation it behaves like a linear macromolecule of greater molar mass. Calibration relationships for the molar mass as a function of the elution volume, which have been established using linear samples, yield too small values of M_t for branched macromolecules, whereas corresponding sedimentation equations yield too high values of M_s measured in the ultracentrifuge. TUNG (1971c) has shown that the true value, M, is the geometric mean of these two apparent values:

$$M = (M_s M_t)^{1/2} \tag{19-35}$$

Here it has been assumed that for branched and linear samples which exhibit equal values of the hydrodynamic volume in a θ solvent these values will be equal in good solvents, too, and that the sedimentation investigations are carried out under θ conditions. Moreover, from the two apparent values for the molar mass it was possible to derive a branching parameter

$$h = (M_t/M_s)^{1/4} \tag{19-36}$$

which represents the ratio of the hydrodynamic radii of branched and linear molecules with equal molar masses. This ratio bears the following relationship to the viscosity ratio g':

$$g' = h^3 \tag{19-37}$$

DIETZ and FRANCIS (1979) investigated PVAC by the method discussed here as well as by viscosimetry/SEC, and found a fairly good agreement. On the basis of their sedimentation measurements in methanol at 6 °C they obtained $g' = 0.876^3 = 0.672$, whereas the values obtained by viscosimetry/SEC were $g' = 0.640$ (in the good solvent THF at 35 °C) and $g'\theta = 0.617$ (in the θ solvent heptane-3-one at 29 °C).

In a subsequent paper, DIETZ (1980) presented a possibility of evaluating branching measurements by SEC and an ultracentrifuge even without the knowledge of the molar mass data.

19.8.5. Branching analysis including the preparative fractionation of the sample

Among the great number of papers of this kind, here we shall only mention those which are closely related to SEC.

OTOCKA et al. (1971) separated three samples of branched PE into approximately 20 fractions, whose viscosity values $[\eta]_{br}$ were measured in 1,2,4-trichlorobenzene at 135 °C. The analytical SEC of the fractions was performed in the same solvent at the same temperature. Using the universal calibration, the peak elution volumes, V_e, were first converted into ($M[\eta]$) values, which, by means of the directly measured values $[\eta]_{br}$, finally yielded the SEC-based molar masses of the fractions. For several fractions the molar masses were also determined directly by light-scattering. The good agreement with the SEC values was considered to confirm the universal calibration. The viscosity ratio $g' = ([\eta]_{br}/[\eta]_l)_M$ was obtained from the measured value $[\eta]_{br}$ and from $[\eta]_l$, which was calculated by the $[\eta]$ vs. M relationship for linear PE. The number of branch points, b was calculated from the g' values by means of eqns. (19-15) (with $\varepsilon = 0.5$) and (19-18). For one of the samples investigated, $\lambda = b/M$ decreased with increasing molar mass, whereas for the two others the λ values obtained were constant, $\lambda = 12 \cdot 10^{-5}$ and $\lambda = 18 \cdot 10^{-5}$, respectively, thus supporting the assumption made for the Drott-Mendelson method (cf., Section 19.8.3.1.).

Wild et al. (1971) investigated low-density polyethylene and used fractions for establishing $a \log \bar{M}_w$ vs. $V_{e,w}$ calibration curve for the evaluation of SEC curves from non-fractionated LDPE. (Whether this curve may indeed be applicable to other samples must be checked in every case because the branching frequency has an influence.) For the fractions prepared by column elution, the viscosity was measured, and the elution curve recorded, in 1,2,4-trichlorobenzene at 140 °C. For each fraction a first approximation to the weight average of the molar mass, $\bar{M}_{w(1)}$, was calculated from the elution curve by means of the SEC calibration curve for linear polyethylene. The averaged value $V_{e,w(1)}$ corresponding to $\bar{M}_{w(1)}$ was read from the same calibration curve. Then this $V_{e,w(1)}$ value was used to obtain $(M[\eta])_1$ from a universal calibration curve. Dividing $(M[\eta])_1$ by the value $[\eta]_{br}$ directly measured for this fraction yielded a slightly improved value, $\bar{M}_{w(2)}$, which partially took account of the effect of branching. Plotting the $\log \bar{M}_{w(2)}$ vs. $V_{e,w(1)}$ data of all fractions yielded a *workable* calibration curve, by means of which the SEC elution curves of the fractions were recomputed. Thus improved approximations, $\bar{M}_{w(3)}$, were obtained, for which the corresponding $V_{e,w(3)}$ were also taken from the workable calibration curve. These $V_{e,w(3)}$ values were employed to find the values, $(M[\eta])_{(3)}$ in the universal calibration curve, which of course had been left unchanged. Again the results were divided by the directly measured values $[\eta]_{br}$ in order to obtain further improved values, $\bar{M}_{w(4)}$. Plotting $\log \bar{M}_{w(4)}$ vs. $V_{e,w(3)}$ yielded a further improved workable calibration curve, by means of which the original elution curves were recomputed in order to find $\bar{M}_{w(5)}$ and $V_{e,w(5)}$. This iteration was repeated until the improvement obtained from one step to the next became insignificant. The relationship between $\bar{M}_{w,\omega}$ and $V_{e,\omega}$ calculated in this way was then used to calculate definite values of $\bar{M}_w$ and $\bar{M}_n$ for the total substance as well as the fractions.

Wagner and McCrackin (1977) fractionated the branched SRM 1476 polyethylene ten times by column elution, using 20 g of substance in each run. By combining the corresponding components, they first prepared twelve main fractions, which were further separated into 122 sub-fractions. The latter were characterized by viscosimetry and partially by osmosis and light-scattering, so that in this case the SEC could be calibrated directly with branched samples. (Naturally the use of such a calibration curve presupposes that all of the products to be investigated will sufficiently correspond with the calibration samples with respect to their branching.)

Wagner and McCrackin additionally investigated the unfractionated sample on a SEC column which was calibrated by linear samples according to

$$\log M = A + BV_e + CV_e^2 \tag{19-38}$$

or:

$$\log (M[\eta]) = \log K_\eta + (1 + a) A + (1 + a) BV_e + (1 + a) CV_e^2 \tag{19-39}$$

Similarly as in Section 19.8.3.2., the following expression was used:

$$\log [\eta]_{br} = P + Q \log M + R \log^2 M \tag{19-40}$$

The V_i values derived from the elugram of the branched sample were converted into $(M[\eta])_i$ values by eqn. (19-39). As the values of the constants P, Q and R in eqn. (19-40) were known from preparatory investigations with characterized fractions, the products $(M[\eta])_i$ could be subdivided to obtain M_i. Finally the mean values were calculated from the M_i values, using the summation rules mentioned in Section 4.2.1. These mean values are listed in Table 19-6 together with the directly measured values and those obtained by the Drott-Mendelson evaluation. The agreement is quite satisfactory.

Table 19-6
Results of the investigation of the branched polystyrene reference sample, type SRM 1476 (according to WAGNER and McCRACKIN, 1977)
All the solutions were filtered before being analyzed (0.45 μm Millipore® filter)

	$\bar{M}_w$ / g · mole^{-1}	$\bar{M}_n$ / g · mole^{-1}	$[\eta]$ ml · g^{-1}	λ
Evaluation of the SEC curve of the total sample by eqns. (19-39, 19-40, 4-3 a, 4-5 a and 4-9)	96 500	22 700	94.0	
Drott-Mendelson evaluation	102 500	23 700		8.8
Light scattering measurement on the unfractionated sample	140 000[1])			
Direct measurements on the fractions and calculation by eqns. (4-3 a and 4-5 a)	105 000	25 000		

[1]) Presumably too high a result (due to the microgel content; cf., Fig. 19-53).

19.9. Special forms of size exclusion chromatography

19.9.1. Vacancy chromatography

The vacancy technique is the counterpart to the ordinary working method (MALONE et al., 1969). Before the injection, a polymer solution is passed through the apparatus until equilibrium is established and the separating bed is saturated. Then a solvent volume is injected and passed through the column by means the solution initially used to flush the column. The sample and the eluent are, so to speak, interchanged. The "vacancy" causes a recorder trace which is similar to the chromatogram of the original sample but with a deflection in the opposite direction (see Fig. 19-50).

Vacant peaks are common in adsorption chromatography. ŠLAIS and KREJČI (1974) investigated this phenomenon which is closely related to the so-called ghost peaks (cf., Section 19.3.).

The *differential SEC technique* (CHUANG and JOHNSON, 1973) likewise uses polymer solutions as an eluent, but investigates the behaviour of a different polymer in this eluent. This method promises advantages in process control and in the search for small differences between similar samples. CHUANG and JOHNSON (1973) demonstrated that the injection of a certain polystyrene sample into its own solution running as an eluent caused no detector response (equal concentration provided). The injection of a different sample, of course, yielded a chromatographic trace which was influenced by the concentration of the eluent solution, but was not such a neat equivalent to the chromatogram of the running polymer as in Fig. 19-50 (OTOCKA and HELLMAN, 1974 b). The flow-rate already has a strong effect in the usual working range.

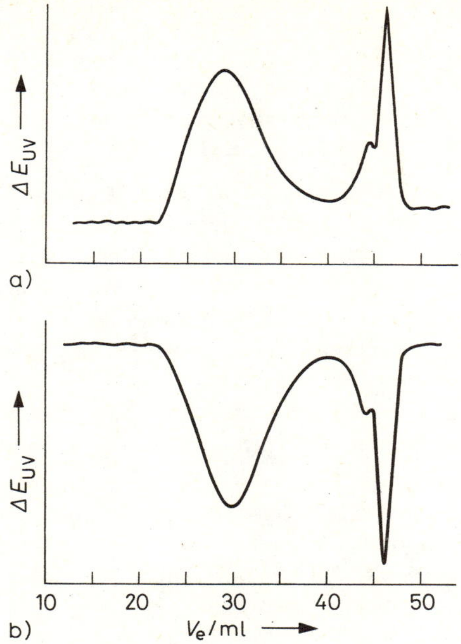

Fig. 19-50
Vacancy technique

a) Standard chromatogram
Injection of a solution of polystyrene with a broad distribution ($c_0 = 1\ g \cdot l^{-1}$) in chloroform as an eluent
b) Vacancy chromatogram
Injection of pure chloroform into the flowing solution $1\ g \cdot l^{-1}$, in chloroform)
(according to MALONE, SUCHAN and YAU, 1969).

It has been suggested that in this way the variation of the hydrodynamic dimensions as a function of concentration should be determined from the variation of the elution volume (BARTICK and JOHNSON, 1976). However, FIGUERUELO et al. (1979) found that the elution volume of macromolecular polystyrene samples also depends on the *molar mass* of the added polymer in the eluent, and not only on its concentration. This unexpected effect on the hydrodynamic volume was confirmed in viscosity measurements.

19.9.2. Column scanning

Column scanning is a variant of column chromatography in which the components of the sample are detected on the column rather than in the eluate.

A device for the identification of substances in the column by their UV absorption has been described by BRUMBAUGH and ACKERS (1968). It was used by WARSHAW and ACKERS (1971) to determine the distribution coefficients of proteins on Sephadex.

19.10. Particle chromatography

A chromatographic analysis of dispersions can be carried out by means of field-flow fractionation (cf., Section 13.1.) or hydrodynamic chromatography (cf., Section 13.2.).

We shall now describe the chromatography of colloidal particles on porous packing material. Compared to investigations by means of an electron microscope, chromatographic techniques have the advantage that they can even be used for soft latex particles. Moreover the particles are investigated in a dispersed condition, so that every variation of the particle size caused by changing the medium can directly be detected.

In most cases the systems to be investigated consist of compact particles and water as a dispersant. The stability of such colloidal solutions is due to the mutual repulsion of the particles rather than to attractive interactions with their dispersant. Consequently the conditions are substantially different from those encountered in macromolecular solutions in good solvents. Therefore in particle chromatography it can be expected that the interactions of the sample with the packing material play a very important rôle. (Such interactions have already been pointed out in another context, cf., Sections 16.6., 19.3. and 19.11.) As particles are mostly investigated in aqueous systems one can especially expect electrostatic interactions here. An undisturbed size exclusion mechanism can result only under carefully balanced conditions.

As electrostatic interactions play a decisive rôle in hydrodynamic chromatography, it is easy to understand that a clear distinction between a flow separation and a size exclusion mechanism will be rather difficult, if it is no longer by several orders of magnitude that the colloidal particles are smaller than the flow channels of the chromatographic bed (Coll and Fague, 1980; Nagy et al., 1981b). The separation efficiency of hydrodynamic chromatography is high for large-sized particles, but it approaches zero as the ratio of the particle diameter to that of the channel becomes very small. On the other hand, SEC is suitable for small particles (cf., Section 19.11.), but its efficiency decreases with increasing molecular size (cf., Fig. 16-14). Thus, strictly speaking, it is no wonder that only low efficiencies were obtained in SEC with colloidal particles. Coll and Fague (1980) worked with different samples of "uniform particles" (polystyrene latices) with graduated diameters ranging between 91 and 500 nm, to which particle weights of $250 \cdot 10^6$ to $40 \cdot 10^9$ g · mole^{-1} would have to be associated. Using samples from the lower part of this range on a 6 m column ($d_C = 5$ mm) packed with controlled-porosity glass ($d_P = 75$—120 μm; $d_0 = 300$, 200, 100 and 50 nm), for $v = 0.7$ ml · min^{-1} they achieved a specific resolution of $R_{sp} = 0.76$, which rather strongly depended on the flow-rate. For $v = 0.067$ ml · min^{-1}, a value of $R_{sp} = 1.09$ was obtained. The calibration curve plotted as log (particle size) vs. V_e exhibited an upward deflection for particle sizes above 200 nm, just as it does at the exclusion limit in the SEC of macromolecules. Below 20 nm the curve deflected downwards, tending to a value of 93 ml for the mobile phase hold-up volume, which correlates well with the dimensions of the column employed ($V_M = 0.8 \times V_C = 0.8 \cdot 118$ ml $= 94.4$ ml).

Nagy et al. (1981b) investigated hydrodynamic separation effects in particle chromatography on porous packing materials. In experiments with Fractosil® ($d_0 = 2500$ nm) and particles occupying from 4 to about 20% of the channel width of this packing material, they found that hydrodynamic effects may occur not only in the channels of the interstitial volume, but for certain sizes also in flowed-through pores of some packing materials ("*hydrodynamic permeation chromatography*").

An SEC of soft particles can be performed rather easily with rigid inorganic packing materials of a suitable pore size. Krebs and Wunderlich (1971) prepared silica gel with pores in the range from 50 to 5000 nm, and used it in the SEC of PS and PMMA latices.

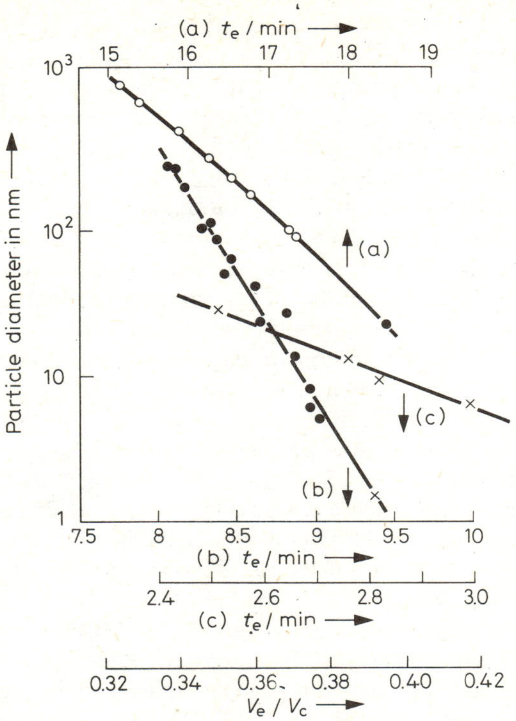

Fig. 19-51
Elution characteristics in particle SEC

a) 7.32 m column of $^3/_8$'' tube (d_C = 12.5 mm; V_C = 354.7 cm^3) packed with Fractosil® and porous glass (d_0: 10 to 3000 nm; d_P: 62—149 µm). Calibration by PS latices (○) and a silica standard (●). Eluent: water with 1 g · l^{-1} Aerosol OT and 1 g · l^{-1} KNO_3. Rate of flow v = 7.6 ml · min^{-1}. (according to Singh and Hamielec, 1978)

b) 2.00 m column (d_C = 7.8 mm; V_C = 95.6 cm^3) with controlled surface porosity support particles (glass beads coated with a 1 µm porous layer of 200 nm silica particles); d_P = 25 µm. Calibration by silica standards (●). Eluent: 0.02 M triethanolamine in water, adjusted to pH = 8 with HNO_3. Rate of flow v = 4.0 ml · min^{-1}. (×: elution value of an unknown polysilicic acid sample). (according to Kirkland, 1979).

c) 0.30 m column (d_C = 7.8 mm; V_C = 14.3 cm^3) packed with porous silica microspheres (d_0 = 75 nm; d_P = 8.9 µm).
Eluent: Na_2HPO_4—NaH_2PO_4, pH = 8.0.
Rate of flow v = 2.0 ml · min^{-1}. Calibration by silica standards. (according to Kirkland, 1979)

Here the elution volumes stated in the original papers refer to the empty volume of the respective column. so that a common representation with the abscissa scale V_e/V_C became possible. The t_e scales were calculated from the V_e data by means of the respective rate of flow.

Gaylor and James (1975) characterized several latices, using porous glasses or hydrophilic polymeric porous gels as packing materials.

Singh and Hamielec (1978) employed six 1.22 m columns connected in series, three of which were packed with Fractosil® (d_0 = 3000, 1400 and 490 nm), while the remaining ones contained different porous glass types with mean pore sizes in the range from 250 to 10 nm. Water with additions of 1 g · l^{-1} Aerosol OT (an anionic surfactant) and 1 g · l^{-1} KNO_3 proved suitable for use as an eluent, by means of which a good resolution was obtained in the particle size range from 20 to 1000 nm. The flow-rate was varied between 0.8 and 7.6 ml · min^{-1}. As the effects on the relationship between the elution volume and the particle size were insignificant, a flow-rate of 7.6 ml · min^{-1} was used, so that the elution was complete after 15—20 min (Fig. 19-51). Owing to this speedy working rate it was possible to utilize the method for monitoring the particle growth in emulsion polymerization. For that purpose samples were taken from the polymerizing system, diluted and a portion containing about 1 mg of particles was injected in each analysis. The growth curves measured at 70 °C for the surfactant-free polymerization of styrene were in excellent agreement with the electron-microscope results obtained by GOODALL et al. (1975), indicating the increase of the particle size to about 650 nm, the almost constant final value, in the course of the first 7 hours. It should be stressed that the method also worked well in the investigation of very soft particles. The emulsion polymerization of vinyl acetate could be watched without any difficulties, although samples were taken already at a conversion of only 5%. In the analysis of latices based on polybutyl acrylate or copolymers of butyl acrylate, styrene and acrylonitrile, there were also no problems caused by clogging of the column.

In these investigations, the packing material was relatively coarse, with d_p = 62—149 µm. Naturally this yielded a good permeability (cf., Section 17.2.), ensuring that even for pores in the µm range the packing particles still had a reasonable geometrical shape rather than being only flat laminae (EISENBEISS et al., 1978). On the other hand, short analysis times can be achieved without a loss of resolution only if the mass transfer between the stationary and the mobile phase proceeds rapidly. In view of the very low diffusion coefficients of colloidal particles this means that especially in this case small- sized packing particles with correspondingly short diffusion distances should be employed. KIRKLAND (1979), using silica microspheres as a packing material, achieved particle separations in the range from 7—30 nm within less than 3 min (curve c in Fig. 19—51). Porous layer beads (cf., Section 10.2.) which, owing to their low capacity, do not exhibit advantages for ordinary

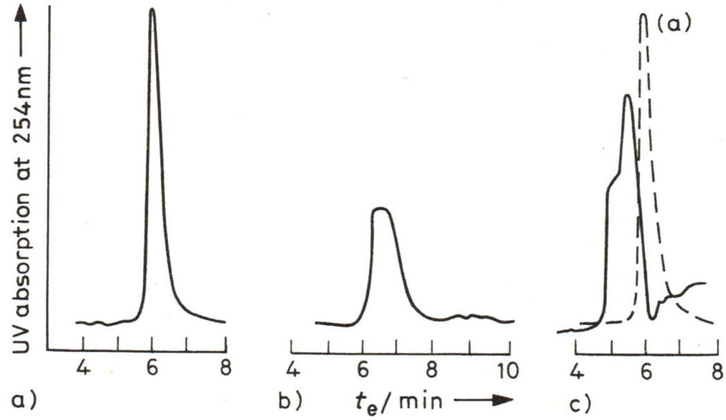

Fig. 19-52
Effect of the composition of the mobile phase on the exclusion chromatography of 8 nm silica particles

0.50 m column (d_c = 6.2 mm) packed with porous silica microspheres (d_p = 6.0 µm; d_0 = 30 nm); rate of flow: v = 1.00 ml · min^{-1}
Eluents:
(a) 0.02 M triethanolamine in water, adjusted to pH = 8 with HNO_3
(b) 0.02 M Na_2HPO_4—NaH_2PO_4, pH = 8
(c) 0.001 M NH_4OH
(according to KIRKLAND, 1979).

exclusion chromatography, may however, in particle chromatography, represent an advantageous compromise between the two contrary requirements for short diffusion distances and large diameters of the packing particles: on 25 μm beads with a 1 μm porous layer of 200 nm silica microparticles, KIRKLAND (1979) achieved separations with a good resolution at high flow-rates (see Fig. 19-51, curve b).

The composition of the eluent has a considerable influence on the behaviour of inorganic colloidal particles on silica packings (see Fig. 19-52). In dilute ammonia solutions with a low ionic strength, both the sol particles and the pores carry a rather high negative charge, so that no adsorption occurs. Even in a 0.02 M phosphate buffer the cation concentration reduces the charge density on the surfaces to such a degree that sol particles 100 nm or more in size are irreversibly retained on the outer surface of porous silica microspheres. In methanol with an addition of 0.5% of 1 M HNO_3, sol particles which are smaller than the pores by at least one order of magnitude are irreversibly adsorbed (KIRKLAND, 1979). Similar phenomena occurred in the exclusion chromatography of polyelectrolytes in aqueous solutions (cf., Section 19.3.3.).

BOOTH et al. (1980) investigated polystyrene-divinylbenzene microgels with particle masses ranging between 18 and 85 · 10^6 g · $mole^{-1}$ and radii of gyration between 25.2 and 40.3 nm (light-scattering data) by exclusion chromatography. They employed columns packed with polystyrene gels, and found an increase of the retention values with increasing content of LiBr in the eluent DMF. The values referred to high salt concentrations (cf., Fig. 16-34) lay together with the experimental results of high-molecular-weight PS standards on a common log ($M[\eta]$) vs. V_e curve. The microgels were irreversibly retained by a column containing a fresh polystyrene gel packing, whenever pure DMF was used for the elution or the LiBr content was above 0.1 mole · l^{-1}, whereas a packing which had been used over several years did not exhibit this phenomenon.

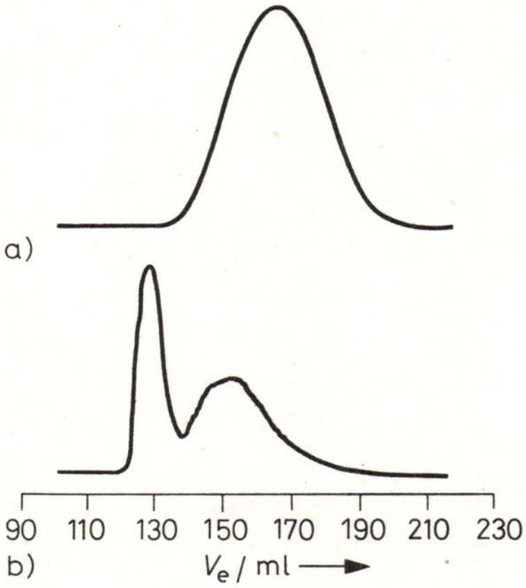

Fig. 19-53
Microgel in the SRM 1476 polyethylene sample
Elution curve obtained by SEC in TCB at 135 °C, v = 2.138 ml/min, recorded (a) with an IR detector (at a wavelength of 3.41 μm) and (b) with the LALLS detector

Column set: L = 5 × 1.22 m; packed with polystyrene gel (nominal porosities 10^7; 10^5; 10^4; 10^3 and 60 Å). (according to MACRURY and MCCONNELL, 1979).

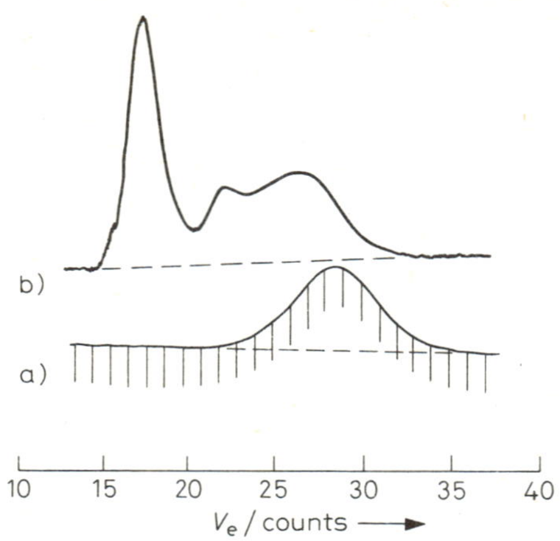

Fig. 19-54
Microgel detection with the LALLS detector

Elution curve of an ethylene-propylene (34 wt.-%) terpolymer (with 5.1 wt.-% 5-ethylidene-2-norbornene), recorded by an R.I. detect or (a) and a LALLS detector (b). While the distribution shown by elugram (a) seems to be quite normal, the light-scattering detector with its high sensitivity in the range of high molar masses clearly reveals the microgel content at the exclusion limit.
SEC in 1,2,4-trichlorobenzene, at 140 °C on polystyrene gel (nominal prosity 10^3, 10^4, 10^5 and 10^6 Å) in a column, $L = 4 \times 1.22$ m.
(according to SCHOLTENS and WEIZEN, 1981).

Even very small additions of microgel may influence some properties of polymers rather strongly. There is reason to suppose that microgels occur much more often than was previously assumed. This is connected with the fact that very small additions are usually difficult to detect and remain unperceived, for instance in sedimentation investigations. They are neither indicated by normal detectors such as R.I. nor by UV units. In such cases the LALLS detector (cf., Section 19.8.3.4.) is of great value, because its reading increases with the molar mass. Fig. 19-53 shows SEC elution curves of the branched NBS polyethylene type SRM 1476: the upper curve from an IR detector seems to prove a common MMD, whereas the record of the LALLS detector (lower curve) indicates a component with a very large molar mass. Although the concentration of this component is so low that it is not perceived in the IR signal, it is most distinctly indicated by a molar-mass-dependent light-scattering detector.

The results of SEC investigations of ethylene-propylene (ter)polymers as carried out by SCHOLTENS and WEIZEN (1981) with a LALLS detector showed: firstly, that only two of eleven samples were free of microgel, and secondly, that the small amounts which could not be perceived in the R.I. signal (see Fig. 19-54), had a considerable influence on the loss angle as observed by dynamic mechanical measurements. The latter approached the theoretical value (90°) at low frequencies only for the microgel-free samples, whereas for samples containing microgel it stopped at a markedly lower value at 10^{-1} rad · s^{-1}. These investigations clearly showed the importance of the microgel problems.

In some cases colloidal particles are of importance for the determination of the interstitial volume of gel packings. Beside the commercial Blue Dextran 2000, a soluble Prussian blue with a particle mass of circ. $3 \cdot 10^6$ g · $mole^{-1}$ is suitable for Sephadex® columns (SAITO and MATSUMOTO, 1979).

19.11. Gel permeation chromatography of small molecules and oligomers

At first sight it appears that investigations of small molecules and oligomers are beyond the scope of this book. In fact, however, there are several reasons to include them here.

For practical reasons the characterization of polymers must also cover the determination of oligomer components, residual monomers and additives such as stabilizers, light-protection substances, etc., because these low-molecular-weight components decisively influence the quality of the polymer. Here the first step is the separation from the macromolecular main product. This can for instance be done by SEC with a separating material which has an exclusion limit lower than the molar mass of the polymer. If concentration conditions permit, the small molecules can be analysed behind this polymer peak in the same analysis (PACCO et al., 1978). Such investigations are of importance for the quality control of finished products, but also for the monitoring of production processes.

Another reason why this section has been included lies in the fact that in the SEC of small molecules a possible *superposition of other mechanisms*, such as adsorption, solvophobic interactions and the like, upon the exclusion principle is easier to interpret than in the chromatography of macromolecular substances. The additional effects in real chromatography, as discussed in Section 16.6. with respect to the SEC of polymers, often lead to an irreversible retention of the sample in the column. Moreover the interactions of the macromolecules with the column packing cannot readily be distinguished from the effect of the solvent on the coil conformation. In this sense the GPC of small molecules can be considered a model which aids the understanding of the more complex situation of the chromatography of polymers.

Further, by SEC it is possible to achieve the base line separation of homologous oligomers, which is comparable with the results of the experimentally more demanding chromatography in supercritical media (cf., Sect. 9.7.) or those of adsorption chromatography (cf., Section 18.3.). Fig. 8-8 already demonstrated the remarkable separation efficiency which can almost automatically be achieved with suitable gels and properly chosen working conditions. Because of the limits given for the distribution coefficient ($0 \leqq K \leqq 1$), the peak capacity in SEC is limited for low-molecular-weight compounds, too, but in many problems this disadvantage is outweighed by the following advantages:

- GPC yields survey chromatograms of unknown substance mixtures under isocratic conditions. Consequently, in most cases the methodical preparatory work can be restricted to testing the solubility of the sample.
- Owing to the limits for K, exclusion chromatograms are so short that the desired survey can readily be obtained. This should not be underestimated, especially for the comparison of similar samples of different origins.
- As exclusion chromatograms are so short, they involve only a relatively low dilution of the samples. Therefore SEC is suitable for the pre-fractionation of complex mixtures, the most interesting components of which will in succession be transferred by column switching (cf., Section 14.3.) into another column for fine separation. For this technique the fact that SEC operates isocratically represents a further advantage. JOHNSON et al. (1978) reported such *sequential analyses* realized by coupling SEC on polystyrene gels (eluent: THF) with a reversed-phase chromatography on RP 18 (with a W/AcN gradient).

- As SEC can be carried out in a purely organic phase and the separations are not based on energetic interactions between the sample and the separating material, it is also possible to investigate unstable compounds, which would be decomposed by water, alcohols or the action of silanol groups, e.g., nickel complexes which are not stable enough for other chromatographic techniques (Tolman and Antle, 1978).
- From the position of the separated peaks one can derive not only qualitative information but even data about the molar mass of the substances. For that purpose, naturally the calibration relationship has to be known for the column in question.

19.11.1. The relationship between the size of small molecules and their elution volume

While in the single-peak resolution of homologous oligomers the peaks can usually be associated to the molar masses in a very simple way by counting them off, the universal calibration in a low-molecular-weight range is problematic. The hydrodynamic volume, i.e., the volume which is actually occupied by a molecule under the respective measuring conditions, cannot simply be set equal to the product $M[\eta]$ for small molecules. Nevertheless Ambler (1976) and Belenkij et al. (1974) empirically found that also in the range of oligomers the data points obtained for different substance classes can be combined into a common curve if the product $M[\eta]$ is plotted vs. V_e. This was confirmed by Ambler and Mate (1977) for oligomers of PE, PS, PBd, polyisoprene and hydrogenated polyisoprene, whereas POP oligomers deviate from the curve because of interactions with the PS separating gel.

To find a correspondence between the elution volume and the molecular dimensions even for substances which do not belong to polymer-homologous series, i.e., for organic molecules with any structure desired, the concept of the *effective number of carbon atoms* was developed (Hendrickson and Moore, 1966; Hendrickson, 1968a, b). It is based on the observation that the elution data for n-alkanes (up to n-$C_{27}H_{56}$) and n-alkenes, if plotted vs. the logarithm of the number of carbon atoms, lie on a common straight line. The equation of this line

$$V_e = C_1 - C_2 \log (\text{number of C atoms}) \tag{19-41}$$

corresponds to the usual calibration relationship (8-2), because M increases linearly with the number of carbon atoms. Other straight-chain compounds can also be associated to the equation if an effective carbon number is allocated to extra groups and heteroatoms. For example, an ether oxygen atom corresponded to 0.67 C, and the halogens Cl, Br and J to 1.08 C, 1.37 C and 1.54 C, respectively. These values were determined using a PS separating gel and THF as an eluent (Hendrickson, 1968a). With benzene as an eluent the same groups exhibited values of 0.67, 0.54, 0.68 and 0.77, respectively (Hendrickson, 1968b). The variations for the halogen atoms show the influence of the solvent. Moreover the value of the effective number of carbon atoms depends on the separating gel, so that a generalization is not possible.

Smith and Kollmansberger (1965) used the *molar volume* of the samples as a calibration quantity. Edwards and Ng (1968) as well as Lambert (1970, 1971) proceeded in the same way. Figueruelo et al. (1980) directly employed the *molar mass*.

The *rectangular volume of projection* was used by Mori (1980) in order to express the elution behaviour of the isomers of phthalate ester and similar compounds. This volume can be calculated from the images of the molecule projected upon the three mutually perpendicular planes of a rectangular XYZ coordinate system.

However, all these calibration relationships fail if other effects are superimposed upon the exclusion mechanism. The introduction of a functional group may change the elution behaviour. In GPC on PS gel columns with THF as an eluent, EDWARDS and NG (1968) found that the semilogarithmic plot of the molar volume vs. the peak elution volume yielded a common curve for alkanes, cycloalkanes, ketones, ethers and esters, whereas (for comparable molar volumes) for compounds with functional groups premature elution resulted in the following order: aliphatic alcohols > carboxylic acids > primary amines > acid anhydrides > secondary amines > alkyl chlorides. The largest deviations occurred for the alcohols; they were interpreted, in accordance with HENDRICKSON and MOORE (1966), as the consequence of complex formation between the OH groups and THF. A rather intriguing fact is the difference between the primary and secondary amines, which continued towards the ternary ones, the latter being eluted even later than the alkanes. Aromatics were also eluted later, probably because of their similarity to the PS gel material.

Of course there are also analogous differences between the various substance classes if $\log M$ is plotted vs. V_e. For n-alkanes, oligostyrene, epoxy resin, p-cresol novolak resin and polyethylene glycol, MORI and YAMAKAWA (1980) stated the equation

$$\log M_1 = \log A + B \log M_2 \tag{19-42}$$

as well as the values of the constants A and B for calculation of the molar mass of type 1 oligomers from values of type 2 (alkanes or oligostyrene) employed in the calibration.

If $B = 1$, this procedure is equal to the method used by KRISHEN and TUCKER (1977) who introduced the quantity A (cf., eqn. (19-42)) as "size factor" and referred all the data to n-alkanes.

The most reliable calibration relationships are those estimated with known substances of the same class of compounds on the same column and with the same eluent. Such calibration curves were established by BRAUN and BAYERSDORF (1980) for oligomeric urea-formaldehyde condensates (with DMF as an eluent and Merckogel® OR 6000 as a separating gel), and by CONCIN et al. (1980) for lignin degradation products (with Dx/W, 7:3, as an eluent and Sephadex® LH 20 or LH 60 as a separating material). Even the elution behaviour of the trimethylsilyl derivatives of silicate anions, e.g., from olivine or laumontite, can thus be represented quite normally (SHIMONO et al., 1980).

19.11.2. Non-exclusion effects in the GPC of small molecules

An ordinary $\log M$ vs. V_e calibration curve not only allows one to state the molar mass associated to each peak, but also enables conclusions to be drawn as to whether or not the separation is brought about by a pure exclusion effect. For the lignin degradation products mentioned above, the calibration curve remained in the standard range of distribution coefficients only for Sephadex LH 20. For Sephadex LH 60 the values observed for the Laurent-Killander distribution coefficient (cf., eqn. (8-30)) ranged up to $K_{av} = 1.5$, which is indicative of *adsorption*. Changing the eluent may completely alter the elution behaviour of the samples. CONCIN et al. (1980) investigated the K_{av} values for substituted phenols, aromatic acids and carbonylic aromatics as functions of the composition of the

Dx/W mixture, and for most of the compounds in pure dioxane they found a strong adsorption, which resulted in K_{av} values of about 2 even when LH 20 gel was used.

FREEMAN and KILION (1977) investigated the distribution coefficients of alkanes (C_2—C_{19}) in a column with poly(80% isodecyl methacrylate-*co*-20% divinylbenzene) and methanol, ethanol or cyclohexane as eluents. The distribution coefficient decrease with increasing molecular size only in the last eluent. In the polar eluent, an additional retention was observed which increased with the molecular size, resulting in a distribution coefficient of more than 2 for $C_{16}H_{34}$ in methanol. This indicates solvophobic interactions between the sample and the gel.

Similar investigations were performed by OZAKI et al. (1979) on Styragel® 60 Å with n-alkanes (C_5—C_{16}) as solutes. Here a great increase in K_{av} was found to occur in acetone, whereas in eight other solvents, including 1,2-dichlorobenzene and ethyl acetate, a normal elution behaviour was observed (see Fig. 16-26).

YANO and JANADO (1980) utilized the *hydrophobic interactions* between n-aliphatic alcohols and Sephadex® G-10 in 2 M NaCl as a basis for a separation of the first eight members of this series. (In 0.1 M NaCl solution the distribution coefficients were smaller and the resolution was much lower.)

Likewise on Sephadex G-10, UJIMOTO et al. (1981) investigated the chromatographic behaviour of tetraalkylammonium ions (ranging from the tetramethyl- to the tetra-n-pentylammonium ion). The process was run at pH 2 in order to suppress the charge interactions. In aqueous NaCl solutions, the distribution coefficients increased with increasing salt concentration, just as they did in the example cited above. At temperatures above 30 °C, the values measured in 0.1 M NaCl increased with increasing length of the aliphatic residue, which is indicative of hydrophobic interactions.

In both cases, increasing the temperature resulted in higher distribution coefficients because of the endothermicity of the distribution process. On the other hand, in the experiments carried out by DUBIN et al. (1977) on Styragel® columns with DMF as an eluent, the retention decreased as the temperature was increased from about 20° to 65 °C. This was the case for the low-molecular-weight solutes benzoic acid, phenol and toluene as well as for the PS and POE standards, i.e., there was a negative enthalpy change induced by the solute-gel interaction.

The temperature dependence indicates that the separation did not follow a pure exclusion mechanism, although the calibration log M vs. V_e exhibited the regular behaviour in the last-mentioned example.

Just as with a change of the solvent, a change in the separating material also influences the elution of functional compounds and polar oligomers. This is illustrated by the data given by DUBIN et al. for the peak elution volume of benzoic acid in DMF: in columns of a comparable geometry it was $V_e = 49.6$ ml on a Styragel® packing, but 61.7 ml on a packing of silanized porous glass. (On non-silanized glass it even reached 98.0 ml. This was accompanied by an extensive tailing.) The influence of the separating gel on the elution behaviour of samples of different polarities is most effectively shown by the almost classical diagram (Fig. 16-15) published by HEITZ and KERN (1967).

To sum up, the following survey of non-exlusion effects in the GPC of small-molecule compounds can be given:

— Variation of the width of the network openings as a result of the altered *swelling* upon a change of the solvent

— Variation of the pore size in rigid, non-swelling materials such as silica gel, caused by eluent adsorption (see Figs. 16-32 and 16-33)
— *Adsorption of the solute* on the surface
— Partition of the solute *in the wall material* of separating gels which are capable of swelling (cf., Section 16.6.3.)
— *Solvophobic interactions*
— Complexing of the solute with solvent molecules. These *solvate complexes* make a solute appear to have different sizes in different eluents.
— *Dimerization* of the solute (e.g., acetic acid in CCl_4)
— Variation of the *effective* molecular volume as a result of intramolecular interactions between donor and acceptor groups in one and the same molecule. As the extent to which such effects occur is different in solvents of different strengths, this again indirectly leads to an influence of the eluent on the effective molecular size.

From this survey, it is easy to understand why hopes for a utilization of GPC as a "liquid phase molecular size spectrometer" for small molecules could not be realized, and also why the universal calibration of GPC for all kinds of low-molecular-weight compounds is so much more complicated than for polymers. The latter substances can, on the one hand, be characterized quite simply by their hydrodynamic volume, $M[\eta]$, while on the other hand their molecules are so large that usually they do not penetrate *into* the wall material of the separating gels. Moreover the size of the pores required for their separation is so large that solvent-induced changes in the network openings or in the available free diameter (items 1 and 2 of the above survey), considered relatively, have a minor effect. The most essential difference is the fact that almost all of the polymers, mainly those of technical importance, are much more similar to one another with respect to their polarity and functionality than the great variety of the low-molecular-weight compounds.

However, the non-exclusion effects must not be considered to be wholly negative factors: for certain difficult separating problems they may even be decisive for success. Thus isomeric carboranes have equal molecular sizes, but their rigid molecules differ widely in their dipole moments. In the GPC of these compounds in THF, Čoupek et al. (1974) found an elution in the order of decreasing dipole moments, which is due to the formation of complexes of different sizes with solvent molecules. Such *solvate complexes* cannot be expected to have the same form in benzene. In fact a much poorer separation with a different elution order was observed in this case.

Pokorný et al. (1978) likewise attributed the separation of the stereoisomers of 2,4-dichloropentane in THF on S-Gel 832 styrene-divinylbenzene copolymer to an interaction of several effects. After a recycling with sixteen cycles, the peak of the *iso* form was separated from the peak of the following *syndio* form almost down to the baseline. The *iso* form travelled faster, because its greater dipole moment resulted in a stronger interaction with the molecules of the eluent, or because the possible conformations *tg'* or *tt* for the two stereoisomers occupy different volumes. On the other hand, the isomers of 2,4,6-trichloroheptane could not be separated from each other in THF even if the number of cycles was increased. The authors attributed this failure to the fact that for this compound an increase in the space requirement due to the conformation is accompanied by a decrease in the dipole moment. Thus obviously the resulting difference in solvation largely cancels the former effect.

19.11.3. Baseline separation of oligomers

A report on the gel chromatographic separation of oligomers was given by HEITZ (1981), who has made essential contributions to the development of this technique. As early as 1969, HEITZ et al. reported the resolution of oligomers, presenting base line separations of oligomers of styrene, butyl methacrylate or ethylene oxide of ether tensides and epoxides. These investigations were carried out using a 2 m column packed with polystyrene-divinylbenzene (2%) gel. Later on, polyvinyl acetate gels were usually employed in very long, coiled columns made of teflon tubing, the separation efficiency of which is indicated in Fig. 8-8.

The durability of these columns is remarkable. In the paper published in 1981, it was mentioned that after 8 years of continuous operation such a 10 m column was still operating with a plate height of $4.1d_P$. A_s SEC at room temperature requires scarcely any attendance, and staggered injections are possible (cf., Section 19.5.), the long waiting times of about 2 days in fact are not so cumbersome as it may appear at first sight. A positive point is that in these high-performance columns the quantity of eluent required is quite small because of the small diameter of 2 mm. The throughput of eluent per hour is about 1 ml. (In the baseline separation on PS gel columns, 2500 ml THF were required for a PS 600 sample (HEITZ et al., 1969; HARMON, 1971).)

Fig. 19-55 shows chromatograms obtained with a polyvinyl acetate gel column for a polytetrahydrofuran, of which samples were taken at different times during the polymeriza-

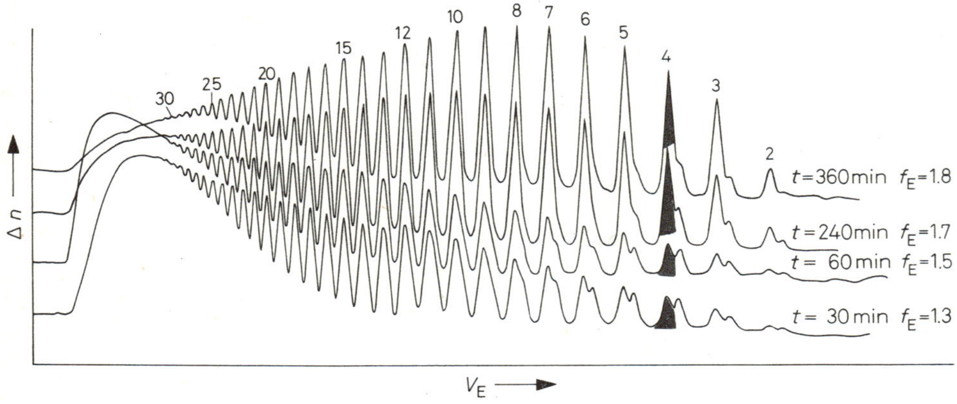

Fig. 19-55
Gel chromatograms of polytetrahydrofuran, with sampling at different times after the start of the polymerization (Initiating system: $HSbF_6/Ac_2O$)

Column: L = 10 m; d_C = 2 mm; packed with Merckogel® OR 6000; d_P = 19 μm (non-swollen) and 27 μm (swollen). Pressure: 8 bar; duration of an analysis appr. 2 days. The figures beside the peaks indicate the number of THF units per molecule.
The polymerization took place at 10 °C in DCM, initially giving molecules with an OH end group and an acetic ester group as well as molecules with two ester groups. After the initiator $HSbF_6$ was exhausted, the acetic anhydride continued to act as a chain-transferring agent: the portion of oligomers with two ester end groups (illustrated by raster for P = 4) increased with time. In this process the initially formed chains with 30 or more THF units were used up. The ester functionality, f_E, determined by titration supported the chromatographic result.
(according to STIX and HEITZ, 1979).

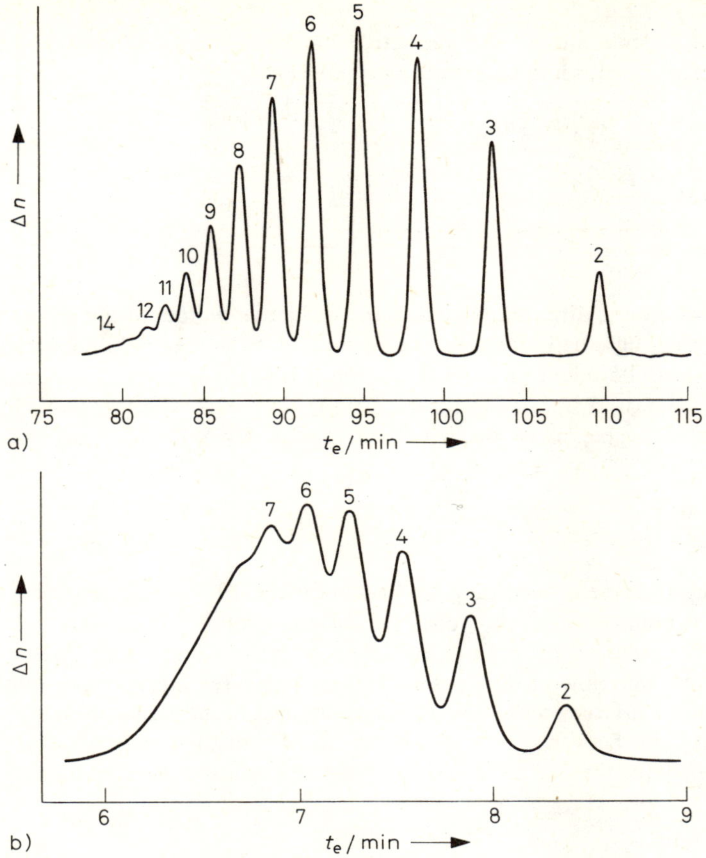

Fig. 19-56
Chromatograms of an oligomeric polystyrene PS 600, obtained by SEC on columns packed with 5 μm polystyrene gel particles; Eluent: tetrahydrofuran

a) $L = 2.44$ m; $d_C = 7.8$ mm; $v = 0.6$ ml · min^{-1}; $c_0 = 40$ g · l^{-1}; $V_0 = 0.05$ ml
b) $L = 0.61$ m; $d_C = 7.8$ mm; $v = 2$ ml · min^{-1}; $c_0 = 20$ g · l^{-1}; $V_0 = 0.05$ ml
(according to KATO, KIDO, WATANABE, YAMAMOTO and HASHIMOTO, 1975).

tion. The high resolution indicates the relatively slow action of acetic acid anhydride as a chain-transfer agent.

Fig. 19-56 demonstrates the possibilities and the limits for the acceleration of high-resolution oligomer separation by exclusion chromatography. The chromatogram shown in Fig. 19-56 (a) was obtained on a 2.44 m column within about 2 hours. As the column was packed with superfine particles, the resolution is comparable with that which can be achieved on polyvinyl acetate gel. On a 0.61 m column with the same separating material it was possible to obtain the chromatogram of the same sample at a nearly three-fold flow-rate within as little as 9 minutes (Fig. 19-56 (b)), but the resolution was much lower.

From chromatograms with base line separation it is possible to calculate the molar mass averages by means of the equations given in Section 4.2.1. Here m_i is represented by the peak area. The variation of the refractive index with the molar mass can be neglected for

Table 19-7
Increments for calculation of the refractive indices of organic compounds from the structural elements (according to VOGEL et al., 1951)

CH_3	17.66	OH	23.57
CH_2	20.64	O	23.73
CH	23.49	(ether)	
C	25.3	CN	36.67
double bond	−6.36		

such systems, where the difference between the refractive indices of the repeat unit and the solvent is great enough, as is the case for PS and THF for example. Generally, however, this influence must be taken into consideration in the investigation of oligomers (HEITZ et al., 1969; CANDAU et al., 1974; MORI, 1978a). For each species, the refractive index can be calculated by means of the increments stated by VOGEL et al. (1950, 1951) (cf., Table 19-7):

$$n = \frac{\Sigma R_v}{M} = \frac{P \Sigma R_{v,0} + \Sigma R_{v,E}}{P \cdot M_0 + M_E} \tag{19-43}$$

where R_v is the increment according to VOGEL (Table 19-6); $R_{v,0}$: R_v for the structural elements of a monomeric unit; $R_{v,E}$ ditto for the end groups.

MORI (1978b) discussed the calculation of $\bar{M}_w$ and $\bar{M}_n$ values from elugrams which are only partially resolved down to the baseline. He investigated a polystyrene with a nominal mass of 600 using different methods of evaluation and found values ranging between 688 and 742 for $\bar{M}_w$ and between 596 and 648 for $\bar{M}_n$. From the comparison of the different methods he concluded that in this case the lowest values should have the highest probability.

20. Experimental parameters and results of precipitation chromatography

The bibliography (Table 20-1) shows that all of the important polymers have been fractionated by means of precipitation chromatography. In the sixties the method was applied in many laboratories. Developed in 1956, it offered advantages so substantial compared to the classical fractionating techniques that it rapidly found a widespread application. However, it has meanwhile been replaced by SEC. If the high prime cost is disregarded, the latter technique is definitely superior to analytical precipitation chromatography with respect to several points:

— the time required for an analysis
— the versatile application without the need for much methodical preparatory work
— the degree of automation
— the possibility of performing analyses with only one milligram of the sample, or even less
— the isocratic conditions of elution, which enable the eluent to be easily recovered.

20.1. Time required for an analysis

In precipitation chromatography, several hours are required for the elution alone. The data shown in Table 20-1, which were either stated in this form by the authors or calculated from the rate of elution and the characteristics of the gradient applied, range between 3 and 228 hours. The amount of time required depends on the system and the quantity of polymer used. For polybutyl methacrylate in acetone-methanol it was found by means of turbidimetric titration of the fractions that the rate of elution can be increased up to 110 ml · h^{-1}. At this rate, 7 hours are required for an analysis (GLÖCKNER and MÜLLER, 1966). In the isothermal elution of a copolymer from styrene and methyl methacrylate, a value of 120 ml · h^{-1} was found to be the limit which must not be exceeded to ensure the establishment of equilibrium (LOVRIĆ et al., 1976).

The time required for the sample preparation and especially for the isolation and characterization of the fractions must be added to the duration of elution. Thus at least three days are required before the result is obtained. SEC is capable of yielding the same result in a much shorter time.

Table 20-1
Bibliography on precipitation chromatograph

Polymer	Authors	m/g	t/h	Precipitant	Solvent	ϑ_o/°C	ϑ_u/°C	Remarks
Polyethylene	GUILLET et al. (1960)	5	24	butoxyethanol	tetralin	152	100	poorer separation for $\vartheta_o = \vartheta_u = 133$ °C; confirmed by means of sedimentation measurements by MOORE et al.
	SLONAKER et al. (1966)	200		butyl carbitol	petroleum hydrocarbons	160	110	2 stages, solvent/precipitant: 220 1
	FERRIER (1967)	3	24	butoxyethanol	tetralin	130	90	2 stages; dissolution at $\vartheta_o = \vartheta_u = 130$ °C
Polypropylene	GUILLET et al. (1962)	1.5		butyl carbitol	tetralin	180	140	
	SLONAKER et al. (1966)	100		butyl carbitol	petroleum hydrocarbons	180	140	2 stages
	FERRIER (1967)	3	24	butoxyethanol	tetralin	165	115	2 stages; dissolution at $\vartheta_o = \vartheta_u = 160$ °C
	BAIJAL et al. (1969)			dimethyl phthalate	1,2,4-trichlorobenzene			$0.130 \leqq [\eta] \leqq 2.780$; with fractionated precipitation $0.348 \leqq [\eta] \leqq 2.608$
Higher α-olefine polymers	JUNGNICKEL and WEISS (1961)	2	>24	ethanol	benzene	60	20	copper column, interior surface gold-plated, $C_{10} \ldots C_{18}$
	FLOWERS et al. (1964)	10	228	ethanol	benzene	73	23	better separation at $\vartheta_o = \vartheta_u = 23$ °C, $C_{12} \ldots C_{18}$ copolymers
Polyisobutylene	M. J. R. CANTOW et al. (1961)	34	192	acetone	benzene	50	28	6 columns in parallel
	M. J. R. CANTOW et al. (1963)	50	200	acetone	benzene	50	28	fractions with $U = 0.013$ and 0.020
	PANTON et al. (1964)	0.3	150	n-propanol	xylene mixture	71	20	on Chromosorb

	M. J. R. Cantow et al. (1967a)	2	72	MEK	benzene	50	18	comparison with SEC
	Johnson et al. (1969)	6.2	(150)	MEK	benzene	30	60	(top-end discharge); d_C 37 mm; L 1200 mm
	Haddon et al. (1964)	10		MEK	benzene	90	40	better separation at $\vartheta_o = \vartheta_u = 30\ °C$
Butyl rubber	Henderson and Hulme (1967)	45	15	methyl isobutyl ketone	isooctane	50	40	
Polybutadiene	Cooper et al. (1962)	0.25	(8)	ethanol	benzene	60	15	inversions for too steep gradients
	Hulme and McLeod (1962)	0.3	3	i-octane	diisobutene	90	40	sample on Chromosorb®
	Poláček et al. (1965)	0.5		methanol	benzene	(35)	(28)	T-cycles
	Ring and Cantow (1965)		(90)	i-propanol	butyl acetate	55	25	
	Uraneck et al. (1965)	1	80	i-propanol	toluene	65	44	
	Henderson and Hulme (1967)	44	15	i-octane	n-heptane	50	40	with 30 l of solvent/precipitant
Polyisoprene	Poláček et al. (1965)	0.5		methanol	benzene	(35)	(28)	T-cycles
Butadiene styrene rubber	Uraneck et al. (1965)	1	80	i-propanol	toluene	65	44	
	Blass and Seide (1966)	4		n-butanol and i-octane	n-heptane and toluene	90	20	ϑ_u is increased step by step
	H. J. Cantow et al. (1968)	0.005	70	methylcyclohexane	methylcyclohexane	θ	$\leqq \theta$	capillary column
Styrene isobutene copolymer	Danon and Jozefonvicz (1969)	2	(72)	i-propanol	benzene	60	25	
Polychloroprene	Poláček et al. (1965)	0.5		methanol	benzene	(35)	(28)	T-cycles

Table 20-1 (continued)

Polymer	Authors	m/g	t/h	Precipitant	Solvent	ϑ_o/°C	ϑ_u/°C	Remarks
Methyl methacrylate on natural rubber	COOPER et al. (1959)			petroleum hydrocarbons	benzene	50	18	
Polystyrene	BAKER and WILLIAMS (1956)	0.3	100	ethanol	MEK	60	10	
	SCHNEIDER et al. (1959b)	0.8		ethanol	MEK	60	10	
	PEPPER and RUTHERFORD (1959)	4	120	ethanol	benzene	65	12	inversion: loading too high
	SCHNEIDER et al. (1960)	0.4		ethanol	MEK	60	10	$c_{max} \cdot M_{max}^{0.5} = const$
	SCHNEIDER et al. (1961)			ethanol	MEK	60	10	better than elution at $\vartheta_o = \vartheta_u = 15$ °C
	JUNGNICKEL and WEISS (1961)	1		ethanol	MEK	64	21	
	STRETCH and ALLEN (1961)							
	ENDO (1961a)	2						
	LANGHAMMER and QUITZSCH (1961)	0.1			toluene	RT	RT	cross-linked PS as a support
	BARONI (1961)							
	SCHULZ et al. (1962)	0.9	(48)	ethanol	benzene	60	20	
	BREITENBACH et al. (1962a, b, c, d)	0.3		ethanol	MEK	65	10	
	HOMMA et al. (1963)	1		ethanol	MEK	60	10	
	JOVANOVIĆ et al. (1965)	0.25	(50)	methanol	benzene	55	20	

	H. J. Cantow et al. (1966)	1		methyl-cyclohexane	methyl-cyclohexane	70	—20	θ-column
	H. J. Cantow et al. (1966)	0.005	30	methyl-cyclohexane	methyl-cyclohexane	32*	8*	capillary 1.6 mm in diam. *after 24 h at 80 °C
	Henderson and Hulme (1967)	43	15	methyl-cyclohexane	cyclohexane	50	40	with 30 l of solvent/precipitant
	Donkai et al. (1968)	10		cyclohexane	cyclohexane			branched PS
	Yamaguchi and Saeda (1969)	1	(14)	methanol	MEK	47	20	better separation than in isothermal extraction
	Casper and Schulz (1971)	0.3	(70)	cyclohexane	cyclohexane	≦25	≦25	phase partition chromatography
	Scholtan and Kwoll (1972)	0.01	3	methanol	benzene			elution gradient chromatography with FI detector
	Kohler et al. (1972)	5		methanol	benzene	$\Delta T = 15$ K		T-cycles; cf., elution
	Ueda (1972)							
	Spatorico et al. (1973)	5		methanol	dioxan	50	25	steel column, gold-plated
	Miyamoto et al. (1973)			cyclohexane	cyclohexane			branched PS
	Böhm et al. (1974)	0.4		methanol	benzene	50	20	
	Alvariño et al. (1978)	0.4—2.0	(12 to 15) × 24	methanol	benzene	50	20	high-grade steel column
PS grafted on PIB	Chapiro et al. (1963)	2	(100)	i-propanol	cyclohexane	60	25	
Poly-α-methylstyrene	Yamamoto et al. (1970)	1.0	(120)	n-hexane	benzene	58	10	comparison with column elution, SEC, and ultracentrifuge
Polymethyl methacrylate	Weakley et al. (1960)	(0.3)	(100)	cyclohexane	MEK	(10)	(60)	top-end discharge
	Poláček (1963a, b)	0.5		methanol	acetone	(40)	(25)	T-cycles

Table 20-1 (continued)

Polymer	Authors	m/g	t/h	Precipitant	Solvent	$\vartheta_o/°C$	$\vartheta_u/°C$	Remarks
	Poláček et al. (1967)	5		methanol	acetone	(36)	(26)	T-cycles
	Daňhelka and Kössler (1970)	1	40	methanol	acetone	(39) (29) (−1)	(31) (21) (−9)	T-cycles
	Dawkins and Peaker (1970)	1	11 × 24	methanol	benzene	40	35	
	Spatorico and Coulter (1973)	5		cyclohexane	dioxan	50	25	steel column, gold-plated
Polybutyl methacrylate	Glöckner and Müller (1968)	0.15	7	methanol	acetone	35	10	
PS grafted on PMMA; PMMA on PS	Acres and Dalton (1963)	0.1	30	petroleum ether	benzene	40	15	
Polyvinyl acetate	M. J. R. Cantow et al. (1963)	40	200	isopropanol	benzene	60	28	
Polyester	M. J. R. Cantow et al. (1964)	90	112	n-heptane	acetone	50	27	
	Dubrowina et al. (1964)	1		ethanol	tetrachloroethane	65	15	polyacrylates
	Hansen and Sather (1964)	5	16	cyclohexane	MEK	70	40	
	Glöckner (1965a)	0.3	12	methanol	dichloromethane	26	1	polycarbonate
	Spatorico and Coulter (1973)	5		cyclohexane	dioxan	40	20	"Pola" polyester
Polyether	Brzezinski (1966)			tetrachloromethane	chloroform	50	24.5	
	Slonaker et al. (1966)	800		water	ethanol	40	30	2 stages; 220 l of solvent/precipitant

	SCHOLTAN and KRANZ (1967)	5	20	water	methanol	20	60	solubility decreases with increasing temperature
Cellulose acetate	TARAKANOV and OKUNEV (1962)		96	heptane	dichloromethane			tube columns without support materials
	OKUNEV and TARAKANOV (1963)	2.0	(160)	heptane	dichloromethane			
Dextran	EBERT and ERNST (1962)	0.1	80	ethanol	water	70	12	
Polysarcosine dimethylamide	CAPLAN (1959)	0.0003	140	water	dioxan	65	15	micro-method
Poly-γ-benzyl-L-glutamate	COSANI et al. (1966a, b)	0.7	220	methanol	dichloro-methane	35	15	
Polypeptides	POPE et al. (1959)	5		cyclohexane	ethanol	60	10	
Polyvinyl chloride	ENDO (1961b) CROOK and WALKER (1963)	2	(700)	methanol	cyclohexanone	60	25	
	FERRIER (1967)	1	24	glycol	cyclohexanone	140	45	2 stages, dissolution at $\vartheta_o = \vartheta_u = 140$ °C
Styrene-p-iodo-styrene copolymers	BRAUN and CHAUDHARI (1970)	2		methanol	benzene	50	20	
Styrene-acryloni-trile copolymers	SCHOLTAN and KWOLL (1972)	0.01	3	cyclohexane	MEK			elution gradient chromato-graphy with FI detector
	WEBER (1976)	0.3	12	methanol	dichloromethane	25	5	
VC-AN copolymers	DING et al. (1965)	0.6		methanol	acetone	40	2	

20.2. Methodical preparatory work for the determination of the separation conditions

While SEC can be used universally after an appropriate calibration, and preliminary or sometimes even sufficient pieces of information already become available through the raw elugrams of the corresponding samples within a few minutes, precipitation chromatography requires rather time-consuming investigations for determining the working conditions. An optimal separation can be achieved only if the solvent and the precipitant, the elution gradient and the temperature gradient are chosen properly (cf., Section 9.5.2.).

Turbidimetric titrations are a good aid in searching for the suitable systems. HULME and McLEOD (1962) recommended the following procedure: portions of the solution containing 0.5% polymer should be titrated to their cloud point by means of all the precipitants in question. Among those precipitants, that of which the largest volume must be added should now be used to search for the most favourable solvent. This should be done by titrating 0.5% solutions in different solvents. The solvent for which the smallest addition of precipitant effects cloudiness is the most suitable and should be used in combination with the precipitant previously selected. Then the compositions between which the optimum gradient will span should be determined by titrations at the specified head (ϑ_0) and bottom (ϑ_u) temperature of the column. (For that purpose the solution must be titrated at ϑ_0 to complete precipitation of the polymer, while for ϑ_u only the cloud point must be determined.)

The evaluation of solvent/precipitant combinations for isothermal column elution of polystyrene according to this recommendation and the intentional selection of a suitable (I) and an unfavourable (II) pair was not, however, reflected by the result of the fractionation (MENCER and KUNST 1979): the resolution obtained by means of system II (cyclohexanone-n-propanol) was almost equivalent to that obtained by means of I (MEK/Eol). This was because the fractionations had been carried out with a temperature programme for the thermostat of the entire column, and the temperature dependence of the solubility was more pronounced in system II than in I. Therefore the authors preferred the criterion stated by SCHIEDERMAIER and KLEIN (1970) for the selection of efficient solvent/precipitant pairs: for all systems in question, $\Delta\varphi_c^*$ must be evaluated for two polymer concentrations (e.g., 0.5 and 0.05%), and $\Delta\varphi_M^*$, for two different molar masses. As φ^* indicates the volume fraction of the precipitant at the cloud point (cf., Section 5.4.3.), $\Delta\varphi^*$ represents the difference observed between titrations carried out under different conditions. The fractionating efficiency of a solvent/precipitant pair increases with decreasing $\Delta\varphi_c^*/\Delta\varphi_M^*$, i.e., in the same sense as the molar mass dependence of the cloud point outweighs its concentration dependence.

In addition to the thermodynamic properties, in column methods the viscosity of the fractionation media must also be taken into account, because sufficiently high diffusion coefficients belong to the preconditions of a rapid establishment of equilibrium.

In Table 20-1 the solvents and precipitants employed have also been listed. These solvents were used in a pure (unmixed) form only in a few exceptional cases. As a rule, the "precipitant" supplied was already mixed with a certain proportion of the other component, and the "solvent" introduced from the supply vessel was blended with the precipitant. In this way it is possible to adjust as flat a *solvent gradient* as desired, which is an important condition for a successful fractionation. In practice this has the additional advantage that the considerable amounts of liquid obtained in each fractionation can be reprocessed with less effort than required for pure components.

The solubility of *copolymers* depends on both the molar mass *and* the composition. In many cases it was possible to find solvent/precipitant combinations which separate mainly according to either the first or the second characteristic. Again suitable pairs can be determined by turbidimetric titration (LITMANOVIČ et al., 1963a, b; KUDRJAVČEVA et al., 1963; JURANIČOVA et al., 1970; GLÖCKNER et al., 1971).

The *temperature gradient* is chosen in the range between the boiling point of the liquids used and their freezing point, in order to avoid unnecessary difficulties.

For practical reasons, POLÁČEK (1963a, b) preferred to use cyclic variations of the column temperature as a whole instead of a stationary gradient. His assumption (1963a) "that the fractionating effect of temperature cycles may be higher than the effect of a temperature gradient, because the temperature variations at every point of the column are so distinctly marked that they practically avoid the possible formation of supersaturated solutions" is confirmed by the good practical results obtained with the method. As is well known, supersaturated solutions and non-stationary, flowing gel phases are problematical in the Baker-Williams technique. The selectivity achieved by the Poláček variant corresponds to that of a well adjusted precipitation chromatography with a linear temperature gradient (GLÖCKNER and KUHNHARDT, 1970). The maximum *permissible loading* can be referred to the mass of the packing material of the sample bed. To a first approximation, this gives a ratio of 1:100—1:40, as stated in Section 11.9. Values taking into account the distribution of the polymer and the molar mass are better. According to PEPPER and RUTHERFORD (1959), c_{max}, the concentration of the polymer in the eluate fraction with the highest content, must not exceed a certain limit. The latter depends on the degree of polymerization, and ranges between 0.3 and 1%. In accordance with FLORY [B 2, p. 341], SCHNEIDER et al. (1960) defined this rule more exactly, stating that the value of the product $c_{max} \cdot M_{max}^{0.5}$ must not exceed 400 (c in %). According to SCHNEIDER et al., M_{max} is the molar mass of the heaviest fraction, whereas YAMAGUCHI and SAEDA (1969) denoted by M_{max} the highest molar mass found.

20.3. Prognosis

As a method for analytical fractionation according to molecular size, the Baker-Williams technique has been replaced by SEC. However, in preparative fractionation, precipitation chromatography is of equal importance when the time required, the amount of solvent and the fractionating efficiency with respect to quality and quantity are compared; it is definitely superior with respect to the cost of the apparatus (cf., Section 17.9.4.).

SEC separates according to the hydrodynamic volume. Heterogeneity in composition, which is possible for copolymers, can be revealed only as far as it is coupled with the hydrodynamic volume. In principle, precipitation chromatography being a solubility method can perform better in this respect. For analytical use, the column must be equipped with adapters having no dead volume, and coupled with a solvent-independent detector. The sample application must be performed by injection of a solution during the operation, as it is generally done in HPLC. With these improvements the method will have a future.

21. Thin-layer chromatography

Since the late sixties, the chromatographic characterization of polymers on thin-layer plates has found increasing interest. The technique itself is customary outside polymer chemistry and has been described in excellent books and reviews [E 1 to E 13]. Its application to polymers has been summarized by INAGAKI (1977a, b) as well as BELENKIJ and GANKINA (1977).

As a rule, sample spots are applied from solution along a starting line parallel to an edge of the thin-layer plate. When the solvent has evaporated, the chromatogram is developed by dipping the starting edge into an eluent to a depth of 5—8 mm (STAHL, 1968). Due to the capillary forces the eluent rises in the layer, flows over the starting spots and arrives at the terminal line after 10—60 minutes. In this form TLC is a *development technique* on a *dry separating bed*. (Almost all of the techniques described in the preceding chapters were wet-bed elution techniques.)

21.1. Flow parameter and speed of migration

In the usual TLC technique, the distance travelled, s_f, increases as the square root of the travelling time:

$$s_f = \sqrt{\varkappa \cdot t} \tag{21-1}$$

The flow parameter, $\varkappa$, is directly proportional to the surface tension of the eluent and inversely proportional to its viscosity. The particle diameter also makes a linear contribution to the value of $\varkappa$. Table 21-1 shows the numerical values for a number of solvents and the travelling times required for 7 cm and 10 cm developments on silica gel H. In reversed-phase TLC the wetting angle may also be of influence.

The commercially available *nanoplates for HPTLC* have very dense and uniform layers, on which the flow constants are smaller than those on ordinary TLC plates. For example, for toluene the value of $\varkappa = 6.3\ \text{cm}^2/\text{min}$ is about 30% smaller than that listed in Table 22-1 for a normal layer. Nanoplates are used for distances ranging from 3—7 cm, because for longer distances not only the development times obtained would be unreasonably long, but also the results of separation would be impaired by the diffusion in the layer. GUIOCHON et al. (1978) concluded from theoretical considerations that this negative effect may reach considerable proportions for particle sizes of less than 7 µm. This was experimentally supported by GUIOCHON and SIOUFFI (1978) and BRINKMAN et al. (1980). The last-mentioned authors compared HPTLC materials obtained from different manufac-

Table 21-1
Flow parameter, $\varkappa$, for calculation of the capillary rise on silica gel *H* (according to GEISS [E 5])

Eluent	$\varkappa$ / $cm^2 \cdot min^{-1}$	Developing time (in min) calculated from $\varkappa$ for a total length of run of	
		(0.5 + 7) cm	(1 + 10) cm
Acetonitrile	11.8	4.7	10.3
Diethyl ether	11.7	4.8	10.3
n-pentane	11.6	4.8	10.4
Acetone	10.9	5.1	11.1
n-hexane	10.9	5.1	11.1
Water	10.3	5.4	11.7
Methylene chloride	9.6	5.8	12.6
Ethyl acetate	8.8	6.4	13.8
Toluene	8.7	6.4	13.9
n-heptane	8.5	6.6	14.3
Benzene	8.2	6.8	14.8
Chloroform	7.2	7.8	16.8
Methanol	6.8	8.2	17.8
Tetrachloromethane	4.4	12.7	27.5
Dioxane	4.3	13.0	28.2
Acetic acid	3.5	16.0	34.6
Ethanol	3.2	17.5	37.9
Bromoform	2.8	20.0	43.2
Formamide	2.6	21.5	46.6
n-propanol	2.0	28.0	60.5
n-butanol	1.5	37.4	80.7

turers. Concerning the grain size, they found that the particles in these layers, ranging between 7 and 10 µm, are not so much smaller than those in normal TLC layers, but that their sizes follow distinctly narrower distributions. The layers are also more homogeneous.

For a full utilization of the separating capacity of nanoplates an extremely precise dosage in the nanolitre range is necessary. Then it is possible to achieve excellent results within a rather short time, above all in the case of radial development ([E 7]; VITEK and KENT, (1978).

C_n-bonded coatings were introduced into TLC by GILPIN and SISCO (1976) (with $n = 1$, 2, 6, 12, and 18). The relatively poor adhesion of these materials to the glass base posed a problem. BRINKMAN and DE VRIES (1980) reviewed the literature about RP-TLC and reported their experience gained with prefabricated RP plates from different manufacturers (E. MERCK, WHATMAN, ANALTECH).

In *overpressured thin-layer chromatography* (OPTLC) the surface of the layer is sealed by a tightly covering foil and the eluent is pressed into the layer by a pump (TYIHÁK et al., 1979; MINCSOVICS et al., 1980). In this technique the front of the eluent advances with a constant velocity.

The optimization of usual TLC was discussed by SIOUFFY et al. (1981). For a given separation problem, the optimum depends on the diffusion coefficient, *D'*, and on the flow parameter $\varkappa$ of the solvent, and hence strongly depends on the particle diameter. In some cases (e.g., dyestuffs, chloroanilines and all substances with low D' values), the best results can be obtained on layers consisting of materials with a particle diameter of 5 µm. In all other cases one can expect better results with a longer development on plates made of coarser materials (d_P = 10 or 20 µm). The latter is mainly true in difficult separation problems.

21.2. The R_f value

The sample components are carried along by the eluent over different distances depending on their respective chromatographic retentions, and at the termination of the development they are located in spots on the separating path. Colourless substances have to be made visible by colouring (e.g., using iodine solution), carbonization or fluorescence. This will be dealt with in Section 21.7.1.

As a measure of the retention, the distance, s, travelled by the substance, compared to the distance, s_f, travelled by the eluent, is indicated as the R_f value (rate of flow):

$$R_f = s/s_f \tag{21-2}$$

If the components in the spots are distributed according to the Gaussian function, then, by analogy with the peak elution volume, s is measured up to the centre of the substance spot, which corresponds to the peak maximum. For asymmetric spots, for a development with "front tailing" (Fig. 21-1b) or "rear tailing" (Fig. 21-1c), one has to proceed as in the case of skewed elution curves: in such a case it is more correct to use the well defined edge rather than an arbitrary central value in the calculation of R_f. Unfortunately, in the thin-layer adsorption chromatography of polymers it is sometimes not possible to avoid such deformed spots, which for small-molecule substances in most cases indicate overloading. Although this makes the R_f determination more difficult, one should make every effort to obtain reproducible data (GEISS, 1968).

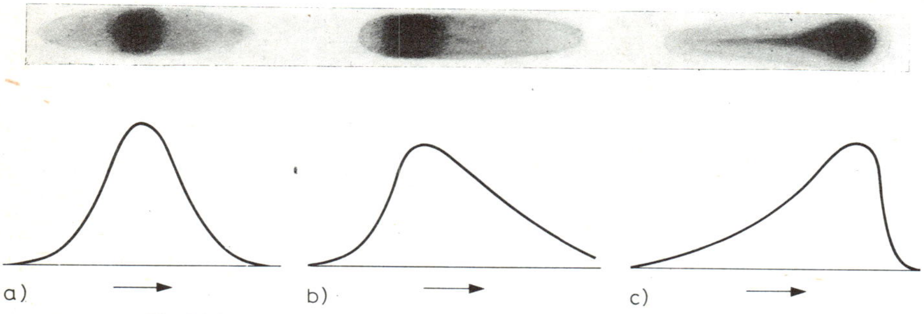

Fig. 21-1
Spot shapes and associated substance distributions on thin-layer chromatograms (development in the direction of the arrows)

a) Symmetric substance distribution (μ_3 = 0)
b) Development with front tailing, substance distribution curve with skewness (μ_3 < 0 as referred to the separating path)
c) Development with rear tailing (μ_3 > 0) (cf., Section 16.3.1.).

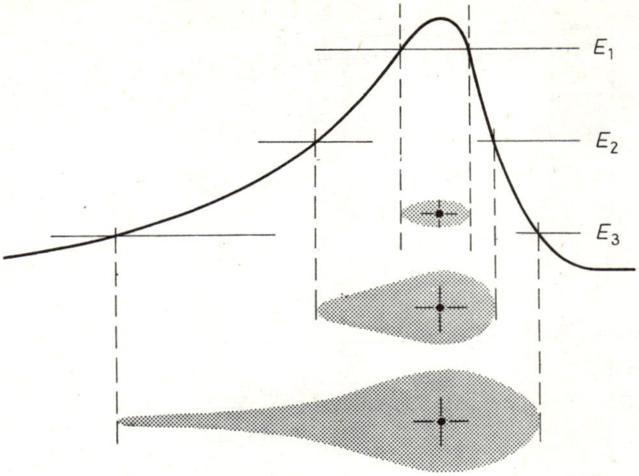

Fig. 21-2
Shape and size of the spot at different detection sensitivities ($E_1 < E_2 < E_3$)

The spot size depends on the starting spot, which should be as small as possible, on the chromatographic spot broadening, the amount of substance applied and the detection sensitivity. What becomes visible is only that part of the spot where the intensity exceeds a certain level; the base of the band remains hidden (see Fig. 21-2). Therefore for skewed bands with a steep slope the intensity maximum and the better defined band edge are indeed almost at the same position.

The R_f value of thin-layer chromatography is related to the *retention rate R* defined by eqns. (3-1) and (3-2), but it is not identical with it. The difference results from the fact that in the conventional technique the development is stopped when the front of the eluent arrives at the terminal line. However, at this time the chromatographic layer is not yet homogeneously filled with solvent. The eluent soaks in due to the capillary forces, which depend on the capillary radius. The narrowest interstitial spaces are filled most rapidly. Here the eluent is ahead of that contained in the other, larger-sized interstitial spaces during the total development, including the time of arrival at the terminal line. Therefore the proportion of the eluent to the adsorbent is constant only for about 80% of the distance travelled. Towards the end of the development distance the saturation of the layer decreases rapidly; the visible eluent front is formed by a few per cent of the saturation quantity. On the other hand the eluent volume profile is slightly raised over a distance of about 10 mm in the neighbourhood of the dipping line (see (Fig. 21-3 a and b). As the distance over which a substance is transported in the chromatographic process is proportional to the eluent flow, the portion lacking at the end of the travelling path causes the R_f values determined by eqn. (21-1) to be too small. For an accurate determination of the correction factor, the *volume profile* would have to be determined for every combination of layer material and eluent. As this is rather difficult, in most cases the following correction is used:

$$R = R_f \cdot 1.1$$

R_f values above 0.9 should be avoided, because they lie in the volume gradient.

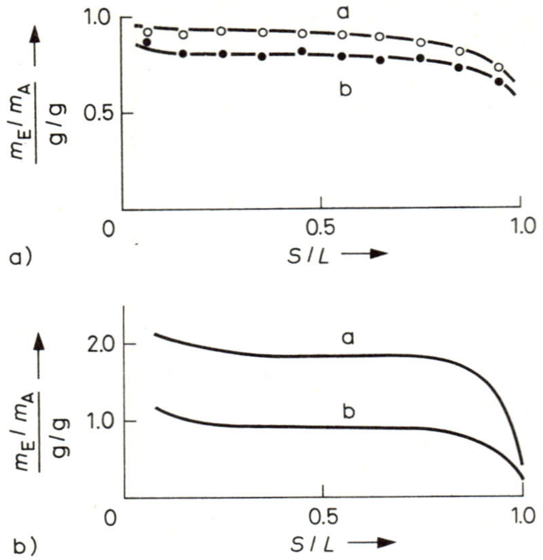

Fig. 21-3
Elution profile upon arrival at the end line, mass of eluent per gram of adsorbent at different points between the start ($s/L = 0$) and the end ($s/L = 1$) of the travelling path.

a) Curve a — upward development; curve b — downward development; eluent: dimethylformamide (according to KAMIYAMA and INAGAKI, 1971)

b) Upward development with dichloromethane-methanol (50:50) on silica gel layers with a thickness of 0.25 (curve a) and 0.73 mm (curve b) (according to KAMIDE, MANABE and OSAFUNE 1973).

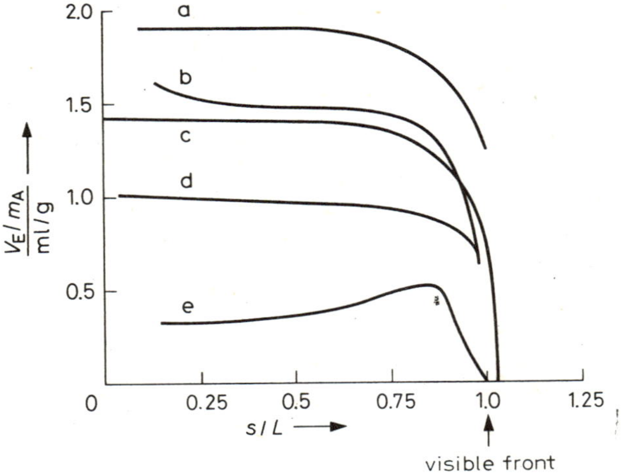

Fig. 21-4
Volume profile of different solvents on silica gel plates (Eluent volume per g of adsorbent at different points s/L between the start and the end of the travelling path)

a benzene with 3 vol.-% ethanol (development in a saturated standard chamber; according to GEISS, SANDRONI and SCHLITT, 1969); b dichloromethane with 9.1 vol.-% methanol; c benzene (according to GEISS [E 5], 1972, p. 13); d dimethylformamide, upward (taken from Fig. 21-3a); e dichloromethane with 50 vol.-% methanol (according to KAMIDE, MANABE and OSAFUNE, 1973; curve b ditto).

Of course the difference between R and R_f can be eliminated by allowing the eluent to continue its migration after passing the terminal line until the total separating path is uniformly saturated. However, apart from the longer time required it is difficult to identify the right moment at which the continued elution should be stopped, so that it is hardly possible to achieve a greater accuracy than that obtained by the usual development and correction by means of eqn. (21-2).

The volume profile depends on the conditions of development (Fig. 21-3a, b), but most of all on the system, even if the density differences of the eluents are taken into account (Fig. 21-4).

21.3. Elimination of activity effects

The R_f value is a complex function of factors which are due partly to the substance, partly to the stationary phase and the elution conditions. It is obvious to attempt to eliminate the contribution of the system-induced effects by means of reference substances. To what extent even the air humidity can effect R_f values is shown in Fig. 21-5.

21.3.1. The R_k value

Brenner et al. (1964) suggested taking the development conditions into account by use of R_k values:

$$R_k = R_{M,S} - R_{M,St} \tag{21-3}$$

$R_{M,S}$ and $R_{M,St}$ are, respectively, the R_M values obtained from eqn. (3-8) for the substance (S) and the reference substance (St) developed together with the former.

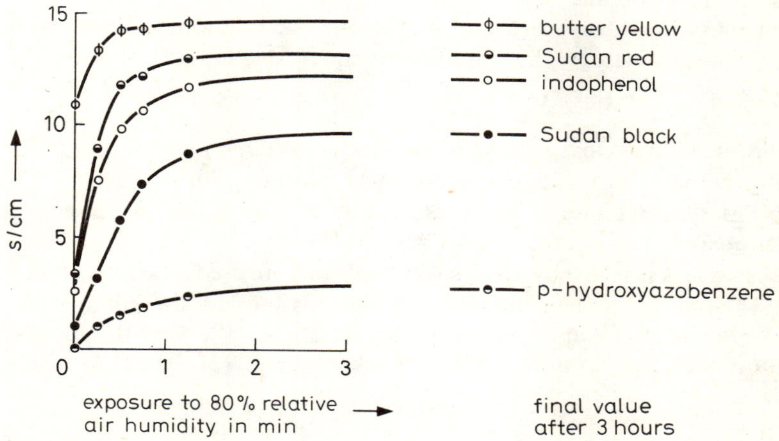

Fig. 21-5
Distance of travel of test dyes (with equal front height L) as a function of the exposure time of the Al_2O_3 layers (conditioned at 15% relative humidity) to air having a relative humidity of 80%

Development by benzene in a sandwich chamber (according to Geiss, 1968).

In view of eqns. (3-7), (3-8) and (21-2), the relationship between R_M and R_f is as follows:

$$R_M = \log\left(\frac{1}{\xi \cdot R_f} - 1\right) \tag{21-4}$$

Here ξ denotes the proportionality constant between R and R_f, for which $\xi = 1.1$ has been assumed in eqn. (21-2).

While R_M bears a simple relationship to the distribution coefficient (cf., Section 3.2.), the relationships between the latter and the R_f value are somewhat more complex. According to systematic investigations, linear dependences on the parameters influencing the distribution coefficients should be expected primarily for the R_M values (SOCZEWIŃSKI, 1969; PERRY, 1979; SOCZEWIŃSKI and JUSIAK, 1981), so that in most cases it is worth the minor trouble of a conversion according to eqn. (21-4).

Taking the difference, as is done in eqn. (21-3), means that the distribution coefficient of the sample is referred to that of a *standard substance*:

$$R_k = \log(K_S^+/K_{St}^+) \tag{21-5}$$

An investigation of one and the same combination S — St in two different systems yields:

$$(R_k)_I = \log(K_S^+/K_{St}^+)_I \quad \text{and} \quad (R_k)_{II} = \log(K_S^+/K_{St}^+)_{II}\,.$$

If the variation of the chromatographic system has proportional effects on *both* of the distribution coefficients, so that the ratio remains constant, then R_k is independent of the system. Then, according to

$$(R_{M,S})_{II} = (R_k)_I + (R_{M,St})_{II} \tag{21-6}$$

the R_M value of a substance on a plate II can be precalculated from the value $(R_{M,St})_{II}$ of the reference substance on this plate.

In the light of the theory developed by SNYDER, the following conclusions can be drawn for adsorption chromatography. Insertion of eqn. (7-11) into (21-5) gives:

$$R_k = \alpha_A[(S^0 - S^0_{St}) - \varepsilon^0(A_S - A_{St})] \tag{21-7}$$

From this it follows that on adsorbents with different activities R_k decreases with decreasing activity, α_A. The use of stronger eluents with a higher ε^0 will have no effect on R_k if the molecules of the sample and the reference substance happen to have equal molecular areas ($A_S \doteq A_{St}$) (cf., Table 7-2).

Based on experience, in most cases stronger eluents are used on an adsorbent with a higher activity. Then α_A and ε^0 have increased in comparison with the initial state. Now, if the "proper" reference substance has been chosen ($A_{St} < A_S$), then the increase of α_A just compensates the variation of the term in brackets in eqn. (21-7), and R_k again remains unchanged.

21.3.2. The vain attempt with the "relative R_f values"

It has been proposed to make the R_f values independent of specific experimental conditions by simply indicating the ratio $R_f/R_{f,St}$ as a relative R_f value. This is indeed simple, but in fact it does not lead to any improvement. As the relative R_f values possibly may show even higher variations than the directly measured ones, SNYDER [A 4] and GEISS [E 5] warned against this method.

21.3.3. The R_f correction using two reference substances

GALANOS and KAPOULAS (1964) observed that the R_f values measured under different conditions (I and II) for the sample S can be related to the simultaneously measured values of two reference substances X and Y as follows:

$$(R_S - R_X)_I = (R_S - R_X)_{II} \frac{(R_X - R_Y)_I}{(R_X - R_Y)_{II}} \tag{21-8}$$

This empirical equation represents an approach to the proposal discussed in Section 21.3.1. For the practical application, the following forms of this equation are very suitable:

$$R_{S,I} = p + q \cdot R_{S,II} \tag{21-9}$$

$$p = R_{X,I} - q \cdot R_{X,II} \tag{21-10}$$

$$q = \frac{(R_X - R_Y)_I}{(R_X - R_Y)_{II}} \tag{21-11}$$

This correction method was tested repeatedly and found useful (DHONT et al., 1970; FREY and ACKERMANN, 1976). All R_f values should range between 0.1 and 0.9, and that of the unknown between those of the substances X and Y. The method fails for eluent mixtures consisting of components of very different polarities (FREY and ACKERMANN, 1980).

VAN WENDEL DE JOODE et al. (1979) derived the theoretically based correction equation

$$\frac{1}{R_{S,I}} = p + \frac{q}{R_{S,II}} \tag{21-12}$$

where

$$p = \frac{1}{R_{X,I}} - \frac{q}{R_{X,II}} \tag{21-13}$$

and

$$q = \frac{(1/R_X - 1/R_Y)_I}{(1/R_X - 1/R_Y)_{II}} \tag{21-14}$$

The derivation starts from

$$R_{M,I} = R_{M,II} + \log C_{I/II} \tag{21-15}$$

This relationship in effect corresponds to eqn. (21-3), and is the mathematical expression of the fact that the distribution coefficients of different substances frequently vary by the same factor if the system is changed.

The authors tested their eqn. (21-12) with experimental data taken from literature and found an even better agreement than with eqn. (21-9). This was confirmed by DHONT (1980).

21.4. Special problems in thin-layer chromatography

The TLC technique is simple; as compared to high-pressure liquid chromatography the equipment cost is almost negligible. This may raise the question why this technique was not discussed in a previous chapter, following the principle of proceeding from simple to more complicated items. There are at least two reasons, why this chapter was intentionally placed at the end. They will be discussed now.

21.4.1. Spontaneous gradients

In almost every case thin-layer chromatography is a gradient technique, even if the experimentalist makes no attempt whatever to establish a gradient: the evaporation from the plate, even favoured by the exothermic heat of wetting, leads to a *flow gradient* (see Fig. 21-6). In circular development, which is also used in "High Performance Thin Layer Chromatography" (HPTLC) [E 7], the geometrical conditions effect a very marked flow gradient.

In planar adsorption chromatography, an *activity gradient* may occur spontaneously due to the absorption of water vapour during the development (cf., Section 10.5.). If multicomponent eluents are used, it is generally impossible to develop without spontaneous gradients. The chromatographic analysis seems to be isocratic, but in fact it follows a gradient

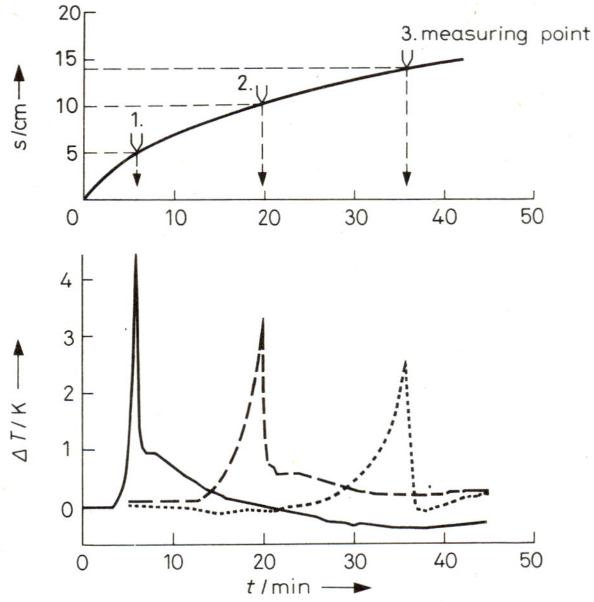

Fig. 21-6
Thermal effect at three different points of a 1.5 mm silica gel layer in the development by benzene

In each case the temperature peaks occurred when the front had arrived at a thermocouple; they reflect the heat of wetting. (according to GEISS and SCHLITT, 1968b).

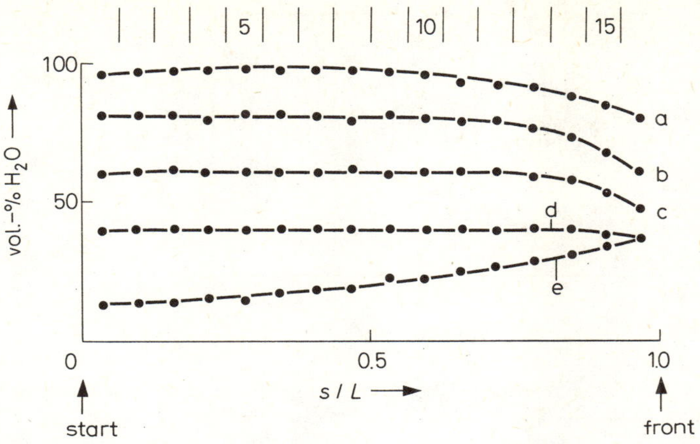

Fig. 21-7
Composition of methanol-water mixtures after passage over a silica gel layer (concentration profile)

Initial value of the water content in the eluent: a 90; b 70; c 50; d 30; e 10 mole-% (according to FRACHE and DADONE, 1973).

technique because of the demixing of eluents. True, the preferential adsorption described in Section 7.4.2. also occurs on the adsorbent in the column, but in this case usually the elution technique is applied with a wet chromatographic bed, which is in equilibrium with the eluent mixture before the sample arrives. In contrast, a dry bed development with eluent components of varying strengths cannot take place at all without exhibiting a gradient due to selective adsorption of the eluent constituents. To this must be added the demixing due to different flow constants and, in the case of an open planar bed, the matter transfer between the chromatographic layer and the gas phase. Fig. 21-7 shows examples of concentration profiles caused by spontaneous eluent demixing.

The possible complications were discussed in detail by GEISS who required "that in TLC, wherever possible, eluent mixtures should be dispensed with" [E 5, p. 150]. However, in the investigation of polymers, the cases where a single solvent is sufficient are even scarcer than in the thin-layer chromatography of low-molecular-weight substances. Apart from a few exceptions, in pure eluents the polymer is either retained at the start or travels together with the front; intermediate R_f values require eluent mixtures. Fig. 21-8 shows the typical transition from $R_f = 0$ to $R_f = 1$ within a relatively narrow range of eluent compositions. With the use of finely graduated mixtures it was also possible to achieve R_f values in the intermediate range (GLÖCKNER and MEISSNER, 1980).

The *vapour pretreatment* of the layer represents a possibility of suppressing the demixing during the development (WAWRZYNOWICZ and SOCZEWIŃSKI, 1979). In this connection it should be mentioned that the fine chromatographic profiles of polymers presented by BELENKIJ and GANKINA were usually obtained with plates pretreated with the eluent vapour. On TLC plates with modified silica (RP 18 and the like), the extent of solvent demixing was observed to be much lower than on bare layers (SIOUFFI et al., 1979).

Spontaneous gradients are difficult to control, and therefore undesirable. However, in macromolecular chromatography there are observations which shed quite a different light on this situation: using methyl acetate/chloroform (3:97) on pre-coated silica gel/glass

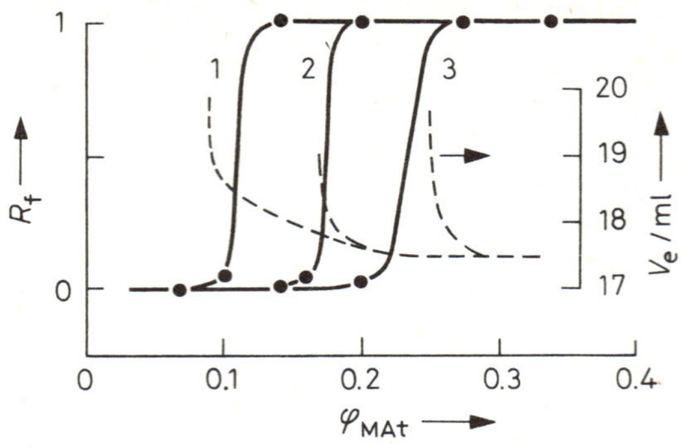

Fig. 21-8
R_f values of styrene-methyl acrylate copolymers in tetrachloromethane/methyl acetate mixtures

TLC on silica gel Replates® 50
1 46.6 mole-% methyl acrylate ($\bar{M}_n$ = 261 000);
2 57.3% (276 000); 3 77.9% (302 000)
The dashed reference curves are taken from Fig. 18-9 showing the behaviour of the three samples in isocratic column chromatography.
(according to TERAMACHI, HASEGAWA, SHIMA, AKATSUKA and NAKAJIMA, 1979).

powder plates, TERAMACHI and ESAKI (1975) succeeded in separating styrene-acrylonitrile copolymers according to their chemical composition. In columns, experiments using the same eluent were unsuccessful.

Similar observations were made by WESSLEN and MÅNSSON (1975) using block copolymers prepared from styrene and ethylene oxide: on cellulose thin-layer plates, ethyl acetate-methanol (90:10) developed homopolystyrene components up to $R_f = 0.8$ and block copolymers up to $R_f = 0.6$, whereas polyethylene oxide was retained at the start. Consequently, the preparative separation of the block copolymer from concomitant homopolymers should also be possible. However, it did not prove feasible to separate the polymers in the column even under strictly the same conditions with respect to support material and eluent. On the other hand, a step-by-step extraction with ethyl acetate (I), ethyl acetate-methanol (4:1; II) and methanol (III) proved successful: after the homopolystyrene had been removed by 170 ml of (I), the block copolymer was obtained free of any additions by the use of (II). The polyethylene oxide was dissolved by methanol. While the gradient-free column elution failed, the thin-layer development, also seemingly isocratic, led to a separation. Apart from other effects (e.g., the loading), a spontaneous gradient certainly played a rôle in this case.

Thus it is not only complications, mainly with respect to reproducibility, which have to be attributed to the spontaneous gradient: in some cases it is no less than the prerequisite to a separation. This especially appears to be the case for polymers. Probably this is also the cause of the remarkable phenomenon that the shape of the TLC chamber is so essential for a success.

21.4.2. Separating mechanisms

In the column chromatography of polymers the principle of separation can be clearly stated in most cases: the vast majority of investigations are based on *size exclusion*, others on *phase partition* and yet others on *adsorption*.

The TLC of small molecules is usually AC. On the other hand, the TLC of polymers may be based on AC, SEC, precipitation or solvophobic effects; frequently several mechanisms interact with one another. On the one hand, this is due to the fact that the pore sizes of some TLC adsorbents are of the order of magnitude as the molecular sizes (see Fig. 21-9). On the other hand, the range of solubility for macromolecules is much narrower than for low-molecular-weight substances. Quite often eluent mixtures are combinations of solvents and precipitants. Thus a phase separation may readily occur (see Fig. 21-10).

This complexity is the other reason why planar chromatography of polymers is discussed only in this chapter, after the column techniques.

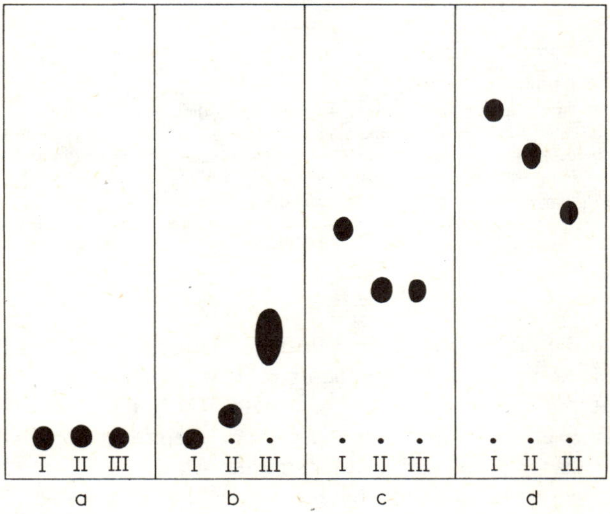

Fig. 21-9
Adsorption and size exclusion
Development of polystyrene on silica gel with a surface area of 35—65 $m^2 \cdot g^{-1}$ (d_0 = 30—70 nm; ${}^1V_p$ = 0.78 $cm^3 \cdot g^{-1}$)

a) Cyclohexane
b) Cyclohexane-benzene (50:50, adsorptive separation)
c) Cyclohexane-benzene (40:60)
d) Benzene (separation by size exclusion)
I — M = 411000; II — M = 19850; III — M = 4800g · $mole^{-1}$.
(according to Heitz, Klatyk, Kraffczyk, Pfitzner and Randau, 1971).

21.4.3. Spot shapes

The gradient-free planar development of low-molecular-weight substances yields elliptic spots, the width of which hardly deviates from that of the starting spot, while their length depends on the number of theoretical plates. For a visual observation, the detection sensitivity has a considerable effect. The spots of polymer samples would certainly disappoint a TLC expert, because they frequently look like the patterns of poorly developed or overloaded plates. This results from the properties of the polymers: macromolecular samples in most cases have an MMD which includes so many similar individuals that there is not the least chance of a subdivision into individual components (cf., Sect. 16.1.). A

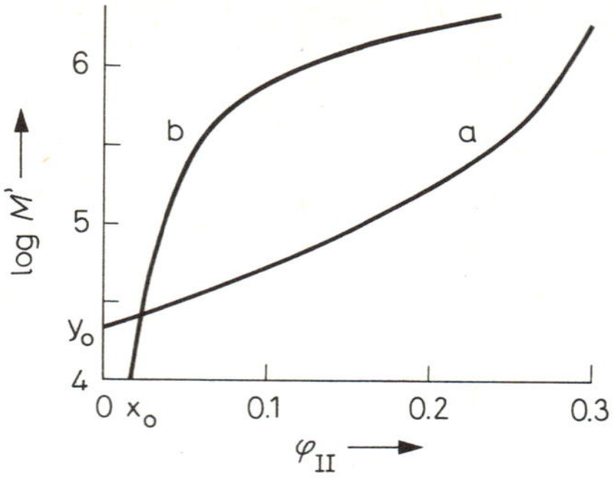

Fig. 21-10
Adsorption and phase separation

Upper mobility limit of polystyrene samples as a function of the eluent composition
a) Addition of chloroform (on silica) and toluene (on alumina), respectively, to acetone: the value y_0 is essentially determined by the insolubility of macromolecular polystyrene in acetone.
b) Addition of tetrahydrofuran to tetrachloromethane (on alumina layers: the value x_0 is determined by the adsorption of polystyrene on alumina, which is overcome only by the addition of the polar solvent.
φ_{II}-volume fraction of the second component in the mixture; M'-upper limit of molar mass for a migrating PS sample (according to Otocka and Hellman, 1970).

separation according to the molar mass at best leads to elongated, ill defined spots. For a complete adsorption of the macromolecules the extremely *non-linear adsorption isotherms* (see Fig. 6-2c) prevent the formation of normal spots. The migration takes place in a substance band with a skewed mass distribution and a very steep front slope. The desorption is highly obstructed. This leads to spots with rear tails which may extend back to the starting point. In this case tailing alone by no means indicates heterogeneity of the sample.

Sometimes even *front tailing* is observed in the chromatography of polymers. In spots of this shape the bulk of the substance is located close to the backward, sharp edge of the spot. The forward end of the spot mostly narrows to a point, sometimes with ill defined contours. According to Inagaki (1977a), such spots indicate a phase separation in the development (*thin-layer precipitation chromatography*). Sometimes the eluent following after drives wedges into the rear of the substance band. Such a *fingering* was exhibited by polystyrene samples without polar end groups in a gradient development using acetone-tetrahydrofuran on silica gel (Min et al., 1975). Polystyrenes with polar groups at both ends of their molecules, which were simultaneously developed on the same plate, yielded normal forward-tailing spots. Very marked nicks occurred in the development of copolymers from α-methyl styrene and acrylonitrile with dimethyl sulphoxide (Meissner, 1977).

In the chromatography of polystyrene ($M = 51\,000$ g · mole^{-1}) on silica gel with benzene-acetone (7:93), Kamiyama and Inagaki (1971) found that the leading edge of the relatively well defined spots always occurred at the same R_f value, irrespective of whether 10, 20, 30 or 40 μg of substance had been applied. However, the higher the loading, the lower was the R_f value of the rear edge. Similar observations were made in our laboratory for α-methylstyrene-acrylonitrile copolymers and acetone-toluene (25:75) for 3—30 μg of substance applied: the lower edges of the spots occurred between $R_f = 0.48$ (30 μg) and

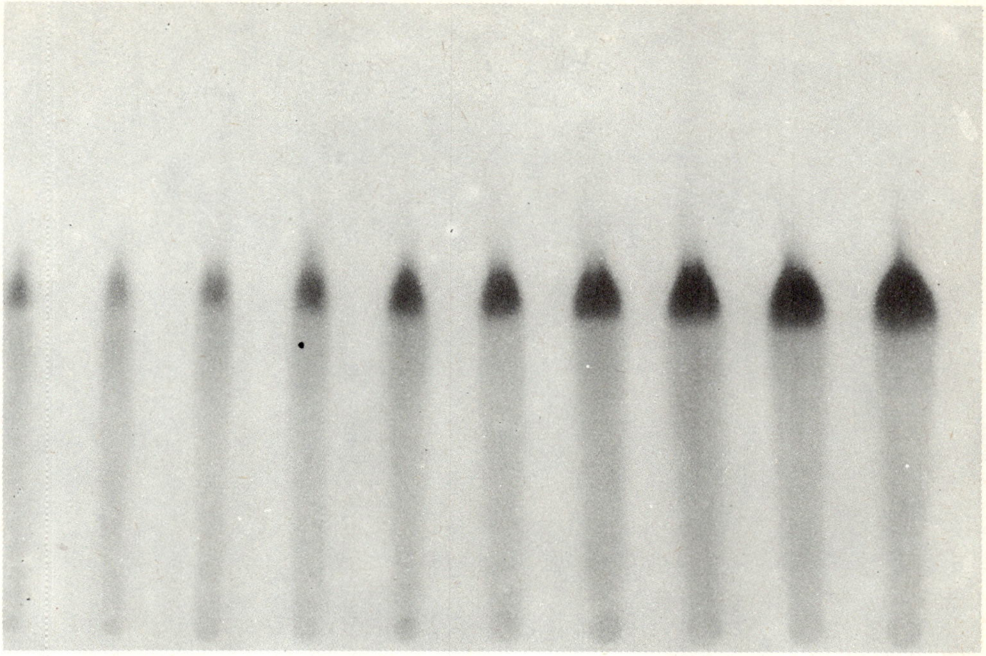

Fig. 21-11
Effect of concentration
Development of an α-methylstyrene acrylonitrile copolymer (43 mole-% AN) by toluene-acetone (75:25) on silica gel in the KN chamber. Substance quantities applied: 3—6—9—12—15—21—30—36—45—60 μg (from left to right).

0.58 (3 μg), but all of the spots extended up to $R_f = 0.62$ (see Fig. 21-11) (MEISSNER, 1977).

Thin-layer chromatography should be carried out with a sample size as small as possible. Table 21-2 shows that the work of the Leningrad team (BELENKIJ et al.) best meets this requirement.

21.5. Results of the TLC of polymers

Table 21-2 shows a bibliography.

21.5.1. Thin-layer exclusion chromatography

In the exclusion mechanism the chromatographic retention decreases with increasing molecular size. This holds true in the column (cf., Chapter 8) as well as on the plate. Consequently:

$$dR_f/dM > 0 \qquad (21\text{-}16)$$

Table 21-2
Separation and characterization of polymers by thin-layer chromatography (Review)

Sample: Polymer	Sample: µg	Layer: Material	Layer: Thickness in mm	Eluent	Gradient	Authors
Chemical heterogeneity of copolymers						
S-MA copolymers	20	SiO_2 (+13% gypsum)	0.25	TCM/EAt; Tetra/MAt	yes	INAGAKI et al. (1968)
S-MMA copolymers	1 ... 2	SiO_2 "S3"		TCM/AC; TCM/E	(yes)	BELENKIJ and GANKINA (1969)
S-MMA block copolymers		SiO_2 "G"	0.3	Bzn/MEK; nitroethane/AC	yes	KAMIYAMA et al. (1972)
S-MMA block copolymers				Tetra/MEK	—	KOTAKA et al. (1975)
S-MA copolymers	20	SiO_2 "Replate 50"		Tetra/MAt	—	TERAMACHI et al. (1979)
CN		SiO_2		AC/EAt/TCM	yes	INAGAKI and KAMIYAMA (1971)
CN	42	SiO_2 "G"	0.25	NM/M	—	KAMIDE et al. (1978)
		kieselguhr	0.25	AC/M/TCM	—	
CA	100	SiO_2 "G"	0.25	DCM/M	yes	KAMIDE et al. (1973)
S-Bd copolymers	25	SiO_2 "G"	0.25	AC; MEK; EAt pyridine; THF; Tol; ether; CHx and mixtures	—	WHITE et al. (1972)
S-Bd copolymers	20 ... 50	SiO_2 "G"	0.25	Tetra/TCM	yes	KOTAKA and WHITE (1974)
S-Bd copolymers	5 ... 10	SiO_2	0.5	CHx/TCM	yes	DONKAI et al. (1975)
S-Chlorobutadiene copolymers		SiO_2 "H"	0.4	CHx/THF	yes	KOHJIYA et al. (1980)
S-AN copolymers	25	SiO_2 + glass powder	0.2	MAt/TCM	—	TERAMACHI and ESAKI (1975)
S-AN copolymers	30	SiO_2 "D"	0.5	Tol/AC	yes	GLÖCKNER and KAHLE (1976a)
S-AN copolymers				Tetra/EAt	—	WÄLCHLI et al. (1978)
Copolyamides		SiO_2, deactivated		formic acid/W formic acid/phenol/M	—	MORI and TAKEUCHI (1972)

VC/VAC copolymers	5 ... 25	SiO_2 "G"	0.2 or 0.3	DCE/Hp/THF	—	JANČA et al. (1979)
(6 ... 15% VAC)						
(15 ... 28% VAC)				DCE/THF	—	
(6 ... 28% VAC)				DCE/Tetra/MEK	yes	
POE-POP copolymers		SiO_2 "KSK"		MEK/W; Hx/MEK/M; Dx/W; EAt(W)/MEK	—	VACHTINA et al. (1978)
Stereoisomers						
PMMA iso-/syndiotactic	10	SiO_2	0.25	EAt	—	INAGAKI et al. (1969)
PMMA a-/syndiotactic		SiO_2 "G"	0.25	EAt/isopropyl acetate AcN/M	yes	INAGAKI and KAMIYAMA (1973)
PMMA iso-/syndiotactic		SiO_2		EAt	—	BUTER et al. (1973)
PMMA iso-/atactic		SiO_2 "KSK"		Bzn		BELENKIJ and GANKINA (1977)
PBD 1,2/1,4 cis/trans	2.5 ... 5	SiO_2 Al_2O_3	0.5	1. Tetra; 2. amyl chloride	—	DONKAI et al. (1974)
Isotopomers						
d-PS/h-PS		SiO_2 "H"		MEK/AC; CHx/Bzn	—	TANAKA et al. (1980a)
Separation by molar masses						
POE derivatives	50	SiO_2 "G"		MEK/W	—	BÜRGER (1963)
POE (esterified)	50 ... 100	SiO_2		MEK/W	—	BÜRGER (1967a)
POE	10 ... 20	SiO_2 Al_2O_3		ethylene glycol/M; M/DMF	yes	OTOCKA and HELLMAN (1970)
POE		SiO_2 "KSK"		pyridine/W	—	BELENKIJ and GANKINA (1977)
POE		Al_2O_3		TCM/Eol	—	ALEKSEEVA et al. (1979)
POP		SiO_2 "KSK"		EAt(W)/MEK	—	BELENKIJ and GANKINA (1977)
POE chloro derivatives		SiO_2		MEK/W	—	FAVRETTO et al. (1970)
Polyhexamethylene adipate		SiO_2 "G"		Bzn/M/glacial acetic acid		HILT et al. (1966)
Polyester oligomers		SiO_2 "D"		E/formic acid; MEK	—	SCHÖLLNER and LÖHNERT (1968)
PETP oligomers		SiO_2		Eol/W/NH_4OH; i-propanol/TCM; Hx/TCM	—	HUDGINS et al. (1978)
Methylene-2-hydroxy benzoic acid oligomers	$3 \cdot 10^5$	SiO_2 "H"	0.75	Bzn/acetic acid/W	—	CIAMPA et al. (1970)

Table 21-2 (continued)

Sample		Layer		Eluent	Gradient	Authors
Polymer	µg	Material	Thickness in mm			
EP oligomers		SiO_2 "GF_{254}"		Bzn/Eat/TCM/E	—	BRAUN and LEE (1975)
Isoprene oligomers		Al_2O_3	1.0	Bzn/AC	—	BRYK et al. (1968)
S-MMA copolymers						INAGAKI et al. (1968)
PMMA; S-MMA copol.	1 ... 2	SiO_2 "S3"		TCM/AC	—	BELENKIJ and GANKINA (1969)
S-Bd-S block copolymers	5 ... 10	SiO_2	0.5	TCM/M	yes	DONKAI et al. (1975)
PS		SiO_2 "KSK"	0.15	CHx/Bzn/AC	—	BELENKIJ and GANKINA (1970a, b)
PS	30 ... 50	SiO_2 "G"	0.30	Bzn/AC/Eol/butanol	yes	KAMIYAMA et al. (1970)
PS	10 ... 20	Al_2O_3		AC/THF	yes	OTOCKA and HELLMAN (1970)
PS	30			AC/isopropyl alcohol/TCM	yes	OTOCKA (1970)
PS	15 ... 20	SiO_2 "G"		CHx/MEK; Bzn/MEK/AC/Eol	yes	KAMIYAMA and INAGAKI (1971)
PMMA				TCM/M	—	INAGAKI et al. (1971)
S-MMA triblock copolymers		SiO_2 "G"	0.3	nitroethane/AC	yes	KAMIYAMA et al. (1972)
PS		SiO_2	0.4	AC/TCM	yes	MIYAMOTO et al. (1973)
CA	100	SiO_2 "G"	0.25	DCM/AC/M	yes	KAMIDE et al. (1973)
CN	42	kieselguhr	0.25	Dx/M/i-propanol	—	KAMIDE et al. (1978)
PMMA				EAt/isopropyl acetate	—	INAGAKI and KAMIYAMA (1973)
Poly-γ-benzyl-L-glutamate		SiO_2		butanol/W/acetic acid	—	COSANI et al. (1966a)
S-Bd-rubber	20 ... 50	SiO_2 "G"	0.25	THF/M	yes	KOTAKA and WHITE (1974)
PMMA				EAt/MAt or isopropyl acetate/methyl formate	—	KAMIYAMA and INAGAKI (1974)
PS		SiO_2 "H"	0.25	THF/AC	yes	MIN et al. (1975)
PS						HIGASHMURA et al. (1975)

α-MS oligomers		SiO_2 "KSK"		Tetra/Hp	—	BELENKIJ and GANKINA
PS (SEC-TLC)		SiO_2 "KSK", 12 nm	0.15	CHx/Bzn/AC	—	BELENKIJ and GANKINA (1970a, b)
PS (SEC-TLC)		SiO_2, d_0 = 0.9 ... 100 nm		CHx/Bzn; TCM; Bzn	—	HALPAAP and KLATYK (1968)
PS (SEC-TLC)		SiO_2, macroporous		THF	—	DONKAI and INAGAKI (1972)
PS (SEC-TLC)		SiO_2, 15 ... 100 nm	0.3	TCM	—	OTOCKA et al. (1972)
PS (SEC-TLC)	40	porous glass	1.0	TCM	—	WAKSMUNDZKI et al. (1979)
Polysil oxan		SiO_2 "KSK"		Bzn/EAt		BELENKIJ et al. (1977)
Polymer architecture						
a) *End groups*						
Polyhexamethylene adipate		Al_2O_3		Bzn/M/glacial acetic acid	—	HILT et al. (1966)
PS		SiO_2 "H"	0.25	1. Bzn 2. TCM	—	MIN et al. (1975)
PS oligomers (with alkyl end groups)		SiO_2		CHx/Bzn	—	BELENKIJ et al. (1978)
PS oligomers with polar terminating groups		SiO_2 (precoated plates)		light petroleum/DCM		MÅNSSON (1980)
PS (hydrolyzed graft product on cellulose)		SiO_2 "G"	0.2	THF	—	TAGA and INAGAKI (1973)
1,2-PBd oligomers (telechelic prepolymers)	10 ... 20	SiO_2 "H"	0.25	Tetra/THF 1. p-xylene } double 2. TCM/THF } development	—	MIN et al. (1977)
Melamine formaldehyde condensate	25	cellulose	0.25	DMF/W/TCM/i-propanol	—	BRAUN and PANDJOJO (1979a, b)
b) *Sequence length*						
S-MMA copolymers	20	SiO_2 + 13% gypsum	0.25	TCM/EAt	yes	KAMIYAMA et al. (1969)
S-MA copolymers	20	SiO_2 + 13% gypsum	0.25	Tetra/MAt		
S-MMA diblock copolymers	50	SiO_2		TCM	—	INAGAKI et al. (1976)
S-MMA block copolymers	20 ... 40	SiO_2	0.25	Tetra/MEK	yes	KOTAKA et al. (1975)
S-Bd copolymers	5 ... 10	SiO_2 Al_2O_3	0.5	CHx/TCM Tetra/n-Hx	—	DONKAI et al. (1975)

Table 21-2 (continued)

Sample		Layer		Eluent	Gradient	Authors
Polymer	µg	Material	Thickness in mm			
c) *Branching*						
PS		SiO_2	0.4	AC/TCM; Tetra/TCM; CHx/Bzn	yes	Miyamoto et al. (1973)
PS		SiO_2 $d_P = 8$ nm	0.2	cyclohexanone/Bzn/AC	—	Belenkij et al. (1973a)
POP (preparative)	$10^3 \ldots 10^4$	SiO_2 "KSK-2"	0.25 to 0.75	EAt(W)/MEK	—	Vachtina et al. (1980)
Polyester oligomers		SiO_2 "KSK-2"		Bzn/Eol	—	Belenkij et al. (1977)
d) *Degree of grafting*						
Graft copolymers: styrene on PVAC, PA, PETP, CA; methyl methacrylate on PVAC	4 ... 40		0.25	backbone polymers: DMC/M; M/W; formic acid; phenol/W graft branchings: TCM; MEK/Tetra	—	Horii et al. (1975)

(HALPAAP and KLATYK, 1968; BELENKIJ and GANKINA, 1970a; JAWOREK, 1970). TLC exclusion chromatography requires solvents with high ε^0 values, which prevent an adsorption. In contrast to the dry bed technique normally used, wet-bed development is applied in this case. The sample is spotted into the flowing eluent. For convenience, the direction of development is usually downward. Size exclusion effects can be expected on layers with a pore size ranging between 3 and 20 nm, depending on the size of the macromolecules.

21.5.2. Thin-layer adsorption chromatography

The statements made about the adsorption behaviour of macromolecules (Chapter 6), about desorption and displacement (Chapter 18) as well as a look at the literature about the column adsorption chromatography of polymers (Table 18-1) are hardly likely to raise great hopes for success in thin-layer chromatography. The fact that the latter is nevertheless possible indicates special circumstances, which allow desorption to be largely achieved in spite of the unfavourable isotherms. In most cases this is due to gradient effects (cf., Section 21.4.1.).

21.5.2.1. Separation by composition

By means of adsorption chromatography, chains with different chemical structures can be separated from each other. This is of importance for the analysis of the heterogeneity of copolymers. Here the first great successes of polymer TLC were achieved (INAGAKI et al., 1968; BELENKIJ and GANKINA, 1969). The previously known techniques required greater efforts and did not allow a separation according to the composition alone, whereas adsorption chromatography allowed a classification according to the chemical structure without a superimposed separation by the molecular size. The separation of binary copolymers may even be possible for single-component eluents, the ε^0 values of which allow the adsorption of one of the primary units, while preventing that of the other one. It is rare that this can be realized, so nevertheless eluent mixtures are used in most cases. In systematic investigations using styrene-methacrylate copolymers, BELENKIJ and GANKINA (1969, 1970b) concluded that combinations of solvents and displacers will separate according to composition. Displacers include substances such as diethyl ether, acetone, methyl ethyl ketone, dioxan and other oxygen-containing compounds which are able to form hydrogen bridges with the surface hydroxyls of silica gel. Only small amounts of displacers are added to the relatively non-polar solvent (e.g., chlorinated hydrocarbons). The separation according to composition is possible without an external gradient, yielding high resolutions within a narrow interval of composition. On the other hand, gradient development must be applied in order to cover wide ranges of composition (cf., Section 21.6.).

21.5.2.2. Separation according to the polymer architecture

The separation of stereoisomeric macromolecules is a most difficult problem of polymer characterization. In Chapter 13, foaming fractionation was referred to as a means of doing this. INAGAKI et al. (1969), using ethyl acetate ($\varepsilon^0 = 0.58$), separated isotactic from syndiotactic PMMA on a thin layer of silica gel. The syndiotactic form reached $R_f = 0.9$, whereas the isotactic one reached only $R_f = 0.1$. In acetone ($\varepsilon^0 = 0.56$) the whole sample travelled with the front; in chloroform ($\varepsilon^0 = 0.40$) both of the isomers were retained at the starting point ($R_f = 0$). Stereo block copolymers ranged between the two limiting forms, depending on the respective isomer contents (BUTER et al., 1973).

Other imposing performances of TLC are the separation of deuterated and normal polystyrene (TANAKA et al., 1980a), the separations according to the end groups, the block structure or the degree of branching or other features of the polymer architecture (see Table 21-2).

21.5.2.3. Separation according to the degree of polymerization

In 1963, BÜRGER reported on the TLC separation of POE according to the molar mass. The adsorption increases with the molecular size, i.e.:

$$dR_f/dM < 0 \tag{21-17}$$

The experimental findings show that the slope of this curve rapidly decreases with increasing molar mass (see Fig. 21-12).

Above $M = 50000$ g · mole^{-1} a differentiation is hardly possible, whereas at 10^4 g × mole^{-1} or less there is a marked dependence. The good separations according to the molar mass, which were achieved relatively early, were all obtained with oligomers (BÜRGER, 1963, 1967a; HILT et al., 1966; BRYK et al., 1968; SCHÖLLNER and LÖHNERT, 1968). In some carefully balanced systems, a molar-mass-dependent migration based on adsorption has also been observed in the macromolecular range. Thus PMMA with $M = 165000$ g · mole^{-1} in ethyl acetate-methyl acetate (81:19) reached an R_f value of 0.95, whereas another fraction with $M = 412000$ g · mole^{-1} only came to $R_f = 0.18$ (KAMIYAMA and INAGAKI, 1974). In pure MAt ($\varepsilon^0 = 0.60$) both samples travelled with the solvent, whereas in EAt ($\varepsilon^0 = 0.58$) they were retained at the start. It is of interest that other, analogous binary mixtures, e.g., benzene-acetone (20:80), although causing the samples to travel up to about the middle of the plate, hardly effected a separation of the two fractions. The authors' finding was that the difference in the dielectric constant of the eluent components determines the separation according to M (see Fig. 21-13).

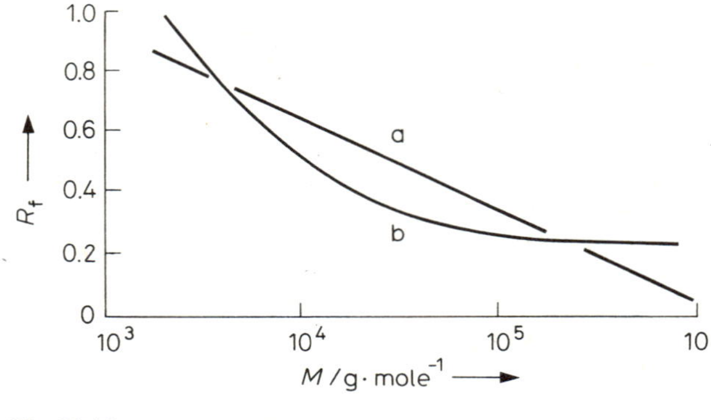

Fig. 21-12
Dependence of the R_f value on the molar mass of the polystyrene samples used

a) in the development using a mixture of solvents and precipitants (benzene/methyl ethyl ketone/acetone/ethanol = 5:3:6:4)

b) in the development using a mixture of 50 ml cyclohexane and 2 ml methyl ethyl ketone, to which another 5 ml of methyl ethyl ketone were added in the course of development (gradient development)

(according to KAMIYAMA and INAGAKI, 1971).

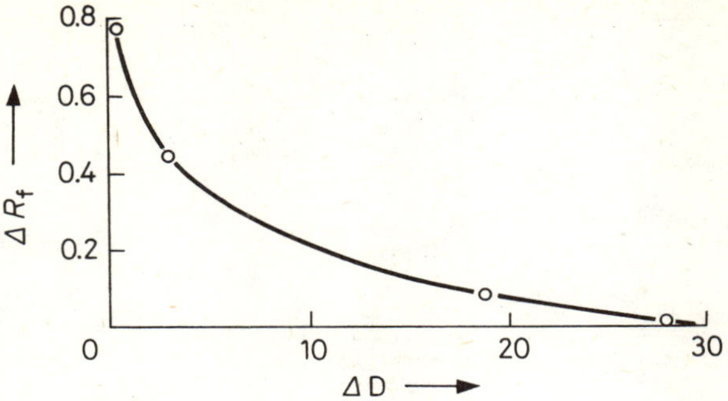

Fig. 21-13
Difference of the R_f values for two fractions of polymethyl methacrylate ($\bar{M}_v = 165000$ and $412000\ g \cdot mole^{-1}$) in the TLC using binary mixtures, as a function of the difference in the dielectric constants, ΔD (according to KAMIYAMA and INAGAKI, 1974)

Another carefully balanced system has been described by BELENKIJ and GANKINA (1970a) for the molar-mass TLC of PS on silica gel, namely $CH_x/Bz_n/Ac$ (12:4:x with $x \approx 1$). Cyclohexane and acetone are non-solvents. This leads us to the next section. The strong effect of the low acetone concentration is shown in Fig. 21-14. See also Fig. 16-17.

21.5.3. Precipitation TLC

Apart from the above-mentioned sophisticated eluents, isocratic separation by molar mass can be achieved by adsorption only in the oligomer range. So what about the mode of action of the numerous TLC methods which can also separate larger-sized molecules?

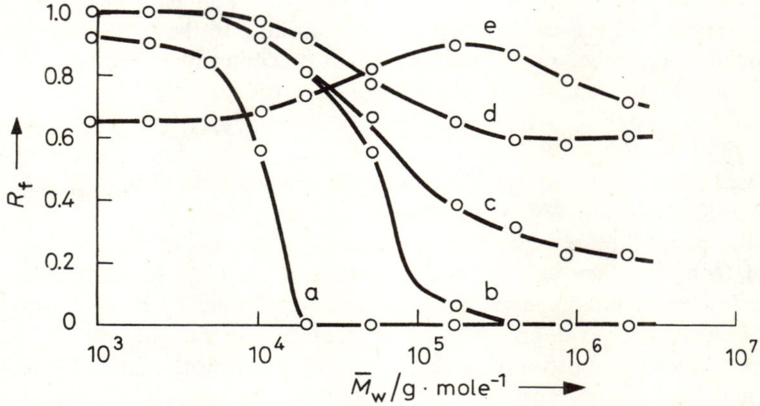

Fig. 21-14
Effect of the acetone proportion in the cyclohexane-benzene-acetone developing system on the R_f value of polystyrene samples, as a function of the molar mass, $\bar{M}_w$
Support material: KSK silica gel; $d_0 = 12$ nm; a 13:3:0.1 (80.8:18.6:0.6); b 12:4:0.2 (74.1:24.7:1.2); c 12:4:0.7 (71.8:24:4.2); d 12:4:1 (70.6:23.5:5.9); e 12:4:2 (66.7:22.2:11.1) (according to BELENKIJ and GANKINA, 1970a).

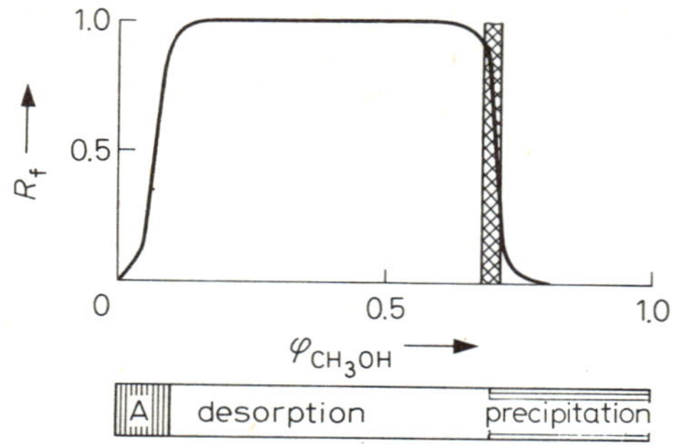

Fig. 21-15
Relationship between the R_f value for an isotactic polymethyl methacrylate ($\bar{M}_v = 412000$ g · mole^{-1}) and the eluent composition

A — immobility due to adsorption; development by M is possible in the slantwise shaded area; chloroform with 70 vol.-% methanol (according to KAMIYAMA and INAGAKI, 1971).

KAMIYAMA and INAGAKI (1971) investigated the behaviour of PMMA in chloroform-methanol mixtures and found the behaviour shown in Fig. 21-15, which is typical of polymers developed in mixtures of a highly polar precipitant and a less polar solvent: in either of the pure media the samples are retained at the start ($R_f = 0$). A few per cent of methanol are sufficient to displace the polymer from the adsorbent, so that it can travel with the front of the solvent ($R_f = 1$). It is not re-adsorbed because all active sites of the surface are occupied by methanol. If the methanol proportion in the mixture increases beyond 70 vol.-%, then the polymer is again retained at the start, because in this case the solvency does not suffice; the critical χ value has been exceeded (cf., Section 5.2.). Naturally the mixtures with $R_f = 0$ or $R_f = 1$ do not effect any separation. Effects may occur at the slopes of the R_f profile; for a homopolymer this may possibly be a separation according to M. However, samples with $43000 \leqq M \leqq 412000$ g · mole^{-1} developed with 5% methanol in chloroform in the range of the displacement slope yielded coincident R_f values. Whereas in this case no separation according to the molecular size occurred, it was obtained at the other slope, in the range of precipitation. Generally, in θ mixtures a separation according to M is possible following the relationship:

$$R_f = A - B \log M \qquad (21\text{-}18)$$

within a wide range (HALPAAP and KLATYK, 1968; KAMIYAMA and INAGAKI 1971; MIYAMOTO et al., 1973; KOTAKA and WHITE, 1974).

From the possibility of separating according to the molar mass at the precipitation slope and from the fact that styrene oligomers in pure acetone are only transported according to their solubility, INAGAKI et al. concluded that the dependence on the molar mass in TLC is due to a precipitation mechanism. In fact almost all of the eluents which separate according to molecular size in the macromolecular range are mixtures of solvents and precipitants and are just on the solvent side. The *Huggins constant*, χ, of these systems is only a little smaller than 0.5; the solutions are close to the θ state. If during the development there are effects which cause the Huggins constant to exceed the critical value, then phase separation will occur. According to KAMIYAMA and INAGAKI (1971), in this case the increase in the polymer concentration as a result of the volume profile (see Figs. 21-3 and 21-4) plays a dominant rôle.

The polymer concentration of the mobile phase is also influenced by a molecular sieve effect: solvent molecules may penetrate into fine-pored silica gel, whereas polymer molecules remain excluded. This effects a further increase in the concentration of the solution in the interstitial volume as compared to the concentration to be observed from the volume profile.

The spontaneous elution gradient can likewise effect a precipitation, but for benzene-acetone (10:90) as a developing agent for polystyrene, KAMIYAMA and INAGAKI showed that no demixing occured. In the diagonal spotting technique, the shifting of the polystyrene spots with respect to the line along which bromocresol green was developed showed that the migration of the polystyrene was not governed by adsorption.

For eluents with a low content of a polar component, demixing phenomena must be expected. If the *solvent* is retained, then phase separation due to a variation in the solvent/precipitant ratio will occur after a certain travelling distance.

The investigations of the adsorption of polymers have shown that from solutions near the θ state large-sized molecules are preferentially adsorbed (cf., Section 6.2.4.). Consequently the adsorption chromatographic separation according to M should also be carried out in poor solvents.

21.6. Generation of gradients

Table 21-2 shows that mainly *gradient developments* were carried out. In addition to the elution gradients, *activity gradients* can be considered in planar chromatography. These activity gradients can be generated by a *vapour pretreatment* in special sandwich chambers for horizontal development. In the Vario-KS chamber® devised by GEISS and SCHLITT (1968a), the plate is arranged on top of a tray, which can be sectionalized longitudinally or transversely in different ways. The VP chamber® devised by DE ZEEUW (1968) is equipped with a transversely sectionalized tray. If the individual partitions contain aqueous solutions in graduated concentrations, then the relative humidity in the vapour space is also graduated. The thin-layer plates are placed on the tray with their coated sides down. Depending on the partial pressure of the water vapour in the respective section, the layer is deactivated to different degrees. For an antiparallel gradient, the vapour pressure in the foremost section must be highest. For the chamber type described by NIEDERWIESER and HONEGGER (1966), the vapour pretreatment is accomplished from a wet cellulose sheet arranged above the thin-layer plate. The gradient can be generated by a non-uniform wetting of the cellulose sheet or by temperature differences. Besides water other highly polar liquids such as methanol can be used to generate activity gradients by a vapour pretreatment.

Possibilities for the generation of *elution gradients* are shown in Fig. 21-16: in sufficiently large development chambers, a gradient can be formed by the dropwise addition of one component to the stirred eluent mixture. To avoid the rise in the eluent level the principle can also be realized in the variant shown in Fig. 21-16b. Both versions require a careful observation of the eluent front and the maintenance of a certain addition regime in order to make the gradient and the development reproducible. For the two-chamber tray shown in Fig. 21-16c, the combination of the components is diffusion-controlled. A close coupling between the development and the shape of the gradient is ensured by the apparatus

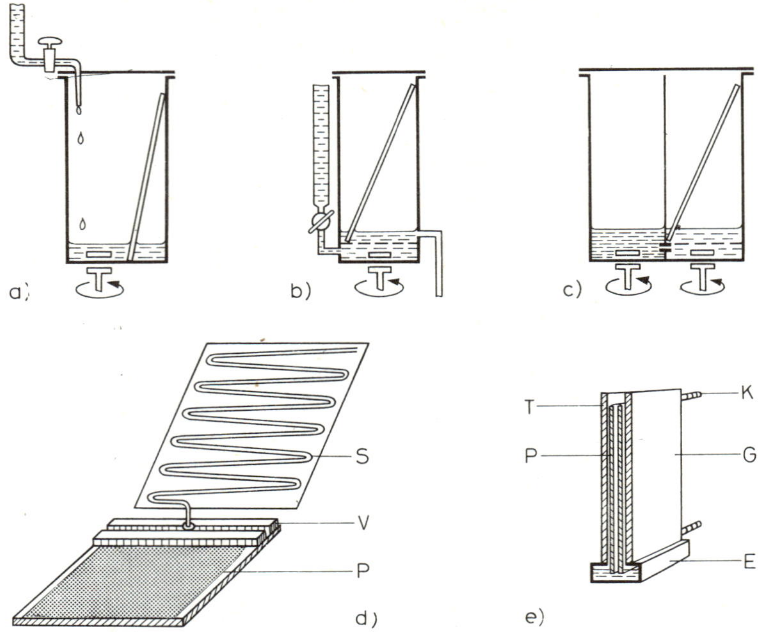

Fig. 21-16
Principle of devices for TLC with gradient elution

a) Dropwise addition of component II
b) Addition of component II without disturbing the vapour space in the chamber
c) Diffusion of the stronger eluent through an opening in the partition wall
d) Gradient development apparatus according to NIEDERWIESER (1969b); P TLC plate; V spreader; S 1 mm tubing with the eluent on an inclined base
e) Thermosandwich chamber
T temperature-controllable metal plate for the generation of a linear temperature gradient; P TLC plate; G counterplate, temperature-controllable metal plate; K pipe connection for the thermostating agent; E tray for the eluent mixture.

described by NIEDERWIESER (1969b) (Fig. 21-16d): before the development, the gradient is built up in a narrow teflon tubing, which is fixed in loops on an inclined base. For an interior diameter of only 1 mm, the axial diffusion will not induce any disturbance in the 10 m tubing. The development is accomplished by means of the connected spreader, which takes up the liquid from the tubing, spreads it by capillary forces and applies it to the layer. The spreader consists of two glass strips (190 mm long, 5 mm wide, 2 mm thick) which are fixed to each other. A gap 0.4 mm wide, 2 mm deep and 190 mm long is left between them, in which the eluent flow is spread to the width of the layer. The outlet flow-rate of the liquid emerging from the tubing can be varied by the inclination of the plane and adapted to the rate at which the layer absorbs the eluent. The development is carried out horizontally. Advantages of this arrangement are that the amount of eluent used is not more than actually required for a development (only 4—6 ml, which fill a length of 5—8 m of the tubing, for a 20×20 cm^2 plate with a 0.25 mm thick layer) and that any gradients desired can be realized in a reproducible way.

The *thermosandwich chamber* (Fig. 21-16e) is easy to operate and reliable with respect to reproducibility: the eluent mixture, rising in an upward-type development, reaches zones of ever-increasing temperature. The component with the higher vapour pressure is stripped, condenses on the counterplate and returns into the tray. This automatically generates an antiparallel gradient, if the more volatile component is at the same time the better eluent (GLÖCKNER and KAHLE, 1976b).

21.7. Quantitative evaluation

While qualitative information about the samples investigated can be obtained from the thin-layer chromatogram in a relatively simple and conclusive way, the quantitative evaluation involves some problems. A bibliography has been given by GÄNSHIRT (1967). Also SHELLARD [E 12], TOUCHSTONE and SHERMA [E 10] as well as HEZEL (1977) discussed these problems in detail.

For colourless samples, which include most of the polymers, there are the following possibilities of a quantitative evaluation:

1. Staining of the substance spots and measurement of their intensity and size
2. UV scanning
3. Analysis after the removal of the separated components from the chromatographic layer

First let us discuss the problem concerning the optical measurements mentioned under 1 and 2: the optical signal obtained from a substance spot is proportional to the concentration in the layer only for very small values (Fig. 21-17). This is true for both *transmission* and *reflectance measurements*. The deviation from the Lambert-Beer law is caused by the scattering in the turbid medium of the layer. This complicated situation was theoretically investigated by KUBELKA and MUNK (1931) and quantitatively described under certain

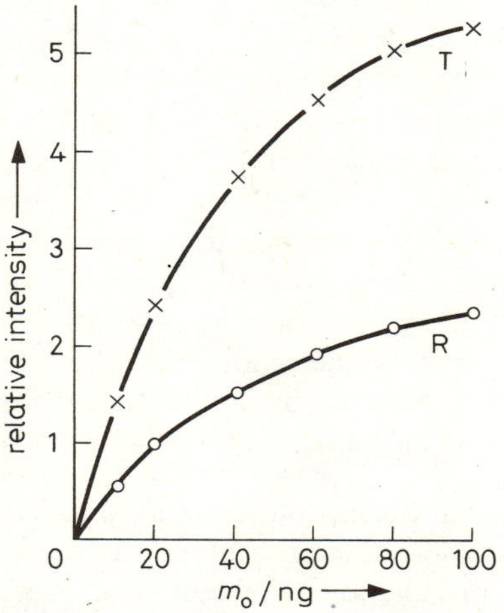

Fig. 21-17
Dependence of the relative intensity of the optical signal in transmission (T) and reflectance (R) measurements on the quantity of substance
Sudan red on silica gel, λ = 500 nm
(according to HEZEL, 1977).

simplifying assumptions. For remission measurements the authors derived the widely used equation

$$\frac{(1 - R)^2}{2R} = \frac{K}{S} \tag{21-19}$$

where R is the reflectance, K denotes the coefficient of absorption, which increases linearly with the *substance concentration*, and S is the coefficient of scattering, which depends on the *layer material*. The shorter the wavelength of the light used in the investigation, the higher is the value of S.

Fig. 21-17 also shows that the transmittance is more sensitive to the substance concentration than does the reflectance. Therefore it was sometimes recommended to carry out TLC evaluations preferably by measuring the transmittance (e.g., ARATANI and MIZUI, 1973). However, on the basis of detailed investigations, HEZEL (1977) concluded that inhomogeneities of the layer are much less disturbing in remittance, so that on the whole a better signal-to-noise ratio is achieved. In addition, silica gel layers permit transmission measurements only down to 325 mm, whereas for remission measurements the total UV range down to 196 nm is accessible. For TLC plates with *RP layers*, SIOUFFI et al. (1979) found that even in the long-wave UV only remission measurements are possible.

POLLAK (1978) worked on the application of the Kubelka-Munk theory, looking for possibilities of expressing the relationships between the optical signal and the coefficients K and S by even simpler approximations, which enable a linearization in certain ranges. This holds for the reflectance, R, with

$$1/R = a_R + b_R K \tag{21-20}$$

and for the transmittance, T, with

$$\ln T = a_T - b_T K \tag{21-21}$$

to an adequate approximation. The coefficients a and b were expressed as functions of S. The linearization of calibration curves was also dealt with by TAUSCH (1971), HEZEL (1978) and MÜLLER (1980). In narrow ranges of medium concentrations, curves like those in Fig. 21-17 can be approximated by straight lines (HEZEL, 1977; MÜLLER, 1980). However, these curves do not pass through the origin and, consequently, must not be extrapolated towards lower concentrations.

The exact dosage of the substance is a general precondition for a quantitative TLC. As at the end it is only possible to detect that part of the separated components which exceeds the detection limit (see Fig. 21-2), the latter must be as low as possible. The maximum concentration in a spot should be at least ten times the detection limit. From this it follows that certain minimum amounts must be applied, while chromatographically the quantity of substance should be as small as possible because of the risk of overloading. These contrary requirements can only be met by a compromise.

The application of the sample in the form of a streak offers the advantage that in the centre the lateral spreading of the substance can be neglected. The width of the streak, or the diameter of the starting spot, d_0, should be as small as possible. Generally it accounts for 20% of the total variance (HALPAAP and RIPPHAHN, 1977). The quality, Q_0, of the application technique can be evaluated by means of its relation to the spot diameter, d, after the development:

$$Q_0 = (d - d_0)/(d + d_0) \tag{21-22}$$

Low d_0 values can be best ensured by using solvents of low polarity (BEREZKIN and BOCKOV, [E 13]).

A promising aid is a plate with a concentrating zone at the bottom of the layer, for instance with a narrow layer of synthetic porous silica of extremely large internal pore diameters (circ. 5000 nm), with a sharp interface leading to the chromatographically active layer (HALPAAP and KREBS, 1977). In the inert concentrating zone, the eluent compresses the sample spots to lines which have become perfectly narrow when they arrive at the interface.

For polymers, a quantitative evaluation is most problematic because of the starting spots, which shall be dealt with in Section 21.7.4.

For a quantitative evaluation, three reference samples should be developed additionally in every experiment, because the calibration curve must be determined for each plate. To eliminate the effects of a non-ideal course of the development, e.g., sagging of the front or boundary effects, it is possible to use the *data pair technique* (BETHKE et al., 1974), in which all the samples as well as the three standards are applied twice to each plate in the same order, i.e., the sequence of samples from the left edge to the middle of the plate is duplicated from the middle to the right edge. The results are taken as the mean values of the two chromatographic traces.

21.7.1. Quantitative evaluation after staining

This method can be considered mainly for samples which do not absorb in the UV range or for which the different components can be differently stained, and thus evaluated most clearly. In some cases, however, it is simply the lack of suitable equipment which calls for the application of this less demanding method. Even with a profound experience and optimum possibilities of comparison, the quantity of substance can be visually estimated to an accuracy of $\pm 10\%$ at best (HEZEL, 1977).

Suitable colouring agents are a 1% solution of iodine in methanol (INAGAKI et al., 1968), concentrated H_2SO_4 containing 3% $KMnO_4$, used with subsequent heating to 150 °C (BELENKIJ and GANKINA, 1970b), or a saturated solution of thymol blue, followed by a treatment with 3 N H_2SO_4 (KAMIYAMA et al., 1970). For oxygen-containing polymers, Dragendorff reagent (basic bismuth nitrate 0.17% and potassium iodide 4%, in an acetic acid solution, with barium chloride solution, 20%, added before use) has proven suitable (BELENKIJ and GANKINA, 1970b). For methylol melamine on cellulose layers, BRAUN and PANDJOJO (1979a, b) used a mixture of equal parts by volume of 0.1 N $AgNO_3$, 5 N NH_4OH and 2 N NaOH. For graft copolymers, HORII et al. (1975) used aqueous 0.05 N iodine solution or 10% $HClO_4$.

These reagents are sprayed onto the plates, which have been dried again after the development. Apart from the work involved (which is sometimes rather unpleasant), this technique entails the risk that the intensity of colour is affected by the method of spraying. Here the requirement for an adequate number of standard samples has double importance.

On RP-TLC plates the staining is more difficult (SIOUFFI et al., 1979). OKUMURA and NAKAOKA (1980) made visible sulphonic amides on RP layers by means of iodine vapour. This method, which is also used for polymers on silica layers, has, however, the disadvantage that the colour fades in the course of time, which is also the case with the previously mentioned application of iodine.

In some cases the possibilities of a chemical conversion of the separated substances are reduced by the limited adhesion and stability of the layers. In this respect plates with *sinter-fused* silica gel (Replate® 50) exhibit great advantages. They can be immersed in the colouring baths and rinsed in running water. After treatment with chromatosulphuric acid (ITOH et al., 1973) they can be re-used.

The intensity of the colouring is measured either directly on the plate or on photographs. In this case the *Sabatier effect*, i.e., an intermediate exposure followed by a reversed development, can be utilized to establish diagrams in which curves are fitted to points of equal optical density. These "*equidensity*" *diagrams* are evaluated by means of an image of a graduated optical wedge, which has been obtained under equal conditions (BELENKIJ and GANKINA, 1977). However, the authors give priority to the densitometric evaluation, because the spot size depends on many factors of the TLC development (WAKSMUNDZKI and RÓZYLO, 1973), and hence does not represent a reliable measure of the quantity of substance. According to INAGAKI (1977b), in densitometry it is possible to obtain a linear relationship between the quantity of substance and the colour intensity for quantities up to 200 $\mu g/cm^2$. However, in many cases the density curve already flattens below this value. This flattening is due to the mentioned optical conditions in a turbid medium. Moreover, for polymer samples there is sometimes an increase in concentration at the interface between the glass base and the silica gel (KOTAKA and WHITE, 1974; INAGAKI, 1977b). Then, if the layers are carefully removed by rinsing, a stronger adhesion of the silica gel is observed even on those places where no colouring is observable on the surface. The non-uniform distribution of the sample in the layer induces errors which are at best likely to be detected by simultaneously measuring the remittance and the transmittance.

In the investigation of copolymers it must be known how the chemically different components respond to the staining.

21.7.2. Quantitative TLC evaluation by UV scanning

A number of manufacturers offer instruments by means of which remission and transmission measurements can be carried out on TLC plates at adjustable wavelengths. Some of the units have a slit-shaped aperture which is adjusted to the width of the spot to be measured. Units which scan the TLC spots by an oscillating light spot 0.25 mm in diameter and determine the total optical density by integration (KOOPMANS and BOUWMEESTER, 1971) yield data which are not affected by the geometrical shape of the spots. For example, this was shown by means of a repetitive measurement of a kidney-shaped TLC spot in different directions [F 42a]. Top quality units allow these measurements at two different wavelengths, which are alternatingly introduced by means of a sector mirror. One of the wavelengths is used for measuring the absorption of the substance, while the other is chosen in such a way that it mainly detects the behaviour of the layer itself. The difference signal yields a curve with a markedly reduced noise level [F 42a]. In another unit a similar result is obtained by simultaneously measuring the remittance and the transmittance. The resulting value is largely free of effects caused by the layer.

For copolymers whose chemical principal units exhibit a different UV absorption, scanning at two different wavelengths opens the possibility of determining the composition as a function of the R_f value even without using any calibration substances (KOTAKA et al., 1975). Naturally the sequence-length dependence of the UV absorption must be taken into account (see Section 19.7.3.1.). The adsorption on the silica gel may also modify the UV spectrum of a substance (ARATANI and MIZUI, 1973).

The detection of colourless components on *plates with a fluorescent indicator* is based on the fact that irradiated UV light is absorbed by the substance, so that on these spots the fluorescence is reduced or even totally absent. Consequently the effect is not due to quenching, but to a filtering action, and does not yield any substantial advantages for a quantitative evaluation. However, if a component itself exhibits fluorescence, then the sensitivity in fluorescence detection increases to 10—1000 times the usual value. Under optimum conditions it is possible to detect as little as 1 ng of substance on TLC plates (JORK, 1968; KEUKER, 1971). Moreover, the component fluorescence enables a selective measurement to be carried out by means of excitation at different wavelengths.

Generally the accuracy of the quantitative optical evaluation ranges between 5 and 7%. Under favourable circumstances 3% (relative) are achieved, but sometimes the accuracy is worse than ±10%. The errors are due to the inhomogeneity and the thickness of the layer (JANAK, 1973).

21.7.3. Quantitative analysis after removal from the layer

In the TLC of small molecules, components which are of special interest are sometimes extracted from the layer material, which is scraped off the chromatografic trace. A device for the isolation of μg quantities by means of not more than 15—30 μl of extraction liquid was described by DEKKER (1979). Apart from the amount of work required in this technique, it is often difficult completely to extract the component from the layer material. Therefore this technique is only chosen if the isolated component is to be used for further reactions or physical investigations.

The quantitative detection of polymers can be carried out by means of pyrolysis and an investigation of the produced gases using a flame ionization detector (FID) (PADLEY, 1969; SZAKASITS et al., 1970; MUKHERJEE et al., 1971; OKUMURA et al., 1975). In a commercial unit used for this detection technique [F 40] the TLC separation is carried out on *quartz rods* (152 mm long, 0.9 mm in diameter) on which a 75 μm thick layer of SiO_2 or Al_2O_3 is sinter-fused. Ten "Chromarods" of this type are supported by a common frame. After the substance has been applied (2—20 μg in 1 μl of solution, in five aliquots), an upward development is carried out, followed by the evaporation of the eluent. Then the dry Chromarods, one by one, are slowly passed through a hydrogen flame, which is part of the FID. The signal yields a curve from which the quantity of substance can be read off for the individual distances along the rod, i.e., as a function of the R_f value (see Fig. 21-18). The hydrogen flame at the same time cleans the rod, so that the latter is re-usable. A Chromarod can be used for up to 100 TLC developments. Using this technique, MIN et al. (1977) investigated telechelic prepolymers (polybutadienes having either COOH or OH groups). The sensitivity is high. GIETZ et al. (1975) detected even 10 pg on TLC rods 0.6 mm in diameter.

21.7.4. Substance immobilization at the start

For the quantitative evaluation the mass of the starting sample must be exactly known. This requires not only an exact operation in the application, but also a complete (and, if possible, instantaneous) elution of the substance from the starting spot. For low-molecular-weight samples the development can be impaired if recrystallisation occurs (STAHL, 1967 in [E 1], p. 64). Then front tailing appears at the starting spot. In most cases the sample portion remaining at the initial spot in the TLC of polymers is clearly separated from the travelling portion. Immobilized spots at the start were observed in the investigation of butadiene-styrene copolymers (WHITE et al., 1972; TAGATA and HOMMA, 1972; KOTAKA and WHITE, 1974), for block copolymers of butadiene and styrene (DONKAI et al., 1974, 1975), for styrene-acrylonitrile copolymers (KAHLE, 1974; WÄLCHLI et al., 1978) and for α-methylstyrene-acrylonitrile copolymers in several eluents (MEISSNER, 1977).

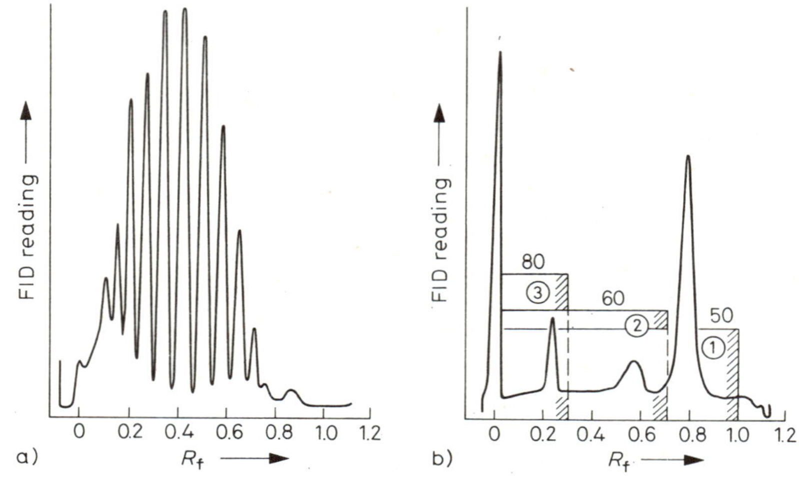

Fig. 21-18
Quantitative TLC evaluation by means of a flame ionization detector after a development on Chromarod SII rods

a) POE oligomers, eluent: ethyl acetate-acetone-water (70:20:4)
b) PBd telechelics with carboxyl or hydroxyl end groups, triple development with chloroform-methanol-acetic acid (x:y:1); (y = 100-x):
1. Development over s = 10 cm with x = 50;
2. s = 7 cm, x = 60; 3. s = 3 cm, x = 80
Stationary phase for both of the TLC investigations: 0.075 mm silica gel, d_p = 5 µm, on quartz rods 0.9 mm in diameter [F 40]
(by courtesy of IATRON Laboratories, Inc., Tokyo).

Whether or not starting spots occur depends on several factors. In the cyclohexane-benzene-acetone developer used by BELENKIJ and GANKINA, anionically prepared polystyrenes travelled without a residue, whereas technical grade polystyrenes left starting spots. Stock solutions in good solvents were more likely to produce residues than those in poor solvents. The eluent had a strong effect: while cyclohexane-benzene-acetone developed without any immobilization, chloroform-carbon tetrachloride mixtures produced remaining spots at the start even with high-purity polystyrenes (BELENKIJ, 1974).

21.8. Importance of the thin-layer chromatography of polymers

Thin-layer chromatography requires only quite simple equipment and takes relatively little time. Chain-length distributions can be determined much better by other methods, but traces of concomitants or components of different molecular sizes, "satellite polymers", can be better detected and analysed by TLC than by other techniques. In macromolecular chromatography, TLC at present has not yet the same importance as in small-molecule separation, but the possibility of clarifying peculiarities of the polymer architecture by means of planar chromatography really deserves attention.

Bibliography

1. Summary presentations, books, firm publications

A Chromatography, general

[A 1] Cvet, M. S.: Chromatographische Adsorptionsanalyse. — Izd. Akad. Nauk, Leningrad 1946

[A 2] Zechmeister, L.; Cholnoky, L.: Die chromatographische Adsorptionsmethode. — Springer-Verlag, Wien 1937

[A 3] Strain, H. H.: Chromatographic Adsorption Analysis. — Wiley-Interscience, New York 1942

[A 4] Snyder, L. R.: Principles of Adsorption Chromatography. — Marcel Dekker, New York 1968

[A 5] Separation and Purification Methods. Ed.: E. S. Perry, C. J. van Oss. — Marcel Dekker, New York 1973ff.

[A 6] Chromatography — A Review of Principles and Applications. Ed.: E. Lederer, M. Lederer, 2nd ed. — Elsevier, Amsterdam 1957

[A 7] Advances in Chromatography. Ed.: A. Zlatkis, L. S. Ettre. — Elsevier, Amsterdam 1974 and following

[A 8] Fundamentals of Chromatography. Ed.: H. G. Cassidy. In: Technique of Organic Chemistry. Ed.: A. Weissberger. Vol. 10. — Wiley-Interscience, New York 1957

[A 9] Giddings, J. C.: Dynamics of Chromatography. Principles and Theory. — Marcel Dekker, New York 1965

[A 10] Guide to Modern Methods of Instrumental Analysis. Ed.: T. H. Gouw. — Wiley-Interscience, New York 1972

[A 11] Chromatographie en Chimie Organique et Biologique. Ed.: E. Lederer. — Masson et Cie Editeurs, Paris 1959

[A 12] Chromatography. Ed.: E. Heftmann. In: Reinhold Chemistry Textbook Series. Ed.: C. A. van der Werf, H. H. Sisler. — Reinhold, New York 1961

[A 13] Bibliography of Paper and Thin-Layer Chromatography, 1966–1969; 1970–1973. J. Chromatogr., Suppl. Vol. 2; 5. — Elsevier, Amsterdam 1972; 1976

[A 14] Bibliography of Column Chromatography, 1967–1970; 1971–1973. J. Chromatogr., Suppl. Vol. 3; 6. — Elsevier, Amsterdam 1973; 1976

[A 15] Advances in Chromatography. Ed.: J. C. Giddings, R. A. Keller. Marcel Dekker, New York 1965 and following

[A 16] Modern Separation Methods of Macromolecules and Particles (Progress in Separation and Purification, Vol. 2). Ed.: E. Gerritsen. — Wiley-Interscience, New York 1969

[A 17] Gas Chromatography 1970. Ed.: R. Stock, S. G. Perry. — The Institute of Petroleum, London 1971

[A 18] Mikeš, O.: Laboratory Handbook of Chromatographic and Allied Methods. — Ellis Horwood, Chichester (England) 1979

[A 19] Walker, J. Q.; Jackson, M. T.; Maynard, J. B.: Chromatographic Systems — Maintenance and Troubleshooting. 2nd ed. — Academic Press, New York/San Francisco/London 1977

[A 20] Zweig, G.; Sherma, J.: Handbook of Chromatography. Vol. 1: Chromatographic Data; Vol. 2: Principles and Techniques; Practical Applications. — CRC Press, Cleveland/Ohio 1972

[A 21] Chromatographic Reviews: Elsevier, Amsterdam, New York

[A 22] Developments in Chromatography. Ed.: C. E. H. Knapman. — Applied Science Publishers, London 1978. Vol. 2: 1980

[A 23] SAKODYNSKI, K. I.: Michail Semenovič Cvet i Chromatografija — k 100-Letiju so dnja roždenija M. S. Cveta. — Naučnyj sovet po Chromatografii AN SSSR 1972
[A 24] Bonded Stationary Phases in Chromatography. Ed.: E. GRUSHKA. — Ann Arbor Science, Ann Arbor 1974
[A 25] UNGER, K.: Porous Silica — Its Properties and Use as Support in Column Liquid Chromatography. — Elsevier, Amsterdam 1979
[A 26] DENNEY, R. C.: A Dictionary of Chromatography. — Macmillan, London 1976

B Polymer sciences

[B 1] HILDEBRAND, J.; SCOTT, R.: The Solubility of Nonelectrolytes. 3rd ed. — Reinhold, New York 1949
[B 2] FLORY, P. J.: Principles of Polymer Chemistry. — Cornell University Press, Ithaca, New York 1953
[B 3] Polymer Fractionation. Ed.: M. J. R. CANTOW. — Academic Press, New York, London 1967
[B 4] PEEBLES, L. H., jr.: Molecular Weight Distribution in Polymers. In: Polymer Reviews. Ed.: H. F. FRANK, E. H. IMMERGUT, Vol. 18. — Wiley-Interscience, New York 1971
[B 5] Physical Methods in Macromolecular Chemistry. Ed.: B. CARROLL. — Marcel Dekker, New York 1969 and following
[B 6] ELIAS, H. G.; BAREISS, R.; WATTERSON, J. G.: Mittelwerte des Molekulargewichtes und anderer Eigenschaften. In: Fortschritte der Hochpolymeren-Forschung **11** (1973), p. 111–204. — Springer-Verlag, Berlin, Heidelberg 1977
[B 7] MORAWETZ, H.: Macromolecules in Solution. — Wiley-Interscience, New York 1975
[B 8] High Polymer Physics. Ed.: H. A. ROBINSON. — Chem. Publ. Co., New York 1948
[B 9] TAGER, A. A.: Physical Chemistry of Polymers. — MIR Publishers, Moscow 1972
[B 10] TOMPA, H.: Polymer Solutions. — Butterworths Scientific Publications, London 1956
[B 11] Novye metody issledovanija polimerov. — Nauk Dumka, Kiev 1975
[B 12] Industrial Polymers: Characterization by Molecular Weight. Ed.: J. H. S. GREEN, R. DIETZ. — Transkripta Books, London 1973
[B 13] Fractionation of Synthetic Polymers. Ed.: L. H. TUNG. — Marcel Dekker, New York 1977

C Adsorption

[C 1] KIPLING, J. J.: Adsorption from Solutions of Nonelectrolytes. — Academic Press, London 1965
[C 2] Proceedings of the International Conference on Colloid and Surface Science, Budapest, 15.–20. Sept. 1975. Ed.: E. WOLFRAM. — Elsevier, Amsterdam 1975
[C 3] LIPATOV, Ju. S.; SERGEEVA, L. M.: Adsorbcija polimerov. — Nauk Dumka, Kiev 1972
[C 4] Methoden der Strukturuntersuchung an hochdispersen und porösen Stoffen. Ed.: H. WITZMANN. — Akademie-Verlag, Berlin 1961
[C 5] SATO, T.; RUCH, R. J.: Stabilization of Colloidal Dispersions by Polymer Adsorption. — Marcel Dekker, New York, Basel 1980

D Liquid column chromatography

[D 1] DETERMANN, H.: Gelchromatographie. — Springer-Verlag, Berlin, Heidelberg 1967
[D 2] Gel Permeation Chromatography. Ed.: K. H. ALTGELT, L. SEGAL. — Marcel Dekker, New York 1971
[D 3] Modern Practice of Liquid Chromatography. Ed.: J. J. KIRKLAND. — Wiley-Interscience, New York 1971
[D 4] HADDEN, N.; BAUMANN, F.; MCDONALD, F.; MUNK, M.; STEVENSON, R.; GERE, D.; ZAMARONI, F.; MAJORS, R.: Basic Liquid Chromatography. — Varian Aerograph, Walnut Creek 1971
[D 5] SNYDER, L. R.; KIRKLAND, J. J.: Introduction to Modern Liquid Chromatography. — Wiley-Interscience, New York 1974
[D 6] DONE, J. N.; KNOX, J. H.; LOHEAC, J.: Applications of High-Speed Liquid Chromatography. — Wiley-Interscience, New York 1974
[D 7] Liquid Column Chromatography. A Survey of Modern Techniques and Applications. Ed.: Z. DEYL, K. MACEK, J. JANÁK. — Elsevier, Amsterdam 1975
[D 8] RAJCSANYI, P. M.; RAJCSANYI, E.: High-Speed Liquid Chromatography. — Marcel Dekker, New York 1975

[D 9] PARRIS, N. A.: Instrumental Liquid Chromatography. A Practical Manual on High-Performance Liquid Chromatographic Methods. — Elsevier, Amsterdam 1976

[D 10] ENGELHARDT, H.: Hochdruckflüssigkeitschromatographie. 2nd ed. — Springer-Verlag, Berlin, Heidelberg 1977
High performance Liquid Chromatography. 2nd ed. — Springer Verlag, Berlin, Heidelberg, New York 1976

[D 11] Chromatography of Synthetic and Biological Polymers. Ed.: R. EPTON.
Vol. 1: Column Packings, GPC, GF and Gradient Elution. Papers of the Birmingham Symposium, July 1976.
Vol. 2: Hydrophobic, ion-exchange and affinity methods.
Ellis Horwood Publishers, Chichester 1978

[D 12] BRISTOW, P. A.: Liquid Chromatography in Practice. HETP, 10 Langley Drive Handforth, Wilmslow Cheshire SK9 3BQ U.K., 1976

[D 13] SCOTT, R. P. W.: Contemporary Liquid Chromatography. — John Wiley & Sons, New York 1976

[D 14] SIMPSON, C. F.: Practical High Performance Liquid Chromatography. — Heyden & Son, London 1976

[D 15] BELENKIJ, B. G.; VILENČIK, L. Z.: Modern Liquid Chromatography of Macromolecules. — Elsevier Science Publishers B.V., Amsterdam, Oxford, New York, Tokyo 1983

[D 16] Size exclusion chromatography (GPC). Ed.: T. PROVDER. — American Chemical Society, Washington 1980 (ACS Symposium Series 138)

[D 17] LOWE, C. R.; DEAN, P. G. D.: Affinity Chromatography. — John Wiley & Sons, New York 1974

[D 18] Process Plant Specifiers Guide Series — Separation and Purification. Ed.: L. H. FORD. — A 4 Publications Ltd. 1975/76

[D 19] YAU, W. W.; KIRKLAND, J. J.; BLY, D. D.: Modern Size — Exclusion Liquid Chromatography: Practice of Gel Permeation and Gel Filtration Chromatography. — John Wiley & Sons, New York, Chichester, Brisbane, Toronto 1979

[D 20] EPPERT, G.: Einführung in die Schnelle Flüssigchromatographie (Hochdruckflüssigchromatographie). — Akademie-Verlag, Berlin 1979

[D 21] FISCHER, L.: Gel Filtration Chromatography. 2nd ed. — Elsevier, North Holland, Biomed. Press, Amsterdam, New York, Oxford 1980

[D 22] Liquid Chromatography of Polymers and Related Materials. — Marcel Dekker, New York
Vol. 1: Ed. by J. CAZES, 1977
Vol. 2: Ed. by J. CAZES, X. DELAMARE, 1980
Vol. 3: Ed. by J. CAZES, 1981

[D 23] MEYER, V.: Praxis der Hochleistungs-Flüssigkeitschromatographie. — Verlag Moritz Diesterweg; Otto Salle Verlag, Frankfurt a. M., Berlin, München; Verlag Sauerländer, Aarau, Frankfurt a. M., Salzburg 1979

[D 24] KAISER, R. E.; OELRICH, E.: Optimierung in der HPLC. — Dr. Alfred Hüthig Verlag, Heidelberg, Basel, New York 1979

[D 25] PRYDE, A.; GILBERT, M. T.: Application of High Performance Liquid Chromatography. — Chapman and Hall, London 1979

[D 26] Instrumentation for High-Performance Liquid Chromatography. Ed.: J. F. K. HUBER. — Elsevier, Amsterdam 1978

[D 27] KNOX, J. H.; DONE, J. N.; FELL, A. F.; GILBERT, M. T.; PRYDE, A.; WALL, R. A.: High-performance liquid chromatography. — Edinburgh University Press, Edinburgh 1978

[D 28] High-performance liquid chromatography — Advances and perspectives. Vol. 1, Ed.: Cs. HORVÁTH. — Academic Press, New York, London 1980

[D 29] KREMMER, T.; BOROSS, L.: Gel Chromatography: Theory, methodology and applications. — John Wiley & Sons, Chichester 1979

E Planar chromatography

[E 1] Dünnschichtchromatographie. Ein Laboratoriumshandbuch. Ed.: E. STAHL. 2nd ed. — Springer, Berlin, Heidelberg 1967

[E 2] Handbuch der Papierchromatographie. Ed.: I. M. HAIS, K. MACEK, 2nd ed. in 3 vols. — Gustav Fischer Verlag, Jena 1963

[E 3] Stationary Phase in Paper and Thin-Layer Chromatography. Ed.: K. MACEK, I. M. HAIS. — Czechoslov. Acad. Sci., Praha 1965

[E 4] Thin-Layer Chromatography. Ed.: J. G. KIRCHNER. In: Technique of Organic Chemistry. Ed.: A. WEISSBERGER. Vol. 12. — Wiley-Interscience, New York 1967

[E 5] GEISS, F.: Die Parameter der Dünnschichtchromatographie. — Friedr. Vieweg & Sohn, Braunschweig 1972

[E 6] Einführung in die Hochleistungs-Dünnschicht-Chromatographie. Ed.: R. E. KAISER. — Inst. für Chromatographie Bad Dürkheim 1976

[E 7] HPTLC — High performance Thin-Layer Chromatography. Ed.: A. ZLATKIS, R. E. KAISER. — Elsevier, Amsterdam 1977

[E 8] BERTSCH, W.; HARA, S.; KAISER, R. E.; ZLATKIS, A.: Instrumental HPTLC. — Dr. Alfred Hüthig Verlag, Heidelberg, Basel, New York 1980

[E 9] TOUCHSTONE, J. C.; DOBBINS, M. F.: Practice of Thin Layer Chromatography. — John Wiley & Sons, New York, Chichester, Brisbane, Toronto 1978

[E 10] TOUCHSTONE, J. C.; SHERMA, J.: Densitometry in Thin Layer Chromatography: Practice and Applications. — John Wiley & Sons, New York, Chichester, Brisbane, Toronto 1979

[E 11] KIRCHNER, J. G.: Thin-layer chromatography. 2nd ed. — Wiley-Interscience, New York 1978

[E 12] SHELLARD, E. J.: Quantitative Paper and Thin-Layer Chromatography. — Academic Press, New York, London 1968

[E 13] BEREŽKIN, V. G.; BOCKOV, A. S.: Količestvennaja tonkoslojnaja chromatografija. — Izd. Nauka, Moskva 1980

F Firm publications

[F 1] WATERS Linear columns

[F 2] SOCIÉTÉ NATIONALE DES PÉTROLES D'AQUITAINE, Courbevoie

[F 3] WATERS Chromatography information 1-1977

[F 4] ELECTRONUCLEONICS INCORPORATION Application sheet. Controlled porous glass-Polyether treatment and silanization

[F 5] VARIAN 973/12000/PMZ. Micro Pak® columns for high performance liquid chromatography

[F 6] REEVE ANGEL Bulletin No. 105

[F 7] CHROMATIX, 8M-11-78. Model 600/200. HPLC constant flow pump

[F 8] BROWNLEE LABS Bulletin 210. MPLC® Guard Column, Brownlee Labs Bulletin 140–260

[F 9] PERKIN ELMER, KHV 57410 Order No. L-358 Perkin-Elmer Sil-X® columns and packing for high performance liquid chromatography

[F 10] SHANDON Publication No. HPLC 3. Shandon Hypersplires

[F 11] REEVE ANGEL Bulletin No. 103, Vol. I

[F 12] E. MERCK High performance liquid chromatography, application of LiChrosorb, LiChrospher, Perisorb, Hibar® pre-packed columns

[F 13] HEWLETT PACKARD, 5952–6234. 1080 Series High performance liquid chromatograph

[F 14] FOXBORO/WILKS Application Report No. 2 (1976). An infrared detector for use with gel permeations and liquid chromatographs. WILKS® Bulletin L-8B-5M-5-78. Solvent functional group correlation chart

[F 15] PYE UNICAM Pye liquid chromatograph systems (83/1,3 1/2 M/9,69) Pye LCM2 liquid chromatograph (1661/7,5m/10,71)

[F 16] HWLETT PACKARD 12/72 (5952–3035) (Data sheet)

[F 17] VARIAN, 773/12000 Variscan® detector system for continuously variable wavelength UV detection

[F 18] HEWLETT PACKARD 12/72 (5952–3036) (Data sheet)

[F 19] WATERS Chromatography information I-1979

[F 20] PHARMACIA FINE CHEMICALS, 1969, 2.9.5. Sephadex® — Gelfiltration in Theorie und Praxis

[F 21] CHROMATIX A-8-4M-3/78. KMX-6 Low angle laser light scattering photometer

[F 22] CHROMATIX application note LS-2 (10 MO 177). Determination of molecular weight distributions by combining low angle light scattering with gel permeation chromatography

[F 23] CHROMATIX application note LS 6. A new data system for the polymer characterization laboratory: application to GPC data treatment

[F 24] WATERS, F 82, 83265. SEP-PAK cartridges for rapid sample preparation

[F 25] ALTEX SCIENTIFIC Inc. Model 110A HPLC solvent metering pump
[F 26] PERKIN ELMER, EP 8/7820. The Sigma 10 chromatography data station
[F 27] E. MERCK, 16/3396/6/1078/RU. Präparate für die HPLC
[F 28] MACHEREY — NAGEL HPLC/e6/01/0/3.79. HPLC: Packings and prepacked columns
[F 29] PERKIN ELMER, LCAS-61. Identification of LC fractions by infrared spectroscopy
[F 30] NATIONAL PHYSICAL LABORATORY, Division of Chemical Standards, Teddington, Middlesex, U.K.
[F 31] WATERS CU 84037 Rev. B, March 1976. μ Styragel® gel permeation chromatography column-Care & use manual
[F 32] BIO-RAD Bulletin 1056 EG. HPLC columns from BIO-RAD laboratories
[F 33] DU PONT Liquid chromatography report E-24012. Biomodal HPSEC-concepts and practice
[F 34] PHASE SEPARATIONS 1980. Packing HPLC columns with the Spherisorb-Range of column packing materials
[F 35] HEWLETT PACKARD Technical paper UV-2 23-5953-4752. Wavelength chromatography: Simultaneous multiwavelength LC monitoring using the HP 8450 A spectrophotometer
[F 36] APPLIED CHROMATOGRAPHY SYSTEMS LIMITED. The ACS 750/14 Mass detector
[F 37] E. MERCK 13/2615/10/478 R. Präparate für die Flüssigkeitschromatographie
[F 38] WATERS Chromatography notes Vol. III, No. 4
[F 39] CHROMATIX Application data. Light-scattering news No. 104
[F 40] IATRON LABORATORIES. Iatroscan TH-10 TLC/FID analyser
[F 41] IATRON LABORATORIES. Analysis of CTPB and HTPB resins
[F 42] SHIMADZU SEISAKUSHO CS-910 SHIMADZU dual-wavelength TLC-scanner. (Data sheet Nr. 1, CA 101–948 and 3001-12605-30ATD-E)
[F 43] CELLOMER ASSOCIATES Characterized polymers
[F 44] HEWLETT PACKARD Technical paper No. 81 (S. A. George and D. R. Baker). Data acquisition and GPC

2. Original papers

The following reference list is alphabetically arranged by the first-mentioned authors; the next-mentioned authors appear at the corresponding places in the alphabetic order with cross-references. This gives the list the character of an *author index*.

The last number put in parentheses indicates the respective chapter of this book in which the reference in question is cited.

ABBOTT, S. R., see WEHR
ABBOUD, J. L. (see also KAMLET)
1977 and KAMLET, M. J.; TAFT, R. W., J. Amer. chem. Soc. **99**, 8325 (7)
1979 and TAFT, R. W., J. phys. Chem. **83**, 412 (7)
ABDEL-ALIM, A. H.
1972 and HAMIELEC, A. E., J. appl. Polymer Sci. **16**, 1093–1101 (19)
1974 and HAMIELEC, A. E., J. appl. Polymer Sci. **18**, 297—300 (19)
ABDELLATIF, E., see GLÖCKNER
ABUAF, N., see LAUFER
ACKERMANN, G., see FREY
ACKERS, G. K.
see BRUMBAUGH
see WARSHAW
1962 and STEEVE, R. L., Biochim. et biophysica Acta **59**, 137 (8)
1964 Biochemistry **3**, 723 (8)
ACRES, G. J. K.
1963 and DALTON, F. L., J. Polymer Sci. A, Gen. Papers **1**, 2419—38 (20)
ADAMS, H. E.
see also ANDERSON
see also BACZEK
1966 and FARHAT, K.; JOHNSON, B. L., Ind. & Engng. Chem. Prod. Res. & Devel. **5**, 126 (19)
1971 Separat. Sci. **6**, 259—73 (19)
1973 and AHAD, E.; CHANG, M. S.; DAVIS, D. B.; FRENCH, D. M.; HYER, H. J.; LAW, R. D.; SIMKINS, R. J. J.; STUCHBURY, J. E.; TREMBLAY, M., J. appl. Polymer Sci. **17**, 269—82 (19)

AGARWAL, P., see PLAZEK
AGHERGHINEI, J., see CRUSOS
AGRELL, C. E., see SVENSSON
AHAD, E., see ADAMS
AIURA, M.
see HASHIMOTO
see KATO
AKATSUKA, M., see TERAMACHI
AKUTSU, S., see FURUKAWA
ALBERTSSON, P.-Å. (see also TISELIUS)
1958a Biochim. et biophysica Acta **27**, 378 (9)
1958b Nature [London] **182**, 709 (9)
1965 Svensk kem. Tidskr. **77**, 439 (9)
ALBRECHT, W.
1972 Beiträge zur Charakterisierung von Acrylnitril-Isobutylen-Copolymerisation durch Gelpermeationschromatographie. — Doctor Thesis, TU Dresden (10)
ALDERLIESTE, E. T., see HUBER
ALDHOUSE, S. T. E.
1968 and STANFORD, D. M., Paper, 5th International GPC-Seminar, London (16)
ALEKSANDROV, M. L.
see BELENKIJ
see VILENČIC
ALEKSEEVA, L. N.
1979 and ULEVA, N. S.; SUCHOTINA, G. N.; BEDINA, Ž. A., Zavodskaja Laboratorija **45**, 702—03 (21)
ALEX, W., see RUMPF
ALEXANDER, W. J. (see also MULLER)
1971 and MULLER, T. E., J. Polymer Sci. C, Polymer Symposia **36**, 87—101 (19)
ALLEN, D. A.
see MUKHERJI
see STRETCH
ALLIET, D. F. (see also GIAMMARE)
1968 and PACCO, J. M., J. Polymer Sci. C, Polymer Symposia **21**, 199—213 (19)
ALLIOT-LUGAZ, M., see PAILHES
ALM, R. S.
1952 and WILLIAMS, R. J. P.; TISELIUS, A., Acta chem. scand. **6**, 826 (14)
ALMIN, K. E.
1957 and STEENBERG, B., Acta chem. scand. **11**, 936 (9)
1959 Acta chem. scand. **13**, 1263, 1274, 1278, 1287, 1293 (9)
ALMQUIST, P. W., see BOMBAUGH
ALTGELT, K. H. (see also BILLMEYER)
1965 Makromol. Chemie **88**, 75—89 (17)
1967 and MOORE, J. C., Gel Permeation Chromatography. In [B 3], 123—79 (12, 19)
1971 Gel Permeation Chromatography with High Loads. In [D 2], 193—201 (16)
ALVARIÑO, J. M.
1978 and HERNÁNDEZ, M. L.; LAIN, L.; DE RENOBALES, M.; TORRE, A., Europ. Polymer J. **14**, 991—94 (11, 20)
AMBLER, M. R.
1974 and MATE, R. D.; PURDON, J. R., J. Polymer Sci. **12**, 1771—80 (19)
1975 and MCINTYRE, D., J. Polymer Sci. B, Polymer Letters **13**, 589—94 (8, 19)
1976 J. Polymer Sci. B, Polymer Letters **14**, 683—88 (8, 19)
1977 J. appl. Polymer Sci. **21**, 1655—65 (19)
1977 and MATE, R. D., The Gel Permeation Chromatography of Oligomers. In [D 22 I], 93—103 (19)
1977 and FETTERS, L. J.; KESTEN, Y., J. appl. Polymer Sci. **21**, 2439—51 (8, 16, 19)
1977 and MCINTYRE, D., J. appl. Polymer Sci. **21**, 2269—82, 3237—50 (8, 19)
1980 J. appl. Polymer Sci. **25**, 901—20 (19)
AMBORSKI, L. E.
1950 and BLACK, C. E.; GOLDFINGER, G., Rubber Chem. Technol. **23**, 417, 803 (18)
AMMUNDSON, N. R., see LAPIDUS
ANDERSON, J. L.
1981 and BRANNON, J. L., J. Polymer Sci. A-2, Polymer Phys. Ed. **19**, 405—21 (19)
ANDERSON, J. N. (see also BACZEK)
1975 and BACZEK, S. K.; ADAMS, H. E.; VESCELIUS, L. E., J. appl. Polymer Sci. **19**, 2255—67 (19)
ANDERSSON, K. B.
1973 and HOLMSTRÖM, A.; SÖRVIK, E. M., Makromol. Chemie **166**, 247—64 (19)
ANDREETTA, H. A.
1981 and FIGINI, R. V., Angew. makromol. Chemie **93**, 143—57 (16)
ANDREEV, A. N., see VACHTINA
ANDREEVA, G. A., see KALNINS
ANDRIANOV, K. A.
1977 and PAVLOVA, S.-S. A.; TVERDOCHLEBOVA, I. I.; PERCIVA, N. V.; TEMNIKOVSKIJ, V. A.; PROMINA, L. N., Vysokomolekularnye Soedinenija Ser. A, **19**, 466—68 (19)
ANGULO-SANCHEZ, J. L., see CAMPOS-LOPEZ
ANSORENA, J., see ECHARRI
ANTHONY, R. G., see COX
ANTLE, P. E.
see KIRKLAND
see TOLMAN
APPELT, B.
1980 and MEYERHOFF, G., Macromolecules **13**, 657—62 (16)
ARATANI, T.
1973 and MIZUI, F., J. Chromatogr. **79**, 173—78 (21)

ARMISTEAD, C. G.
1969 and TYLER, A. J.; HAMBLETON, F. H.; MITCHELL, S. A.; HOCKEY, J. A., J. phys. Chem. **73**, 3947—53 (11)
ARMONAS, J. E., see PIERCE
ARNDT, J., see BRAUN
ARNDT, K. F., see SCHRÖDER
ARPIN, M.
1976 and STRAZIELLE, C., Makromol. Chemie **177**, 293—98 (19)
ARRIGHETTI, S., see DE CHIRICO
ASAI, K., see ISHII
ASKILL, I. N., see GILDING
ASSHAUER, J.
1974 and HALÁSZ, I., J. Chromatogr. Sci. **12**, 139 to 47 (17)
ASTM D-20.70.04
1971 Gel Permeation Chromatography Minutes. Atlantic City, New Jersey, June 30 (cit. from ADAMS et al., 1973) (19)
ATKINSON, C. M. L.
1976 and DIETZ, R., Makromol. Chemie **177**, 213—31 (8, 19)
1978 and DIETZ, R.; FRANCIS, M. A., Whole Polymer Reference Materials in Analytical GPC. In [D 11/1], 297—303 (8)
1979 and DIETZ, R., Europ. Polymer J. **15**, 21—26 (8, 16, 19)
ATWOOD, J. G.
1979 and SCHMIDT, G. J.; SLAVIN, W., J. Chromatogr. **171**, 109—15 (11)
AUDEBERT, R.
see also BOUÉ
see also LECOURTIER
1979 Polymer [London] **20**, 1561—66 (16)
AUDU, T. O. K., see STOJANOV
AUE, W. A.
see also DANIEWSKI
see also WICKRAMANAYAKE
1969 and HASTINGS, C. R., J. Chromatogr. **42**, 319—35 (8, 11)
1973 and HASTINGS, C. R.; KAPILA, S., Analyt. Chem. **45**, 725—28 (11)
AXELSON, D. E.
1980 and KNAPP, W. C., J. appl. Polymer Sci. **25**, 119—23 (19)

BACZEK, S. K. (see also ANDERSON)
1975 and ANDERSON, J. N.; ADAMS, H. E., J. appl. Polymer Sci. **19**, 2269—77 (19)
BADGER, R. M., see BROOKS
BAHR, H., see PASTUSKA
BAIJAL, M. D.
1969 and DILER, R. M.; POOL, F. R., Macromolecules **2**, 679—81 (20)
BAISAL, M. D.
1971 and KAUPPILA, K. M., Polymer Engng. Sci. **11**, 182—86 (19)
BAKALYAR, S. R.
1977 and MCILWRICK, R.; ROGGENDORF, E., J. Chromatogr. **142**, 353—65 (7)
BAKER, C. A.
1956 and WILLIAMS, R. J. P., J. chem. Soc. [London], 2352—62 (9, 11, 20)
BAKER, G. K.
1971 J. Chromatogr. **55**, 173 (19)
BAKOŠ, D. (see also BEREK)
1976 and BEREK, D.; BLEHA, T., Europ. Polymer J. **12**, 801—04 (19)
1979 and BLEHA, T.; OZIMÁ, A.; BEREK, D., J. appl. Polymer Sci. **23**, 2233—44 (8, 16)
BALDRIAN, J.
1978 and PEŠKA, J.; ŠTAMBERG, J.; LUDWIG, J.; PAUL, D., Cellulose Packings in Bead Form — Porous Structure. In [D 11/1], 109—21 (12)
BALKE, S. T.
1969 and HAMIELEC, A. E., J. appl. Polymer Sci. **13**, 1381—420 (16, 17)
1969 and HAMIELEC, A. E.; LE CLAIR, B. P., PEARCE, S. L., Ind. & Engng. Chem. Prod. Res. & Devel. **8**, 54 (8)
1973 and HAMIELEC, A. E., J. appl. Polymer Sci. **17**, 905—49 (19)
1980 and PATEL, R. D., J. Polymer Sci. B, Polymer Letters **18**, 453—56 (19)
BALL, W.
1978 and SCHOLTE, Th. G., Pure & Appl. Chem. **50**, 1715—23 (19)
BANNER, B., see LOCKE
BANNISTER, D. W.
1954 and PHILLIPS, C. S. G.; WILLIAMS, R. J. P., Analyt. Chem. **26**, 1451 (18)
BAO, Y. T.
1980 and BOSE, A.; LADISCH, M. R.; TSAO, G. T., J. appl. Polymer Sci. **25**, 263—75 (12, 19)
BARKER, M. G., see URANECK
BARKER, P. E.
1978 Developments in continuous chromatographic refining. In [D 16], 41—85 (17)
1978a and ELLISON, F. J.; HATT, B. W., Continuous chromatography of macromolecular solutes. In [D 11/1], 219—39 (17)
1978b ELLISON, J. F.; HATT, B. W., Ind. & Engng. Chem. Prod. Res. & Devel. **17**, 302—09 (17, 19)
1979 and HATT, B. W.; HOLDING, S. R., J. Chromatogr. **174**, 143—51 (19)
1981a and HATT, B. W.; HOLDING, S. R., J. Chromatogr. **206**, 27—34 (11)
1981b and HATT, B. W.; VLACHOGIANNIS, G. J., J. Chromatogr. **208**, 74—77 (19)

BARLOW, A. (see also WILD)
1971 and WILD, L.; ROBERTS, T., J. Chromatogr. **55**, 155—64 (17, 19)
1977 and WILD, L.; RANGANATH, R., J. appl. Polymer Sci. **21**, 3319—29 (8, 17, 19)

BARMAKOVA, T. V.
1974 and KISELEV, A. V.; KOVALEVA, N. V., Kolloidnyi Žurnal **36**, 133 (11)

BARNES, A. D., see ROBINSON, J.

BARNES, D. E., see RUNYON

BARONI, A. T.
1961 Materie plast. **27**, 1050—53 (20)

BARRALL, E. M., see also COOPER
see also JOHNSON, J. F.
see also MIRABELLA
see also OUANO
1968 and CAIN, J. H., J. Polymer Sci. C, Polymer Symposia **21**, 253—65 (11)

BARTHOLIN, M., see JACOBELLI

BARTICK, E. G. (see also MOEBUS)
1976 and JOHNSON, J. F., Polymer [London] **17**, 455 (19)

BARTMANN, D.
1972 Z. Elektrochemie, Ber. Bunsenges. phys. Chemie **76**, 336—39 (14)

BASEDOW, A. M. (see also HALLER)
1976 and EBERT, K. H.; EDERER, H.; HUNGER, H., Makromol. Chemie **177**, 1501—24 (8, 11, 16, 19)
1979 and EBERT, K. H., J. Polymer Sci. C, Polymer Symposia **66**, 101—15 (19)
1980 and EBERT, K. H.; EDERER, H. J.; FOSSHAG, E., J. Chromatogr. **192**, 259—74 (19)

BASILA, M. R.
1961 J. chem. Phys. **35**, 1151—58 (11)

BATA, G. L.
1970 and HAZELL, J. E.; PRINCE, L. A., J. Polymer Sci. C, Polymer Symposia **30**, 157—62 (19)

BATE-SMITH, E. C.
1950 and WESTALL, R. G., Biochim. et biophysica Acta **4**, 427 (3)

BATHER, J. M.
1976 and GRAY, R. A. C., J. Chromatogr. **122**, 159—69 (11)
1978 and GRAY, R. A. C., J. Chromatogr. **156**, 21—26 (11)

BATZER, H.
1975 and ZAHIR, S. A., J. appl. Polymer Sci. **19**, 585—600; 609—17 (19)

BAUER, P.
1981 and RICHTER, M., J. Chromatogr. **206**, 343—52 (19)

BAUMANN, H., see KRANZ

BAUMANN, W. C., see WHEATON

BAYER, E.
1968 and CHOVIN, P.; CREMER, E.; DEANS, D. R.; GEISS, F.; GUIOCHON, G.; HALÁSZ, I.; JENTZSCH, D.; KAISER, R.; LEBBE, J.; MACHATA, G.; MUNIER, R. L.; OSTER, H.; PREVOT, A.; ROHRSCHNEIDER, L.; SCHAY, G.; SCHOMBURG, G.; STAHL, E.; SZEPESY, L.; TRANCHANT, J., Chromatographia **1**, 153—59 (3)

BAYERSDORF, F., see BRAUN

BEAU, R. (see also LE PAGE)
1969 and LE PAGE, M.; DE VRIES, A. J., Appl. Polymer Symposia **8**, 137 (19)

BECK, W.
1978 and HALÁSZ, I., Fresenius Z. Anal. Chem. **291**, 312—18 (18)

BECKER, N., see UNGER

BEDINA, Ž. A., see ALEKSEEVA

BELENKIJ, B. G.
see also JANČA
see also KALNINS
see also KEVER
see also NEFEDOV
see also NESTEROV
see also SKVORZOV
see also VILENČIK
see also ŽDANOV
1969 and GANKINA, E. S., Doklady Akademii Nauk SSSR **186**, 857 (21)
1970a and GANKINA, E. S., Doklady Akademii Nauk SSSR **194**, 573—76 (21)
1970b and GANKINA, E. S., J. Chromatogr. **53**, 3—25 (21)
1972 and NEFEDOV, P. P., Vysokomolekularnye Soedinenija, Ser. A **14**, 1568 (8)
1973 and GANKINA, E. S.; NEFEDOV, P. P.; KUZNECOVA, M. A.; VALČICHINA, M. D., J. Chromatogr. **77**, 209—17 (19, 20, 21)
1973b and VILENČIK, L. Z.; NESTEROV, V. V.; ŠAŠINA, T. I., Vysokomolekularnye Soedinenija Ser. A **15**, 2614—15 (16)
1974 private communication, May (21)
1974 and VACHTINA, I. A.; TARAKANOV, O. G., Vysokomolekularnye Soedinenija, Ser. B **16**, 507—09 (19)
1975a and GANKINA, E. S.; NEFEDOV, P. P.; LAZAREVA, M. A.; SAVITSKAYA, T. S.; VOLCHIKHINA, M. D., J. Chromatogr. **108**, 61—78 (19)
1975b and VILENCHIK, L. Z.; NESTEROV, V. V.; KOLEGOV, V. J.; FRENKEL, S. Ja., J. Chromatogr. **109**, 233—38 (16)
1976 and GANKINA, E. S., TENNIKOV, M. B.; VILENČIK, L. Z., Doklady Akademii Nauk SSSR **231**, 1147—49 (16)
1977 and GANKINA, E. S., J. Chromatogr. **141**, 13—90 (21)

1977 and KOLEGOV, V. I.; NEFEDOV, P. P.; ALEKSANDROV, M. L.; MELAS, V. B., Vysokomolekularnye Soedinenija, Ser. A **19**, 907—09 (8, 19)
1978a and GANKINA, E. S.; TENNIKOV, M. B.; VILENCHIK, L. Z., J. Chromatogr. **147**, 99—110 (16)
1978b and GANKINA, E. S.; EGONNIK, V. N.; KRASNOVA, T. P.; MELENEVSKAJA, E. Ju.; NEFEDOV, P. P., Vysokomolekularnye Soedinenija, Ser. B **20**, 574—77 (21)
1979 Pure & appl. Chem. **51**, 1519—35 (16, 18)

BELLENS, G.
1973 and HELME, J. P.; Leautey, J., Double Liaison **20**, 223—29 (19)

BELYAKOVA, L. D.
1966 and KISELEV, A. V.; KOVALEVA, N. V., Zh. Fiz. Khim. **40**, 1494 (7)

BENEDETTI, E., see PINO

BENNICH, H., see PORATH

BENOIT, H.
see also DONDOS
see also FRANCOIS
see also GRUBIŠIĆ
see also GRUBIŠIĆ-GALLOT
see also GUENET
see also PEYROUSET
see also PRECHNER
1966 and GRUBIŠIĆ, Z.; REMPP, P.; DECKER, D.; ZILLIOX, J.-G., J. chim. Phys. **63**, 1507—14 (8)

BENSON, S. W., see KING

BEREK, D.
see also BAKOŠ
see also BŁEDZKI
see also BLEHA
see also JURANIČOVA
see also ŠOLTÉS
see also SPYCHAJ
1974 and BAKOŠ, D.; ŠOLTÉS, L.; BLEHA, T., J. Polymer Sci. B, Polymer Letters **12**, 277—80 (19)
1975 and BAKOŠ, D.; BLEHA, T.; ŠOLTÉS, L., Makromol. Chemie **176**, 391—98 (16, 19)
1976 and NOVÁK, L.; PEVNÁ, Z., J. Polymer Sci. B, Polymer Letters **14**, 323—27 (19)
1976 and BLEHA, T.; PEVNÁ, Z., J. Chromatogr. Sci. **14**, 560—563 (19)
1977 and KATUŠČÁKOVÁ, G.: NOVÁK, L., Brit. Polymer J. 62—65 (11, 19)

BERENDSEN, G. E.
1980 Preparation and Characterization of Well-defined Chemically Bonded Stationary Phases for High Pressure Liquid Chromatography. Thesis, Delft (11, 16)

BERG, H. C.
1967 and PURCELL, E. M., Proc. Nat. Acad. Sci. USA **58**, 1821 (13)

BERG, K., see UNGER, K.

BERGER, H. L.
1965 and SHULTZ, A. R., J. Polymer Sci. A, Gen. Papers **3**, 3643—48 (8)

BERGER, K.
see KUNATH
see SCHULZ

BERGER, K. C. (see also SCHULZ, G. V.)
1970 and SCHULZ, G. V., Makromol. Chemie **136**, 221—39 (16, 19)
1971 and SCHULZ, G. V., Apparent and Real Distribution in GPC (Experiments with PMMA Samples). In [D 2], 325—421 (17)
1974 Makromol. Chemie **175**, 2121—32 (16, 19)
1975 Makromol. Chemie **176**, 399—410 (16, 19)
1978 Makromol. Chemie **179**, 719—32 (16)
1979a Progr. Colloid & Polymer Sci. **66**, 159—68 (16)
1979b Makromol. Chemie **180**, 1257—75 (16)
1979c Makromol. Chemie **180**, 2567—80 (16)

BERGER, R., see MINDNER
BERGES, J., see URIEL
BERGMANN, J. G., see PASCHKE
BERLIN, Al. Al., see POZNJAK

BERRY, G. C.
1971 J. Polymer Sci. A-2, Polymer Phys. **9**, 687—715 (19)

BERRY, V. N.
1980 J. Chromatogr. **199**, 219—38 (14)

BERTICAT, P., see SAGE
BERTRAND, J. C., see MUKHERJI

BETHKE, H.
1974 and SANTI, W.; FREI, R. W., J. Chromatogr. Sci. **12**, 392—97 (21)

BETTGER, T., see VACHTINA

BEYER, G. L., see SPATORICO
BIANCHI, U., see COPPOLA
BIDLINGMEYER, B. A., see PASCHKE
BIER, P., see HEITZ
BIERI, V., see v. TAVEL
BIERINGER, H. F., see WOLF

BIESENBERGER, J. A.
see also OUANO
see also DUVDEVANI
1971a and TAN, M.; DUVDEVANI, I.; MAURER, T., Gov. Rep. Announce (US) **71**, 53 (19)
1971b and TAN, M.; DUVDEVANI, I.; MAURER, T., J. Polymer Sci. B, Polymer Letters **9**, 353—57 (17)
1971c and TAN, M.; DUVDEVANI, I., J. appl. Polymer Sci. **15**, 1549—52 (17)

BIJ, K. E.
1981 and HORVÁTH, Cs.; MELANDER, W. R.; NAHUM, A., J. Chromatogr. **203**, 65—84 (11)
BILLIET, H. A. H.
see SCHOENMAKERS
see TIJSSEN
BILLMEYER, F. W. (see also KELLEY)
1968 and KELLEY, R. N., J. Chromatogr. **34**, 322—31 (16)
1969 and KATZ, J., Macromolecules **2**, 105 (19)
1971 and ALTGELT, K. H.: The Sizes of Polymer Molecules and the GPC Separation. In [D 2], 3—12 (8)
BINFORD, J. S.
1959 and GESSLER, E. M., J. phys. Chem. **63**, 1376—78 (18)
BIRLEY, A. W.
1980 and DAWKINS, J. V.; KYRIACOS, D., Polymer **21**, 632—38 (19)
BIRSHTEIN, T. M.
1979 Macromolecules **12**, 715—21 (6)
BLACK, C. E., see AMBORSKI
BLACKBURN, D. W. J., see SCOTT
BLAIR, D. E.
US Patent 3392 158 (9)
1970 J. appl. Polymer Sci. **14**, 2469—79 (9)
BLANCHARD, L. P., see CHEN
BLANKS, R. F.
1964 and PRAUSNITZ, J. M., Ind. & Engng. Chem. Fundamentals **3**, 1—8 (5)
BLASS, E.
1966 and SEIDE, H., Plaste u. Kautschuk **13**, 405—08 (20)
BLAZEJEWICZ, L.
1968 Z. analyt. Chemie **234**, 121—24 (18)
BŁEDZKI, A.
1978 and PRUSINSKA, M. I.; BEREK, D., J. Polymer Sci. A-1, Polymer Chemistry **16**, 107—14 (19)
BLEHA, T.
see also BAKOŠ
see also BEREK
1980 and MLYNEK, J.; BEREK, D., Polymer [London] **21**, 798—804 (19)
1981 and BEREK, D., Chromatographia **14**, 163 to 68 (16)
BLY, D. D. (see also YAU)
1968a J. Polymer Sci. C, Polymer Symposia **21**, 13—21 (16)
1968b J. Polymer Sci. A-1, Polymer Chem. **6**, 2085—89 (16)
1971 and BONI, K. A.; CANTOW, M. J. R.; CAZES, J.; HARMON, D. J.; LITTLE, J. N.; WEIR, E. D., J. Polymer Sci. B, Polymer Letters **9**, 401—11 (8, 16)
1980 J. Liqu. Chromatogr. **3**, 465—70 (8, 19)
BOBLETER, O., see CONCIN
BOCHKOV, A. S., see DAVANKOV
BOCK, R. M.
1954 and LING, N.-S., Analyt. Chem. **26**, 1543 to 46 (14)
BOGACHEVA, E. K.
1969 and KISELEV, A. V.; CHOCHLOVA, T. D.; ELTEKOV, Ju. A., Vysokomolekularnye Soedinenija, Ser. A **11**, 2180—85 (18)
BOHDANECKÝ, M.
1977 Macromolecules **10**, 971—75 (19)
BÖHM, L. L.
1974 and CASPER, R. H.; SCHULZ, G. V., J. Polymer Sci. A-2, Polymer Phys. **12**, 239 to 54 (9, 20)
BOLLINGER, W., see v. TAVEL
BOMBAUGH, K. J.
see also DARK
see also LITTLE
1968 and DARK, W. A.; KING, R. N., J. Polymer Sci. C, Polymer Symposia **21**, 131—42 (17, 19)
1969a and DARK, W. A.; LITTLE, J. N., Analyt. Chem. **41**, 1337—42 (16)
1969b and KING, R. N.; COHEN, A. J., J. Chromatogr. **43**, 332—38 (14)
1970a and LEVANGIE, R. F., Amer. chem. Soc., Div. Petrol. Chem., Prepr. **15** (2), A-105 (17)
1970b and LEVANGIE, R. F., Separat. Sci. **5**, 751—63 (17)
1970c and LEVANGIE, R. F., J. Chromatogr. Sci. **8**, 560—66 (17)
1975 and ALMQUIST, P. W., Chromatographia **8**, 109 (17)
BÖMER, B.
see HEITZ
see WALZ
BONI, K. A. (see also BLY)
1968 and SLIEMERS, F. A.; STICKNEY, P. B., J. Polymer Sci. A-2, Polymer Phys. **6**, 1567 (19)
BONN, R.
1972 and VAN AARTSEN, J. J., Europ. Polymer J. **8**, 1055—66 (5)
BOOTH, C. (see also PRICE)
1980 and FORGET, J.-L.; GEORGII, I.; LI, W. S.; PRICE, C., Europ. Polymer J. **16**, 255—59 (16, 19)
BORÁK, J.
1978 and CADERSKY, I.; KISS, F.; SMRZ, M.; VISKA, J., Poly(hydroxyethylmethacrylate) Gels (Spheron®). In [D 11/1], 91—108 (12)
BOSCHETTI, E. (see also URIEL)
1974 and TIXIER, R.; GARELLE, R., Sci. Tools, LKB Instr. J. **21**, 35—38 (16)
BOSE, A., see BAO

BOSWORTH, P., see MONTAGUE

BOTHAM, R.
1970 and THIES, C., J. Polymer Sci. C, Polymer Symposia **30**, 369—80 (6, 18)

BÖTTGER, J.
1976 and KRAUSE, T.; SCHURZ, J., Holzforschung **30**, 41—44 (19)

BOUÉ, J.
1980 and AUDEBERT, R.; QUIVORON, C., Direct resolution of α-Aminoacids by Ligand Exchange: Use of Silica Packings Coated by a Chiral Polymer. Contribution C 25 at the 13th International Symposium on Chromatography, Cannes, 30 June to 4 July (11)

BOULTON, A. A., see GIETZ

BOUNINE, J. P.
see GUILLEMIN
see THOMAS

BOUWMEESTER, P. C., see KOOPMANS

BOWDEN, M. J.
1975 and THOMPSON, L. F., J. appl. Polymer Sci. **19**, 905—06 (19)

BOWEN, J. N.
1967 and BOWREY, R.; MALIN, A. S., J. Catal. **1**, 209 (11)

BOWMAN, L. M.
see TALLEY
see GIDDINGS

BOWMAN, R. L.
see ITO
see TANIMURA

BOWREY, R., see BOWEN

BRADLEY, W. A., see HANCOCK

BRAKS, J. G.
1978 and HUANG, R. Y. M., J. appl. Polymer Sci. **22**, 3111—20 (19)

BRANNON, J. L., see ANDERSON

BRASH, J. L., see OMORODION

BRAUER, E. (see also ECKHARDT)
1977a and WIEGLEB, H.; HELMSTEDT, M.; EIFERT, W., Plaste und Kautschuk **24**, 463—68 (19)
1977b and WIEGLEB, H.; GEYER, U.; MISCHE, W., Plaste u. Kautschuk **24**, 630—35 (19)

BRAUN, D.
1970 and CHAUDHARI, D., Makromol. Chemie **133**, 241—50 (20)
1972 and ARNDT, J.; KÄMMERER, H., Angew. makromol. Chemie **26**, 181—85 (19)
1972 and LEGRADIC, V., Angew. makromol. Chemie **25**, 193—96 (19)
1975 and LEE, D. W., Angew. makromol. Chemie **48**, 161—73 (19)
1976 and LEE, D. W., Angew. makromol. Chemie **51**, 11—24 (19)
1978 and ARNDT, J., Angew. makromol. Chem. **73**, 143—51 (19)
1979 and BAYERSDORF, F., Angew. makromol. Chem. **81**, 147—70 (19)
1979a PANDJOJO, W., Fresenius Z. Anal. Chem. **294**, 375—78 (19, 21)
1979b and PANDJOJO, W., Angew. makromol. Chemie **80**, 195—205 (19, 21)
1980 and BAYERSDORF, F., Angew. makromol. Chem. **85**, 1—13 (19)

BREBNER, K. I.
1980 and CHAHAL, R. S.; St.-PIERRE, L. E., Polymer [London] **21**, 533—40 (6)

BREITENBACH, J. W. (see also WOLF)
1962a and BURGER, H. G., Makromol. Chemie **54**, 60—69 (20)
1962b and BURGER, H. G., Monatsh. Chemie **93**, 733—45 (20)
1962c and BURGER, H. G.; SCHINDLER, A., Monatsh. Chemie **93**, 160—72 (20)
1962d and STREICHSBIER, O., Kolloid-Z. u. Z. Polym. **182**, 35—40 (20)
1967 and WOLF, B. A., Makromol. Chem. **108**, 263—80 (9)

BRENNER, M. (see also NIEDERWIESER)
1964 and PATAKI, G.; NIEDERWIESER, A. In: Thin Layer Chromatography. Ed.: G. B. MARINI-BETTOLO. — Elsevier, Amsterdam, p. 117 (21)

BRESLER, S. E.
1963 and DOBYČIN, D. P.; POPOV, A. G., Ž. prikl. chim. **36**, 66—74 (11)

BRESSOLLE, F., see SIOUFFI

BRETTON, R. H., see CARBERRY

BREWER, P. I.
1965 Polymer [London] **6**, 603—13 (10)

BRIDGMAN, A. L., see SHULTZ

BRINKMAN, U. A. Th.
1980 and DE VRIES, G., J. Chromatogr. **192**, 331—40 (21)
1980 and DE VRIES, G.; CUPERUS, R., J. Chromatogr. **198**, 421—28 (21)

BRISTOW, P. A.
1977 and BRITTAIN, P. N.; RILEY, C. M.; WILLIAMSON, B. F., J. Chromatogr. **131**, 57—64 (17)
1978 J. Chromatogr. **149**, 13—28 (17)

BRITTAIN, P. N., see BRISTOW

BROCKMANN, H.
1941 and SCHODDER, H., Ber. dt. chem. Ges. **74B**, 73 (10)

BRODILOVÁ, J.
1979 and ROTSCHOVÁ, J.; POSPÍŠIL, J., J. Chromatogr. **168**, 530—32 (18)

BROIDO, A., see OUANO

BROOKS, M. C.
1950 and BADGER, R. M., J. Amer. chem. Soc. **72**, 4384—88 (18)

BROQUAIRE, M.
1979 J. Chromatogr. **170**, 43—52 (17)
BROUGH, A. W. J.
1977 and EPTON, R.; MCLAREN, J. V.; MARR, G., Polymer [London] **18**, 1208—12 (12)
1978 and EPTON, R.; MARR, G.; SHACKLEY, A. T.; SNIEZKO-BLOCKI, G. A., Poly(acryloyl morpholine) (Enzacryl® Gels) Packings. In [D 11/1], 70—90 (12)
BROWN, D. R., see KENYON
BROWN, D. W.
1979 and LOWRY, R. E., J. Polymer Sci. A-1, Polymer Chem. **17**, 1039—46 (19)
BRUCK, S. D.
1958 J. Polymer Sci. **32**, 519—21 (18)
BRUGMAN, W. J. T., see SLAATS
BRUMBAUGH, E. E.
1968 and ACKERS, G. K., J. Biol. Chem. **243**, 6315—24 (19)
BRUN, A., see THOMAS
BRUNAUER, S.
1938 and EMMETT, P. H.; TELLER, E., J. Amer. chem. Soc. **60**, 309—19 (10)
BRÜSSAU, R. J.
1970 and STEIN, D. J., Angew. makromol. Chemie **12**, 59—72 (19)
1977 Erste Erfahrungen mit dem Laserlichtstreugerät als Molgrößendetektor bei der GPC. Paper, read on Königsteiner Chromatography Symposium (19)
BRUZZONE, A. R.
see COOPER
see JOHNSON
BRUZZONE, M., see DE CHIRICO
BRYK, M. T.
1968 and SEVLJAKOV, A. S.; KRESUB, O. B., Vysokomolekularnye Soedinenija, Ser. A **10**, 893—95 (21)
BRZEZIŃSKI, J. (see also DOBKOWSKI)
1966 Polimery Tworzywa, Warschau **11**, 423 to 29, 476 (20)
1980 and DOBKOWSKI, Z., Europ. Polymer J. **16**, 85—87 (19)
BUDTOV, V. P.
1976 and PODOSENOVA, N. G.; GAL'PERIN, V. M.; OTRADINA, G. A.; RUPYŠEV, V. G.; DOMAREVA, N. M.; KRIVČENKO, E. I., Vysokomolekularnye Soedinenija, Ser. A **18**, 1522—27 (19)
BUHLER, D. R., see RAMSDELL
BULAS, R., see ROWLAND
BUNGENBERG, DE JONG, H. G.
1960 and HOOGEVEEN, J. T., Proc. Kon. nederl. Akad. Wetensch. **B 63**, 228, 243, 383 (14)
1961 and HOOGEVEEN, J. T., Proc. Kon. nederl. Akad. Wetensch. **B 64**, 1, 18, 167, 183 (14)
BURGER, H. G., see BREITENBACH
BÜRGER, K.
1963 Z. analyt. Chemie **196**, 259—68 (21)
1967a Z. analyt. Chemie **224**, 421—25 (21)
1967b Z. analyt. Chemie **224**, 425—35 (18)
BURKAT, T. M., see DOBYČIN
BURNS, C. M., see NARASIMHAN
BURNS, H.
1968 and CARPENTER, D. K., Macromolecules **1**, 384—90 (6, 18)
BURTSCHER, E., see CONCIN
BUTER, R.
1973 and TAN, Y. Y.; CHALLA, G., Polymer [London] **14**, 171—72 (21)
BUTNARU, R., see SIMIONESCU
BUYTENHUYS, F. A.
1978 and VAN DER MAEDEN, F. P. B., J. Chromatogr. **149**, 489—500 (19)
BYRNE, S. H., see HENRY

CACACE, M. G., see SADA
CADERSKY, I., see BORÁK
CAIN, J. H.
see BARRALL
see COOPER, A. R.
CALDWELL, K. D.
see also GIDDINGS
see also GRUSHKA
1972 and KESNER, L. F.; MYERS, M. D.; GIDDINGS, J. C., Science [Washington] **176**, 296 (13)
CALHOUN, R. C., see FOX
CALUGARU, E.-M., see THEIL
CALZOLARI, C.
1971 and FAVRETTO, L.; STAUCHER, B., Chromatographia **3**, 313 (18)
CAMPBELL, D. E., see MELANDER
CAMPOS, A.
see also FIGUERUELO
see also KATIME
1979 and SORIA, V.; FIGUERUELO, J. E., Makromol. Chemie **180**, 1961—74 (16)
CAMPOS-LOPEZ, E.
1976 and ANGULO-SANCHEZ, J. L., J. Polymer Sci. B, Polymer Letters **14**, 649—52 (19)
CANE, F.
1978 and CAPACCIOLI, T., Europ. Polymer J. **14**, 185—88 (19)
CANTOW, H. J.
see also MIYAMOTO
see also PROBST
see also RING
1966 and SIEFERT, E.; KUHN, R., Chemie-Ing.-Techn. **38**, 1032—38 (9, 20)
1968 and PROBST, J.; STOJANOW, C., Rev. gén. Caoutchoucs et Plastiques **45**, 1233—59; Kautschuk—Gummi—Kunstst. **21**, 609 (9, 11, 19, 20)

CANTOW, M. J. R.
see also BLY
see also PICKETT
1961 and PORTER, R. S.; JOHNSON, J. F., Nature [London] **192**, 752—53 (9, 17, 20)
1963 and PORTER, R. S.; JOHNSON, J. F., J. Polymer Sci. C, Polymer Symposia **1**, 187—94 (9, 17, 20)
1964 and PORTER, R. S.; JOHNSON, J. F., J. appl. Polymer Sci. **8**, 2963—69 (9, 17, 20)
1967a and PORTER, R. S.; JOHNSON, J. F., J. Polymer Sci. C, Polymer Symposia **16**, 13—21 (19, 20)
1967b and PORTER, R. S.; JOHNSON, J. F., J. Polymer Sci. A-1, Polymer Chem. **5**, 987—91 (8, 19)
1967 and JOHNSON, J. F., J. Polymer Sci. A-1, Polymer Chem. **5**, 2835—41 (8)
1967 and JOHNSON, J. F., J. appl. Polymer Sci. **11**, 1851—54 (8, 11, 19)
CAPACCIOLI, T., see CANE
CAPLAN, S. R.
1959 J. Polymer Sci. **35**, 409—21 (9, 20)
CAPLE, M. B., see CHOW
CAPP, M., see OHARE
CAPRA, J. D., see HANCOCK
CARBERRY, J. J.
1958 and BRETTON, R. H., Amer. Inst. Chem. Engng. J. **4**, 367 (3)
CARPENTER, D. K., see BURNS
CARR, P. W.
1980 J. Chromatogr. **194**, 105—19 (7)
CARROLL, W. R.
1966 and EISENBERG, H., J. Polymer Sci. A-2, Polymer Phys. **4**, 599—610 (19)
CARSON, L. M., see STERNBERG
CARTER, J. D., see DARK
CASASSA, E. F.
1967 J. Polymer Sci. B, Polymer Letters **5**, 773—78 (8)
1971 Gel Permeation Chromatography and Thermodynamic Equilibrium. In [D 2], 119—33 (8)
1976 Macromolecules **9**, 182—85 (8)
CASE, L. C.
1960 Makromol. Chemie **41**, 61—76 (9)
CASPER, R. H. (see also BÖHM)
1970 J. Polymer Sci. A-2, Polymer Phys. **8**, 833—39 (9)
1971a and SCHULZ, G. V., Separat. Sci. **6**, 321 (9, 20)
1971b and SCHULZ, G. V.: Phase Distribution Chromatography. In [D 2], 225—33 (9, 20)
CATTERALL, E., see LYONS
CAUDE, M.
see GAREIL
see HENNION
CAZES, J., see BLY
CEAUŞESCU, E.
1972 and LEBEDA, P.; MINTI, M., Mater. Plast [Bukarest] **9**, 463—66 (19)
CERVENKA, A. (see also WILLIAMSON)
1973 and WILLIAMSON, G. R., Characterization of low-density polyethylene by GPC. In [B 12], 102—110 (19)
1974 and WILLIAMSON, G. R., Europ. Polymer J. **10**, 305—08 (19)
CHA, C. Y.
1969 J. Polymer Sci. B, Polymer Letters **7**, 343—48 (16, 19)
CHADWICK, V. S., see MURPHY
CHAHAL, R. S., see BREBNER
CHALLA, G., see BUTER
CHAMPTIER, G., see CHAUFER
CHAN, R. K. S.
1971 Polymer Engng. & Sci. **11**, 187—93 (19)
1972 and WORMAN, C., Polymer Engng. & Sci. **12**, 437—43 (19)
CHANG, F. S. C.
1971 J. Chromatogr. **55**, 67—71 (19)
CHANG, K. S.
1969 and HUANG, R. Y. M., J. appl. Polymer Sci. **13**, 1459—71 (16)
1972 and HUANG, R. Y. M., J. appl. Polymer Sci. **16**, 329—35 (16)
1973 and POUND, T. C.; MANLEY, R. S. J., J. Polymer Sci. A-2, Polymer Phys. **11**, 399—411 (19)
CHANG, M. S., see ADAMS
CHANG, P.
1955 and WILKE, C. R., J. phys. Chem. **59**, 592—96 (15)
CHANG, S.-H.
1976 and NOEL, R.; REGNIER, F. E., Analytical Chem. **48**, 1839—45 (11)
CHAPIRO, A.
1963 and CORDIER, P.; JOZEFOWICZ, J.; SEBAN-DANON, J., J. Polymer Sci. C, Polymer Symposia **4**, 491—506 (20)
CHAUDHARI, D., see BRAUN
CHAUFER, B.
1977 and LESEC, J.; QUIVORON, C.; CHAMPTIER, G., C. R. hebd. Séances Acad. Sci. [Paris] **284**, Ser. C, 881—84 (19)
CHEN, B.-K. (see also MELANDER)
1979 and HORVÁTH, Cs., J. Chromatogr. **171**, 15—28 (7)
CHEN, H. R.
1972 and BLANCHARD, L. P., J. appl. Polymer Sci. **16**, 603—13 (19)
CHOCHLOVA, T. D.
see BOGACHEVA
see ELTEKOV
CHOPINA, V. V., see ELTEKOV

CHOVIN, P., see BAYER
CHOW, C. D.
1976 J. app. Polymer Sci. **20**, 1619—26 (19)
CHOW, H.-C.
1978 and CAPLE, M. B.; STROUSE, C. E., J. Chromatogr. **151**, 357—62 (11)
CHRAMOVA, T. S., see LIPATOV
CHRISTENSEN, G.
1973 and FINK-JENSEN, P., Farbe u. Lack **79**, 301—06 (19)
CHRISTOPHE, A. B.
1971 Chromatographia **4**, 455—58 (3)
CHRISTOPHER, P. C.
1976 J. appl. Polymer Sci. **20**, 2989—3003 (19)
CHUA, M. G. S.
1979 and WAYMAN, M., Can. J. Chem. **57**, 1141 (19)
CHUANG, J.-Y.
1973 and JOHNSON, J. F., J. appl. Polymer Sci. **17**, 2123—29 (8, 19)
1973 and COOPER, A. R.; JOHNSON, J. F., J. Polymer Sci. C, Polymer Symposia **43**, 291—97 (16)
CHUBAROV, V. A., see VILENČIK
CHURÁČEK, J., see JANDERA
CIAMPA, G.
1970 and GRIECO, C.; SILIPO, C., J. Chromatogr. **46**, 132—33 (21)
CIARDELLI, F., see PINO
CICCIOLI, P.
1981 and TAPPA, R.; DI CORCIA, A.; LIBERTI, A., J. Chromatogr. **206**, 35—42 (11)
CILIPOTKINA, M. V.
1976 and TAGER, A. A.; DAVANKOV, V. A.; CJURUPA, M. P.; LIROVA, B. I.; KOLČINA, G. M.; ŽIGUNOVA, O. A., Vysokomolekularnye Soedinenija, Ser. B **19**, 874—77 (10)
CJURUPA (TSYURUPA), M. P.
see also CILIPOTKINA
see also DAVANKOV
1980 and DAVANKOV, V. A., J. Polymer Sci. A-1, Polymer Chem. **18**, 1399—1406 (10)
CLAESSON, J.
1944/45 and CLAESSON, S., Ark. Kem. Mineralog. Geol., Ser. A, **19**, Nr. 1 (18)
CLAESSON, S. (see also CLAESSON, J.)
1949a Ark. Kem. Mineralog. Geol. Ser. A, **26**, Nr. 24 (18)
1949b Discuss. Faraday Soc. **7**, 321 (18)
CLARK, J. C., see WESTERMAN
CLARK, J. H., see PROVDER
CLARK, O. K., see RODRIGUEZ
CLUFF, J. R.
1976 and HAWKES, S. J., J. Chromatogr. Sci. **14**, 248—55 (15)
CLUSIUS, K.
1938 and DICKEL, G., Naturwissenschaften **26**, 546 (13)
COHEN, A. J., see BOMBAUGH
COHN-GINSBERG, E., see WEISS
COLIN, H.
1976 and EON, C.; GUIOCHON, G., J. Chromatogr. **122**, 223—42 (7,14)
1977 and GUIOCHON, G., J. Chromatogr. **141**, 289—312 (11)
1978a and WARD, N.; GUIOCHON, G., J. Chromatogr. **149**, 169 (11)
1978b and DIEZ-MASA, J. C.; GUIOCHON, G.; CZAJKOWSKA, T.; MIEDZIAK, I., J. Chromatogr. **167**, 41—65 (11)
1979 and MARTIN, M.; GUIOCHON, G., J. Chromatogr. **185**, 79—95 (17)
COLL, H.
1970 and GILDING, D. K., J. Polymer Sci. A-2, Polymer Phys. **8**, 89—103 (8, 19)
1971 Calibration of GPC Columns. In [D 2], 135—44 (8)
1980 and FAGUE, G. R., J. Colloid & Interface Sci. **76**, 116—23 (19)
COLLET, G., see SAGE
COMBELLAS, C., see HENNION
COMBS, R. L.
see GUILLET
see SLONAKER
COMPAAN, H., see DHONT
CONCIN, R.
1980 and BURTSCHER, E.; BOBLETER, O., J. Chromatogr. **198**, 131—41 (19)
CONSTANTIN, D.
1977 Europ. Polymer J. **13**, 907—13 (19)
COOKE, R. C., see FRANCIS
COOPER, A. R.
see also CHUANG
see also JOHNSON
1969 and JOHNSON, J. F., J. appl. Polymer Sci. **13**, 1487—92 (10,11)
1969 and BUZZONE, A. R.; JOHNSON, J. F., J. appl. Polymer Sci. **13**, 2029—30 (17)
1971 and BRUZZONE, A. R.; CAIN, J. H.; BARRALL, E. M., J. appl. Polymer Sci. **15**, 571—88 (11)
1971 and JOHNSON, J. F., J. appl. Polymer Sci. **15**, 2293—2301 (8, 19)
1973 and JOHNSON, J. F.; BRUZZONE, A. R., Europ. Polymer J. **9**, 1381—91, 1393—1403 (19)
1973 and KISS, J., Brit. Polymer J. **5**, 433—441 (16, 17, 19)
1974 J. Polymer Sci. A-2, Polymer Phys. **12**, 1969—77 (15, 16, 19)
1975a and HUGHES, J. J.; JOHNSON, J. F., J. appl. Polymer Sci. **19**, 435—50 (8, 17, 19)

1975b and HUGHES, A. J.; JOHNSON, J. F., Chromatographia **8**, 136—42 (17)
1977 and MATZINGER, D. P., American Laboratory **9**, 15—23 (19)
1978 and VAN DERVEER, D. S., J. Liq. Chromatogr. **1**, 693—726 (19)
1978 Column Fractionation of Macromolecules by Gradient Elution. In [D 11/1], 240—65 (9, 17)
1979 and MATZINGER, D. P., J. appl. Polymer Sci. **23**, 419—27 (11, 19)
1980 J. Liq. Chromatogr. **3**, 393—402 (19)

COOPER, W.
1959 and VAUGHAN, G.; MADDEN, R. W., J. appl. Polymer Sci. **1**, 329—35 (20)
1962 and VAUGHAN, G.; YARDLEY, J., J. Polymer Sci. **59**, 2—4 (9, 20)

COOVER, H. W.
see GUILLET
see SLONAKER

COPPOLA, G.
1972 and FABBRI, P.; PALLESI, B.; BIANCHI, U., J. appl. Polymer Sci. **16**, 2829—34 (16, 19)

COQ, B.
1979 and CRETIER, G.; ROCCA, J. L., J. Chromatogr. **186**, 457—73 (17)

CORBINO, O. M.
1910 Phys. Z. **11**, 756—61 (6)

CORDIER, P., see CHAPIRO

CORGIER, M., see MONSIGNY

CORNET, M., see MONSIGNY

CORRADINI, P., see NATTA

COSANI, A.
1966a and PEGGION, E.; SCOFFONE, E., Makromol. Chemie **97**, 113—27 (20, 21)
1966b and PEGGION, E.; SCOFFONE, E.; VERDINI, A. S., J. Polymer Sci. B, Polymer Letters **4**, 55—58 (20)

COTE, J. A.
1971 and SHIDA, M., J. Polymer Sci. A-2, Polymer Phys. **9**, 421—30 (19)

COULOMBE, S.
1978 and SCHAUWECKER, P.; MARCHESSAULT, R. H.; HAUTTECOEUR, B., Macromolecules **11**, 279—81 (18)

COULTER, B., see SPATORICO

COUPE, D., see DAWKINS

ČOUPEK, J.
see also HEITZ
see also MINÁRIK
1968 and HEITZ, W., Makromol. Chemie **112**, 286—92 (19)
1973 and KŘIVÁKOVÁ, M.; POKORNY, S., J. Polymer Sci. C, Polymer Symposia **42**, 185 (12)
1974 and POKORNÝ, S.; GREGOR, V.; JEHLIČKA, V., J. Chromatogr. **88**, 323 (19)

COX, J. H.
1975 and ANTHONY, R. G., J. appl. Polymer Sci. **19**, 821—30 (17)

CRAIG, L. C.
1944 J. biol. Chem. **155**, 519 (9)

CRAUBNER, H.
1965 Makromol. Chemie **89**, 105—26 (6)

CREMER, E. (see also BAYER)

CRESSWELL, W. T., see VOGEL

CRETIER, G., see COQ

CRIGHTON, T. J.
1978 and PEAKER, F. W., Composite Polystyrene Packings for GPC. In [D 11/1], 146—55 (12)

CRONE, H. D.
1975 J. Chromatogr. **107**, 25—33 (19)

CRONIN, P. J., see RYBICKA

CROOK, M. A.
1963 and WALKER, F. S., Nature [London] **198**, 1163—65 (20)

CROUZET, P.
1975 and MARTENS, A. J., Chromatographia **8**, 675—81 (19)

CROWTHER, J. A., see MOEBUS

CRUSOS, A.
1972 and UGLEA, C.; AGHERGHINEI, J., Rev. roum. Chim. **17**, 565—70 (19)

ČUBAROVA, E. V., see NESTEROV

CUPERUS, R., see BRINKMAN

CURTHOYS, G.
1974 and DAVIDOV, V. Y.; KISELEV, A. V.; KISELEV, S. A.; KUZNECOV, B., J. Colloid & Interface Sci. **48**, 58—72 (6)

CVET (TSWETT), M. S.
1906 Ber. dt. bot. Ges. **24**, 316—23 (10)

CVETKOVSKIJ, I. B.
1977 and VALUEV, V. I.; ŠLICHTER, R. A., Vysokomolekularnye Soedinenija, Ser. A **19**, 2637—39 (8)

CYPEROVIC, A. S.
1962 and GALIC, I. P., Ukrainskij biochimičnij Žurnal **34**, 666 (18)

CZAJKOWSKA, T., see COLIN

CZARNECKA, B.
1979 and DOBKOWSKI, Z., Polimery **24**, 151—55 (19)

DADONE, A., see FRACHE

DALE, A. D., see LITTLE

DALLAS, M. S. J.
1965 J. Chromatogr. **17**, 267—77 (10)

DALTON, F. L., see ACRES

DAŇHELKA, J.
1970 and KÖSSLER, I., Collect. czechoslov. chem. Commun. **35**, 2870—77 (20)

DANIELEWICZ, M.
1977 and VOZKA, S.; KUBIN, M., Collect. czechoslov. chem. Commun. **42**, 2498—2510 (16)
1981 and KUBIN, M., J. appl. Polymer Sci. **26**, 951—56 (18)
DANIEWSKI, M. M.
1978 and AUE, W. A., J. Chromatogr. **147**, 119—25 (11)
DANON, J.
1969 and JOZEFONVICZ, J., Europ. Polymer J. **5**, 405—17 (20)
DAOUST, H.
see GOURDENNE
see HO-DUC
DARK, W. A. (see also BOMBAUGH)
1968 and LEVANGIE, R. F.; BOMBAUGH, K. J., 5th Internat. Seminar on Gel Permeation Chromatography, London, Seminar Proc., 414 (17, 21)
1975 and LIMPERT, R. J.; CARTER, J. D., Polymer Engng. & Sci. **15**, 831—33 (19)
DAVANKOV, V. A.
see also CILIPOTKINA
see also CJURUPA (TSYURUPA)
see also ROGOŽIN
1973 and ROGOŽIN, S. V.; CJURUPA, M. P., Vysokomolekularnye Soedinenija, Ser. B, **15**, 463—65 (10)
1974a and ROGOŽIN, S. V.; CJURUPA, M. P.; PANKRATOV, E. A., Ž. fiz. chimii **48**, 2964 (10)
1974b and ROGOŽIN, S. V.; CJURUPA, M. P., J. Polymer Sci. C, Polymer Symposia **47** 95—101 (10)
1976 and CJURUPA, M. P.; ROGOŽIN, S. V., Angew. makromol. Chemie **53**, 19—27 (10)
1977 and SEMECHKIN, A. V., J. Chromatogr. **141**, 313—53 (11, 18)
1980 Resolution of Racemates by Ligand-Exchange Chromatography. In [A15], **18**, 139—95 (18)
1980 and BOCHKOV, A. S.; KURGANOV, A. A.; ROUMELIOTIS, P.; UNGER, K., Chromatographia **13**, 677—85 (11)
1980 and CJURUPA, M. P., Angew. makromol. Chem. **91** 127—42 (10)
DAVE, S. B.
1975 Ind. & Engng. Chem., Prod. Res. & Devel. **14**, 85—89 (11)
DAVIDOV, V. Y., see CURTHOYS
DAVIS, B. E., see DICKSON
DAVIS, D. B., see ADAMS
DAWIDOWICZ, A., see WAKSMUNDZKI
DAWKINS, J. V. (see also BIRLEY)
1968 J. Macromol. Sci. **B2**, 623 (8)
1969 and DENYER, R.; MADDOCK, J. W., Polymer [London] **10**, 154—57 (8)
1970 Europ. Polymer. J. **6**, 831—40 (8)
1970 and PEAKER, F. W., Europ. Polymer J. **6**, 209—18 (20)
1971 and MADDOCK, J. W.; COUPE, D., J. Chromatogr. **55**, 65 (21)
1972b and HEMMING, M., Polymer [London] **13**, 553—54 (1)
1973 and MADDOCK, J. W.; NEVIN, A., Europ. Polymer J. **9**, 327—43 (21)
1975 and TAYLOR, G., J. Polymer Sci. B, Polymer Letters **13**, 29—33 (21)
1975a and HEMMING, M., Makromol. Chemie **176**, 1777—93 (16, 19)
1975b and HEMMING, M., Makromol. Chemie **176**, 1795—1813 (21)
1975c and HEMMING, M., Makromol. Chemie **176**, 1815—28 (16, 19)
1975d and HEMMING, M., J. appl. Polymer Sci. **19**, 3107—18 (19)
1975e and HEMMING, M., Polymer **16**, 554—60 (19)
1977 and STONE, T.: YEADON, G., Polymer [London] **18**, 1179—84 (12, 19)
1979 Pure & appl. Chem. **51**, 1473—81 (16)
1979 and STONE, T.; YEADON, G.; WARNER, F. P., Polymer [London] **20**, 1164—66 (12)
1979 and YEADON, G., Polymer [London], **20**, S. 981—89 (15, 19)
1980 and YEADON, G., J. Chromatogr. **188**, 333—45 (19)
1981 and GABBOTT, N. P., Polymer **22**, 291—92 (12)
DAY, J. C.
1980 and ROBB, I. D., Polymer [London] **21**, 408—12 (6)
DEANS, D. R., see BAYER
DE BRUILLE, R., see SERVOTTE
DEBYE, P.
1948 and BUECHE, A. M., in [B 8], 497 (13)
DE CHIRICO, A.
1979 and ARRIGHETTI, S., Preprints IUPAC Mainz 758—61 (19)
1981 and ARRIGHETTI, S.; BRUZZONE, M., Polymer [London], **22**, 529—33 (19)
DECKER, D., see BENOIT
DE GALAN, K., see SCHOENMAKERS
DE GENNES, P. G., see JOANNY
DEHLÉN, S. O., see SVENSSON
DE JONG, A. W. J.
1978 and POPPE, H.; KRAAK, J. C., J. Chromatogr. **148**, 127—41 (17)
1980 and KRAAK, J. C.; POPPE, H.; NOOITGEDACHT, F., J. Chromatogr. **193**, 181—95 (17)

1981 and POPPE, H.; KRAAK, J. C., J. Chromatogr. **209**, 432—36 (17)

DEKKER, D.
1979 J. Chromatogr. **168**, 508—11 (21)

DELLWEG, H., see JOHN

DEMENT'EVA, M. N.
1962 and DOBYČIN, D. P.; ŠEFTER, V. E., Ž. fizičeskoi chim. **36**, 228—29 (11)

DEMING, S. N., see MORGAN

DENYER, R., see DAWKINS

DE RENOBALES, M., see ALVARINO

DERHAM, K. W. (see also GORDON)
1974 and GOLDSBROUGH, J.; GORDON, M., Pure & appl. Chem. **38**, 97—116 (5)

DERUVO, A., see LINDSTRÖM

DESBRIERES, J., see RINAUDO

DESREUX, V. (see also HUGGINS)
1949 Recueil Trav. chim. Pays-Bas **68**, 789 (9)
1950 and SPIEGELS, M. C., Bull. Soc. chim. Belgique **59**, 476—89 (9)
1952 and OTH, A., Chem. Weekbl. **48**, 247—59 (9)

DETERMANN, H.
1966 and MICHEL, W., J. Chromatogr. **25**, 303 (21)
1968 and REHNER, H.; WIELAND, T., Makromol. Chemie **114**, 263—74 (12)
1971 and LAMBERT, K., J. Chromatogr. **56**, 140—42 (12)

DEUSSEN, P., see SCHULZ, G. V.

DE VRIES, A. J.
see BEAU
see LE PAGE

DE VRIES, G., see BRINKMAN

DE ZEEUW, R. A. (see also DHONT)
1968 Analyt. Chem. **40**, 2134—43 (21)

DHONT, J. H.
1970 and VINKENBORG, C.; COMPAAN, H.; RITTER, J. F.; LABADIE, R. P.; VERWEIJ, A DE ZEEUW, R. A., J. Chromatogr. **47**, 376—81 (21)
1980 J. Chromatogr. **202**, 15—20 (21)

DICKEL, G., see CLUSIUS

DICKSON, F. E.
1971 and WIRKKALA, R. A.; DAVIS, B. E.: Combined Gel Permeation Chromatography — NMR Techniques in the Characterization of Petroleum Residuals. In [D 2], 579—91 (17)

DI CORCIA, A., see CICCIOLI

DIECKMAN, J. F., see SCHMIT

DIETZ, E.
1976 Makromol. Chemie **177**, 2113—37 (6, 18)

DIETZ, R.
see also ATKINSON
see also VAUGHAN
1979 and FRANCIS, M. A., Polymer [London] **20**, 450—454 (19)
1980 J. appl. Polymer Sci. **25**, 951—54 (19)

DIEZ-MASA, J. C., see COLIN

DILER, R. M., see BAIJAL

DIMARZIO, E. A.
1965 J. chem. Phys. **42**, 2101—06 (6)
1969 and GUTTMAN, C. M., J. Polymer Sci. B, Polymer Letters **7**, 267—72 (8)
1970 and GUTTMAN, C. M., Macromolecules **3**, 681—91 (8)
1971 and RUBIN, R. J., J. Chem. Phys. **55**, 4318—36 (16)

DING, H.-C.
1965 and PAN, H.-C.; YU, T.-K., Gaofenzi Tongxun **7**, 260—63 (19)

DI PASCALE, G., see SADA

DIRHEIMER, G., see FELTER

D'JAČKOVSKIJ, F. S., see POZNJAK

DOBBIN, Ch. J. B.
1980 and RUDIN, A.; TCHIR, M. F., J. appl. Polymer Sci. **25**, 2985—92 (8)

DOBKOWSKI, Z.
see also BRZEZIŃSKI
see also CZARNECKA
1981 and BRZEZIŃSKI, J., Europ. Polymer J. **17**, 537—40 (19)

DOBRY, A.
1956 Makromol. Chemie **18/19**, 317—21 (9)

DOBYČIN, D. P.
see also BRESLER
see also DEMENT'EVA
1962 and BURKAT, T. M.; KISELEVA, N. N., In: Synthetische Zeolithe, Nauka, Moskau, p. 75 (11)

DODGSON, K.
1978 and SYMPSON, D.; SEMLYEN, J. A., Polymer [London], **19**, 1285—89 (19)

DOLAN, J. W.
see also GANT
see also SNYDER
1979 and GANT, J. R.; SNYDER, L. R., J. Chromatogr. **165**, 31—58 (14)

DOI, M.
1975 J. chem. Soc. [London], Faraday Trans. **71**, 1720—29 (8)

DOLL, H., see SCHULZ, G. V.

DOMARD, A.
1979 and RINAUDO, M.; ROCHAS, C., J. Polymer Sci., A-2, Polymer Phys. **17**, 673—81 (16, 19)

DOMAREVA, N. M., see BUDTOV

DONDOS, A.
1974 and REMPP, P.; BENOIT, H., Makromol. Chemie **175**, 1659—63 (8, 19)

DONE, J. N.
1973 and KENNEDY, G. J.; KNOX, J. H.: In: Gas Chromatography 1972. Ed.: Perry. — Applied Science Publishers, London, p. 145 (15)
DONKAI, N.
see also INAGAKI
see also TANAKA
see also MIYAMOTO
1968 and KOTAKA, T., Kagaku kogaku (Japan) **71**, 1039 (20)
1972 and INAGAKI, H., J. Chromatogr. **71**, 473 to 80 (21)
1974 and MURAYAMA, N.; MIYAMOTO, T.; INAGAKI, H., Makromol. Chemie **175**, 187—95 (21)
1975 and MIYAMATO, T.; INAGAKI, H., Polymer J. **7**, 577—83 (21)
DÖRFFEL, J.
1976 and RÜTER, J.; HOLTRUP, W.; FEINAUER, R., Farbe u. Lack **82**, 796—800 (21)
DOERNER, W. A., see KIRKLAND
DOUSEK, F. P., see SMOLKOVA
DRAKE, L. C., see RITTER
DRAUGLIS, E., see HASSELL
DREHER, T. W.
1979 and HAWTHORNE, D. B.; GRANT, B. R., J. Chromatogr. **174**, 443—46 (19)
DROTT, E. E.
1970a and MENDELSON, R. A., J. Polymer Sci. A-2, Polymer Phys. **8**, 1361—71 (19)
1970b and MENDELSON, R. A., J. Polymer Sci. A-2, Polymer Phys. **8**, 1373—85 (19)
1971 and MENDELSON, R. A., Amer. chem Soc. Polymer Prepr. **12**, 277—81 (21)
1977a Use of Hexafluoro-2-propanol as a GPC solvent. In [D 22 I], 41—50 (19)
1977b Determination of Long-Chain Branching in Polymers by GPC: A Review, in [D 22 I], 161—66 (19)
DRUDE, P.
1889 Ann. Physik u. Chemie **36**, 865 (6)
1890 Ann. Physik u. Chemie **39**, 481 (6)
DUBIN, P. L.
1977 and KOONTZ, S.; WRIGHT, K. L., J. Polymer Sci. A-1, Polymer Chem. **15**, 2047—57 (16, 19)
1977 and MILLER, V. P., Amer. chem. Soc. Polymer Prepr. **18**/2, 243—48 (19)
DUBININ, M. M.
1961 Adsorption von Gasen und Dämpfen und die Struktur der Adsorbenzien. In [C 4], 107—31 (10)
DUBROVINA, L. W.
1964 and TIMOFEEVA, G. I.: KORŠAK, V. V.; PAVLOVA, S. A., Vysokomolekularnye Soedinenija **6**, 2011—14 (20)
DUDLEY, M. A.
1972 J. appl. Polymer Sci. **16**, 493—504 (21)
DUGONE, J., see KRAUS
DUMONT, B., see REVILLON
DUMONT, E., see EISENBEISS
DUNKY, M.
1981 and LEDERER, K.; ZIMMER, E.: Österr. Kunststoff-Z. **12**, 36—39 (19)
DUERKSEN, J. H.
1968a and HAMIELEC, A. E., J. Polymer Sci. C, Polymer Symposia **21**, 83—103 (16)
1968b and HAMIELEC, A. E., J. appl. Polymer Sci. **12**, 2225—55 (19)
1971 Comparison of Different Techniques of Correcting for Band Broadening in GPC. In [D 2], 81—102 (16)
DUŠEK, K., see SEIDL
DUVAL, M.
1970 Off. Nat. Etud. Rech. Aerosp. Note Techn. **171**, 43
DUVDEVANI, I. (see also BIESENBERGER)
1971 and BIESENBERGER, J. A.; TAN, M., J. Polymer Sci. B, Polymer Letters **9**, 419—34
DUVEAU, N., see PAILHES

EBEL, J. P., see FELTER
EBERT, K. H. (see also BASEDOW)
1962 and ERNST, E., Makromol. Chemie **56**, 88—103 (20)
ECHARRI, J.
1979 and IRUIN, J. J.; GUZMAN, G. M.; ANSORENA, J., Makromol. Chem. **180**, 2749—51 (19)
ECKHARDT, G.
1980 and BRAUER, E.; WÜNSCHE, P.; WIEGLEB, H.; VOIGT, D., Acta Polymerica **31**, 382—87 (19)
EDE, P. S.
1971 J. Chromatogr. Sci. **9**, 275—80 (19)
EDERER, H., see BASEDOW
EDWARDS, G. D.
1968 and NG, Q. Y., J. Polymer Sci. C, Polymer Symposia **21**, 105—17 (19)
EGGERS, E. A.
1971 and HUMPHREY, J. S. J., J. Chromatogr. **55**, 33—44 (19)
EGLINTON, W. J., see FOX
EGONNIK, V. N., see BELENKIJ
EIFERT, W., see BRAUER
EIRICH, F. R.
see FRISCH
see KORAL
see ROWLAND
see SIMHA
EISENBEISS, F.
1976 Kontakte (E. MERCK, Darmstadt) **1**, 19—28; **2**, 17—29 (14)

1978 and DUMONT, E.; HENKE, H., Angew. makromol. Chemie **71**, 67—89 (18, 19)

EISENBERG, H., see CARROLL

EJZNER, Ju. E. (see also PTICYN)
1961 Vysokomolekularnye Soedinenija **3**, 748 to 57 (19)

EK, L.
1968 Process Biochem. **3**, 25 (17)

EL-AWADY, N. I.
1979 and GHANEM, N. A.; PEDERSEN, W. B.; SINGER, K., Europ. Polymer J. **15**, 1017—24 (19)

ELGERT, K.-F.
1977 and WOHLSCHIESS, R., Angew. makromol. Chemie **57**, 87—97 (19)

ELLERSTEIN, S.
1961 and ULLMAN, R., J. Polymer Sci. **55**, 123—35 (6, 18)

ELLIOTT, J. H., see FRANCIS

ELLIS, E. J., see SALAMONE

ELLISON, F. J., see BARKER

ELLISON, J. F., see BARKER

ELTEKOV, Ju. A.
see also ŽDANOV
see also BOGACHEVA
see also KISELEV
1973 and KISELEV, A. V.; CHOCHLOVA, T. D.; NIKITIN, Ju. S., Chromatographia **6**, 187 to 89 (11)
1974 and CHOPINA, V. V.; KISELEV, A. V.; KOVALEVA, N. V., J. Colloid & Interface Sci. **47**, 795 (6, 18)

EMERY, A. H.
1967 Thermal Diffusion. In [B 3], 181—90 (13)

EMMETT, P. H., see BRUNAUER

ENDO, R.
1961a KOBUNSHI KAGAKU (Japan) **18**, S. 148—52 (20) (ref. Makromol. Chemie **47**, 247)
1961b KOBUNSHI KAGAKU (Japan) **18**, 477—81 (20)

ENDOH, H., see HATTORI

ENGELHARDT, H.
see also GUIOCHON
see also KARCH
1975 Z. Analyt. Chem. **277**, 267—74 (14)
1977 and MATHES, D., J. Chromatogr. **142**, 311—20 (11, 19)
1979 and MATHES, D., J. Chromatogr. **185**, 305—19 (8, 11, 19)

ENGLERT, A.
1970 and TOMPA, H., Polymer [London] **11**, 507—32 (9)

ENOMOTO, J., see SHIONO

ENTELIS, S. G., see KUZAEV

EON, C.
see COLIN
see KARGER
see MARTIN

EON, C. H.
1978 J. Chromatogr. **149**, 29—42 (17)

EPPERT, G., see PEKER

EPTON, R. (see also BROUGH)
1974 and HOLDING, S. R.; MCLAREN, J. V., Polymer [London] **15**, 466—67 (12)
1974b and HOLLOWAY, C.; MCLAREN, J. V., J. Chromatogr. **90**, 249—58 (12)
1974c and HOLLOWAY, C.; MCLAREN, J. V., J. appl. Polymer Sci. **18**, 179—92 (12)
1975 and HOLDING, S. R.; MCLAREN, J. V., J. Chromatogr. **110**, 327—34 (12)
1976 and HOLLOWAY, C.; MCLAREN, J. V., J. Chromatogr. **117**, 245—55 (19)

ERNST, E., see EBERT

ERNST, O., see REHAGE

ESAKI, H., see TERAMACHI

ESTUPINAN, L., see PATAT

ETTRE, L. S.
1981 J. Chromatogr. **220**, 29—63 (3, 8)

EVANS, D. L., see PACCO

EVANS, M. B., see LITTLE

EYRING, H., see GIDDINGS

FABBRI, P., see COPPOLA

FAGAN, E. A., see MURPHY

FAGUE, G. R., see COLL

FALLICK, G. L.
1979 and RAUSCH, C. W., American Lab. **11**, 89—90 (15)

FANG, P., see WILSON

FARHAT, K., see ADAMS

FARKAS, D. F., see HEFTMANN

FATOU, J. M., see GUZMÁN

FAVRETTO, L. (see also CALZOLARI)
1970 and PERTOLDI-MARLETTA, G.; GABRIELLI, F. L., J. Chromatogr. **50**, 304—10 (21)

FEDYANTSEVA, A. A., see LOSIKOV

FEHER, I., see SZEPESY

FEINAUER, R., see DÖRFFEL

FEKETE, J., see SLAATS

FELDMAN, D., see THEIL

FELTER, R. E.
1970 and RAY, L. N., J. Colloid & Interface Sci. **32**, 349—60 (6, 18, 19)
1971 J. Polymer Sci. C, Polymer Symposia **34**, 227—41 (6, 18)

FELTER, S.
1968 and DIRHEIMER, G.; EBEL, J. P., J. Chromatogr. **35**, 207—10 (19)

FELTL, L.
1974 and LUTOVSKÝ, P.; SOSNOVÁ, L.; SMOLKOVÁ, E., J. Chromatogr. **91**, 321—32 (11)

FENDLER, H. G.
1955 and ROHLEDER, H.; STUART, H. A., Makromol. Chemie **18**, 383—96 (6, 18)
FERAUD, J. P., see GAREIL
FERGUSON, R. C.
1978 and STOKLOSA, H. J.; YAU, W. W.; HOEHN, H. H., J. appl. Polymer Sci. **34**, 119—38 (19)
FERLAUTO, E. C.
1971 J. Chromatogr. **55**, 45—54 (19)
FERRIER, J.
1967 J. Polymer Sci. C, Polymer Symposia **16**, 1937—45 (9, 20)
FETTERS, K. L., see SLAGOWSKI
FETTERS, L. J., see also AMBLER
1974 and MORTON, M., Macromolecules **7**, 552—59 (19)
1976 J. appl. Polymer Sci. **20**, 3437—42 (19)
1979 Macromolecules **12**, 344—46 (19)
FIGINI, R. V.
see also ANDREETTA
see also SCHULZ, G. V.
1979 Polymer Bull. **1**, 619—23 (16)
FIGUERUELO, J. E., see also CAMPOS
1979 and SORIA, V.; CAMPOS, A., Makromol. Chemie **180**, 1069—72 (19)
1980 and SORIA, V.; CAMPOS, A., J. Liq. Chromatogr. **3**, 367—80 (16, 19)
FILISKO, F. E., see MARON
FINK-JENSEN, P., see CHRISTENSEN
FINLAYSON, J. S.
1963 and WOSESSON, M. W., Biochemistry **2**, 42 (18)
FINLEY, J. W.
1978 and KROCHTA, J. M.; HEFTMANN, E., J. Chromatogr. **157**, 435—39 (17)
FISHER, M., see MURPHY
FISHER, N. E., see MOTTLAU
FLEER, G. J.
1972 and KOOPAL, L. K.; LYKLEMA, J., Kolloid-Z. u. Z. Polym. **250**, 689—702 (18)
1976 and SMITH, L. E.: The Adsorption Mechanism of Polyvinyl Alcohol on Silica, as studied by Ellipsometric Measurements of Oxidized Silicon. Prepr. C 27, Conference "Phases and Interfaces in Macromolecular Systems", Prag, 12.—16. 7. 76 (6, 18)
FLEMING, S. W., see YAN
FLODIN, P., see PORATH
FLORIAN, S., see JURANIČOVA
FLORY, P. J.
1936 J. Amer. chem. Soc. **58**, 1877—85 (4)
FLOWERS, D. L.
1964 and HEWETT, W. A.; MULLINEAUX, R. D., J. Polymer Sci. A, Gen. Papers **2**, 2305—26 (9, 20)
FOKSOVÁ, A., see MANDIK
FOLTYN, J., see MANDIK
FONTANA, B. J.
1958 and THOMAS, J. R., J. phys. Chem. **62**, 54 (7)
1961 and THOMAS, J. R., J. phys. Chem. **65**, 480 (6, 18)
FORGET, J.-L.
see BOOTH
see PRICE
FORSMAN, W. C.
see also HAMORI
see also PODDAR
1963 and HUGHES, R. E., J. chem. Phys. **38**, 2130—55 (6)
FORSS, K. G.
1973 and STENLUND, B. G., J. Polymer Sci. C, Polymer Symposia **42**, 951—63 (19)
FOSSHAG, E., see BASEDOW
FOX, J. B. (see also NICHOLAS)
1969a and CALHOUN, R. C.; EGLINTON, W. J., J. Chromatogr. **43**, 48—54 (17)
1969b J. Chromatogr. **43**, 55—60 (17)
FOX, K. K.
1974 and ROBB, I. D.; SMITH, R., J. chem. Soc. **70**, 1186 (6)
FRACHE, R.
1973 and DADONE, A., Chromatographia **6**, 274—76 (19)
FRANCIS, M. A.
see ATKINSON
see DIETZ
see VAUGHAN
FRANCIS, P. S.
1958 and COOKE, R. C.; ELLIOTT, J. H., J. Polymer Sci. **31**, 453 (9)
FRANCOIS, J.
1978 and JACOB, M.; GRUBIŠIĆ-GALLOT, Z.; BENOIT, H., J. appl. Polymer Sci. **22**, 1159 to 62 (19)
FRANCUSKIEWICZ, F., see GLÖCKNER
FRANK, F. C.
1968 and WARD, I. M.; WILLIAMS, T., J. Polymer Sci. A-2, Polymer Phys. **6**, 1357—69 (8, 16)
FRANKENBERG, C. A. v., see HUGHES
FREEMAN, D. H.
1977 and POINESCU, I. C., Analytical Chem. **49**, 1183—88 (10)
1977 and KILLION, D., J. Polymer Sci. A-2, Polymer Phys. **15**, 2047—49 (19)
FREI, R. W., see BETHKE
FRENCH, D. M., see ADAMS
FRENKEL, S. Ja.
see BELENKIJ
see NEFEDOV
see TENNIKOV

FREY, H.-P.
1976 and ACKERMANN, G., Paper read on 5th Symposium "Chromatography in clinical Biochemistry", Leipzig, 6/7 May 1976 (21)
1980 and ACKERMANN, G., J. Chromatogr. **198**, 357—62 (21)
FRIEDEL, H., see KLEIN
FRISCH, H. L., see also SIMHA
see also YEH
1953 and SIMHA, R.; EIRICH, F. R., J. chem. Phys. **21**, 365 (6)
1959 and HELLMAN, M. Y.; LUNDBERG, J. L., J. Polymer Sci. **38**, 441—49 (6, 18)
FRITZSCHE, P.
1967 Faserforsch. u. Textiltechnik **18**, 21—24 (17, 19)
FROLEN, S., see ROSS
FRÖLICH, P.
1979 and SCHWACHULA, G.; SARODNIK, E., Angew. makromol. Chemie **79**, 55—65 (10)
FROLOVA, G. S., see PROKOVSKAJA
FUJIMOTO, T.,
see KATO, TADAYA
see KATO, TOSHIO
FUJISHIGE, S.
1975 and OHGURI, N., Makromol. Chemie **176**, 233 (19)
FUJITA, H., see HOMMA
FUJITA, T., see HANSCH
FUKANO, K.
1978 and KOMIYA, K.; SASAKI, H.; HASHIMOTO, T., J. Chromatogr. **166**, 47—54 (11, 19)
FUKOTOMI, M., see KATO, Y.
FUKUDA, M., see KATO, Y.
FUNKE, W., see HILT
FURUKAWA, J.
1966 and AKUTSU, S.; SEAGUSA, T., Makromol. Chemie **94**, 68—73 (18)
FURUKAWA, K., see KATO, Y.
FURUSAWA, K.
see also NORDE
see also ONDA
1975 and NAKANISHI, H.; KOTERA, A.: Polymer Adsorption into Porous Glass form Theta Solvent. In [C 2], 221—28 (6, 18)
FÜZES, L., see SAMAY

GABBOTT, N. P., see DAWKINS
GABRIELLI, F. L., see FAVRETTO
GAESSER, G. L., see SLAGOWSKI
GAHNER, E.
1967 Experimentelle Untersuchungen über die Trennleistung bei der Baker-Williams-Fraktionierung. Theses, TU Dresden (9)
GALANOS, D. S.
1964 and KAPOULAS, V. M., J. Chromatogr. **13**, 128—38 (21)
GALIC, I. P., see CYPEROVIC
GALIMORE, J. M., see HIATT
GALLEI, E., see UNGER
GALLOT, Z.
see GUENET
see MEUNIER
GAL'PERIN, V. M., see BUDTOV
GANKINA, E. S.
see BELENKIJ
see JANČA
see KEVER
see NESTEROV
see SKVORZOV
GÄNSHIRT, H.
1967 Quantitative Auswertung von Dünnschicht-Chromatogrammen. In [E 1], 133 to 55 (21)
GANT, J. R.
see also DOLAN
see also KARGER
see also SNYDER
1979 and DOLAN, J. W.; SNYDER, L. R., J. Chromatogr. **185**, 153—77 (15)
GARCIA-RUBIO, L. H.
1979 and MACGREGOR, J. F.; HAMIELEC, A. E., ACS Analytical Chemistry, 178th National Meeting, September 10—14, Washington, Paper No. 78 (19)
GAREIL, P.
1980 and PERSONNAZ, L.; FERAUD, J. P.; CAUDE, M., J. Chromatogr. **192**, 53—74 (17)
GARELLE, R., see BOSCHETTI
GAYLOR, V. F.
1975 and JAMES, H. L., Preprints, Pittsburgh Conference on Analytical Chemistry, Cleveland, Ohio, March (19)
GEBAUER, R. C., see SLAGOWSKI
GEHRKE, K., see POPOV
GEISS, F.
see also BAYER
1965 and SCHLITT, H.; KLOSE, A., Z. analyt. Chemie **213**, 331 (10)
1968 J. Chromatogr. **33**, 9—24 (21)
1968a and SCHLITT, H., Chromatographia **1**, 392 to 402 (21)
1968b and SCHLITT, H., J. Chromatogr. **33**, 208 (21)
1969 and SANDRONI, S.; SCHLITT, H., J. Chromatogr. **44**, 290—99 (14, 21)
1972 and SCHLITT, H.; In [E 5], 127 (10)
GELOTTE, B.
1960 J. Chromatogr. **3**, 330—42 (12)
GEMMILL, F. Q., see KROL
GEORGII, I., see BOOTH
GESSLER, E. M., see BINFORD
GEYER, R., see PEKER
GEYER, U., see BRAUER

GHAEMI, Y.
1979 and WALL, R. A., J. Chromatogr. **174**, 51 (11)
GHANEM, N. A., see EL-AWADY
GIAMMARSE, A. T.
1968 and ALLIET, D. F.; PACCO, J. M., J. Polymer Sci. B, Polymer Letters **6**, 499 (17)
GIDDINGS, J. C.
see also CALDWELL
see also GRUSHKA
see also HOVINGH
see also THOMPSON
1955 and EYRING, H., J. phys. Chem. **59**, 419 (3, 8)
1957 J. chem. Phys. **26**, 169—73 (3, 8)
1959 J. chem. Phys. **31**, 1462—67 (15)
1961 Analyt. Chem. **33**, 962—63 (15)
1966 Separat. Sci. **1**, 123—25 (13)
1966 and MALLIK, K. L., Analyt. Chem. **38**, 997—1000 (15, 19)
1967a Analyt. Chem. **39**, 1027—28 (13, 16)
1967b J. Gas Chromatogr. 413—19 (14)
1969 and MYERS, M. N.; KING, J. W., J. Chromatogr. Sci. **7**, 276 (15)
1973 J. Chem. Educ. **50**, 667—69 (13)
1974 and YANG, F. J.: MYERS, M. N., Analyt. Chem. **46**, 1917—24 (13)
1975 and YOON, Y. H.; MYERS, M. N., Analyt. Chem. **47**, 126—31 (13)
1976 and YANG, F. J.; MYERS, M. N., Analyt. Chem. **48**, 1126—32 (13)
1977a and YANG, F. J.; MYERS, M. N., Science [Washington] (13)
1977b and BOWMAN, L. M.; MYERS, M. N., Macromolecules **10**, 443—49 (15)
1978a and MYERS, M. N.; MOELLMER, J. F., J. Chromatogr. **149**, 501—17 (13)
1978b and MARTIN, M.; MYERS, M. N., J. Chromatogr. **158**, 419—35 (13)
1979 Pure & appl. Chem. **51**, 1459—71 (13)
1979a and MYERS, M. N.; CALDWELL, K. D.; PAV, J. W., J. Chromatogr. **185**, 261—71 (10, 13)
1979b and MYERS, M. N.; JANČA, J., J. Chromatogr. **186**, 37—44 (13)
GIERLINGER, W., see TREIBER
GIERRA, A., see LEVI
GIETZ, W.
1975 and BOULTON, A. A.; POLLAK, V., J. Chromatogr. **107**, 81—89 (21)
GILBERT, J.
1980 and SHEPHERD, M. J.; WALLWORK, M. A., J. Chromatogr. **193**, 235—42 (19)
GILDING, D. K. (see also COLL)
1981 and REED, A. M.; ASKILL, I. N., Polymer [London], **22**, 505—12 (19)
GILON, C., see GRUSHKA
GILPIN, R. K.
1976 and SISCO, W. R., J. Chromatogr. **124**, 257—64 (21)
GIMPEL, M. G., see UNGER
GINNARD, C. R., see YAU
GIROT, P., see MONSIGNY
GLADKOVSKIJ, G. A., see VACHTINA
GLAJCH, J. L.
1980 and KIRKLAND, J. J.; SQUIRE, K. M.; MINOR, J. M., J. Chromatogr. **199**, 57—79 (15)
GLASSER, W. G., see MOROHOSHI
GLÖCKNER, G.
1964 Über das Verhalten von Polycarbonat auf Basis 4,4'-Dioxydiphenyl-2,2-propan in Lösung. — Habilitation Thesis, TU Dresden (4, 9)
1965a Plaste u. Kautschuk **12**, 96—102 (9, 11, 20)
1965b Z. phys. Chemie **229**, 98—109 (5, 9)
1966 Plaste u. Kautschuk **13**, 340—41 (9)
1966 and MÜLLER, K.-D., Plaste u. Kautschuk **13**, 396—97 (20)
1968 and MÜLLER, K.-D., Plaste u. Kautschuk **15**, 812—16 (20)
1970 and KUHNHARDT, R., Faserforsch. u. Textiltechnik **21**, 523—28 (20)
1971 and FRANCUSKIEWICZ, F.; MÜLLER, K.-D., Plaste u. Kautschuk **18**, 654—56 (20)
1975 and KAUFMANN, W.; ABDELLATIF, E.; RIEDEL, S., Faserforsch. u. Textiltechnik **26**, 606—13 (5)
1976a and KAHLE, D., Plaste u. Kautschuk **23**, 338—45 (21)
1976b and KAHLE, D., Plaste u. Kautschuk **23**, 577—80 (21)
1977 and KAUFMANN, W., Faserforsch. u. Textiltechnik **28**, 409—16 (9)
1980 and MEISSNER, R., Acta Polymerica **31**, 191—92 (21)
1980a J. Polymer Sci. C, Polymer Symposia **68**, S. 179—83 (7)
1980b J. Chromatogr. **191**, 279—86 (16)
GLOOR, R., see JOHNSON
GNAUCK, R., see SCHULZ, G.
GOLAY, M. J. E.
1959 In: Gas Chromatography 1958. Ed.: D. H. Desty. Butterworths, London 1959, 36 (15)
GOLDFINGER, G. (see also AMBORSKI)
1945 Rubber Chem. & Technol. **18**, 286 (18)
1945 Rubber Chem. & Technol. **19**, 614 (18)
GOLDSBROUGH, J.
see GORDON
see DERHAM
GOLKIEWICZ, W., see SOCZEWIŃSKI

GOLUB, M. A.
1953 J. Polymer Sci. **11**, 583—99 (18)
GOODALL, A. R.
1975 and WILKINSON, M. Č.; HEARN, J., J. Colloid & Interface Sci. **53**, 327—31 (19)
GOODBAN, A. E.
1957 and OWERS, H. S., J. Polymer Sci. **23**, 825—29 (18)
GOODELL, H., see TANAKA
GORDIJENKO, A.
1953 Faserforsch. u. Textiltechnik **4**, 499—507 (9)
GORDON, M., (see also DERHAM)
1973 and GOLDSBROUGH, J.; READY, B. W.; DERHAM, K. W., Pulse induced critical scattering. In [B 12], 45—51 (5)
1977 and IRVINE, P.; KENNEDY, J. W., J. Polymer Sci. C, Polymer Symposia **61**, 199—220 (5)
GOTO, Y., see NAKANO
GOTTLIEB, M.
1960 J. phys. Chem. **64**, 427—32 (6, 18)
GÖTZ, H., see UNGER
GÖTZE, T., see UEBERREITER
GOURDENNE, A., (see also HO-DUC)
1972 and HO-DUC, N.; DAOUST, H., J. Chromatogr. **74**, 225—31 (19)
GOUW, T. H. (see also JENTOFT)
1972 and JENTOFT, R. E., J. Chromatogr. **68**, 303—23 (15)
GRACHEVA, L. I., see VACHTINA
GRAMAIN, Ph., see GRUBIŠIĆ-GALLOT
GRANATH, K. A.
1967 and KVIST, B. E., J. Chromatogr. **28**, 69—81 (19)
GRANT, B. R., see DREHER
GRANT, J. W., see KIRKLAND
GRANT, W. H.
see STROMBERG
see TUTAS
GRAESSLEY, W. W., see PARK
GRAY, R. A. C., see BATHER
GREEAR, G. R., see MOORE, L. D.
GREEN, J. H. S., see VAUGHAN
GREFF, R. J., see WAGNER
GREGOR, V., see ČOUPEK
GRESCHNER, G. S.
1979a IUPAC-Symposium, Mainz, 746—49 (9)
1979b Makromol. Chem. **180**, 2551—66 (9)
GRIECO, C., see CIAMPA
GROH, R.
1980 and HALÁSZ, I., J. Chromatogr. **199**, 23—34 (16)
GRUBIŠIĆ, Z. (see also BENOIT)
1967 and REMPP, P.; BENOIT, H., J. Polymer Sci. B, Polymer Letters **5**, 753—59 (8, 19)
1968 and BENOIT, H., C. R. hebd. Séances Acad. Sci. [Paris] **C266** 1275—77 (8)
GRUBIŠIĆ-GALLOT, Z.
see also FRANCOIS
see also LOVRIĆ
see also MENCER
1972 and PICOT, M.; GRAMAIN, Ph.; BENOIT, H., J. appl. Polymer Sci. **16**, 2731—45 (19)
1976 and MARAIS, L.; BENOIT, H., J. Polymer Sci. A-2, Polymer Phys. **14**, 959—61 (16,19)
GRUSHKA, E.
1974 and CALDWELL, K. D.; MYERS, M. N.; GIDDINGS, J. C.: In [A 5], Bd. 2, 127 (13)
1977 and KIKTA, E. J., Analyt. Chem. **49**, 1004A—1014A (11)
1980 and LESHEM, R.; GILON, C., The Reversed-Phase Separation of Amino Acid Enantiomers. Lecture 2 B 6 at the 13th International Symposium on Chromatography, Cannes, 30 June—4 July (11)
GUENET, J. M.
1977 and GALLOT, Z.; PICOT, C.; BENOIT, H., J. appl. Polymer Sci. **21**, 2181—90 (19)
GUILLEMIN, C. L.
1977 and THOMAS, J. P.; THIAULT, S.; BOUNINE, J. P., J. Chromatogr. **142**, 321—37 (15)
GUILLET, J. E. (see also SLONAKER)
1960 and COMBS, R. L.; SLONAKER, D. F.; COOVER, H. W., J. Polymer Sci. **47**, 307—20 (9, 20)
1962 and COMBS, R. L.; SLONAKER, D. F.; SUMMERS, J. T.; COOVER, H. W., Soc. Plastic Engin. Trans. **2**, 164—69 (20)
GUIOCHON, G.
see also BAYER
see also COLIN
see also MARTIN
see also SIOUFFI
1976 Optimization of Liquid-Chromatography Separation and Analysis. — Paper read on Meeting of Analyzers, Leipzig, 16—19 November (15)
1978 and SIOUFFI, A. M., J. Chromatogr. Sci. **16**, 470—81 (21)
1978 and SIOUFFI, A. M.; ENGELHARDT, H.; HALÁSZ, I., J. Chromatogr. Sci. **16**, 152—57 (21)
1979 J. Chromatogr. **185**, 3—26 (15)
GULIANA, R., see WILD
GUTFINGER, C., see LAUFER
GUTTMAN, C. M., see DIMARZIO
GUYOT, A.
see JACOBELLI
see REVILLON
GUZMAN, G. M., (see also ECHARRI)
1958 and FATOU, J. M., An. Real Soc. españ. Física Quím., Ser. B **54**, 609 (13)

HAAHTI, E., see MUKHERJEE
HADDON, W. F.
1974 and PORTER, R. S.; JOHNSON, J. F., J. appl. Polymer Sci. **8**, 1371—78 (20)
HADSELL, E. M., see SHULTZ
HAGDAHL, L. (see also SVENSSON)
1952 and WILLIAMS, R. J. P.; TISELIUS, A., Ark. Kemi **4**, 193 (14)
HAGEL, L.
1978 J. Chromatogr. **160**, 59—71 (19)
HAGNAUER, G. L.
1978 and SETTON, I., J. Liq. Chromatogr. **1**, S. 55—73 (18)
1980 Polymer Composites **1**, 81—87 (18, 19)
HAIR, M. L.
1969 and HERTL, W., J. phys. Chem. **73**, 2372 to 78; 4269—76 (11)
HALÁSZ, I.
see also ASSHAUER
see also BECK
see also GROH
see also GUIOCHON
see also KARCH
see also NIKOLOV
1975 Z. Analyt. Chem. **277**, 257—66 (16)
1975 and MARTIN, K., Z. Elektrochemie, Ber. Bunsenges. phys. Chemie **79**, 731—32 (10)
1978 and MARTIN, K., Angew. Chemie **90**, 954 (10)
1979 J. Chromatogr. **173**, 229—47 (15)
1980 and VOGTEL, P., Angew. Chemie **92**, 25—29; (Internat. Ed. **19**, 24—28) (10)
HALÁSZ, J.
see also BAYER
see also RÖSSLER
1966 J. Gas Chromatogr., 8—11 (16)
1972 and NAEFE, M., Analyt. Chem. **44**, 76—84 (15)
HALL, R. R.
1959 In: Techniques of Polymer Characterization. Ed.: P. W. Allen. Butterworths Science Publishers, London, Chapt. 2 (11)
HALLER, W.
1965 J. chem. Phys. **42**, 686—93 (11)
1965b Nature [London] 206, 693—97 (11)
1968 J. Chromatogr. **32**, 676—84 (8)
1977 and BASEDOW, A. M.; KÖNIG, B., J. Chromatogr. **132**, 387—97 (11)
1977 Macromolecules **10**, 83—86 (8, 11, 19)
HALLING, M. S., see INGELMAN
HALPAAP, H.
1968 and KLATYK, K., J. Chromatogr. **33**, 80—89 (21)
1968 and REICH, W., J. Chromatogr. **33**, 70—79 (10, 11)
1977 and KREBS, K.-F., J. Chromatogr. **142**, 823—53 (14, 21)
1977 and RIPPHAHN, R., Chromatographia **10**, 613—23 (21)
HAMA, T.
1972 and YAMAGUCHI, K.; SUZUKI, T., Makromol. Chemie **155**, 283—98 (19)
HAMANN, K., see HILT
HAMASHIMA, M., see HATTORI
HAMBLETON, F. H. (see also ARMISTEAD)
1966 and HOCKEY, J. A., Trans. Faraday Soc. **62**, 1694—1701 (11)
HAMIELEC, A. E.
see also ABDEL-ALIM
see also BALKE
see also DUERKSEN
see also GARCIA-RUBIO
see also ISHIGE
see also OMORODION
see also SINGH
1969 and RAY, W. H., J. appl. Polymer Sci. **13**, 1319—21 (16)
1970 J. appl. Polymer Sci. **14**, 1519 (16)
1978 and OUANO, A. C., J. Liq. Chromatogr. **1**, 111—20 (8)
1979 and OMORODION, S. N. E., Recent Advances in Size Exclusion Chromatography. ACS Symposium Proceedings, Washington (16)
1981 Characterization of complex polymer systems by SEC: Homopolymers with long chain branching and copolymers with composition drift. Paper read on IUPAC-Symposium "Advances in Polymer Characterization", Durham, 13—17 July (19)
HAMMOND, J. M.
1971 and HOOPER, J. F.; ROBERTSON, W. G. P., J. Polymer Sci. A-1, Polymer Chem. **9**, 281—94
HAMORI, E.
1971 and FORSMAN, W. C.; HUGHES, R. E., Macromolecules **4**, 193 (18)
HANCOCK, W. S.
1981 and CAPRA, J. D.; BRADLEY, W. A.; SPARROW, J. T., J. Chromatogr. **206**, 59—70 (15)
HANN, N. D.
1977 J. Polymer Sci. A-1, Polymer Chem. **6**, 1331—39 (16, 19)
HANSCH, C.
1964 and FUJITA, T., J. Amer. chem. Soc. **86**, 1616 (7)
HANSEN, C. M.
1964 and SATHER, G. A., J. appl. Polymer Sci. **8**, 2479—84 (18, 20)
1965 Off. Digest Federat. Soc. Paint Technol. **37**, No. 480 (5)
1967 Off. Digest Federat. Soc. Paint Technol. **39**, 104—17, 505—10, 511—14 (5)

1969 Ind. & Engng. Chem. Prod. Res. & Devel. **8**, 2—11 (5)

HANSEN, S. H.
1981 J. Chromatogr. **209**, 203—10 (11)

HANSSON, L.
1981 and TROJER, L., J. Chromatogr. **207**, 1—11 (11)

HARA, K.
1970 and IMOTO, T., KOLLOID.-Z. u. Z. Polym. **237**, 297—302 (6, 18)

HARMON, D. J.
see also BLY
see also LATTIMER
1965 J. Polymer Sci. C, Polymer Symposia **8**, 243—51 (19)
1971 Gel Permeation Chromatography, Column Packings: Types and Uses. In [D 2], 13—23 (19)
1978 The Role of Polymer Structure in Fractionation by GPC. In [D 11/1], 122—45 (19)

HARREN, H., see HUBER

HARTKOPF, A., see KARGER

HARTLEY, R. W.
1962 and PETERSON, E. A.; SOBER, H. A., Biochemistry **1**, 60 (18)

HARTMANN, W. (see also KLESPER)
1977 and KLESPER, E., J. Polymer Sci. B, Polymer Letters **15**, 713—19 (8, 9)

HASEGAWA, A., see TERAMACHI

HASHIMOTO, T.
see also FUKANO
see also KATO, TOSHIO
see also KATO, Y.
1978 and SASAKI, H.; AIURA, M.; KATO, Y., J. Polymer Sci. A-2, Polymer Phys. **16**, 1789 to 1800 (12, 17, 19)

HASSELL, J. A.
1979 and SLIEMERS, F. A.; DRAUGLIS, E.; NANCE, G. P., J. Polymer Sci. B, Polymer Letters **17**, 111—13 (16, 19)

HASTINGS, C. R., see AUE

HATANO, H., see ROKUSHIKA

HATT, B. W., see BARKER

HATTORI, S.
1970 and HAMASHIMA, M., Kobunshi Kagaku (Japan) **27**, 775—84 (19)
1978 and ENDOH, H.; NAKAHARA, H.; KAMATA, T.; HAMASHIMA, M., Polymer J. **10**, 173—80 (17, 19)

HÄUPKE, K.
1968 and HOFFMANN, H., Z. Chemie **8**, 463—64 (12)
1974 and PIENTKA, V., J. Chromatogr. **102**, 117—21 (12)
1980 private communication, September (12)

HAUTTECOEUR, B., see COULOMBE

HAVLIK, A. J., see KAYE

HAWKES, S. J., see CLUFF

HAWTHORNE, D. B., see DREHER

HAYWOOD, P. A.
1980 and MUNRO, G., Preparative scale liquid chromatography. In [A 22], 33—68 (17)

HAZELL, J. E. (see also BATA)
1968 and PRINCE, L. A.; STAPELFELDT, H. E., J. Polymer Sci. C, Polymer Symposia **21**, 43—65 (8, 17, 19)

HEARN, J., see GOODALL

HECKLES, J. S., see QUINN

HEFTMANN, E. (see also FINLEY)
1972 and KROCHTA, J. M.; FARKAS, D. F.; SCHWIMMER, S., J. Chromatogr. **66**, 365—69 (17)

HEITZ, W.
see also ČOUPEK
see also SEIDL
see also STIX
see also WALZ
1967 and ČOUPEK, J., Makromol. Chemie **105**, 280—84 (19)
1967 and KERN, W., Angew. makromol. Chemie **1**, 150—73 (16, 19)
1967 and PLATT, K.-L.; ULLNER, H.; WINAU, H., Makromol. Chemie **102**, 63 (16)
1968 and ULLNER, H., Makromol. Chemie **120**, 58—67 (17)
1969 and BÖMER, B.; ULLNER, H., Makromol. Chemie **121**, 102—16 (19)
1969 and PLATT, K.-L., Makromol. Chemie **127**, 113—40 (12, 19)
1970 J. Chromatogr. **53**, 37—49 (12, 19)
1970 and KRAFFCZYK, F.; PFITZNER, K.; RANDAU, D., Chimia **24**, 126—27 (12)
1970 and WINAU, H., Makromol. Chemie **131**, 75—88 (12, 19)
1971 and KLATYK, K.; KRAFFCZYK, F.; PFITZNER, K.; RANDAU, D., J. Chromatogr. Sci. **9**, 521—25 (19, 21)
1973a Z. Elektrochemie, Ber. Bunsenges. phys. Chemie **77**, 210—17 (8)
1973b J. Chromatogr. **83**, 223—31 (17, 19)
1975 Z. analyt. Chemie **277**, 323—33 (8, 16, 19)
1977 and HENKE, G.; PLATT, K.-L.; BIER, P., Angew. makromol. Chem. **63**, 203—14 (12)
1977a Polymeric Reagents Polymer Design, Scope, and Limitations. In Adv. in Polymer Sci. **23**, 1—23 (12)
1977b Kontakte (MERCK) **2/77**, 29—36 (12)
1979 Angew. makromol. Chemie **76/77**, 273—97 (12, 19)
1981 Characterization of Oligomers by GPC. In [D 22 III], 137—56 (19)

HELLER, W.
1951 and TANAKA, W., Phys. Rev. **82**, 302 (18)
1966 Pure & appl. Chem. **12**, 249—74 (6)
HELLMAN, M. Y.
see FRISCH
see OTOCKA
see SALOVEY
HELLWIG, J.
1973 and SCHÖLLNER, R., Plaste u. Kautschuk **20**, 216—21 (19)
HELME, J. P., see BELLENS
HELMSTEDT, M., see BRAUER
HEMETSBERGER, H.
1977 and KELLERMANN, M.; RICKEN, H., Chromatographia **10**, 726—30 (11)
HENDERSON, J. F.
1967 and HULME, J. M., J. appl. Polymer Sci. **11**, 2349—60 (17, 20)
HENDRICKSON, J. G.
1966 and MOORE, J. C., J. Polymer Sci. A-1, Polymer Chem. **4**, 167—88 (19)
1968 J. Polymer Sci. A-2, Polymer Phys. **6**, 1903—17 (16)
1968a Analyt. Chem. **40**, 49—53 (19)
1968b J. Chromatogr. **32**, 543—58 (19)
HENKE, G., see HEITZ
HENKE, H., see EISENBEISS
HENKENS, W. C. N., see VAN DIJK
HENNION, M. C.
1981 and PICARD, C.; COMBELLAS, C.; CAUDE, M.; ROSSET, R., J. Chromatogr. **210**, 211 to 228 (7)
HENRY, M. P.
1959 J. Polymer Sci. **36**, 3—19 (11)
HENRY, R. A. (see also SCHMIT)
1973 and SCHMIT, J. A.; WILLIAMS, R. C., J. Chromatogr. Sci. **11**, 358 (11)
1974 and BYRNE, S. H.; HUDSON, D. R., J. Chromatogr. Sci. **12**, 197—99 (17)
HERD, M. J.
1971 and HOPKINS, A. J.; HOWARD, G. J., J. Polymer Sci. C, Polymer Symposia **34**, 211—26 (6, 18)
HERMANS, J. J. (see also TUIJNMAN)
1968 J. Polymer Sci. A-2, Polymer Phys. **6**, 1217—26 (8, 15)
HERNÁNDEZ, M. L., see ALVARIÑO
HEROLD, J.
1980 and MEYERHOFF, G., Makromol. Chemie **181**, 2625—36 (19)
HERTL, W., see HAIR
HERZEL, U.
1978 Internat. Labor., May—June, 73 (21)
HES, J., see MARTIN
HESS, M.
1966 and KRATZ, R. F., J. Polymer Sci. A-2, Polymer Phys. **4**, 731—43 (16)
HESSELINK, F. T.
1975 J. Colloid & Interface Sci. **50**, 606—08 (6)
HEUBLEIN, G.
1980 and MEISSNER, H.; MIKOLAJCZAK, E.; WONDRACZEK, R., Acta Polymerica **31**, 145—46 (19)
HEWETT, W. A., see FLOWERS
HEYER, W.
1977 and JANOWSKI, F.; WOLF, F., Z. Chemie **17**, 212—15 (11)
1980 Chem. Technik **32**, 86—90 (11)
1981 Private communication (11)
HEZEL, U. B.
1977 Potential and experience in quantitative "High performance thin-layer chromatography". In [E 7], 147—88 (21)
HIATT, C. W.
1971 and SHELOKOV, A.; ROSENTHAL, E. J.; GALIMORE, J. M., J. Chromatogr. **56**, 362—64 (11)
HIBY, J. W.
1968 J. Inst. Chem. Eng. [London], June, 317 (15)
HICKS, G. E., see KROL
HIGASHIMURA, T.
1975 and KISHIRO, O.; MATZUZAKI, K.; URYU, T., J. Polymer Sci. A-1, Polymer Chem. **13**, 1393—99 (21)
HILD, G.
1980 and REMPP, P., Mechanism of Network Formation by Radical Copolymerization. Paper read on CL 9, 7th Discussion Conference: Polymer Networks. Karlovy Vary, 15—19 September (12)
HILLMAN, D. E., see HOARE
HILT, A.
1966 and FUNKE, W.; HAMANN, R., Dt. Farben-Z. **20**, 567—72 (18, 21)
HINDRIKS, H., see VAN WENDEL DE JOODE
HIRAI, Y., see ONDA
HIROSE, A., see ISHII
HJERTÉN, S.
1962 and MOSBACH, R., Analyt. Biochem. **3**, 109 (12)
1964 Biochim et biophysica Acta [Amsterdam] **79**, 393—98 (12)
HOARE, H. C.
1971 and HILLMAN, D. E., Brit. Polymer J. **3**, 259—62 (19)
HOBDEN, J. F.
1953 and JELLINEK, H. H. G., J. Polymer Sci. **11**, 365—78 (18)
HOCKEY, J. A., see HAMBLETON
HOCKEY, J. A., see ARMISTEAD

HO-DUC, N. (see also GOURDENNE)
1971 and DAOUST, H.; GOURDENNE, A., Amer. Chem. Soc. Polym. Prepr. **12**, 639—44 (19)
1973 and PRUD'HOMME, J., Macromolecules **6**, 472—74 (19)
HOFFMANN, H., see HÄUPKE
HOFFMANN, M.
1977 and URBAN, H., Makromol. Chem. **178**, 2683—96 (19)
HOFSTEE, B. H. J.
1979 Pure & appl. Chem. **51**, 1537—48 (12)
HOEGY, H. L. W., see RUDIN
HOEHN, H. H., see FERGUSON
HOLDING, S. R.
see BARKER
see EPTON
HOLLOWAY, C., see EPTON
HOLMES, L. G., see SCHNEIDER
HOLMSTRÖM, A. (see also ANDERSSON)
1970 and SÖRVIK, E., J. Chromatogr. **53**, 95—108 (19)
HOLT, C.
1978 and MACKIE, W.; SELLEN, D. B., Polymer **19**, 1421—26 (19)
HOLTRUP, W.
see DÖRFFEL
see RING
HOMMA, T. (see also TAGATA)
1963 and KAWAHARA, K.; FUJITA, H.; UEDA, M., Makromol. Chemie **67**, 132—48 (20)
HOOGEVEEN, J. T., see BUNGENBERG DE JONG
HOOPER, J. F., see HAMMOND
HOPKINS, A. J., see HERD
HORIE, K.
1980 PMM-Discussions-Conference "Polymer Networks", Karlovy Vary, 15—19 Sept. (12)
HORII, F.
1974 and IKADA, Y., J. Polymer Sci. B, Polymer Letters **12**, 27—29 (18)
1975 and IKADA, Y.; SAKURADA, J., J. Polymer Sci. A-1, Polymer Chem. **13**, 755—60 (21)
HORIUCHI, Y., see UENO
HORVATH, C. G.
1967 and PREISS, B. A.; LIPSKY, S. R., Analyt. Chem. **39**, 1422—28 (15)
HORVÁTH, Cs.
see also BIJ
see also CHEN
see also MELANDER
see also NAHUM
1976 and LIN, H.-J., J. Chromatogr. **126**, 401—20 (15)
1976 and MELANDER, W.; MOLNAR, I., J. Chromatogr. **125**, 129—56 (7)
1977 and MELANDER, W., J. Chromatogr. Sci. **15**, 393—404 (11)
1978 and MELANDER, W., International Laboratory, Nov.—Dec., 11—35 (7)
1978 and LIN, H.-J., J. Chromatogr. **149**, 43—70 (15)
HOSHINO, S., see OGAWA
HOSTETTMANN, K.
1979a and HOSTETTMANN-KALDAS, M.; NAKANISHI, K., J. Chromatogr. **170**, 355—61 (9)
1979b and HOSTETTMANN-KALDAS, M.; STICHER, O., J. Chromatogr. **186**, 529—34 (9)
HOSTETTMANN-KALDAS, M., see HOSTETTMANN
HOEVE, C. A. J.
1965 J. chem. Phys. **43**, 3007—08 (6)
1970 J. Polymer Sci. C, Polymer Symposia **30**, 361—67 (6)
1976 Paper L 12.C, read on Conference "Phases and Interfaces in Macromolecular Systems", Prague, 12—16 July (6)
HOVINGH, M. E.
1970 and THOMPSON, G. H.; GIDDINGS, J. C., Analyt. Chem. **42**, 195—203 (13)
HOWARD, G. A.
1950 and MARTIN, A. J. P., Biochem. J. **46**, 532—38 (7)
HOWARD, G. J. (see also HERD)
1967 and MCCOUNELL, P., J. phys. Chem. **71**, 2974—81 (6, 19)
1972 and WOODS, S. J., J. Polymer Sci. A-2, Polymer Phys. **10**, 1023—28 (6, 18)
HOY, K. L.
1969 Tables of Solubility Parameters. — Publication of Union Carbide, July, 21 (5)
HRADIL, J., see PEŠKA
HSU, C. C., see LOO
HUANG, R. Y. M.
see BRAKS
see CHANG
see NARASIMHAN
HUBER, J. F. K.
1967 and HULSMAN, J. A. R. J., Analytica chim. Acta [Amsterdam] **38**, 305 (15)
1969 J. Chromatogr. Sci. **7**, 85 (15)
1972a and KOLDER, F. F. M.; MILLER, J. M., Analyt. Chem. **44**, 105—10 (17)
1972b and MEIJERS, C. A. M.; HULSMAN, J. A. R. J., Analyt. Chem. **44**, 111—16 (7)
1973 and ALDERLIESTE, E. T.; HARREN, H.; POPPE, H., Analyt. Chem. **45**, 1337—43 (7)
1973 Z. Elektrochemie, Ber. Bunsenges. phys. Chemie **77**, 179—84 (7, 15)
1976 Ausnutzung der Phasensystemselektivität in der Säulen-Flüssigkeits-Chromatographie. Paper read on Meeting of Analyzers, Leipzig, 16—19 Nov. (11, 14)

HUDGINS, W. R.
1978 and THEURER, K.; MARIANI, T., J. appl. Polymer Sci. **34**, 145—55 (18)
HUDSON, B. E.
1971 J. Chromatogr. **55**, 185—92 (19)
HUDSON, D. R., see HENRY
HUGGINS, M. L.
1962 and NATTA, G.; DESREUX, V.; MARK, H., J. Polymer Sci. **56**, 153—61 (4)
1967 and OKAMOTO, H.: Theoretical Considerations. In: Polymer Fractionation. Ed.: M. J. R. CANTOW. — Academic Press, New York, London. p. 1—42 (5)
HUGHES, A. J., see COOPER, A. R.
HUGHES, R. E.
see also FORSMAN
see also HAMORI
1963 and FRANKENBERG, C. A. v., Ann. Rev. phys. Chem. **14**, 291—313 (6)
HUISMAN, T. H. J.
1960 and VAN DE BRANDE, J.; MEYERING, C. A., Clin. chim. Acta **5**, 375 (18)
HULME, J. M. (see also HENDERSON)
1962 and MCLEOD, L. A., Polymer [London] **3**, 153—65 (9, 11, 20)
HULSMAN, J. A. R. J., see HUBER
HUMPHREY, J. S. J., see EGGERS
HUNGER, H., see BASEDOW
HUPPENTHAL, L., see ULIŃSKA
HSU, W. W.
1971 HUA HSUEH 116—25 (19)
HYER, H. J., see ADAMS

IKADA, Y., see HORII
ILAVSKÝ, M., see PESKA
IMAI, K.
1963 and MATSUMOTO, M., Bull. chem. Soc. Japan **36**, 455 (13, 15)
IMOTO, T., see HARA
INABA, T., see OGAWA
INAGAKI, H.
see also DONKAI
see also KAMIYAMA
see also KOHJIYA
see also KOTAKA
see also MIN
see also MIYAMOTO
see also OMOTO
see also TAGA
see also TANAKA
see also WÄLCHLI
1968 and MATSUDA, H.; KAMIYAMA, F., Macromolecules **1**, 520—25 (21)
1969 Bull. Inst. chem. Res., Kyoto Univ. **47**, 196—208 (21)
1969 and MIYAMOTO, T.; KAMIYAMA, F., J. Polymer Sci. B, Polymer Letters **7**, 329—33 (21)
1971 and KAMIYAMA, F., Paper read on the 25th Anniversary Meeting of the Japanese Chemical Society, Tokyo, 1971, October (21)
1971 and KAMIYAMA, F.; YAGI, T., Macromolecules **4**, 133—34 (21)
1973 and KAMIYAMA, F., Macromolecules **6**, 107—11 (21)
1976 and KOTAKA, T.; MIN, T.-I., Pure & appl. Chem. **46**, 61—70 (21)
1977a Thin Layer Chromatography. In: Fractionation of Synthetic Polymers. Ed.: L. H. TUNG. — Marcel Dekker, New York (21)
1977b Fortschr. Hochpolymeren-Forsch. **24**, 189 bis 244 (21)
1979 and DONKAI, N., Preprints IUPAC Mainz, 738—41 (18)
1981 and TANAKA, T., Characterization of Copolymers by TLC, GPC and Other Methods. Paper read on IUPAC-Symposium "Advances in Polymer Characterization", Durham, 13—17 July (19)
INGELMAN, B.
1949/50 and HALLING, M. S., Ark. Kemi **1**, 61 (18)
INOUE, H.
1981 and KOHAMA, S., J. Polymer Sci. A-1, Polymer Chem. **19**, 759—71 (19)
INOUE, O., see OGIHARA
IRUIN, J. J., see ECHARRI
IRVINE, P., see GORDON
ISHIDA, Y.
1972 and KAWAI, K., Shimadzu Hyoron 29112, 89—92 (19)
ISHIGE, T.
1971 and LEE, S. I.; HAMIELEC, A. E., J. appl. Polymer Sci. **15**, 1607—22 (16)
ISHIGURO, S., see NAKAMURA
ISHII, D.
1978a and HIROSE, A.; HIBI, K.; IWASAKI, Y., J. Chromatogr. **157**, 43—50 (17)
1978b and HIBI, K.; ASAI, K.; JONOKUCHI, T., J. Chromatogr. **151**, 147—54 (17,19)
ISHIMURA, A., see SHIONO
ISOBE, T., see SHIMONO
ISO'O, A., see OKUMURA
ITO, Y.
see also TANIMURA
1970 and BOWMAN, R. L., J. Chromatogr. Sci. **8**, 315 (9)
1971 and BOWMAN, R. L., Analyt. Chem. **43**, 69A (9)
1972 and BOWMAN, R. L.; NOBLE, F. W., Analyt. Biochem. **49**, 1 (9)

1978 and BOWMAN, R. L., J. Chromatogr. **147**, 221—31 (9, 17)

ITOH, T.
1973 and TANAKA, M.; KANEKO, H., Lipids **8**, 259 (21)

ITSUBO, A., see KATO, TADAYA

IWAEDA, T., see KATO

IWAMA, M.
1972 Kobunshi Kagaku (Japan) **29**, 367—72 (19)

IWANOV, W., see NITSCH

IWASAKI, H.
1980 and SHIMADA, K.; TARUTANI, T., J. Chromatogr. **198**, 429—434 (19)

IWASAKI, Y., see ISHII

IWATA, K., see KOHJIYA

JACKSON, R. A.
1973 Discussion remark. — In: [B 12], 109 (19)

JACOB, M., see FRANCOIS

JACOBELLI, H.
1979 and BARTHOLIN, M.; GUYOT, A., Angew. makromol. Chemie **80**, 31—51 (12)

JACOBI, E.
1980 and SCHUTTENBERG, H.; SCHULZ, R. C., Macromol. Chem., Rapid Commun. **1**, 397—402 (19)

JAHANGIR, L. M.
1980 and SAMUELSON, O., J. Chromatogr. **193**, 197—206 (12)

JAMES, H. L., see GAYLOR

JAMES, P. M.
1973 and OUANO, A. C., J. appl. Polymer Sci. **17**, 1455—66 (8, 19)

JAMIESON, B. S.
1973 Discussion remark. In [B 12], 102 (8)

JANAK, J.
1973 J. Chromatogr. **78**, 117—22 (21)

JANČA, J.
see also GIDDINGS
see also JIŠOVÁ
see also KOLÍNSKÝ
see also LUKÁŠ
see also MRKVICKOVÁ-VACULOVÁ
see also POKORNY
1975 and VLČEK, P.; TREKOVAL, J.; KOLÍNSKÝ, M., J. Polymer Sci. A-1, Polymer Chem. **13**, 1471—73 (19)
1977 and KOLÍNSKÝ, M., J. appl. Polymer Sci. **21**, 83—90 (8, 19)
1977 J. Chromatogr. **134**, 263—72 (19)
1978 and MRKVIČKOVA, L.; KOLÍNSKÝ, M., J. appl. Polymer Sci. **22**, 2661—68 (17, 19)
1978a and POKORNÝ, S., J. Chromatogr. **148**, 31—36 (19)
1978b and POKORNÝ, S., J. Chromatogr. **156**, 27—33 (19)
1979 and POKORNÝ, S., J. Chromatogr. **170**, 319—24 (19)
1979 J. Chromatogr. **170**, 309—18 (19)
1979a and POKORNÝ, S.; KALAL, J., Plasty kauc. **16**, 334—37 (19)
1979b and KROUPOVÁ, J.; POKORNÝ, S.; STEHLIČEK, J.; KÁLAL, J.; GANKINA, E. S.; KEVER, E. E.; BELENKIJ, B. G., J. Chromatogr. **171**, 81—88 (21)
1979c POKORNÝ, S.; KOLINSKÝ, M., J. appl. Polymer Sci. **23**, 1811—17 (19)
1980a J. Chromatogr. **187**, 21—26 (19)
1980b Polymer J. **12**, 405—06 (19)
1981 and POKORNÝ, S.; VILENČIK (VILENCHIK), L. Z.; BELENKIJ, B. G., J. Chromatogr. **211**, 39—44 (19)

JANDERA, P.
1979 and CHURÁČEK, J.; SVOBODA, L., J. Chromatogr. **174**, 35—50 (14)

JANOWSKI, F., see HEYER

JANSON, J.-C. (see also PORATH)
1971 J. Agr. Food Chem. **19**, 581 (17)

JANSTA, J., see SMOLKOVA

JASTREBOVA, L. S., see ŽDANOV

JAVIER-SON, A. C., see OUANO

JAWOREK, D.
1970 Chromatographia **3**, 414—17 (21)

JEFFERY, G. H., see VOGEL

JEHLIČKA, V., see ČOUPEK

JEHRING, H.
1974 Elektrosorptionsanalyse mit der Wechselstrompolarographie. — Akademie-Verlag, Berlin. p. 131 (6, 18)

JELLINEK, H. H. G. (see also HOBDEN)
1954 and NORTHEY, H. L., J. Polymer Sci. **14**, 583 (18)

JENCKEL, E. (see also SONNTAG)
1951 and RUMBACH, B., Z. Elektrochemie u. angew. phys. Chemie **55**, 612—18 (6, 18)

JENKINS, R.
1980 and PORTER, R. S., J. Polymer Sci. B, Polymer Letters **18**, 743—50 (19)

JENTOFT, R. E. (see also GOUW)
1969 and GOUW, T. H., J. Polymer Sci. B, Polymer Letters **7**, 811—13 (9, 15)

JENTZSCH, D., see BAYER

JERMYN, M. A.
1957 Australien J. Chem. **10**, 55; cit. from [A 4], 198 (7)

JIDA, N., see KUWAHARA

JIŠOVÁ, V.
1977 and JANČA, J.; KOLINSKÝ, M., J. Polymer Sci. A-1, Polymer Chem. **5**, 533—36 (19)

JOANNY, J. F.
1979 and LEIBLER, L.; DE GENNES, P. G., J. Polymer Sci. A-2, Polymer Phys. **17**, 1073 to 1084 (6)

JOHANSSON, E., see LINDBERG
JOHANSSON, K., see LINDBERG
JOHN, M.
1973 and DELLWEG, H., Gel Chromatographic Separation of Oligosaccharides. In [A 5], Vol. 2, 231—57 (19)
JOHNE, R., see RUMPF
JOHNSON, B. L., see ADAMS
JOHNSON, E. L.
1978 and GLOOR, R.; MAJORS, R. E., J. Chromatogr. **149**, 571—85 (14)
JOHNSON, J. F.
see also BARTICK
see also CANTOW
see also CHUANG
see also COOPER
see also HADDON
see also MARTIN
see also MOEBUS
see also MIRABELLA
see also PICKETT
1969 and BRUZZONE, A. R.; BARRALL, E. M., J. Colloid & Interface Sci. **31**, 253—56 (9, 20)
1973 and COOPER, A. R.; PORTER, R. S., J. Chromatogr. Sci. **11**, 292—94 (16)
JOHNSON, K. J., see ROBINSON
JOHNSON, M. F. L.
1967 and MOOI, J., ACS Meeting, Chicago, Sept. In: Petroleum Div. Prepr. **12**, 3 (11)
JOHNSON, W. D., see URANECK
JONOKUCHI, T., see ISHII
JOVANOVIĆ, S. (see also MEYERHOFF)
1965 and ROMATOWSKI, J.; SCHULZ, G. V., Makromol. Chemie **85**, 187—94 (20)
JOZEFOWICZ, J.
see CHAPIRO
see DANON
JUNGNICKEL, J. L.
1961 and WEISS, F. T., J. Polymer Sci. **49**, 437—54 (9, 11, 20)
JURANIČOVA, V.
1970 and FLORIAN, Š.; BEREK, D., Europ. Polymer J. **6**, 52—62 (20)
JURŽENKO, A. I.
1958 and MALEEV, I. I., J. Polymer Sci. **31**, 301—07 (18)
JUSIAK, J., see SOCZEWIŃSKI
JUST, U., see PASTUSKA

KADOKURA, S., see OMOTO
KADONO, T., see OKUMURA
KAHLE, D. (see also GLÖCKNER)
1974 Über die dünnschichtchromatographische Bestimmung der chemischen Zusammensetzung von Styrol-Acrylnitril-Copolymeren. Thesis, TU Dresden (21)
KAHN, A., see KOLTHOFF
KAISER, R. (see also BAYER)
1970 Chromatographia **3**, 38—40 (11)
KALAL, J., see JANČA
KALÁSZ, H.
see also MINCSOVICS
see also TYIHAK
1975 and NAGY, J.; KNOLL, J., J. Chromatogr. **107**, 35—42 (17)
KALNINŠ, K. K.
1976 and KRASOVSKIJ, A. N.; BELENKIJ, B. G.; ANDREEVA, G. A., Vysokomolekularnye Soedinenija, Ser. A **18**, 2304—10 (6)
KAMATA, T., see HATTORI
KAMETANI, T., see KATO, Y.
KAMIDE, K.
1973 and MANABE, S.-I.; OSAFUNE, E., Makromol. Chemie **168**, 173—93 (21)
1978 and OKADA, T.; TERAKAWA, T.; KANEKO, K., Polymer J. **10**, 547—56 (21)
KAMIYAMA, F. (see also INAGAKI)
1969 and MATSUDA, H.; INAGAKI, H., Makromol. Chemie **125**, 286—90 (7, 21)
1970 and MATSUDA, H.; INAGAKI, H., Polymer J. **1**, 518—23 (21)
1971 and INAGAKI, H., Bull. Inst. chem. Res., Kyoto Univ. **49**, 53—68 (21)
1972 and INAGAKI, H.; KOTAKA, T., Polymer J. **3**, 470—75 (21)
1974 and INAGAKI, H., Bull. Inst. chem. Res., Kyoto Univ. **52**, 393—402 (7, 21)
KAMLET, M. J. (see also ABBOUD)
1977 and ABBOUD, J. L.; TAFT, R. W., J. Amer. chem. Soc. **99**, 6027—38 (7)
KÄMMERER, H., see BRAUN
KANDA, A., see KATO
KANEKO, H., see ITOH, T.
KANEKO, K., see KAMIDE
KANEKO, M., see KUWAHARA
KANGLE, P. J.
1961 and PASCU, E., J. Polymer Sci. **54**, 301—20 (18)
KAPILA, S., see AUE
KAPOULAS, V. M., see GALANOS
KARCH, K.
1976 and SEBESTIAN, I.; HALÁSZ, I., J. Chromatogr. **122**, 3—16 (10, 11)
1976a and SEBESTIAN, I.; HALÁSZ, I.; ENGELHARDT, H., J. Chromatogr. **122**, 171—84 (7)
KARGER, B. L.
see also KELLER
see also TANAKA
1976 and SNYDER, L. R.; EON, C., J. Chromatogr. **125**, 71—88 (5, 7)
1976b and GANT, J. R.; HARTKOPF, A.; WEINER, P. H., J. Chromatogr. **128**, 65—78 (7)

1978 and SNYDER, L. R.; EON, C., Analyt. Chem. **50**, 2126—36 (5, 7)

1980 and TAPUHI, Y.; LINDNER, W.; LE PAGE, J. N., Ligand Exchange in High Performance Liquid Chromatography. Lecture 2P1 at the 13th International Symposium on Chromatography, Cannes, 30 June to 4 July (11)

KARINO, I., see SHIONO

KATIME, I.

1979 and CAMPOS, A.; RIVERA, J. M. T., Europ. Polymer J. **15**, 291—93 (19)

KATO, TADAYA

1975 and ITSUBO, A.; YAMAMOTO, Y.; FUJIMOTO, T.; NAGASAWA, M., Polymer J. **7**, 123—29 (8, 19)

1979 and KANDA, A.; TAKAHASHI, A.; NODA, I.; MAKI, S.; NAGASAWA, M., Polymer J. **11**, 575—81 (19)

KATO, TOSHIO

1975 and HASHIMOTO, T.; FUJIMOTO, T.; NAGASAWA, M., J. Polymer Sci. A-2, Polymer Phys. **13**, 1849—54 (8, 19)

KATO, Y. (see also HASHIMOTO)

1973 and KIDO, S.; HASHIMOTO, T., J. Polymer Sci. A-2, Polymer Phys. **11**, 2329—37 (9, 19)

1974a and HASHIMOTO, T., J. appl. Polymer Sci. **18**, 1239—41 (8, 19)

1974b and HASHIMOTO, T., J. Polymer Sci. A-2, Polymer Phys. **12**, 813—14 (8, 19)

1974c and HASHIMOTO, T., Kobunshi Ronbunshu (Japan) **31**, 224—26 (16)

1974 and KIDO, S.; YAMAMOTO, M.; HASHIMOTO, T., J. Polymer Sci. A-2, Polymer Phys. **12**, 1339—45 (8, 16, 19)

1975 and KAMETANI, T.; FURUKAWA, K.; HASHIMOTO, T., J. Polymer Sci. B, Polymer Letters **13**, 1695—1703 (17, 19)

1975 and KIDO, S.; WATANABE, H.; YAMAMOTO, M.; HASHIMOTO, T., J. appl. Polymer Sci. **19**, 629—31 (19)

1976 and KAMETANI, T.; HASHIMOTO, T., J. Polymer Sci. A-2, Polymer Phys. **14**, 2105 to 2108 (19)

1977 and TAKAMATSU, T.; FUKOTOMI, M.; FUKUDA, M.; HASHIMOTO, T., J. appl. Polymer Sci. **21**, 577—79 (19)

1978 and SASAKI, H.; AIURA, M.; HASHIMOTO, T., J. Chromatogr. **153**, 546—49 (19)

1980a and KOMIYA, K.; SASAKI, H.; HASHIMOTO, T., J. Chromatogr. **190**, 297—303 (11, 19)

1980b and KOMIYA, K.; SAWADA, Y.; SASAKI, H.; HASHIMOTO, T., J. Chromatogr. **190**, 305 to 310 (19)

1980c and KOMIYA, K.; SASAKI, H.; HASHIMOTO, T., J. Chromatogr. **193**, 29—36 (19)

1980d and KOMIYA, K.; SASAKI, H.; HASHIMOTO, T., J. Chromatogr. **193**, 311—15 (19)

1980e and KOMIYA, K.; SASAKI, H.; HASHIMOTO, T., J. Chromatogr. **193**, 458—63 (19)

1981 and KOMIYA, K.; IWAEDA, T.; SASAKI, H.; HASHIMOTO, T., J. Chromatogr. **205**, 185 to 188; **206**, 135—38; **208**, 71—73; **211**, 383—87 (12, 17)

KATUŠČÁKOVÁ, G., see BEREK

KATZ, J., see BILLMEYER

KAUFMANN, W., see GLÖCKNER

KAUPPILA, K. M., see BAISAL

KAVERINA, N. I., see LOSIKOV

KAWAHARA, K., see HOMMA

KAWAI, K.

see OGIHARA

see ISHIDA

KAYE, W.

1973 HAVLIK, A. J., Appl. Opt. **12**, 541 (19)

KAZANSKIJ, K. S.

1969 and SOLOVJANOV, A. A., Kinet. Mechn. Polyreactions, Internat. Symp. Macromol. Prepr. **2**, 103—06 (19)

KELLER, R. A.

1970 and KARGER, B. L.; SNYDER, L. R., Use of the Solubility Parameter in Predicting Chromatographic Retention and Eluotropic Strength. In [A 18], 125—40 (5, 7)

KELLERMANN, M., see HEMETSBERGER

KELLEY, R. N. (see also BILLMEYER)

1971 and BILLMEYER, F. W.: A Review of Peak Broadening in Gel Chromatography. In [D 2], 47—72 (15)

KENDRICK, T. C.

1969 J. Polymer Sci. A-2, Polymer Phys. **7**, 297—307 (19)

KENNEDY, G. J., see DONE

KENNEDY, J. W., see GORDON

KENT, D. N., see VITEK

KENYON, A. S.

1960 and SALYER, I. O., J. Polymer Sci. **43**, 427—44 (9, 11)

1965 and SALYER, I. O.; KURZ, J. E.; BROWN, D. R., J. Polymer Sci. C, Polymer Symposia **8**, 205—16 (11)

KERN, W. (see also HEITZ)

1955 and SCHMIDT, H.; v. STEINWEHR, H. E., Makromol. Chemie **16**, 74—76 (18)

KESNER, L. F., see CALDWELL

KESTEN, Y., see AMBLER

KEUKER, H.

1971 Chromatographia **4**, 40—47 (21)

KEVER, E. E. (see also JANČA)

1981a and GANKINA, E. S.; BELENKIJ, B. G., Vysokomolekularnye Soedinenija, Ser. A, **23**, 234—36 (19)

1981 b and Belenkij, B. G.; Gankina, E. S.; Vilenčik, L. Z.; Kurenbin, O. I.; Žmakina, T. P., J. Chromatogr. **207**, 145—47 (19)

Kho, B. T., see Krol

Kido, S., see Kato, Y.

Kikta, E. J., see Grushka

Kilb, R. W., see Zimm

Killander, J., see Laurent

Killion, D., see Freeman

Killmann, E. (see also Patat)

1962 and Schneider, G., Makromol. Chemie **57**, 212—19 (18)

1974 and v. Kuzenko, M., Angew. makromol. Chemie **35**, 39—55 (6, 18)

1976 Polymer [London] **17**, 864—68 (6, 18)

Kim, C. W., see Smith, W. B.

King, J.

1966 and Benson, S. W., J. chem. Phys. **44**, 1007—14 (11)

King, J. W., see Giddings

King, R. N., see Bombaugh

Kipling, J. J.

1965 Adsorption from Solutions of Nonelectrolytes. Academic Press, London, Chapt. 8 (6)

Kirkland, J. J.

see also Glajch

see also Yau

1972 J. Chromatogr. Sci. **10**, 593 (15, 17)

1973 J. Chromatogr. **83**, 149—67 (10)

1975 Chromatographia **8**, 661—68 (11, 17, 18)

1976 J. Chromatogr. **125**, 231—50 (11)

1977 and Antle, P. E., J. Chromatogr. Sci. **15**, 137 (11)

1979 J. Chromatogr. **185**, 273—88 (10, 16, 19)

1980 and Yau, W. W.; Doerner, W. A.; Grant, J. W., Analytical Chem. **52**, 1944—54 (13)

Kirkwood, J. G.

1934 J. Chem. Phys. **2**, 351—61 (7)

Kiselev, A. V.

see also Barmakova

see also Belyakova

see also Bogacheva

see also Curthoys

see also Eltekov

see also Ždanov

1961 Spezifische Oberfläche von Adsorbenzien verschiedener Struktur. Absolute Isothermen und Adsorptionswärmen. In [C 4], 132—76 (10)

1963 and Novikova, V. N.; Eltekov, Ju. A., Doklady Akademii Nauk SSSR **149**, 131 to 134 (6, 18)

1967 Advances Chromatogr. **4**, 113 (11)

1968 and Lygin, V. I.; Solomonova, I. N.; Usmanova, D. O.; Eltekov, Ju. A., Kolloidnyj Ž. **30**, 386 (6,18)

1971 and Ryabinina, T. I.; Eltekov, Ju. A., Doklady Akademii Nauk SSSR **200**, 1132 to 1135 (18)

1976 Paper read on Meeting of Analyzers, Leipzig, November (11)

Kiselev, S. A., see Curthoys

Kiseleva, N. N., see Dobyčin

Kishiro, O., see Higashimura

Kiss, F., see Borák

Kiss, J., see Cooper

Kisselew, A. V., see Kiselev, A. V.

Klatyk, K.

see Halpaap

see Heitz

Klein, J. (see also Schiedermaier)

1970 and Friedel, H., J. appl. Polymer Sci. **14**, 1927—38 (9)

1970 and Weinhold, G., Angew. makromol. Chemie **10**, 49—68 (9)

1979 and Leidigkeit, G., Makromol. Chemie **180**, 2753—56 (18)

Kleintjens, L. A.

1976 and Koningsveld, R.; Stockmayer, W. H., Brit. Polymer J. **8**, 144 (9)

Klesper, E.

see also Hartmann

see also Schmitz

1977 a and Hartmann, W., J. Polymer Sci. B, Polymer Letters **15**, 9—16 (9, 14, 15)

1977 b and Hartmann, W., J. Polymer Sci. B, Polymer Letters **15**, 707—12 (9)

1978 and Hartmann, W., Europ. Polymer J. **14**, 77—88 (9)

1978 Angew. Chemie **90**, 785—93 (9)

Kline, G. M., see Stromberg

Klinkenberg, A., see van Deemter

Klose, A., see Geiss

Knapp, W. C., see Axelson

Knight, G. W.

see May

see Tung

Knoll, J., see Kalász

Knopp, Ch., see Vachtina

Knox, J. H. (see also Dohne)

1969 and Parcher, J. F., Analyt. Chem. **41**, 1599—1606 (15)

1975 and Pryde, A., J. Chromatogr. **112**, 171 to 188 (11)

1979 and McLennan, F., J. Chromatogr. **185**, 289—304 (16, 18, 19)

Kohama, S., see Inoue

Kohjiya, S.

1980 and Iwata, K.; Yamashita, S.; Miyamoto,

T.; Inagaki, H., Colloid & Polymer Sci. **258**, 1192—93 (21)
Kohler, A.
1972 and Zilliox, J.-G.; Rempp, P.; Poláček, J.; Koessler, I., Europ. Polymer J. **8**, 627—39 (20)
Kohlschütter, H. W.
1966 and Unger, K.; Vogel, K., Makromol. Chemie **93**, 1—12 (10)
Kolčina, G. M., see Cilipotkina
Kolder, F. F. M., see Huber
Kolegov, V. I., see Belenkij
Koliňský, M.
see also Janča
see also Jišova
see also Lukáš
see also Pokorný
1974 and Janča, J., J. Polymer Sci. A-1, Polymer Chem. 1181—91 (8)
Kollmansberger, A., see Smith
Kolthoff, I. M. (see also Shapiro)
1950 and Kahn, A., J. phys. & Colloid Chem. **54**, 251—56 (18)
Komiya, K.
see Fukano
see Kato
Komuro, S., see Onda
König, B., see Haller
Koningsveld, R. (see also Kleintjens)
1967 and Staverman, A., J. Polymer Sci. C, Polymer Symposia **16**, 1775—86 (11)
1969 Pure & appl. Chem. **20**, 271—307 (5)
1970a Discuss. Faraday Soc. **49**, 144—61 (5)
1970b Fortschr. Hochpolymeren-Forsch. **7**, 1—69 (9)
Konton, M. M., see Nefedov
Konuma, M., see Kuwahara
Koontz, S., see Dubin
Koopal, L. K., see Fleer
Koopmans, H. J.
1971 and Bouwmeester, P. C., Chromatographia **4**, 83—84 (21)
Koral, J.
1958 and Ullman, R.; Eirich, F. R., J. phys. Chem. **62**, 541—50 (6, 18)
Koromal'di, E. V.
see Ždanov
see Ždanov (Zhdanov)
Koršak, V. V., see Dubrovina
Korus, R. A.
1978 and Rosser, R. W., Analyt. Chem. **50**, 249—51 (19)
Kössler, I.
see also Daňhelka
see also Kohler
see also Poláček
1959 and Krejsa, J., J. Polymer Sci. **35**, 308—09 (13)
1962 and Krejsa, J., J. Polymer Sci. **57**, 509—15 (13)
Kotaka, T.
see also Donkai
see also Inagaki
see also Kamiyama
1974 and White, J. L., Macromolecules **7**, 106—16 (21)
1975 and Uda, T.; Tanaka, T.; Inagaki, H., Makromol. Chemie **176**, 1273—88 (21)
Kotera, A., see Furusawa
Kotliar, A.
1964 J. Polymer Sci. A., Gen. Papers, **2**, 4327—37 (4)
Kouchiyama, K.
1978 and Yoza, N.; Ohashi, S., J. Chromatogr. **147**, 271—79 (19)
Kouřilova, D., see Krejči
Kovaleva, N. V.
see Barmakova
see Belyakova
see Eltekov
Kraak, J. C.
see de Jong
see Paanakker
see Slaats
Kraemer, E. O., see Lansing
Kraffczyk, F., see Heitz
Kranz, D. (see also Scholtan)
1972 and Pohl, H. U.; Baumann, H., Angew. makromol. Chemie **26**, 67—84 (8, 19)
Krasikov, V. D., see Nesterov
Krasnova, T. P., see Belenkij
Krasovskij, A. N., see Kalnins
Kratochvil, P.
see Mrkvičková-Vaculová
see Stejskal
see Tuzar
Kratz, R. F., see Hess
Kraus, G.
1955 and Dugone, J., Ind. & Engng. Chem. **47**, 1809—16 (18)
1972 and Stacy, C. J., J. Polymer Sci. A-2, Polymer Phys. **10**, 657—72 (19)
Krause, T., see Böttger
Krebs, K. F.
see also Unger
see also Halpaap
see also Unger
1971 and Wunderlich, W., Angew. makromol. Chemie **20**, 203—05 (19)
Krejči, M. (see also Šlais)
1980a and Kóuřilova, D.; Vespalec, R.; Šlais, K., J. Chromatogr. **191**, 3—7 (16)
Krejsa, J., see Kössler

KRESSMAN, T. R. E., see MILLAR
KRESUB, O. B., see BRYK
KRIGBAUM, W. R.
1959 and KURZ, J. E., J. Polymer Sci. **41**, 275—93 (9, 11)
KRISHEN, A.
1977 and TUCKER, R. G., Analyt. Chem. **49**, 898 (19)
KŘIVÁKOVÁ, M., see ČOUPEK
KRIVČENKO, E. I., see BUDTOV
KROCHTA, J. M.
see FINLEY
see HEFTMANN
KROL, G. J.
1972 and MANNAN, C. A.; GEMMILL, F. Q.; HICKS, G. E.; KHO, B. T., J. Chromatogr. **74**, 43—49 (15)
KROUPOVÁ, J., see JANČA
KUBELKA, P.
1931 and MUNK, F., Z. Techn. Phys. **12**, 593 (21)
KUBIN, M.
see also DANIELEWICZ
see also SAMAY
see also VOZKA
1965 Collect. czechoslov. chem. Commun. **30**, 1104—18, 2900 (3, 8)
1967 Collect. czechoslov. chem. Commun. **32**, 1505—17 (4)
1975 Gel Permeation Chromatography. In [D 7], 57—67 (9)
1978a and VOZKA, S., Series MACRO, Inst. Macromol. Chem. Czechoslovak Academy of Sciences, K-28; C 34/1-C 34-4 (10)
1978b and VOZKA, S., J. Chromatogr. **147**, 85—98 (17)
KUĆERA, P. (see also SCOTT)
1981 and MANIUS, G., J. Chromatogr. **216**, 9—21 (19)
KUDRJAVCEVA, L. G.
1963 and LITMANOVIČ, A. D.; TOPČIEV, A. V.; ŠTERN, V. Ja., Neftechimija **3**, 343—47 (20)
KUGA, S.
1981 J. Chromatogr. **206**, 449—61 (10)
KUHN, R., see CANTOW, H. J.
KUHNHARDT, R., see GLÖCKNER
KULAKOWSKI, R. A., see RODRIGUEZ
KUMANOTANI, J., see MIYATAKE
KUMLIN, K.
1981 and SIMONSON, R., Angew. makromol. Chemie **93**, 43—54 (19)
KUN, K. A.
1968 and KUNIN, R., J. Polymer Sci. A-1, Polymer Chem. **6**, 2689—2701 (12)
KUNATH, D. (see also SCHULZ, D.)
1978 and SCHULZ, D.; BERGER, K., Faserforsch. und Textiltechn. **29**, 16—18 (6)
KUNIN, R., see KUN
KUNST, B.
see LOVRIĆ
see MENCER
KUO, C.-Y., see PROVDER
KURAMOTO, K., see NAKAO
KURENBIN, O. I., see KEVER
KURGANOV, A. A., see DAVANKOV
KURIHARA, H., see UJIMOTO
KURZ, J. E.,
see KENYON
see KRIGBAUM
KUWAHARA, N.
1966 and OGINO, K.; KONUMA, M.; JIDA, N.; KANEKO, M., J. Polymer Sci. A-2, Polymer Physics **4**, 173—81 (4)
KUZAEV, A. I.
1973 and SUSLOVA, E. N.; ENTELIS, S. G., Doklady Akademii Nauk SSSR **208**, 142—44 (18)
KUZENKO, M. v., see KILLMANN
KUZNECOVA, M. A.
see BELENKIJ
see ŽDANOV
KUZNETSOV, B., see CURTHOYS
KVIST, B. E., see GRANATH
KWOLL, F. J., see SCHOLTAN
KYRIACOS, D., see BIRLEY
KYRITSOS, J., see PAILHES

LÅÅS, T., see PORATH
LABADIE, R. P., see DHONT
LADISCH, M. R., see BAO
LAGAREVA, M. A., see TENNIKOV
LAIN, L., see ALVARIÑO
LAKEMAN, J., see VAN WENDEL DE JOODE
LAMBERT, A.
1970 J. appl. Chem. **20**, 305 (19)
1971 Analytica chim. Acta [Amsterdam] **53**, 63 (19)
LAMBERT, K., see DETERMANN
LAMMENS, J., see VERZELE
LANDLER, I.
1947 C. R. hebd. Séances Acad. Sci. [Paris] **225**, 629 (18)
LANGHAMMER, G. (see also SCHWEIGER)
1954 Naturwissenschaften **41**, 552 (13)
1955 and QUITZSCH, K., Makromol. Chemie **17**, 74—76 (13)
1958 and PFENNIG, H.; QUITZSCH, K., Z. Elektrochemie, Ber. Bunsenges. phys. Chemie **62**, 458—80 (13)
1961 Z. Elektrochemie, Ber. Bunsenges. phys. Chemie **65**, 706 (13)

1961 and QUITZSCH, K., Makromol. Chemie **43**, 160—63 (9, 11, 20)
1967 and SEIDE, H., Kolloid-Z. u. Z. Polym. **216/217**, 264—68 (10, 11, 16)
1970 and SEIDE, H., Plaste u. Kautschuk **7**, 486—88 (19)

LANSING, W. D.
1935 and KRAEMER, E. O., J. Amer. chem. Soc. **57**, 1369—77 (4)

LAPIDUS, L.
1955 and AMUNDSON, N. R., J. phys. Chem. **59**, 416 (15)

LARSEN, F. N.
1968 6th International Gel Permeation Chromatography Seminar, Seminar Proc., 111 (19)
1971 J. Chromatogr. **55**, 220 (19)

LATTIMER, R. P.
1979 and HARMON, D. J.; WELCH, K. R., Analyt. Chem. **51**, 1293—96 (8, 18)

LAUB, R. J.
1975 and PURNELL, J. H., J. Chromatogr. **112**, 71—79 (15)
1978 and PURNELL, J. H., J. Chromatogr. **161**, 49—57 (15)

LAUFER, G.
1976 and GUTFINGER, C.; ABUAF, N., Ind. & Engng. Chem. Fundamentals **15**, 74—77 (17)

LAURENT, T. C.
1964 and KILLANDER, J., J. Chromatogr. **14**, 317—30 (8)

LAVERTU, R., see PERRAULT

LAW, R. D. (see also ADAMS)
1969 J. Polymer Sci. A-1, Polymer Chem. **7**, 2097—2116 (17)
1971 J. Polymer Sci. A-1, Polymer Chem. **9**, 589—615 (19)

LAZAREVA, M. A.
see BELENKIJ
see NEFEDOV

LEA, A. G. H.
1980 J. Chromatogr. **194**, 62—68 (14)

LEA, D. J.
1962 and SEHON, A. H., Can. J. Chem. **40**, 159—60 (12)

LEAUTEY, J., see BELLENS

LEBBE, J., see BAYER

LEBEDA, P., see CEAUŞESCU

LE CLAIR, B. P., see BALKE

LECOURTIER, J.
1978a and AUDEBERT, R.; QUIVORON, C., Theoretical and Experimental Aspects of Fractionation. Mechanism on Cross-Linked Organic Polymers. In [D 11/1], 156—68 (11)
1978b and AUDEBERT, R.; QUIVORON, C., J. Liq. Chromatogr. **1**, 367—84 (9, 11)
1978c and AUDEBERT, R.; QUIVORON, C., J. Liq. Chromatogr. **1**, 479—90 (9, 11)
1979a and AUDEBERT, R.; QUIVORON, C., Pure & appl. Chem. **51**, 1483—1501 (11, 16)
1979b and AUDEBERT, R.; QUIVORON, C., Macromolecules **12**, 141—46 (16)

LEDERER, K.
see DUNKY
see NITSCH

LEE, D. W., see BRAUN

LEE, S. I., see ISHIGE

LEE, W. Y.
1978 J. appl. Polymer Sci. **22**, 3343—44 (19)

LEGRADIC, V., see BRAUN

LEIBLER, L., see JOANNY

LEICESTER, J., see VOGEL

LEIDIGKEIT, G., see KLEIN

LEMON, P., see MONTAGUE

LENSCHONSKI, K., see RUMPF

LENZ, R. W., see SIEBOURG

LÉONARD, J. (see also MALHOTRA)
1977 and MALHOTRA, S. L., J. Macromol. Sci., Chem. A 11, **10**, 1867—1906 (19)

LEOPAIRAT, P.
1978 and MERRILL, E. W., J. Liq. Chromatogr. **1**, 21—32 (16)

LEOPOLD, H. (see also KRATKY)

LE PAGE, J. N., see KARGER

LE PAGE, M. (see also BEAU)
1967 and BEAU, M.; DE VRIES, A. J., Amer. chem. Soc. Polym. Prepr. **8**, 1211 (11)
1968 and BEAU, M.; DE VRIES, A. J., J. Polymer Sci. C, Polymer Symposia **21**, 119—30 (10, 17, 19)

LERMAN, L. S.
1953 Proc. nat. Acad. Sci. USA **39**, 232 (18)

LESEC, J., see CHAUFER

LESHEM, R., see GRUSHKA

LEVANGIE, R. F.,
see BOMBAUGH
see DARK

LEVI, G. R.
1937 and GIERRA, A., Gazz. chim. ital. **67**, 719 (18)

LI, W. S., see BOOTH

LIBERTI, A., see CICCIOLI

LIGHTBODY, B., see VIVILECCHIA

LIMPERT, R. J., see DARK

LIN, H.-J., see HORVÁTH

LINDBERG, W.
1981 and JOHANSSON, E.; JOHANSSON, K., J. Chromatogr. **211**, 201—12 (15)

LINDNER, W., see KARGER

LINDSTRÖM, T.
1977 and DERUVO, A.; SÖREMARK, Ch., J. Polymer Sci. A-1, Polymer Chem. **15**, 2029—32 (19)

LING, N.-S., see BOCK
LIPATOV, Ju. S.
1975 and SERGEEVA, L. M.; TODOSLIČUK, T. T.; CHRAMOVA, T. S., Kolloidnyj Žurnal **37**, 280—84 (6, 18)
LIPSKY, S. R., see HORVATH
LIROVA, B. I., see CILIPOTKINA
LI-RU, D.
1979 J. Chromatogr. **186**, 317—38 (11, 12)
LISICYN, D. M., see POZNJAK
LITMANOVIČ, A. D. (see also KUDRJAVCEVA)
1963 and ŠTERN, V. Ja.; TOPČIEV, A. V., Neftechimija **3**, 217 (20)
1963 and TOPČIEV, A. V., Neftechimija **3**, 336 to 342 (20)
1967 and ŠTERN (SHTERŃ), V. Ja., J. Polymer Sci. C, Polymer Symposia **16**, 1375—82 (19)
LITTLE, C. J.
1978a and DALE, A. D.; EVANS, M. B., J. Chromatogr. **153**, 381—89 (11)
1978b and DALE, A. D.; EVANS, M. B., J. Chromatogr. **153**, 543—45 (11)
1979a and DALE, A. D.; WHATLEY, J. A.; EVANS, M. B., J. Chromatogr. **171**, 431—34 (11)
1979b and WHATLEY, J. A.; DALE, A. D.; EVANS, M. B., J. Chromatogr. **171**, 435—38 (11)
LITTLE, J. N.
see also BLY
see also BOMBAUGH
1969 and WATERS, J. L.; BOMBAUGH, K. J.; PAUPLIS, W. J., J. Polymer Sci. A-2, Polymer Physics **7**, 1775—83 (17, 19)
1970 and WATERS, J. L.; BOMBAUGH, K. J.; PAUPLIS, W. J., Separat. Sci. **5**, 765—75 (19)
1971 and PAUPLIS, W. J., J. Chromatogr. **55**, 211 (15)
LLOYD, D. R., see NARASIMHAN
LOCHMÜLLER, C. H.
1979 and WILDER, D. R., J. Chromatogr. Sci. **17**, 17 (11)
LOCKE, D. C.
1972 and SCHMERMUND, J. T.; BANNER, B., Analyt. Chem. **44**, S. 90—92 (8)
1973 J. Chromatogr. Sci. **11**, 120 (11)
1974 J. Chromatogr. Sci. **12**, 433 (7)
LOCONTI, J. D., see SCHNEIDER
LÖNNERT, P., see SCHÖLLNER
LOMBANA, L., see SOSA
LOO, C. C.
1971 and HSU, C. C., J. clin. Inst. chem. Engng. **2**, 57—64 (19)
LOPOUR, P., see MRKVIČKOVÁ-VACULOVÁ
LORA, J. H.
1980 and WAYMAN, M., J. appl. Polymer Sci. **25**, 589—96 (19)
LORENZI, G. P., see PINO
LOSIKOV, B. V.
1956 and KAVERINA, N. I.; FEDYANTSEVA, A. A., Chimija i Technologija Topliv i Masel **3**, 51 (18)
LOVRIĆ, L.
1969 J. Polymer Sci. A-2, Polymer Physics **7**, 1357—66 (11)
1976 and GRUBIŠIĆ-GALLOT, Z.; KUNST, B., Europ. Polymer J. **12**, 189—93 (19, 20)
LOWRY, R. E., see BROWN
LOY, B. R.
1976 J. Polymer Sci. A-1, Polymer Chem. **14**, 2321 (8)
LUCE, I. E.
1961 and ROBERTSON, A. A., J. Polymer Sci. **51**, 317 (18)
LUDWIG, J., see BALDRIAN
LUKÁŠ, R. (see also POKORNÝ)
1978 and JANĆA, J.; MRKVIČKOVÁ, L.; POKORNÝ, S.; KOLINSKÝ, M., J. Polymer Sci. A-1, Polymer Chem. **16**, 2539—44 (19)
LUNDBERG, J. L., see FRISCH
LUNDBERG, R. D., see SIEBOURG
LÜTJE, H.
1970 Makromol. Chemie **133**, 295—98 (5)
LUTOVSKY, P., see FELTL
LYGIN, V. I., see KISELEV
LYKLEMA, J.
see also FLEER
see also NORDE
1973 and NORDE, W., Croat. chem. Acta **45**, 67—84 (18)
LYNGAAE-JORGENSEN, J.
1971 J. Chromatogr. Sci. **9**, 331—40 (19)
LYONS, A. R.
1971 and CATTERALL, E., Makromol. Chemie **146**, 141—44 (19)

MACGREGOR, J. F., see GARCIA-RUBIO
MACHATA, G., see BAYER
MACKIE, W., see HOLT
MACRURY, T. B.
1979 and MCCONNELL, M. L., J. appl. Polymer Sci. **24**, 651—62 (19)
MADDEN, R. W., see COOPER, W.
MADDOCK, J. W., see DAWKINS
MAHABADI, H. K.
1977 and O'DRISCOLL, K. F., J. appl. Polymer Sci. **21**, 1283—87 (8, 19)
1979 and RUDIN, A., Polymer J. **11**, 123—31 (19)
MAJORS, R. E. (see also JOHNSON)
1972 Analyt. Chem. **44**, 1722—26 (17)
1973 J. Chromatogr. Sci. **11**, 88 (15)
1976 Bonded Phase Chromatography. In [D 14], 121—32 (11)

1980 J. Chromatogr. Sci. **18**, 488—511 (10, 11)
MAKI, S., see KATO
MALAWER, E. G.
1980 and MONTANA, A. J., J. Polymer Sci. A-2, Polymer Phys. **18**, 2303—05 (8)
MALEEV, I. I., see JURŽENKO
MALEY, L. E.
1965 J. Polymer Sci. C, Polymer Symposia **8**, 253—68 (19)
MALHOTRA, S. L. (see also LEONARD)
1977 and LEONARD, J., J. Macromol. Sci., Chem. A 11, **10**, 1907—48 (19)
MALIN, A. S., see BOWEN
MALINSKY, J., see SEIDL
MALLIK, K. L., see GIDDINGS
MALONE, C. P. (see also YAN)
1969 and SUCHAN, H. L.; YAU, W. W., J. Polymer Sci. B, Polymer Letters **7**, 781—84 (8, 19)
MALONE, W. M., see ALBERT
MANABE, S.-I., see KAMIDE
MANDEL, J., see ROTHMAN
MANDEL, M.
1981 Polyelectrolytes: A Survey, in Molecular Electro-Optics. Ed.: S. Krause. Plenum Publishing Corporation, New York, pp. 285 to 307 (19)
MANDÍK, L.
1979 and FOKSOVÁ, A.; FOLTYN, J., J. appl. Polymer Sci. **24**, 395—404 (19)
MANIUS, G., see KUČERA
MANLEY, R. S. J., see CHANG
MANNAN, C. A., see KROL
MÅNSSON, P. (see also WESSLEN)
1980 J. Polymer Sci. A-1, Polymer Chem. **18**, 1945—56 (18, 19, 21)
MARC, R.
1913 Z. phys. Chemie **81**, 641—94 (6, 18)
MARCHESSAULT, R. H., see COULOMBE
MARIANI, T., see HUDGINS
MARK, H. (see also HUGGINS)
1936 and SAITO, G., Monatsh. Chemie **68**, 237 to 243 (18)
MARKOVSKI, W., see SLAATS
MARON, S. H.
1972 and FILISKO, F. E., J. Macromol. Sci. **B6**, 79—90 (18)
MARR, G., see BROUGH
MARSHALL, M. A., see ROBINSON
MARTENS, A. J., see CROUZET
MARTIN, A. J. P.
see also HOWARD
1941 and SYNGE, R. L. M., Biochem. J. **35**, 1358 (3, 9)
1949 Biochem. Soc. Symposia [London] **3**, 4 (7, 16)
MARTIN, J. R.
1973 and JOHNSON, J. F., Separation Sci. **8**, 619 (17)
MARTIN, K., see HALÁSZ
MARTIN, M.
see also COLIN
see also GIDDINGS
1974 and EON, C.; GUIOCHON, G., J. Chromatogr. **99**, 357—76 (15)
1975a and EON, C.; GUIOCHON, G., J. Chromatogr. **108**, 229—41 (15)
1975b and EON, C.; GUIOCHON, G., J. Chromatogr. **110**, 213—32 (15)
1976 VERILLON, F.; EON, C.; GUIOCHON, G., J. Chromatogr. **125**, 17—41 (17)
1980 and HES, J., Thermal Field Flow Fractionation — Laser Light Scattering: A Powerful Tandem for Polymer Analysis. Lecture 1P5 at the 13th International Symposium on Chromatography, Cannes, 30 June—4 July (13)
MATE, R. D., see AMBLER
MATHES, D., see ENGELHARDT
MATSUDA, H.
see INAGAKI
see KAMIYAMA
MATSUMOTO, M., see IMAI
MATSUMOTO, Y., see SAITO
MATSUZAKI, K., see HIGASHIMURA
MATYSKA, B., see POLÁČEK
MATZINGER, D. P., see COOPER
MAURER, T., see BIESENBERGER
MAY, J. A. (see also SMITH, W. B.)
1971 and KNIGHT, G. W., J. Chromatogr. **55**, 111—19 (8, 16, 19)
MCCANN, F. R., see ROTHMAN
MCLENDON, J. H.
1979 J. Chromatogr. **171**, 460—61 (8, 19)
MCCONNELL, M. L. (see also MACRURY)
1979 Recent Advances in Size Exclusion Chromatography. ACS Symposium Proceedings, Washington (16)
MCCORMICK, Ch. L.
1981 and PARK, L. S., J. appl. Polymer Sci. **26**, 1705—17 (19)
MCCOUNELL, P., see HOWARD
MCCRACKIN, F. L. (see also WAGNER)
1977 J. appl. Polymer Sci. **21**, 191—98 (8)
1980 and WAGNER, H. L., Macromolecules **13**, 685—90 (16, 17, 19)
MCCULLOUGH, C. R., see SHULTZ
MCGINNIS, G. D., see WILSON
MCHUGH, A. J., see NAGY
MCILWRICK, R., see BAKALYAR
MCINTYRE, D.
see AMBLER
see SLAGOWSKI

McLAREN, J. V.
see BROUGH
see DOUGLAS
see EPTON
see TAYLOR
McLEAN, D. L.
1972 and WHITE, J. L., Polymer [London], **13**, 124—32 (11)
McLENNAN, F., see KNOX
McLEOD, L. A., see HULME
McQUARRIE, D. A.
1963 J. chem. Phys. **38**, 437—45 (3, 8)
MEDA, E., see RIGAMONTI
MEIER, D. J.
1967 J. phys. Chem. **71**, 1861—68 (6)
MEIJERINK, N. L. J., see SCHOLTE
MEIJERS, C. A. M., see HUBER
MEISSNER, H., see HEUBLEIN
MEISSNER, R. (see also GLÖCKNER)
1977 Dünnschicht-Chromatographie von Copolymeren aus Acrylnitril/Styrol/Alphamethylstyrol. Thesis, TU Dresden (21)
MELANDER, W., (see also HORVÁTH)
1978 and CAMPBELL, D. E.; HORVÁTH, Cs., J. Chromatogr. **158**, 215—25 (7)
1979a and CHEN, B.-K.; HORVÁTH, Cs., J. Chromatogr. **185**, 99—109 (7)
1979b and NAHUM, A.; HORVÁTH, Cs., J. Chromatogr. **185**, 129—52 (18)
MELANDER, W. R., see BIJ
MELAS, V. B., see BELENKIJ
MELENEVSKAJA, E. Ju., see BELENKIJ
MENCER, H. J.
1978 and KUNST, B., Colloid & Polymer Sci. **256**, 696—98 (9)
1979 and KUNST, B., Makromol. Chemie **180**, 2463—71 (20)
1979 and GRUBIŠIČ-GALLOT, Z., J. Liq. Chromatogr. **2**, 649—62 (19)
MENDELSON, R. A., see DROTT
MENIN, J. P.
1972 and ROUX, R., J. Polymer Sci. A-1, Polymer Chem. **10**, 855—66 (19)
MERCIER, P., see OUANO
MERRILL, E. W., see LEOPAIRAT
MESSER, W., see UNGER
MEUNIER, J. C.
1972 and GALLOT, Z., Makromol. Chemie **156**, 117—20 (2, 19)
MEYERHOFF, G.
see also APPELT
see also HEROLD
see also MOHITE
1965a Makromol. Chemie **89**, 282—84 (8)
1965b Z. Elektrochemie, Ber. Bunsenges. phys. Chemie **69**, 866—74 (8, 19)
1967 and JOVANOVIČ, S., J. Polymer Sci. B, Polymer Letters **5**, 495—99 (17, 19)
1967 and SHIMOTSUMA, S., Makromol. Chemie **109**, 263—65 (13)
1968 J. Polymer Sci. C, Polymer Symposia, **21**, 31—41 (19)
1971 Separat. Sci. **6**, 239—48 (17, 19)
MEYERING, C. A., see HUISMAN
MICHEL, W., see DETERMANN
MIEDZIAK, I., see COLIN
MIJAL, C. F., see SCHNEIDER
MIKEŠ, O.
1975 and VESPALEC, R.: Mobile Phases. In [D 7], 233—80 (14)
MIKOLAJCZAK, E., see HEUBLEIN
MIKULÁŠOVÁ, D., see ŠOLTÉS
MILES, B. H.
1965 Amer. chem. Soc., Div. Org. Coatings and Plast. Chem. **24/2**, 273 (19)
MILLAR, J. R.
1965 and SMITH, D. G.; KRESSMAN, T. R. E., J. chem. Soc. [London] 304—10 (12)
MILLER, J. M., see HUBER
MILLER, R. L.
1980 and VANDEMARK, F. L., Chromatogr. Newslett. **8**, 69—72 (12, 19)
MILLER, V. P. see DUBIN
MILTZ, J. (see also RAM)
1971 and RAM, A., Polymer [London] **12**, 685 to 696 (19)
MIN, T.-I. (see also INAGAKI)
1975 and MIYAMOTO, T.; INAGAKI, H., Bull. Inst. chem. Res., Kyoto Univ. **53**, 381—86 (21)
1977 and MIYAMOTO, T.; INAGAKI, H., Rubber Chem. & Technol. **50**, 63—70 (21)
MINÁRIK, M.
1977 and ŠIR, Z.; ČOUPEK, J., Angew. Makromol. Chem. **64**, 147—56 (19)
MINCSOVICS, E. (see also TYIHÁK)
1980 and TYIHÁK, E.; KALÁSZ, H., J. Chromatogr. **191**, 293—300 (21)
MINDNER, K.
1969 and BERGER, R., Plaste u. Kautschuk **16**, 405—11, 882—88 (8)
1979 and BERGER, R., Plaste und Kautschuk **26**, 251—54 (19)
MINOR, J. M., see GLAJCH
MINSKER, K. S., see POZNJAK
MINTI, M., see CEAUŞESCU
MIRABELLA, F. M.
1975a and BARRALL, E. M.; JOHNSON, J. F., J. appl. Polymer Sci. **19**, 2131—41 (17, 19)
1975b and BARRALL, E. M.; JOHNSON, J. F., IBM Research July **17** RJ 1618, Chemistry (19)
1976 and BARRALL, E. M.; JOHNSON, J. F., Polymer [London] **17**, 17—20 (17, 19)

1980 J. appl. Polymer Sci. **25**, 2265—71 (19)
MISCHE, W., see BRAUER
MITCHELL, S. A., see ARMISTEAD
MIYAMOTO, T.
see also DONKAI
see also INAGAKI
see also KOHJIYA
see also MIN
see also WÄLCHLI
1972 and CANTOW, H.-J., Makromol. Chemie **162**, 43—51 (18)
1973 and DONKAI, N.; TAKIMI, N.; INAGAKI, H., Nippon Kagakusen'i Kenkyusho Koenshu (Ann. Rep. Res. Inst. Chem. Fibers, Japan) **30**, 27—34 (20, 21)
1974 and TOMOSHIGE, S.; INAGAKI, H., Polymer J. **6**, 564—70 (6, 18)
MIYATAKE, R.
1976 and KUMANOTANI, J., Makromol. Chemie **177**, 2749—57 (19)
MIZUI, F., see ARATANI
MIZUTANI, T.
1980 J. Chromatogr. **196**, 485—88 (11)
MLYNEK, J., see BLEHA
MOEBUS, G. A.
1979 and CROWTHER, J. A.; BARTICK, E. G.; JOHNSON, J. F., J. appl. Polymer Sci. **23**, 3501—04 (17)
MOELLMER, J. F., see GIDDINGS
MOHITE, R. B.
1972 and MEYERHOFF, G., Angew. makromol. Chemie **25**, 41—50 (19)
MÖLLER, D., see REHAGE
MOLNAR, I., see HORVÁTH
MONSIGNY, M.
1978 and CORNET, M.; TIXIER, R.; CORGIER, M.; GIROT, P., Polyacrylamide-Agarose Gels — Structure and Applications. In [D 11/1], 57—69 (12)
MONTAGNOLI, G., see PINO
MONTAGUE, P. G.
1971 and PEAKER, F. W.; BOSWORTH, P.; Lemon, P., Brit. Polymer J. **3**, 93—94 (19)
1973 and PEAKER, F. W., J. Polymer Sci. C, Polymer Symposia **43**, 277—89 (17)
MONTANA, A. J., see MALAWER
MOOI, J., see JOHNSON, M. F. L.
MOORE, J. C.
see also ALTGELT
see also HENDRICKSON
see also TUNG
1964 J. Polymer Sci. A., Gen. Papers **2**, 835—43 (8, 10, 12, 13, 19)
1968 unpublished, cit. from BOYER, R. F., Paper read on 6th International GPC-Seminar, Miami, October (19)
1970 Separat. Sci. **5**, 723, and: The Overload Effect in Gel Permeation Chromatography. In [D 2], 157—64 (19)
MOORE, L. D. (see also OVERTON)
1962 and GREEAR, G. R.; SHARP, J. O., J. Polymer Sci. **59**, 339—55 (9)
MORGAN, S. L.
1976 and DEMING, S. N., Sep. Purif. Methods **5**, 333 (15)
MORI, S.
1972 and TAKEUCHI, T., Kobunshi Kagaku (Japan) **29**, 383—87 (21)
1977 J. appl. Polymer Sci. **21**, 1921—32 (19)
1978a Analyt. Chem. **50**, 1639—43 (19)
1978b J. Chromatogr. **157**, 75—84 (19)
1978c J. Chromatogr. **156**, 111—20 (19)
1979 J. Chromatogr. **174**, 23—33 (8)
1980 and YAMAKAWA, A., J. Liq. Chromatogr. **3**, 329—42 (19)
1980a J. Chromatogr. **192**, 295—305 (8, 19)
1980b J. Chromatogr. **194**, 163—73 (19)
MORIIZUMI, S., see NAKAMURA
MORIMOTO, K.
1972 and SUZUKI, S., J. appl. Polymer Sci. **16**, 2947—61 (19)
MOROHOSHI, N.
1979 and GLASSER, W. G., Wood Sci. Technol. **13**, 249—64 (19)
MORRIS, M. C.
1971 J. Chromatogr. **55**, 203—10 (19)
MORRIS, R. J.
1963 and PERSINGER, E. H., J. Polymer Sci. A, Gen. Papers **1**, 1041—48 (18)
MORTON, M. (see also FETTERS)
1974 J. appl. Polymer Sci. **18**, 2391—97 (19)
1976 J. appl. Polymer Sci. **20**, 2157—64 (19)
MOSBACH, R., see HJERTEN
MOTTLAU, A. Y.
1962 and FISHER, N. E., Analyt. Chem. **34**, 714—15 (10)
MOTTWEILER, R.
1981 and SCHEINERT, W., Angew. makromol. Chemie **94**, 147—59 (19)
MRKVIČKOVÁ, L.
see JANČA
see LUKÁŠ
MRKVIČKOVÁ-VACULOVÁ, L.
1972 and KRATOCHVIL, P., Collect. czechoslov. chem. Commun. **37**, 2015—28; 2029—38 (19)
1980 and LOPOUR, P.; POKORNÝ, S.; JANČA, J., Angew. makromol. Chemie **90**, 217—21 (19)
MUGLIA, P. M., see OTOCKA
MUKHERJEE, K. D.
1971 and SPAANS, H.; HAAHTI, E., J. Chromatogr. **61**, 317—21 (21)

MUKHERJI, A. K. (see also PACCO)
1978 and BERTRAND, J. C.; ALLEN, D. A., Angew. makromol. Chemie **72**, 213—19 (17)
MUELLER, H., see UNGER
MÜLLER, K.-D., see GLÖCKNER
MÜLLER, M.
1980 Chromatographia **13**, 557 (21)
MULLER, T. E. (see also ALEXANDER)
1968 and ALEXANDER, W. J., J. Polymer Sci. C, Polymer Symposia **21**, 283—95 (19)
MULLINEAUX, R. D., see FLOWERS
MUNIER, R. L., see BAYER
MUNK, F., see KUBELKA
MUNRO, G., see HAYWOOD
MURANAKA, M., see UENO
MURAYAMA, N., see DONKAI
MURPHY, R.
1981 and SELDEN, A. C.; FISHER, M.; FAGAN, E. A.; CHADWICK, V. S., J. Chromatogr. **211**, 160—65 (18, 19)
MUSSCHE, P., see VERZELE
MUUS, L. T., see STOCKMAYER
MYERS, M. N.
see CALDWELL
see GIDDINGS
see GRUSHKA
see THOMPSON

NAEFE, M., see HALÁSZ
NAGAOKA, T., see OKUMURA
NAGASAWA, M.
see KATO, TADAYA
see KATO, TOSHIO
see YAMAMOTO
NAGY, D. J.
1981 a and SILEBI, C. A.; MCHUGH, A. J., J. appl. Polymer Sci. **26**, 1555—65 (13)
1981 b and SILEBI, C. A.; MCHUGH, A. J., J. appl. Polymer Sci. **26**, 1567—78 (19)
NAGY, J., see KALÁSZ
NAGY, Z., see SZEPESY
NAHUM, A.
see also BIJ
see also MELANDER
1981 and HORVÁTH, Cs., J. Chromatogr. **203**, 53—63 (7, 11)
NAKAGAWA, G., see TSUDA
NAKAHARA, H., see HATTORI
NAKAJIMA, M., see TERAMACHI
NAKAJIMA, N.
1966 J. Polymer Sci. A-2, Polymer Physics **4**, 101—11 (19)
1968 J. Polymer Sci. C, Polymer Symposia **21**, 153—58 (19)
1971 a Separat. Sci. **6**, 275—86 (19)
1971 b Amer. chem. Soc. Polym. Prepr. **12**, 804 to 11 (19)
1971 c J. appl. Polymer Sci. **15**, 3089—99 (19)
1972 J. appl. Polymer Sci. **16**, 2417—23 (19)
NAKAMURA, S.
1973 and ISHIGURO, S.; YAMADA, T.; MORIIZUMI, S., J. Chromatogr. **83**, 279—88 (17)
NAKANISHI, H., see FURUSAWA
NAKANISHI, K., see HOSTETTMANN
NAKANO, S.
1975 and GOTO, Y., J. appl. Polymer Sci. **19**, 2655—65 (16, 19)
1976 and GOTO, Y., J. appl. Polymer Sci. **20**, 3313—20 (19)
NAKAO, K.
1972 and KURAMOTO, K., Nippon Secchaku Kyokai Shi (Japan) **8**, 186—92 (19)
NANCE, G. P., see HASSELL
NARASIMHAN, V.
1979 and LLOYD, D. R.; BURNS, C. M., J. appl. Polymer Sci. **23**, 749—54 (19)
1981 and HUANG, R. Y. M.; BURNS, C. M., J. appl. Polymer Sci. **26**, 1295—1300 (19)
NATTA, G.
see also HUGGINS
see also PINO
1958 and PEGORARO, M.; PERALDO, M., Ric. Sci. **28**, 1473—79 (9, 11, 18)
1960 and PASQUON, I.; CORRADINI, P.; PERALDO, M.; PEGORARO, M.; ZAMBELLI, A., Atti Accad. naz. Lincei, Rend., Cl. Sci. fisiche, mat. natur. **28/8**, 539 (18)
NAZANSKIJ, A. S., see ŽDANOV
NEDDERMEYER, P. A.
1968 and ROGERS, L. B., Analyt. Chem. **40**, 755—62 (19)
1969 and ROGERS, L. B., Analyt. Chem. **41**, 94—102 (19)
NEFEDOV, P. P.
see also BELENKIJ
see also TENNIKOV
see also ŽDANOV
1979 and LAZAREVA, M. A.; BELENKIJ, B. G.; FRENKEL, S. Ya.; KONTON, M. M., J. Chromatogr. **170**, 11—24 (16, 19)
NENTWIG, W.
1980 and SINN, H., Makromol. Chem., Rapid Commun. **1**, 59—63 (19)
NESTEROV, V. V. (see also BELENKIJ)
1978 and KRASIKOV, V. D.; ČUBAROVA, E. V.; TURKOVA, L. D.; GANKINA, E. S.; BELENKIJ, B. G., Vysokomolekularnye Soedinenija, Ser. A **20**, 2320—25 (19)
NEVIN, A., see DAWKINS
NG, Q. Y., see EDWARDS
NICE, E. C. (see also O'HARE)
1978 and O'HARE, M. J., J. Chromatogr. **166**, 263—67 (11)

NICHOLAS, R. A.
1969 and FOX, J. B., J. Chromatogr. **43**, 61—65 (17)
NIEDERWIESER, A. (see also BRENNER)
1965 and BRENNER, M., Experientia [Basel] **21**, 50, 105 (21)
1966a and HONEGGER, C. G., Advances Chromatogr. **2**, 123 (21)
1966b and HONEGGER, C. G., in [A 11], Bd. 2, 123—67 (21)
1969a Chromatographia **2**, 23—30 (14, 21)
1969b Chromatographia **2**, 362—75 (21)
NIEMAN, J. A.
1975 and ROGERS, L. B., Separat. Sci. **10**, 517 (9)
NIKOLOV, R.
1980 and WERNER, W.; HALÁSZ, I., J. Chromatogr. Sci. **18**, 207—16 (10)
NISHIKAWA, M., see UENO
NITSCH, E.
1979 and IWANOV, W.; LEDERER, K., Carbohydrate Res. **72**, 1—12 (19)
NITSCHMANN, R., see PATAT
NOACK, D.
1971 Wiss. Z. Techn. Univ. Dresden **20**, 693—97 (19)
NOBLE, F. W., see ITO
NODA, H.
1979 and SAITOH, K.; SUZUKI, N., J. Chromatogr. **168**, 250—54 (19)
NODA, I.
see KATO
see YAMAMOTO
NOEL, R.
see CHANG
see REGNIER
NONDEK, L.
1981 and VYSKOČIL, V., J. Chromatogr. **206**, 581—85 (11)
NOOITGEDACHT, F., see DE JONG
NORDE, W. (see also LYKLEMA)
1972 and FURUSAWA, K.; LYKLEMA, J.: Adsorption of Human Serum Albumin on Well-defined Polystyrene Latices. Paper read on 6th International Congress „Grenzflächenaktive Stoffe“, Zurich, 11 to 15 September. In: Tagungsberichte. Hanser Verlag München, 1973, 209—19 (18)
1975 and LYKLEMA, J.: Adsorption of Human Plasma Albumin and Bovine Ribonuclease at Charged Polystyrene-Water Interfaces. In [C 2], 205—11 (18)
NORDT, E., see SCHULZ, G. V.
NORTHEY, H. L., see JELLINEK
NORRIS, J. F.
1973 and PEAKER, F. W., Fractionation and Analysis of Methyl-Methacrylate-Styrene Copolymers. In [B 12], 118—27 (19)
NOVAK, L., see BEREK
NOVIKOV, D. D., see POZNJAK
NOVIKOVA, V. N., see KISELEV
NOVOŽILOVA, O. S., see VACHTINA
NOZAWA, A.
1980 and OHNUMA, T., J. Chromatogr. **187**, 261—63 (18)

OBIAGA, T. I.
1974 and WAYMAN, M., J. appl. Polymer Sci. **18**, 1943 (19)
O'DRISCOLL, K. F., see MAHABADI
OGAWA, T.
1972a and TANAKA, S.; HOSHINO, S., Kobunshi Kagaku (Japan) **29**, 6—9 (19)
1972b and SUZUKI, Y.; INABA, T., J. Polymer Sci. A-1, Polymer Chem. **10**, 737—50 (19)
1973 and TANAKA, S.; INABA, T., J. appl. Polymer Sci. **17**, 779—88 (9)
1977 and INABA, T., J. appl. Polymer Sci. **21**, 2979—90 (19)
1979 J. appl. Polymer Sci. **23**, 3515—23 (19)
OGIHARA, Y.
1976 and INOUE, O.; OTSUKA, H.; KAWAI, K.; TANIMURA, T.; SHIBATA, S., J. Chromatogr. **128**, 218—23 (9)
OGINO, K., see KUWAHARA
O'HARE, M. J. (see also NICE)
1980 and NICE, E. C.; CAPP, M., J. Chromatogr. **198**, 23—39 (11)
OHASHI, S. (see also KOUCHIYAMA)
1966 and YOZA, N.; UENO, Y., J. Chromatogr. **24**, 300—01 (19)
OHGURI, N., see FUJISHIGE
OHKAWA, T., see ROKUSHIKA
OHNUMA, T., see NOZAWA
ÖHRN, O. E.
1956 J. Polymer Sci. **19**, 199—200 (6)
OKADA, T., see KAMIDE
OKUMURA, T.
1975 and KADONO, T.; ISO'O, A., J. Chromatogr. **108**, 329—36 (21)
1980 and NAGAOKA, T., J. Liq. Chromatogr. **3**, 1947—60 (21)
OKUNEV, P. A. (see also TARAKANOV)
1963 and TARAKANOV, O. G., Chimičeskie Volokna **6**, 44—46 (20)
OLIEMAN, C.
1981 and SEDLICK, E.; VOSKAMP, D., J. Chromatogr. **207**, 421—24 (11)
OLLEY, R. H., see RYBICKA
OMORODION, S. N. E., see HAMIELEC
1980 and HAMIELEC, A. E.; BRASH, J. L., ACS Symposium Series, Size Exclusion Chromatogr. **138**, 267—84 (19)

OMOTO, M. (see also TANAKA)
1979 and TANAKA, T.; KADOKURA, S.; INAGAKI, H., Polymer [London], **20**, 129—30 (19)
ONDA, N.
1979 and FURUSAWA, K.; YAMAGUCHI, N.; KOMURO, S., J. appl. Polymer Sci. **23**, 3631 to 3638 (19)
1980 and FURUSAWA, K.; YAMAGUCHI, N.; TOKIWA, M.; HIRAI, Y., J. appl. Polymer Sci. **25**, 2363—72 (19)
OPREA, S. P., see SIMIONESCU
OSAFUNE, E., see KAMIDE
OŠCIK, J.
1971 and RÓZYLO, J. K., Chromatographia **4**, 516—23 (7)
OSTER, H., see BAYER
OSTERHOUDT, H. W. (see also QUINN)
1967 and RAY, L. N., J. Polymer Sci. A-2, Polymer Phys. **5**, 569—81 (17, 19)
OTH, A., see DESREUX
OTHMER, D. F.
1953 and THAKER, M. S., Ind. & Engng. Chem., 589—93 (15)
OTOCKA, E. P.
1970 and HELLMAN, M. Y., Macromolecules **3**, 362—65 (21)
1970 Macromolecules **3**, 691—94 (21)
1971 and ROE, R. J.; HELLMAN, M. Y.; MUGLIA, P. M., Macromolecules **4**, 507—12 (19)
1972 and HELLMAN, M. Y., MUGLIA, P. M., Macromolecules **5**, 227—28 (21)
1973 J. Chromatogr. **76**, 149—57 (11)
1974a and HELLMAN, M. Y., J. Polymer Sci. B, Polymer Letters **12**, 331—38 (8, 19)
1974b and HELLMAN, M. Y., J. Polymer Sci. B, Polymer Letters **12**, 439—45 (8, 19)
OTRADINA, G. A., see BUDTOV
OTSUKA, H., see OGIHARA
OUANO, A. C.
see also HAMIELEC
see also JAMES
1968 and MERCIER, P., J. Polymer Sci. C, Polymer Symposia **21**, 309—15 (19)
1971 and BIESENBERGER, J. A., J. Chromatogr. **55**, 145—54 (17)
1972a J. Polymer Sci. A-1, Polymer Chem. **10**, 2169—80 (19)
1972b Ind. Engng. Chem., Fundam., **11**, 268 (17)
1973 and BARRALL, E. M.; BROIDO, A.; Javier-Son, A. C., Amer. chem. Soc., Advances Chem. Ser. **125**, 187—93 (8, 19)
1976 J. Chromatogr. **118**, 303—12 (19)
1978 J. Colloid & Interface Sci. **63**, 275—81 (19)
OVERTON, J. R.
1968 and RASH, J.; MOORE, L. D., 6. Internat. GPC-Seminar, Seminar Proc., 422 (19)
OWERS, H. S., see GOODBAN
OZAKI, E.
1979 and SAITOH, K.; SUZUKI, N., J. Chromatogr. **177**, 122—26 (16)
OZIMÁ, A., see BAKOŠ

PAANAKKER, J. E.
1978 and KRAAK, J. C.; POPPE, H., J. Chromatogr. **149**, 111—26 (7)
PACCO, J. M.
see also ALLIET
see also GIAMMARSE
1971 J. Chromatogr. **55**, 99—109 (19)
1978 and MUKHERJI, A. K.; EVANS, D. L., Sep. Sci. Technol. **13**, 277 (19)
PADLEY, F. B.
1969 J. Chromatogr. **39**, 37 (21)
PAILHES, F.
1967 and ALLIOT-LUGAZ, M.; DUVEAU, N.; KYRITSOS, J., J. Polymer Sci. C, Polymer Symposia **16**, 1177—90 (9)
PAJUREK, J., see KREJČI
PALLAS, G., see PANARIS
PALLESI, B., see COPPOLA
PAN, H.-C., see DING
PANARIS, R.
see also PEYROUSET
see also PRECHNER
1970 and PALLAS, G., J. Polymer Sci. B, Polymer Letters **8**, 441—46 (19)
PANDJOJO, W., see BRAUN
PANKRATOV, E. A., see DAVANKOV
PANNELL, J.
1972 Polymer [London] **13**, 277—82 (8, 19)
PANTON, C. J.
1964 and PLESCH, P. H.; RUTHERFORD, P. P., J. chem. Soc. **495**, 2586—91 (9, 20)
PARCHER, J. F., s. KNOX
PARK, L. S., see MCCORMICK
PARK, W. S.
1977a and GRAESSLEY, W. W., J. Polymer Sci. A-2, Polymer Phys. **15**, 71—83 (16, 19)
1977b and GRAESSLEY, W. W., J. Polymer Sci. A-2, Polymer Phys. **15**, 85—95 (17, 19)
PARKER, M. S., see STROMBERG
PARRIS, N. A.
1978 J. Chromatogr. **157**, 161—70 (18)
PARSONS, J. L., see ZINBO
PASCHKE, E. E.
1977 and BIDLINGMEYER, B. A.; BERGMANN, J. G., J. Polymer Sci. A-1, Polymer Chem. **15**, 983—89 (19)
PASCU, E., see KANGLE
PASQUON, I., see NATTA
PASSAGLIA, R., see STROMBERG

PASTUSKA, G.
1979 and WÜNDRICH, K.; BAHR, H., Kautschuk u. Gummi, Kunststoffe **32**, 157—60 (19)
1979 and JUST, U., Angew. makromol. Chemie **81**, 11—18 (11, 19)

PATAKI, G., see BRENNER

PATAT, F.
1961 and ESTUPINAN, L., Makromol. Chemie **49**, 182—99 (18)
1961 and SCHLIEBENER, C., Makromol. Chemie **44—46**, 643—68 (18)
1964 and NITSCHMANN, R., Monatsh. Chemie **95**, 716—32 (6)
1964 and KILLMANN, E.; SCHLIEBENER, C., Fortschr. Hochpolymeren-Forsch. **3**, 332—93 (6)

PATEL, C. K.
see VYAS
see PATEL, V. M.

PATEL, G. N.
1974 J. appl. Polymer Sci. **18**, 3537—42 (19)
1974 and STEJNY, J., J. appl. Polymer Sci. **18**, 2069 (8)

PATEL, R. D.
see BALKE
see PATEL, V. M.
see TRIVEDI
see VYAS

PATEL, V. M. (see also TRIVEDI)
1970 and PATEL, C. K.; PATEL, R. D., Angew. makromol. Chemie **13**, 195—98 (18)
1971 and PATEL, C. K.; PATEL, R. D., Angew. makromol. Chemie **18**, 39—47 (18)

PATNAIK, S., see VARMA

PÄTZ, R., see SCHRÖDER

PAUL, D., see BALDRIAN

PAUPLIS, W. J., see LITTLE

PAV, J. W., see GIDDINGS

PAVLOVA, S.-S. A., see ANDRIANOV
see DUBROVINA

PAVLU, B., see WICHMAN

PEAKER, F. W.
see CRIGHTON
see DAWKINS
see MONTAGUE
see NORRIS

PEARCE, S. L., see BALKE

PEDERSEN, K. O.
1962 Arch. Biochem. & Biophysics **1**, 157 (8)

PEDERSEN, W. B., see EL-AWADY

PEGGION, E., see COSANI

PEGORARO, M., see NATTA

PEKER, J.
1972 and GEYER, R.; EPPERT, G., J. Chromatogr. **67**, 329—34 (15)
1976 and GEYER, R.; EPPERT, G.: Paper read on 5th Symposium "Chromatography in clinical Biochemistry". Leipzig, 6/7 May (15)

PENCEL, U., see VACHTINA

PEPPER, D. C.
1959 and RUTHERFORD, P. P., J. appl. Polymer Sci. **2**, 100—07 (9, 11, 20)

PERALDO, M., see NATTA

PERCIVA, N. V., see ANDRIANOV

PERKEL, R.
1961 and ULLMAN, R., J. Polymer Sci. **54**, 127—48 (6, 18)

PERRAULT, G.
1971 and TREMBLAY, M.; LAVERTU, R.; TREMBLAY, R., J. Chromatogr. **55**, 121—36

PERRY, J. A.
1979 J. Chromatogr. **165**, 117—40 (21)

PERSINGER, E. H., see MORRIS

PERSONNAZ, L., see GAREIL

PERTOLDI-MARLETTA, G., see FAVRETTO

PEŠKA, J.
see also BALDRIAN
see also ŠTAMBERG
1976 and ŠTAMBERG, J.; HRADIL, J.; ILAVSKÝ, M., J. Chromatogr. **125**, 455—69 (12)

PETERSON, E. A. (see also HARTLEY)

PETIT, L., see SOSA

PETRAKOVA, E. A., see VACHTINA

PEURIFOY, P. V., see SZAKASITS

PEYROUSET, A.
1972 and PANARIS, R., J. appl. Polymer Sci. **16**, 315—28 (17, 19)
1974 and PANARIS, R., Bull. Soc. chim. France **5**, 2279—84 (19)
1975 and PRECHNER, R.; PANARIS, R.; BENOIT, H., J. appl. Polymer Sci. **19**, 1363—71 (8, 19)

PEYSER, P.
1967 and TUTAS, D. J.; STROMBERG, R. R., J. Polymer Sci. A-1, Polymer, Chem. **5**, 651 to 663 (6, 9, 18)

PFEIFFER, D. G.
1980 J. appl. Polymer Sci. **25**, 369—80 (5)

PFENNIG, H., see LANGHAMMER

PFITZNER, K., see HEITZ

PHILLIPS, C. S. G., see BANNISTER

PICARD, C., see HENNION

PICKETT, H. E.
1966 and CANTOW, M. J. R.; JOHNSON, J. F., J. appl. Polymer Sci. **10**, 917—24 (16)
1968 and CANTOW, M. J. R.; JOHNSON, J. F., J. Polymer Sci. C, Polymer Symposia **21**, 67—81 (16)

PICOT, C., see GUENET

PICOT, M., see GRUBIŠIĆ-GALLOT

PIENTKA, V., see HÄUPKE

PIERCE, P. E.
1968 and ARMONAS, J. E., J. Polymer Sci. C, Polymer Symposia **21**, 23—29 (16)
PINO, P.
1962 and CIARDELLI, F.; LORENZI, G. P.; NATTA, G., J. Amer. chem. Soc. **84**, 1487—88 (9, 11, 19)
1966 and MONTAGNOLI, G.; CIARDELLI, F.; BENEDETTI, E., Makromol. Chemie **93**, 158 to 179 (9, 11, 18)
PISANO, J. J., see TANIMURA
PLATT, K.-L., see HEITZ
PLAZEK, D. J.
1978 and AGARWAL, P., J. appl. Polymer Sci. **22**, 2127—35 (8)
PLESCH, P. H., see PANTON
PLZAK, Z., see SMOLKOVA
PODDAR, S. K.
1976 and FORSMAN, W. C., Ind. & Engng. Chem. Prod. Res. & Devel. **15**, 274—78 (19)
PODEŠVA, J., see SAMAY
PODOSENOVA, N. G. (see also BUDTOV)
1980 and ROŽKOVA, V. F., Vysokomolekularnye Soedinenija, Ser. A **22**, 947—50 (19)
POHL, H. U., see KRANZ
POILE, A.
1980 Chromatogr. Newslett. **8**, 65—68 (17)
POINESCU, I. C., see FREEMAN
POKORNÝ, S.
see also ČOUPEK
see also JANČA
see also LUKÁŠ
see also MRKVIČKOVÁ-VACULOVÁ
1978 and LUKÁS, R.; JANČA, J.; KOLINSKÝ, M., J. Chromatogr. **148**, 183—87 (19)
POLÁČEK, J. (see also KOHLER)
1962 and MATYSKA, B., Collect. czechoslov. chem. Commun. **27**, 816—22 (18)
1963a Collect. czechoslov. chem. Commun. **28**, 1838—47 (20)
1963b Collect. czechoslov. chem. Commun. **28**, 3011—18 (20)
1965 and KÖSSLER, I.; VODEHNAL, J., J. Polymer Sci. A. Gen. Papers **3**, 2511—26 (20)
1967 and SCHULZ, L.; KÖSSLER, I., J. Polymer Sci. C, Polymer Symposia **16**, 1327—32 (20)
POLLAK, V. (see also GIETZ)
1978 J. Chromatogr. **152**, 201—07 (21)
POLSON, A.
1966 and RUSSELL, B. W., Biochim. et biophysica Acta [Amsterdam] **117**, 477—79 (14, 17)
POOL, F. R., see BAIJAL
Pope, N. T.
1959 and WEAKLEY, T. J. R.; WILLIAMS, R. J. P., J. chem. Soc. [London] 3442—46 (20)
POPOV, A. G., see BRESLER
POPOV, G.
1980 and SCHWACHULA, G.; GEHRKE, K., Plaste u. Kautschuk **27**, 65—68 (19)
POPPE, H.
see also DE JONG
see also HUBER
see also PAANAKKER
see also SLAATS
see also SNYDER
1980 and SLAATS, E. H., Some remarks on the derivation of solvent polarity factors from gas-liquid partition coefficients. Part IV (p. 89—111) of Slaats thesis 1980 (7)
PORATH, J. (see also TISELIUS)
1959 and FLODIN, P., Nature **183**, 1657—59 (12)
1962 Nature **196**, 47—48 (19)
1962 and BENNICH, H., Arch. Biochem. & Biophysics **1**, 152 (17)
1963 Pure & appl. Chem. **6**, 233 (8)
1971 and JANSON, J.-C.; LÅÅS, T., J. Chromatogr. **60**, 167—77 (12)
1978 Polysaccharide Supports: Synthesis, Derivatization and Application. In [D 11, 1], 9—29 (12)
1979 Pure & Appl. Chem. **51**, 1549—59 (10, 12)
POROD, G., see TREIBER
PORSCH, B., see VOZKA
PORTER, R. S.
see CANTOW, M. J. R.
see HADDON
see JENKINS
see JOHNSON, J. F.
POSPIŠIL, J., see BRODILOVÁ
POUND, T. C., see CHANG
POZNJAK, T. I.
1977 and LISICYN, D. M.; NOVIKOV, D. D.; D'JAČKOVSKIJ, F. S., Vysokomolekularnye Soedinenja, Ser. A **19**, 1168—70 (19)
1980 and LISICYN, D. M., NOVIKOV, D. D.; BERLIN, Al. Al.; DJAČKOVSKIJ, F. S.; PROČUCHAN, Ju. A.; SANGALOV, Ju. A.; MINSKER, K. S., Vysokomolekularnye Soedinenija, Ser. A **22**, 1424—27 (19)
PRAGNELL, R. J., see TAYLOR
PRAUSNITZ, J. M., see BLANKS
PRECHNER, R. (see also PEYROUSET)
1972 and PANARIS, R.; BENOIT, H., Makromol. Chemie **156**, 39—54 (19)
PREISS, B. A., see HORVATH
PREVNÁ, Z., see BEREK
PREVOT, A., see BAYER
PRICE, C. (see also BOOTH)
1977 and FORGET, J.-L.; BOOTH, C., Polymer [London] **18**, 526—27 (19)

PRIEL, Z.
1978 and SILBERBERG, A., J. Polymer Sci. A-2, Polymer Phys. **16**, 1917—25 (6)
PRINCE, L. A.
see also BATA
see also HAZELL
1968 and STAPELFELDT, H. E., 6th International GPC-Seminar. Seminar Proc., 78 (19)
PŘISTOUPIL, T. I.
1970 J. Chromatogr. (Chromatogr. Rev.) **12**, 109—25
PROBST, J. (see also CANTOW, H. J.)
1974 and UNGER, K.; CANTOW, H.-J., Angew. makromol. Chemie **35**, 177—202 (16, 19)
PROČUCHAN, Ju. A., see POZNJAK
PROKOVSKAJA, L. A.
1969 and FROLOVA, G. S., Fiz. Chim. i Chim. Technol. 201—06 (19)
PROMINA, L. N., see ANDRIANOV
PROVDER, T. (see also ROSEN)
1970 and ROSEN, E. M., Separat. Sci. **5**, 437—84 (17)
1971 and ROSEN, E. M.: The Instrument Spreading Correction in GPC. I. The General Shape Function Using a Linear Calibration Curve. In [D 2], 243—90 (8, 16)
1971 and WOODBREY, J. C.; CLARK, J. H.: Gel Permeation Chromatography Calibration. I. Use of Calibration Curves Based on Polystyrene in THF and Integral Distributation Curves of Elution Volume to Generate Calibration Curves for Polymers in 2,2,2-Trifluorethanol. In [D 2], 493—528 (8, 17)
1976 and KUO, C.-Y., Applications of Multiple Detector GPC to the Compositional and Molecular Weight Distribution Analysis of Multicomponent Polymers. ACS Organic Coat. Plastics Chem., Prepr. **36**, (2), p. 7 (19)
PRUD'HOMME, J., see HO-DUC
PRUSINSKA, M. I., see BŁEDZKI
PRYDE, A., see KNOX
PRYKE, P., see WAKSMUNDZKI
PTICYN, O. B.
1959 and EJZNER, Ju. E., Žurnal Techničeskoj Fiziki **29**, 1118—34 (8)
PURCELL, E. M., see BERG
PURDON, J. R., see AMBLER
PURNELL, J. H., see LAUB

QUASIUS, A. R., see STROMBERG
QUINN, E. J.
1968 and OSTERHOUDT, H. W.; HECKLES, J. S.; ZIEGLER, D. C., Analyt. Chem. **40**, 547 (19)
QUINN, H., see VIVILECCHIA
QUISENBERRY, D. O., see WHITE
QUITZSCH, K., see LANGHAMMER
QUIVORON, C.
see BOUÉ
see CHAUFER
see LECOURTIER

RACHOW, J. W., see TIMM
RAFLER, G., see REINISCH
RAM, A. (see also MILTZ)
1971 and MILTZ, J., J. appl. Polymer Sci. **15**, 2639—44 (19)
RAMSDELL, H. S.
1981 and BUHLER, D. R., J. Chromatogr. **210**, 154—58 (12)
RANDAU, D., see HEITZ
RANGANATH, R.
see BARLOW
see WILD
RASH, J., see OVERTON
RATHORE, R. N. S., see SUSSMAN
RAUSCH, C. W., see FALLICK
RAY, L. N.
see FELTER
see OSTERHOUDT
READY, B. W., see GORDON
REED, A. M., see GILDING
REGNIER, F. E. (see also CHANG)
1976 and NOEL, R., J. Chromatogr. Sci. **14**, 316—20 (11, 19)
REHAGE, G.
1965 and MÖLLER, D.; ERNST, O., Makromol. Chemie **88**, 232—55 (5)
REHNER, H., see DETERMANN
REICH, W., see HALPAAP
REINER, R. H.
1971 and WALCH, A., Chromatographia **4**, 578 to 587 (18)
REINISCH, G.
1969 and RAFLER, G.; TIMOFEEVA, G., Angew. makromol. Chemie **7**, 110—20 (9)
REJFMAN, L. S., s. VILENČIK
REMPP, P.
see BENOIT
see DONDOS
see GRUBIŠIĆ
see HILD
see KOHLER
REVILLON, A.
1976 and DUMONT, B.; GUYOT, A., J. Polymer Sci. A-1, Polymer Chem. **14**, 2263—73 (19)
RICHTER, M., see BAUER
RICKEN, H., see HEMETSBERGER
RIEDEL, S., see GLÖCKNER

RIGAMONTI, R.
1955 and MEDA, E., Ricerca sci. **25**, 457 (9)
RIJNDERS, G. W. A., see SIE
RILEY, C. M., see BRISTOW
RINAUDO, M.
see also DOMARD
1980 and DESBRIERES, J., Europ. Polymer J. **16**, 849—54 (19)
RING, W.
1965 and CANTOW, H.-J., Makromol. Chemie **89**, 138—55 (9)
1966 and CANTOW, H.-J.; HOLTRUP, W., Europ. Polymer J. **2**, 151—62 (11)
1967 and HOLTRUP, W., Makromol. Chemie **103**, 83—90 (19)
RIPPHAHN, R., see HALPAAP
RITTER, H. L.
1945 and DRAKE, L. C., Ind. Engng. Chem., Anal. Ed. **17**, 782 (10)
RITTER, J. F., see DHONT
RIVERA, J. M. T., see KATIME
ROBB, I. D.
see also DAY
see also FOX
1974 and SMITH, R., Europ. Polymer J. **10**, 1005—10 (6, 18)
1977 and SMITH, R., Polymer [London] **18**, 500—07 (18)
ROBERTS, T., see BARLOW
ROBERTSON, A. A., see LUCE
ROBERTSON, W. G. P., see HAMMOND
ROBINSON, J. L.
1980 and ROBINSON, W. J.; MARSHALL, M. A.; BARNES, A. D.; JOHNSON, K. J.; SALAS, D. S., J. Chromatogr. **189**, 145—67 (7)
ROBINSON, W. J., see ROBINSON
ROCCA, J. L.
see COQ
see SAGE
ROCHAS, C., see DOMARD
RODRIGUEZ, F.
see also SHAW
see also TERRY
1966 and KULAKOWSKI, R. A.; CLARK, O. K., Ind. Engng. Chem., Prod. Res. & Devel. **5**, 121—25 (19)
ROE, R. J. (see also OTOCKA)
1974 J. chem. Phys. **60**, 4192—4207 (6)
ROELS, J. P. M.
see SOETEMAN
see VAN DIJK
ROGERS, L. B.
see NEDDERMEYER
see NIEMAN
ROGGENDORF, E., see BAKALYAR
ROGOŽIN (ROGOSHIN), S. V. (see also DAVANKOV)
1970 and DAVANKOV, V. A., Dokl. Akad. Nauk SSSR **192**, 1288—90 (18)
ROHLEDER, H., see FENDLER
ROHN, C. L.
1967 J. Polymer Sci. A-2, Polymer Phys. **5**, 547—54 (19)
ROHRSCHNEIDER, L. (see also BAYER)
1968 J. Gas Chromatogr. **6**, 5 (7)
1973 Analyt. Chem. **45**, 1241—47 (7)
ROKUSHIKA, S.
1979 and OHKAWA, T., HATANO, H., J. Chromatogr. **176**, 456—61 (15, 19)
ROMATOWSKI, J., see JOVANOVIĆ
ROSEN, E. M. (see also PROVDER)
1971 and PROVDER, T.: The Instrument Spreading Correction in GPC. II. The General Shape Function Using the Fourier Transform Method with a Nonlinear Calibration Curve. In [D 2], 291—327 (8, 16)
ROSENTHAL, E. J., see HIATT
ROSS, G.
1972 and FROLEN, S., J. Res. nat. Bur. Standards **76** A, January—December, 163 (19)
ROSSER, R. W., see KORUS
ROSSET, R., see HENNION
RÖSSLER, G.
1973 and SCHNEIDER, W.; HALÁSZ, I., Chromatographia **6**, 237—38 (17)
ROTHMAN, S.
1953 and SCHWEBEL, Á.; WEISSBERG, S. G., J. Polymer Sci. **11**, 381—82 (18)
1955 and MANDEL, J.; MCCANN, F. R.; WEISSBERG, S. G., J. Colloid Sci. **10**, 338—50 (18)
ROTHSTEIN, E., see ROWLAND
ROTSCHOVÁ, J., see BRODILOVÁ
ROUMELIOTIS, P., see also DAVANKOV
see also UNGER
1979 and UNGER, K., J. Chromatogr. **185**, 445 to 452 (17, 19)
ROUX, R., see MENIN
ROWLAND, F. W.
1965 and BULAS, R.; ROTHSTEIN, E.; EIRICH, F. R., Ind. & Engng. Chem. **57**, 46—52 (6)
1966a and EIRICH, F. R., J. Polymer Sci. A-1, Polymer Chem. **4**, 2033—40 (6)
1966b and EIRICH, F. R., J. Polymer Sci. A-1, Polymer Chem. **4**, 2401—21 (6, 18)
ROŽKOVA, V. F., see PODOSENOVÁ
RÓZYŁO, J. K.
see WAKSMUNDZKI
see OŚCIK
RUBIN, R. J., see DI MARZIO
RUDD, J. F., see RUNYON

RUDIN, A.
see also DOBBIN
see also MAHABADI
1971 J. Polymer Sci. A-1, Polymer Chem. **9**, 2587—2604 (8, 19)
1972 and HOEGY, H. L. W., J. Polymer Sci. A-1, Polymer Chem. **10**, 217—35 (8)
1976 and WAGNER, R. A., J. appl. Polymer Sci. **20**, 1483—90 (8, 19)
RÜFENACHT, H. J. (see also v. TAVEL)
1976 and v. TAVEL, P., Makromol. Chemie **177**, 2449—56 (9)
RUMBACH, B., see JENCKEL
RUMPF, H.
1967 and ALEX, W.; JOHNE, R.; LENSCHONSKI, K., Z. Elektrochemie, Ber. Bunsenges. phys. Chemie **71**, 253—70 (10)
RUNYON, J. R. (see also TUNG)
1969 and BARNES, D. E.; RUDD, J. F.; TUNG, L. H., J. appl. Polymer Sci. **13**, 2359—69 (19)
RUPYŠEV, V. G., see BUDTOV
RUSSELL, B. W., see POLSON
RUTER, J., see DÖRFFEL
RUTHERFORD, P. P.
see PANTON
see PEPPER
RYABININA, T. I., see KISELEV
RYBICKA, S. M.
1971 and OLLEY, R. H.; CRONIN, P. J., J. Chromatogr. Sci. **9**, 424—27 (19)
RYLE, T., see WILD

SADA, A.
1979 and DI PASCALE, G.; CACACE, M. G., J. Chromatogr. **177**, 353—56 (12)
SAEDA, S., see YAMAGUCHI
SAGE, D.
1976 and BERTICAT, P.; COLLET, G.; ROCCA, J.-L., Angew. makromol. Chemie **54**, 167 to 86 (10,11)
SAITO, G., see MARK
SAITO, H.
1979 and MATSUMOTO, Y., J. Chromatogr. **168**, 227—29 (19)
SAITOH, K.
see NODA
see OZAKI
SAKURADA, J., see HORII
SALAMONE, J. C.
1975 and WISNIEWSKI, A. W.; ELLIS, E., J. Amer. chem. Soc., Polym. Prepr. **16**, 727—30 (19)
SALAS, D. S., see ROBINSON
SALLADAY, D. G., see WHITE
SALOVEY, R.
1967 and HELLMAN, M. Y., J. Polymer Sci. A-2, Polymer Phys. **5**, 333 (19)
SALYER, I. O., see KENYON
SAMAY, G. (see also VOZKA)
1978 and KUBIN, M.; PODEŠVA, J., Angew. makromol. Chemie **72**, 185—98 (8, 19)
1979 and KUBIN, M., J. appl. Polymer Sci. **23**, 1879—81 (19)
1980 and FÜZES, L., J. Polymer Sci. C, Polymer Symposia **68**, 185—90 (16, 17)
SAMUELSON, O., see JAHANGIR
SANDRA, P., see VERZELE
SANDRONI, S., see GEISS
SANGALOV, Ju. A., see POZNJAK
SANTI, W., see BETHKE
SARODNIK, E., see FRÖLICH
SASAKI, H.
see FUKANO
see HASHIMOTO
see KATO
ŠAŠINA, T. I., see BELENKIJ
SATHER, G. A., see HANSEN
SAUNDERS, D. L. (see also SNYDER)
1976 J. Chromatogr. **125**, 163—77 (7)
SAUNDERS, F. L., see SMALL
SAVITSKAYA, T. S., see BELENKIJ
SAWADA, Y., see KATO
SCHAUWECKER, P., see COULOMBE
SCHAY, G., see BAYER
SCHEINERT, W., see MOTTWEILER
SCHICK-KALB, J., see UNGER
SCHIEDERMAIER, E.
1970 and KLEIN, J., Angew. makromol. Chemie **10**, 169—88 (20)
SCHIMMEL, K.-H., see SCHRÖDER
SCHINDLER, A., see BREITENBACH
SCHLIEBENER, C., see PATAT
SCHLITT, H., see GEISS
SCHMERMUND, J. T., see LOCKE
SCHMIDT, G. J., see ATWOOD
SCHMIDT, H., see KERN
SCHMIDT, J. A.
see also HENRY
see also WILLIAMS
1971 and HENRY, R. A.; WILLIAMS, R. C.; DIECKMAN, J. F., J. Chromatogr. Sci. **9**, 645 (11)
SCHMIDT-KUSTER, W. J., see SCHULTZE
SCHMITZ, F. P.
1981 and KLESPER, E., Polymer Bull. **5**, 603—08 (9)
SCHNABEL, W.
1976 and SOTOBAYASHI, H., Polymer J. **8**, 423—27 (19)
SCHNEIDER, G., see KILLMANN

SCHNEIDER, N. S.
1959 and HOLMES, L. G., J. Polymer Sci. **38**, 552—54 (9, 20)
1959 and HOLMES, L. G.; MIJAL, C. F.; LOCONTI, J. D., J. Polymer Sci. **37**, 551—53 (9, 20)
1960 and LOCONTI, J. D.; HOLMES, L. G., J. appl. Polymer Sci. **3**, 251 (20)
1961 and LOCONTI, J. D.; HOLMES, L. G., J. appl. Polymer Sci. **5**, 354—63 (9, 20)
1961 Analyt. Chemie **33**, 1829—31 (9)
SCHNEIDER, W., see RÖSSLER
SCHODDER, H., see BROCKMANN
SCHÖLLNER, R. (see also HELLWIG)
1967 Plaste u. Kautschuk **14**, 188—98 (9)
1968 and LÖHNERT, P., Plaste u. Kautschuk **15**, 436—39 (9, 21)
SCHOENMAKERS, P. J. (see also TIJSSEN)
1978 and BILLIET, H. A. H.; TIJSSEN, R.; DE GALAN, L., J. Chromatogr. **149**, 519—37 (14)
1979 and BILLIET, H. A. H.; DE GALAN, L., J. Chromatogr. **185**, 179—95 (7, 14)
1981 and BILLIET, H. A. H.; DE GALAN, L., J. Chromatogr. **205**, 13—30 (14)
SCHOLTAN, W.
1967 and KRANZ, D., Makromol. Chemie **110**, 150—72 (16, 19, 20)
1972 and KWOLL, F. J., Makromol. Chemie **151**, 33—48 (9, 20)
SCHOLTE, Th. G. (see also BALL)
1977 and MEIJERINK, N. L. J., British Polymer J., June 133—39 (19)
SCHOLTEN, P. C.
1978 Discuss. Faraday Soc. **65**, 242—51 (6)
SCHOLTENS, B. J. R.
1981 and WEIZEN, T. L., Makromol. Chemie **182**, 269—72 (19)
SCHOLZ, A. G. R., see SCHULZ, G. V.
SCHOMBURG, G. (see also BAYER)
SCHRÖDER, E.
1977 Studien zur Selektivtrennung von Polymeren durch adsorptive Blasentrennung. — Paper read on Symposium in Bratislava, May 11
SCHULZ, D. (see also KUNATH)
1977 and KUNATH, D.; BERGER, K., Paper read on Microsymposium "Polymer compositions". Berlin, 25/26 April (6)
SCHULZ, G.
1976 Plaste u. Kautschuk **23**, 398—401 (19)
1982 and GNAUCK, R.; ZIEBARTH, G., Plaste u. Kautschuk **29**, 398—401 (16)
SCHULZ, G. V.
see also BERGER
see also BÖHM
see also CASPER
see also JOVANOVIĆ
1935 Z. phys. Chemie B **30**, 379—98 (4)
1939 Z. phys. Chemie B **43**, 25—46 (4)
1940 and NORDT, E., J. prakt. Chemie **155**, 115 (9)
1952 and DOLL, H., Z. Elektrochemie, Ber. Bunsenges. phys. Chemie **56**, 248—57 (5)
1962 and SCHOLZ, A.; FIGINI, R. V., Makromol. Chemie **57**, 220—40 (9, 20)
1963 and DEUSSEN, P.; SCHOLZ, A. G. R., Makromol. Chemie **69**, 47—73 (9)
1965 and BERGER, K. C.; SCHOLZ, A. G. R., Z. Elektrochemie, Ber. Bunsenges. phys. Chemie **69**, 856—66 (9, 11, 18)
SCHULZ, L., see POLÁČEK
SCHULZ, R. C., see JACOBI
SCHURZ, J.
see BÖTTGER
see TREIBER
SCHUTTENBERG, H., see JACOBI
SCHWACHULA, G.
see FRÖLICH
see POPOV
SCHWEBEL, A., see ROTHMANN
SCHWEIGER, M.
1977 and LANGHAMMER, G., Plaste u. Kautschuk **24**, 101—04 (19)
SCHWENKE, K.-D.
1964 Faserforsch. u. Textiltechnik **15**, 266—72 (9)
1965 Z. Chemie **5**, 322—33 (9)
SCHWIMMER, S., see HEFTMANN
SCOFFONE, E., see COSANI
SCOTT, R. P. W.
1967 and BLACKBURN, D. W. J.; WILKINS, T., J. Gas Chromatogr. **5**, 183 (17)
1973 and KUČERA, P., Analyt. Chem. **45**, 749 to 754 (7)
1976 and KUČERA, P., J. Chromatogr. **125**, 251—63 (17)
1977 and KUČERA, P., J. Chromatogr. **142**, 213—32 (7, 11)
1979 and KUČERA, P., J. Chromatogr. **171**, 37—48 (11, 18)
1980 and SIMPSON, C. F., J. Chromatogr. **197**, 11—20 (11)
1980a J. Chromatogr. Sci. **18**, 49—54 (19)
1980b Faraday Disc. Royal Soc. Chem. **15**, 49—68 (18)
SEAGUSA, T., see FURUKAWA
SEBAN-DANON, J., see CHAPIRO
SEBESTIAN, I., see KARCH
SEBESTYÉN, Zs., see SZEPESY
SEDLICK, E., see OLIEMAN
ŠEFTER, V. E., see DEMENT'EVA
SEGAL, L.
1968 J. Polymer Sci. C., Polymer Symposia **21**, 267—82 (19)

1970 and TIMPA, J. D.; WADSWORTH, J. I., J. Polymer Sci. A-1, Polymer Chem. 8, 25—35 (19)
SEHON, A. H., see LEA
SEIDE, H., see BLASS
see LANGHAMMER
SEIDL, J.
1967 and MALINSKY, J.; DUŠEK, K.; HEITZ, W., Makroporöse Styrol-Divinylbenzol-Copolymere und ihre Verwendung in der Chromatographie und zur Darstellung von Ionenaustauschern. Adv. Polymer Sci. **5**, 113—213 (12)
SELDEN, A. C., see MURPHY
SELLEN, D. B., see HOLT
SEMECHKIN, A. V., see DAVANKOV
SEMLYEN, J. A., see DODGSON
SERGEEVA, L. M., see LIPATOV
SERVOTTE, A.
1975 and DE BRUILLE, R., Makromol. Chemie **176**, 203—12 (16, 19)
SETTON, I., see HAGNAUER
ŠEVLJAKOV, A. S., see BRYK
SHACKLEY, A. T., see BROUGH
SHAPIRO, I.
1950 and KOLTHOFF, I. M., J. Amer. chem. Soc. **72**, 776—82 (11)
SHARP, J. O., see MOORE, L. D.
SHASHIKANTS, S., see VYAS
SHAW, M. T.
1967 and RODRIGUEZ, F., J. appl. Polymer Sci. **11**, 991 (19)
SHDANOW, S. P., see ŽDANOV
SHELOKOV, A., see HIATT
SHEPHERD, M. J., see GILBERT
SHIBATA, S., see OGIHARA
SHIDA, M., see COTE
SHIMA, Y., see TERAMACHI
SHIMADA, K. (see also IWASAKI)
1979 and TARUTANI, T., J. Chromatogr. **168**, 401—06 (19)
SHIMONO, T.
1979 and ISOBE, T.; TARUTANI, T., J. Chromatogr. **179**, 323—29 (19)
1980 and TAKAGI, H.; ISOBE, T.; TARUTANI, T., J. Chromatogr. **197**, 59—70 (19)
SHIMOTSUMA, S., see MEYERHOFF
SHIONO, S.
1980 and KARINO, I.; ISHIMURA, A.; ENOMOTO, J., J. Chromatogr. **193**, 243—53 (18)
SHIRADZI, Z. H., see STOJANOV
SHRIRAZI, Z. H., see STOJANOV
SHTERN, V. Ja., see ŠTERN
SHULTZ, A. R. (see also BERGER, H. L.)
1972 and BRIDGMAN, A. L.; HADSELL, E. M.; MCCULLOUGH, C. R., J. Polymer Sci. A-2, Polymer Phys. **10**, 273—82 (19)
SHYLUK, S.
1962 J. Polymer Sci. **62**, 317—29 (9)
SIE, S. T.
1967 and RIJNDERS, G. W. A., Separat. Sci. **2**, 699, 729, 755 (15)
SIEBOURG, W.
1980 and LUNDBERG, R. D.; LENZ, R. W., Macromolecules **13**, 1013—16 (16, 19)
SILBERBERG, A.
see also PRIEL
1962a J. phys. Chem. **66**, 1872—83 (6)
1962b J. phys. Chem. **66**, 1884—1907 (6)
1970 J. Polymer Sci. C, Polymer Symposia **30**, 393—97 (6)
SILEBI, C. A. (see also NAGY)
1979 J. appl. Polymer Sci. **23**, 1699—1721 (13)
SILIPO, C., see CIAMPA
SIMHA, R. (see also FRISCH)
1953 and FRISCH, H. L.; EIRICH, F. R., J. phys. Chem. **57**, 584—89 (6)
SIMIONESCU, C. R.
1972 and OPREA, S. P.; BUTNARU, R., Cellul. Chem. & Technol. **6**, 365—77 (19)
SIMKINS, R. J. J., see ADAMS
SIMONSON, R., see KUMLIN
SIMPSON, C. F. (see also SCOTT)
1976 Column packing techniques. In [D 14], 291—97 (17)
SINGER, K., see EL-AWADY
SINGH, S.
1978 and HAMIELEC, A. E., J. appl. Polymer Sci. **22**, 577—84 (11, 19)
SINN, H., see NENTWIG
SIOUFFI, A. M. (see also GUIOCHON)
1979 and WAWRZYNOWICZ, T.; BRESSOLLE, F.; GUIOCHON, G., J. Chromatogr. **186**, 563 to 574 (21)
1981 and BRESSOLLE, F.; GUIOCHON, G., J. Chromatogr. **209**, 129—47 (21)
ŠIR, Z., see MINÁRIK
SISCO, W. R., see GILPIN
SKVORZOV, A. M.
1978 and BELENKIJ, B. G.; GANKINA, E. S.; TENNIKOV, M. B., Vysokomolekularnye Soedinenija, Ser. A **20**, 678—86 (16)
SLAATS, E. H. (see also POPPE)
1978 and KRAAK, J. C.; BRUGMAN, W. J. T.; POPPE, H., J. Chromatogr. **149**, 255—70 (7)
1980 Influence of Mobile Phase and Stationary Phase Compositions on the Retention in HPLC. Thesis, Amsterdam 1980 (16)
1981 and MARKOVSKI, W., FEKETE, J.; POPPE, H., J. Chromatogr. **207**, 299—323 (7, 16)
SLAGOWSKI, E. L.
1974 and FETTERS, K. L.; MCINTYRE, D., Macromolecules **7**, 394—96 (16, 19)

1977 and GEBAUER, R. C.; GAESSER, G. J., J. appl. Polymer Sci. **21** 2293—95 (19)

ŠLAIS, K. (see also KREJČI)

1974 and KREJČI, M., J. Chromatogr. **91**, 161 to 66 (19)

SLAVIN, W., see ATWOOD

ŠLICHTER, R. A., see CVETKOVSKIJ

SLIEMERS, F. A.
see BONI
see HASSELL

SLONAKER, D. F. (see also GUILLET)

1966 and COMBS, R. L.; GUILLET, J. E.; COOVER, H. W., J. Polymer Sci. A-2, Polymer Phys. **4**, 523—26 (11, 20)

SMALL, H.

1974 J. Colloid & Interface Sci. **48**, 147—61 (13)

1977 and SAUNDERS, F. L.; SOLC, J., Hydrodynamic Chromatography. A New Approach to Particle Size Analysis. — Phys. Research Laboratory, Dow Chem. Comp. Midland, Michigan (8, 13)

SMALL, P. A.

1953 J. appl. Chem. **3**, 71 (5)

1975 Long-Chain Branching in Polymers. Advances in Polymer Science **18**, 1—64 (19)

SMIT, J. A. M.
see SOETEMAN
see VAN DIJK
see VRIJBERGEN

SMITH, D. G., see MILLAR

SMITH, L. E., see FLEER

SMITH, R.
see FOX
see ROBB

SMITH, W. B.

1965 and KOLLMANSBERGER, A., J. phys. Chem. **69**, 4157—61 (19)

1966 and MAY, J. A.; KIM, C. W., J. Polymer Sci. A-2, Polymer Phys. **4**, 365—374 (19)

SMITH, W. N.

1967 J. appl. Polymer Sci. **11**, 639—57 (16)

SMITH, W. V.

1970 and THIRUVENGADA, S., Rubber Chem. & Technol. **43**, 1439—50 (19)

1974 J. appl. Polymer Sci. **18**, 3685—97 (8, 19)

SMOLKOVÁ, E. (see also FELTL)

1980 and ZIMA, J.; DOUŠEK, F. P.; JANSTA, J.; PLZAK, Z., J. Chromatogr. **191**, 61—69 (11)

SMRZ, M., see BORÁK

SNIEZKO-BLOCKI, G. A., see BROUGH

SNYDER, L. R., see also DOLAN
see also GANT
see also KARGER
see also KELLER

1963 J. Chromatogr. **12**, 488—509 (11)

1965 Principles of Gradient Elution. In [A 6], Vol. 7, 1—51 (14)

1966 and WARD, J. W., J. phys. Chem. **70**, 3941—52 (11)

1966a Separat. Sci. **1**, 191 (11)

1966b J. Chromatogr. **25**, 274—93 (11)

1968 In [A 4], 85 (10, 11)

1969 J. Chromatogr. Sci. **7**, 352 (15, 18)

1969a and SAUNDERS, D. L., J. Chromatogr. Sci. **7**, 195—208 (14)

1969b and SAUNDERS, D. L., J. Chromatogr. **44**, 1—13 (3, 14)

1970 J. Chromatogr. Sci. **8**, 692—706 (10, 14)

1971a Coupled Columns. In [D 3], 154—56 (14)

1971b In [D 3], 214 (10, 14)

1974a Analyt. Chem. **46**, 1384—93 (7)

1974b Techniques for Solving the General Elution Problem. In [D 5], 462—82 (14)

1974c J. Chromatogr. **92**, 223—30 (7)

1978 J. Chromatogr. Sci. **16**, 223—34 (7)

1979 and DOLAN, J. W.; GANT, J. R., J. Chromatogr. **165**, 3—30 (3, 14)

1980 and POPPE, H., J. Chromatogr. **184**, 363 to 413 (7)

1980a A New Route to Large Column Plate Numbers with Reasonable Per-Sample Separation Time. Lecture 1P3 at the 13th International Symposium on Chromatography, Cannes, 30 June—4 July (14)

1980b High Performance Liquid Chromatography **1**, 207—316 (14)

SOCZEWIŃSKI, E. (see also WAWRZYNOWICZ)

1969 Analyt. Chem. **41**, 179—82 (7, 21)

1973 and GOŁKIEWICZ, W., Chromatographia **6**, 269—73 (7)

1981 and JUSIAK, J., Chromatographia **14**, 23 to 31 (21)

SOETEMAN, A. A. (see also VRIJBERGEN)

1978 and ROELS, J. P. M.; VAN DIJK, J. A. P. P.; SMIT, J. A. M., J. Polymer Sci. A-2, Polymer Physics **16**, 2147—55 (19)

SOLOMONOVA, J. N., see KISELEV

SOLOV'JANOV, A. A., see KAZANSKIJ

ŠOLTÉS, L. (see also BEREK)

1980 and BEREK, D.; MIKULÁŠOVÁ, D., Colloid Polym. Sci. **258**, 702—10 (19)

SONNTAG, F.

1954 and JENCKEL, E., Kolloid-Z. u. Z. Polym. **135**, 1 (18)

SOREMARK, Ch., see LINDSTRÖM

SORIA, V.
see CAMPOS
see FIGUERUELO

SÖRVIK, E. M. (see also ANDERSON)

1977 J. appl. Polymer Sci. **21**, 2769—90 (19)

SOSA, J. M.
1978 and LOMBANA, L.; PETIT, L., J. Chromatogr. **157**, 371—77 (19)
SOSNOVÁ, L., see FELTL
SOTOBAYASHI, H., see SCHNABEL
SPAANS, H., see MUKHERJEE
ŠPAČEK, P., see VOZKA
SPARROW, J. T., see HANCOCK
SPATORICO, A. L.
1973 and COULTER, B., J. Chromatogr. **79**, 121—28 (9, 11, 20)
1975 J. appl. Polymer Sci. **19**, 1601—10 (11, 16, 17, 19)
1975 and BEYER, G. L., J. appl. Polymer Sci. **19**, 2933—45 (11, 19)
SPIEGELS, M. C., see DESREUX
SPORER, A. H.
1959 and TRUEBLOOD, K. N., J. Chromatogr. **2**, 499 (7)
SPYCHAJ, T.
1979 and BEREK, D., Polymer [London], **20**, 1108—14 (19)
SQUIRE, K. M., see GLAJCH
ST.-PIERRE, L. E., see BREBNER
STABINGER, H., see KRATKY
STACY, C. J., see KRAUS
STAHL, E.
see also BAYER
1968 Chromatographia **1**, 338—42 (21)
ŠTAMBERG, J.
see also BALDRIAN
see also PEŠKA
1978 and PEŠKA, J., Inst. Macromol. Chem. Prague, Series Macro **1—5**, 1—23 (12)
STANFORD, D. M., see ALDHOUSE
STAPELFELDT, H. E.
see PRINCE
see HAZELL
STAUCHER, B., see CALZOLARI
STAVERMAN, A., see KONINGSVELD
STEIN, D. J., see BRÜSSAU
Steinwehr, H. E. v., see KERN
STEHLIČEK, J., see JANCA
STEIN, D. J., see BRÜSSAU
STEINWEHR, H. E. v., see KERN
STEJNY, J., see PATEL
STEJSKAL, J.
1978 and KRATOCHVIL, P., J. appl. Polymer Sci. **22**, 2925—39 (20)
1981 and KRATOCHVIL, P.; STRAKOVÁ, D., Macromolecules **14**, 150—53 (19)
STENLUND, B. G., see FORSS
ŠTERN, V. Ja.
see KUDRJAVCEVA
see LITMANOVIČ
STICHER, O., see HOSTETTMANN
STICKNEY, P. B., see BONI
STIX, W.
1979 and HEITZ, W., Makromol. Chemie **180**, 1367—69 (19)
STOCKMAYER, W. H.
see also KLEINTJENS
see also ZIMM
1945 J. Chem. Phys. **13**, 199 (19)
1959 and MUUS, L. T., unpublished, cit. from BILLMEYER, F. W., in: Textbook of Polymer Chemistry. Interscience, New York, London, 1962 (4)
STOJANOV, Ch. (see also CANTOW, H. J.)
1977 and SHIRADZI, Z. H.; AUDU, T. O. K., Z. Elektrochemie, Ber. Bunsenges. phys. Chemie **81**, 767—73 (19)
1978 and AUDU, T. O. K.; SHRIRAZI, Z. H., Z. anal. Chemie **293**, 122—26 (19)
STOKLOSA, H. J.
see FERGUSON
see YAU
STONE, T., see DAWKINS
STRAKOVÁ, D., see STEJSKAL
STRAZIELLE, C. (see also ARPIN)
1976 IUPAC, Macromol. Div. 615—25 (19)
STREICHSBIER, O., see BREITENBACH
STRETCH, C.
1961 and ALLEN, G., Polymer [London] **2**, 151 to 160 (20)
STROMBERG, R. R. (see also PEYSER)
1959 and QUASIUS, A. R.; TONER, S. C.; PARKER, M. S., J. Res. nat. Bur. Standards **62**A, 71 (18)
1961 and KLINE, G. M., Mod. Plastics **38**, 123 to 124, 129—30, 133, 192—96 (18)
1963 and GRANT, W. H., J. Res. nat. Bur. Standards **67**A, 601—06 (6, 18)
1964 and GRANT, W. H.; PASSAGLIA, R., J. Res. nat. Bur. Standards **68**A, 4 (6, 18)
1965 and TUTAS, D. J.; PASSAGLIA, E., J. phys. Chem. **69**, 3955—63 (6, 18)
1967 In: Treatise on Adhesion and Adhesives. Ed.: R. L. PATRICK. Edward Arnold, London, Vol. 1, 69 (6)
STROUSE, C. E., see CHOW
STRUBERT, W.
1973 Chromatographia **6**, 50—52 (17)
STUART, H. A., see FENDLER
STUCHBURY, J. E., see ADAMS
STÜTZEL, B.
1971 Thesis, Freiburg (cit. from PROBST et al., 1974) (19)
SUBRAMAMIAN, A.
1972 Rubber Chem. & Technol. **45**, 346—58 (19)
SUCHAN, H. L.
see MALONE

see WILLIAMS
see YAU
SUCHOTINA, G. N., see ALEKSEEVA
SUMMERS, J. T., see GUILLET
SUSLOVA, E. N., see KUZAEV
SUSSMAN, M. V.
1975 and RATHORE, R. N. S., Chromatographia **8**, 55—59 (17)
SUZUKI, K., see UJIMOTO
SUZUKI, N.
see NODA
see OZAKI
SUZUKI, S., see MORIMOTO
SUZUKI, T., see HAMA
SUZUKI, Y., see OGAWA
SVENSSON, H.
1955 and AGRELL, C.-E.; DEHLÉN, S.-O.; HAGDAHL, L., Science Tools **2**, 17 (17)
SVOBODA, L., see JANDERA
SWENSON, H. A.
1955 Acta chem. scand. **9**, 572 (18)
SYLVESTER, N. D., see VERHOFF
SYMPSON, D., see DODGSON
SZAKASITS, J. J.
1970 and PEURIFOY, P. V.; WOODS, L. A., Analyt. Chem. **42**, 351 (21)
SZEPESY, L., (see also BAYER)
1975 and SEBESTYÉN, Zs.; FEHÉR, I.; NAGY, Z., J. Chromatogr. **108**, 285—97 (17)
SZESZTAY, M.
1981 and TÜDÖS, F., Polymer Bull. **5**, 429—35 (19)
SZEWCZYK, P.
1976 Polymer [London] **17**, 90—91 (8)
1980 J. Polymer Sci. C, Polymer Symposia **68**, 191—97 (8)

TAFT, R. W.
see ABBOUD
see KAMLET
TAGA, T.
1973 and INAGAKI, H., Angew. makromol. Chemie **33**, 129—42 (21)
TAGATA, N.
1972 and HOMMA, T., Nippon Kagaku Zassi (Japan) **8**, 1330—38 (18, 21)
TAGER, A. A., see CILIPOTKINA
TAKAGI, H., see SHIMONO
TAKAHASHI, A., see KATO
TAKAMATSU, T., see KATO, Y.
TAKEMOTO, N., see TERAMACHI
TAKEUCHI, T.
see also MORI
TAKIMI, N., see MIYAMOTO
TALEB-BENDIAB, S. A.
1980 and VERGNAUD, J.-M., J. appl. Polymer Sci. **25**, 499—510 (19)
TALLEY, C. P.
1979 and BOWMAN, L. M., Analyt. Chem. **51**, 2239—42 (19)
TAN, M.
see BIESENBERGER
see DUVDEVANI
TAN, Y. Y., see BUTER
TANAKA, M., see ITOH
TANAKA, N.
1978 and GOODELL, H.; KARGER, B. L., J. Chromatogr. **158**, 233—48 (7, 11)
TANAKA, S., see OGAWA
TANAKA, T.
see also INAGAKI
see also KOTAKA
see also OMOTO
1980 a and DONKAI, N.; INAGAKI, H., Macromolecules **13**, 1021—23 (21)
1980 b and OMOTO, M.; DONKAI, N.; INAGAKI, H., J. Macromol. Sci. — Phys. **B 17** (2); 211 to 228 (18, 19)
TANAKA, W., see HELLER
TANIMURA, T. (see also OGIHARA)
1970 and PISANO, J. J.; ITO, Y.; BOWMAN, R. L., Science (Japan) **169**, 54 (9)
TAPPA, R., see CICCIOLI
TAPUHI, Y., see KARGER
TARAKANOV, O. G.
see also BELENKIJ
see also OKUNEV
see also VACHTINA
1962 and OKUNEV, P. A., Vysokomol. Soed. **4**, 683—89 (20)
TARUTANI, T.
see also IWASAKI
see also SHIMADA
see also SHIMONO
1970 J. Chromatogr. **50**, 523—26 (19)
TAUSCH, W.
1971 Meßtechnik **80**, 38 (21)
TAVEL, P. v. (see also RÜFENACHT)
1968 and BOLLINGER, W., Helv. chim. Acta **51**, 278 (9)
1971 and BIERI, V., Makromol. Chemie **149**, 63—74 (9)
1976 and RÜFENACHT, H. J., Makromol. Chemie **177**, 2431—48 (9)
TAYLOR, G. B.
1947 J. Amer. chem. Soc. **69**, 635—38 (9)
TAYLOR, R.
1980 and PRAGNELL, R. J.; MCLAREN, J. V., J. Chromatogr. **195**, 154—57 (19)
TCHIR, M. F., see DOBBIN
TELLER, E., see BRUNAUER
TEMNIKOVSKIJ, V. A., see ANDRIANOV

TENNIKOV, M. B.
see also BELENKIJ
see also SKVORZOV
1977 and NEFEDOV, P. P.; LAGAREVA, M. A.; FRENKEL, S. Ja., Vysokomol. Soed., Ser. A **19**, 657—60 (16)

TERAKAWA, T., see KAMIDE

TERAMACHI, S.
1975 and ESAKI, H., Polymer J. **7**, 593—99 (18, 21)
1978 and HASEGAWA, A.; AKATSUKA, M.; YAMASHITA, A.; TAKEMOTO, N., Macromolecules **11**, 1206—10 (19)
1979 and HASEGAWA, A.; SHIMA, Y.; AKATSUKA, M.; NAKAJIMA, M., Macromolecules **12**, 992—96 (18, 21)

TERRY, S. L.
1968 and RODRIGUEZ, F., J. Polymer Sci. C., Polymer Symposia **21**, 191—97 (19)

THAKER, M. S., see OTHMER

THEIL, I.
1972 and CALUGARU, E.-M.; FELDMAN, D., Faserforsch. u. Textiltechnik **23**, 123—27 (9)

THEURER, K., see HUDGINS

THIAULT, S., see GUILLEMIN

THIES, C. (see also BOTHAM)
1966 J. phys. Chem. **70**, 3783—90 (6, 19)
1968 J. Colloid & Interface Sci. **27**, 734—42
1971 J. Polymer Sci. C, Polymer Symposia **34**, 201—10 (18)

THIMOT, N., see VIVILECCHIA

THIRUVENGADA, S., see SMITH, W. V.

THOMAS, J. P. (see also GUILLEMIN)
1979 and BRUN, A.; BOUNINE, J.-P., J. Chromatogr. **172**, 107—30 (7)

THOMAS, J. R., see FONTANA

THOMPSON, G. H. (see also HOVINGH)
1967 and MYERS, M. N.; GIDDINGS, J. C., Separat. Sci. **2**, 797 (13)
1969 and MYERS, M. N.; GIDDINGS, J. C., Analyt. Chem. **41**, 1219—22 (13)

THOMPSON, L. F., see BOWDEN

TIJSSEN, R. (see also SCHOENMAKERS)
1976 and BILLIET, H. A. H.; SCHOENMAKERS, P. J., J. Chromatogr. **122**, 185—203 (5, 7)

TIMM, D. C.
1975 and RACHOW, J. W., J. Polymer Sci. A-1, Polymer Chem. **13**, 1401—16 (16, 19)

TIMOFEEVA, G. I.
see DUBROVINA
see REINISCH

TIMPA, J. D., see SEGAL

TISELIUS, A.
see also ALM
see also HAGDAHL
1952 Endeavour **11**, 5 (14)
1963 and PORATH, J.; ALBERTSSON, P.-Å., Science [Washington] **141**, 13 (9)

TIXIER, R., see BOSCHETTI
see MONSIGNY
see URIEL

TODOSLIČUK, T. T., see LIPATOV

TOKIWA, M., see ONDA

TOLMAN, C. A.
1978 and ANTLE, P. E., J. Organomet. Chem. **159**, C5 (19)

TOMOSHIGE, S., see MIYAMOTO

TOMPA, C. R.
1949 Trans. Faraday Soc. **45**, 1142 (5)

TOMPA, H., see ENGLERT

TONER, S. C., see STROMBERG

TOPČIEV, A. V.
see KUDRJAVCEVA
see LITMANOVIČ

TORRE, A., see ALVARIÑO

TRANCHANT, J., see BEYER

TRAPPE, W.
1940 Biochem. Z. **305**/**306**, 316 (7)

TREIBER, E.
1953 and POROD, G.; GIERLINGER, W.; SCHURZ, J., Makromol. Chemie **9**, 241—43 (18)

TREKOVAL, J., see JANČA

TREMBLAY, M.
see ADAMS
see PERRAULT

TREMBLAY, R., see PERRAULT

TRIVEDI, H. C.
1973 and PATEL, V. M.; PATEL, R. D., Europ. Polymer J. **9**, 525—31 (18)

TROJER, L., see HANSSON

TRUEBLOOD, K. N., see SPORER

TSAO, G. T., see BAO

TSWETT, M. S., see CVET

TSYURUPA, M. P., see CJURUPA

TUCKER, R. G., see KRISHEN

TÜDÖS, F., see SZESZTAY

TUIJNMAN, C. A. F.
1957 and HERMANS, J. J., J. Polymer Sci. **25**, 385—401 (18)

TUNG, L. H. (see also RUNYON)
1956 J. Polymer Sci. **20**, 495—506 (4)
1966a J. appl. Polymer Sci. **10**, 375—85 (16)
1966b and MOORE, J. C.; KNIGHT, G. W., J. appl. Polymer Sci. **10**, 1261—70 (16)
1966c J. appl. Polymer Sci. **10**, 1271—83 (4, 16)
1969 J. appl. Polymer Sci. **13**, 775—84 (16)
1969 and RUNYON, J. R., J. appl. Polymer Sci. **13**, 2397—2409 (16, 19)
1971a Mathematical Methods of Correcting Instrumental Spreading in GPC. In [D 2], 73—80 (8, 16)
1971b Data Treatment in GPC. In [D 2], 145—53 (9, 16)

1971c J. Polymer Sci. A-2, Polymer Phys. **9**, 759 to 62 (19)
1979 J. appl. Polymer Sci. **24**, 953—63 (19)
TURKOVA, L. D., see NESTEROV
TUTAS, D. J.
see PEYSER
see STROMBERG
TUZAR, Z.
1967 and KRATOCHVIL, P., Collect. Czechoslov. chem. Commun. **32**, 3358—70 (5)
TVERDOCHLEBOVA, I. I., see ANDRIANOV
TYIHÁK, E.
see also MINCSOVICS
1979 and MINCSOVICS, E.; KALÁSZ, H., J. Chromatogr. **174**, 75—81 (21)
TYLER, A. J., see ARMISTEAD

UDA, T., see KOTAKA
UEBERREITER, K.
1959 and GÖTZE, T., Makromol. Chemie **29**, 61—69 (9)
UEDA, M. (see also HOMMA)
1972 Polymer. J. **3**, 431—41 (20)
UENO, H.
1981 and NISHIKAWA, M.; MURANAKA, M.; HORIUCHI, Y., J. Chromatogr. **207**, 425 to 429 (19)
UENO, Y., see OHASHI
UGLEA, C. V.
1973 Makromol. Chemie **166**, 275—82 (19)
UJIMOTO, K.
1981 and SUZUKI, K.; KURIHARA, H., J. Chromatogr. **210**, 1—10 (19)
UL'EVA, N. S., see ALEKSEEVA
ULIŃSKA, A.
1960 and HUPPENTHAL, L., Roczniki Chem. [Ann. Soc. chim. Polonorum] **34**, 1089 to 1101 (18)
ULLMAN, R.
see also ELLERSTEIN
see also KORAL
see also PERKEL
1964 In: Encyclopedia of Polymer Science and Technology. Ed.: H. F. MARK; N. G. GAYLORD; N. BIKALES Vol. 1. — Interscience, New York, 551 (6)
ULLNER, H., see HEITZ
UNGER, K.
see also DAVANKOV
see also KOHLSCHÜTTER
see also PROBST
see also ROUMELIOTIS
1969 and BERG, K.; GALLEI, E., Kolloid-Z. u. Z. Polym. **234**, 1108—14 (11)
1973 and SCHICK-KALB, J.; KREBS, K.-F., J. Chromatogr. **83**, 5—9 (10, 11)
1974 Chimia [Zurich] **28**, 679—89 (9, 10)
1976 and BECKER, N.; ROUMELIOTIS, P., J. Chromatogr. **125**, 115—27 (11)
1978 and MESSER, W., KREBS, K. F., J. Chromatogr. **149**, 1—12 (10, 15)
1979 and GIMPEL, M. G., J. Chromatogr. **180**, 93—102 (10)
1979 In [A 25], 130—133 (19)
1980a and ROUMELIOTIS, P., MUELLER, H.; GOETZ, H., J. Chromatogr. **202**, 3—14 (11)
1980b ROUMELIOTIS, P., MUELLER, H., Porous carbon packings as supports in reverse phase ion-pair chromatography. Paper 4 A2-1 at the 13th International Symposium on Chromatography, Cannes, 30 June to 4 July (11)
URANECK, C. A.
1965 and BARKER, M. G.; JOHNSON, W. D., Rubber Chem. & Technol. **38**, 802—16 (20)
URBAN, H., see HOFFMANN
URIEL, J.
1966 Bull. Soc. Chim. Biol. **48**, 969 (12)
1966 and BERGES, J., C. R. hebd. Séances Acad. Sci. [Paris] **262**, 164—67 (12)
1971 and BERGES, J.; BOSCHETTI, E.; TIXIER, R., C. R. hebd. Séances Acad. Sci. [Paris] **273**, 2358 (12)
URYU, T., see HIGASHIMURA
USMANOVA, D. O., see KISELEV

VACHTINA, I. A. (see also BELENKIJ)
1970 and TARAKANOV, O. G.; GRACHEVA, L. I., Plastičeskije Massy **10**, 58—59 (19)
1976 and BETTGER, T.; ANDREEV, A. N.; NOVOŽILOVA, O. S.; TARAKANOV, O. G., Vysokomol. Soed., Ser. A **18**, 2138—40 (19)
1978 and PETRAKOVA, E. A.; KNOPP, Ch.; GLADKOVSKIJ, G. A., Vysokomol. Soed., Ser. A **20**, 2388—91 (21)
1980 and PETRAKOVA, E. A.; PENCEL, U.; TARAKANOV, O. G., Vysokomol. Soed., Ser. A **22**, 1671—74 (19, 21)
VALČICHINA, M. D., see BELENKIJ
VALUEV, V. I., see CVETKOVSKIJ
VAN AARTSEN, J. J., see BONN
VAN DE BRANDE, J., see HUISMAN
VAN DEEMTER, J. J.
1956 and ZUIDERWEG, F.; KLINKENBERG, A., Chem. Engng. Sci. **5**, 271 (15)
VAN DEN HOED, N., see VAN KREVELD
VAN DER LINDEN, C.
1980 Polymer [London] **21**, 171—76 (8)
VAN DER MAEDEN, F. P. B., see BUYTENHUYS
VAN DERVEER, D. S., see COOPER

van Dijk, J. A. P. P. (see also Soeteman)
1976 and Henkens, W. C. M.; Smit, J. A. M., J. Polymer Sci. A-2, Polymer Phys. **14**, 1485—93 (19)
1980 and Roels, J. P. M.; Smit, J. A. M., Aqueous Gel Permeation Chromatography: The Separation of Neutral Polymers on Silicagel. In [D 22 II], 95—111 (8, 11, 19)
van Kreveld, M. E.
1978 and Van Den Hoed, N., J. Chromatogr. **149**, 71—91 (15, 17)
van Roelenbosch, M., see Verzele
van Wendel de Joode, M. D.
1979 and Hindriks, H.; Lakeman, J., J. Chromatogr. **170**, 412—18 (21)
Vandemark, F. L., see Miller
Varma, I. K.
1979 and Patnaik, S., J. Polymer Sci. A-1, Polymer Chem. **17**, 3279—89 (19)
Vaughan, G., see Cooper, W.
Vaughan, M. F.
1963 and Green, J. H. S., Techniques of Polymer Sciences. Monograph 17: Soc. Chem. Ind. [London], 81—95 (9, 11)
1977 and Francis, M. A., J. appl. Polymer Sci. **21**, 2409—18 (17, 19)
1978 and Dietz, R., Preparative gel permeation chromatography. In [D 11/1], 199—217 (17)
Verdini, A. S., see Cosani
Verhoff, F. H.
1970 and Sylvester, N. D., J. Macromol. Sci. A **4**, 979 (8)
Vergnaud, J.-M., see Taleb-Bendiab
Verillon, F., see Martin
Verweij, A., see Dhont
Verzele, M.
1979 and Lammens, J.; Van Roelenbosch, M., J. Chromatogr. **186**, 435—41 (10, 11)
1980 and Mussche, P.; Sandra, P., J. Chromatogr. **190**, 331—37 (11)
Vescelius, L. E., see Anderson
Vespalec, R.
see also Krejči
see also Mikeš
Vesper, W., see Wulff
Vilenčik (Vilenchik), L. Z.
see also Belenkij
see also Janča
see also Kever
1971 and Belenkij, B. G., J. Chromatogr. **56**, 13—22 (19)
1974 and Belenkij, B. G.; Aleksandrov, M. L.; Rejfman, L. S.; Chubarov, V. A., Ž. fiz. Chimii **48**, 2086 (8)
Vinkenborg, C., see Dhont
Viska, J., see Borák
Vitek, R. K.
1978 and Kent, D. N., International Laboratory, March—April, 73—76 (21)
Vivilecchia, R.
1977 and Lightbody, B.; Thimot, N.; Quinn, H., Exclusion Chromatography of Hydrophilic and other Polymers. In [D 22/I], 11—27 (19)
Vlachogiannis, G. J., see Barker
Vlček, P., see Janča
Vogel, A. I.
1950 and Cresswell, W. T.; Jeffery, G. H.; Leicester, J., Chemistry & Industry (Chem. Week) **1950**, 358 (19)
1951 and Cresswell, W. T.; Jeffery, G. H.; Leicester, J., Chemistry & Industry (Chem. Week) **1951**, 376 (19)
Vogel, K., see Kohlschütter
Vogtel, P., see Halász
Voigt, D., see Eckhardt
Volchikhina, M. D., see Belenkij
Voskamp, D., see Olieman
Vozka, S.
see also Danielewicz
see also Kubin
1977 and Kubin, M., J. Chromatogr. **139**, 225 to 235 (16)
1980b and Kubin, M.; Samay, G., J. Polymer Sci. C, Polymer Symposia **68**, 199—208 (16)
Vrijbergen, R. R.
1978 and Soeteman, A. A.; Smit, J. A. M., J. appl. Polymer Sci. **22**, 1267—76 (8, 19)
Vyas, N. G.
1979 and Shashikants, S.; Patel, C. K.; Patel, R. D., J. Polymer Sci. A-2, Polymer Phys. **17**, 2021—29 (19)
Vyskočil, V., see Nondek

Wadsworth, J. I., see Segal
Wagner, E. P.
1971 and Greff, R. J., J. Polymer. Sci. A-1, Polymer Chem. **9**, 2193—2207 (21)
1972 and Greff, R. J., J. Polymer Sci. A-1, Polymer Chem. **10**, 1195—1205 (21)
Wagner, H. L. (see also McCrackin)
1977 and McCrackin, F. L., J. appl. Polymer Sci. **21**, 2833—45 (19)
Wagner, R. A., see Rudin
Waksmundzki, A.
1973 and Różyło, J. K., J. Chromatogr. **78**, 55—62 (21)
1979 and Pryke, P.; Dawidowicz, A., J. Chromatogr. **168**, 234—40 (11, 21)
Walch, A., see Reiner

WÄLCHLI, J.
1978 and MIYAMOTO, T.; INAGAKI, H., Bull. Inst. Chem. Res., Kyoto Univ. **56**, 80—88 (21)
WALKER, F. S., see CROOK
WALL, R. A., see GHAEMI, Y.
WALLWORK, M. A., see GILBERT
WALSH, E. K.
1971 J. Chromatogr. **55**, 193—202 (21)
WALZ, R.
1977 and BÖMER, B.; HEITZ, W., Makromol. Chemie **178**, 2527—34 (19)
1978 and HEITZ, W., J. Polymer Sci. A-1, Polymer Chem. **16**, 1807—14 (19)
WARD, I. M., see FRANK
WARD, J. M.
1971 and WILLIAMS, T., J. Macromol. Sci., B **5**, 693—719 (21)
WARD, J. W., see SNYDER
WARD, N., see COLIN
WARNER, F. P., see DAWKINS
WARSHAW, H. S.
1971 and ACKERS, G. K., Analyt. Biochem. **42**, 405—21 (19)
WATANABE, H., see KATO
WATERS, J. L. (see also LITTLE)
1970 J. Polymer Sci. A-2, Polymer Phys. **8**, 411—15 (17)
1971 J. Chromatogr. **55**, 213—19 (17)
WAWRZYNOWICZ, T. (see also SIOUFFI)
1979 and SOCZEWIŃSKI, E., J. Chromatogr. **169**, 191—203 (21)
WAYMAN, M.
see CHUA
see LORA
see OBIAGA
WEAKLEY, T. J. R. (see also POPE)
1960 and WILLIAMS, R. J. P.; WILSON, J. D., J. chem. Soc. [London], 3963—73 (20)
WEBER, U.
1976 Herstellung eines azeotropen SAN-Copolymerisats von hoher chemischer Einheitlichkeit und Charakterisierung durch Baker-Williams-Fraktionierung mit und ohne Temperaturgradient. Thesis, TU Dresden (9, 20)
WEHR, C. T.
1979 and ABBOTT, S. R., J. Chromatogr. **176**, 453—62 (11, 19)
WEINER, P. H., see KARGER
WEINHOLD, G., see KLEIN
WEIR, E. D., see BLY
WEISS, A. R.
1969 and COHN-GINSBERG, E., J. Polymer Sci. B, Polymer Letters **7**, 379 (8)
1970 and COHN-GINSBERG, E., J. Polymer Sci. A-2, Polymer Phys. **8**, 148—52 (8)
WEISS, F. T., see JUNGNICKEL
WEISSBERG, S. G., see ROTHMAN
WEIZEN, T. L., see SCHOLTENS
WELCH, K. R., see LATTIMER
WERNER, W., see NIKOLOV
WESSLAU, H.
1956 Makromol. Chemie **20**, 111—42 (4)
WESSLEN, B.
1975 and MÅNSSON, P., J. Polymer Sci. A-1, Polymer Chem. **13**, 2545—49 (18, 21)
WESTALL, R. G., see BATE-SMITH
WESTERMAN, L.
1974 and CLARK, J. C., J. Polymer Sci. A-2, Polymer Phys. **11**, 559—69
WHATLEY, J. A., see LITTLE
WHEALS, B. B.
1975 J. Chromatogr. **107**, 402—06 (11)
WHEATON, R. M.
1953 and BAUMANN, W. C., Ann. New York Acad. Sci. **57**, 159 (8)
WHITE, J. L.
see also KOTAKA
see also MCLEAN
1972 and SALLADAY, D. G.; QUISENBERRY, D. O.; MCLEAN, D. L., J. appl. Polymer Sci. **16**, 2811—27 (19, 21)
WICHMAN, A.
1980 and PAVLU, B., High Speed Gel Filtration of Proteins. Contribution C12-1 at the 13th International Symposium on Chromatography, Cannes, 30 June—July (11)
WICKE, E.
1965 Z. Elektrochemie, Ber. Bunsenges. phys. Chemie **69**, 761 (17)
WICKRAMANAYAKE, P. P.
1980 and AUE, W. A., J. Chromatogr. **195**, 25 34 (11)
WIEGLEB, H.
see BRAUER
see ECKHARDT
WIELAND, T., see DETERMANN
WILD, L. (see also BARLOW)
1967 and GULIANA, R., J. Polymer Sci. A-2, Polymer Phys. **5**, 1087—1101 (19)
1971 and RANGANATH, R.; RYLE, T., J. Polymer Sci. A-2, Polymer Phys. **9**, 2137—50 (21)
1977 and RANGANATH, R.; BARLOW, A., J. appl. Polymer Sci. **21**, 3331—43 (19)
WILDER, D. R., see LOCHMÜLLER
WILKE, C. R., see CHANG
WILKINS, T., see SCOTT
WILKINSON, M. C., see GOODALL
WILLIAMS, R. C.
see also HENRY
see also SCHMIT

1971 and SCHMIT, J. A.; SUCHAN, H. L., J. Polymer Sci. B, Polymer Letters **9**, 413—17 (8)

WILLIAMS, R. J. P.
see ALM
see BAKER
see BANNISTER
see HAGDAHL
see POPE
see WEAKLEY

WILLIAMS, T.
see FRANK
see WARD

WILLIAMSON, B. F., see BRISTOW

WILLIAMSON, G. R. (see also CERVENKA)
1972 and CERVENKA, A., Europ. Polymer J. **8**, 1009—17 (8, 19)

WILSON, J. D., see WEAKLEY

WILSON, W. W.
1979 and FANG, P.; McGINNIS, G. D., J. appl. Polymer Sci. **24**, 2195—98 (19)

WINAU, H., see HEITZ

WIRKKALA, R. A., see DICKSON

WISNIEWSKI, A. W., see SALAMONE

WOHLSCHIESS, R., see ELGERT

WOJTANIA, W.
1979 Polimery **24**, 170—73 (19)

WOLF, B. A. (see also BREITENBACH)
1978 and BIERINGER, H. F.; BREITENBACH, J. W., Brit. Polymer J. **10**, 156—60 (9)

WOLF, F., see HEYER

WONDRACZEK, R., see HEUBLEIN

WOODBREY, J. C., see PROVDER

WOODS, L. A., see SZAKASITS

WOODS, S. J., see HOWARD

WORMAN, C., see CHAN

WOSESSON, M. W., see FINLAYSON

WULFF, G.
1978 and VESPER, W., J. Chromatogr. **167**, 171—86 (12)

WUNDERLICH, W., see KREBS

WÜNDRICH, K., see PASTUSKA

WÜNSCHE, P., see ECKHARDT

YAGI, T., see INAGAKI

YAMADA, T., see NAKAMURA

YAMAGUCHI, K. (see also HAMA)
1969 and SAEDA, S., J. Polymer Sci. A-2, Polymer Phys. **7**, 1303—14 (9, 20)

YAMAGUCHI, N., see ONDA

YAMAKAWA, A., see MORI

YAMAKAWA, H.
1965 J. Chem. Phys. **43**, 1334—44 (19)

YAMAMOTO, A.
1970 and NODA, I.; NAGASAWA, M., Polymer J. **1**, 304—11 (9)

YAMAMOTO, M., see KATO, Y.

YAMAMOTO, Y., see KATO, TADAYA

YAMASHITA, A., see TERAMACHI

YAMASHITA, S., see KOHJIYA

YAU, W. W.
see also FERGUSON
see also KIRKLAND
see also MALONE
1967 and MALONE, C. P., J. Polymer Sci. B, Polymer Letters **5**, 663—69 (8)
1968 and SUCHAN, H. L.; MALONE, C. P., J. Polymer Sci. A-2, Polymer Phys. **6**, 1349 to 1355 (17, 19, 21)
1968 and FLEMING, S. W., J. appl. Polymer Sci. **12**, 2111—16 (19)
1971 and MALONE, C. P.; SUCHAN, H. L.: Separation Mechanisms in Gel Permeation Chromatography. In [D 2], 105—17 (8, 11, 15, 16)
1976 and KIRKLAND, J. J.; BLY, D. D.; STOKLOSA, H. J., J. Chromatogr. **125**, 219 to 230 (16)
1977 and STOKLOSA, H. J., BLY, D. D., J. appl. Polymer Sci. **21**, 1911—20 (16, 19)
1978a and GINNARD, C. R.; KIRKLAND, J. J., J. Chromatogr. **149**, 465—87 (8, 10)
1978b and STOKLOSA, H. J.; GINNARD, C. R.; BLY, D. D., Paper PO 13, 12th Middle Atlantic Regional Meeting, American Chemical Society, 5—7 April (16)
1980 and KIRKLAND, J. J., Programmed Sedimentation Field Flow Fractionation (SFFF): A New Method for Particle-Size Distribution Analyses. Lecture 1P4 at the 13th International Symposium on Chromatography, Cannes, 30 June—4 July (13)

YANG, F. J., see GIDDINGS

YARDLEY, J., see COOPER, W.

YEADON, G., see DAWKINS

YEH, S. J.
1958 and FRISCH, H. L., J. Polymer Sci. **27**, 149—66 (18)

YOON, Y. H., see GIDDINGS

YOZA, N.
see KOUCHIYAMA
see OHASHI

YU, T.-K., see DING

ZABORSKY, L. M.
1977 Analyt. Chem. **49**, 1166—68 (18)

ZAHIR, S. A., see BATZER

ZAMBELLI, A., see NATTA

ŽDANOV (SHDANOW), S. P.
1961 Strukturuntersuchung poröser Adsorbenzien nach der Kapillarkondensationstheorie. In [C 4], 177—205 (10)

1962 and KOROMAL'DI, E. V.; JASTREBOVA, L. S. In: Synthetische Zeolithe. Nauka, Moskva, p. 68 (11)

1973 and BELENKIJ, B. G.; NEFEDOV, P. P.; KOROMAL'DI, E. V.; KUZNECOVA (KUZNETSOVA), M. A., J. Chromatogr. **77**, 149 to 151 (16, 19)

1977 and KISELEV, A. V.; KOROMAL'DI, E. V.; NAZANSKIJ, A. S.; ELTEKOV, Ju. A., Kolloidnyj Žurnal **39**, 345—57 (11)

ZHDANOV, S. P., see ŽDANOV

ZHMAKINA, T. P., see ŽMAKINA, T. P.

ZIEBARTH, G., see SCHULZ

ZIEGLER, D. C., see QUINN

ŽIGUNOVA, O. A., see CILIPOTKINA

ZILLIOX, J.-G.
see BENOIT
see KOHLER

ZIMA, J., see SMOLKOVA

ZIMM, B. H.

1948 J. chem. Phys. **16**, 1099 (4)

1949 and STOCKMAYER, W. H., J. Chem. Phys. **17**, 1301 (19)

1959 and KILB, R. W., J. Polymer Sci. **37**, 19—42 (19)

ZIMMER, E., see DUNKY

ZINBO, M.

1971 and PARSONS, J. L., J. Chromatogr. **55**, 55—64 (21)

ŽMAKINA (ZHMAKINA), T. P., see KEVER

ZUIDERWEG, F., see VAN DEEMTER

ZWET, M. S., see CVET

Sources

We gratefully acknowledge the kind permission given for the reproduction of figures or data by the publishing houses, firms, or institutions listed.

Academic Press, Inc., New York/London:
Fig. 14-11 [D 28]

American Chemical Society, Washington:
Fig. 13-4 (Giddings et al., 1975)
Fig. 16-31 (Lecourtier et al., 1979 b)
Fig. 18-6 (Lattimer et al., 1979)
Fig. 18-7 (Lattimer et al., 1979)
Fig. 19-15 (Dubin and Miller, 1977)
Fig. 19-38 (Teramachi et al., 1978)
Fig. 19-43 (Teramachi et al., 1978)

Elsevier, Amsterdam:
Table 7-5 b (Robinson et al., 1980)
Table 10-2 (Kirkland 1973)
Fig. 8-5 (May and Knight, 1971)
Fig. 11-1 (Bather and Gray, 1976)
Fig. 11-3 (Bather and Gray, 1978)
Fig. 12-2, 12-4 b (Heitz, 1970)
Fig. 13-5, 13-6 (Giddings, 1978 a)
Fig. 14-9, 14-10 (Schoenmakers et al., 1979)
Fig. 15-9 (Unger et al., 1978)
Fig. 16-19 (Knox and McLennan, 1979)
Fig. 16-26 (Ozaki et al., 1979)
Fig. 16-30 (Belenkij et al., 1975 b)
Fig. 16-32 (Groh and Halász, 1980)
Fig. 17-6 (Martin et al., 1976)
Fig. 17-9 (Nakamura et al., 1973)
Fig. 17-10 (Martin et al., 1976)
Fig. 18-4 (Melander et al., 1979)
Fig. 18-6 (Knox and McLennan, 1979)
Fig. 19-1, 19-2 (Nefedov et al., 1979)
Fig. 19-3 (Janča and Pokorný, 1978 a)
Fig. 19-4 (Janča and Pokorný, 1979)
Fig. 19-17 (Epton et al., 1976)
Fig. 19-19 (Crone, 1975)
Fig. 19-20, 19-21, 19-22 (Buytenhuys et al., 1978)
Fig. 19-23 (Roumeliotis and Unger, 1979)
Fig. 19-24 (Rokushika et al., 1979)
Fig. 19-25 (Kato et al., 1980)
Fig. 19-32, 20-33 (May and Knight, 1971)
Fig. 19-34 (Ishii et al., 1978)
Fig. 19-52 (Kirkland, 1979)

Friedr. Vieweg & Sohn, Verlagsgesellschaft mbH, Wiesbaden:
Fig. 21-5, 21-6 [E 5]

Hüthig & Wepf Verlag, Basel/Heidelberg:
Fig. 9-3 (Greschner, 1979)
Fig. 10-1 (Davankov and Cjurupa, 1980)
Fig. 16-3, 16-4 (Berger, 1976)
Fig. 16-11 (Berger, 1979 c)
Fig. 16-23 (Dawkins and Hemming, 1975 a)
Fig. 16-33 (Campos et al., 1979)
Fig. 18-5, 18-6 (Eisenbeiss et al., 1978)
Fig. 18-8 (Klein and Leidigkeit, 1979)
Fig. 19-9 (Berger, 1976)
Fig. 19-40 (Hoffmann and Urban, 1977)
Fig. 19-54 (Scholtens and Weizen, 1981)
Fig. 19-55 (Stix and Heitz, 1979)

Iatron Laboratories, Inc., Tokyo:
Fig. 21-18 [F 40]

Institut für Chromatographie, Bad Dürkheim
Fig. 21-17 [E 7]

Institute for Chemical Research Kyoto University, Kyoto:
Fig. 19-41 (Inagaki and Tanaka, 1981)
Fig. 19-42 (Tanaka et al., 1980 b)
Fig. 21-12, 21-15 (Kamiyama and Inagaki, 1971)

International Union of Pure and Applied Chemistry, Oxford:
Fig. 5-4 (Derham et al., 1974)
Fig. 5-6 (Koningsveld, 1969)
Fig. 13-1 (Giddings, 1979)
Fig. 16-22 (Lecourtier et al., 1979)
Fig. 18-10 (Belenkij, 1979)

IPC Business Press, Ltd., Guildford:
Fig. 19-7, 19-8 (Bleha et al., 1980)

John Wiley & Sons, Ltd., Chichester/New York/ Sydney/Toronto:
Fig. 8-2 (GRUBIŠIĆ et al., 1967)
Fig. 9-14 (MOORE et al., 1962)
Fig. 9-18 (JENTOFT and GOUW, 1969)
Fig. 9-19 (KLESPER and HARTMANN, 1977a)
Fig. 9-20 (KLESPER and HARTMANN, 1977b)
Fig. 15-7, Table 5-2 [D 6]
Fig. 16-25, 16-27 (DUBIN et al., 1977)
Fig. 16-28 (HANN, 1977)
Fig. 16-35 (DOMARD et al., 1979)
Fig. 17-2 (OSTERHOUDT and RAY, 1967)
Fig. 17-11 (KATO et al., 1975)
Fig. 19-1, 19-2 (SAMAY and KUBIN, 1979)
Fig. 19-12 (CHA, 1969)
Fig. 19-13 (DOMARD et al., 1979)
Fig. 19-14 (HANN, 1977)
Fig. 19-16 (DOMARD et al., 1979)
Fig. 19-44 (WAGNER and MCCRACKIN, 1977)
Fig. 19-46, 19-47, 19-48 (PARK and GRAESSLEY, 1977a/b)
Fig. 19-49 (WAGNER and MCCRACKIN, 1977)
Fig. 19-50 (MALONE et al., 1969)
Fig. 19-53 (MACRURY and MCCONNELL, 1979)
Fig. 19-56 (KATO et al., 1975)
Table 19-6 (WAGNER and MCCRACKIN, 1977)

Marcel Dekker, Inc., New York:
Fig. 7-1, 7-3, 7-4, 10-3, 10-4; Table 7-1, 7-2, 7-3 [A 4]
Fig. 16-2, 16-6 [D 2]
Fig. 17-7, 17-8 (BOMBAUGH and LEVANGIE, 1970b)

Nauka, Moskau:
Fig. 16-18 (SKVORZOV et al., 1978)
Fig. 16-20, 16-21 (TENNIKOV et al., 1977)

Pergamon Press, Ltd., Oxford:
Fig. 9-21 (KLESPER and HARTMANN, 1978)
Fig. 16-29, 16-34 (BOOTH et al., 1980)
Fig. 19-18 (RINAUDO, 1980)
Fig. 19-45 (CONSTANTIN, 1977)

Springer-Verlag GmbH, Berlin/Heidelberg/ New York/Tokyo:
Fig. 8-8 (HEITZ, 1975)

Subject Index

Compiling the index it has been tried to cover also synonymous terms for a notion. So the entries are not in every case literally.
Boldface numbers refer to essential information. *Italic numbers* indicate figures.